Plant Hormones

Biosynthesis, Signal Transduction, Action!

Edited by

Peter J. Davies
*Cornell University,
Ithaca, NY, U.S.A.*

KLUWER ACADEMIC PUBLISHERS
DORDRECHT / BOSTON / LONDON

A C.I.P. Catalogue record for this book is available from the Library of Congress.

ISBN 1-4020-2685-4 (PB)
ISBN 1-4020-2684-6 (HB)
ISBN 1-4020-2686-2 (e-book)

Published by Kluwer Academic Publishers,
P.O. Box 17, 3300 AA Dordrecht, The Netherlands.

Sold and distributed in North, Central and South America
by Kluwer Academic Publishers,
101 Philip Drive, Norwell, MA 02061, U.S.A.

In all other countries, sold and distributed
by Kluwer Academic Publishers,
P.O. Box 322, 3300 AH Dordrecht, The Netherlands.

The cover picture shows Mendel's dwarf (*le-1*, left) and tall (*LE*, right) peas. The tall, wild-type peas possess a gene encoding gibberellin 3â-hydroxylase (GA 3-oxidase) that converts GA_{20} to GA_1. GA_1 (inset) promotes stem elongation whereas GA_{20} is inactive. Tall plants possess a relatively high level of GA_1 but the level in dwarf plants is much lower. In the mutant dwarf plants the gene differs by one base and the protein by one amino acid from the wild-type tall, and the enzyme activity is $1/20^{th}$ of the level in the tall plants (see Chapters A2, B2 and B7).

Figures with a literature citation to a journal (referenced in the list at the end of every chapter) are copyright of that prospective journal and are reprinted by kind permission.

Printed on acid-free paper

All Rights Reserved
© 2004 Kluwer Academic Publishers
No part of this work may be reproduced, stored in a retrieval system, or transmitted in any form or by any means, electronic, mechanical, photocopying, microfilming, recording or otherwise, without written permission from the Publisher, with the exception of any material supplied specifically for the purpose of being entered and executed on a computer system, for exclusive use by the purchaser of the work.

Printed in the Netherlands.

Dedicated to
Arthur W. Galston, Richard P. Pharis
and to the memory of Kenneth. V. Thimann.
Gentlemen researchers in the field of plant hormones.

PLANT HORMONES:
Biosynthesis, Signal Transduction, Action!

3rd Edition 2004

Edited By Peter J. Davies

Contents

Preface xi
Color Plates xiii

A. INTRODUCTION

1. The plant hormones: Their nature, occurrence and function
   Peter J. Davies 1-15
2. Regulatory factors in hormone action: Level, location and signal transduction
   Peter J. Davies 16-35

B. HORMONE BIOSYNTHESIS, METABOLISM, AND ITS REGULATION

1. Auxin biosynthesis and metabolism
   Jennifer Normanly, Janet P. Slovin and Jerry D. Cohen 36-62
2. Gibberellin biosynthesis and inactivation
   Valerie M. Sponsel and Peter Hedden 63-94
3. Cytokinin biosynthesis and metabolism
   Hitoshi Sakakibara 95-114
4. Ethylene biosynthesis
   Jean-Claude Pech, Mondher Bouzayen and Alain Latché 115-136
5. Abscisic acid biosynthesis and metabolism
   Steven H. Schwartz and Jan A.D. Zeevaart 137-155
6. Brassinosteroids biosynthesis and metabolism
   Sunghwa Choe 156-178
7. Regulation of gibberellin and brassinosteroid biosynthesis by genetic, environmental and hormonal factors
   James B. Reid, Gregory M. Symons and John J. Ross 179-203

C. THE FINAL ACTION OF HORMONES

1 Auxin and cell elongation
   Robert E. Cleland 204-220
2 Gibberellin action in germinating cereal grains
   Fiona Woodger, John V. Jacobsen, and Frank Gubler 221-240
3 Cytokinin regulation of the cell division cycle
   Luc Roef and Harry Van Onckelen 241-261
4 Expansins as agents of hormone action
   Hyung-Taeg Cho and Daniel Cosgrove 262-281

D. HORMONE SIGNAL TRANSDUCTION

1 Auxin signal transduction
   Gretchen Hagen, Tom J. Guilfoyle and William M. Gray 282-303
2 Gibberellin signal transduction in stem elongation and leaf growth
   Tai-Ping Sun 304-320
3 Cytokinin signal transduction
   Bridey B. Maxwell and Joseph J. Kieber 321-349
4 Ethylene signal transduction in stem elongation
   Ramlah Nehring and Joseph R. Ecker 350-368
5 Ethylene signal transduction in flowers and fruits
   Harry J. Klee and David G. Clark 369-390
6 Abscisic acid signal transduction in stomatal responses
   Sarah M. Assmann 391-412
7 Brassinosteroid signal transduction
   Steven D. Clouse 413-436

E. THE FUNCTIONING OF HORMONES IN PLANT GROWTH AND DEVELOPMENT

1 Auxin transport
   David A. Morris, Jiří Friml and Eva Zažímalová 437-470
2 The induction of vascular tissues by auxin
   Roni Aloni 471-492
3 Hormones and the regulation of water balance
   Ian C. Dodd and William J. Davies 493-512
4 The role of hormones during seed development and germination
   Ruth R. Finkelstein 513-537
5 Hormonal and daylength control of potato tuberization
   Salomé Prat 538-560
6 The hormonal regulation of senescence
   Susheng Gan 561-581
7 Genetic and transgenic approaches to improving crop performance via hormones
   Andy L. Phillips 582-609

F. THE ROLES OF HORMONES IN DEFENSE AGAINST INSECTS AND DISEASE

1. Jasmonates
   Gregg A. Howe 610-634
2. Salicylic acid
   Terrence P Delaney 635-653
3. Peptide hormones
   Clarence A. Ryan and Gregory Pearce 654-670

G. HORMONE ANALYSIS

1. Methods of plant hormone analysis
   Karin Ljung, Göran Sandberg and Thomas Moritz 671-694

   TABLE OF GENES 695-716
   INDEX 717

PREFACE

Plant hormones play a crucial role in controlling the way in which plants grow and develop. While metabolism provides the power and building blocks for plant life, it is the hormones that regulate the speed of growth of the individual parts and integrate these parts to produce the form that we recognize as a plant. In addition, they play a controlling role in the processes of reproduction. This book is a description of these natural chemicals: how they are synthesized and metabolized; how they work; what we know of the molecular aspects of the transduction of the hormonal signal; a description of some of the roles they play in regulating plant growth and development; their role in stimulating defensive responses; and how we measure them. Emphasis has also been placed on the new findings on plant hormones deriving from the expanding use of molecular biology as a tool to understand these fascinating regulatory molecules. Even at the present time, when the role of genes in regulating all aspects of growth and development is considered of prime importance, it is still clear that the path of development is nonetheless very much under hormonal control, either via changes in hormone levels in response to changes in gene transcription, or with the hormones themselves as regulators of gene transcription.

This is not a conference proceedings, but a selected collection of newly written, integrated, illustrated reviews describing our knowledge of plant hormones, and the experimental work which is the foundation of this knowledge. The aim of this book is to *tell a story* of what is known at the present time about plant hormones. This volume forms the third edition of a book originally published in 1987 under the title *Plant Hormones and Their Role in Plant Growth and Development,* with the second edition, published in 1995, being titled *Plant Hormones: Physiology, Biochemistry and Molecular Biology.* The title has been changed again from the previous editions in order to reflect the changing nature of the field of plant hormones. The changes that have taken place in the nine years since the last edition are particularly striking: genetics and mutations have moved from one chapter to almost all, Arabidopsis has gone from a mention to central stage in almost every chapter, signal transduction has changed from one example to a major section, the concept of negative regulators has emerged and the role of signal destruction has become an underlying theme. The new edition bears only a superficial resemblance to its first forebear!

The information in these pages is directed at advanced students and professionals in the plant sciences: molecular biologists, botanists, biochemists, or those in the horticultural, agricultural and forestry sciences. It should also form an invaluable reference to molecular biologists from other disciplines who have become aware of the fact that plants form an exiting class of organisms for the study of development, and who need information on the regulators of development that are exclusive to plants. It is intended that the book should serve as a text and guide to the literature for graduate level courses in the plant hormones, or as a part of courses in plant, comparative, or molecular aspects of development. Scientists in other disciplines who wish to know more about the plant hormones and their role in plants should also find this volume a valuable

Preface

resource. It is hoped that anyone with a reasonable scientific background can find valuable information in this book expounded in an understandable fashion.

Gone are the days when one person could write a comprehensive book in an area such as plant hormones. I have thus drawn together a team of fifty one experts who have individually or jointly written about their own area. At my direction they have attempted to tell a story in a way that will be both informative and interesting. Their styles and approaches vary, because they each have a tale to tell from their own perspective. The choice of topics has been my own. Within each topic the coverage and approach has been decided by the authors. While the opinions expressed by the authors are their own, they are, in general, also mine, because I knew their perspective before I invited them to join the project.

Where appropriate the reader will find cross references between chapters. In addition the extensive, sub-divided index at the end of the volume should allow this book to be used as a reference to find individual pieces of information. Sometimes the same information can be found in more than one location, though usually from a different perspective. Rather than edit out such duplication, I have chosen to let it remain so that the complete story on any topic can be obtained without having to excessively transfer between chapters.

A volume such as this cannot be encyclopedic. Nevertheless, we have covered the majority of topics in which active research is taking place. The author of each chapter has provided a set of references that will guide the reader to further information in the area. So as not to disrupt the narrative excessively, and to maintain the book at a reasonable length, the chapter authors were asked to severely limit the number of citations to those papers of greatest or most recent significance, and to also bring to the attention of readers other reviews in which other papers are cited and discussed, so that you, the reader, can find the original information. As such a reference citation at any point within the chapters may be to a later paper in a series, or to a review covering the material, from which the original citation(s) can be obtained if desired; it does not imply that this is the original reference to the work discussed. The authors wish to apologize to their colleagues whose significant work is not directly cited for their lack of inclusion due to the strictures placed on them by the editor.

I would like to thank all the authors who made this volume possible, and produced their chapters not only in timely fashion, but in line with the many restrictions placed on them by this editor. Finally I would like to thank my wife, Linda, who put up with my many hours of absences during which I edited, produced and formatted this book, and to the continuing support and cooperation of Jacco Flipsen and Noeline Gibson from Kluwer Academic Publishers.

Peter J. Davies
Ithaca, New York
June 2004

Color Plates   *For captions see individual chapters*

Chapter A2, Figure 4.

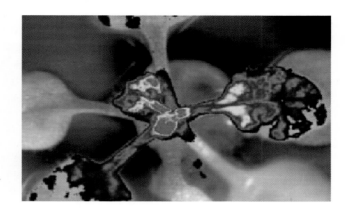

Chapter A2, Figure 6.

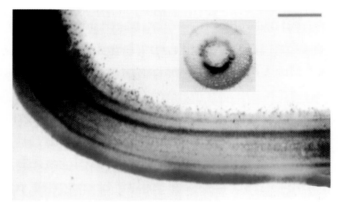

Chapter A2, Figure 13.

WT   WT+ACC   *gin1(aba2)*   *gin4(ctr1)*   *ein2*   *gin1 ein2*

Chapter E7, Figure 12.

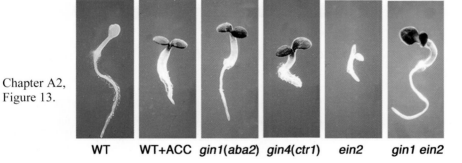

Color Plates *For captions see individual chapters*

Arabidopsis seedlings grown on IAA-Ala

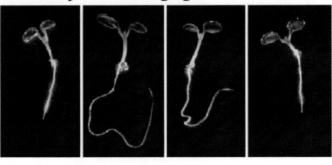

Wild type   *iar1*   *iar3*   *ilr1*

Maize *orp* kernels   *Lemna* plants

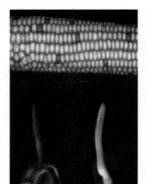 

Normal   *orp*   10 uM   5 uM
Seedlings under UV   alpha-MT treated

Wild type   jsR1
Lemna gibba G3   mutant

Chapter B1, Figure 4.

Color Plates   For captions see individual chapters

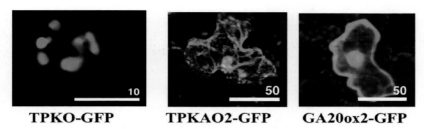

Chapter B2, Figure 8.

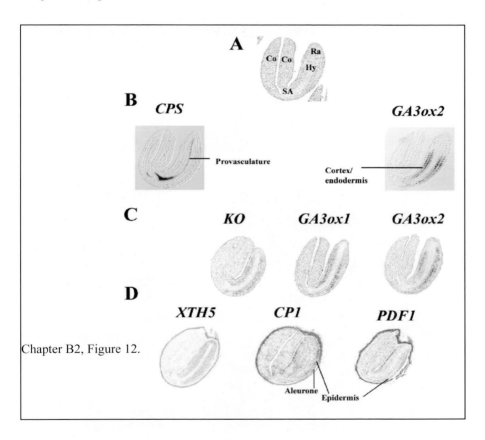

Chapter B2, Figure 12.

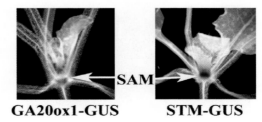

Chapter B2, Figure 13.

CP3

Color Plates   *For captions see individual chapters*

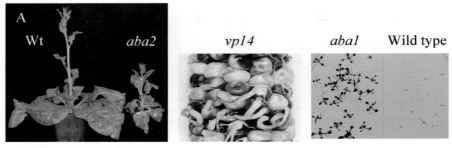

Chapter B5, Figure 7.

Chapter B5, Figure 11.

Control   *GVG::PvNCED1*

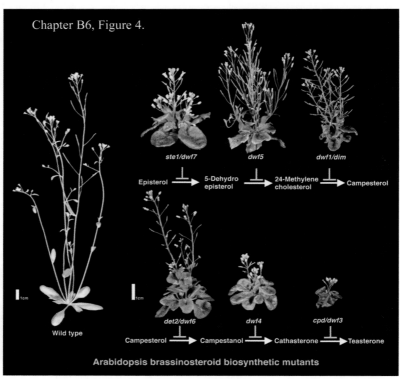

CP4

Color Plates *For captions see individual chapters*

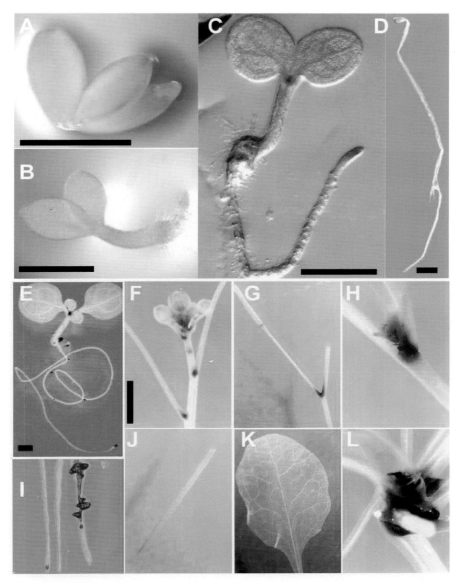

Chapter B6, Figure 5.

Color Plates   *For captions see individual chapters*

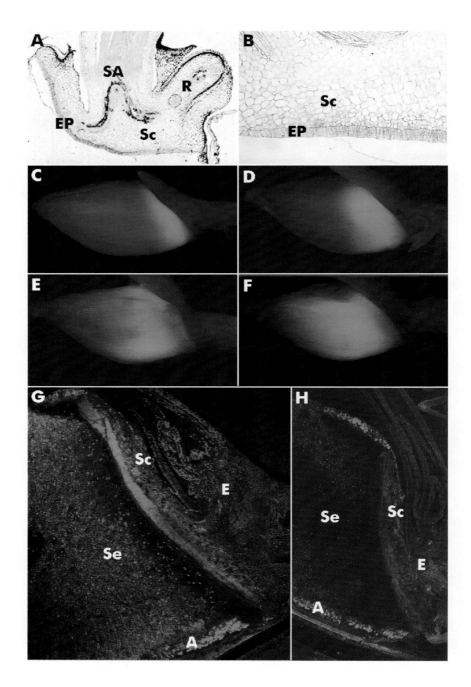

Chapter C2, Figure 3.

Color Plates   For captions see individual chapters

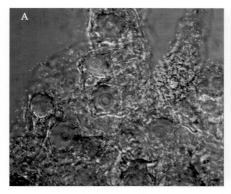

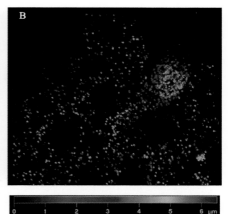

Chapter C3, Figure 8.

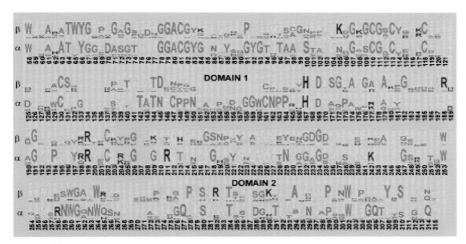

Chapter C4, Figure 2.

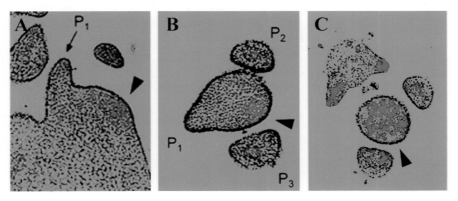

Chapter C4, Figure 7.

CP7

Color Plates   *For captions see individual chapters*

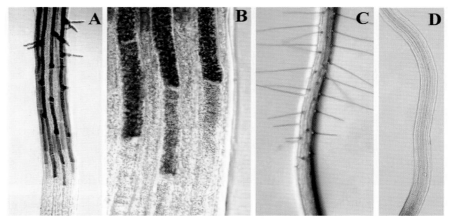

Chapter C4, Figure 11A-D

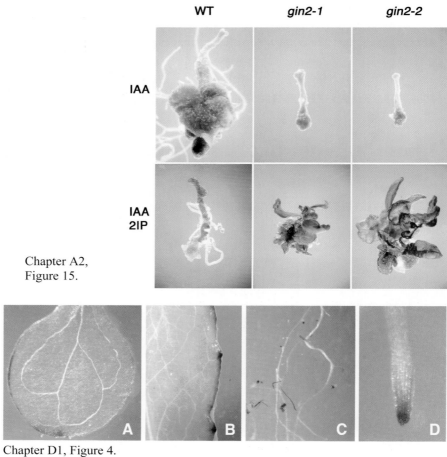

Chapter A2, Figure 15.

Chapter D1, Figure 4.

CP8

Color Plates   For captions see individual chapters

Chapter D2, Figure 5.

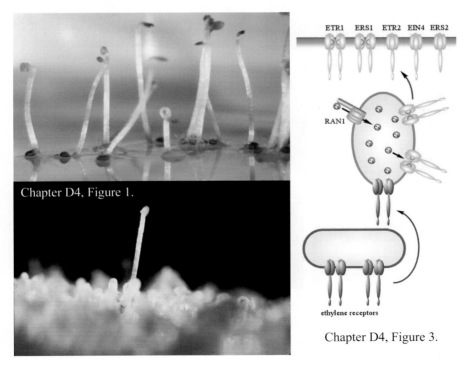

Chapter D4, Figure 3.

Color Plates   For captions see individual chapters

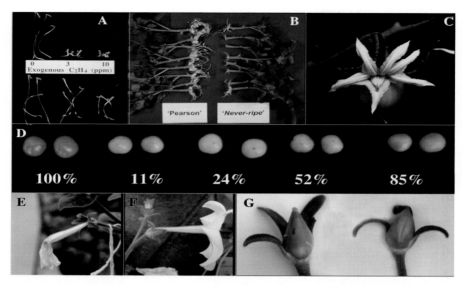

Chapter D5, Figure 2.

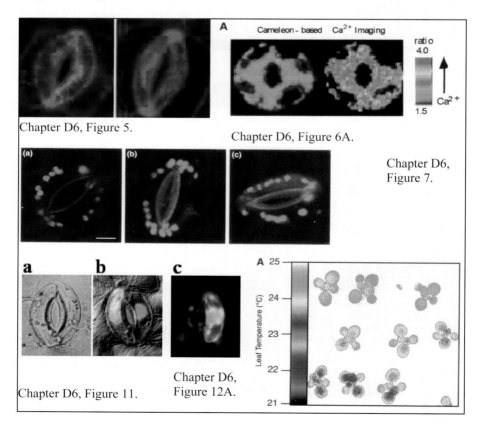

Color Plates — For captions see individual chapters

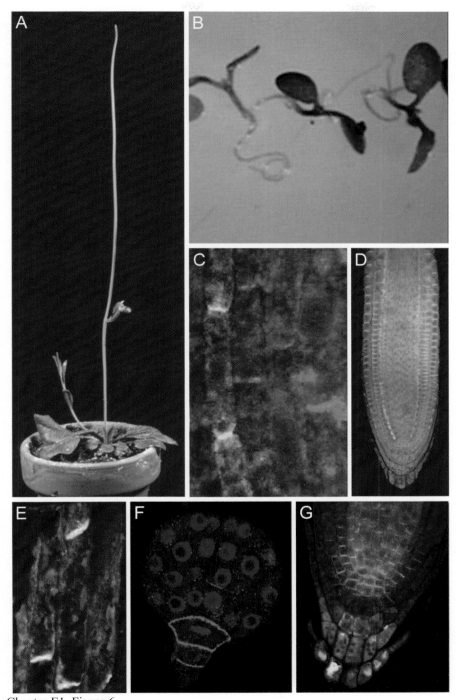

Chapter E1, Figure 6.

Color Plates   *For captions see individual chapters*

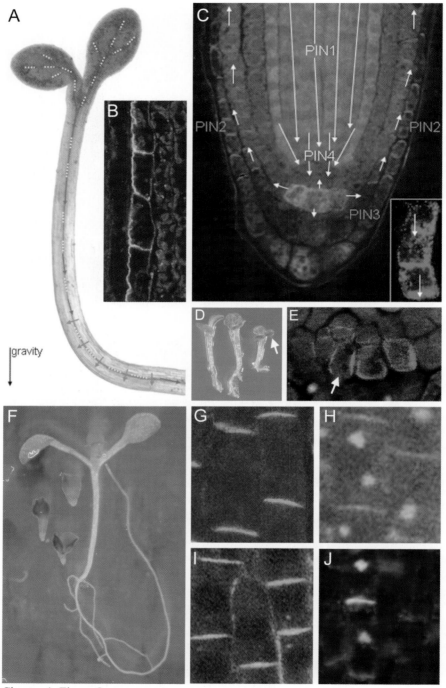

Chapter 1, Figure 8

Color Plates   *For captions see individual chapters*

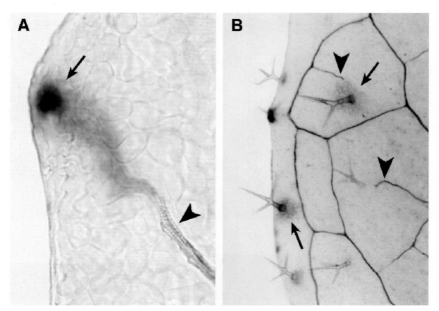

Chapter E2, Figure 2.

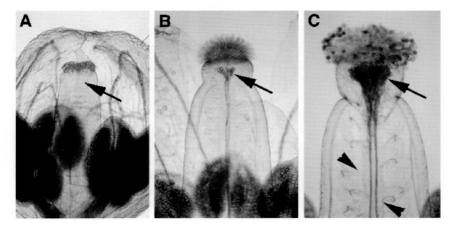

Chapter E2, Figure 3.

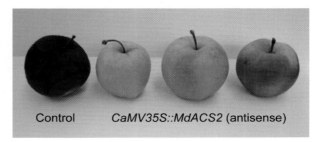

Chapter E7, Figure 14.

Color Plates   *For captions see individual chapters*

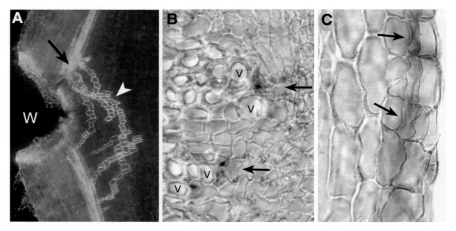

Chapter E2, Figure 5

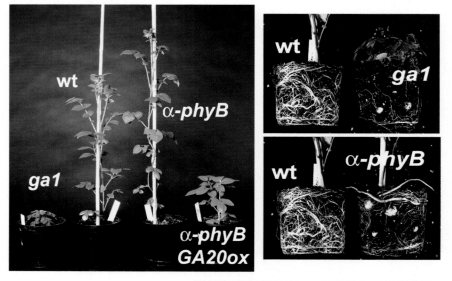

Chapter E5, Figure 1.

Chapter E7, Figure 13.

Color Plates   *For captions see individual chapters*

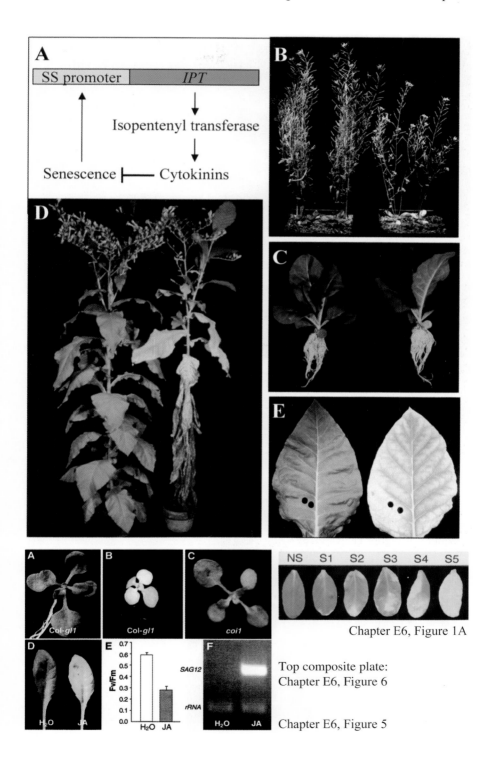

Chapter E6, Figure 1A

Top composite plate:
Chapter E6, Figure 6

Chapter E6, Figure 5

Color Plates *For captions see individual chapters*

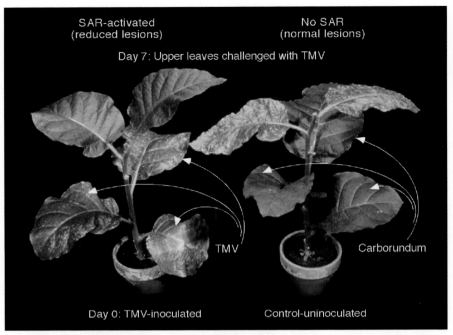

Chapter F2, Figure 3.

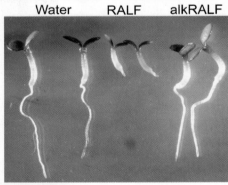

Chapter G1, Figure 11.

Chapter F3, Figure 9.

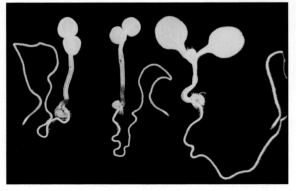

A. INTRODUCTION

A1. The Plant Hormones: Their Nature, Occurrence, and Functions

Peter J. Davies
Department of Plant Biology, Cornell University, Ithaca, New York 14853, USA. E-mail: pjd2@cornell.edu

INTRODUCTION

The Meaning of a Plant Hormone

Plant hormones are *a group of naturally occurring, organic substances which influence physiological processes at low concentrations*. The processes influenced consist mainly of growth, differentiation and development, though other processes, such as stomatal movement, may also be affected. Plant hormones[1] have also been referred to as 'phytohormones' though this term is infrequently used.

In their book *Phytohormones* Went and Thimann (10) in 1937 define a hormone as a substance which is transferred from one part of an organism to another. Its original use in plant physiology was derived from the mammalian concept of a hormone. This involves a localized site of synthesis, transport in the bloodstream to a target tissue, and the control of a physiological response in the target tissue via the concentration of the hormone. Auxin, the first-identified plant hormone, produces a growth response at a distance from its site of synthesis, and thus fits the definition of a *transported* chemical messenger. However this was before the full range of what we now consider plant hormones was known. It is now clear that plant hormones do not fulfill the requirements of a hormone in the mammalian sense. The synthesis of plant hormones may be localized (as occurs for animal hormones), but it may also occur in a wide range of tissues, or cells within tissues. While they may be transported and have their action at a distance this is not always the case. At one extreme we find the transport of

---

[1] The following abbreviations are used throughout this book with no further definition: ABA, abscisic acid; BR, brassinosteroid; CK, cytokinin; GA gibberellin; IAA, indole-3-acetic acid

cytokinins from roots to leaves where they prevent senescence and maintain metabolic activity, while at the other extreme the production of the gas ethylene may bring about changes within the same tissue, or within the same cell, where it is synthesized. Thus, transport is not an essential property of a plant hormone.

The term 'hormone' was first used in medicine about 100 years ago for a stimulatory factor, though it has come to mean a transported chemical message. The word in fact comes from the Greek, where its meaning is 'to stimulate' or 'to set in motion'. Thus the origin of word itself does not require the notion of transport *per se*, and the above definition of a plant hormone is much closer to the meaning of the Greek origin of the word than is the current meaning of hormone used in the context of animal physiology.

Plant hormones[2] are a unique set of compounds, with unique metabolism and properties, that form the subject of this book. Their only universal characteristics are that they are natural compounds in plants with an ability to affect physiological processes at concentrations far below those where either nutrients or vitamins would affect these processes.

THE DISCOVERY, IDENTIFICATION AND QUANTITATION OF PLANT HORMONES.

The Development of the Plant Hormone Concept and Early Work.

The plant hormone concept probably derives from observations of morphogenic and developmental correlations by Sachs between 1880 and 1893. He suggested that "Morphological differences between plant organs are due to differences in their material composition" and postulated the existence of root-forming, flower forming and other substances that move in different directions through the plant (10).

At about the same time Darwin (3) was making his original observations on the phototropism of grass coleoptiles that led him to postulate the existence of a signal that was transported from the tip of the coleoptile to the bending regions lower down. After further characterizations by several workers of the way in which the signal was moved, Went in the Netherlands was finally able to isolate the chemical by diffusion from coleoptile tips into agar blocks, which, when replaced on the tips of decapitated coleoptiles, resulted in the stimulation of the growth of the decapitated coleoptiles, and their bending when placed asymmetrically on these tips. This thus demonstrated the existence of a growth promoting chemical that was

---

[2] The term "plant growth substance" is also used for plant hormones but this is a rather vague term and does not describe fully what these natural regulators do - growth is only one of the many processes influenced. The international society for the study of plant hormones is named the "International Plant Growth Substance Association" (IPGSA). While the term plant growth regulator is a little more precise this term has been mainly used by the agrichemical industry to denote synthetic plant growth regulators as distinct from endogenous growth regulators.

synthesized in the coleoptile tips, moved basipetally, and when distributed asymmetrically resulted in a bending of the coleoptile away from the side with the higher concentration. This substance was originally named *Wuchsstoff* by Went, and later this was changed to *auxin*. After some false identifications the material was finally identified as the simple compound indoleacetic acid, universally known as IAA (11).

Discovery of Other Hormones

Other lines of investigation led to the discovery of the other hormones: research in plant pathogenesis led to gibberellins (GA); efforts to culture tissues led to cytokinins (CK); the control of abscission and dormancy led to abscisic acid (ABA); and the effects of illuminating gas and smoke led to ethylene. These accounts are told in virtually every elementary plant physiology textbook, and further elaborated in either personal accounts (9, 11) or advanced treatises devoted to individual hormones (see book list at the end of the chapter) so that they need not be repeated here. More recently other compounds, namely brassinosteroids (Chapters B7 and D7), jasmonates (Chapter F1) (including tuberonic acid, Chapter E5), salicylic acid (Chapter F2), and the peptides (Chapter F3) have been added to the list of plant hormones, and these are fully covered in this book for the first time. Polyamines, which are essential compounds for all life forms and important in DNA structure, have also been categorized as plant hormones as they can modulate growth and development, though typically their levels are higher than the other plant hormones. However, as little further understanding of their exact function in plants at the cellular and molecular levels has been added in the last few years, no individual chapter has been devoted to polyamines in this edition (a chapter on polyamines can be found in the previous edition (4): 2E Chapter C1).

It is interesting to note that, of all the original established group of plant hormones, only the chemical identification of abscisic acid was made from higher plant tissue. The original identification of the others came from extracts that produced hormone-like effects in plants: auxin from urine and the fungal cultures of *Rhizopus*, gibberellins from culture filtrates of the fungus *Gibberella*, cytokinins from autoclaved herring sperm DNA, and ethylene from illuminating gas. Today we have at our disposal methods of purification (such as high performance liquid chromatography: HPLC, following solid phase extraction: SPE cartridges) and characterization (gas chromatography-mass spectrometry: GC-MS, and high performance liquid chromatography-mass spectrometry: HPLC-MS) that can operate at levels undreamed of by early investigators (Chapter G1). Thus while early purifications from plant material utilized tens or even hundreds of kilograms of tissues, modern analyses can be performed on a few milligrams of tissue, making the characterization of hormone levels in individual leaves, buds, or even from tissues within the organs much more feasible. Thus it is not surprising to see the more-recently discovered hormones being originally identified within plant tissues. Nonetheless only brassinosteroids were identified following investigations of plant growth effects, with the discovery of jasmonates, salicylic acid and peptide hormones deriving from work on insect and disease resistance.

Immunoassay (see 2nd edition, Chapter F2) is also used for hormone quantitation, though is considered much less precise because of interfering effects of other compounds and cross reactivity. Immunoassay columns can, however, permit the very precise isolation of plant hormones prior to more rigorous physico-chemical characterization. While the exact level and location of the hormones within the individual tissues and cells is still largely elusive (Chapter G1), huge strides have been made in analyzing and localizing the expression of genes for hormone biosynthesis using sensitive techniques such as PCR (polymerase chain reaction), or the expression, in transgenic plants, of marker genes driven by promoters of one or more steps in the biosynthetic process. The location of hormone action in tissues and cells has also been investigated by examining the location of marker gene expression driven by promoters of genes known to be induced by the presence of hormone (e.g. Chapter A2).

THE NATURE, OCCURRENCE, AND EFFECTS OF THE PLANT HORMONES

Before we become involved in the various subsequent chapters covering aspects of hormone biochemistry and action it is necessary to review what hormones do. In subsequent chapters some or most of these effects will be described in more detail, whereas others will not be referred to again. It is impossible to give detailed coverage of every hormonal effect, and the reader is referred to the book list at the end of this chapter. The choice of topics for subsequent chapters has been determined largely by whether there is active research in progress in that area. Over the last few years there has been active progress in elucidating the biosynthesis, signal transduction and action of almost every hormone. Thus whereas previously the progress in understanding the action of one hormone was much better than that of another we now find increased understanding of hormone action across the board. A good case in point is cytokinin, where we now know much more about perception, signal transduction (Chapter D3) and action (Chapter C3) than just a few years ago. In fact progress on understanding one hormone as opposed to another has been leapfrogging: whereas the action of auxin at the physiological level was one of the first to be understood (Chapter C1) we still do not understand the connection between auxin signal transduction (Chapter D1) and its final action in inducing cell elongation, and while the identification of the auxin receptor was previously regarded as established, this is now regarded as far less certain. By contrast, after two decades of relatively little advance in the understanding of brassinosteroids, or even much interest in these compounds, following their discovery by extraction from *Brassica* pollen and the demonstration of growth activity in a bean petiole bioassay, the entire biosynthetic pathway has been elucidated (Chapter B6), receptors identified (Chapter D7), mutants characterized and crosstalk with other hormones investigated (Chapter B7).

The effects produced by each hormone were initially elucidated largely from exogenous applications. However in more and more cases we have evidence that the endogenous hormone also fulfills the originally designated roles, and new functions are being discovered. Such more recent evidence derives from correlations between hormone levels and growth of defined genotypes or mutants, particularly of the model plant Arabidopsis, or from transgenic plants. In other cases it has not yet been conclusively proved that the endogenous hormone functions in the same manner.

The nature, occurrence, transport and effects of each hormone (or hormone group) are given below. (Where there is no specific chapter on the topic in this edition but a reference in the second edition of this book (4) this is indicated with the notation '2E'.) It should, however, be emphasized that hormones do not act alone but in conjunction, or in opposition, to each other such that the final condition of growth or development represents the net effect of a hormonal balance (Chapter A2) (5).

Auxin

Nature
Indole-3-acetic acid (IAA) is the main auxin in most plants.

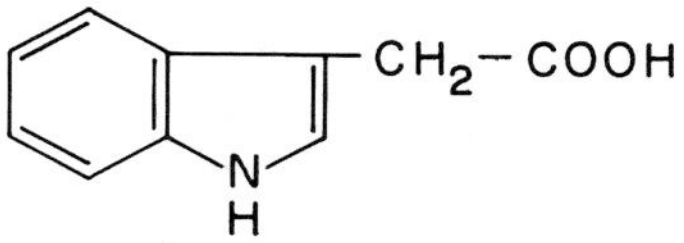

INDOLEACETIC ACID

Compounds which serve as IAA precursors may also have auxin activity (e.g., indoleacetaldehyde). Some plants contain other compounds that display weak auxin activity (e.g., phenylacetic acid). IAA may also be present as various conjugates such as indoleacetyl aspartate (Chapter B1)). 4-chloro-IAA has also been reported in several species though it is not clear to what extent the endogenous auxin activity in plants can be accounted for by 4-Cl-IAA. Several synthetic auxins are also used in commercial applications (2E: G13).

Sites of biosynthesis
IAA is synthesized from tryptophan or indole (Chapter B1) primarily in leaf primordia and young leaves, and in developing seeds.

Nature, occurrence and functions

Transport
IAA transport is cell to cell (Chapters E1 and E2), mainly in the vascular cambium and the procambial strands, but probably also in epidermal cells (Chapter E2). Transport to the root probably also involves the phloem.

Effects
- Cell enlargement - auxin stimulates cell enlargement and stem growth (Chapter D1).
- Cell division - auxin stimulates cell division in the cambium and, in combination with cytokinin, in tissue culture (Chapter E2 and 2E: G14).
- Vascular tissue differentiation - auxin stimulates differentiation of phloem and xylem (Chapter E2).
- Root initiation - auxin stimulates root initiation on stem cuttings, and also the development of branch roots and the differentiation of roots in tissue culture (2E: G14).
- Tropistic responses - auxin mediates the tropistic (bending) response of shoots and roots to gravity and light (2E: G5 and G3).
- Apical dominance - the auxin supply from the apical bud represses the growth of lateral buds (2E: G6).
- Leaf senescence - auxin delays leaf senescence.
- Leaf and fruit abscission - auxin may inhibit or promote (via ethylene) leaf and fruit abscission depending on the timing and position of the source (2E: G2, G6 and G13).
- Fruit setting and growth - auxin induces these processes in some fruit (2E: G13)
- Assimilate partitioning - assimilate movement is enhanced towards an auxin source possibly by an effect on phloem transport (2E: G9).
- Fruit ripening - auxin delays ripening (2E: G2 & 2E:G12).
- Flowering - auxin promotes flowering in Bromeliads (2E: G8).
- Growth of flower parts - stimulated by auxin (2E: G2).
- Promotes femaleness in dioecious flowers (via ethylene) (2E: G2 & 2E: G8).

In several systems (e.g., root growth) auxin, particularly at high concentrations, is inhibitory. Almost invariably this has been shown to be mediated by auxin-produced ethylene (2, 7) (2E: G2). If the ethylene synthesis is prevented by various ethylene synthesis inhibitors, the ethylene removed by hypobaric conditions, or the action of ethylene opposed by silver salts (Ag+), then auxin is no longer inhibitory.

Gibberellins (GAs)

Nature
The gibberellins (GAs) are a family of compounds based on the *ent*-gibberellane structure; over 125 members exist and their structures can be found on the web (Chapter B2). While the most widely available compound is GA_3 or gibberellic acid, which is a fungal product, the most important GA

in plants is GA_1, which is the GA primarily responsible for stem elongation (Chapters A2, B2, and B7). Many of the other GAs are precursors of the growth-active GA_1.

GIBBERELLIN A_1 or GA_1

Sites of biosynthesis.
GAs are synthesized from glyceraldehyde-3-phosphate, via isopentenyl diphosphate, in young tissues of the shoot and developing seed. Their biosynthesis starts in the chloroplast and subsequently involves membrane and cytoplasmic steps (Chapter B2).

Transport
Some GAs are probably transported in the phloem and xylem. However the transport of the main bioactive polar GA_1 seems restricted (Chapters A2 and E5).

Effects
- Stem growth - GA_1 causes hyperelongation of stems by stimulating both cell division and cell elongation (Chapters A2, B7 and D2). This produces tall, as opposed to dwarf, plants.
- Bolting in long day plants - GAs cause stem elongation in response to long days (Chapter B2, 2E: G8).
- Induction of seed germination - GAs can cause seed germination in some seeds that normally require cold (stratification) or light to induce germination (Chapter B2).
- Enzyme production during germination - GA stimulates the production of numerous enzymes, notably α-amylase, in germinating cereal grains (Chapter C3).
- Fruit setting and growth - This can be induced by exogenous applications in some fruit (e.g., grapes) (2E: G13). The endogenous role is uncertain.
- Induction of maleness in dioecious flowers (2E: G8).

Cytokinins (CKs)

Nature
CKs are adenine derivatives characterized by an ability to induce cell division in tissue culture (in the presence of auxin). The most common

cytokinin base in plants is zeatin. Cytokinins also occur as ribosides and ribotides (Chapter B3).

ZEATIN

Sites of biosynthesis
CK biosynthesis is through the biochemical modification of adenine (Chapter B3). It occurs in root tips and developing seeds.

Transport
CK transport is via the xylem from roots to shoots.

Effects
- Cell division - exogenous applications of CKs induce cell division in tissue culture in the presence of auxin (Chapter C3; 2E: G14). This also occurs endogenously in crown gall tumors on plants (2E: E1). The presence of CKs in tissues with actively dividing cells (e.g., fruits, shoot tips) indicates that CKs may naturally perform this function in the plant.
- Morphogenesis - in tissue culture (2E: G14) and crown gall (2E: E1) CKs promote shoot initiation. In moss, CKs induce bud formation (2E: G1 & G6).
- Growth of lateral buds - CK applications, or the increase in CK levels in transgenic plants with genes for enhanced CK synthesis, can cause the release of lateral buds from apical dominance (2E: E2 & G6).
- Leaf expansion (6), resulting solely from cell enlargement. This is probably the mechanism by which the total leaf area is adjusted to compensate for the extent of root growth, as the amount of CKs reaching the shoot will reflect the extent of the root system. However this has not been observed in transgenic plants with genes for increased CK biosynthesis, possibly because of a common the lack of control in these systems.
- CKs delay leaf senescence (Chapter E6).
- CKs may enhance stomatal opening in some species (Chapter E3).
- Chloroplast development - the application of CK leads to an accumulation of chlorophyll and promotes the conversion of etioplasts into chloroplasts (8).

Ethylene

Nature
The gas ethylene (C_2H_4) is synthesized from methionine (Chapter B4) in many tissues in response to stress, and is the fruit ripening hormone. It does not seem to be essential for normal mature vegetative growth, as ethylene-deficient transgenic plants grow normally. However they cannot, as seedlings, penetrate the soil because they lack the stem thickening and apical hook responses to ethylene, and they are susceptible to diseases because they lack the ethylene-induced disease resistance responses. It is the only hydrocarbon with a pronounced effect on plants.

Sites of synthesis
Ethylene is synthesized by most tissues in response to stress. In particular, it is synthesized in tissues undergoing senescence or ripening (Chapters B4 and E5).

Transport
Being a gas, ethylene moves by diffusion from its site of synthesis. A crucial intermediate in its production, 1-aminocyclopropane-1-carboxylic acid (ACC) can, however, be transported and may account for ethylene effects at a distance from the causal stimulus (2E: G2).

Effects
The effects of ethylene are fully described in 2E: G2. They include:
- The so called *triple response*, when, prior to soil emergence, dark grown seedlings display a decrease in tem elongation, a thickening of the stem and a transition to lateral growth as might occur during the encounter of a stone in the soil.
- Maintenance of the apical hook in seedlings.
- Stimulation of numerous defense responses in response to injury or disease.
- Release from dormancy.
- Shoot and root growth and differentiation.
- Adventitious root formation.
- Leaf and fruit abscission.
- Flower induction in some plants (2E: G8).
- Induction of femaleness in dioecious flowers (2E: 8).
- Flower opening.
- Flower and leaf senescence.
- Fruit ripening (Chapters B4 and E5).

Abscisic acid (ABA)

Nature
Abscisic acid is a single compound with the following formula:

Nature, occurrence and functions

ABSCISIC ACID

Its name is rather unfortunate. The first name given was "abscisin II" because it was thought to control the abscission of cotton bolls. At almost the same time another group named it "dormin" for a purported role in bud dormancy. By a compromise the name abscisic acid was coined (1). It now appears to have little role in either abscission (which is regulated by ethylene; 2E: G2) or bud dormancy, but we are stuck with this name. As a result of the original association with abscission and dormancy, ABA has become thought of as an inhibitor. While exogenous applications can inhibit growth in the plant, ABA appears to act as much as a promoter, such as in the promotion of storage protein synthesis in seeds (Chapter E4), as an inhibitor, and a more open attitude towards its overall role in plant development is warranted. One of the main functions is the regulation of stomatal closure (Chapters D6 and E3)

Sites of synthesis
ABA is synthesized from glyceraldehyde-3-phosphate via isopentenyl diphosphate and carotenoids (Chapter B5) in roots and mature leaves, particularly in response to water stress (Chapters B5 and E3). Seeds are also rich in ABA which may be imported from the leaves or synthesized in situ (Chapter E4).

Transport
ABA is exported from roots in the xylem and from leaves in the phloem. There is some evidence that ABA may circulate to the roots in the phloem and then return to the shoots in the xylem (Chapters A2 and E4).

Effects
- Stomatal closure - water shortage brings about an increase in ABA which leads to stomatal closure (Chapters D6 and E3).
- ABA inhibits shoot growth (but has less effect on, or may promote, root growth). This may represent a response to water stress (Chapter E3; 2E: 2).
- ABA induces storage protein synthesis in seeds (Chapter E4).
- ABA counteracts the effect of gibberellin on α-amylase synthesis in germinating cereal grains (Chapter C2).
- ABA affects the induction and maintenance of some aspects of dormancy in seeds (Chapters B5 and E4). It does not, however, appear

to be the controlling factor in 'true dormancy' or 'rest,' which is dormancy that needs to be broken by low temperature or light.
- Increase in ABA in response to wounding induces gene transcription, notably for proteinase inhibitors, so it may be involved in defense against insect attack (2E: E5).

Polyamines

$$H_2N-(CH_2)_3-NH-(CH_2)_4-NH_2$$

SPERMIDINE

Polyamines are a group of aliphatic amines. The main compounds are putrescine, spermidine and spermine. They are derived from the decarboxylation of the amino acids arginine or ornithine. The conversion of the diamine putrescine to the triamine spermidine and the quaternaryamine spermine involves the decarboxylation of S-adenosylmethionine, which also is on the pathway for the biosynthesis of ethylene. As a result there are some complex interactions between the levels and effects of ethylene and the polyamines.

The classification of polyamines as hormones is justified on the following grounds:
- They are widespread in all cells and can exert regulatory control over growth and development at micromolar concentrations.
- In plants where the content of polyamines is genetically altered, development is affected. (E.g., in tissue cultures of carrot or *Vigna*, when the polyamine level is low only callus growth occurs; when polyamines are high, embryoid formation occurs. In tobacco plants that are overproducers of spermidine, anthers are produced in place of ovaries.)

Such developmental control is more characteristic of hormonal compounds than nutrients such as amino acids or vitamins.

Polyamines have a wide range of effects on plants and appear to be essential for plant growth, particularly cell division and normal morphologies. At present it is not possible to make an easy, distinct list of their effects as for the other hormones. Their biosynthesis and a variety of cellular and organismal effects is discussed in 2E Chapter C1. It appears that polyamines are present in all cells rather than having a specific site of synthesis.

Brassinosteroids

Brassinosteroids (Chapters B6 and D7) are a range of over 60 steroidal compounds, typified by the compound brassinolide that was first isolated from *Brassica* pollen. At first they were regarded as somewhat of an oddity but they are probably universal in plants. They produce effects on growth and development at very low concentrations and play a role in the endogenous regulation of these processes.

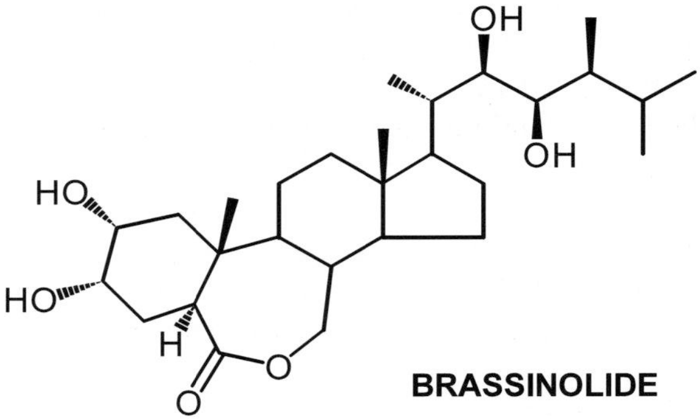

Nature, occurrence and functions

BRASSINOLIDE

Effects
- Cell Division, possibly by increasing transcription of the gene encoding cyclinD3 which regulates a step in the cell cycle (Chapter D7).
- Cell elongation, where BRs promote the transcription of genes encoding xyloglucanases and expansins and promote wall loosening (Chapter D7). This leads to stem elongation.
- Vascular differentiation (Chapter D7).
- BRs are needed for fertility: BR mutants have reduced fertility and delayed senescence probably as a consequence of the delayed fertility (Chapter D7).
- Inhibition of root growth and development
- Promotion of ethylene biosynthesis and epinasty.

Jasmonates
Jasmonates (Chapter F1) are represented by jasmonic acid (JA) and its methyl ester.

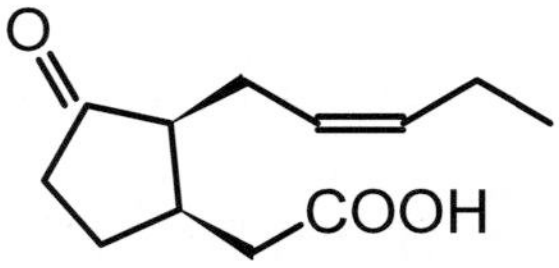

JASMONIC ACID

They are named after the jasmine plant in which the methyl ester is an important scent component. As such they have been known for some time in the perfume industry. There is also a related hydroxylated compound that has been named tuberonic acid which, with its methyl ester and glycosides, induces potato tuberization (Chapter E5). Jasmonic acid is synthesized from linolenic acid (Chapter F1), while jasmonic acid is most likely the precursor of tuberonic acid.

Effects
- Jasminates play an important role in plant defense, where they induce

the synthesis of proteinase inhibitors which deter insect feeding, and, in this regard, act as intermediates in the response pathway induced by the peptide systemin.
- Jamonates inhibit many plant processes such as growth and seed germination.
- They promote senescence, abscission, tuber formation, fruit ripening, pigment formation and tendril coiling.
- JA is essential for male reproductive development of Arabidopsis. The role in other species remains to be determined.

Salicylic Acid (SA)

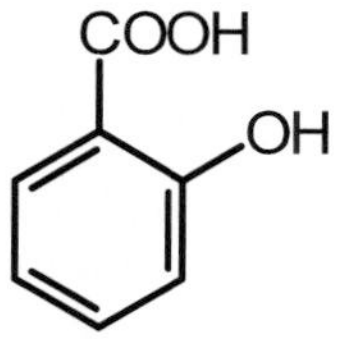

SALICYLIC ACID

Salicylates have been known for a long time to be present in willow bark, but have only recently been recognized as potential regulatory compounds. Salicylic acid is biosynthesized from the amino acid phenylalanine.

Effects
- Salicylic acid (Chapter F2) plays a main role in the resistance to pathogens by inducing the production of 'pathogenesis-related proteins'. It is involved in the systemic acquired resistance response (SAR) in which a pathogenic attack on older leaves causes the development of resistance in younger leaves, though whether SA is the transmitted signal is debatable.
- SA is the calorigenic substance that causes thermogenesis in *Arum* flowers.
- It has also been reported to enhance flower longevity, inhibit ethylene biosynthesis and seed germination, block the wound response, and reverse the effects of ABA.

Signal Peptides

The discovery that small peptides could have regulatory properties in plants started with the discovery of systemin, an 18 amino acid peptide that travels in the phloem from leaves under herbivore insect attack to increase the content of jasmonic acid and proteinase inhibitors in distant leaves, so protecting them from attack (Chapters F1 and F3). Since then, over a dozen peptide hormones that regulate various processes involved in defense, cell

division, growth and development and reproduction have been isolated from plants, or identified by genetic approaches (Chapter F3). Among these effects caused by specific peptides are:
- The activation of defense responses.
- The promotion of cell proliferation of suspension cultured plant cells.
- The determination of cell fate during development of the shoot apical meristem
- The modulation of root growth and leaf patterning in the presence of auxin and cytokinin
- Peptide signals for self-incompatability.
- Nodule formation in response to bacterial signals involved in nodulation in legumes.

Are the More-Recently-Discovered Compounds Plant Hormones?

Two decades ago there was a heated discussion as to whether a compound had to be transported to be a plant hormone, and could ethylene therefore be a plant hormone. To this Carl Price responded: "Whether or not we regard ethylene as a plant hormone is unimportant; bananas do…"[3]. Hormones are a human classification and organisms care naught for human classifications. Natural chemical compounds affect growth and development in various ways, or they do not do so. Clearly brassinosteroids fit the definition of a plant hormone, and likely polyamines, jasmonates salicylic acid and signal peptides also can be so classified. Whether other compounds should be regarded as plant hormones in the future will depend on whether, in the long run, these compounds are shown to be endogenous regulators of growth and development in plants in general.

A Selection of Books on Plant Hormones Detailing their Discovery and Effects

Abeles FB, Morgan PW, Saltveit ME (1992) Ethylene in Plant Biology. Academic Press, San Diego

Addicott FT (ed) (1983) Abscisic acid. Praeger, New York

Arteca RN (1995) Plant Growth Substances, Principles and Applications. Chapman and Hall, New York

Audus LJ (1959) Plant Growth Substances (2E) L. Hill, London; Interscience Publishers New York. (Editors note: the 2nd edition of Audus contains a lot of information on auxins that was cut out of the later, broader, 3rd edition and it is therefore still a valuable reference.)

Audus LJ (1972) Plant Growth Substances (3E). Barnes & Noble, New York

Crozier A (ed) (1983) The Biochemistry and Physiology of Gibberellins. Praeger, New York

Davies PJ (ed) (1995) Plant Hormones: Physiology, Biochemistry and Molecular Biology. Kluwer Academic, Dordrecht, Boston

Davies WJ, Jones HG (1991) Abscisic Acid: Physiology and Biochemistry. Bios Scientific Publishers, Oxford, UK

Hayat S, Ahmad A (eds) (2003) Brassinosteroids: Bioactivity and Crop Productivity. Kluwer Academic, Dordrecht , Boston

---

[3] Carl A. Price, in Molecular Approaches to Plant Physiology

Jacobs WP (1979) Plant Hormones and Plant Development. Cambridge University Press
Khripach VA, Zhabinskii VN, de Groot AE (1999) Brassinosteroids: A New Class of Plant Hormones. Academic Press, San Diego
Krishnamoorthy HN (1975) Gibberellins and Plant Growth. Wiley, New York
Mattoo A, Suttle J (1991) The Plant Hormone Ethylene. CRC Press Boca Raton FL
Mok DWS, Mok MC (1994) Cytokinins: Chemistry, Activity and Function. CRC Press Boca Raton FL
Sakurai A, Yokota T, Clouse SD (eds) (1999) Brassinosteroids: Steroidal Plant Hormones Springer, Tokyo, New York
Slocum RD, Flores HE (eds) (1991) Biochemistry and Physiology of Polyamines in Plants. Boca Raton CRC Press
Thimann KV (1977) Hormone Action in the Whole Life of Plants. University of Massachusetts Press, Amherst
Takahashi N, Phinney BO, MacMillan J (eds) (1991) Gibberellins. Springer-Verlag, New York
Yopp JH, Aung, LH, Steffens GL (1986) Bioassays and Other Special Techniques for Plant Hormones and Plant Growth Regulators. Plant Growth Regulator Society of America Lake Alfred FL, USA

References

1. Addicott FT, Carns HR, Cornforth JW, Lyon JL, Milborrow BV, Ohkuma K, Ryback G, Thiessen WE, Wareing PF (1968) Abscisic acid: a proposal for the redesignation of abscisin II (dormin). In F Wightman, G Setterfield, eds, Biochemistry and Physiology of Plant Growth Substances, Runge Press, Ottawa, pp 1527-1529
2. Burg SP, Burg EA (1966) The interaction between auxin and ethylene and its role in plant growth. Proc Natl Acad Sci USA 55: 262-269
3. Darwin C (1880) The power of movement in plants. John Murray, London,
4. Davies PJ (1995) Plant Hormones: Physiology, Biochemistry and Molecular Biology. Kluwer Academic Publishers, Dordrecht, The Netherlands; Norwell, MA, USA,
5. Leopold AC (1980) Hormonal regulating systems in plants. In SP Sen, ed, Recent Developments in Plant Sciences, Today and Tomorrow Publishers, New Delhi, pp 33-41
6. Letham DS (1971) Regulators of cell division in plant tissues. XII. A cytokinin bioassay using excised radish cotyledons. Physiol Plant 25: 391-396
7. Mulkey TJ, Kuzmanoff KM, Evans ML (1982) Promotion of growth and hydrogen ion efflux by auxin in roots of maize pretreated with ethylene biosynthesis inhibitors. Plant Physiol 70: 186-188
8. Parthier B (2004) Phytohormones and chloroplast development. Biochem Physiol Pflanz 174: 173-214
9. Phinney BO (1983) The history of gibberellins. In A Crozier, ed, The Biochemistry and Physiology of Gibberellins, Vol 1. Praeger, New York, pp 15-52
10. Went FW, Thimann KV (1937) Phytohormones. Macmillan, New York,
11. Wildman SG (1997) The auxin-A, B enigma: Scientific fraud or scientific ineptitude? Plant Growth Regul 22: 37-68

A2. Regulatory Factors in Hormone Action: Level, Location and Signal Transduction

Peter J. Davies
Department of Plant Biology, Cornell University, Ithaca, New York 14853, USA. Email: pjd2@cornell.edu

INTRODUCTION

The way in which a plant hormone influences growth and development depends on:
1) The amount present: this is regulated by biosynthesis, degradation and conjugation.
2) The location of the hormone: this is affected by movement or transport.
3) The sensitivity (or responsiveness) of the tissue: this involves the presence of receptors and signal-transduction chain components.

All of the above are active areas of current research that will be considered in this volume. Examples of each will be introduced below and some will be considered in more detail in later chapters.

EFFECTS DETERMINED BY THE AMOUNT OF HORMONE PRESENT: BIOSYNTHESIS, DEGRADATION AND CONJUGATION

Hormone Quantitation and Its Interpretation

Hormone concentration clearly has a major role to play in hormonal regulation. Even if sensitivity changed, there would still be a response to a change in concentration, though of a different magnitude. In addition we should ask what we are measuring when we calculate "concentration," and where we are measuring it, in relation to the tissues or cells that respond. If the accuracy of measurement or localization is vague then so is our knowledge of hormone concentration at the active site.

One of the difficulties in correlating endogenous hormone concentration with differences in growth and development is the problem of what is measured when the hormone is assayed, even if this is done by highly accurate physico-chemical means (Chapter G1). Many hormonal extractions are done of whole shoots, roots or fruits. Quantitation is normally done by measuring total extractable hormone. Often this is of little relevance because the total hormonal amount often tells us little about the hormonal concentration in the tissue in question, let alone in the cell, cell compartment, or at the hormone receptor site. Studies investigating hormone metabolism in different tissues have indicated that there are often specific qualitative and quantitative differences between even adjacent tissues. For example the embryo of pea has a different ABA and gibberellin content and metabolism than the seed coat (2E G9). Applied GA_{20} is metabolized to GA_{29} in the embryo, and this GA_{29} is then further metabolized to GA_{29} catabolite in the seed coat (34). Hormone biosynthesis, and almost certainly level, also differs in different parts of young embryos within seeds (Chapters B2 and E4). Differences probably also exist within seemingly uniform tissues.

While the intracellular site of biosynthesis is now, in many cases, well understood (e.g., Chapter B2) we have little idea of the differences in hormonal contents between or within cells. While the receptors for some hormones, such as ethylene (Chapters D 4 and D5), are located on the plasma membrane, the location of the auxin receptor is still uncertain, with evidence for both a plasma membrane location and an intracellular location. It was noted many years ago that growth of stem or coleoptile sections correlates with the amount of auxin that will diffuse out of the tissue rather than the total extractable auxin (26). This tells us that much of the hormone in the tissue is probably compartmentalized away from the growth active site, possibly within an organelle. A demonstration of this compartmentation can be seen if a stem segment is loaded with radiolabelled auxin, and then washed over an extended period to remove the auxin effluxing from the tissue. Growth initially increases in response to the applied IAA, but then declines to a very low rate over 3-4 hours, even though about 70% of the radiolabel is still in the tissue, and can be chromatographically identified as IAA. If the tissue is then put in an anaerobic environment, under which growth ceases, the rate of auxin efflux increases. A return to aerobic conditions is accompanied by a burst in the growth rate such that the total growth more than makes up for the loss of growth during anaerobiosis, and the rate of auxin efflux gradually returns to its previous condition (20) (Fig. 2). This indicates that the IAA is sequestered within an inner membrane-bound compartment, from which it leaks at a slower rate than when it exits the cell through the plasma membrane. The growth-active site, with which auxin interacts, is exterior to the cellular compartment in which it is contained. Once the auxin originally in the cytoplasm leaves the cell, the auxin concentration at the growth active site remains very low (under normal aerobic conditions) because the auxin exits through the plasma membrane faster than it leaves the internal compartment. Under anaerobic conditions the compartment membrane becomes leaky, possibly because an inwardly-directed, ATP-requiring IAA carrier becomes inactivated, allowing outward passive diffusion to predominate. This allows more IAA to leak into the rest of the cell and to the exterior. On the return of oxygen the high concentration of auxin at the growth active site allows ATP-requiring growth to take place, while the compartment membrane ceases being leaky. We do not know the location of the IAA-storage compartment, or the location of the

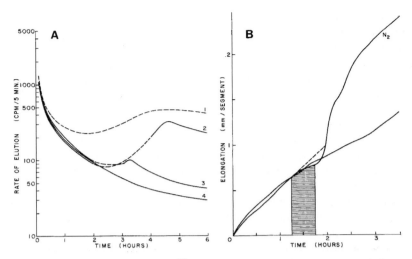

Figure 1. A) The elution of label from [^{14}C]IAA-treated pea stem segments as influenced by anaerobiosis and subsequent oxygenation. Under aerobic conditions (bubbled with oxygen-- solid line) the rate of efflux steadily decreases (curve 4), but if the sections become anaerobic (bubbled with nitrogen), as indicated by the dashed line (curves 2 and 3), the rate of efflux increases. The increase can subsequently be reversed by the return of aerobic conditions. Segments in curve 1 were anaerobic throughout. B) In segments that become anaerobic for a short period, and then return to aerobic conditions, growth is stimulated beyond that of those that remain in aerobic conditions. From (20).

growth active site (see Chapters C1 and C4 for the latter). One possibility for the storage compartment is the chloroplast (or proplastid) which, because of its relatively high pH, would tend to trap acidic molecules (see Chapter E1). In fact 30-40% of the cellular IAA has been reported to be localized in the chloroplast (25). The vacuole seems unlikely because its acidic pH would dictate an auxin concentration only about 1/10th that in the cytoplasm, though its vast bulk in parenchymatous cells might make it a hormone reservoir.

Correlations of Growth with Hormonal Concentration

Gibberellins
The correlation of growth with hormonal levels has been well demonstrated by the work of Reid and co-workers in peas (Chapter B7) (9), and by Phinney and co-workers in maize (27). They have shown that tall plants have GA_1, while in dwarf plants GA_1 is absent or present at a very low level. In sorghum genotypes, the genotype with the greatest height and weight also showed a 2-6 fold higher GA_1 concentration over two shorter, related genotypes (3). Another well-characterized response is that of bolting in spinach (*Spinacea oleracea*) where GA_1 naturally increases in response to long days to produce bolting (33).

In many plants GA_1 is the main growth-active GA, while the other GAs are not active in their own right, but only after conversion to GA_1. The difference between Mendel's tall and dwarf peas can be accounted for in terms of the level of GA_1 present, rather than the total GA content. Mendel's tallness gene has been shown to code for the enzyme GA 3β-hydroxylase (3-oxidase), which catalyzes the conversion of GA_{20} to GA_1 by the addition of a hydroxyl group to carbon 3 of GA_{20} (Fig. 2). Mendel's dwarf (*le-1*) mutant gene was shown to differ from the tall wild type (*LE*) by having a single base change (G to A), which led to a single amino acid change (alanine to threonine) in the enzyme (12). When the activities of *LE* and *le-1* recombinant GA 3β-hydroxylases in E. coli were assayed the le mutant enzyme was shown to possess only 1/20 of the activity found in the *LE* wild-type enzyme in converting radiolabelled GA_{20} into GA_1; this would therefore account for the decrease in GA_1 content and the shorter stature of the dwarf plants. The mutation is close to iron-binding motif of other 2-oxoglutarate-dependent dioxygenases so a change in iron binding may be the cause of the decreased enzyme activity. The levels of GA_1 in genotypes possessing different alleles of the *LE* gene that result in a range of heights shows a good relationship between internode length in the different genotypes and the GA_1 concentration (Fig. 3) (24).

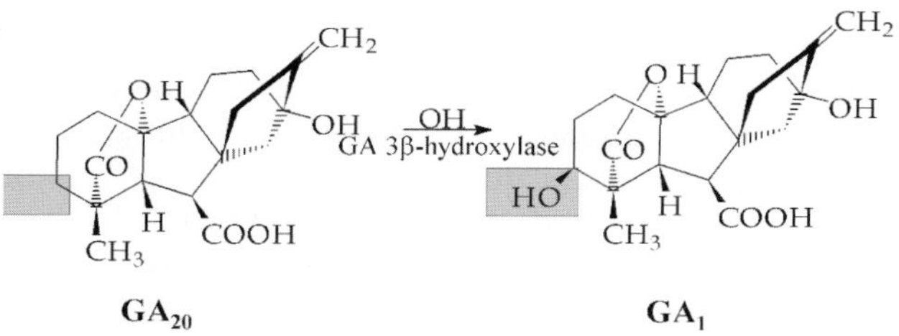

Figure 2. Conversion of GA_{20} to GA_1 by GA 3b-hydroxylase, which adds a hydroxyl group (OH) to carbon 3 of GA_{20}

If the gibberellin content is measured by bioassay, without prior chromatographic separation, then tall and dwarf plants are found to contain the same amount of "gibberellin activity". This is because the vast majority of bioassays will show no difference between GA_{20} and GA_1, since the bioassay plant has the ability to convert inactive GA_{20} to the active GA_1. However when a plant lacking GA 3β-hydroxylase (3-oxidase), such as Waito-C rice, is used as the bioassay plant, then such a difference can be seen (18). It should also be noted that GA_1 is mainly present in the youngest internodes of the stem, so that mature internodes display a reduced difference between tall and dwarf genotypes.

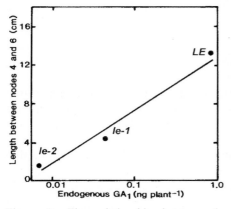

Figure 3. The relationship between the endogenous GA_1 content and internode length in pea plants with different alleles of the *LE* gene. From (24)

Analyses of plant parts for hormones sometimes yield some seemingly contradictory results on first inspection. In wild-type andigena potatoes tuberization occurs in short days. High levels of bioactive GAs inhibit tuberization. But in short days the shoots have a high level of GA_1 biosynthesis. This is seemingly in contradiction to that which is expected, until it is realized that GA_{20}, the GA_1 precursor, is transported, whereas GA_1 itself is not, so that the formation of GA_1 in the shoots decreases the level of GA_{20} available for transport to the stolon tips where tuberization takes place (Chapter E5).

The relationship between stem length and GA content beaks down, however, when one considers signal transduction chain mutants (Chapter D2), such as 'slender' peas (Fig. 6). These pea plants, which possess the alleles *la cry^s*, are ultratall regardless of their GA content, and appear as if they are GA-saturated. However, the concentration of IAA shows a good relationship to tallness across a wide range of pea tallness genotypes, including slender (see Fig. 10), pointing towards a complex multifaceted hormonal control of growth (see later and Chapter B7).

Indoleacetic acid
Early work seemed to indicate that IAA caused elongation in stem segments, but not in intact plants. However, continuous application of auxin solutions via a cotton wick wrapped around the upper stem of light-grown pea seedlings results in a 6 fold increase in growth rate in dwarf plants, and a 2 fold increase in growth rate in tall plants (32). This difference is most likely due to the differing hormonal content of the two height types: tall plants already contain more auxin (10) and GA_1 (9) than do dwarf plants, and are already growing close to the maximal elongation rate. Even at saturating concentrations of applied auxin, tall plants still grow faster than dwarf plants, probably because of their enhanced content of GA_1. The application of auxin to intact plants induces a growth increase with similar kinetics to that seen in isolated segments (Chapter D1) i.e., a lag of about 15 min followed by a rise in the growth rate to a maximum, a drop in growth rate, and then a rise to a more steady rate slightly lower than the initial growth rate peak (see Fig. 12 Chapter D1). Over time the growth rate measured at any particular point slowly declines over 1-2 days as the tissues age and move out of the growing zone. Growth, of course, continues in the younger tissues that are being continually produced. The two peaks in growth rate appear to be related to two responses to auxin: an initial, short term, rapid growth response (IGR), followed by a more prolonged growth response (PGR) with a longer induction time.

Hormone Levels in Plants Transformed by Hormone Biosynthesis Genes

The generation of transgenic plants with enhanced or decreased hormone levels is a powerful tool to demonstrate the effects of changing hormone concentrations. The effects of decreasing hormone concentration, either by the antisense constructs, or more recently the use of RNAi (interference RNA), for hormone biosynthesis genes (e.g., Chapter B4), or the constitutive expression of genes for hormone degradation, especially gibberellins (Chapter B2), are, in general, more satisfactory demonstrations than the overexpression of hormone biosynthesis genes. The reason for this is that the unregulated expression of high hormone levels in unusual tissues can have unexpected consequences. Nevertheless, increases in cytokinin or IAA synthesis, achieved through the incorporation of the genes for their synthesis from *Agrobacterium*, have demonstrated the internal regulation of some hormone-mediated phenomena. For example the incorporation of the gene for cytokinin synthesis into tobacco plants shows the effect of cytokinin concentration on lateral branching and leaf senescence. The transformed plants, which contained a 3-23 fold increase in zeatin riboside level, displayed increased axillary branching, an underdeveloped root system and delayed leaf senescence (15). Particularly striking is the prevention of leaf senescence in tobacco plants when the gene for cytokinin biosynthesis is driven by a promoter for the induction of gene transcription of senescence-associated genes (Chapter E5). The relationship between IAA content and stem height in transgenic plants is less clear, possibly related to the morphological distortions observed when the IAA content is elevated throughout the entire plant. Ethylene, which is induced by high levels of IAA, may play a part in this altered morphology.

EFFECTS DETERMINED BY THE LOCATION OF THE HORMONE

The location of a hormone within a tissue is a most important regulator of the way in which that tissue responds. This is particularly true of auxin whose asymmetric presence in growing stems and roots determines the lateral differences in cellular extension rate, leading to tropistic curvatures. We still have no reliable way of determining the location of native auxin within tissues. However the powerful technique of using marker genes regulated by a) promoters of genes whose transcription is involved in the biosynthesis of the hormone in question, or b) the promoter of a gene that is known to be turned on in response to the hormone, have given a method of visualizing the location of biosynthesis or the differences in likely hormone concentrations in tissues.

The Sites of Biosynthesis or Conversion

The site of biosynthesis also has a bearing on where the higher concentrations of hormone are present. This can be visualized by coupling the promoter of a biosynthesis gene to a reporter gene and transferring the construct into plants. Figure 4 shows an Arabidopsis plant transformed with the *AtGA20ox1* promoter coupled to the luciferase reporter gene and then the light output by the plants recorded. This demonstrates the apical and young leaf location of GA 20-oxidase activity, which catalyzes the steps in the biosynthesis of GAs involving the conversion of GA_{53} to GA_{20}.

Hormone level, location and signal transduction

The response to hormones depends on the tissue. In peas the control of tallness resides in GA_1, It is found mainly in the youngest internodes and is not transmitted through a graft junction. The transmission of tallness can, however, be seen if the correct system is chosen. GA_1 is the final product of a long biosynthetic pathway which produces the biologically active compound. The genotype *NA le* is dwarf as it has the ability to produce GA_{20} (through the gene *NA*), but it cannot convert GA_{20} to GA_1, because it lacks the dominant gene *LE*. The genotype *na LE* is

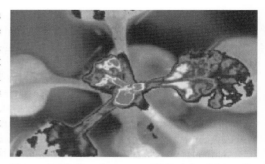

Figure 4. (Color plate page CP1). The false-color image of light emitted by transgenic Arabidopsis plants containing the firefly luciferase (LUC) reporter gene coupled to the *AtGA20ox1* promoter demonstrates that the *GA20ox1* gene is expressed most strongly in the shoot apex and in young developing leaves. (From P. Hedden, IAACR-BBSRC annual report, 2000.)

ultradwarf (*nana*) as it lacks the ability to synthesize GAs (because of *na*), but it has the ability to convert GA_{20} to GA_1 (because of *LE*), though normally it does not do so because GA_{20} is lacking. Now if a *nana* scion is grafted onto a dwarf stock the resulting plant is tall. The stock synthesizes the GA_{20} and passes it to the scion, which converts the GA_{20} to GA_1, giving tall growth (22)

Hormone Movement or Transport

The polar movement of auxin enables it to regulate growth in the subapical regions. IAA is synthesized in young apical tissues and then is transported basipetally to the growing zones of the stem, and more distantly, the root. The major stream is via the procambial strands and then the cells of the vascular cambium. In the root tip the downward transport in the central stele reverses in the cortical tissues (4, 23) to give what has been termed an "inverted umbrella" (Fig. 5). The mechanism of IAA transport involves a pH difference across the cell membrane allowing the movement of un-ionized IAA into the cell, an uptake carrier and an efflux carrier for ionized IAA at the lower side of the cell (Chapter E1).

Recent advances have now localized several of these carriers, most notably the efflux carrier(s). This has been achieved through the use of Arabidopsis *pin* mutants, so named because they often have an inflorescence stem resembling a pinhead, rather than

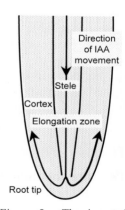

Figure 5. The inverted umbrella model of IAA transport in roots. IAA moves downward in the stele and then returns a short distance to the growing zone in the outer cortical tissues.

the normal inflorescence. *Atpin1* mutants have a contorted inflorescence stem and abnormal vascular tissue patterns in the stem. This is correlated with a greatly reduced level of basipetal auxin transport in the inflorescence tissues. *AtPIN1* gene analysis revealed that the gene product is membrane localized. When the gene product was made fluorescently labeled it was shown to occur in continuous vertical cell strands in vascular bundles and to be located at the basal end of elongated, parenchymatous xylem cells (6) (Chapter E1). This is the same as predicted and previously found for the auxin efflux carrier. The *Atpin2* mutant influences root gravitropism so that the roots of these mutants fail to respond to gravity. The AtPIN2 protein functions as a transmembrane component of the auxin carrier complex and regulates direction of auxin efflux in cortical and epidermal root cells. Localization of the *AtPIN2* gene product with a fluorescent antibody showed it to be located in the membranes of root cortical and epidermal cell files just behind the root apex, with a basal and outer lateral distribution towards the elongation zones (17), again matching the previously-determined pathways of basipetal auxin transport at the root tip, consistent with the "inverted umbrella" (Chapter E1). *Atpin3* mutant hypocotyls are defective in gravitropic as well as phototropic responses and root gravitropism. Expression of the *PIN3* gene in *PIN3:GUS* transgenic seedlings shows the PIN3 protein to also be located in the cell membranes of the auxin transporting tissues. What was unique about PIN3 is that it was found to be relocated in root columella cells in response to a change in the gravity vector; one hour after roots were turned sideways the PIN3 protein relocalized from the basipetal cell membrane to the upper cell membrane with respect to gravity (5), so perhaps PIN3 is important in redirecting auxin laterally in response to gravity and thus leading to auxin regulation of the gravitropic response. The possible mechanism of PIN protein relocalization has been revealed by the location of the fluorescently-labeled protein in the presence of metabolic inhibitors. Treatment with brefeldin (BFA), a vesicle trafficking inhibitor, leads to the detection of PIN1 in vesicles in 2 hours. When the BFA is washed out the PIN1 returns to its previous membrane localization (7). This implies a constant recycling of the PIN proteins between the correct localization in the cell membrane and internal vesicles, with the gravity vector possibly determining the site of reintegration of the auxin transporting PIN auxin efflux carriers.

Hormone Location Following Transport, and its Effects

Tropistic responses to light or gravity are mediated by changes in IAA concentration. If the promoter of an auxin-responsive gene is fused to the *GUS* (β-glucuronidase) gene and used to transform tobacco plants this results in the production of a blue color in the presence of auxin when tissues are exposed to the GUS substrate. This technique nicely demonstrates that auxin is redistributed in response to gravitropic stimuli. When a young shoot of tobacco is placed on its side the blue color, indicating the presence of auxin, is found on the lower side of the stem (Fig. 6). This side elongates faster causing a bending. This is evident prior to the appearance of any gravitropic curvature (13)

EFFECTS DETERMINED BY THE SENSITIVITY (OR RESPONSIVENESS) OF PLANT TISSUES TO HORMONES

The concept of negative regulators

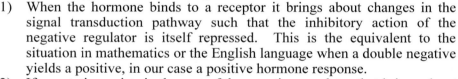

Figure 6. (Color plate page CP1). The location of auxin can be determined from the expression of GUS resulting in the presence of a blue color when a plant is transformed with the gene for GUS driven by the promoter for an early auxin activated gene (*GH3*). In a gravistimulated shoot auxin is located on the lower side of the stem, as seen in the longitudinal section and inset transverse section of a seedling tobacco shoot. The higher concentration in the vascular parenchyma is also evident in the transverse section. Bar = 1mm. From (13).

In one plant hormone system after another we are seeing that the native situation with no restraints gives the same growth or response as it does in the presence of the hormone, even though there is no hormone present. What maintains the hormone-absent phenotype is the presence of a repressor or negative regulator. This is represented in diagrams by an inverted **T**. There are then two ways in which the hormone-type response can be generated:

1) When the hormone binds to a receptor it brings about changes in the signal transduction pathway such that the inhibitory action of the negative regulator is itself repressed. This is the equivalent to the situation in mathematics or the English language when a double negative yields a positive, in our case a positive hormone response.
2) If a mutation exists in the part of the negative regulator that brings about the response inhibition then the inhibition becomes lost, so that a hormone-type response occurs even in the absence of the hormone.

Protein Breakdown: Ubiquitin and Proteasomes

Protein breakdown is turning out to be very important in hormonal regulation in that the repression of negative regulators appears to generally involve the actual destruction of the protein negative regulator in the presence of the hormone.

Protein breakdown is a natural and necessary part of cellular metabolism. It occurs in a small organelle termed a proteasome. Proteasomes deal primarily with endogenous proteins; that is proteins that were synthesized within the cell such as cyclins (which must be destroyed to prepare for the next step in cell division (Chapter C3)), proteins encoded by viruses and other intracellular parasites, and proteins that are folded incorrectly because of errors or that have been damaged by other molecules in the cytosol.

Proteins destined for destruction first become conjugated to molecules of ubiquitin. Ubiquitin is a small protein (76 amino acids), conserved throughout all the kingdoms of life, that is used to target proteins for destruction. Ubiquitin molecules look like little balls and chains attached to the protein. The protein-ubiquitin complex binds to ubiquitin-recognizing sites on the regulatory particle of a proteasome. The protein is unfolded by ATPases (using ATP) and degraded (Chapters D1 and D2). The regulatory particle then releases the ubiquitins for reuse.

The Role of Responsiveness to Hormones in Growth and Development

Responsiveness to hormones can change during development, in response to environmental changes, or as the result of genetic changes.

Changes in hormone responsiveness (sensitivity) during development
The sensitivity of stem elongation in gibberellin-deficient *nana* (*na*) peas to gibberellin is different in dark-grown and light-grown seedlings with the light-grown plants being less sensitive, so that dark-grown seedlings elongate more in response to applied GA_1 than do light-grown seedlings (Fig. 7) (21). During development under constant conditions sensitivity can also change.

Gibberellin Signal Transduction in Stem Elongation Shown by mutants in Arabidopsis
Progress is now being made in the elucidation of the gibberellin signal transduction pathway through the use of Arabidopsis mutants (Chapter D2). In a GA-insensitive mutant (*gai-1*) growth is much reduced. In other mutants lacking GA because of a mutation blocking GA biosynthesis, a second mutation (*rga*) partially restores growth. Some mutants are also extra tall without gibberellin: the *spy* mutant behaves as if treated with GA. The phenotypes of the *spy* and *rga* mutants show that stem elongation is normally repressed and that what GA does is to negate the repression. The SPY and RGA proteins are therefore negative regulators.

GAI and RGA, which are very similar, act in the absence of GA to suppress growth. Whether mutations in these genes result in the promotion of growth or an insensitivity to GA depends on the position of the mutation in the protein -

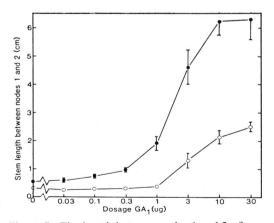

Figure 7. The length between nodes 1 and 2 of *nana* (*na*) pea seedlings resulting from the application of differing amounts of GA_1 to plants grown in light (○) or darkness (●). From (21).

whether it is in the functional domain (the action part of the protein) or the regulatory domain (the part responding to the presence of the hormone) respectively. The wheat dwarfing gene of the "green revolution" (*rht*) represents a mutation in the regulatory domain of an RGA/GAI homologous protein (Chapter D2) so that the plant has a much reduced response to GA and thus possesses a reduced stature. This in turn helps to prevent lodging (the collapse of the stem) when bearing the high-yielding heads.

SPY, which is also a negative regulator, enhances the effect of GAI and RGA. GA acts to block the actions of SPY, GAI and RGA. GAI and RGA turn out to be transcription factors, whereas SPY functions as a second messenger.

The negative regulator RGA protein is found in the nucleus. The level and location of RGA has been visualized by the production of a green color in Arabidopsis transformed with the gene encoding the green fluorescent protein attached to the gene for RGA (Chapter D2). In the absence of GA, experimentally brought about by the GA biosynthesis inhibitor paclobutrazol, the amount of RGA in the nucleus increases, whereas it is severely decreased in the presence of GA. This occurs because the RGA protein is degraded in the presence of GA; in other words, GA negates an inhibition of growth.

Cytokinin Signal Transduction

Cytokinins have long been a hormone for which plant biologists have searched for an action product or pathway; now this long search is producing results (Chapters C3 and D3). The promoter of an Arabidopsis gene switched on by cytokinin, was fused to the firefly luciferase (*LUC*) gene. This was then used to transform protoplasts, so that the luciferase, which is detected by its light emission, was produced in the presence of cytokinin. Sequences encoding suspected signaling intermediates were transfected into protoplasts and the luciferase production measured. CKI1, a membrane localized histidine protein kinase, was found to induce luciferase even in the absence of cytokinin. Another cytokinin signaling intermediate AHP1 moves to the nucleus in the presence of cytokinin (8). Ectopic expression of CKI1 (and other signaling intermediates) is sufficient to promote cytokinin responses in transgenic tissues so that tissues expressing these genes develop as if they had been treated with cytokinins (Fig. 8).

Regulation by Both Changing Hormone Levels and Changes in Sensitivity

As a dark-grown seedling emerges

Figure 8. Ectopic expression of CKI1 (and other signaling intermediates) is sufficient to promote cytokinin responses in transgenic tissues. Tissue culture with: left: vector alone; center: IAA alone; right: tissues transformed with CKI1 growing in the presence of IAA. From (8).

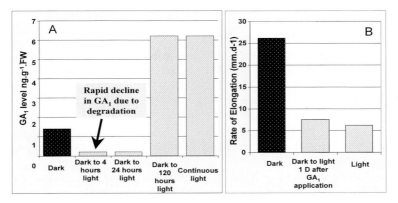

Figure 9. The decrease in growth of seedlings emerging into the light results from a change in both the level and the responsiveness to GA. A) When dark-grown pea seedlings transferred to light the GA_1 level drops rapidly due to metabolism of GA_1. It then increases to a higher level, similar to light grown plants, over the next 4 days. B) Sensitivity to GA of pea seedlings transferred to the light rapidly falls, so that the elongation rate of plants in the light is lower than in the dark even though their GA1 content is higher. Redrawn from (19).

into the light its growth rate slows down substantially, despite the fact that plants grown continuously in the light have higher levels of hormones. This turns out to be regulated by a change in both hormone level and sensitivity. When dark-grown pea seedlings are transferred to light their GA_1 level drops rapidly, due to metabolism of GA_1. The GA_1 level then increases to a higher level, similar to that in plants grown in continuous light, over the next 4 days (Fig. 9A) (19). At the same time sensitivity to GA of the seedlings transferred to the light rapidly falls, so that the elongation rate of plants in the light is lower than in the dark even though their GA_1 content is higher (Fig. 9B).

Another example of changing levels and sensitivity is the response of fruits to ethylene. Immature flower or fruit tissue show no response to ethylene, but mature tissue responds to ethylene with ripening or senescence (Chapter D5). This occurs at about the same time as a rise in natural ethylene production occurs at the start of ripening (Chapter B4). The increasing responsiveness of flowers and fruit to ethylene as they mature represents a change in the number of ethylene receptors (Chapter D5). In addition a shift may occur in the hormonal balance so that an inhibitor of ethylene action, such as IAA (28), may decrease with advancing maturity.

Integrated Effects of Multiple Hormones

We tend to examine the effects of plant hormones on an individual basis, but the growth of a plant is in response to all signals, internal and external, including the net effect of multiple hormones. These hormones may vary in level and responsiveness and also produce numerous, often complicated, interactions.

The effect of any one hormone is often dependent upon the presence of one or more other hormones. A notable case in point is stem elongation. GA application to dwarf plants results in the production of the tall phenotype. However, it is now equally clear that the appropriate treatment of dwarf plants with IAA also increases the growth rate, though not to the same rate as is found in tall plants. Tall plants not only have a higher content of GA_1, but a high content of IAA. This all points towards both GA and IAA being needed for stem elongation.

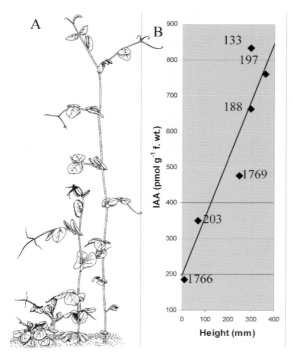

Figure 10. Four genetic lines of peas (A) differing in internode length, and gibberellin and IAA content (B) in their vegetative tissue. Left to right: *Nana* (1766, *na*): an ultradwarf containing no detectable GAs and a very low IAA content; Dwarf (203, *NA le*) containing GA_{20}, but very little GA_1, and low IAA; Tall (1769, *NA LE*) containing GA_1, and a medium IAA content; Slender (188, *na LE la cry^s*): an ultratall with no detectable GAs, but a high IAA content. The IAA content of slender plants with high (197, *LE*) or low (133, *le*) GA1 levels reflects the ultratall characteristic and not the GA level. B from (10)

As an example, pea height varies because of differences in GA_1 content and gibberellin sensitivity (signal transduction components): *nana* has very low GA_1 because an early step in GA biosynthesis is blocked; dwarf has a low level of GA_1; wild-type tall has a high level of GA_1; whereas "slender" is ultratall regardless of GA_1 content and probably represents a mutation in a signal transduction component (Fig. 10A). However auxin level also tracks height, including in slender plants that are tall regardless of GA_1 content (Fig. 10B) (10). Auxin levels also rise by three fold following treatment of dwarf pea stem segments with GA_3 (1).

Not only does gibberellin promote IAA biosynthesis, but auxin promotes gibberellin biosynthesis. Decapitation of pea shoots inhibits, and IAA promotes, the step between GA_{20} to GA_1 in subapical internodes, as detected by the metabolism of applied radiolabelled GA_{20}. Decapitation also reduces, and IAA restores, endogenous GA_1 content. This is because IAA up-regulates GA 3β-hydroxylase transcription, and down-regulates that of GA 2-oxidase, which degrades GA_1 (Chapter B7). We can conclude that

IAA coming from the apical bud promotes, and is required for, GA_1 biosynthesis in subtending internodes. Thus the growth of young stems is the net result of both IAA and GA_1, with considerable mutual effects on the levels of the two hormones.

The correlation of both IAA and GA_1 content with tallness clearly shows that stem elongation is governed both by IAA and GA_1, and that plant height has a clear relationship to hormonal concentration. Notwithstanding the promotive effect of IAA on the biosynthesis of GA_1 (Chapter B7), and vice versa, each has a distinct effect of its own as indicated by the difference in the speed of action following endogenous application (see Fig. 12 Chapter D1), and also by the stem location at which the effects can be observed. The effect of GA on elongation is greater in internodes that are less than 25% expanded, whereas IAA produces a more growth in the relatively more mature, subtending internodes (Fig. 11) (31). The effect of endogenous GA is seen both on cell number and cell length, whereas the effect of IAA is exclusively on cell length.

The Influence of Other Hormones
Complex hormone interactions are more difficult to unravel. For example the brassinosteroid-deficient pea genotype *lkb* (Chapter B7) appears dwarf, but does not respond to applications of GA. It does, however, respond strongly to IAA applications, and, surprisingly, in the presence of IAA it responds to GA applications (Fig. 12) (31). These plants thus behave as if 1) auxin is needed for the response to GA_3, and 2) they are deficient in IAA. This has in fact been shown to be the case, with *lkb* plants having about 1/3 the level of IAA present in the normal *Lkb* plants (14), even though the principal lesion is in Br biosynthesis (Chapter B7). Recent evidence indicates that there is crosstalk between the auxin and brassinosteroid signal transduction pathways (Chapter D7).

In the GA-deficient dwarf plant containing the allele *ls*, which responds well to

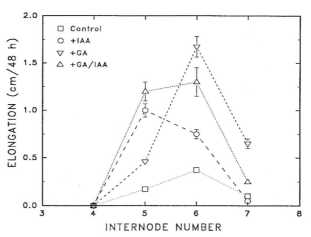

Figure 11. Comparison of the distribution of the induced elongation growth in individual internodes in dwarf pea seedlings 48h following application of IAA, GA_3 or a combination of both. Treatments were administrated when the fourth, fifth and sixth internodes were about 100%, 75% and 25% fully expanded while the seventh internode was still enclosed in the apical bud. GA_3 was applied to the leaf at node 6 whereas IAA was applied around both the fifth and sixth internodes. From (31)

Hormone level, location and signal transduction

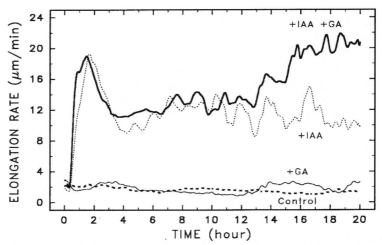

Figure 12. Dwarf *lkb* plants, which are brassinolide deficient and have a sub-normal auxin level, do not respond to GA, but will do so in the presence of added IAA. The figure shows *lkb* stem elongation responses to exogenous IAA (light dots), GA_3 (light solid line) or both (dark solid line), as compared to a control treatment (heavy dashed line). From (31)

GA_3, the GA-response is enhanced in the presence of IAA and vice versa. Thus the level one hormone can probably influence the sensitivity of the plant to changing levels of another hormone.

Sugar-Hormone interactions

Glucose can also act as a signaling molecule in plants in addition to being a metabolic substrate. High levels of glucose (6%) in the growth medium inhibit Arabidopsis seedling growth (Fig. 13), causing a decrease in photosynthesis and in the transcription of photosynthesis-associated genes. This is associated with the loss of the expression of hexokinase (which uses ATP to phosphorylate hexoses), and can be negated by expression of the gene for hexokinase, though it appears to be the enzyme or the process and not the product that is important, as glucose-6-phosphate is without effect (29).

A glucose insensitive mutant *gin1* turned out to be the ABA-deficient mutant *aba2* (11) (Chapter B5). While unaffected by glucose (Fig. 13) it is less vigorous in the absence of glucose than is the wild-type plant (Fig. 14). An ABA insensitive mutant similarly displays glucose tolerance. When ABA is provided to the ABA-deficient mutant the glucose sensitivity is restored. Glucose up-regulates many of the genes in the ABA biosynthesis and signaling pathways, and ABA levels increase in response to glucose. The *gin1* (*aba2*) mutant reveals that abscisic acid may also have a crucial role as a growth-promoting hormone in addition to its growth-inhibitory effects.

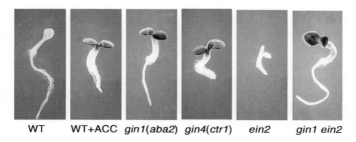

WT WT+ACC gin1(aba2) gin4(ctr1) ein2 gin1 ein2

Figure 13. (Color plate page CP1). Glucose in the medium is detrimental to Arabidopsis (WT, wild type) seedling growth. Here seedlings were grown on 6% glucose in the light. Some mutants are glucose insensitive (*gin* mutants). The glucose-insensitive phenotype can be mimicked by treatment with the ethylene precursor 1-aminocyclopropane-1-carboxylate (ACC) (b). (c) Abscisic acid-deficient *gin1* (*aba2*) and (d) constitutive ethylene-signaling *gin4* (*ctr1*) mutants display a similar glucose-insensitive phenotype except for root elongation. (e) Endogenous abscisic acid is required for the glucose-oversensitive phenotype of the ethylene-insensitive mutant ein2 because (f) the *gin1 ein2* double mutant exhibits the glucose-insensitive phenotype. From (11)

Further interactions occur with ethylene (11). Another glucose-insensitive mutant (*gin4*) is the same as the mutation in the ethylene negative signal-transduction component *ctr1*, which confers a constitutive ethylene response even in the absence of ethylene (Fig. 13) (Chapter D4). Interestingly, in light of the above, treating wild-type seedlings with ethylene overcomes the glucose inhibition. There may be a signaling network connecting endogenous glucose, abscisic acid and ethylene signals. By contrast a positive signal transduction component when mutated to bring about ethylene insensitivity (*ein2*) confers a glucose oversensitive phenotype, so that seedlings are even more inhibited on a glucose-containing medium than are wild-type seedlings (Fig 13), yet in the absence of glucose *ein2* is more vigorous than wild-type (Fig 14). The over-sensitivity of *ein2* to glucose can be negated in the double mutant with *gin1* (Fig. 13). When protoplasts are transformed with *EIN3* combined with *GFP* (green fluorescent protein) the protoplasts appear green because of the transcription

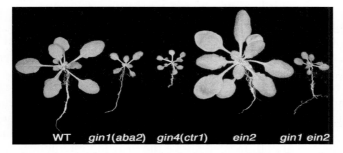

Figure 14. Glucose insensitive *gin* mutants display a retardation of growth as compared to wild-type (a), in the absence of glucose. (b) abscisic acid deficient *gin1* (*aba2*), (c) constitutive ethylene-signaling *gin4* (*ctr1*), (d) ethylene insensitive *ein2*, (e) *gin1 ein2* double mutant. From (11)

of *EIN3*. When glucose is added the protein is destroyed by the proteasome (and the fluorescence is no longer seen) (30), whereas if EIN3 is constitutively produced under the control of the 35S promoter the protein is stabilized.

The glucose resistant mutant *gin2* is a mutant in hexokinase, and displays a delay in senescence. It is unable to proliferate in response to auxin in tissue culture, but if CK (isopentenyl adenine) is also added the explants not only grow but are more vigorous and produce prolific shoots (Fig. 15) (16).

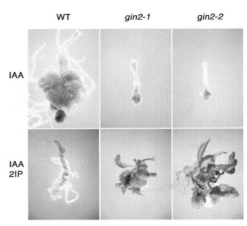

Figure 15. (Color plate page CP8). The glucose insensitive *gin2* mutants in culture are inhibited by auxin (upper panels) but this is negated when cytokinin (isopentenyladenine, 2IP) is added to the medium (lower panels). From (16).

The nature or regulation of these interactions is not fully understood but it shows that we cannot regard hormone responses or interactions as isolated systems. We need to consider complicated interactions between the many factors and pathways that influence growth and development.

THE ROLE OF TRANSPORT AND REDISTRIBUTION IN PLANT HORMONE FUNCTION

A prime function of hormones in plants, which lack a nervous system, is to convey information from one part of the plant to another. The idea that transport was an essential part of the role of plant hormones originally came from experiments on the phototropic control of coleoptile growth. The hypothesis was that the IAA was synthesized in the tip, transported basipetally and was then redistributed laterally to give differential growth and bending.

Auxin transport occurs in a very specific manner in a basipetal direction in stems (Chapter G3). The auxin undergoing transport is clearly involved in vascular differentiation (Chapter G4), lateral root initiation, and the regulation of stem elongation. The fact that transported auxin does regulate growth can be seen from the results of IAA applications to different parts of the stem of growing pea plants. When application is directly to the growing internodes, the classical growth response results. If, however, the IAA is applied to the apical bud, located above the growing internodes, then two or three initial peaks in the growth rate are observed before the PGR plateau occurs (Fig. 16) (32). This in fact represents the auxin being transported down the stem. When a transducer is also located at the tip of the second

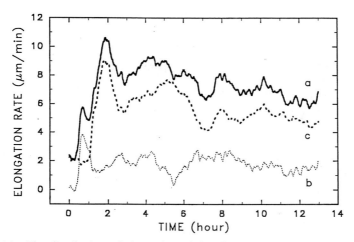

Figure 16. The distribution of elongation of dwarf pea seedlings induced by 10^{-4} M IAA applied to the uppermost internode, starting at time zero, as detected by transducers positioned at nodes 1 (a) or 2 (c) below the apical bud. Line (a) represents the total growth; (c) the growth of the second internode (internodes below the second have ceased growth); and (b) represents the growth of the uppermost internode as determined by the difference between lines (a) and (c). Note that the growth of internode two lags behind that of node one, representing the time taken for the applied IAA to transport from the first to the second internode. Thus IAA in transport from the apical bud promotes growth as it moves down the stem. From (32)

internode it can be seen that this internode responds later than the uppermost internode, corresponding to the time the basipetally-transported IAA reaches that internode. The multiple growth rate peaks observed by the upper transducer in fact correspond to the summed growth of the two or three internodes in which growth is taking place.

The role of transported auxin in the phototropic curvature of maize coleoptiles has been demonstrated by the fact that the basipetal migration of the response from the tip occurs at the same rate as the growth stimulation caused by exogenous auxin (2). Thus there is good evidence for a role of transported auxin in growth.

A most notable case of inter-organ communication via long-distance hormone transport is root to shoot signaling via ABA. When the soil dries out ABA is synthesized in the root and is transported to the shoot in the xylem stream, causing stomatal closure (Chapter E3). This occurs without a change in the water status of the shoot. For example, apple trees with the roots divided into two containers, one moist and one dry, showed restricted leaf expansion and leaf initiation, even though the water status of the shoot was unaffected. When the roots in the dry container were severed, leaf growth recovered. The xylem ABA level is usually found to be a sensitive indicator of water status, and there is a good relationship between the xylem ABA and the stomatal conductance of the leaves. The leaves appear to respond to the amount of ABA arriving rather than the concentration, as during the day the concentration in the xylem may fluctuate with the flux of water through the xylem. Other factors, such as water potential and temperature, affect the sensitivity to the ABA. For example, a decrease in water potential leads to increased sensitivity. Thus at midday there is increased stomatal sensitivity to ABA, leading to stomatal closure. From solely physical (water potential) considerations stomata should reopen in the late afternoon though this is seldom observed in drying soil conditions because of the ABA reaching the leaves.

In other cases plant hormones operate in or near the tissue in which they are produced. Ethylene is the prime example here. In almost all ethylene controlled phenomena, the ethylene is produced in the responding tissue (Chapter B4).

Although transport is not an *essential* part of the definition of a plant hormone, this does not mean that transport plays no part in hormone functioning. Indeed, we find that transport is important in the role of plant hormones in most, but not all, systems.

Acknowledgments

I thank Dr Jen Sheen for the photographs on glucose hormone interaction, and John Ross and Jim Reid for constructive criticism.

References

1. Barratt NM, Davies PJ (1997) Developmental changes in the gibberellin-induced growth response in stem segments of light-grown pea genotypes. Plant Growth Regul 21: 127-134
2. Baskin TI, Briggs WR, Iino M (1986) Can lateral redistribution of auxin account for phototropism of maize coleoptiles? Plant Physiol 81: 306-309
3. Beall FD, Morgan PW, Mander LN, Miller FR, Babb KH (1991) Genetic regulation of development of *Sorghum bicolor*. V. The *ma- 3* allele results in gibberellin enrichment. Plant Physiol 95: 116-125
4. Davies PJ, Mitchell EK (1972) Transport of indoleacetic acid in intact roots of *Phaseolus coccineus*. Planta 105: 139-154
5. Friml J, Wisniewska J, Benkova E, Mendgen K, Palme K (2002) Lateral relocation of auxin efflux regulator PIN3 mediates tropism in Arabidopsis. Nature 415: 806-809
6. Gälweiler L, Guan C, Müller A, Wisman E, Mendgen K, Yephremov A, Palme K (1998) Regulation of polar auxin transport by AtPIN1 in Arabidopsis vascular tissue. Science 282: 2226-2230
7. Geldner N, Friml J, Stierhof YD, Juergens G, Palme K (2001) Auxin transport inhibitors block PIN1 cycling and vesicle trafficking. Nature 413: 425-428
8. Hwang I, Sheen J (2001) Two-component circuitry in Arabidopsis cytokinin signal transduction. Nature London 413: 383-389
9. Ingram TJ, Reid JB, Murfet IC, Gaskin P, Willis CL (1984) Internode length in *Pisum*. The *Le* gene controls the 3-β-hydroxylation of gibberellin A_{20} to gibberellin A_1. Planta 160: 455-463
10. Law DM, Davies PJ (1990) Comparative indole-3-acetic acid levels in the slender pea and other pea phenotypes. Plant Physiol 93: 1539-1543
11. Leon P, Sheen J (2003) Sugar and hormone connections. Trends in Plant Science 8: 110-116
12. Lester DR, Ross JJ, Davies PJ, Reid JB (1997) Mendel's stem length gene (*Le*) encodes a gibberellin 3β-hydroxylase. Plant Cell 9: 1435-1443
13. Li Y, Wu YH, Hagen G, Guilfoyle T (1999) Expression of the auxin-inducible GH3

promoter/GUS fusion gene as a useful molecular marker for auxin physiology. Plant Cell Physiol 40: 675-682
14. McKay MJ, Ross JJ, Lawrence NL, Cramp RE, Beveridge CA, Reid JB (1994) Control of internode length in *Pisum sativum*: further evidence for the involvement of indole-3-acetic acid. Plant Physiol 106: 1521-1526
15. Medford JI, Horgan R, El-Sawi Z, Klee HJ (1989) Alterations of endogenous cytokinins in transgenic plants using a chimeric isopentenyl transferase gene. Plant Cell 1: 403-413
16. Moore B, Zhou L, Rolland F, Hall Q, Cheng WH, Liu YX, Hwang I, Jones T, Sheen J (2003) Role of the Arabidopsis glucose sensor HXK1 in nutrient, light, and hormonal signaling. Science (Washington D C) 300: 332-336
17. Müller A, Guan C, Gälweiler L, Taenzler P, Huijser P, Marchant A, Parry G, Bennett M, Wisman E, Palme K (1998) AtPIN2 defines a locus of Arabidopsis for root gravitropism control. EMBO J 17: 6903-6911
18. Nishijima T, Koshioka M, Yamazaki H (1993) A highly-sensitive rice seedling bioassay for the detection of femtomole quantities of 3-β-hydroxylated gibberellins. Plant Growth Regul 13: 241-247
19. O'Neill DP, Ross JJ, Reid JB (2000) Changes in gibberellin A_1 levels and response during de-etiolation of pea seedlings. Plant Physiol 124: 805-812
20. Parrish DJ, Davies PJ (1977) Emergent growth - an auxin-mediated response. Plant Physiol 59: 745-749
21. Reid JB (1988) Internode length in *Pisum*. Comparison of genotypes in the light and dark. Physiol Plant 74: 83-88
22. Reid JB, Murfet IC, Potts WC (1983) Internode length in *Pisum*. II. Additional information on the relationship and action of loci *Le La Cry Na* and *Lm*. J Exp Bot 34: 349-364
23. Reid JB, Potts WC (1986) Internode length in *Pisum*. Two further mutants, *lh* and *ls*, with reduced gibberellin synthesis, and a gibberellin insensitive mutant, *lk*. Physiol Plant 66: 417-426
24. Ross JJ, Reid JB, Gaskin P, Macmillan J (1989) Internode length in *Pisum*. Estimation of GA_1 levels in genotypes *Le, le* and *le*^{d}. Physiol Plant 76: 173-176
25. Sandberg G, Gardestrom P, Sitbon F, Olsson O (1990) Presence of IAA in chloroplasts of *Nicotiana tabacum* and *Pinus sylvestris*. Planta 180: 562-568
26. Scott TK, Briggs WR (1962) Recovery of native and applied auxin from the light grown Alaska pea seedlings. Amer J Bot 49: 1056-1083
27. Spray C, Phinney BO, Gaskin P, Gilmour SI, Macmillan J (1984) Internode length in *Zea mays* L. The dwarf-1 mutant controls the 3β-hydroxylation of gibberellin A_{20} to gibberellin A_1. Planta 160: 464-468
28. Vendrell M (1985) Dual effect of 2,4-D on ethylene production and ripening of tomato fruit tissue. Physiol Plant 64: 559-563
29. Xiao W, Sheen J, Jang JC (2000) The role of hexokinase in plant sugar signal transduction and growth and development. Plant Mol Biol 44: 451-461
30. Yanagisawa S, Yoo SD, Sheen J (2003) Differential regulation of EIN3 stability by glucose and ethylene signalling in plants. Nature (London) 425: 521-525
31. Yang T, Davies PJ, Reid JB (1996) Genetic dissection of the relative roles of auxin and gibberellin in the regulation of stem elongation in intact light-grown peas. Plant Physiol 110: 1029-1034
32. Yang T, Law DM, Davies PJ (1993) Magnitude and kinetics of stem elongation induced by exogenous indole-3-acetic acid in intact light-grown pea seedlings. Plant Physiol 102: 717-724
33. Zeevaart JAD, Gage DA, Talon M (1993) Gibberellin A_1 is required for stem elongation in spinach. Proc Natl Acad Sci USA 90: 7401-7405
34. Zhu Y-X, Davies PJ, Halinska A (1991) Metabolism of gibberellin A_{12} and A_{12}-aldehyde in developing seeds of *Pisum sativum* L. Plant Physiol 97: 26-33

B. HORMONE BIOSYNTHESIS, METABOLISM AND ITS REGULATION

B1. Auxin Biosynthesis and Metabolism

Jennifer Normanly[a], Janet P. Slovin[b] and Jerry D. Cohen[c]
[a]Department of Biochemistry and Molecular Biology, University of Massachusetts, Amherst, MA 01003, USA; [b]Beltsville Agricultural Research Center, Agricultural Research Service, U.S. Department of Agriculture, Beltsville, MD 20705-2350, USA; [c]Department of Horticultural Science, University of Minnesota, Saint Paul, MN 55108, USA.
E-mail: normanly@biochem.umass.edu

INTRODUCTION
Auxins function at the intersection between environmental and developmental cues and the response pathways that they trigger (Fig. 1).

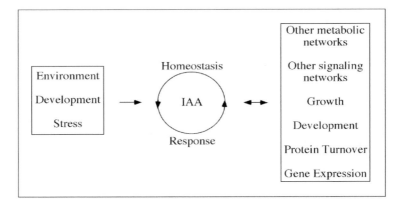

Fig. 1. Auxin homeostasis, the maintenance of an IAA level in the plant cell that is optimal for growth and development, is dependent on abiotic and biotic sources that impinge on the system, and results in a variety of often interconnected responses that themselves impinge on the system. Some of the intricate regulatory networks have evolved with considerable redundancy and adaptive plasticity to maintain optimal auxin levels in response to changing environmental and developmental conditions.

Auxin levels vary dramatically throughout the body and life of the plant, forming gradients that are a central component of its action (4, 5, 14, 20, 34). Accordingly, plants have evolved intricate regulatory networks with considerable redundancy and adaptive plasticity to maintain auxin levels in response to changing environmental and developmental conditions. We refer to this phenomenon as auxin homeostasis; specifically the biosynthesis, inactivation, transport, and inter-conversion pathways that regulate and maintain auxin levels.

The study of auxin homeostasis is aimed at understanding the mechanisms by which plants manage to have the hormone available in the required amount at the right time and place, and at determining how the developmental and environmental signals impact these processes. A combination of molecular genetic and analytical approaches in the past ten years has resulted in an increased understanding of auxin homeostasis; primarily that it involves a highly interactive network of redundant pathways, the complexity of which are only beginning to be discerned. In this chapter, we will focus our attention on the naturally occurring auxins that have been described to date: IAA, IBA [1], PAA and 4-Cl-IAA (Fig. 2). We will summarize the growing body of knowledge concerning the enzymes and the genes involved in IAA synthesis and metabolism. Several other reviews provide a more detailed analysis of particular aspects of auxin synthesis and metabolism (3, 7, 12, 13, 29, 31, 40, 56).

Chemical Forms of Auxins in Plants

Auxins are defined as organic substances that promote cell elongation growth when applied in low concentrations to plant tissue segments in a bioassay. In addition to the most often studied auxin, IAA, there are several other native auxins that have been reported to occur in plants. All natural auxins are found in plants as the free acid and in conjugated forms (Fig. 2).

The halogenated derivative 4-Cl-IAA has been found in peas, several other members of the Fabaceae and in seeds of pine (2). In many biological tests, 4-Cl-IAA is more active than IAA; for example, it has approximately 10 times the activity of IAA on oat coleoptile elongation (38). Most of the 4-Cl-IAA occurs as the methyl ester in many of the plants examined, however, 4-Cl-IAAsp and its monomethyl ester have also been described (2). Studies into the physiological role of 4-Cl-IAA have focused on its activity in pea pod development (44). Normal pod development requires seeds, but in their absence, 4-Cl-IAA and GAs act synergistically to promote pea pod development, specifically cell elongation and division (44). Expression of PsGA3ox1, the gene encoding the oxidase that converts GA_{20} to biologically active GA_1 is developmentally and spatially regulated during pod

---

[1] Abbreviations: Asp, aspartate; Gln, glutamine; Glu, glutamate; Gluc, glucose; IAOx, indole-3-acetaldoxime; IBA, indole-3-butyric acid; IG, indole methyl glucosinolate; Inos, inositol; Lys, lysine; OxIAA, 2-oxindole-3-acetic acid; PAA, phenylacetic acid; Trm, tryptamine; Trp, tryptophan; Trp-D, tryptophan dependent; Trp-I, tryptophan independent.

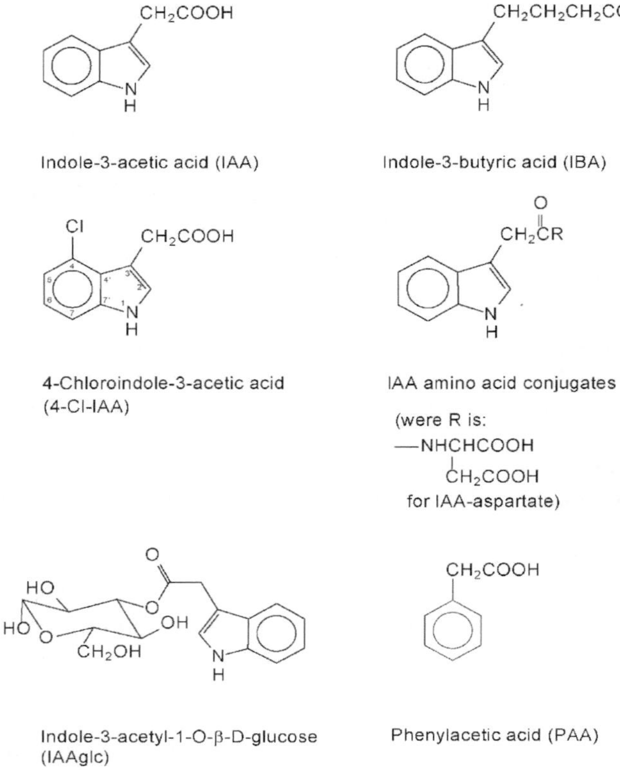

Auxin biosynthesis and metabolism

Fig. 2. Naturally occurring auxins and some examples of the lower molecular weight conjugates found in plants. Related low molecular weight con-jugates (such as IAA-Inos, IAA-Inos-arabinose, and conjugates with several other amino acids) and higher molecular weight conjugates (such as the IAA protein IAP1, IAA-peptides, IAA glycoprotein, and IAA-glucans) have also been isolated from plant materials.

development (44). Additionally, it is dramatically up-regulated by exogenous 4-Cl-IAA suggesting that 4-Cl-IAA is necessary for GA synthesis during pea pod development. PsGA3ox1 expression is not up-regulated by IAA and IAA has very little effect on stimulating GA synthesis or pod development (36). However, IAA clearly affects expression of GA biosynthesis genes in pea shoots and in tobacco (52), although the primary targets of IAA regulation are different in the two species.

IAA and its proposed precursors undergo metabolic conversions to indole-3-lactic acid, indole-3-ethanol and IBA (10). IBA has been used commercially for plant propagation for decades because of its efficacy in the stimulation of adventitious roots (2). The observation that IBA is more efficient than IAA at inducing rooting may be partially explained by the observation that IBA is more stable than IAA against *in vivo* catabolism and inactivation by conjugation (3). IBA was definitively shown to occur naturally in plants in 1989 (31). Feeding studies and analysis of metabolic mutants show that plants are able to convert IBA to IAA by a mechanism that parallels fatty acid β-oxidation (3) and to convert IAA to IBA (31). It is still not clear whether IBA is an auxin *per se*, or a precursor to IAA. Arabidopsis mutants that were resistant to the inhibitory effects of exogenous IBA on root elongation revealed a number of loci that are involved in β-oxidation of fatty acids in the peroxisome, peroxisome protein import and possibly IBA signaling and response (3). The functions of these corresponding genes strongly suggest that IBA metabolism is centered in the peroxisome (3).

In addition to the indolic auxins, various phenolic acids in plants (such as PAA) have low auxin activity. In *Tropaeolum majus* (nasturtium), PAA is present at levels 10- to 100-fold lower than those of both IAA and IBA and it appears to be distinctly localized (32). The low biological activity, the low levels found in some plants, and its unique distribution within the plant, suggest that PAA either does not act *in planta* as an auxin or has an as yet unknown specialized function.

In most tissues, the majority of the auxin is found conjugated to a variety of sugars, sugar alcohols, amino acids and proteins (Fig. 2) (2, 56). Enzyme activities for synthesis and hydrolysis of various conjugated forms are known to exist, suggesting that these forms function to provide a readily accessible and easily regulated source of free IAA without *de novo* synthesis. One general class of conjugated forms consists of those linked through carbon-oxygen-carbon bridges. These compounds have been referred to generically as "ester-linked", although some 1-*O* sugar conjugates such as 1-*O*-IAA-Gluc are actually linked by acyl alkyl acetal bonds. Typical ester-linked moieties include 6-*O*-IAGluc, IAA-Inos, IAA-glycoproteins, IAA-glucans and simple methyl and ethyl esters. The other type of conjugates present in plants are linked through carbon-nitrogen-carbon amide bonds (referred to as "amide-linked"), as in the IAA-amino acid and protein and peptide conjugates (see Fig. 2). IAA conjugates have been identified in all plant species examined for them (56). The conjugate pool in endosperm tissues of monocots and dicots is generally comprised of ester-linked moieties, whereas conjugate moieties in most dicot seeds seem to be predominantly amide-linked. Light grown vegetative tissues of most plants, both monocots and dicots, appear to contain primarily amide-linked conjugates.

Historically, our knowledge of IAA conjugate function comes mainly from tissue culture and bioassays. Various conjugates of IAA, both ester- and amide-linked, act as "slow release" forms of IAA in tissue culture (2) and when used to stimulate rooting of cuttings. Early studies showed that the biological effect of the conjugate was related to its ability to release IAA when applied to tissue (2). Differential growth induced by IAA-conjugates applied to stems of bean (*Phaseolus vulgaris*) was quantitatively correlated to the degree of hydrolysis of the conjugate by the tissue (56). Conjugated auxins are thought to be either storage or long distance mobile forms of the active hormone, or, in the case of IAA-Asp, to act as an intermediate in the degradation of IAA. There are, however, some indications that conjugates themselves can have specialized auxin activities (1, 3, 43, 60, 63).

Auxin biosynthesis and metabolism

Structural determinants of IAA function

The search for a simple structure-activity correlation model valid for IAA and a series of alkylated and halogenated derivatives has not defined a simple or straightforward relationship, although detailed analysis of their structural parameters and their physico-chemical properties has evolved our concepts of activity/structural determinants (37). Contrary to expectations, a simple relationship based on the relative lipophilicity of the compounds does not account for differences in biological activity nor does weak π-complexing ability or reduced indole ring (NH) acidity. These factors had been suggested to explain the differences in auxin activity. The ability of modified structures to function as active auxins may be defined by the relationship between the acidic side chain and unspecified properties of the ring system (37).

IAA HOMEOSTASIS PATHWAYS

Biochemical pathways that result in IAA production within a plant tissue are illustrated in Fig. 3. They include: (A) *de novo* synthesis, whether from tryptophan [referred to as Trp-dependent (Trp-D) IAA synthesis], or from indolic precursors of Trp [referred to as Trp-independent (Trp-I) IAA synthesis, since these pathways bypass Trp]; (B) hydrolysis of both amide- and ester-linked IAA conjugates; (C) transport from one site in the plant to another site; and (D) conversion of IBA to IAA. IAA turnover mechanisms include: (E) oxidative catabolism; (F) conjugate synthesis; (G) transport away from a given site; and (H) conversion of IAA to IBA. This model for inputs and outputs to the IAA pool has remained essentially the same for the past twenty years (11).

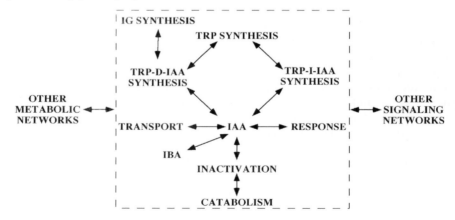

Figure 3. A diagram of our current understanding of metabolic and signaling interactions shows the major interactions discussed. There still remain a number of gaps in our understanding of the complete pathways and broader interactions between auxin regulation and other signaling networks, but hypotheses generated from the interactions shown should serve as a guide to future research efforts.

Classical genetic approaches, such as screening for auxin-deficient mutants, have yielded little information about IAA homeostasis. This has been attributed to the likelihood that auxin auxotrophs would be lethal and that IAA homeostatic pathways are redundant. Thus, a mutation in one pathway would be compensated by another functional pathway. Fortunately, the manner in which IAA metabolism is now studied has yielded many more details about the pathways and genes involved. Early characterization of IAA synthesis and metabolism involved the use of radiolabeled precursors (usually Trp) applied to cut surfaces or extracts of plants. Radiolabeled intermediates were then identified, IAA and IAA metabolites quantified, and the enzymes that catalyze these reactions isolated and characterized. This type of approach has been widely useful in delineating many metabolic pathways, with the caveat that only the capacity of the plant to carry out the metabolic reaction has been demonstrated, not the physiological relevance of the reaction. Recent advances in mass spectrometry have allowed researchers to replace radiolabeled tracer compounds with stable isotope labeled precursors that are monitored by mass spectrometry. Thus, the endogenous, unlabeled compounds of interest could be monitored as well (39). Most of these experiments are done with intact plants, providing a better biological context than previous experiments with extracts or cut sections of plants. Mutants and transgenic plants with altered IAA homeostasis have been integral to these studies, as they reveal the compensatory mechanisms that plants utilize to achieve IAA homeostasis (Fig. 4). Several major findings have emerged (29). They are that: 1) plants utilize multiple pathways for the homeostasis of IAA; 2) these pathways are differentially regulated; and 3) numerous environmental factors including light, temperature and biotic or abiotic stress regulate the activity of these pathways.

De novo synthesis

De novo synthesis refers to the synthesis of the heterocyclic indole ring from non-aromatic precursors. The only known source of indole ring compounds found in nature is the shikimic acid pathway, which is exclusive to bacteria, fungi and plants (Fig. 5). One of these compounds, the abundant amino acid Trp, contains all the carbon and structural features necessary to make IAA.

IAA Synthesis from Tryptophan

Trp-D-IAA synthesis is induced by wounding, temperature and developmental cues

An enormous body of knowledge demonstrates that IAA can be synthesized from Trp, and a summary of some of the known reactions is presented in Fig. 5. These pathways were largely elucidated by radiotracer labeling studies and biochemical assays. While they were proposed decades ago, there have always been some significant aspects of these pathways to regard with caution (42). The pool size of Trp is three or more orders of magnitude

Auxin biosynthesis and metabolism

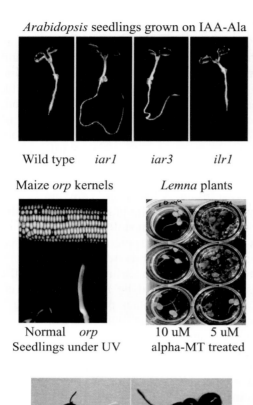

Fig. 4 (Color plate page CP2). Tryptophan or auxin metabolism mutants have served as important tools for understanding IAA metabolism. This figure illustrates some of the mutants that have enhanced our knowledge of IAA conjugate hydrolysis, the redundancy of IAA biosynthesis, and the complexity of IAA homeostasis. Shown are: [Top row] Arabidopsis mutants resistant to high levels of IAA-alanine were selected by failure of the conjugate to suppress root growth. These mutants identified genes involved in IAA conjugate hydrolysis (3). [Middle row, left] Kernels of maize *orange pericarp* (*orp*) segregate 15:1 as expected for a recessive trait encoded the two genes for tryptophan synthase β in maize. Seedlings from *orp* fluoresce brightly under UV illumination due to the accumulation of the glucoside of anthranilic acid, while normal seedlings do not. The *orp* mutant demonstrated the existence of a Trp-I pathway for IAA biosynthesis (56). [Middle row, right] Selection of lines of *Lemna gibba* resistant to α-methyltryptophan yielded plants that produced elevated levels of tryptophan due to loss of feedback inhibition of anthranilate synthase. These lines were used to show that the rate of IAA production/turnover could change without altering the steady state levels of IAA (29). [Bottom row] A large variant, jsR$_1$, of *Lemna gibba* was shown to have high levels of free IAA, and provided the first quantitative data indicating that a Trp-I pathway to IAA was active in plants (56).

larger than that of IAA. Trp is readily converted to IAA by microorganisms and even a minute amount of microbial conversion of Trp would produce the picomolar amounts of IAA commonly found in plants. Radiolabeled Trp can be non-enzymatically converted to IAA, presumably by decomposition mediated by peroxides and free radicals that accumulate in the radioactive solution. In addition, many of the early studies demonstrating Trp-D-IAA synthesis employed cut sections (39) and we now know that wounding induces Trp-D-IAA synthesis (61). Therefore, many of the earlier studies with tissue sections probably overestimated the contribution of Trp-D-IAA synthesis. Stable isotope labeling studies performed with intact plants have confirmed the existence and utilization of Trp-D-IAA synthetic pathways in a variety of species (29, 56). Trp-D-IAA synthesis is also induced by environmental factors such as light and temperature (49). While a direct correlation to Trp-D-IAA synthesis has yet to be made, there are several

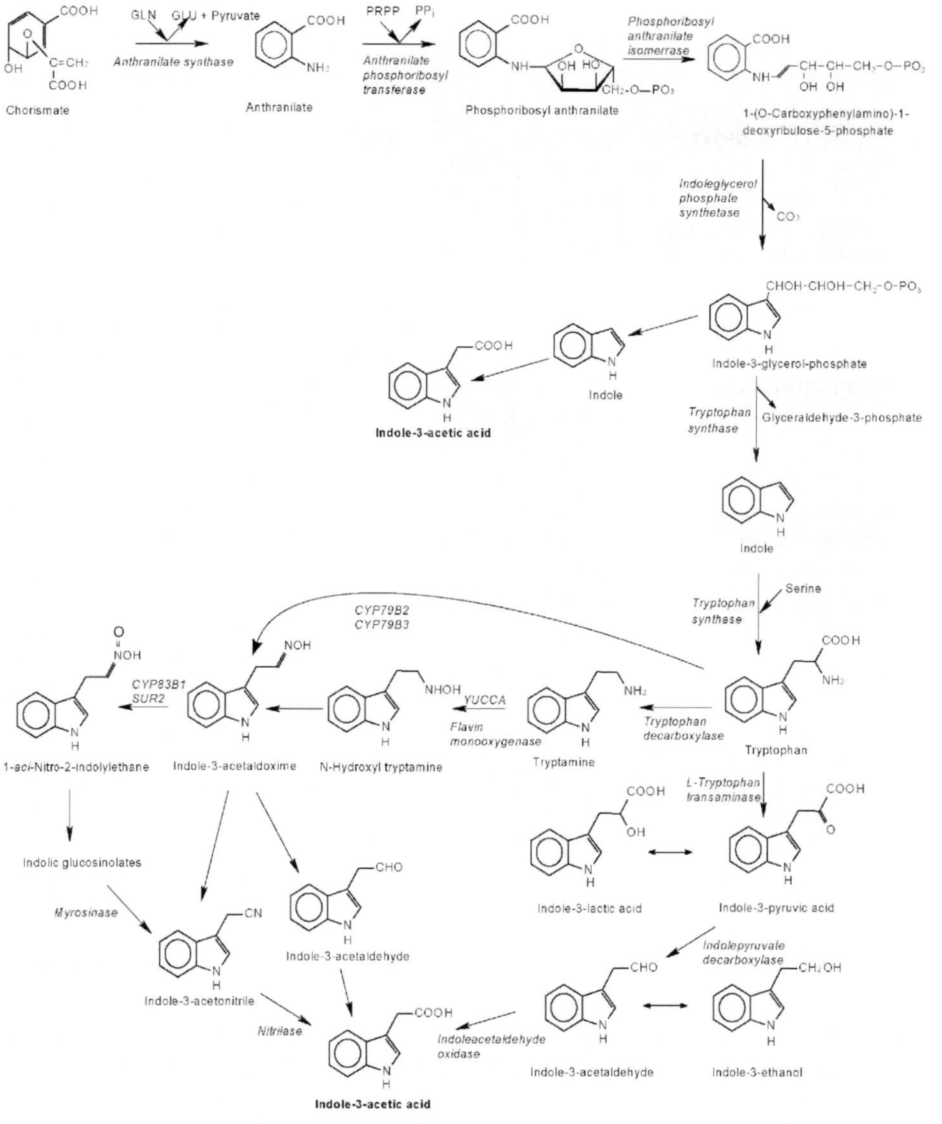

Figure 5. Summary of the reactions leading from chorismate (the first committed step of indolic metabolism) to IAA and tryptophan. In maize, the production of free indole, independent of tryptophan synthase β, has been shown to by catalyzed by two genes, *BX1* and *IGL* (17). In addition to YUCCA, the gene FLOOZY from petunia has been shown to encode a similar flavin monooxygenase (12, 64). As noted in the text, indolic glucosinolates are present only in a limited number of plant species and thus reactions involving these compounds are likely different in other plants. In fact, most of what we know on this scheme is derived from studies of only a small number of plant species.

examples of surges in auxin levels that correspond with developmental or environmental stimuli. Such surges follow zygotic fertilization (51) and changes in temperature (29). IAA metabolism is also perturbed by pathogen infection (40).

Identification of genes involved in Trp-D-IAA synthesis
Most of the Trp-D-IAA synthesis pathways shown in Fig. 5 were postulated as many as 55 years ago, yet identification of genes that encode enzymes responsible for some of the reactions has only come about within the past decade. (see Fig. 5). Genes encoding nitrilases that are able to convert IAN to IAA *in vitro* were isolated in a screen that was completely unrelated to IAA biosynthesis, and their identification was based on homology to microbial nitrilases (3). There are 4 nitrilase genes in Arabidopsis with varying degrees of affinity for IAN as a substrate (3, 48). Demonstrating a precise role for nitrilases in IAA synthesis is complicated by the redundancy of the genes. The *nit1* mutant of Arabidopsis is resistant to high levels of exogenous IAN, because the NIT1 protein is not available to convert IAN to IAA (41). However, IAA levels are normal in this mutant. The challenge in using genetic approaches to assign function to a member of a gene family is to discover those conditions under which gene redundancy does not provide adequate compensation for a defect in one member of a gene family. For example, Arabidopsis NIT3 may in fact be specifically involved in IAA synthesis in cases of sulfur starvation, as this nitrilase gene is the only one that is up-regulated under such conditions (27). In another case, nitrilase activity is up-regulated during pathogen attack, specifically by infection of Arabidopsis by *Plasmodiophora brassicae*, the causal agent of clubroot disease (40, 41). It remains to be seen just how ubiquitous the nitrilase pathway is, as IGs, which are presumed to be the likely precursors to IAN, are limited in the plant kingdom to the order Capparales (54). Tobacco and maize both encode nitrilases, but IAN levels are not believed to be significant in these species.

Genes have been identified for steps in the metabolism of Trp to IAOx (7), which in Arabidopsis is a branch point compound between IAA and IG synthesis pathways (12, 69, see Fig. 5). Several of the genes involved in Trp metabolism to IAA or IG encode enzymes that belong to the cytochrome P450 family of monooxygenases, a large group of heme-containing enzymes that primarily catalyze hydroxylation reactions. P450 enzymes are involved in biosynthesis of a number of plant hormones (26) and secondary metabolites such as the glucosinolates, which are a class of amino acid-derived plant defense compounds with anti-cancer properties (67).

Mutant analysis has revealed cross talk between IAA synthesis and IG synthesis. Specifically, altered flux through the IG pathway perturbs IAA levels (12), which is not surprising given that the two pathways share a common intermediate. What is surprising it that the biosynthetic pathways for glucosinolates derived from amino acids other than Trp can also feed into IAA homeostasis pathways in a more global regulatory network (8, 12, 21, 50). For example, *CYP79F1* is involved in synthesis of short-chain methionine-derived aliphatic glucosinolates. The bushy phenotype of the *CYP79F1* null mutant *(bus1-1f)* is indicative of a perturbation in IAA homeostasis and indeed IAA, IAN and IG levels are altered in this mutant (50). Mutations in *CYP83B1*, which directs IAOx toward IG synthesis, result in elevated IAA and IAN levels (12) as well as increased transcription of Trp biosynthetic genes and a myb transcription factor gene, ATR1 (57). The physiological role of this transcription factor has yet to be determined; however the corresponding increase in Trp synthetic genes in CYP83B1 mutants could be explained by up-regulation of a positive regulatory factor. Ethylene, jasmonic acid and salicylic acid, elicitors of defense signaling pathways, stimulate increased CYP79B2 and CYP79B3 expression along with IG accumulation (6, 33), while non-indole glucosinolates were not affected. IAA levels have not been measured under conditions of elicitation, so the degree to which IAOx is directed between IG and IAA synthesis pathways in a defense response is not yet known.

At least one of the flavin monooxygenase-like YUCCA proteins of Arabidopsis is able to convert Trm to N-hydroxyl Trm *in vitro*, and overexpression of three of the nine related YUCCA genes in Arabidopsis results in over-production of IAA *in planta* (12, 68). Thus, members of this gene family are implicated in Trp-D-IAA synthesis, and by a novel pathway that had not been postulated from earlier biochemical approaches (Fig. 5). In a screen of activation tagged Arabidopsis mutants, the YUCCA mutants stood out as having phenotypes consistent with altered IAA levels (68). Knockouts of individual YUCCA genes were not affected in IAA homeostasis, which is not surprising given the redundancy of this gene family and of IAA synthesis pathways in general. Conventional loss of function genetic approaches are not likely to reveal redundant pathways, while overexpression approaches apparently do. The *FZY* gene of petunia appears to be an ortholog of the YUCCA family, and the *fzy* mutant exhibits blockage in the early stages of floral organ primordia formation and aberrant leaf vein patterning (12, 64). While *FZY* appears to be a single gene in petunia, auxin levels are normal in the *fzy* mutant, indicating that if these genes normally play a role in IAA synthesis, there are alternate pathways available.

De Novo Synthesis of IAA Not Involving Tryptophan

Trp-I-IAA synthesis
Since the first studies in the 1940s showing that plants have the capacity to convert exogenous Trp to IAA, many have equated IAA synthesis from Trp with *de novo* biosynthesis. Although doubts about this general concept have been raised since the early 1950s (29), only more recently have the methods for exacting studies on this question been developed. Two advances have been critical to the demonstration that plants have one or more Trp–I-IAA synthesis routes; the use of stable isotope tracer methods with mass spectrometric analysis and the availability of Trp auxotrophic mutants of higher plants (Fig 4). These methods have also resulted in critical reevaluations of when in a plant's life IAA is made *de novo*, and these studies have yielded surprising results.

The measurement of *de novo* synthesis of aromatic ring compounds can be followed by allowing plants to grow in the presence of water enriched in deuterium oxide. Under these conditions, any newly formed aromatic rings have deuterium locked into non-exchangeable positions on the ring. Such labeling techniques provide at least two advantages. First, since the "labeled precursor" is water, such an approach does not require exact knowledge of precursors or pathways in order to accurately ascertain the extent of *de novo* synthesis. Second, since all cell compartments are freely permeable to water, problems of compartmentation and uptake are not an issue. Experiments from several laboratories in which young plants of *Zea mays*, Arabidopsis, or pea, and cell cultures of carrot were grown on 30% deuterium oxide clearly showed that IAA is made in such a way that deuterium is incorporated into non-exchangeable positions of the indole ring of IAA to a greater extent than that found in the indole ring of Trp. These results indicated that Trp is not the only precursor to IAA (2, 42).

A more exacting procedure for isotopic labeling is possible by using specific labeling with stable isotopes of carbon or nitrogen. This allows quantitative evaluations of the rates of metabolism and precise analysis of both the labeled and unlabeled compounds (2, 42). Perhaps the most striking of these isotopic labeling studies used the *orange pericarp* (*orp*) mutant of maize (2, 42 56, see Fig. 4). *Orp* carries a double recessive trait caused by mutations in the two genes encoding the enzyme for Trp synthase β. Despite this metabolic block in the terminal step for Trp biosynthesis, mutant seedlings produced IAA *de novo* and, in fact, accumulated up to 50 times the level of IAA as do non-mutant seedlings. Labeling studies established that the *orp* mutants were able to convert [^{15}N]anthranilate to [^{15}N]IAA but did not convert [^{15}N]anthranilate to [^{15}N]Trp. Neither *orp* seedlings nor control seedlings converted [^{2}H$_5$]Trp to [^{2}H$_5$]IAA in significant amounts even when the orp seedling were fed levels of [^{2}H$_5$]Trp high enough to reverse the lethal effects of the mutation (2, 42, 56). These results were the first to clearly establish that Trp-I-IAA biosynthesis does occur, and they suggested that in these plants, Trp-I-IAA biosynthesis actually predominates over the Trp-D-IAA biosynthesis pathway.

Despite the demonstration that IAA biosynthesis can occur without the amino acid Trp as a precursor, the exact pathway for Trp-I-IAA biosynthesis remains elusive. *In vivo* labeling techniques using Arabidopsis mutants confirmed and extended the *orp* findings, and suggested that the branch point for IAA production occurs at indole or its precursor, indole-3-glycerol phosphate (2, 40, 42). The pathway has also been directly demonstrated in an *in vitro* system using enzyme(s) from light grown seedlings of normal or *orp* maize. This system catalyzed the direct conversion of [^{14}C]indole, but not [^{14}C]Trp, to [^{14}C]IAA, and the addition of unlabeled Trp to the reaction mixture did not alter the rate of conversion of [^{14}C]indole to [^{14}C]IAA (3, 29, 40).

IAA is not the only compound made by plants from precursors that can lead directly to tryptophan. Camalexin, an Arabidopsis defense compound, appears to be produced from indole directly and not through Trp (70). In maize, which does not produce camalexin, a remodeling of indolic metabolism occurs as part of defense responses and two genes, *bx1* and *igl*, produce free indole independent of Trp biosynthesis. The first gene yields the chemical protectant DIMBOA and the second gene responds to a herbivore elicitor (17).

Evidence showing the existence of a Trp-I-IAA biosynthetic pathway must be viewed in the context of compelling evidence that many plant species have been shown to convert Trp to IAA and in some cases at rates that make it important for the auxin economy of the plant. With this in mind, some of the strongest evidence supporting the importance of Trp-I-IAA biosynthesis comes from studies where, in the same plant and/or tissue, a switch occurs from one biosynthetic pathway to the other. In embryogenic carrot suspension cultures, the conversion of Trp to IAA is the primary route (29). However, following induction of these cells to form somatic embryos, the Trp-D-IAA pathway was diminished and the Trp-I-IAA pathway predominated. Similarly, IAA production in expanding green fruit of tomato occurred via Trp-D-IAA routes, but continued IAA production in red-ripe fruit was predominantly by the Trp-I-IAA biosynthesis pathway (16).

The change from one pathway to the other is not only controlled by developmental programs, but is also impacted by external signals. To study these changes, specialized techniques capable of measuring metabolic events over short time periods were required. Early experiments using stable isotope labeling over several days following the surgical removal of the cotyledons showed that in germinating axes of bean seedlings, Trp-D-IAA biosynthesis accounted for essentially all of the IAA production. When reexamined in a study using rapid labeling techniques (61) Trp-D-IAA synthesis was found to be operative only in the first hours after cotyledon removal, followed by several days when the Trp-I-IAA biosynthesis pathway predominated. Clearly then, bean uses both the Trp-I-IAA pathway and the Trp-D-IAA routes, and in this case Trp-D-IAA synthesis is activated as a consequence of the wounding resulting from cotyledon removal.

Growth temperature also impacts regulation of the IAA synthesis pathways (49) and also changes the amount of IAA that accumulates in the plant (29). In *Lemna*, low temperature growth ($\leq 15°C$) resulted in a measured increase in IAA levels, greatly reduced the rate of growth, and initiated the early events leading to the development of resting structures known as turions. Growth at 30°C resulted in normal growth rate and IAA levels (49). When *Lemna gibba* was grown at 30°C in the presence of [^{15}N]anthranilic acid and [^{2}H$_5$]Trp, the IAA pool became enriched in [^{15}N] from [^{15}N]anthranilic acid well before conversion of [^{2}H$_5$]Trp to [^{2}H$_5$]IAA could be observed. These data show that during normal growth at 30°C, the Trp-I-IAA biosynthesis pathway predominates in these plants. However, when *Lemna* plants are grown at 15°C, label from exogenous [^{2}H$_5$]Trp labels the IAA pool at the same time as, and to a greater extent, than label from [^{15}N]anthranilic acid, indicating that the Trp-D-IAA synthesis pathway is primarily active. Thus, the utilization of the Trp-I-IAA and Trp-D-IAA synthesis pathways differ at the two growth temperatures (49). In general, a variety of stress situations appear to induce the Trp-D-IAA biosynthesis system.

CONJUGATE BIOSYNTHESIS

All the higher plants tested, as well as many lower plants, conjugate exogenous IAA, primarily to the amino acid aspartate (11, 13). The ability to make IAA-Asp can be enhanced by pretreatment with IAA, and RNA or protein synthesis inhibitors abolish this induction (2). The production of conjugates from applied IAA proved to be a useful assay to establish a relationship between evolutionary development and the complexity of hormone metabolism in land plants (13). These results, together with measurements of the steady state levels of free IAA and IAA metabolites in developmentally staged axenic tissues, revealed that lower plants use a strategy for regulating free IAA that is dependent on *de novo* synthesis of IAA and degradation of existing hormone. As evolution progressed and plants developed more complex vascularization, a strategy for regulating free IAA levels involving conjugate synthesis and hydrolysis became established (Fig. 6). Temporal and spatial patterns of hormone conjugation and hydrolysis, along with differences in conjugate identity and complexity, may reflect the more precise regulation required for growth and development of diverse or complex plant forms (13).

High molecular weight conjugates account for the majority of IAA in many of the tissues examined so far (Fig. 6). A primarily storage role can be envisioned for the considerable amount of IAA-glucan in maize seeds and the IAA-protein or peptide conjugates in bean seeds. Only 2% of the amide conjugates in Arabidopsis seedlings are represented by IAA-Asp and IAA-Glu, the major low molecular weight conjugates, while the majority of the amide conjugates are small peptides of as yet unknown function (29, 62). Antibody to a 3.6 kDa IAA-peptide conjugate found in bean seed cross-reacts with a larger IAA-modified seed protein, IAP1 (66). Cross-reacting proteins were also found in seeds of many other plants, and were found in bean leaf too. The presence of specific peptides or proteins to which IAA is attached may indicate that higher molecular weight conjugates have significance beyond their role in the modulation of IAA levels.

Developmental Controls of IAA Conjugate Synthesis and Hydrolysis

Conjugation of free IAA in maturing seeds serves as a mechanism for storage of IAA in a form that is readily hydrolyzed to provide hormone to the growing shoot upon germination. This strategy is present in gymnosperms as well as monocots and dicots (3), although the particular ester or amide conjugate moieties that are stored in the seed appear to be division specific. In maize and pine, formation of conjugates resumes when the seedling runs out of stored conjugates and the seedling is capable of *de novo* IAA synthesis

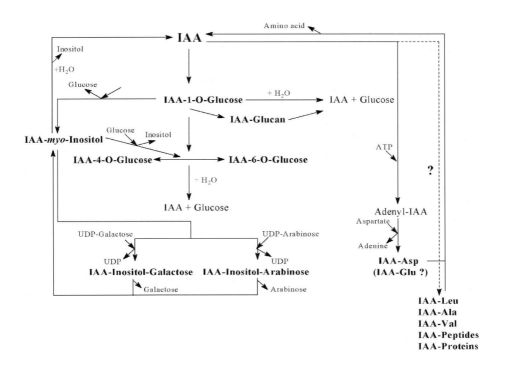

Figure 6. Metabolic transformations between IAA and conjugate pools are shown. Reactions shown have mostly been derived from studies of maize, Arabidopsis and bean, although there is ample evidence for many of these reactions occurring in a diverse array of plant materials (29, 56). The biosynthetic routes, hydrolysis reactions and known inter-conversions are shown. Transacylation from IAA-Inos back to glucose to form the 4-O or 6-O-IAA-gluc are reversible reactions that can lead to the subsequent hydrolysis of either form. The biosynthesis of IAA-Asp by formation of the adenyl-IAA intermediate following auxin induction of the process now appears to be established (58): however, it remains to be determined if the biosynthesis of other IAA-amino acids and higher molecular weight amide conjugates proceed via a similar reaction scheme.

(2, 56). In bean, conjugated forms decrease dramatically only during the first few hours following imbibition (56, 66). *De novo* IAA and IAA-peptide biosynthesis begins in the growing axes while conjugate hydrolysis is still ongoing in the cotyledonary tissue (3, 56, 61, 66). Within a given plant, the type of conjugate formed may be developmentally specific: in pine, ester conjugates predominate in the seed while IAA-Asp, an intermediate in IAA catabolism, is made in the growing shoot (2, 56).

The best-studied systems pertaining to developmental regulation of IAA conjugate formation are the germinating maize kernel (Fig. 7), the germinating pine seed, and developing and germinating bean seed (2, 29, 56). The ester conjugate IAA-Inos is one of the major conjugates in both the kernel and the shoot of maize (2, 56). [^{3}H]IAA-Inos applied to the kernel is transported to the shoot and is, in part, hydrolyzed to yield free [^{3}H]IAA (2, 56). [^{3}H]IAA-Inos-[^{14}C]galactoside applied to the endosperm is also

Auxin biosynthesis and metabolism

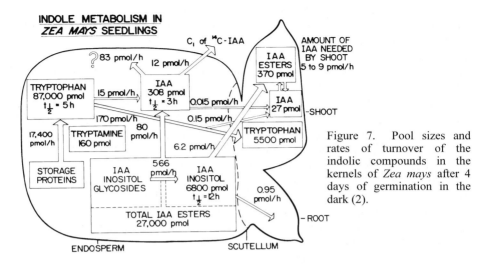

Figure 7. Pool sizes and rates of turnover of the indolic compounds in the kernels of *Zea mays* after 4 days of germination in the dark (2).

transported to the shoot. In these experiments, the double-labeled compound was used to show that hydrolysis of the galactose from the conjugate occurred after the conjugate left the endosperm but before it entered the shoot, presumably in the scutellum. Studies with [^{3}H]IAA-Inos indicated that between 2 to 6 pmol/hr of IAA (Fig . 7) can be supplied to a maize shoot from the kernel and this would be a major source of the estimated 10 pmol/hr required by each shoot (2, 56). As in maize, ester conjugates form the majority of conjugate in the pine seed, and their hydrolysis results in a dramatic increase in free IAA within 48h of imbibition (2, 56). In bean, IAA amide conjugates accumulate throughout seed development, with the major IAA-protein conjugate, IAP1, accumulating late during seed development and rapidly disappearing from both axes and cotyledons during the first day of imbibition (29, 66). Carrot cell cultures differentiating into somatic embryos exhibit a dramatic increase in the rate of conjugate utilization. Order of magnitude decreases in the amide conjugate pool occur over a period of only a few days and this increase in conjugate utilization correlates with a 3-4 times increase in the activity of an IAA-amino acid hydrolase (2, 56).

Biosynthesis of IAA Ester Conjugates

Early biochemical studies in *Zea mays* showed that plants make the IAA-*myo*-inositol family of conjugates by first synthesizing 1-*0*-D-IAA-Gluc from IAA and UDP Gluc, then transferring the IAA to inositol by transacylation (56). IAA-Inos may then be glycosylated to form IAA-Inos-galactoside or IAA-Inos-arabinoside by reaction with the appropriate uridine diphosphosugar. All of the low molecular weight IAA conjugates of *Zea mays* have been shown to be synthesized *in vitro* using enzyme extracts at different levels of purification (56).

A 50.6 kDa UDP-glucosyl transferase catalyzing the formation of IAA-Gluc was purified and characterized from maize, and antibodies raised to the protein were used to select a cDNA clone for the transferase from a maize library (56). Transgenic tomato seedlings expressing the antisense maize sequence had, as might be expected, more roots as well as reduced levels of IAA-Gluc and esterified IAA (22). A 75 kDa tomato protein that cross reacted with antiserum to the maize IAGLU protein was undetectable in these antisense tomato seedlings. By contrast, overexpressing the maize gene in Arabidopsis resulted in a phenotype (diminished root growth and small and/or curly leaves) consistent with reduced free IAA levels. Unexpectedly, the measured IAA levels in these plants were comparable to wild type even though IAA-Gluc levels were increased (J. Ludwig-Muller, personal communication). Arabidopsis UDP glucosyl transferase was identified by an impressive brute force approach combining genomics and biochemistry (23). Putative glucosyltransferase genes encoding enzymes capable of forming glucosyl esters were identified in the Arabidopsis genome. The clones for these genes were expressed in *E. coli* and the products were assayed for activity with a variety of substrates. One clone, UGT84B1, produced the correct products, including the 2-0-, 4-0-, and 6-0-isomers of IAA-Gluc, as identified by mass spectrometry. The expressed protein could use indole proprionic acid and IBA as substrates, but competition studies showed that IAA was preferred. Both IAA and IBA glucose conjugates are formed in plants (62). RT-PCR analyses showed that UGT84B1 is primarily expressed in the siliques, but low expression was detected in inflorescences and roots.

IAA-Gluc is also the precursor to the high molecular weight glucans that form the majority of IAA ester conjugates in the maize kernel (2, 56). In addition, IAA-Gluc undergoes non-enzymatic conversion to 4- and 6-*0*-IAA-Gluc, and enzymatic conversion to di-0-IAA-Gluc in the kernel (2, 56).

Biosynthesis of IAA Amide Conjugates

Mysteriously, and despite numerous attempts, an *in vitro* enzymatic system for synthesis of IAA amide conjugates has never been successfully reconstituted from plants. IAA-Gluc was suggested as a precursor for IAA amide conjugates in plants (23), however transgenic plants overexpressing IAA glucosyl transferase genes either did not accumulate higher levels of amide conjugates or exhibited reduced levels of IAA-Asp or IAA-Glu (22 and J. Ludwig-Muller, personal communication). The recent discovery that at least seven members of the JAR1 family of genes in Arabidopsis were capable of adenylating IAA pointed to a likely pathway for IAA amide conjugation (58). JAR1 proteins belong to the acyl adenylate-forming firefly luciferase superfamily and activate substrates for subsequent biochemical modification by adding AMP to their carboxyl group. One of the Arabidopsis JAR1 family members shows homology to a soybean early auxin inducible protein, GH3, that had been previously described as being involved in auxin signal transduction based on the phenotype of an activation tagged mutant (35). While it is therefore likely that the mechanism by which inducible IAA-Asp formation proceeds involves adenylation of IAA by a JAR1 family member, we cannot be sure the same mechanism is employed for lower level, constitutive, amide conjugate formation. The redundancy of JAR1-like genes is a possible reason for failure to associate phenotypes with defects in IAA amide conjugate synthesis in the past; however we are now in a position to test the role(s) of these genes in IAA conjugation and in auxin responses.

Hydrolysis of IAA-Conjugates

In 1935, Cholodny (11) showed that application of a water-moistened piece of endosperm to a seedling led to an auxin response whereas alcohol moistened endosperm elicited no response. Thus, Cholodny was the first person to observe enzyme-catalyzed hydrolysis of the IAA conjugates stored in the endosperm. A few years later, Skoog (11) made the important observation that a "seed auxin precursor" moved from the seed to the shoot and that this precursor could be converted by the plant to an active auxin. Extraction of corn shoots or roots with ether for 3 hr at 4°C yielded IAA whereas an extraction with 80% ethanol yielded no IAA. Cold ether does not, in general, inactivate enzymes whereas 80% ethanol does. From these results it was concluded that the tissue was autolyzing in the wet ether, with consequent enzymatic hydrolysis of IAA conjugates to yield free IAA (11). Subsequent work established that the "bound" IAA of the seed could also be released by mild alkaline hydrolysis (11), and that the IAA was not the product of degradation of Trp or storage proteins.

While conjugate hydrolysis in ether-induced autolyzing tissues occurs fairly readily, it has been difficult to obtain *in vitro* hydrolysis of IAA conjugates with purified enzymes. Many commercial proteases and esterases that might be expected to hydrolyze IAA-amino acid conjugates fail to do so (11). *In vitro* hydrolysis has been reported (2, 56), but the hydrolytic enzymes have proven difficult to extract and purify. IAA-inositol hydrolase activity, as well as the co-fractionation of IAGlu synthetase and two enzymes for IAGlu hydrolysis, a 1-*O*-IAGluc hydrolase and a 6-*O*-IAGluc hydrolase, have been reported from maize (2, 56). IAA-Gluc hydrolases have also been found in other plants, including oat, potato and bean (2, 56).

Recently, a genetic approach was used to identify genes involved in hydrolyzing IAA conjugates. The selection based on resistance to the inhibition of root growth by high levels of the IAA conjugates. Those candidates that were resistant to the inhibitory effects of IAA conjugates, but still sensitive to the inhibitory effects of excess IAA were subjected to positional cloning. In this manner, a gene family encoding amidohydrolases was identified. Nineteen IAA-L-amino acid conjugates were tested against the amidohydrolases encoded by this gene family and differing specificities were identified (3, 28). The existence of amidohydrolases with differing specificities gave impetus to determining the identities of all the IAA conjugates in Arabidopsis (29), and indicates that the less abundant conjugates may play as yet unknown roles in auxin regulation.

Not all the Arabidopsis conjugate resistant mutants identified to date have corresponded to genes encoding amidohydrolases 3). Therefore, these mutants could be useful in determining if the activities of conjugates are solely due to their hydrolysis with release free IAA. Thus, they may ultimately prove to be pivotal in revealing any conjugate functions that do not involve hydrolysis.

The role of IAA-Asp as an intermediate in the catabolism of IAA in some plants (Fig.8) has led to IAA-Asp formation being referred to as an "irreversible" conversion process. This may indeed be true in some plant tissues, and under some conditions. However, a subset of the Arabidopsis amide conjugate hydrolase family has low-level IAA-Asp hydrolase activity (28). In soybean, which stores IAA-Asp and IAA-Glu in its seeds, the

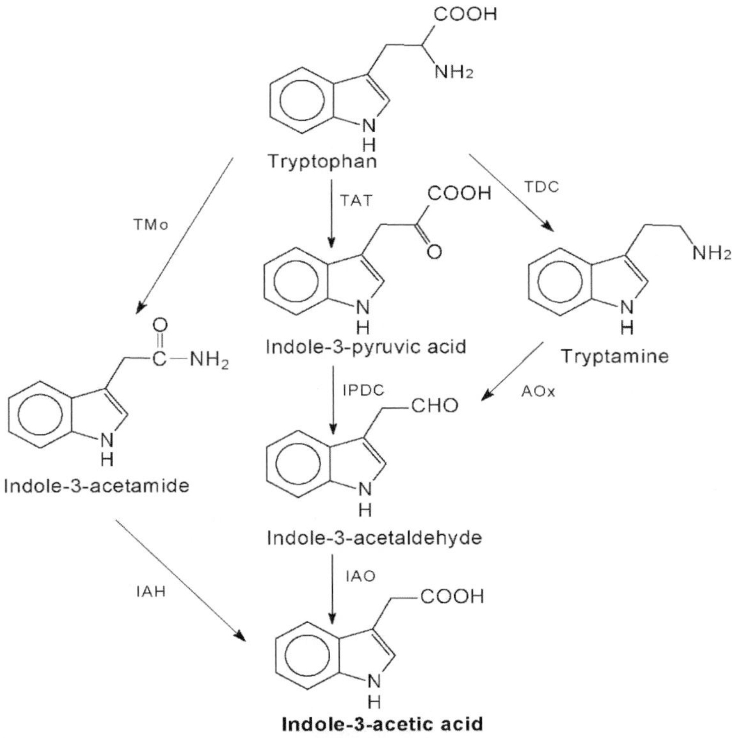

Figure 8. Pathways for the microbial biosynthesis of IAA from tryptophan. Although a number of different enzymes have been proposed for several of the steps, representative ones that have been described are labeled: Tmo: Trp monooxygenase; IAH: indoleacetamide hydrolase; TAT: Trp aminotransferase; IPDC: indole-3-pyruvic acid decarboxylase; IAO: indole-3-acetaldehyde oxidase; IAO; TDC: Trp decarboxylase; Aox: amine oxidase. Little attention has been focused on the Trm pathway in bacteria, although the enzymes are present and bacteria can both make Trm and convert it to IAA added to the culture medium (18).

hypocotyls respond to IAA-L-Asp, but not to IAA-D-Asp, by cell elongation; the growth-inducing activity of IAA-Asp is significant, about 70% of that seen with IAA (2, 56). IAA-Asp is also hydrolyzed, albeit slowly, by bean stems (2, 56).

Microbial Pathways for IAA Biosynthesis

IAA production by plant-associated bacteria has been an important aspect of research on IAA metabolism. It is estimated that as much as 80% of all soil bacteria have the capacity to make IAA, yet this function is generally not essential (18). IAA producing bacteria can be either phytopathogenic or beneficial to plant growth, so the role of IAA production by the microbe isn't completely clear. Wide diversity exists in the specific interactions between plants and bacteria. Within a given bacterial species, there are many variations in metabolic capacity between strains, suggesting a myriad of regulatory mechanisms (18).

The IAA biosynthetic genes encoded by the crown gall forming bacteria *Agrobacterium tumefaciens* have proven very useful in refining hypotheses about auxin action in plants (25). *Agrobacterium tumefaciens* transfers genetic material (a fragment, termed T-DNA, of the tumor inducing Ti plasmid) from the infecting bacteria into the host plant. Encoded within the transferred DNA are genes for the enzymes Trp monooxygenase and indoleacetamide hydrolase (18), which carry out the conversion of Trp to indoleacetamide and the hydrolysis of the indoleacetamide to IAA by the infected plant cell (Fig. 8). Ectopic overexpression of these genes in a plant independent of other Ti plasmid genes that induce tumor formation, has been useful in characterizing the effect of increased IAA production *in planta*. In addition to confirming many of the observations that were made in classic hormone application studies (25), stable isotope labeling studies in such lines have revealed cross talk between IAA and cytokinin homeostasis pathways (15).

Another gall forming bacterium, *Pseudomonas syringae* pv. *Savastanoi*, uses the identical pathway for IAA production (2); however, in this case no genetic material is transferred and the bacteria themselves produce high levels of IAA. Pathogenic *Pseudomonas* also have the capacity to form the novel conjugate of IAA, IAA-Lys, as well as its α-N-acetyl derivative (2). Although IAA-Lys formation reduces the pool of free IAA produced by the bacteria by about 30%, the role of these conjugates in gall formation has not been established.

Nonpathogenic, plant-associated bacteria that inhabit the rhizosphere appear to enhance plant growth by a variety of means, including secretion of IAA into the rhizosphere. Mutations in the *Pseudomonas putida* gene encoding indole pyruvate decarboxylase, *ipdc* (Fig. 8), render the strain unable to stimulate root growth compared to wild type strains (46). Some strains of *Erwinia herbicola* cause crown and root galls on *Gypsophila paniculata* (2). The strains of *Erwinia* that are pathogenic and form such galls have the capacity to make most of their IAA by the same indoleacetamide pathway as *Pseudomonas* and *Agrobacterium* T-DNA transformed plant cells. Mutation analysis indicated that the capacity for gall formation requires the indoleacetamide pathway (18). *Erwinia* mutants defective in the indoleacetamide pathway were unable to form galls, but were still capable of producing IAA via the indole-3-pyruvic acid pathway, albeit at lower levels (18). The *ipdc* promoter in *Pseudomonas putida* strains appears to be regulated by both Trp and the stationary phase sigma factor RpoS (47) while the indoleacetamide pathway appears to be constitutive in most systems examined (18). Overexpression of rpoS in *Enterobacter* strains results in accumulation of IAA and enhanced root growth when the strain is incubated with seedlings (47).

Some soil bacteria have the capacity to hydrolyze IAA conjugates from plants (9). Indeed, the gene for a specific IAA-Asp hydrolase (9) has provided a useful tool for studies of the compound's role *in planta* (63). Many of the nonpathogenic strains of IAA-producing bacteria are saprophytic epiphytes that are widespread in nature. Their presence in plants grown in non-sterile conditions is a potential complication for IAA biosynthesis studies because bacterial IAA production by the endophyte could be considerably greater than plant IAA production rates.

Recently, cyanobacteria, the photosynthetic bacteria that are the progenitors of chloroplasts, have been shown to produce IAA by a Trp-D-IAA synthesis pathway, probably via indole-3-pyruvic acid (Figs. 4 and 8) (55). In higher plants, Trp biosynthetic enzymes are chloroplast localized, as are some enzymes in Trp-D IAA synthesis pathways (e.g. CYP79B2 and CYP79B3) (12).

IAA CATABOLISM

Oxidative catabolism of IAA is generally considered to be the chemical modification of the indole nucleus or side chain that results in loss of auxin activity. As far as we know, it is the only truly irreversible step regulating IAA levels. Studies of IAA catabolism predate even the identification of IAA as a ubiquitous auxin in plants, and reports of peroxidase enzymes that catalyze the oxidative decarboxylation of IAA are numerous (2). For many years this reaction was generally assumed to be the physiological route for oxidative catabolism. Although a copious amount of information is available on the peroxidation of IAA *in vitro*, the products formed are not usually found in plants in significant amounts. Also, the rate of peroxidation of IAA applied to plants, as measured by CO_2 evolution, is usually only a small fraction of the measured rate of IAA turnover. These same problems, low rates of CO_2 evolution and failure to identify the oxidative products in normal plant materials, also apply to recent studies of ascorbic acid oxidase catabolism of IAA in maize roots (24). These considerations, coupled with studies of plants over and under expressing specific peroxidases (39), suggest

Auxin biosynthesis and metabolism

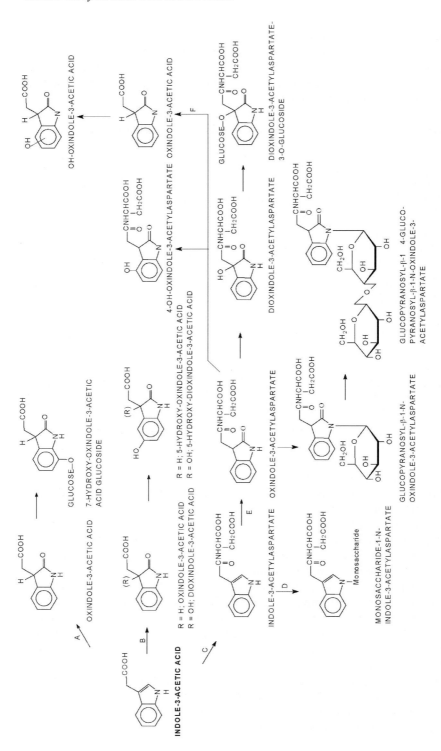

Figure 9. Oxidation routes for the catabolism of IAA known to occur in plants, based on the identification of the oxidation products from untreated plant materials as well as from isotopic labeling studies (39). In some cases, the responsible enzyme(s) have also been partially purified (2). Each of these pathways have been primarily studied in only one or a few species: A: *Zea mays*; B: *Oryza sativa*; C: most plants, including *Zea mays*, *Pisum sativum* and Arabidopsis; D: *Pinus sylvestris*; E: *Vicia faba, Dalbergia dolechopetala*; F: *Lycopersicon esculentum*.

that this is not likely a major route for IAA catabolism *in vivo* under most conditions. More recently a new pathway of IAA catabolism, non-decarboxylative oxidation with retention of the carboxyl side chain and oxidation of the indole nucleus, has been identified (2, 39) and shown to occur in a number of plant species.

The Oxindole-3-Acetic Acid/Dioxindole-3-Acetic Acid Pathway

The direct ring oxidation pathway for IAA catabolism has been the subject of detailed study in several plant species including rice, corn, and broad bean (29). Based on simple colorimetric assays, the first product in this pathway (Fig. 9), OxIAA, also occurs in germinating seeds of *Brassica rapa*, and developing seeds of *Ribes rubrum* (2). *Oryza sativa*, rice, is interesting in that it is the only plant known to contain both OxIAA and DiOxIAA (2), and was also shown to have the 5-hydroxy analogs of OxIAA and DiOxIAA. OxIAA has been shown not to be an intermediate or a substrate for the peroxidase pathway (2), so the pathways are independent.

The first report of IAA oxidation to oxindole-3-acetic acid (OxIAA) was in the basidiomycete *Hygrophorus conicus* (2). OxIAA and DiOxIAA were later found to be synthesized by *Zea mays* and *Vicia faba*, respectively, following feeding of 1-[^{14}C]-IAA (2). Isotope dilution experiments (2) showed that OxIAA was a naturally occurring compound in *Zea mays* endosperm and shoot tissues, occurring in amounts of 357 pmol per endosperm and 47 pmol per shoot, about the level of free IAA in these tissues. In *Vicia faba*, DiOxIAA was estimated by UV measurements to be 1 µM in roots (2).

OxIAA is further metabolized in maize by hydroxylation at the 7 position, and by glucose addition to form 7-OH-OxIAA-glucoside (2). Isotope dilution assays have shown that the 7-OH-OxIAA-glucoside is a naturally occurring compound in corn in amounts of 62 pmol per shoot and 4,800 pmol per endosperm. *Vicia faba* is also reported to form a glucose derivative of DiOx-IAA-aspartate (2). 7-OH-OxIAA has also been identified as a catabolite of IAA in germinating kernels of *Zea mays*. It was found to be present at 3,100 pmol in endosperm and has been shown to be an intermediate in the synthesis of 7-OH-OxIAA-glucoside (2). Since the glucoside is present in much higher amounts than IAA, OxIAA, and somewhat higher than 7-OH-OxIAA, it has been speculated that it might be the form that accumulates in vacuoles (2). The further metabolism of 7-OH-OxIAA and 7-OH-OxIAA-glucoside has not been studied except that 5-[^{3}H]-7-OH-OxIAA loses tritium to water upon further enzymatic oxidation (2). This implies a second oxidation of the benzenoid ring leading to a highly unstable dioxindole.

Much less is known about the enzymology of the OxIAA and DiOxIAA pathways than has accumulated about the action of peroxidase on IAA. This is due to its more recent discovery, the limited availability of enzymes and substrates, and the difficulty of establishing robust *in vitro* assays. In *Zea mays*, the rate of oxidation of IAA to OxIAA has been measured in shoot, root, scutellar, and endosperm tissues at 1-10 pmol h^{-1} mg protein^{-1}. The enzyme that carries out this reaction is soluble, of high molecular weight and clearly different from lipoxygenase or peroxidase (2). Enzyme activity was reduced by 90 per cent when assayed under argon, indicating an oxygen requirement. Enzyme activity was stimulated up to ten fold by addition of an ionic detergent extract of corn tissue. A heat stable lipophylic component of these extracts was identified as the factor that increased enzyme activity when added to buffer extracted enzyme. This heat stable factor could be replaced by linolenic, linoleic, or arachidonic acid (2), which could be acting as co-substrates.

IAA-Asp Oxidation

A specific route of IAA-conjugate oxidation was demonstrated in work that showed that in *Vicia* seedlings, IAA-Asp is oxidized to di-Ox-IAA-aspartate without prior hydrolysis to the free acid (2). As with IAA oxidation, once IAA-Asp oxidation occurs, the product is glycosylated to form the 3-(*O*-β-glucosyl) derivative. IAA-Asp could be oxidized by peroxidase, but onl when peroxide was added to the reaction mixture (2). The product of this oxidation was 2-OH-Ox-IAAsp, thus, the reaction with peroxidase and H_2O_2 yields a different product from that isolated from the plant. In *Populus*, IAAsp, OxIAAsp and ring hydroxylated products of OxIAAsp are the major products formed after feeding with IAA (2). In Scots pine, IAA is oxidized to OxIAA as well as conjugated to IAAsp. These compounds are further metabolized to glucopyranosyl-1-N-IAAsp and glucosyl-1-N-IAA. Several other species have been studied (2, 29) and, in general, two basic strategies appear to be followed with individual differences dependent on species: IAA is either oxidized and glycosylated directly, or these processes are carried out following conjugation with aspartate.

SUMMARY

Through the marriage of molecular genetics to highly sensitive analytical methods, a wealth of information about IAA homeostasis pathways has recently been uncovered. Redundancy and plasticity are becoming the two overriding themes to the network of pathways that determine how much auxin is present in a plant at any given time (12, 40). There are still more players to be identified, from the enzymes that carry out the individual steps in a pathway to the regulatory components that render these pathways so responsive. Genome level approaches should yield more new information in this regard and, activation tagging[2] is a new method that holds promise for

---

[2] Activation tagging refers to transgenic plants in which a DNA sequence that activates transcription of downstream sequences has been randomly inserted into the genome, thus randomly activating nearby genes. Mutant phenotypes may result from the activation of said gene and can provide clues to the function of these genes.

identifying genes in families that would be masked in a conventional genetic screen. Rapid and ultrasensitive assays for high throughput auxin analyses are in the pipeline, and should greatly facilitate genetic discoveries.

It is now well established that a plant's auxin metabolic response is intricately tied to other pathways that respond to a wide variety of signals, including temperature, light, jasmonic acid, other plant hormones and second messengers. The molecular intermediaries in these interactions are just beginning to be discovered and protein phosphorylation is emerging as a central component (59). Genetic approaches and global profiling strategies will be particularly useful in assessing the effect of mutations in one pathway upon interacting pathways, both during normal plant growth and development as well as during periods of pathogen infection or abiotic stress.

Acknowledgements

We thank Bonnie Bartel and Allen Wright for providing photographic images used as part of Fig. 4. Work in the laboratories of the authors was funded by grants from the US National Science Foundation (DBI0077769, IBN0111530), US Department of Energy (DE-FG02-00ER15079), US Department of Agriculture National Research Initiative Competitive Grants Program (USDA-2002-03555), and by funds from the Gordon and Margaret Bailey Endowment for Environmental Horticulture, the Minnesota Agricultural Experiment Station and the USDA Agricultural Research Service.

References

1. Aharoni N, Cohen JD (1986) Identification of IAA conjugates from IAA-treated tobacco leaves and their role in the induction of ethylene. Plant Physiol 80(S):34
2. Bandurski R, Cohen J, Slovin J, and Reinecke D (1995) Auxin biosynthesis and metabolism, in Plant Hormones: Physiology, Biochemistry and Molecular Biology, P Davies, Editor, Kluwer Academic Publishers: Dordrecht, Boston, London. ISBN 0-7923-2984-8. p. 39-65
3. Bartel B, LeClere S, Magidin M, Zolman BK (2001) Inputs to the active indole-3-acetic acid pool: *de novo* synthesis, conjugate hydrolysis and indole-3-butyric acid β-oxidation. J. Plant Growth Regulation. 20:198-216
4. Benfey P (2002) Auxin action: Slogging out of the swamp. Current Biology 12: R389-R390.
5. Berleth T, Mattsson J, Hardtke C (2000) Vascular continuity and auxin signals. Trends in Plant Science 5:387-393
6. Brader G, Tas E, Palva E, (2001) Jasmonate-dependent induction of indole glucosinolates in Arabidopsis by culture filtrates of the nonspecific pathogen Erwinia cartovora. Plant Physiol 126:849-860
7. Celenza J, (2001) Metabolism of tyrosine and tryptophan-new genes for old pathways. Current Opinion in Plant Biology 4:234-240
8. Chen S and Glawischnig E (2003) CYP79F1 and CYP79F2 have distinct functions in the biosynthesis of aliphatic glucosinolates in Arabidopsis. Plant J 33:923-937
9. Chou J-C, Mulbry WW, Cohen JD (1998) The gene for indole-3-acetyl-L-aspartic acid hydrolase from *Enterobacter agglomerans*: molecular cloning, nucleotide sequence and expression in *Escherichia coli*. Molecular and General Genetics 259:172-178
10. Cohen JD, Bialek K (1984) The biosynthesis of indole-3-acetic acid in higher plants. In: A. Crozier and J.R. Hillman, eds., The biosynthesis and metabolism of plant hormones. Society for Experimental Biology Seminar 23. Cambridge Univ. Press pp. 165-181
11. Cohen JD, Bandurski, RS (1982) Chemistry and physiology of the bound auxins. Annu

Rev Plant Physiol 33:403-430
12. Cohen JD, Slovin JP, Hendrickson A (2003) Two genetically discrete pathways convert tryptophan to auxin: more redundancy in auxin biosynthesis. Trends in Plant Science 8:197-199
13. Cooke TJ, Poli DB, Sztein AE, Cohen, JD (2002) Evolutionary patterns in auxin action. Plant Molecular Biology 49:319-338
14. Doerner P (2000) Root patterning: Does auxin provide positional clues? Current Biology 10:R201-R203
15. Eklöf S, Åstot C, Sitbon F, Moritz T, Olsson O, and Sandberg G (2000) Transgenic tobacco plants co-expressing Agrobacterium iaa and ipt genes have wild-type hormone levels but display both auxin- and cytokinin-overproducing phenotypes. Plant J 23:279-284.
16. Epstein E, Cohen JD, Slovin JP (2002) The biosynthetic pathway for indole-3-acetic acid changes during tomato fruit development. Plant Growth Regulation 38:15-20
17. Gierl A, Frey M (2001) Evolution of benzoxazinone biosynthesis and indole production in maize. Planta 213:493-498
18. Glick B, Patten C, Holguin G, Penrose D (1999) Biochemical and genetic mechanisms used by plant growth promoting bacteria. Ontario, Canada: Imperial College Press, 267 pp.
19. Glick B, Saleh S (2001) Involvement of gacS and rpoS in enhancement of the plant growth-promoting capabilities of Enterobacter cloacae CAL2 and UW4. Canadian J Microbiology 47:698-705
20. Hamann T (2001) The role of auxin in apical-basal pattern formation during Arabidopsis embryogenesis. J Plant Growth Regulation 20:292-299
21. Hemm M, Reugger M, Chapple C (2003) The Arabidopsis ref2 mutant is defective in the gene encoding CYP83A1 and shows both phenylpropanoid and glucosinolate phenotypes. Plant Cell 15:179-194
22. Iyer M, Cohen JD, Epstein E, Slovin JP (1999) An unexpected change in free IAA levels and alteration of fruit ripening in tomatoes transformed with the *iaglu* gene (Abstract). Amer Soc Plant Physiol. 1999:150
23. Jackson RG, Lim EK, Li Y, Kowalczyk M, Sandberg G, Hoggett J, Ashford DA, Bowles DJ. (2001) Identification and biochemical characterization of an Arabidopsis indole-3-acetic acid glucosyltransferase. J Biol Chem.276:4350-4356
24. Kerk N, Jiang K, Feldmann L (2000) Auxin metabolism in the root apical meristem. Plant Physiol 122:925-932
25. Klee HJ, Lanahan MB (1995) Transgenic plants in hormone biology, in Plant Hormones: Physiology, Biochemistry and Molecular Biology, PJ Davies, Editor, Kluwer Academic Publishers: Dordrecht, Boston, London. ISBN 0-7923-2984-8. p. 340-353
26. Kim G, Tsukaya H (2002) Regulation of the biosynthesis of plant hormones by P450s. Journal of Plant Research 115:169-177
27. Kutz A, Muller A, Hennig P, Kaiser W, Piotrowski M, Weiler E (2002) A role for nitrilase 3 in the regulation of root morphology in sulphur-starving Arabidopsis thaliana. Plant J 30:95-106
28. LeClere S, Tellez R, Rampey R, Matsuda S, Bartel B (2002) Characterization of a family of IAA-amino acid conjugate hydrolases from Arabidopsis. J Biological Chemistry 277:20446-20452
29. Ljung K, Hull AK, Kowalczyk M, Marchant A, Celenza J, Cohen JD, Sandberg G (2002) Biosynthesis, conjugation, catabolism and homeostasis of indole-3-acetic acid in *Arabidopsis thaliana*. Plant Molecular Biology 50:309-332
30. Ljung K, Östin A, Lioussanne L, Sandberg G (2001) Developmental regulation of indole-3-acetic acid turnover in Scots Pine seedlings. Plant Physiol 125:464-475
31. Ludwig-Müller J (2000) Indole-3-butyric acid in plant growth and development. Plant

Growth Regulation 32:219-230
32. Ludwig-Müller J, Cohen JD (2002) Identification and quantification of three active auxins in different tissues of *Tropaeolum majus*. Physiol Plant 115:320-329
33. Mikkelsen M, Petersen B, Glawischnig E, Jensen A, Andreasson E, Halkier B (2003) Modulation of CYP79 genes and glucosinolate profiles in Arabidopsis by defense signaling pathways. Plant Physiol 131:298-308
34. Muday G (2001) Auxin and tropisms. J Plant Growth Regulation 20:226-243
35. Nakazawa M, Yabe N, Ichikawa T, Yamamoto YY, Yoshizumi T, Hasunuma K, Matsui M. (2001) DFL1, an auxin-responsive GH3 gene homologue, negatively regulates shoot cell elongation and lateral root formation, and positively regulates the light response of hypocotyl length. Plant J 25:213-221
36. Ngo P, Ozga J, Reinecke D (2002) Specificity of auxin regulation of gibberellin 20-oxidase gene expression in pea pericarp. Plant Molecular Biology 49:439-448
37. Nigovic G, Antolic S, Kojic-Prodic B, Kiralj R, Magnus V, Salopek-Sondi B (2000) Correlation of structural and physico-chemical parameters with bioactivity of alkylated derivatives of indole-3-acetic acid, a phytohormone (auxin). Acta Cryst B56:94-111
38. Nigovic B, Kojic-Prodic B, Antolic S, Tomic S, Puntarec V, Cohen JD (1996) Structural studies on monohalogenated derivatives of the phytohormone indole-3-acetic acid (auxin). Acta Cryst B52:332-343
39. Normanly J (1997) Auxin metabolism. Physiol Plant 100:431-442
40. Normanly J, Bartel B (1999) Redundancy as a way of life-IAA metabolism. Current Opinion in Plant Biology 2:207-213
41. Normanly J, Grisafi P, Fink GR, Bartel B (1997) Arabidopsis mutants resistant to the auxin effects of indole-3-acetonitrile are defective in the nitrilase encoded by the NIT1 gene. Plant Cell 9:1781-1790
42. Normanly J, Slovin JP, Cohen JP (1995) Rethinking auxin biosynthesis and metabolism. Plant Physiol 107:323-329
43. Oetiker J, Aeschbacher G (1997) Temperature-sensitive plant cells with shunted indole-3-acetic acid conjugation. Plant Physiol 114:1385-1395
44. Ozga J, Reinecke D (2003) Hormonal interactions in fruit development. J Plant Growth Regulation (in press).
45. Ozga J, Yu J, Reinecke D (2003) Pollination-, development-, and auxin-specific regulation of gibberellin 3β-hydroxylase gene expression in pea fruit and seeds. Plant Physiol 131:1137-1146
46. Patten C, Glick B (2002) Role of *Pseudomonas putida* indoleacetic acid in development of the host plant root system. Applied Environ Microbiology 68:3795-3801
47. Patten C, Glick B (2002) Regulation of indoleacetic acid production in *Pseudomonas putida* GR12-2 by tryptophan and the stationary-phase sigma factor RpoS. Canadian J Microbiology 48:635-642
48. Piotrowski M, Schonfelder S, and Weiler EW (2001) The *Arabidopsis thaliana* isogene NIT4 and its orthologs in tobacco encode β-cyano-L-alanine hydratase/nitrilase. J Biological Chemistry 276:2616-2621
49. Rapparini F, Tam Y, Cohen JD, Slovin JP (2002) IAA metabolism in *Lemna gibba* undergoes dynamic changes in response to growth temperature. Plant Physiol 128:1410-1416
50. Reintanz B, Lehnen M, Reichelt M, Gershenzon J, Kowalczyk M, Sandberg G, Godde M, Uhl R, Palme K (2001) bus, a bushy Arabidopsis CYP79F1 knockout mutant with abolished synthesis of short-chain aliphatic glucosinolates. Plant Cell 13:351-367
51. Ribnicky D, Cohen JD, Hu W, Cooke TJ (2002) An auxin surge following fertilization in carrots: A mechanism for regulating plant totipotency. Planta 214: 505-509
52. Ross J, O'Neill D, Wolbang C, Symons G, Reid, J (2002) Auxin-gibberellin interactions and their role in plant growth. J Plant Growth Regulation 20:346-353
53. Scherer GF (2002) Secondary messengers and phospholipase A2 in auxin signal

transduction. Plant Molecular Biology 49:357-372
54. Selmar D (1999) Biosynthesis of cyanogenic glycosides, glucosinolates and non protein amino acids, in Biochemistry of Plant Secondary Metabolism, M Wink, Editor, CRC Press: Boca Raton, FL. p. 79-150
55. Sergeeva E, Liaimer A, Bergman B (2002) Evidence for production of the phytohormone indole-3-acetic acid by cyanobacteria. Planta 215:229-238
56. Slovin JP, Bandurski RS, Cohen JD (1999) Auxin, in Biochemistry and Molecular Biology of Plant Hormones, P Hooykaas, M Hall, and K Libbenga, Editors, Elsevier Science: Oxford. p. 115-140
57. Smolen G, Bender J (2002) Arabidopsis cytochrome P450 cyp83B1 mutations activate the tryptophan biosynthetic pathway. Genetics 160:323-332
58. Staswick PE, Tiryaki I, Rowe ML (2002) Jasmonate response locus JAR1 and several related Arabidopsis genes encode enzymes of the firefly luciferase superfamily that show activity on jasmonic, salicylic, and indole-3-acetic acids in an assay for adenylation. Plant Cell 14:1405-1415
59. Swarup R, Parry G, Graham N, Allen T, Bennett M (2002) Auxin cross-talk: integration of signalling pathways to control plant development. Plant Molecular Biology 49:411-426
60. Szmidt-Jaworska A, Kesy J, Kocewicz, J (1997) Transformation of 1-O-(indole-3-acetyl)-β-D-glucose into di-O-(indole-3-acetyl)-D-glucose catalysed by enzyme preparations from corn seedlings. Acta Biochmica Polonica 44:215-220
61. Sztein A, Ilic N, Cohen, JD, Cooke TJ (2002) Indole-3-acetic acid biosynthesis in isolated axes from germinating bean seeds: The effect of wounding on the biosynthetic pathway. Plant Growth Regulation 36:201-207
62. Tam Y, Epstein E, Normanly J (2000) Characterization of auxin conjugates in *Arabidopsis thaliana*: low steady state levels of indole-3-acetyl-aspartate, indole-3-acetyl-glutamate, and indole-3-acetyl glucose. Plant Physiol 123:589-595
63. Tam Y, Normanly J, (2002) Overexpression of a bacterial indole-3-acetyl-L-aspartic acid hydrolase in Arabidopsis thaliana. Physiol Plant 115:513-522
64. Tobena-Santamaria R, Bliek M, Ljung K, Sandberg G, Mol J, Souer E, Koes R (2002) FLOOZY of petunia is a flavin mono-oxygenase-like protein required for the specification of leaf and flower architecture. Genes and Development 16:753-763
65. Vorwerk S, Biernacki S, Hillebrand H, Janzik IAM, Weiler EW, Piotrowski M (2001) Enzymatic characterization of the recombinant *Arabidopsis thaliana* nitrilase subfamily encoded by the NIT2/NIT1/NIT3-gene cluster. Planta 212:508-516
66. Walz A, Park S, Slovin JP, Ludwig-Müller J, Momonoki YS, Cohen JD (2002) A gene encoding a protein modified by the phytohormone indoleacetic acid. Proc. Natl. Acad. USA 99:1718-1723.
67. Wittstock U, Halkier B (2002) Glucosinolate research in the Arabidopsis era. Trends in Plant Science 7:263-270.
68. Zhao Y, Christensen S, Fankhauser C, Cashman J, Cohen, JD, Weigel D, and Chory, J (2001) A role for flavin monooxygenase-like enzymes in auxin biosynthesis. Science 291:306-309
69. Zhao Y, Hull A, Gupta N, Goss K, Alonso J, Ecker J, Normanly J, Chory J, Celenza J (2002) Trp-dependent auxin biosynthesis in Arabidopsis: involvement of cytochrome P450s CYP79B2 and CYP79B3. Genes and Development 16:3100-3112
70. Zook M (1998) Biosynthesis of camalexin from tryptophan pathway intermediates in cell-suspension cultures of Arabidopsis. Plant Physiol 118:1389-1398

B2. Gibberellin Biosynthesis and Inactivation

Valerie M. Sponsel[a] and Peter Hedden[b]
[a] Biology Department, University of Texas at San Antonio, San Antonio, TX 78249, USA. [b]Rothamsted Research, Harpenden, Herts AL5 2JQ, UK.
vsponsel@utsa.edu

INTRODUCTION

The gibberellins (GAs[1]) are defined by chemical structure. Naturally-occurring tetracyclic diterpenoid acids with structures based on the *ent*-gibberellane carbon skeleton (Fig. 1) are assigned gibberellin "A numbers" in chronological order of their identification (45) (http://www.plant-hormones.info/gibberellin_nomenclature.htm). At the present time there are

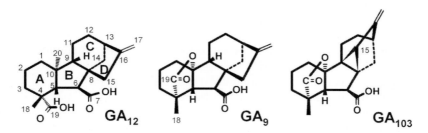

Figure 1. Structure of GA_{12}, a C_{20}-GA with the *ent*-gibberellane skeleton, showing the carbon atom numbering system and the assignment of the four rings. Also shown are GA_9, the simplest C_{19}-GA, which has an *ent*-20-norgibberellane skeleton, and GA_{103}, which has an extra cyclopropane ring.

---

[1] Abbreviations: BR, brassinosteroid; CDP-ME, 4-diphosphocytidyl-methylerythritol; CDP-MEP, CDP-ME 2-phosphate; CMK, CDP-ME kinase; CMS, CDP-ME synthase; CPS, *ent*-copalyl diphosphate synthase; DMAPP, dimethylallyl diphosphate; DXP, deoxyxylulose 5-phosphate; DXR, DXP reductoisomerase; DXS, DXP synthase; FPP, farnesyl diphosphate; GA, gibberellin; GA-Glc ester, GA-glucosyl ester; GA-*O*-Glc, GA-*O*-glucosyl ether; GA2ox, GA 2-oxidase; GA20ox, GA 20-oxidase; GA3ox, GA 3-oxidase; GA_n, gibberellin A_n; GFP, green fluorescent protein; GGPP, geranylgeranyl diphosphate; GPP, geranyl diphosphate; GUS, β-glucuronidase; HDS, HMBPP-synthase; HMBPP, hydroxymethyl-butenyl 4-diphosphate; IAA, indole 3-acetic acid; IDS, IPP/DMAPP synthase; IPP, *iso*pentenyl diphosphate; KAO, *ent*-kaurenoic acid oxidase; KO, *ent*-kaurene oxidase; KS, *ent*-kaurene synthase; MCS, ME-cPP synthase; MEP, methylerythritol 4-phosphate; ME-cPP, ME 2,4-cyclodiphosphate; MVA, mevalonic acid.

Gibberellin biosynthesis and inactivation

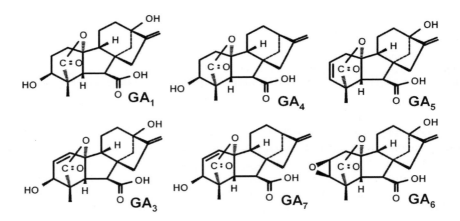

Figure 2. The structures of the main growth-active GAs.

136 fully characterized GAs, designated gibberellin A_1 (GA_1) through GA_{136}, that have been identified from 128 different species of vascular plants, and also from seven bacteria and seven fungi (44) (http://www.plant-hormones.info/ga1info.htm).

Gibberellins were first isolated from the pathogenic fungus *Gibberella fujikuroi* from which they derive their name (50). The presence of large quantities of GAs as secondary metabolites in this fungus leads to the extensive overgrowth of infected rice plants. Thus, from the time of their discovery GAs were known to be effective in promoting stem elongation and their characterization from the fungus was followed shortly by their identification as natural components of non-infected plants.

Gibberellins are now known to be regulators of many phases of higher plant development, including seed germination, stem growth, induction of flowering, pollen development and fruit growth. The concentration of bioactive GAs in plants is in the range $10^{-11} - 10^{-9}$ g/g fresh weight, depending on the tissue and species, and is closely regulated. This chapter describes the biosynthesis and catabolism of GAs, and examines the regulatory processes that optimize the levels of bioactive GAs within plant tissues.

Only a few of the 136 known GAs have intrinsic biological activity. Not surprisingly, many of the GAs that were identified in the earliest years of GA research are the ones which possess the highest biological activity, and are candidates for the role of an active hormone. These include GA_1, GA_3, GA_4, GA_5, GA_6 and GA_7 (Fig. 2). The bioactive GA(s) present in a particular plant species is/are accompanied by a dozen or more GAs that are likely to be inactive precursors or deactivation products of the active forms. Experiments utilizing single gene dwarf mutants have provided convincing evidence that GA_1 is the major active GA for stem elongation in *Zea mays* and *Pisum sativum*. Moreover, GA_1 has been identified in 86 plants, more than any other GA, implying that it has an important and widespread role as a

GA "hormone." However, GA$_4$ co-occurs with GA$_1$ in many species and in some members of the Cucurbitaceae and in Arabidopsis, it is the major bioactive GA. Gibberellin A$_3$, which is also known as gibberellic acid, has been identified in 45 plants. It is the major GA accumulating in *G. fujikuroi*, from which it is produced commercially. Gibberellic acid is used to promote seed germination, stem elongation, and fruit growth in a variety of agronomically and horticulturally important plants.

By definition, GAs possess tetracyclic *ent*-gibberellane (C$_{20}$) or 20-nor-*ent*-gibberellane (C$_{19}$) skeletons, with the rings designated A through D as shown in Fig. 1. The C$_{20}$-GAs (e.g., GA$_{12}$, Fig. 1) have the full complement of 20 carbon atoms, whereas the C$_{19}$-GAs (e.g., GA$_9$, Fig. 1) possess only 19 carbon atoms, having lost carbon-20 by metabolism. In almost all the C$_{19}$-GAs the carboxyl at C-4 forms a lactone at C-10. In some cases, there is a bond between C-9 and C-15 to form an additional cyclopropane ring (e.g., GA$_{103}$, Fig. 1). Other structural modifications can be made to the *ent*-gibberellane skeleton of both C$_{20}$- and C$_{19}$-GAs, such as the insertion of additional functional groups. Both the position and stereochemistry of these substituents can have a profound effect on the biological activity of the GA. For example, a hydroxyl (OH) group in the 3β-position is required for growth-promoting activity, whereas the insertion of an OH in the 2β-position will substantially reduce the bioactivity of an active GA. In either instance the insertion of an OH in α-orientation has little effect.

The C$_{20}$-GAs do not normally have biological activity *per se*, but can be

Figure 3. Gibberellin-responsive dwarf mutants of Arabidopsis (ecotype Landsberg *erecta*) showing the position of the lesion in the GA-biosynthetic pathway for each locus. (From 65) (©Academic Press).

metabolized to C_{19}-GAs that may be bioactive. Many assays determine bioactivity in terms of stem elongation in dwarf seedlings (rice, maize), or hypocotyl elongation in seedlings whose growth is inhibited by light (lettuce, cucumber). However, because an applied GA may be metabolized in the assay plant, the observation that it has bioactivity does not establish that it is active *per se*. More sophisticated studies, most often using mutants in which specific steps in GA metabolism are blocked, are required to assess intrinsic activity. For example, in mutant seedlings in which 3β-hydroxylation is blocked, applied GA_9 may be inactive, but applied GA_4 (3β-hydroxyGA_9) may elicit a growth response. From this result one can infer that GA_9 does not have activity *per se*, but must be metabolized to GA_4. A series of GA-biosynthetic mutants of Arabidopsis with lesions at different points of the pathway (Fig. 3) has been extremely valuable in elucidating the genetics of GA-biosynthesis in this species. As discussed later, the different severities of dwarfism for mutations at each locus is due to the different sizes of gene families for each enzyme.

The requirements for intrinsic biological activity for growth stimulation are that it is a C_{19}-GA, with a 4,10 lactone and a carboxylic acid group on C-6, that it possesses a 3β-hydroxyl group, or some other functionalization at C-3, such as a 2, 3-double bond (as in GA_5) and that it does not possess a 2β-hydroxyl group. All the GAs shown in Fig. 2 fulfill these requirements. Since 2β-hydroxylation is a deactivating mechanism, a GA that has some functionality at C-2 that prevents 2β-hydroxylation may have enhanced activity. Thus GA_3 and GA_7, which possess a 1, 2 double bond, are not substrates for the 2β-hydroxylating enzyme and have longer lasting activity than their 1, 2-dihydro counterparts (GA_1 and GA_4). There is evidence that the structural requirements for florigenic activity may be subtly different than those for stem elongation. For example, GA_5 and GA_6 are more active than GA_1 and GA_4 in enhancing flowering in *Lolium*, whereas GA_1 and GA_4 show enhanced activity over GA_5 and GA_6 in promoting stem growth in this genus (37).

THE BIOSYNTHETIC PATHWAY

As diterpenoids, GAs are synthesized from geranylgeranyl diphosphate (GGPP) *via* isopentenyl diphosphate (IPP), which is the 5-carbon building block for all terpenoid/isoprenoid compounds (Fig. 4). The GA-biosynthetic pathway can be divided into three parts. The first part, which occurs in plastids, leads to the synthesis of the tetracyclic hydrocarbon, *ent*-kaurene (Figs. 5 and 6). In the second part of the pathway, which occurs in the endoplasmic reticulum, *ent*-kaurene is sequentially oxidized to yield the first-formed GA, GA_{12} and its 13-hydroxylated analog GA_{53} (Fig. 7). In the third part of the pathway, which occurs in the cytosol, GA_{12} and GA_{53} are further oxidized to other C_{20}-GAs, and C_{19}-GAs (Fig. 9).

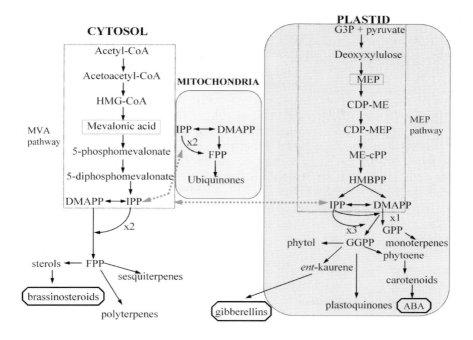

Figure 4. Pathways for isoprenoid biosynthesis and their compartmentalization in plant cells, showing the origin of plant hormone and other products.

The Pathway to *ent*-Kaurene.

For four decades IPP was thought to be formed exclusively *via* the acetate/mevalonate (MVA) pathway. Since the last edition of this book a novel biosynthetic route to IPP has been elucidated (55). The new pathway has been referred to in the literature by several names: the non-mevalonate, the Rohmer (after one of its discoverers), the deoxyxylulose phosphate (DXP) and the methylerythritol phosphate (MEP) pathway. The latter name has gained most recent acceptance, since MEP is considered to be the first committed intermediate in the pathway. The MEP pathway was first characterized in Eubacteria, and is now known to occur in plastids of eukaryotic organisms.

While most organisms possess either the MVA pathway (fungi and animals) or the MEP pathway (most eubacteria and many algae), plants and their immediate algal ancestors, the Charophyceae, possess both pathways. The two pathways exist in different cellular compartments and the resulting IPP has different fates (Fig. 4). In plants the MVA pathway is cytosolic, and IPP formed by this route is further metabolized to sesqui- (C_{15}), and tri-terpenoids (C_{30}), including sterols. In contrast, the MEP pathway is plastidic, and IPP formed in plastids is converted to monoterpenes (C_{10}), diterpenes

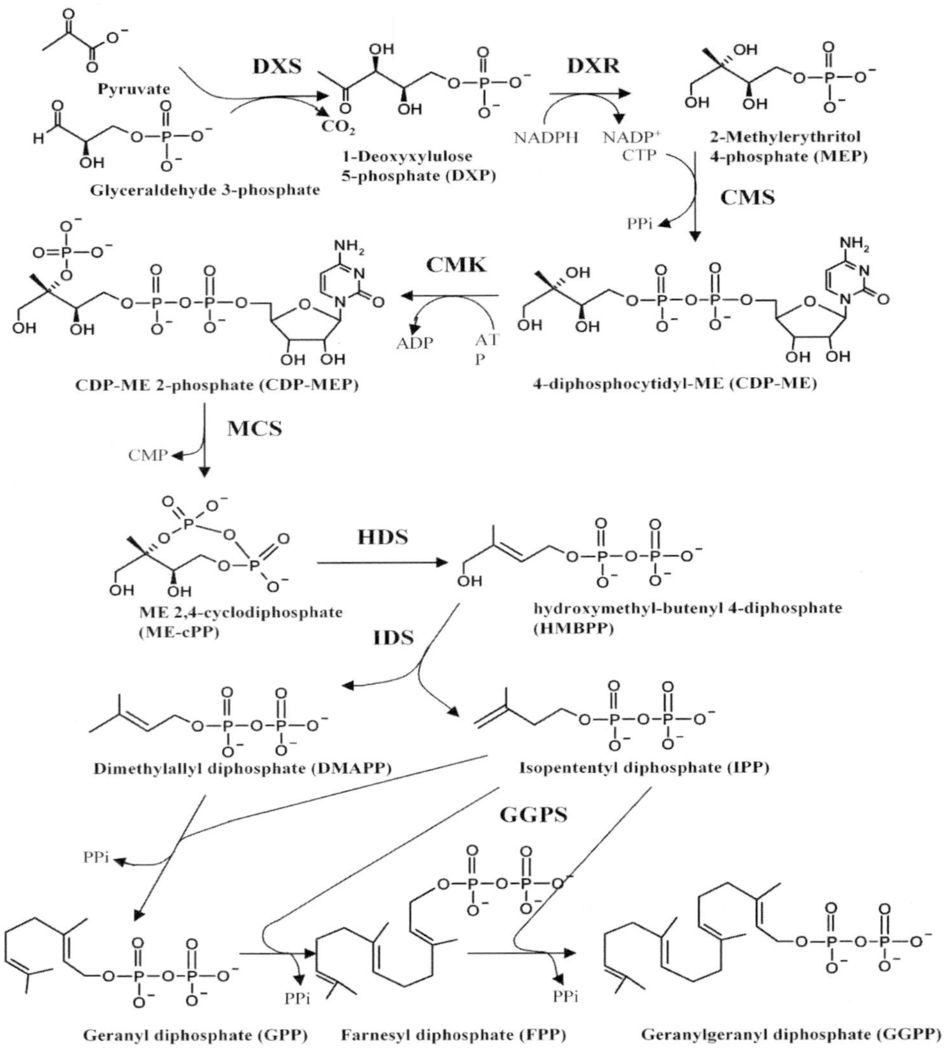

Gibberellin biosynthesis and inactivation

Figure 5. Biosynthesis of geranylgeranyl diphosphate (GGPP) in plastids *via* the MEP pathway. Enzymes in biosynthetic sequence are: DXS, deoxyxylulose 5-phosphate synthase; DXR, deoxyxylulose 5-phosphate reductoisomerase; CMS, 4-diphosphocytidyl methylerythritol synthase; CMK, 4-diphosphocytidyl methylerythritol kinase; MCS, methylerythritol 2,4-cyclodiphosphate synthase; HDS, hydroxymethyl-butenyl 4-diphosphate synthase; IDS, isopentenyl diphosphate/dimethylallyl diphosphate synthase; GGPS, GGPP synthase.

(C_{20}), including GAs and the phytyl side-chain of chlorophyll, and tetraterpenes (C_{40}), including carotenoids. The co-occurrence and compartmentation of the two pathways is unique to plants and their immediate ancestors. Thus in fungi the MVA pathway provides IPP for synthesis of all terpenoids, including GAs. Although there was considerable indirect evidence to suggest that the MEP pathway provides the IPP for GA biosynthesis in plants (47, 64), this has now been confirmed by feeding studies (34). This latter study also addressed the question of whether MEP is the only source of IPP for GA biosynthesis, or whether there is cross-over of one or more intermediates from the MVA pathway.

The MEP pathway for the production of *trans*-geranylgeranyl diphosphate (GGPP) in plastids is shown in Fig. 5, and its elucidation in Eubacteria is described in several reviews (12, 55). Briefly, a two-carbon unit (derived by decarboxylation of pyruvate) is added to glyceraldehyde 3-phosphate to give deoxyxylulose 5-phosphate (DXP). The condensation is catalyzed by DXP synthase (DXS), which requires thiamin diphosphate, indicating a transketolase type reaction. Although this is not the first committed reaction in the pathway, for DXP is also a precursor of thiamine, it is known to be a regulated step (see below). The next reaction, catalyzed by DXP reductoisomerase (DXR), involves the reduction and intramolecular rearrangement of the linear DXP to give the branched structure, methylerythritol 4-phosphate (MEP), which is subsequently converted to ME 2,4-cyclodiphosphate (ME-cPP) in three stages: the formation of 4-diphosphocytidyl-ME (CDP-ME), catalyzed by CDP-ME synthase (CMS), ATP- dependent phosphorylation of CDP-ME to CDP-ME 2-phosphate (CDP-MEP), catalyzed by CDP-ME kinase (CMK), and then conversion of CDP-MEP to ME-cPP with elimination of CMP in a reaction catalyzed by ME-cPP synthase (MCS). Reductive ring opening of ME-cPP to give hydroxymethyl-butenyl 4-diphosphate (HMBPP) is catalyzed by HMBPP-synthase (HDS). The final step in the pathway yields both IPP and its isomer dimethylallyl diphosphate (DMAPP), in a reaction catalyzed by IPP/DMAPP synthase (IDS). In *E.coli* the products, both of which are required for subsequent isoprenoid synthesis, are formed in a 5 to 1 ratio (56). This branch-point in the MEP pathway is in contrast to the MVA pathway in which DMAPP is formed from IPP, by the action of IPP isomerase. Analysis of this final step in the MEP pathway in plants is eagerly awaited.

Although the enzymes in the MEP pathway, and genes encoding them, were first identified in *E. coli*, orthologous genes and expressed sequence tags for all the enzymes in this pathway have now been found in Arabidopsis databases (e.g., DXS: (14); DXR: (62); MCS: (57); HDS: (51)). The enzymes have also been studied in peppermint leaves, which synthesize high levels of monoterpenes, and fruits of tomato and pepper, which are excellent systems for studying carotenoid biosynthesis. All the plant enzymes are encoded by nuclear genes, possess N-terminal sequences for putative plastid targeting, and conserved cleavage motifs. Direct plastid targeting has been demonstrated for DXS (3, 43), DXR (9) and HDS (formerly known as GCPE) (51). Cleavage of the signal peptides give enzymes similar to the bacterial enzymes except in the case of HDS, which contains a large plant-specific domain of unknown function (51).

DXP synthase is rate-limiting for IPP synthesis in all systems studied (13, 43). In Arabidopsis it is encoded by *CLA1* (chloroplastos alterados or altered chloroplasts), mutations in which result in an albino phenotype with very low levels of chlorophyll and carotenoids. Feeding of ^{13}C DXP to *cla1* mutants led to the identification of labeled *ent*-kaurene and GA_{12}, with very little dilution, confirming, for the first time, that the MEP pathway is the major source of IPP for GA biosynthesis (34). Despite the severe phenotype of *cla1* mutants there are two other *CLA* homologs in Arabidopsis. Both have plastid targeting sequences, but whether they encode functional proteins is not yet known. Other enzymes in the MEP pathway are encoded by single-copy genes in the Arabidopsis genome. T-DNA insertional mutants of DXR and CMS have seedling-lethal albino phenotypes (8). DXR is competitively inhibited by fosmidomycin, an antibiotic from *Streptomyces lavedulae* that acts as an effective herbicide (38).

The conversion of MEP-derived IPP and DMAPP to isoprenoids proceeds in the plastids, and is catalyzed by prenyl transferases. Repetitive addition of IPP to DMAPP in a head-to-tail fashion gives geranyl diphosphate (GPP), the C_{10} precursor of all monoterpenes, and GGPP. Both GPP synthase and GGPP synthase have been characterized from several plant sources, and shown to be plastid localized (7). In Arabidopsis GGPP synthases are encoded by a small family of five nuclear genes, the products of which are targeted to specific subcellular locations (48). Two of them, GGPS-1 and GGPS-3, become plastid localized and catalyze the formation of GGPP for use in plastid pathways (see below).

At GGPP the pathway in plastids branches in several directions, with separate pathways leading to the *ent*-kaurenoids and GAs, the phytyl side-chain of chlorophyll, phytoene and carotenoids, and the nonaprenyl (C_{45}) side-chain of plastoquinone (Fig. 4). It should be noted any regulation in the MEP pathway before GGPP will affect all branches, which is why *cla1*, *dxr* and *cms* mutants have albino, seedling-lethal phenotypes, and why fosmidomycin, which blocks DXS, is a herbicide. Also it should be remembered that since there is a shared precursor pool of GGPP, manipulation of one of the branch pathways can have significant effects on the flux through other branches. Thus, up- and down-regulation of phytoene synthase (*PSY*) to increase and decrease carotenoid biosynthesis causes reciprocal changes in GA levels (16).

Cyclization of the linear GGPP to the tetracyclic *ent*-kaurene occurs in two stages (Fig. 6). GGPP is converted first to the bicyclic compound, *ent*-copalyl diphosphate, by *ent*-copalyl diphosphate synthase (CPS), which was

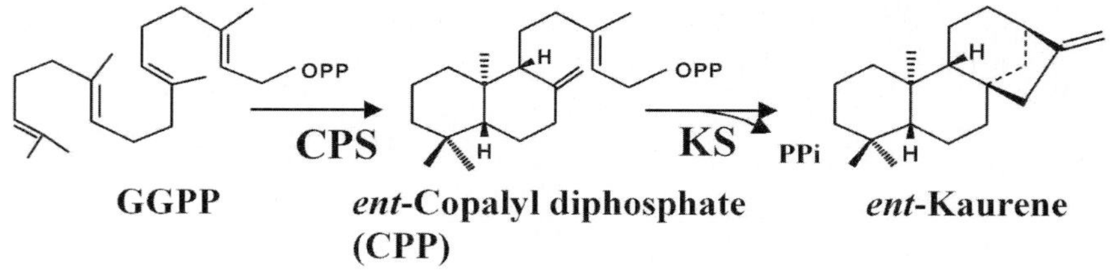

GGPP → CPS → ***ent*-Copalyl diphosphate (CPP)** → KS, PPi → ***ent*-Kaurene**

Figure 6. Two-step cyclization of GGPP to *ent*-kaurene *via ent*-copalyl diphosphate, catalyzed by *ent*-copalyl diphosphate synthase (CPS) and *ent*-kaurene synthase (KS).

previously called *ent*-kaurene synthase A. In Arabidopsis CPS is encoded by the *GA1* gene, which is present as a single-copy. However, *ga1* null mutants, although very severely dwarfed (Fig. 3), contain traces of GA, suggesting either the presence of an additional unrecognized *CPS* gene in the Arabidopsis genome, or an additional minor pathway to GA. *ent*-Copalyl diphosphate is converted to *ent*-kaurene by *ent*-kaurene synthase (KS) previously called *ent*-kaurene synthase B. KS, like CPS, is encoded by a single-copy gene (*GA2*) in Arabidopsis, mutations in which cause severe dwarfism (Fig. 3). The enzyme has been shown to be localized in plastids in Arabidopsis (30). *Stevia rebaudiana*, which produces large quantities of stevioside (glycosides of 13-hydroxykaurenoic acid) in mature leaves, has two *KS* gene copies (54). It has been suggested that gene duplication and differential regulation allows for tightly-regulated GA biosynthesis in rapidly dividing tissues, and the much greater flux to stevioside in mature leaves.

In the fungi, *G. fujikuroi* and *Phaeosphaeria* spp., CPS and KS activities reside in the same protein, though at separate catalytic sites (35). In plants, the occurrence of two separate proteins gives more opportunity for independent regulation of the two activities than exists in the fungi. Thus *CPS* transcript levels are much lower than those of *KS* in Arabidopsis, indicating that *CPS* may have a "gate-keeper" function on the branch of the terpenoid pathway that is committed to *ent*-kaurenoids and GAs.

Conversion of *ent*-Kaurene to GA_{12} and GA_{53}

All metabolic steps after *ent*-kaurene are oxidative (26). Initially, *ent*-kaurene is converted by a membrane-associated cytochrome P450 monooxygenase, *ent*-kaurene oxidase (KO), to *ent*-kaurenoic acid, which is oxidized by a second P450, *ent*-kaurenoic acid oxidase (KAO), to GA_{12}. The formation of GA_{12} from *ent*-kaurene requires six steps, the two enzymes involved each catalyzing three reactions (Fig. 7): KO catalyzes the sequential oxidation of the C-19 methyl group of *ent*-kaurene *via* the alcohol and aldehyde to the carboxylic acid, while KAO oxidizes C-7 of *ent*-kaurenoic acid to produce *ent*-7α-hydroxykaurenoic acid[2], which is then oxidized by this enzyme on C-6 to form GA_{12}-aldehyde. Finally, KAO oxidizes GA_{12}-aldehyde on C-7 to produce GA_{12}. The conversion of *ent*-7α-hydroxykaurenoic acid to GA_{12}-aldehyde involves the contraction of ring B from six C atoms to five, thereby transforming the *ent*-kaurane carbon skeleton to the *ent*-gibberellane structure. Arabidopsis contains a single *KO* gene (*GA3*) and two *KAO* genes. Whereas loss-of-function *ga3* mutants are severely dwarfed (Fig. 3) and, in common with the *ga1* and *ga2* null mutants, are male sterile, and need GA treatment to produce seeds that themselves do not germinate unless treated with GA, no *kao* mutants have been identified in

---

[2] Using IUPAC nomenclature the precursor of GAs is the enantiomeric form of kaurene, designated *ent*-kaurene. By convention α-substituents are designated *ent*-β, and β-substituents are designated *ent*-α.

Gibberellin biosynthesis and inactivation

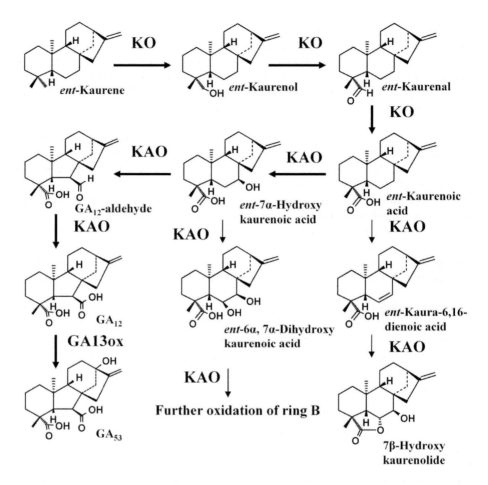

Figure 7. Reactions catalyzed by cytochrome P450-dependent mono-oxygenases. GA_{12} is formed from *ent*-kaurene by the sequential action of *ent*-kaurene oxidase (KO) and *ent*-kaurenoic acid oxidase (KAO). GA_{12} is converted to GA_{53} by 13-hydroxylases (GA13ox). Also shown are side reactions of KAO noted in some systems that result in the formation of kaurenolides and seco ring B compounds (in which ring B has been oxidatively cleaved between C-6 and C-7).

Arabidopsis because the two genes are fully redundant with the same expression patterns (29). However, *kao* mutants of barley (29) and pea (11) are GA-deficient dwarfs since most of the KAO activity present in vegetative tissues of these plants is due to expression of single genes. Pea has a second *KAO* gene, which is expressed in developing seeds (11).

GA_{12} is the common precursor for all GAs in higher plants. It lies at a branch-point in the pathway, undergoing either oxidation at C-20, or hydroxylation on C-13 to produce GA_{53}. GA_{12} and GA_{53} are precursors for the so-called non-13-hydroxylation and 13-hydroxylation pathways, respectively, which will be discussed in the following section. Genes encoding 13-hydroxylases have not yet been identified and there is still some uncertainty about the nature of this enzyme, with evidence suggesting the involvement of P450s in developing and germinating seed tissues (32, 39), but of a soluble dioxygenase in vegetative tissues of spinach (19).

Experiments using subcellular fractionation (22) and transient expression of enzyme fused to green fluorescent protein (GFP) (30)(see later) show that KAO is located in the endoplasmic reticulum (Fig. 8). Work with KO-GFP fusions transiently expressed in tobacco leaves indicate that KO is located on the outer membrane of the plastid (Fig. 8) and may thus participate in the translocation of *ent*-kaurene from its site of synthesis in plastids to the endoplasmic reticulum (30). However, as is discussed in more detail later, other experiments based on the use of reporter genes and *in situ* hybridisation suggest that *ent*-kaurene synthesis and oxidation may occur in different cell types in young germinating Arabidopsis embryos (73). Movement of the highly hydrophobic *ent*-kaurene between cells would require the assistance of a carrier protein. The discrepancy between these results may be due to the different experimental systems used, but clearly more work is required to resolve this disparity.

The fungus *G. fujikuroi* contains P450 monooxygenases that are functionally equivalent to the plant KO and KAO, although they have very low sequence similarity with the plant enzymes and are probably not closely related in evolutionary terms (28). The fungal *ent*-kaurene oxidase (P450-4) catalyzes the same reactions as KO, while the KAO equivalent possesses 3β-hydroxylase activity in addition to the other activities, such that it converts *ent*-kaurenoic acid to GA_{14} (3β-hydroxyGA_{12}). The substrate for 3β-hydroxylation is GA_{12}-aldehyde, which is converted to GA_{14} via GA_{14}-aldehyde. This remarkably multifunctional enzyme is also responsible for the formation of kaurenolides, via *ent*-kaura-6, 16-dienoic acid, and of seco-ring B compounds, such as fujenal, via *ent*-6α, 7α-dihydroxykaurenoic acid. Kaurenolides and *ent*-6α, 7α-dihydroxykaurenoic acid are also by-products of KAOs from pumpkin (23) and pea (11), but these enzymes do not have 3β-hydroxylase activity.

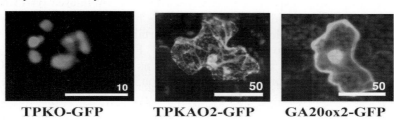

TPKO-GFP　　　　TPKAO2-GFP　　　GA20ox2-GFP

Fig. 8 (Color plate page CP3). Confocal images of tobacco leaves after microprojectile bombardment with the following constructs: TPKO-GFP, N-terminal region of Arabidopsis *ent*-kaurene oxidase (AtKO) fused to green fluorescent protein (GFP); TPKAO2-GFP, N-terminal region of *ent*-kaurenoic acid oxidase AtKAO2 fused to GFP; GA20ox2-GFP, full coding region of an Arabidopsis GA 20-oxidase (AtGA20ox2) fused to GFP. TPKO-GFP is associated with plastids, while TPKAO2-GFP is associated with the endoplasmic reticulum and GA20ox2-GFP is in the cytosol. Images are modified from (30).

Conversion of GA_{12} and GA_{53} to C_{19}-GAs

Reactions in the third part of the biosynthetic pathway, illustrated in Fig. 9, are catalyzed by soluble 2-oxoglutarate-dependent dioxygenases. The first enzyme, GA 20-oxidase (GA20ox), is responsible for the removal of C-20 in the formation of the C_{19}-GA skeleton. GA_{12} and GA_{53} are converted by this enzyme in parallel pathways to GA_9 and GA_{20}, respectively, by sequential oxidation of C-20 to the alcohol and aldehyde, and then removal of this C atom with formation of the 4, 10-lactone. Each of the C_{20}-GA intermediates in the reaction sequence is converted by GA20ox, although the alcohol intermediate must be present as the free alcohol. These alcohol intermediates form δ-lactones with the 19-carboxyl group when extracted from plant tissues and are then no longer oxidized by GA20ox. It is unclear whether the δ-lactones form naturally *in planta*, but it is of interest that vegetative tissues, but apparently not seeds, contain an enzyme capable of converted the δ-lactones to the aldehydes (71). This enzyme may serve to ensure formation of C_{19}-GAs when GA20ox activity is low, as in vegetative tissues, when lactone formation may compete with further oxidation of the alcohol intermediate. In most systems C_{20}-GAs containing a 20-carboxylic acid group are formed by GA20ox as minor biologically inactive by-products, which are not converted to C_{19}-GAs. However, these tricarboxylic acid GAs are major products of the GA20ox present in endosperm and immature embryos of *C. maxima* that was the first *GA20ox* to be cloned (40). The function of this pumpkin enzyme is considered abnormal as this type of activity has not been encountered in other species or, indeed, in vegetative tissues of pumpkin. The chemical mechanism for the loss of C-20 has not been elucidated. There is evidence that it is lost as CO_2 and that both O atoms in the lactone function originate from the 19-carboxyl group. Direct removal of C-20 as CO_2 requires the formation of an intermediate between the aldehyde and final C_{19}-GA product, but none has been identified and it may remain bound to the enzyme.

The growth-active GAs GA_4 and GA_1 are formed by 3β-hydroxylation of GA_9 and GA_{20}, respectively, catalyzed by GA 3-oxidases (GA3ox). The major GA3ox of Arabidopsis (AtGA3ox1) is highly regiospecific, producing a single product, while enzymes from certain other species also oxidize neighboring C atoms to a small extent (26). For example, oxidation of both C-2 and C-3 produces a 2, 3-double bond, as in the conversion of GA_{20} to GA_5. Further oxidation of GA_5, initially on C-1 and then on C-3, by the same enzyme results in the formation of GA_3 (2, 66). While most GA 3-oxidases are specific for C_{19}-GAs, some plants, and particularly seeds, produce 3β-hydroxylated C_{20}-GAs. For example, an enzyme from pumpkin endosperm 3β-hydroxylates C_{20}-GAs more readily than C_{19}-GAs (41).

A third class of dioxygenase, GA 2-oxidase (GA2ox), is responsible for the irreversible deactivation of GAs by 2β-hydroxylation, so ensuring GA turnover, which is necessary for effective regulation of GA concentration. In some tissues, such as the cotyledons and, particularly, the testae of

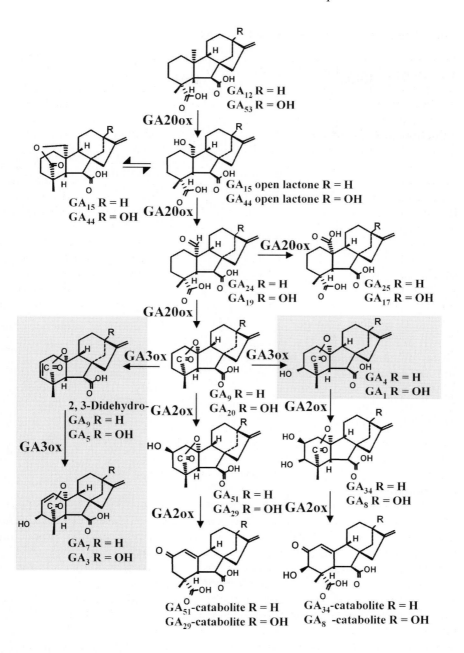

Figure 9. Reactions catalyzed by soluble 2-oxoglutarate-dependent dioxygenases. GA_{12} and GA_{53} are converted in parallel pathways to the active GAs, GA_4 and GA_1, respectively, by the sequential action of GA 20-oxidases (GA20ox) and GA 3-oxidases (GA3ox). The tricarboxylic acid GAs GA_{25} and GA_{17} are bi-products of GA 20-oxidases, while 2,3-didehydroGA$_9$/GA$_5$ and GA_7/GA_3 are by-products of some GA 3-oxidases. Gibberellin 2-oxidases (GA2ox) act mainly on C_{19}-GAs to form inactive products. The shaded regions indicate GAs with biological activity.

developing pea seeds, C-2 is oxidized further to the ketone, giving rise to the so-called GA-catabolites. As detected by GC-MS, the catabolites appear as dicarboxylic acids in which the lactone has opened (Fig. 9), although this rearrangement is possibly an artefact occurring during the analytical procedure. Studies of GA2ox function using recombinant protein prepared in *E. coli* reveal that these enzymes are capable of both reactions, 2β-hydroxylation and ketone formation (67). However, ketone production is relatively inefficient and occurs only when enzyme levels are very high, as in pea seeds. Most 2-oxidases are specific for C_{19}-GAs, and will accept the 3β-hydroxy bioactive GAs and their non-3β-hydroxylated precursors as substrates (Fig. 9). C_{20}-GAs may also be 2β-hydroxylated and Arabidopsis has been shown to contain GA 2-oxidases that are specific for these compounds (61). Such enzymes may be important for maintenance of GA homeostasis (see later) when levels of C_{20}-GA precursors become very high.

In contrast to the enzymes responsible for the earlier steps of the pathway that are encoded by single or small numbers of genes, the dioxygenases are encoded by multigene families, members of which differ in their positional and temporal patterns of expression. The phylogenetic relationship between the Arabidopsis *GA20ox*, *GA3ox* and *GA2ox* genes that have been identified on the basis of their derived amino acid sequences are shown in Fig. 10. Five *GA20ox* and four *GA3ox* genes have been identified. So far mutant phenotypes have been recognized only for *AtGA20ox1* (*ga5*) and *AtGA3ox1* (*ga4*), in both cases the mutants growing as semi-dwarfs (Fig. 3) with normal seed germination and flower fertility. There is, therefore, partial redundancy between the gene family members, which may have overlapping expression or, alternatively, GA products from other family members may move between tissues leading to partial rescue.

In Arabidopsis there are two classes of *GA2ox* genes (Fig. 10). One class contains six genes (*AtGA2ox1–6*), although one of these (*AtGA2ox5*) contains a large DNA insert and is apparently not expressed. Functional analysis of the expressed genes indicate that they are all C_{19}-GA 2-oxidases, whereas two further genes (*AtGA2ox7* and

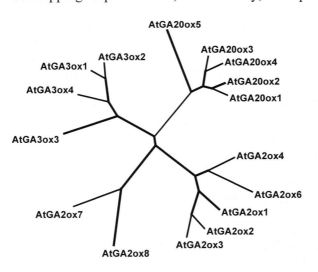

Figure 10. Unrooted phylogenetic tree of the Arabidopsis GA dioxygenases, produced using the Phylip programs PROTDIST and NEIGHBOR and displayed using TREEVIEW.

-8) encode enzymes that are specific for C_{20}-GAs (61). These last two genes, which were identified by activation tagging, are not closely related to the other 2-oxidases and their function could not have been predicted on the basis of sequence. It is, therefore, possible that further dioxygenses involved in GA biosynthesis in Arabidopsis have yet to be discovered.

Dioxygenases are not utilized for GA biosynthesis by *G. fujikuroi*, in which the GA 20-oxidase that converts GA_{14} to GA_4 is a P450 monooxygenase (28). Although the potential C_{20}-GA intermediates in this conversion are present in low levels in fungal cultures they are not metabolised to C_{19}-GAs so, in contrast to plants, the reaction sequence is unclear. In the next step in the fungal pathway, GA_4 is converted to GA_7 by a desaturase that has little homology to known enzymes (68). Finally, another P450 catalyzes the 13-hydroxylation of GA_7 to GA_3 (68). In contrast to plants, *G. fujikuroi* does not deactivate GAs by 2β-hydroxylation or conjugate formation (see below). Deactivation would be unnecessary since GAs have no physiological function in the fungus, but may act on the host by inducing the production of hydrolytic enzymes to aid infection or facilitate the acquisition of nutrients (28). A physiological role for the fungal GAs on the host rather than the fungus itself is supported by the large amounts of active GAs produced and their efficient secretion by the fungus.

Gibberellin Conjugates

In order to reduce the pool of bioactive GA, GAs can either be deactivated by 2β-hydroxylation and further catabolism, as described above, or they can be converted into conjugates (60). Conjugation to glucose is found most commonly for GAs and this can occur either *via* a hydroxyl group to give a GA-*O*-glucosyl ether (GA-*O*-Glc), or *via* the 7-carboxyl group to give a GA-glucosyl ester (GA-Glc ester). The most common sites within the GA molecule for –*O*-glc conjugation are C-2, C-3 and C-13. When applied to bioassay plants GA-*O*-Glcs show little or no activity, whereas GA-glc esters can exhibit bioactivity in certain assays, although this is unlikely to be activity of the conjugate *per se*. Instead it appears that if a bioassay plant, or microbial contaminant of the plant, possesses the requisite hydrolytic enzyme to cleave the glucose moiety and if the resulting aglycone is a potentially active GA, then the GA-conjugate will appear to have bioactivity. Feeding studies suggest that GA-Glc-esters sequester bioactive GAs, often quite rapidly, and release the free GA as required. On the other hand, the fate of GA-*O*-Glcs appears to be determined by the nature of the parent GA. GA-2-*O*-glcs, upon hydrolysis, will yield inactive GAs, whereas GA-3-*O*-glcs are hydrolyzed to bioactive GAs. Thus the enzymes for the synthesis and hydrolysis of GA-3-*O*-glcs have higher specificity than those catalyzing the hydrolysis of GA 2-*O*-glcs, reflecting their more direct role in maintaining the pool size of active GA.

Gibberellin biosynthesis and inactivation

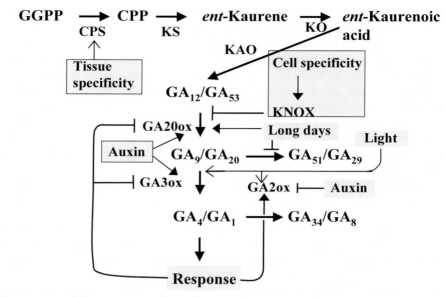

Figure 11. Major points of regulation of GA biosynthesis. Environmental or endogenous stimuli that modify expression of GA-biosynthetic genes are shaded. Solid arrows (→) indicate positive regulation (stimulation of gene expression), while bars (⊣) show negative regulation (suppression of gene expression). Open arrows (→) indicate that regulation can be positive or negative depending on the tissue or the nature of the signal.

REGULATION OF GA BIOSYNTHESIS AND CATABOLISM

With the identification of most of the genes involved in GA biosynthesis and catabolism, considerable advances are being made in our understanding of their tissue/cell specificity and their regulation by environmental factors, such as light and temperature. Furthermore, it is becoming evident that not only is GA metabolism moderated by its own action in order to achieve homeostasis, but that GAs participate in complex interactions with other hormones. The main sites of regulation in the GA-biosynthetic pathway are indicated in Fig. 11.

Developmental Regulation – Sites of GA Biosynthesis

Many different methods of assessing sites of GA biosynthesis have been used. The demonstration that all reactions in GA biosynthesis can be demonstrated in cell-free systems from seeds or seed parts (for instance liquid endosperm of Cucurbits) provided definitive evidence that developing seeds are sites of GA biosynthesis. However, GA biosynthesis in seeds is stage-specific. A surge in GA biosynthesis accompanies and is necessary for seed and early fruit growth, whereas studies with inhibitors of GA biosynthesis has shown that GAs that accumulate to high levels later in seed maturation may not have a physiological role (17).

For vegetative tissues, studies using cell-free systems have provided evidence that rapidly growing regions of the plant are sites of GA biosynthesis. For instance, *ent*-kaurene synthesizing activity was demonstrated only in immature plastids in plant parts undergoing active growth, rather than in mature chloroplasts (1). These and other studies have raised the possibility that GA biosynthesis and action occur in the same tissues, and perhaps even in the same cells. This contention is contrary to the classical definition of a hormone, which states that its synthesis and action are remote from one another. Certainly, long distance transport of GAs has been reported in some systems, but a wealth of new information is providing strong evidence for GA biosynthesis and action occurring in close proximity.

The advent of reporter genes and their use to localize the expression of genes encoding enzymes in GA biosynthesis has provided the technology to study GA metabolism in individual tissues and cells. In these experiments a reporter gene encoding, for example, β-glucuronidase (GUS) or GFP is fused to the promoter and regulatory sequences of the gene of interest. Use of promoter-reporter gene fusions has been instrumental in not only identifying sequences necessary for tissue- or developmentally-specific expression, but in defining the sites and stages that the expression occurs. In several instances it appears that the sites of GA synthesis and action are close-by each other, or may even be coincident. Two comprehensive series of studies, one using Arabidopsis and the other rice, have been selected from the considerable literature on this subject, and are described below.

In Arabidopsis, expression of *GA1*, encoding CPS, was observed in shoot apices, root tips, anthers and immature seeds (63). Importantly, these sites of *GA1* expression in wild-type plants coincide with sites where the *ga1* mutant phenotype is evident. Thus, Silverstone *et al.* (63) concluded that CPS is present in tissues and cells that are sites of GA action. This work has now been extended to look at the sites of expression of several other genes coding for enzymes of GA biosynthesis (73). In addition to CPS this study includes AtKO, encoded by the *GA3* gene in Arabidopsis, and two different 3β-hydroxylases, AtGA3ox1 (encoded by *GA4*) and AtGA3ox2. Interestingly within germinating seeds the gene encoding CPS is expressed in the provasculature, whereas those encoding KO, AtGA3ox1 and AtGA3ox2 are expressed in the cortex and endodermis of the embryonic axis (Fig. 12). The results imply that an intermediate, most likely *ent*-kaurene is transported intercellularly from the provasculature to the endodermis and cortex during GA biosynthesis, which, given the hydrophobic nature of *ent*-kaurene, would presumably require the assistance of a carrier (as previously discussed). Since cortical cells expand on GA treatment it was assumed that the *ent*-kaurene oxidation and 3β-hydroxylation occur at or close to the site of GA action. Following on from this work Ogawa *et al.* (46) used *in situ* hybridization to study the site(s) of GA response, by examining the cellular

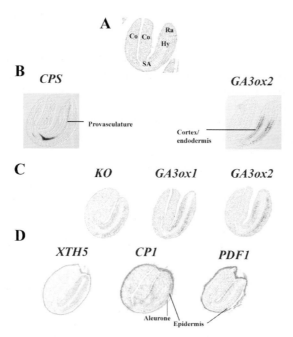

distribution of transcripts of three genes whose expression is dramatically up-regulated by endogenous GA_4 during germination. The genes *AtXTH5*, encoding a xyloglucan endotransglycosylase / hydrolase, *AtCP1*, which encodes a putative cysteine proteinase, and *PDF1*, PROTODERMAL FACTOR 1, showed unique expression profiles (Fig. 12). *AtXTH5*, whose gene product is likely to be involved in the wall loosening that is a prerequisite for cell expansion, was co-expressed with *AtGA3ox1* in the cortex of the embryonic axis, but was also expressed in the radicle. *AtCP1* transcripts were abundant in the epidermis and provasculature of the embryonic axis, and in the aleurone layer where *AtGA3ox1* transcripts were undetectable. The expression of *PDF1* was also in the epidermal cells of the axis. Thus GA_4 can up-regulate the expression of genes in cellular locations both at and separate from its site of synthesis, suggesting

Figure 12. (Color plate page CP3). Cellular localization of GA-biosynthesis and GA-regulated genes in germinating Arabidopsis (ecotype Landsberg *erecta*) seeds. A, longitudinal section showing component organs: Co, cotyledon; Hy, hypocotyl; Ra, radicle; SA, shoot apex. B, X-gluc staining of GUS activity in transgenic seeds expressing the reporter genes *AtCPS-GUS* and *AtGA3ox2-GUS*, showing expression in the provasculature and cortex/endodermis of the embryonic axis, respectively. C and D, *in situ* hybridization of (C) the biosynthetic genes *AtKO*, *AtGA3ox1* and *AtGA3ox2* and (D) the GA-regulated genes *AtXTH5* (encoding xyloglucan endotransglycosylase/ hydrolase), *AtCP1* (encoding a putative cysteine proteinase) and *PDF1* (PROTODERMAL FACTOR1). The GA-biosynthetic genes are expressed in the cortex/endodermis of the embryonic axis, while the GA-regulated genes show different expression patterns: *XTH5* is expressed in the cortex/endodermis of the embryonic axis, including the radicle, *CP1* is expressed in the epidermis and provasculature of the embryonic axis, and in the aleurone, while *PDF1* is expressed mainly in the epidermis. The images are from (73)(A-C) and (46)(D).

that it or a downstream signal must move between different cell types in the germinating seed.

The potential co-localization of GA biosynthesis and response has been studied further in rice, using reporter genes, and *in situ* hybridization. The expression patterns of two GA 20-oxidase genes (*OsGA20ox1* and

OsGA20ox2), and two 3-oxidase genes (*OsGA3ox1* and *OsGA3ox2*), and the expression of two genes involved in GA signaling have been monitored throughout development (33). The GA-signaling genes are *Gα*, which encodes a subunit of a heterotrimeric G protein and is a positive regulator of GA signaling in rice, and *SLR1*, which encodes a negative regulator. The results provide a comprehensive picture of the timing and precise locations of GA metabolism, in relation to the sites of action. The two 20-oxidase and 3β-hydroxylase genes show tissue- and cell-specific patterns of expression in rice. *OsGA20ox1* and *OsGA3ox1* are expressed in just two cell layers- the epithelium of the embryo and the tapetum of the pollen sacs, suggesting that they have very specific functions in GA metabolism. In germinating grain, *OsGA20ox2* and *OsGA3ox2* are expressed in both the epithelium (like *OsGA20ox1*) and in the developing shoot of the embryo, whereas the response genes, *Gα* and *SLR* are expressed at both these locations, and also in the aleurone. Therefore the embryo seems to be a site of GA biosynthesis and response, whereas the aleurone is a site of response only. From early work with cereal half-seeds and isolated aleurone layers the movement of active GA from embryo to aleurone has long been implicated in the *de novo* synthesis of amylase and other hydrolytic enzymes in this cell layer, for use in mobilizing endosperm reserves.

OsGA20ox2 and *OsGA3ox2* are also expressed in young leaf primordia of developing seedlings and elongating stems, and in seedling root tips, whereas expression of the other genes encoding each enzyme is low to absent in these sites. *Gα* and *SLR* are expressed in both locations inferring that young leaf primordia and root tips are sites of GA synthesis and action. During the development of floral organs *OsGA20ox2* and *OsGA3ox2*, and the response genes, are expressed in stamen primordia. Each of the genes, including *OsGA20ox1* and *OsGA3ox1* and the response genes are expressed in the tapetal layer. However the observation that *Gα* and *SLR1* are also expressed in other floral organs, suggests that the tapetum is a source of bioactive GAs whose site of action includes floral tissues separate from the site of their biosynthesis.

In summary, it appears that together OsGA20ox2 and OsGA3ox2 catalyze the production of bioactive GA(s) that have a site of action within the cells of the vegetative tissues in which they are formed. The fact that the loss of function *OsGA3ox2* mutant *d18* is a severe dwarf with normal levels of GA_4 in its reproductive tissue supports a role for this gene only in vegetative growth. In contrast *OsGA20ox1* and *OsGA3ox1* show high expression only in cell layers (embryo epithelium and anther tapetum) that provide bioactive GA for nearby, but separate locations (aleurone layer and other floral tissues, respectively). This and other work suggest that for genes encoded by small families, the time and site of expression of individual members is tightly regulated. Furthermore, the expression of these genes appears to be located in cells and tissues which respond to GA, or in close proximity, suggesting that GAs are produced at or very near to their site of action.

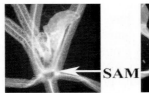

GA20ox1-GUS   **STM-GUS**

Fig. 13 (Color pate page CP3). The GA 20-oxidase (*AtGA20ox1*) and KNOX (*STM*) genes have non-overlapping expression patterns at the Arabidopsis shoot apex. Transgenic plants were produced containing the reporter genes *AtGA20ox1-GUS*, in which the *AtGA20ox1* promoter including part of the coding region was fused to the β-glucuronidase (*GUS*) coding region, and *STM-GUS*, in which the *STM* promoter was fused to *GUS*. GUS expression (blue staining) in the *AtGA20ox1-GUS* line is in the leaf primordia and sub-apical region, but is excluded from the shoot apical meristem (SAM), where *STM-GUS* is expressed. The images are modified from (25).

The molecular mechanisms underlying cell-specific expression of the GA-biosynthetic genes are beginning to emerge. For example, Chang and Sun (10) identified one *cis*-regulatory region in the Arabidopsis *CPS* promoter (between −997 and −796 with respect to the translation initiation site) that is required for its expression in developing seeds. Other regions provided either positive regulation in all tissues, or regulation in all tissues except seeds, as well as negative regulation.

A number of transcription factors have been described that alter expression of GA-biosynthesis genes and may be involved in their regulation and/or cell-specific expression. The most fully characterized of these in terms of their involvement with GA biosynthesis are the KNOTTED-1-like homeobox (KNOX) transcription factors. Ectopic expression of KNOX genes in tobacco causes dwarfism due, at least in part, to suppression of *GA20ox* gene expression (58). This effect was specific to the *GA20ox* gene with *GA3ox* expression apparently not affected. The tobacco KNOX protein NTH-15 was shown to bind to a *cis* element within the first intron of the *GA20ox* gene thereby inhibiting its expression. Consequently, *KNOX* and *GA20ox* genes have non-overlapping expression patterns in the shoot apex, with *KNOX* expressed in the meristem and *GA20ox* in the leaf primordia and sub-meristem region (25, 58) (Fig. 13). It is proposed that KNOX functions to exclude GA biosynthesis from the meristem so that the cells in this region remain indeterminate. It has also been suggested that leaf morphology may be determined by the presence or absence of KNOX expression in the developing leaf resulting in uneven GA distribution (25). However, GA movement to meristematic regions would compromise this control mechanism so that GAs from external sources would need to be removed from the meristems by, for example, deactivation. In rice, expression of a GA 2-oxidase gene (*OsGA2ox1*) is localized to a ring of cells at the base of the shoot apical meristem (59). Expression of *OsGA2ox1* was reduced on flower initiation leading Sakamoto *et al.* (59) to suggest that this enzyme controls the amount of active GAs reaching the shoot apical meristem; its reduction would allow more GA to reach the meristem and thereby contribute to its transition to a meristem inflorescence.

Gibberellin Homeostasis

Observations that certain GA-insensitive dwarf mutants contain abnormally high amounts of C_{19}-GAs provided the first clues for a link between GA biosynthesis and response (27). It is now clear that part of a plant's response to GAs is to depress GA biosynthesis and stimulate catabolism in order to establish GA homeostasis. This is achieved through down-regulation of GA 20-oxidase and 3-oxidase gene expression (negative feedback regulation) and up-regulation of GA 2-oxidase expression (positive feedforward regulation)(Fig. 11), the relative importance of each gene to this process varying with species and tissue. Mutations that disrupt GA response (as in GA-insensitive dwarfs) or severely reduce levels of endogenous GAs such that the GA response pathway does not occur (as in GA-deficient dwarfs) have similar effects on expression of genes encoding enzymes of GA metabolism. Numerous studies have demonstrated that regulation occurs, at least partly, at the level of mRNA accumulation so that GA-deficient plants accumulate transcript for *GA20ox* and *GA3ox* genes, but have lower levels of *GA2ox* transcript, as for example in Arabidopsis (67)(Fig. 14). However, effects on enzyme levels have been little investigated. The homeostatic mechanism allows changes in GA content in response to developmental or environmental signals, but probably serves to restore GA concentrations to normal levels, particularly after gross changes (70). The existence of this close association between GA metabolism and response provides further evidence that at least the final stages of GA biosynthesis must occur at a site of GA action.

Regulation by other hormones

Recently it has been shown that other hormones also regulate GA biosynthesis, with the most compelling evidence being provided for auxin. Most of this evidence has come from work with pea and is discussed in detail by Reid et al. (Chapter B7). In summary, auxin acts as a long-distance signal, between pea seeds and the pod or between the shoot apex and the internodes, to promote cell enlargement in the target tissues *via* enhanced GA production. In the first example, 4-chloro-IAA up-regulates GA 20-oxidase and 3-

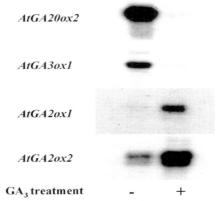

Fig. 14. Northern blot analysis of the GA-biosynthetic genes *AtGA20ox2* and *AtGA3ox1*, and the GA-catabolic genes *AtGA2ox1* and *AtGA2ox2* in the shoot apex of the Arabidopsis GA-deficient mutant *ga1-2* with and without treatment with GA_3. Expression of the biosynthetic genes is strongly depressed by GA treatment while expression of the GA 2-oxidase genes is stimulated by GA. From (67). (© 1999, National Academy of Sciences, U.S.A).

oxidase gene expression in the pod (49, 69), while, in pea internodes, IAA stimulates GA 3-oxidase expression and suppresses GA 2-oxidase expression. Removal of the source of auxin prevents growth of the target tissues unless auxin or GA is supplied. The interaction between auxin and GA in stem elongation was also demonstrated in tobacco, in which removal of the apical bud resulted in reduced GA_1 concentration in internodes (72). Application of IAA to the cut end of the stem restored GA_1 levels. In this case the major effect of IAA was to promote GA 20-oxidase activity and to reduce 2-oxidase activity; although 3-oxidase activity was also reduced by removal of the apical bud it could not be restored by IAA. Thus, although regulation of GA biosynthesis by auxin is probably a general phenomenon, the target enzymes may vary from species to species. In Arabidopsis, expression of the GA 20-oxidase gene *AtGA20ox1* (*GA5*) was shown to be up-regulated by application of epibrassinolide to a BR-deficient mutant, indicating that BR may also promote GA biosynthesis (6).

Environmental Regulation

Plants are extremely sensitive to their environment, reacting to external stimuli by changing their patterns of growth and development. Gibberellins are intermediaries for a number of environmental signals, which may induce changes in GA concentration and/or sensitivity. In particular, light quality, quantity and duration (photoperiod) can all influence the rate of GA biosynthesis and catabolism through effects on expression of specific genes. Examples of developmental processes that are regulated particularly by light-induced changes in GA biosynthesis will be discussed below (31). Further examples from pea are provided in chapter B7, while the role of light and GAs in potato tuberization is discussed in chapter E5.

Seed germination
In dicotyledonous species such as lettuce and Arabidopsis, seed germination is stimulated by exposure to light, which acts to promote production of GAs. Germination in such species has an absolute requirement for *de novo* biosynthesis of GAs, which induce the production of enzymes to digest the endosperm that would otherwise form a mechanical barrier to radicle emergence. Red light acts to promote specifically the 3β-hydroxylation step in the pathway; in Arabidopsis expression of two genes, *AtGA3ox1* (*GA4*) and *AtGA3ox2* (*GA4H*), is induced in imbibed seeds from about 4 hours following exposure to red light (74) (Fig. 15). Regulation of *AtGA3ox2* by light is mediated by phytochrome B, whereas *AtGA3ox1* expression is regulated by a different photoreceptor, which has not been identified. The two genes differ also in their developmental patterns of expression, with *AtGA3ox2* being expressed only during seed germination up to about 2 days following imbibition, in contrast to *AtGA3ox1*, which is induced more rapidly, but less strongly and is involved in most of the other GA-regulated developmental processes. Furthermore, *AtGA3ox2*, unlike *AtGA3ox1*, is not

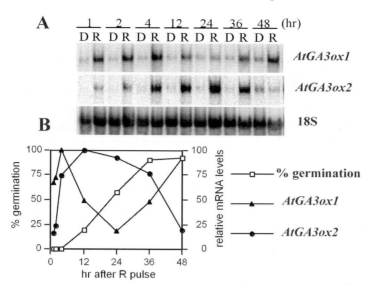

Fig. 15. Red light-induced expression of *AtGA3ox1* (*GA4*) and *AtGA3ox2* (*GA4H*) genes and germination in Arabidopsis (ecotype Landsberg *erecta*) seeds. Seeds were imbibed in the dark and irradiated with a pulse of far-red light one hour after imbibition. They were then either irradiated with a pulse of red light (R) 24 hours later, or kept in the dark (D). A, northern blot analysis of RNA from germinating seeds at different times following the red-light pulse or the equivalent time in the dark. B, germination frequency of red light-irradiated seeds (non-irradiated seeds do not germinate) at different times following irradiation, and the mRNA levels for *AtGA3ox1* and *AtGA3ox2* relative to the 18S rDNA loading control. After (74).

subject to feedback regulation. Expression of *AtGA3ox1* is also enhanced by cold treatment of the seeds prior to imbibition, a process known as stratification that stimulates germination in many species, including Arabidopsis (75).

De-etiolation
Plants growing in low light conditions are characterized by an etiolated growth habit with long, thin stems and small, chlorotic cotyledons and leaves. On exposure to normal light conditions, the growth rate of the stem is reduced, the stems thicken and leaves expand and green, in a process known as de-etiolation. After many years of controversy about whether or not de-etiolation involved changes in GA content it has now been clearly established by several groups working with etiolated pea seedlings that exposure to light results in a rapid, but temporary reduction in GA_1 content. The light, which is detected by phytochrome A and possibly a blue light receptor, causes down-regulation of a *GA3ox* gene (*PsGA3ox1*) and up-regulation of a *GA2ox* gene (*PsGA2ox2*) within 0.5 hours after exposure (53). After about four hours there is a strong increase in expression of the *GA3ox* gene and the *PsGA20ox1* gene, presumably due to feedback regulation, leading eventually to recovery of the GA_1 concentration. The lower growth rate in light in the longer term is thought to be due to reduced responsiveness

to GA in the light compared with the dark. Related mechanisms may explain the control of plant growth by phytochrome in the light, such as in the shade avoidance response in which plants respond to a change in the light spectrum. Plants also increase their GA content in low light conditions without changes in light quality, for example in the shade of a building. There is evidence that pea seedlings respond to reduced light by enhancing production of C_{19}-GAs (18), presumably due to higher GA 20-oxidase activity, but the detailed mechanism has not been determined.

The induction of flowering
Plants use changes in day-length (photoperiod) as cues for the initiation of numerous development changes. Gibberellins have been shown to act as secondary messengers in processes induced by long photoperiods, such as flowering in long-day rosette plants, the breaking of bud dormancy in woody plants (24), and the induction of stolons rather than tubers in potato (Chapter E5).

The interaction of photoperiod and GAs in reproductive development is complex, although significant progress in dissecting the various signaling pathways in Arabidopsis, a facultative long day plant, has been made utilizing mutants affecting flowering time, floral meristem identity and floral organ identity. In Arabidopsis a facultative long-day pathway exists which is separate from a GA-dependent pathway that promotes flowering in short days. Thus the *ga1-3* mutant will not flower in SD unless treated with GA, whereas this GA-deficient mutant will flower in LD without GA treatment, albeit after a slight delay compared to wildtype plants. Mutations in *CONSTANS* (*CO*), a component of the long-day pathway, together with *ga1-3* prevent flowering in LDs too. However, there is an interaction between the long-day and GA pathways, so that LDs do in fact enhance GA 20-oxidase expression in Arabidopsis, leading to rapid stem extension known as bolting and accelerated flowering. In spinach (*Spinacia oleracea*) too, which has an absolute requirement for long days, bolting requires the presence of bioactive GA(s), synthesis of which is induced when plants are transferred from short days to long days. Enhanced GA biosynthesis in long days is associated with higher levels of *ent*-kaurene production and of GA 20-oxidase activity, the latter resulting from greatly increased transcription of the *SoGA20ox1* gene in shoot tips (42). Exposure to long days also depressed expression of *SoGA2ox1* but had little effect on GA 3-oxidase gene expression.

In *Lolium temulentum*, ryegrass, which requires only 1 long photoperiod of at least 16 h duration for induction, flowering occurs without much stem extension, allowing the efficacy of GAs for both processes to be measured separately. GA_5 and GA_6 have high florigenic activity, a property that may be related to their failure to be 2β-hydroxylated and hence deactivated. GA_5 levels have been shown to increase five-fold in illuminated leaves at the end of a single 16 h photoperiod, and to double in the shoot apex, along with GA_6, on the following day (36). King et al. (37) have suggested that these values reflect the transport of GA_5 from leaf to apex at a rate consistent with the proposed movement of the hypothetical flowering stimulus, florigen, from the site of illumination to the site of floral initiation. Changes in expression of a MADS-box-containing transcription factor, which is functionally related to the Arabidopsis *AP1* gene, within hours after the purported arrival of GA_5 at the shoot apex is considered as strong evidence for a relationship between GA_5 and floral induction (36). Interestingly GA_1 and GA_4 appear to be excluded from the apical meristem at this time perhaps, by analogy with rice, because of a transient zone of high 2β-hydroxylating activity surrounding the apical meristem. Thus these GAs are not thought to be associated with the earliest events of floral evocation in *Lolium*, although they can promote stem elongation, and may also be involved in inflorescence initiation 2 or 3 d later when they gain access and accumulate in the shoot meristem.

In Arabidopsis, for which the most complete data are available, both the photoperiod and GA pathways to flowering converge with the activation of *LFY*, a floral meristem identity gene. *LFY* expression is only observed in SD-grown *ga1-3* mutants that have been treated with GA, though constitutive *LFY* expression from a 35S promoter will allow SD-grown *ga1-3* to flower without GA treatment (4). However, the up-regulation of *LFY* expression in *ga1-3* plants transferred from SD to LD is quite weak, and occurs more slowly than in WT plants, providing additional evidence for the biosynthesis of bioactive GA being important even in the long-day pathway to flowering. Using deletion analysis, Blasquez and Weigel (5) identified a 8 bp motif in the *LFY* promoter which is necessary for GA responsiveness, but not for response to daylength. This sequence is a potential target for GAMYB transcription factors, and candidate MYBs, whose expression increases in the shoot apex in response to exogenous GA_4, or to elevated levels of native GA_4, have been identified in both Arabidopsis and *L. temulentum* (20, 21).

CHEMICAL CONTROL OF GIBBERELLIN BIOSYNTHESIS

Growth retardants, synthetic chemicals that inhibit GA biosynthesis, have been used for several decades for the manipulation of agronomic and horticultural crops (52). The importance of these retardants may depend on their selectivity to inhibit only GA biosynthesis. Most often their commercial use is to reduce vegetative growth, without toxicity, and without decreasing crop yield. Since much of the GA-biosynthetic pathway is common to other diterpenoids, and tetraterpenoids, a chemical which blocks the pathway before GGPP (for example fosmidomycin which blocks DXR in the MEP pathway) will most likely be lethal, and thus be an effective herbicide. For inhibitors that block later in the pathway, there is the greater potential to inhibit GA biosynthesis selectively. However, many of the inhibitors target a particular category of enzyme (terpene cyclases/synthases, monooxygenases, or dioxygenases), and do in fact also affect other pathways

Gibberellin biosynthesis and inactivation

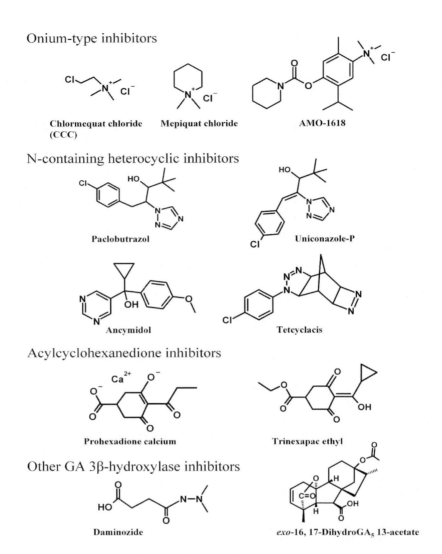

Figure. 16. Structures of growth retardants that act as inhibitors of GA biosynthesis.

that utilize these types of enzymes. The most selective action comes from inhibitors whose chemical structure is modeled on that of GAs, and are therefore likely to block specifically the active site of GA-metabolizing enzymes. There are four main types of GA biosynthesis inhibitors, which are discussed briefly below. The most recently developed retardants are considered in more detail. Selected structures for each type of retardant are shown in Fig. 16.

The type exemplified by chlormequat chloride, mepiquat chloride and AMO-1618 may exert their growth retarding activity by primarily blocking CPS activity, though there is a minor inhibition of KS too. These quaternary ammonium compounds, which are positively charged at cellular pH, may mimic cationic intermediates in the cyclization of GGPP to CPP. They have been used extensively for reducing stem length in wheat, thereby preventing lodging and subsequent decrease in yield. They are also used to reduce vegetative growth in cotton, and to control the growth habit of ornamental plants.

The second class of GA biosynthesis inhibitors include the triazoles, exemplified by paclobutrazol and uniconazol, and other heterocyclic compounds such as ancymidol and tetcyclasis (Fig. 16). These retardants are inhibitors of KO, which catalyzes the oxidation of *ent*-kaurene to *ent*-kaurenoic acid, and other monooxygenases. There is some overlap of activities with triazole-type fungicides such as triadimenol and metconazole whose main function is to inhibit sterol biosynthesis. Thus the growth retardants also have some fungicidal activity, and the fungicides have some growth-retarding properties. Of the two diastereoisomers in commercial samples of paclobutrazol, the 2S, 3S form is the more effective growth retardant whereas the 2R, 3R-enantiomer is more active in blocking ergosterol biosynthesis in fungi, by inhibiting lanosterol C-24 demethylation. The commercial use of these growth retardants is, amongst other things, for the reduction of stem growth in rice seedlings, and the restriction in vegetative growth of fruit trees and ornamentals.

The acylcyclohexanedione growth retardants, for example prohexadione-calcium and trinexapac, block the dioxygenase enzymes. These inhibitors have some structural features in common with 2-oxoglutarate, the co-substrate for this class of enzymes, and have been shown to inhibit the dioxygenases competitively. However, since these compounds inhibit both the biosynthetic enzyme GA 3-oxidase and the deactivating enzyme GA 2-oxidase, their activity might be somewhat unpredictable. Although on most plants these compounds act as growth retardants, there are a few examples where they may enhance growth though inhibition of GA-catabolism. The mode of action of daminozide, which was developed many years before the acylcyclohexanediones and has been used to control the stature of potted ornamentals such as chrysanthemum, has only recently been demonstrated. It has been shown to inhibit 2-oxogutarate-dependent dioxygenases in the same manner as prohexadione-Ca.

The most recently developed growth retardants are 16, 17-dihydro-GAs. These compounds were evaluated initially for activity as inducers of flowering in *Lolium temulentum*, but were found to act as effective growth retardants on Graminaceous species (15), with *exo*-16, 17-dihydro-GA$_5$ (applied as the 13-acetate) being one of the most effective examples. Like the acylcyclohexanediones these compounds also inhibit 2-oxoglutarate-dependent dioxygenases, but are more specific for the GA-metabolizing enzymes since the retardant mimics the GA substrate, rather than 2-oxoglutarate, which is common for all reactions. The activity of these compounds is somewhat puzzling, for their impressive efficacy on grasses is accompanied often by inactivity in dicot species.

CONCLUSIONS

Since the previous edition of this book, published almost ten years ago, progress in our understanding of GA biosynthesis and its regulation has been spectacular. This has occurred largely because advances in molecular genetics have enabled the isolation of most of the genes encoding the enzymes of GA biosynthesis and catabolism. In particular, the completion of the Arabidopsis genome sequence has allowed almost all the genes of GA metabolism to be identified in this species. It should be pointed out, however, that this rapid progress was possible to a large extent because of the earlier detailed metabolism studies that delineated the pathways and determined the nature of the enzymes involved. An important recent advance has been the recognition that GAs are biosynthesized predominantly *via* the MEP, rather than the MVA, pathway in plants, coupled with the identification of the Arabidopsis genes for the enzymes in this pathway. Notable omissions from the identified GA-biosynthesis genes are those encoding the 13-hydroxylase. It has, therefore, not been possible to assess the significance of this step in the production of the active hormones. In addition, little is known about the other enzymes that modify GAs on rings C and D, for example, those that hydroxylate at C-11, C-12 and C-15. Although GAs oxidized at these positions are found mainly in developing seeds, it is possible that they have roles in specific developmental processes, such as flowering, in some species.

The availability of the GA-metabolic genes is allowing the molecular mechanisms underlying the regulation of GA metabolism to be deciphered. Such studies have reinforced the importance of GAs as mediators of environmental signals, and are also revealing previously unsuspected cross-talk between the GA signaling pathways and those of other hormones. Furthermore, GA signal transduction has been shown to impinge directly on GA metabolism in the establishment of homeostasis for this hormone. Progress in understanding this process is likely to accelerate with the continuing impressive advances in our knowledge of GA signal transduction (Chapter D2).

Another benefit accruing from the availability of the GA-metabolic genes is the ability to manipulate GA biosynthesis and catabolism by altering expression of these genes (Chapter E7). Genetic manipulation of GA metabolism could provide an alternative to chemical treatments for control of plant growth and development and, furthermore, can be targeted to specific tissues or stages of development. It will also provide an important experimental tool to further understanding of GA biosynthesis and the function of GAs in plant development.

Acknowledgements

We thank Drs Chris Helliwell, CSIRO Plant Industry, Canberra, Australia, and Shinjiro Yamaguchi, Plant Science Center, RIKEN, Yokohama, Japan, for providing images used in

Figs 8 (CH), and 12 and 15 (SY). Work in the authors' laboratories is supported by the National Science Foundation (IBN 0080934 to VMS) and by the Biotechnology and Biological Sciences Research Council of the United Kingdom (PH).

References

1. Aach, H., Bose, G., Graebe, J. E. (1995) *ent*-Kaurene biosynthesis in a cell-free system from wheat (*Triticum aestivum* L) seedlings and the localization of *ent*-kaurene synthetase in plastids of three species. Planta 197, 333-342.
2. Albone, K. S., Gaskin, P., MacMillan, J., Phinney, B. O., Willis, C. L. (1990) Biosynthetic origin of gibberellin A_3 and gibberellin A_7 in cell-free preparations from seeds of *Marah macrocarpus* and *Malus domestica*. Plant Physiol. 94, 132-142.
3. Araki, N., Kusumi, K., Masamoto, K., Niwa, Y., Iba, K. (2000) Temperature-sensitive Arabidopsis mutant defective in 1-deoxy-*D*-xylulose 5-phosphate synthase within the plastid non- mevalonate pathway of isoprenoid biosynthesis. Physiol. Plant. 108, 19-24.
4. Blazquez, M. A., Green, R., Nilsson, O., Sussman, M. R., Weigel, D. (1998) Gibberellins promote flowering of Arabidopsis by activating the LEAFY promoter. Plant Cell 10, 791-800.
5. Blazquez, M. A., Weigel, D. (2000) Integration of floral inductive signals in Arabidopsis. Nature 404, 889-892.
6. Bouquin, T., Meier, C., Foster, R., Nielsen, M. E., Mundy, J. (2001) Control of specific gene expression by gibberellin and brassinosteroid. Plant Physiol. 127, 450-458.
7. Bouvier, F., Suire, C., d'Harlingue, A., Backhaus, R. A., Camara, B. (2000) Molecular cloning of geranyl diphosphate synthase and compartmentation of monoterpene synthesis in plant cells. Plant J. 24, 241-252.
8. Budziszewski, G. J., Lewis, S. P., Glover, L. W., Reineke, J., Jones, G., Ziemnik, L. S., Lonowski, J., Nyfeler, B., Aux, G., Zhou, Q., McElver, J., Patton, D. A., Martienssen, R., Grossniklaus, U., Ma, H., Law, M., Levin, J. Z. (2001) Arabidopsis genes essential for seedling viability: Isolation of insertional mutants and molecular cloning. Genetics 159, 1765-1778.
9. Carretero-Paulet, L., Ahumada, I., Cunillera, N., Rodriguez-Concepcion, M., Ferrer, A., Boronat, A., Campos, N. (2002) Expression and molecular analysis of the Arabidopsis DXR gene encoding 1-deoxy-*D*-xylulose 5-phosphate reductoisomerase, the first committed enzyme of the 2-C-methyl-*D*-erythritol 4-phosphate pathway. Plant Physiol. 129, 1581-1591.
10. Chang, C. W., Sun, T. P. (2002) Characterization of *cis*-regulatory regions responsible for developmental regulation of the gibberellin biosynthetic gene *GA1* in *Arabidopsis thaliana*. Plant Mol. Biol. 49, 579-589.
11. Davidson, S. E., Elliott, R. C., Helliwell, C. A., Poole, A. T., Reid, J. B. (2003) The pea gene *NA* encodes *ent*-kaurenoic acid oxidase. Plant Physiol. 131, 335-344.
12. Eisenreich, W., Rohdich, F., Bacher, A. (2001) Deoxyxylulose phosphate pathway to terpenoids. Trends Plant Sci. 6, 78-84.
13. Estevez, J. M., Cantero, A., Reindl, A., Reichler, S., Leon, P. (2001) 1-deoxy-*D*-xylulose-5-phosphate synthase, a limiting enzyme for plastidic isoprenoid biosynthesis in plants. J. Biol. Chem. 276, 22901-22909.
14. Estevez, J. M., Cantero, A., Romero, C., Kawaide, H., Jimenez, L. F., Kuzuyama, T., Seto, H., Kamiya, Y., Leon, P. (2000) Analysis of the expression of *CLA1*, a gene that encodes the 1- deoxyxylulose 5-phosphate synthase of the 2-C-methyl-*D*-erythritol-4-phosphate pathway in Arabidopsis. Plant Physiol. 124, 95-103.
15. Evans, L. T., King, R. W., Mander, L. N., Pharis, R. P., Duncan, K. A. (1994) The differential effects of C-16,17-dihydro gibberellins and related compounds on stem elongation and flowering in *Lolium temulentum*. Planta 193, 107-114.
16. Fray, R. G., Wallace, A., Fraser, P. D., Valero, D., Hedden, P., Bramley, P. M., Grierson, D. (1995) Constitutive expression of a fruit phytoene synthase gene in transgenic tomatoes causes dwarfism by redirecting metabolites from the gibberellin pathway. Plant J. 8, 693-701.

17. García-Martínez, J. L., Sponsel, V. M., Gaskin, P. (1987) Gibberellins in developing fruits of *Pisum sativum* cv. Alaska: studies on their role in pod growth and seed development. Planta 170, 130-137.
18. Gawronska, H., Yang, Y. Y., Furukawa, K., Kendrick, R. E., Takahashi, N., Kamiya, Y. (1995) Effects of low irradiance stress on gibberellin levels in pea seedlings. Plant Cell Physiol. 36, 1361-1367.
19. Gilmour, S. J., Zeevaart, J. A. D., Schwenen, L., Graebe, J. E. (1986) Gibberellin metabolism in cell-free-extracts from spinach leaves in relation to photoperiod. Plant Physiol. 82, 190-195.
20. Gocal, G. F. W., Poole, A. T., Gubler, F., Watts, R. J., Blundell, C., King, R. W. (1999) Long-day up-regulation of a *GAMYB* gene during *Lolium temulentum* inflorescence formation. Plant Physiol. 119, 1271-1278.
21. Gocal, G. F. W., Sheldon, C. C., Gubler, F., Moritz, T., Bagnall, D. J., MacMillan, C. P., Li, S. F., Parish, R. W., Dennis, E. S., Weigel, D., King, R. W. (2001) *GAMYB*-like genes, flowering, and gibberellin signaling in Arabidopsis. Plant Physiol. 127, 1682-1693.
22. Graebe, J. E. (1980) GA-biosynthesis: The development and application of cell-free systems for biosynthetic studies. *In*: Plant Growth Substances 1979, ed. Skoog, F., pp. 180-187. Berlin-Heidelberg-New York: Springer Verlag.
23. Graebe, J. E. (1987) Gibberellin biosynthesis and control. Annu. Rev. Plant Physiol. Plant Mol. Biol. 38, 419-465.
24. Hansen, E., Olsen, J. E., Junttila, O. (1999) Gibberellins and subapical cell divisions in relation to bud set and bud break in *Salix pentandra*. J. Plant Growth Regul. 18, 167-170.
25. Hay, A., Kaur, H., Phillips, A. L., Hedden, P., Hake, S., Tsiantis, M. (2002) The gibberellin pathway mediates KNOTTED1-type homeobox function in plants with different body plans. Curr. Biol. 12, 1557-1565.
26. Hedden, P. (1997) The oxidases of gibberellin biosynthesis: Their function and mechanism. Physiol. Plant. 101, 709-719.
27. Hedden, P., Kamiya, Y. (1997) Gibberellin biosynthesis: enzymes, genes and their regulation. Annu. Rev. Plant Physiol. Plant Mol. Biol. 48, 431-460.
28. Hedden, P., Phillips, A. L., Rojas, M. C., Carrera, E., Tudzynski, B. (2002) Gibberellin biosynthesis in plants and fungi: A case of convergent evolution? J. Plant Growth Regul. 20, 319-331.
29. Helliwell, C. A., Chandler, P. M., Poole, A., Dennis, E. S., Peacock, W. J. (2001) The CYP88A cytochrome P450, *ent*-kaurenoic acid oxidase, catalyzes three steps of the gibberellin biosynthesis pathway. Proc. Natl. Acad. Sci. U. S. A. 98, 2065-2070.
30. Helliwell, C. A., Sullivan, J. A., Mould, R. M., Gray, J. C., Peacock, W. J., Dennis, E. S. (2001) A plastid envelope location of Arabidopsis *ent*-kaurene oxidase links the plastid and endoplasmic reticulum steps of the gibberellin biosynthesis pathway. Plant J. 28, 201-208.
31. Kamiya, Y., Garcia-Martinez, J. L. (1999) Regulation of gibberellin biosynthesis by light. Curr. Opin. Plant Biol. 2, 398-403.
32. Kamiya, Y., Graebe, J. E. (1983) The biosynthesis of all major pea gibberellins in a cell-free system from *Pisum sativum*. Phytochemistry 22, 681-689.
33. Kaneko, M., Itoh, H., Inukai, Y., Sakamoto, T., Ueguchi-Tanaka, M., Ashikari, M., Matsuoka, M. (2003) Where do gibberellin biosynthesis and gibberellin signaling occur in rice plants? Plant J. 35, 104-115.
34. Kasahara, H., Hanada, A., Kuzuyama, T., Takagi, M., Kamiya, Y., Yamaguchi, S. (2002) Contribution of the mevalonate and methylerythritol phosphate pathways to the biosynthesis of gibberellins in Arabidopsis. J. Biol. Chem. 277, 45188-45194.
35. Kawaide, H., Sassa, T., Kamiya, Y. (2000) Functional analysis of the two interacting cyclase domains in *ent*-kaurene synthase from the fungus *Phaeosphaeria* sp L487 and a comparison with cyclases from higher plants. J. Biol. Chem. 275, 2276-2280.
36. King, R. W., Evans, L. T. (2003) Gibberellins and flowering of grasses and cereals: Prizing open the lid of the "florigen" black box. Annu. Rev. Plant Biol. 54, 307-328.

37. King, R. W., Evans, L. T., Mander, L. N., Moritz, T., Pharis, R. P., Twitchin, B. (2003) Synthesis of gibberellin GA_6 and its role in flowering of *Lolium temulentum*. Phytochemistry 62, 77-82.
38. Kuzuyama, T., Shimizu, T., Takahashi, S., Seto, H. (1998) Fosmidomycin, a specific inhibitor of 1-deoxy-*D*-xylulose 5-phosphate reductoisomerase in the nonmevalonate pathway for terpenoid biosynthesis. Tetrahedron Lett. 39, 7913-7916.
39. Lange, T., Hedden, P., Graebe, J. E. (1993) Biosynthesis of 12a-hydroxylated and 13-hydroxylated gibberellins in a cell-free system from *Cucurbita maxima* endosperm and the identification of new endogenous gibberellins. Planta 189, 340-349.
40. Lange, T., Hedden, P., Graebe, J. E. (1994) Expression cloning of a gibberellin 20-oxidase, a multifunctional enzyme involved in gibberellin biosynthesis. Proc. Natl. Acad. Sci. U. S. A. 91, 8552-8556.
41. Lange, T., Robatzek, S., Frisse, A. (1997) Cloning and expression of a gibberellin 2b, 3b-hydroxylase cDNA from pumpkin endosperm. Plant Cell 9, 1459-1467.
42. Lee, D. J., Zeevaart, J. A. D. (2002) Differential regulation of RNA levels of gibberellin dioxygenases by photoperiod in spinach. Plant Physiol. 130, 2085-2094.
43. Lois, L. M., Rodriguez-Concepcion, M., Gallego, F., Campos, N., Boronat, A. (2000) Carotenoid biosynthesis during tomato fruit development: regulatory role of 1-deoxy-*D*-xylulose 5-phosphate synthase. Plant J. 22, 503-513.
44. MacMillan, J. (2002) Occurrence of gibberellins in vascular plants, fungi and bacteria. J. Plant Growth Regul. 20, 387-442.
45. MacMillan, J., Takahashi, N. (1968) Proposed procedure for the allocation of trivial names to the gibberellins. Nature 217, 170-171.
46. Ogawa, M., Hanada, A., Yamauchi, Y., Kuwahara, A., Kamiya, Y., Yamaguchi, S. (2003) Gibberellin biosynthesis and response during Arabidopsis seed germination. Plant Cell 15, 1591-1604.
47. Okada, K., Kawaide, H., Kuzuyama, T., Seto, H., Curtis, I. S., Kamiya, Y. (2002) Antisense and chemical suppression of the nonmevalonate pathway affects *ent*-kaurene biosynthesis in *Arabidopsis*. Planta 215, 339-344.
48. Okada, K., Saito, T., Nakagawa, T., Kawamukai, M., Kamiya, Y. (2000) Five geranylgeranyl diphosphate synthases expressed in different organs are localized into three subcellular compartments in Arabidopsis. Plant Physiol. 122, 1045-1056.
49. Ozga, J. A., Ju, J., Reinecke, D. M. (2003) Pollination-, development-, and auxin-specific regulation of gibberellin 3b-hydroxylase gene expression in pea fruit and seeds. Plant Physiol. 131, 1137-1146.
50. Phinney, B. O. (1983) The history of gibberellins. *In*: The Biochemistry and Physiology of Gibberellins, ed. Crozier, A., pp. 19-52. New York: Praeger Publishers.
51. Querol, J., Campos, N., Imperial, S., Boronat, A., Rodriguez-Concepcion, M. (2002) Functional analysis of the *Arabidopsis thaliana* GCPE protein involved in plastid isoprenoid biosynthesis. FEBS Lett. 514, 343-346.
52. Rademacher, W. (2000) Growth retardants: Effects on gibberellin biosynthesis and other metabolic pathways. Annu. Rev. Plant Physiol. Plant Mol. Biol. 51, 501-531.
53. Reid, J. B., Botwright, N. A., Smith, J. J., O'Neill, D. P., Kerckhoffs, L. H. J. (2002) Control of gibberellin levels and gene expression during de- etiolation in pea. Plant Physiol. 128, 734-741.
54. Richman, A. S., Gijzen, M., Starratt, A. N., Yang, Z. Y., Brandle, J. E. (1999) Diterpene synthesis in *Stevia rebaudiana*: recruitment and up-regulation of key enzymes from the gibberellin biosynthetic pathway. Plant J. 19, 411-421.
55. Rodriguez-Concepcion, M., Boronat, A. (2002) Elucidation of the methylerythritol phosphate pathway for isoprenoid biosynthesis in bacteria and plastids. A metabolic milestone achieved through genomics. Plant Physiol. 130, 1079-1089.
56. Rodriguez-Concepcion, M., Campos, N., Lois, L. M., Maldonado, C., Hoeffler, J. F., Grosdemange-Billiard, C., Rohmer, M., Boronat, A. (2000) Genetic evidence of branching in the isoprenoid pathway for the production of isopentenyl diphosphate and dimethylallyl diphosphate in *Escherichia coli*. FEBS Lett. 473, 328-332.

57. Rohdich, F., Wungsintaweekul, J., Eisenreich, W., Richter, G., Schuhr, C. A., Hecht, S., Zenk, M. H., Bacher, A. (2000) Biosynthesis of terpenoids: 4-Diphosphocytidyl-2C-methyl-*D*-erythritol synthase of *Arabidopsis thaliana*. Proc. Natl. Acad. Sci. U. S. A. 97, 6451-6456.
58. Sakamoto, T., Kamiya, N., Ueguchi-Tanaka, M., Iwahori, S., Matsuoka, M. (2001) KNOX homeodomain protein directly suppresses the expression of a gibberellin biosynthetic gene in the tobacco shoot apical meristem. Genes Dev. 15, 581-590.
59. Sakamoto, T., Kobayashi, M., Itoh, H., Tagiri, A., Kayano, T., Tanaka, H., Iwahori, S., Matsuoka, M. (2001) Expression of a gibberellin 2-oxidase gene around the shoot apex is related to phase transition in rice. Plant Physiol. 125, 1508-1516.
60. Schneider, G., Schliemann, W. (1994) Gibberellin conjugates - an overview. Plant Growth Regul. 15, 247-260.
61. Schomburg, F. M., Bizzell, C. M., Lee, D. J., Zeevaart, J. A. D., Amasino, R. M. (2003) Overexpression of a novel class of gibberellin 2-oxidases decreases gibberellin levels and creates dwarf plants. Plant J. 14, 1-14.
62. Schwender, J., Muller, C., Zeidler, J., Lichlenthaler, H. K. (1999) Cloning and heterologous expression of a cDNA encoding 1-deoxy-*D*-xylulose-5-phosphate reductoisomerase of *Arabidopsis thaliana*. FEBS Lett. 455, 140-144.
63. Silverstone, A. L., Chang, C.-W., Krol, E., Sun, T.-p. (1997) Developmental regulation of the gibberellin biosynthetic gene *GA1* in *Arabidopsis thaliana*. Plant J. 12, 9-19.
64. Sponsel, V. M. (2002) The deoxyxylulose phosphate pathway for the biosynthesis of plastidic isoprenoids: early days in our understanding of the early stages of gibberellin biosynthesis. J. Plant Growth Regul. 20, 332-345.
65. Sponsel, V. M. (2003) Gibberellins. *In*: Encyclopedia of Hormones, ed. Henry, H. L., Norman, A. W., pp. 29-40: Academic Press.
66. Spray, C. R., Kobayashi, M., Suzuki, Y., Phinney, B. O., Gaskin, P., Macmillan, J. (1996) The *dwarf-1* (*d1*) mutant of *Zea mays* blocks 3 steps in the gibberellin-biosynthetic pathway. Proc. Natl. Acad. Sci. U. S. A. 93, 10515-10518.
67. Thomas, S. G., Phillips, A. L., Hedden, P. (1999) Molecular cloning and functional expression of gibberellin 2-oxidases, multifunctional enzymes involved in gibberellin deactivation. Proc. Natl. Acad. Sci. U. S. A. 96, 4698-4703.
68. Tudzynski, B., Mihlan, M., Rojas, M. C., Linnemannstons, P., Gaskin, P., Hedden, P. (2003) Characterization of the final two genes of the gibberellin biosynthesis gene cluster of *Gibberella fujikuroi* - *des* and *P450-3* encode GA_4 desaturase and the 13-hydroxylase, respectively. J. Biol. Chem. 278, 28635-28643.
69. van Huizen, R., Ozga, J. A., Reinecke, D. M. (1997) Seed and hormonal regulation of gibberellin 20-oxidase expression in pea pericarp. Plant Physiol. 115, 123-128.
70. Vidal, A. M., Ben-Cheikh, W., Talon, M., Garcia-Martinez, J. L. (2003) Regulation of gibberellin 20-oxidase gene expression and gibberellin content in citrus by temperature and citrus exocortis viroid. Planta 217, 442-448.
71. Ward, J. L., Jackson, G. J., Beale, M. H., Gaskin, P., Hedden, P., Mander, L. N., Phillips, A. L., Seto, H., Talon, M., Willis, C. L., Wilson, T. M., Zeevaart, J. A. D. (1997) Stereochemistry of the oxidation of gibberellin 20-alcohols, GA_{15} and GA_{44}, to 20-aldehydes by gibberellin 20-oxidases. Chem. Commun., 13-14.
72. Wolbang, C. M., Ross, J. J. (2001) Auxin promotes gibberellin biosynthesis in decapitated tobacco plants. Planta 214, 153-157.
73. Yamaguchi, S., Kamiya, Y., Sun, T. P. (2001) Distinct cell-specific expression patterns of early and late gibberellin biosynthetic genes during *Arabidopsis* seed germination. Plant J. 28, 443-453.
74. Yamaguchi, S., Smith, M. W., Brown, R. G. S., Kamiya, Y., Sun, T. P. (1998) Phytochrome regulation and differential expression of gibberellin 3b-hydroxylase genes in germinating Arabidopsis seeds. Plant Cell 10, 2115-2126.
75. Yamauchi, Y., Ogawa, M., Kuwahara, A., Hanada, A., Kamiya, Y., Yamaguchi, S. (2004) Activation of gibberellin biosynthesis and response pathways by low temperature during imbibition of *Arabidopsis thaliana* seeds. Plant Cell 16, 367-378.

B3. Cytokinin Biosynthesis and Metabolism

Hitoshi Sakakibara
Plant Science Center, RIKEN, 1-7-22, Tsurumi, Yokohama 230-0045, Japan.
E-mail: sakaki@postman.riken.go.jp

INTRODUCTION

Since the discovery of cytokinins in the 1950s, it has been clearly established that they play an important role in various processes in the growth and development of plants, including the promotion of cell division, the counteraction of senescence, the regulation of apical dominance and the transmission of nutritional signals. Kinetin (Fig. 1a) was the first substance to be identified as a cytokinin, and although it was isolated from autoclaved herring sperm DNA (34) it is not naturally produced and has not been found in living plants. The naturally occurring cytokinin *trans*-zeatin (tZ, Fig. 1b) was first isolated from immature maize endosperm in the early 1960s (26). In the following 40 years, several cytokinin species have been identified from various plant species (35, 43). These studies demonstrated that natural plant cytokinins are adenines which have substituted at the N^6 terminal either an isoprene-derived side chain (isoprenoid cytokinins), or an aromatic derivative side chain (aromatic cytokinins). In most cases, nucleosides, nucleotides and other sugar-conjugates have also been found, implying that there is a metabolic network for their interconversion. Another well-known type of cytokinin is the phenylurea-type species (Fig. 2). Although they were first reported to be isolated from coconut milk, we now understand that the phenylurea-type cytokinins are synthetic, and that there is no evidence that any of them occur naturally in plant tissues (35). Progress in chemical research

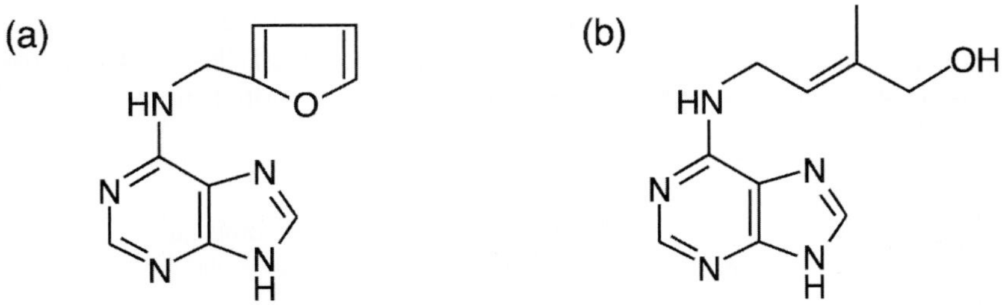

Figure 1. Structures of kinetin (a) and *trans*-zeatin (b). Kinetin was the first synthetic cytokinin to be discovered and *trans*-zeatin was the first naturally occurring one to be discovered.

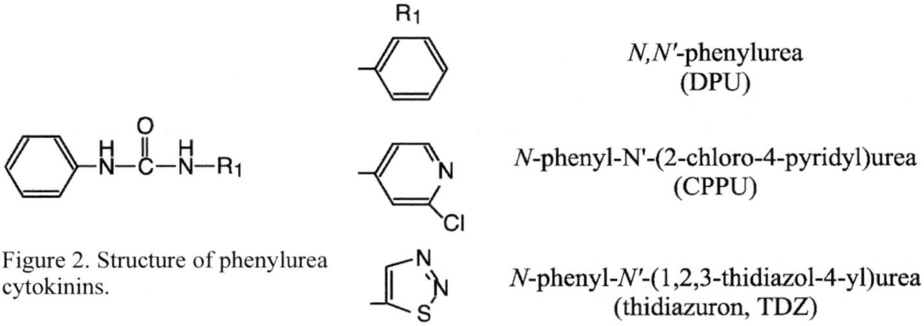

Figure 2. Structure of phenylurea cytokinins.

N,N'-phenylurea (DPU)

N-phenyl-N'-(2-chloro-4-pyridyl)urea (CPPU)

N-phenyl-*N'*-(1,2,3-thidiazol-4-yl)urea (thidiazuron, TDZ)

techniques has allowed several cytokinin metabolic enzymes to be characterized; some have also been purified and their corresponding genes cloned. These extensive studies have produced a broad picture of cytokinin biosynthetic and metabolic pathways in plants, although some elements are still putative. In this chapter, we outline the synthesis and metabolism of cytokinins in terms of their structural diversity, the biochemical nature of the metabolic enzymes involved and the regulation of gene expression.

STRUCTURAL DIVERSITY AND ACTIVITY OF NATURAL CYTOKININS

Skoog and Armstrong, pioneers of cytokinin research, defined cytokinin as 'a generic name for substances which promote cell division and exert other growth regulatory functions in the same manner as kinetin' (41). Figures 3 and 4 summarize the structural diversity of naturally occurring cytokinins and the derivatives. The variation and distribution of cytokinin species are not ubiquitous and depend on the plant species, tissue and developmental stage. Although several cytokinins have been identified, the physiological function of individual species is not yet completely understood. However, previous studies have proposed that cytokinins could be classified into three major groups: active forms, translocation forms, and storage and inactivated forms.

A bioassay with *Funaria hygrometrica* suggested that free-base cytokinins such as tZ[1] and iP were active, whereas tZR only exhibits a low level of activity (42). The identification of cytokinin receptor genes provides more concrete evidence. An *in vitro* binding assay of ^{3}H-labeled cytokinin with AHK4 (CRE1), a cytokinin receptor, clearly demonstrated that free-base

---

[1] Abbreviations: BA, benzyladenine; CKX, cytokinin oxidase/dehydrogenase; cZ, *cis*-zeatin; cZR, *cis*-zeatin riboside; cZRMP, *cis*-zeatin riboside monophosphate; DMAPP, dimethylallyl diphosphate; DZ, dihydrozeatin; HMBDP, 4-hydroxy-3- methyl-2-(*E*)-butenyl diphosphate; iP, isopentenyladenine; iPR, isopentenyladenine riboside; iPRMP, isopentenyladenine riboside 5'-monophosphate; IPT, isopentenyl- transferase; MEP, methylerythritol phosphate; mT, meta-topolin; oT, ortho-topolin; tZ, *trans*-zeatin; tZR, *trans*-zeatin riboside; tZRMP, *trans*-zeatin riboside 5'-monophosphate

R₁	Base	Nucleoside	Nucleotide
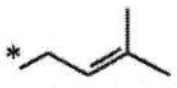	isopentenyladenine (iP)	isopentenyladenine riboside (iPR)	isopentenyladenine riboside 5'-(mono, di, tri) phosphate (iPRMP, iPRDP, iPRTP)
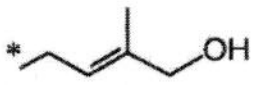	trans-zeatin (tZ)	trans-zeatin riboside (tZR)	trans-zeatin riboside 5'-(mono, di, tri) phosphate (tZRMP, tZRDP, tZRTP)
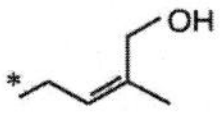	cis-zeatin (cZ)	cis-zeatin riboside (cZR)	cis-zeatin riboside 5'-monophosphate (cZRMP)
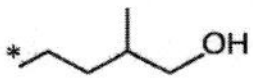	dihydrozeatin (DZ)	dihydrozeatin riboside (DZR)	dihydrozeatin riboside 5'-monophosphate (DZRMP)
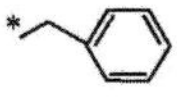	benzyladenine (BA)	benzyladenine riboside (BAR)	benzyladenine riboside 5'-monophosphate (BARMP)
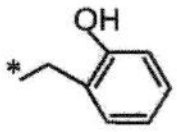	ortho-topolin (oT)	o-topolin riboside (oTR)	
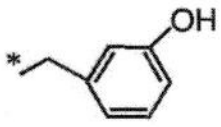	meta-topolin (mT)	m-topolin riboside (mTR)	
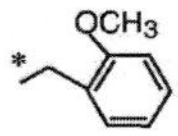	ortho-methoxytopolin (MeoT)	o-methoxytopolin riboside (MeoTR)	
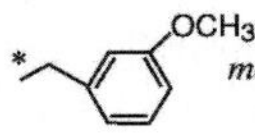	meta-methoxytopolin (MemT)	m-methoxytopolin riboside (MemTR)	

Figure 3. Structures and names of naturally occurring cytokinins. Conjugation site of the R_1 with adenine ring is asterisked (*).

cytokinins such as tZ and iP could bind the receptor but not the nucleosides (53). Thus, free-base cytokinins are now regarded as active forms.

The relative abundance of nucleosides in the xylem and phloem sap of various plant species, suggests that they are translocation forms. It is not clear whether cytokinin nucleotides, primary products of the biosynthetic reaction, have their own specific function or not.

On the other hand, reciprocal changes in the accumulation level of the cytokinin O-glucosides versus free bases in response to environmental factors such as low temperature (9) suggest that the sugar conjugates are the storage and inactivated cytokinin forms. The observed metabolic stability of the O-glucosides against cytokinin oxidase/dehydrogenase (CKX), a cytokinin

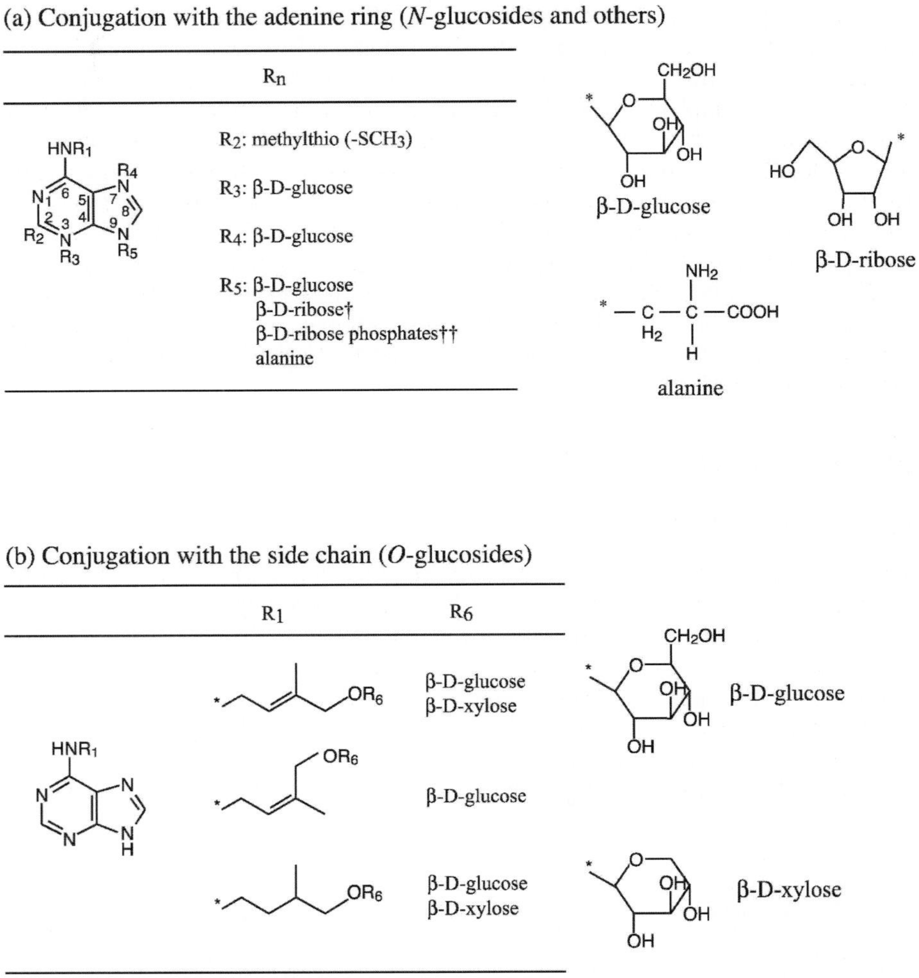

Cytokinin biosynthesis and metabolism

(a) Conjugation with the adenine ring (*N*-glucosides and others)

R_n	
R_2: methylthio (-SCH3)	
R_3: β-D-glucose	
R_4: β-D-glucose	
R_5: β-D-glucose	
β-D-ribose†	
β-D-ribose phosphates††	
alanine	

β-D-glucose

β-D-ribose

alanine

(b) Conjugation with the side chain (*O*-glucosides)

R_1	R_6
	β-D-glucose
	β-D-xylose
	β-D-glucose
	β-D-glucose
	β-D-xylose

β-D-glucose

β-D-xylose

Figure 4. Cytokinin derivatives conjugating with sugars and amino acid. (a) Conjugation with the adenine ring (*N*-glucosides and others). Variation in R_1 and †, †† are also shown in Fig. 3 (b) Conjugation with the side chain (*O*-glucosides). R_1 side chain is conjugated at asterisked site.

degradation enzyme (33), also supports this hypothesis. Other conjugates, such as *N*-glucosyl and *N*-alanyl derivatives (Fig. 4a) are also more stable metabolically than the free bases (33). In terms of their physiological activity, the *O*-glucosides, the *N*7- and *N*9-glucosides and the *N*9-alanyl cytokinins exhibit little or no activity in cytokinin bioassays, suggesting that they are inactive forms (28). In IPT overexpressing plants, tZ-*O*-glucoside and/or iP-*N*7- and iP-*N*9-glucosides hyper-accumulate (18, 55), probably due to the inactivation and storage of the excess cytokinins.

Cytokinin species also have small side chain variations: the presence or absence and position of a hydroxyl group (Fig. 3). For example, there are four isoprenoid cytokinins, namely iP, tZ, cZ and DZ. Although there is currently no widely accepted explanation for the physiological significance of these variants, some differences have been found in terms of their activity, abundance and stability. iP and tZ are active and their derivatives are the most abundant species in plants, but they are susceptible to CKX (7). DZ appears to be a biologically stable species because of its resistance to CKX, is generally found in small quantities. cZ has been thought to be a less active and relatively more stable species than tZ and iP because of its low affinity to CKX (7). On the other hand, the hydroxylated side chain of tZ, cZ and DZ can be modified by O-glucosylation, which converts them to storage forms (Fig. 4b). Although Skoog and Armstrong defined cytokinins on the basis of their physiological activity, we usually call the various derivatives 'cytokinins' as well as the more active forms.

BIOSYNTHESIS

Isoprenoid cytokinins

Two possible pathways are proposed for the biosynthesis of isoprenoid cytokinins: one is derived from tRNA degradation and another is derived from the isopentenylation of free adenine nucleotides (Fig. 5). The first identification of substrates for the latter pathway was reported in a slime mold, *Dictyostelium discoideum* (50). This study found that dimethylallyldiphosphate (DMAPP) and AMP were the biosynthetic reaction substrates of two cytokinins, iPR and iP, and that the resulting primary product was iPRMP. This reaction is catalyzed by isopentenyltransferase (IPT).

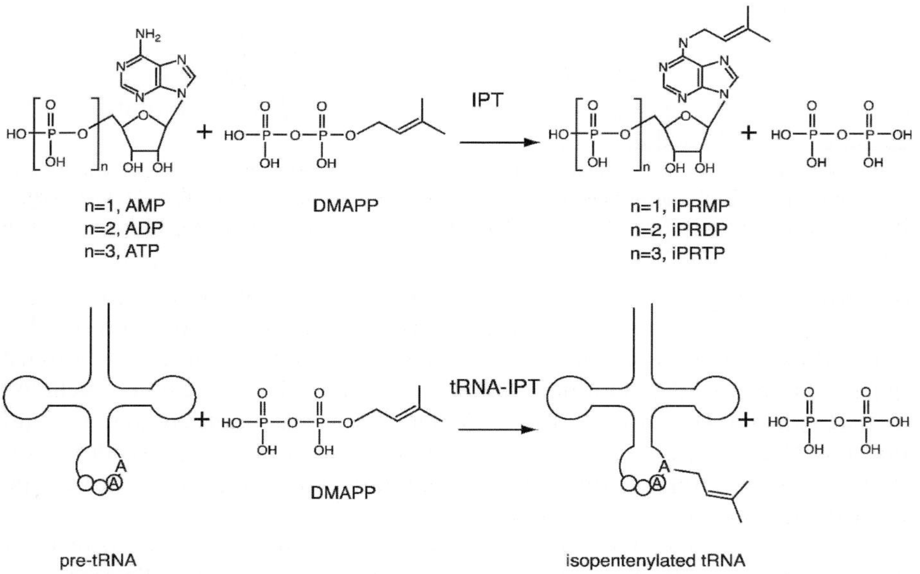

Figure 5. Primary reactions of cytokinin biosynthesis catalyzed by IPT and tRNA-IPT.

Biochemical analysis of the purified enzyme has shown that the *D. discoideum* IPT can utilize both AMP and ADP as an isoprenoid-acceptor (21).

The first identification of a gene encoding an IPT, *tmr*, was carried out in *Agrobacterium tumefaciens*, a crown gall-forming bacterium (1, 4). It is now known that *A. tumefaciens* has two IPT genes, *tmr* and *tzs*. *tmr* is encoded on the T-DNA region of the Ti-plasmid and *tzs* is located on the virulence region. After infection, *tmr* is integrated into the genome of host plants, whereas *tzs* functions within the bacterial cells themselves. The IPT activity of the recombinant tmr was identified *in vitro*, and it was revealed that tmr exclusively utilizes AMP as the acceptor.

The reaction pathway of iPRMP synthesis in higher plants has been assumed to be similar to that found in *D. discoideum* and *A. tumefaciens*. Although multiple attempts have been made to purify and characterize plant IPT from cytokinin-autotrophic tobacco pith tissue culture and developing maize kernels, only a few biochemical properties have been reported perhaps due to low content or enzyme instability. However, a breakthrough in the identification of a cytokinin biosynthesis gene in higher plants was brought about by a reverse genetic approach using the *Arabidopsis* genome sequence.

IPT genes in plants

Plant IPT genes were first identified in *Arabidopsis* (22, 46). The *Arabidopsis* genome encodes nine genes (*AtIPT1* to *AtIPT9*) which show similarity to Agrobacterial IPT. Phylogenetic analysis has classified them into two types: one (containing *AtIPT2* and *AtIPT9*) is similar to the gene encoding a tRNA modifying enzyme, known as tRNA-isopentenyltransferase (tRNA-IPT; EC 2.5.1.8) and the other (containing *AtIPT1*, *AtIPT3* to *AtIPT8*) is similar to the gene encoding the iPRMP-forming enzyme, IPT (EC 2.5.1.27). tRNA-IPT and IPT are structurally related, suggesting that these transferases diverged from a common ancestral gene. *AtIPT6* appears to be inactive in the ecotype WS and some other ecotypes because they contain a frameshift mutation (22, Takei, K. unpublished).

The enzyme activity of the AtIPTs was identified both *in vitro* and *in vivo*. Extracts of *E. coli* transformants expressing *AtIPT1* and *AtIPT3* to *AtIPT8* demonstrated IPT activity (46). In addition, the overexpression of *AtIPT4* enabled calli to regenerate shoots even in the absence of externally applied cytokinins (22).

In *Petunia hybrida*, an activation tagging screening identified an IPT gene (*Sho*) (55). The *Sho* overexpressing line exhibits several phenotypes including enhanced shooting, reduced apical dominance and delayed senescence and flowering. In particular, *Sho* expression in petunia and tobacco enhances the accumulation levels of iP-type cytokinins, in contrast to the IPT gene from *A. tumefaciens*, which primarily increases tZ-type cytokinins (18).

Biochemical nature of IPTs

AtIPT1 and AtIPT4 have been purified as recombinant enzymes and their cytokinin biosynthetic activity established (Table 1) (22, 46, Sakakibara et al. unpublished). These studies imply a difference in substrate preference between the two isoenzymes. AtIPT1 was found to be capable of synthesizing iPMP from AMP and DMAPP *in vitro*. From the respective Kms, AtIPT1 can utilize all three adenine nucleotides, although it prefers ADP and ATP, whereas AtIPT4 uses ATP and ADP exclusively. These results demonstrate that plant IPTs have different substrate preferences, at least in terms of adenine nucleotides. At present, we do not have a convincing explanation for the broad specificity of plant IPTs for the adenine nucleotides. *AtIPTs* are expressed in various non-photosynthetic cells and tissues including the root stele, lateral root primodium and phloem companion cells (48). This broad specificity for adenine nucleotides in higher plants might be efficient for the utilization of the adenine skeleton in the reaction, regardless of the energy charge status.

Table 1. Kinetic parameters of IPTs

enzyme	species	substrates	K_m (substrate)	ref.
tmr	*A. tumefaciens*	AMP	85 nM (AMP)	(8)
		DMAPP	8.3 µM (DMAPP)	(8)
IPT	*D. discoideum*	AMP, ADP	100 nM (AMP)	(21)
		DMAPP	2.2 µM (DMAPP)	(21)
AtIPT1	*Arabidopsis*	ATP, ADP, AMP	185 µM (AMP)	(46)
		DMAPP	14.6 µM (ADP)*	
			11.4 µM (ATP)*	
			8.3 µM (DMAPP)*	
AtIPT4	*Arabidopsis*	ATP, ADP	9.1 µM (ADP)*	(22)
		DMAPP	3.4 µM (ATP)*	
			11.6 µM (DMAPP)*	

*Sakakibara H. et al. manuscript in preparation.

Spatial expression patterns of the AtIPTs

Although several studies have implied that cytokinin could be synthesized in various sites in plant body (27), the idea is still controversial and it is generally thought that the main site for cytokinin production is in the roots. Analyses of the spatial expression pattern of the *AtIPTs* have provided new insight into our understanding of the cytokinin biosynthetic tissues. In mature *Arabidopsis* plants, *AtIPTs* are expressed not only in the roots, but also in various tissues of the shoots, such as the leaves, stem, flowers and siliques (48). Analysis of transgenic plant seedlings expressing a gene for the green fluorescent protein, driven by *AtIPT* promoters, showed that *AtIPT1* was predominantly expressed in the vascular stele of roots, *AtIPT3* was expressed in the phloem companion cells throughout the entire seedling, *AtIPT5* was expressed in the lateral root primordium, and *AtIPT7* was expressed in both the vascular stele and the root phloem companion cells (48) (Fig. 6). This evidence indicates that the expression of the *AtIPTs* is spatially differentiated, and that cytokinin can be synthesized at various sites within the plant body. The specific expression of *AtIPT5* in the lateral root primordium implies the induction of cell division within the primordium and neighboring cells at an early stage of the plant's development. On the other hand, the expression of *AtIPT3* and *AtIPT7* in the phloem companion cells implies a contribution to long-distance signaling via the vascular systems. Spatial variation in individual *AtIPT* expressions and their cooperative actions must be important for the normal growth and development of plants.

Regulation of *IPT* expression

Because the use of molecular approaches for determining the regulation of *IPT* expression is in its infancy, only limited information is currently available. The environmentally-responsive accumulation of cytokinins has been reported in response to nitrogen availability (47), decapitation (40), long-daylight treatment (6) and so on. In *Arabidopsis*, the accumulation levels of the *AtIPT3* transcripts were correlated with the availability of nitrate ions (48). *AtIPT3* expression was induced by nitrate ions accompanying the accumulation of tZRMP. Nitrate-dependent accumulation both of the *AtIPT3* transcript and the cytokinins was greatly reduced in a Ds transposon-insertion mutant of *AtIPT3*. *AtIPT3* expression was also correlated with the availability of sulfate and phosphate ions. These results imply that *AtIPT3* is a key determinant of cytokinin biosynthesis in response to the availability of inorganic macronutrients (48).

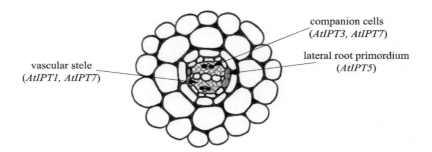

Figure. 6. Schematic representation of the spatial expression pattern of *AtIPTs* in Arabidopsis roots. Phloem companion cells in which *AtIPT3* and *AtIPT7* express are marked in black. Lateral root primordium in which *AtIPT5* exclusively expresses are marked in grey. Vascular stele in which *AtIPT1* and *AtIPT7* express are hatched.

tRNA-IPT

Mature tRNA molecules contain many modified nucleosides. Some tRNA molecules contain an iPR or a derivative residue at the site adjacent to the anticodon. tRNA-IPT catalyzes the first step of the isopentenylation of the adenine residue. tRNA-IPT genes appear to exist ubiquitously in living organisms: they have been identified in *Homo sapiens*, *Saccharomyces cerevisiae* and *E. coli*. In *Arabidopsis*, *AtIPT2* functionally complements the *MOD5* mutation in *S. cerevisiae* and AtIPT2 exhibits tRNA-IPT activity *in vitro* (19).

Because the degradation of tRNA could produce active cytokinins, it has been suggested to be a possible source of the hormone. Several different compounds, iPR, methylthio-iPR, cZR, tZR, and *cis*- and *trans*-methylthio-ZR have been found in plant tRNAs (Figs. 3 and 4), and cZR is the most abundant (38). The ratio of *cis*- to *trans*- isomer in tRNA is approximately 40:1. It has been proposed that this heavily skewed distribution suggests that *cis*-type cytokinins originated from tRNA degradation. Although interconversion between cZ and tZ can be catalyzed zeatin isomerase, calculations of the rates of tRNA turnover and cytokinin production suggest that the tRNA degradation pathway is not a major source of cytokinins, at least not the tZ-type (23). Further modifying enzymes producing cZR and other cytokinin species have not yet been characterized.

Cis-type cytokinins

Although *cis*-cytokinins have exhibited weak activity in several bioassays, substantial quantities of cZ and their derivatives are found in some plant species, such as chickpea (15), maize (51) and potato tuber (44). The physiological function of cZ has not been elucidated yet. One study suggests that cZ-type cytokinins act as transient storage forms, which are transported to the shoot and converted to the *trans* isomers (more active species) by zeatin isomerase in the presence of light (5). If the major origin of cZ is tRNA degradation even in such species, we cannot ignore the physiological contribution of metabolic flow from tRNA degradation. More recently, it has been reported that a maize cytokinin receptor, ZmHK1, could respond to cZ with a comparable sensitivity to tZ (54). This report leads us to reconsider the physiological significance of cis-type cytokinins, as they might have a specialized function in some plant species.

Alternative pathway for cytokinin biosynthesis

It was formerly believed that the primary product of cytokinin biosynthesis was the iP-type species, and that tZ-type cytokinins were produced only by hydroxylation of the side chain, brought about by microsomal hydroxylase activity (13) (the iPRMP-dependent pathway). However, recent progress in cytokinin metabolism research has provided us with a new picture of the pathway. From *in vivo* deuterium labeling experiments in *Arabidopsis*, Åstot et al. (3) have suggested the existence of an alternative cytokinin biosynthetic pathway whose initial product is not iPRMP, but instead is tZRMP (i.e. an iPRMP-independent pathway). They first found that the overexpression of Agrobacterial IPT in *Arabidopsis* caused labeling of tZRMP with deuterium oxide to a much greater extent than of iPRMP. The rate of tZRMP biosynthesis in the transgenic plants was calculated to be 66-fold higher than that of iPRMP. In wild type plants, the existence of the iPRMP-independent pathway was also confirmed by the labeling approach. Thus, in addition to tRNA degradation, there are at least two pathways for the *de novo* biosynthesis of cytokinin, both of which are catalyzed by IPT. One substrate is DMAPP, whose product is iPRMP, and another substrate is a hydroxylated precursor originating from a terpenoid. In this case, a possible substrate is a hydroxylated derivative of DMAPP, such as 4-hydroxy-3-methyl-2-(*E*)-butenyl diphosphate (HMBDP, Fig. 7). HMBDP is a metabolic intermediate of the methylerythritol phosphate (MEP) pathway, which occurs in bacteria and plastids. In fact, an *Agrobacterium* IPT, tzs, can catalyze the direct formation of tZRMP from AMP and HMBDP (24). When several AtIPTs (AtIPT1, AtIPT4, AtIPT7 and AtIPT8) were expressed in *E. coli*, significant amounts of tZ were secreted into the culture medium, in addition to isopentenyladenine (iP) (46). These results imply that AtIPTs can

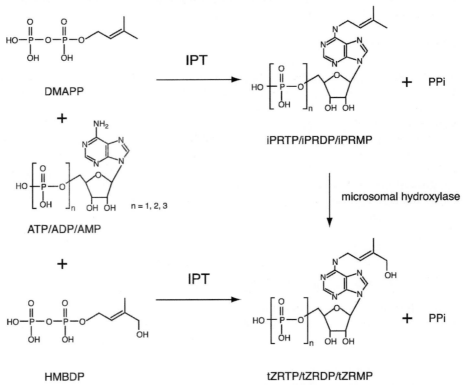

Figure 7. Alternative pathway for cytokinin biosynthesis. PPi, pyrophosphate.

at least partially catalyze the formation of tZ-type cytokinins by utilizing HMBDP, a hydroxylated side chain substrate. However, in *in vitro* experiments, the reaction efficiency of purified AtIPT7 for HMBDP was much lower than for DMAPP (about $1/100^{th}$ of the efficiency) (45). Furthermore, the overexpression of *Agrobacterium* IPT causes the dominant accumulation of tZ-type cytokinins, whereas that of plant IPT causes the dominant accumulation of iP-type species (18, 55). Thus, the extent of the physiological contribution of the iPRMP-independent pathway in higher plants has not yet been established.

Aromatic cytokinins

Aromatic cytokinins have been identified in several plant species, such as poplar (43) and *Arabidopsis* (49). Among the cytokinins, meta-topolin (mT) has the highest activity: in plant species, not only the free-base species but the nucleosides, nucleotides and glucosides have also been found (43). This implies that the at least part of the metabolic pathway is shared with that of the isoprenoid cytokinins. However, the catalytic enzymes used in the biosynthetic reaction and interconversion among mT, oT and BA aromatic cytokinins have not yet been identified. The influence of the AtIPTs, which have already been identified as isoprenoid cytokinin biosynthetic enzymes, on the aromatic cytokinins has not been examined.

METABOLISM

The proposed cytokinin metabolic pathway has been based on the results of experiments in which exogenous application of radiolabeled cytokinins is followed by determination of the metabolites. When such radiolabeled cytokinins are applied to plant tissue, they are usually rapidly distributed among the respective nucleotide, nucleoside, and base forms, and are then further metabolized into degraded products or sugar conjugates (27). The cytokinin metabolic pathway can be broadly classified into two types: modifications of adenine moiety and enzymes affecting the side chain. Significant advances during the 1980s in characterizing the metabolic enzymes suggested that the processes leading to the modification of adenine moiety are shared with the purine metabolic pathway. However, to date there is no evidence for the existence of an enzyme with absolute specificity for the cytokinins. So far as they have been studied, the reaction efficiency ($Vmax$: Km ratio) of the purine metabolic enzymes favors adenine over cytokinins. However, taking the difference in the abundance of the cytokinins (nM order) and other adenine compounds (mM order), they might nevertheless be sufficient to serve as cytokinin metabolic enzymes. On the other hand, enzymes for modifications of the side chain appear to be specific for cytokinins.

Modification of adenine moiety

5'-ribonucleotide phosphohydrolase (5'-nucleotidase)
Dephosphorylation from cytokinin nucleotides to nucleosides is catalyzed by 5'-ribonucleotide phosphohydrolase (5'-nucleotidase, EC 3.1.3.5) (Fig. 8, reaction 2). This enzyme has been partially purified and characterized in wheat germ (12). Column chromatography was used to separate the activity into two forms, F-I (110 kD) and F-II (57 kD). To our current knowledge, both forms can catalyze AMP, iPRMP, GMP and IMP, and require divalent metal ions such as Mg^{2+}. The Km values of F-I and F-II for iPRMP were found to be 3.5 µM and 12.8 µM, respectively, and the $Vmax : Km$ ratios for iPRMP were 70% and 65% of those for AMP, respectively.

In addition to the 5'-nucleotidase, non-specific phosphatases should participate in the dephosphorylating step of cytokinin nucleotides to the nucleosides. However, a comparison of the extent of the physiological contribution of the two enzymes to this step has not yet been undertaken.

Adenosine nucleosidase
The deribosylation step from cytokinin nucleosides to free bases is catalyzed by adenosine nucleosidase (EC 3.2.2.7) (Fig. 8, reaction 3). Because free base cytokinins are physiologically active, this step must be important for the regulation of cytokinin action. This enzyme has been characterized in several plant species and it has been partially purified in wheat germ (56). The Km for iPR is 2.4 µM and for adenine it is 1.4 µM at an optimum pH of 4.7. The $Vmax : Km$ ratio for iPR was approximately one-fifth of that for adenosine. In the wheat germ study, the enzyme's activity did not require orthophosphate, indicating that the reaction is not mediated by purine nucleoside phosphorylase.

Purine nucleoside phosphorylase
Conversion from free base cytokinins to nucleosides could be catalyzed by purine nucleoside phosphorylase (EC 2.4.2.1) (Fig. 8, reaction 5). Characterization of the partially purified enzyme of wheat germ demonstrated that the Km values for iP, BA and adenine are 57 µM, 46.5 µM 32 µM, respectively (14). The $Vmax : Km$ ratios for iP and BA are 70% and 71% of that for adenine, respectively. The reaction mechanism is essentially reversible. However, the equilibrium constant values show that the reaction strongly favors nucleoside formation (14). Based on the differences in enzymatic properties, this enzyme is distinct from inosine-guanosine phosphorylase, another enzyme involved in purine metabolism.

Adenosine kinase
Adenosine kinase (EC 2.7.1.20) could catalyze the formation of cytokinin nucleotides from the corresponding nucleoside (Fig. 8, reaction 4). The wheat germ form of the enzyme has been partially characterized: the Km value for iPR was 31 µM whereas that for adenosine was 8.7 µM (11). The cDNA was

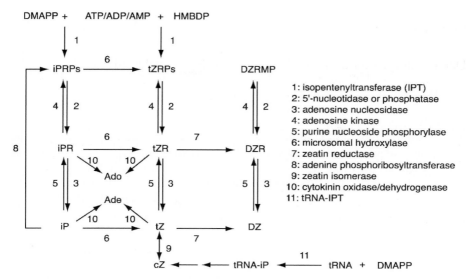

Figure 8. Current model of the cytokinin metabolic pathway and catalytic enzymes. iPRPs, isopentenyladenine riboside mono-, di-, tri-phosphates; tZRPs, trans-zeatin riboside mono-, di-, tri-phosphates; tRNA-iP, isopentenylated tRNA

cloned from *Physcomitrella patens* by functional complementation of an *E. coli* purine auxotrophic strain (52).

Adenine phosphoribosyltransferase
There is another route for converting free base cytokinins into nucleotides, which is catalyzed by adenine phosphoribosyltransferase (EC 2.4.2.7) (Fig. 8, reaction 8). In *Arabidopsis*, five relevant genes (*APT1* to *APT5*) have been found on the nuclear genome and the three of them (APT1 to APT3) have been biochemically identified (2, 39). The biochemical characterization by Schnorr et al revealed that APT2 (ATapt2) has a higher affinity to BA ($Km = 3$ μM) than adenine ($Km = 10$ μM) although APT1 (ATapt1) has an opposite affinity to these substrates (39). Another study reported that APT3 has more similar properties to APT2 than APT1 (2), suggesting that APT2 and APT3 both participate in cytokinin metabolism.

The extent of the physiological contribution of the two pathways, namely the purine nucleoside phosphorylase/adenosine kinase pathway and the adenine phosphoribosyltransferase pathway, has not yet been fully understood.

N-Glucosylation
Glucosylation of cytokinins occurs at the 3, 7, and 9 positions of the purine moiety and the modification of 7- and 9-positions makes cytokinins inactive. A glucosyltransferase (EC 2.4.1.118) catalyzes the formation both of the 7- and 9-glucosides using UDP- or TDP-glucose (17). The formation of 7-glucosides takes place preferentially, at a rate 10-fold greater than 9-glucosides in the case of tZ and 21-fold greater in that of cZ (17). The Km

values for tZ and UDP-glucose are 150 µM and 190 µM, respectively. The biochemical nature of glucosyltransferase for the 3-position has not been characterized. So far as examined, the glucosides cannot be hydrolyzed by β-glucosidase. Thus, the modification is thought to be an irreversible inactivation.

N-Alanylation
β-(9-cytokinin)-alanine (lupinic acid) synthase (EC 4.2.99.13) catalyzes the conversion of tZ, cZ, iP and DZ into 9-alanyl derivatives. The enzyme has been partially purified from lupin seeds (16). Although the affinity for *O*-acetyl-L-serine, a possible donor of alanine moiety, is high (Km = 47 µM), that for tZ is relatively low (Km = 880 µM). The physiological significance of alanylation is not well understood in this context.

Modification of the side chain

Zeatin isomerase
The conversion of cZ to tZ can be catalyzed by zeatin isomerase (Fig. 8, reaction 9). Although low non-enzymatic isomeric conversion occurs *in vitro*, the presence of the enzyme substantially enhances the reaction. The enzyme has been partially purified from the endosperm of *Phaseolus vulgaris* (5) and appears to be a 68 kD glycoprotein. The reaction favors conversion from the *cis* to the *trans* form and requires flavin (FAD or FMN), light and dithiothreitol. At present, the physiological significance of isomeration in cytokinin metabolism is not well understood, but it possibly provides a link between tRNA metabolism and tZ synthesis.

Zeatin reductase
Zeatin reductase (EC 1.3.1.69) catalyzes the irreversible conversion of tZ to DZ (Fig. 8, reaction 7). Because DZ and its derivatives are not substrates for cytokinin oxidase, this reductive conversion stabilizes the physiological activity of cytokinins. Zeatin reductase has been partially purified from the immature embryos of *Phaseolus vulgaris* (32). Elution on chromatographic columns revealed that there are two forms, a high molecular weight isozyme (HMW, 55 kD) and a low molecular weight isozyme (LMW, 25 kD), which have distinct electrical charges. Both of the activities require NADPH as a cofactor and specifically catalyze tZ but not cZ, iP and the nucleoside derivatives. In spite of the high substrate specificity, the Km values of the HMW and LMW isozymes for tZ were estimated at 70 µM and 100-230 µM, respectively, which are much higher than the concentration of tZ *in vivo*.

Hydroxylase
Although conversion of iP-type cytokinin to a tZ-type one is thought to be catalyzed by a hydroxylase (Fig. 8, reaction 6), the biochemical nature of the enzyme is limited. Partially characterized hydroxylase from cauliflower microsomal fractions could hydroxylate both iP and iPR to tZ and tZR,

respectively (13). Sensitivity to metyrapone, an inhibitor of cytochrome P450, and its dependency on the presence of NADPH strongly suggests that the reaction is catalyzed by a P450 enzyme. In spite of the physiological importance of this step in the cytokinin metabolism pathway, further characterization has not yet been carried out.

As described above, an alternative pathway whose initial product is tZRMP has been identified in *Arabidopsis* (3). However, several lines of evidence have not confirmed the existence, or the importance, of this hydroxylation step.

O-Glucosylation
Glucosylation of the isoprenoid side chain to form the O-β-D-glucopyranosyl derivatives of tZ and/or cZ appears to occur in a wide variety of plant species. As far as is currently known, most zeatin O-glucosyltransferases are specific for certain tZ-type cytokinins. The enzyme has been purified and its kinetic parameters characterized: the *Km* values for tZ and UDP-glucose are 28 μM and 200 μM, respectively at an optimal pH 8.0 (36). The gene for the transferase was first identified in *Phaseolus lunatus* (31). ZOG1 utilizes tZ and UDP-glucose and the gene predominantly expresses in immature seeds but only faintly in roots and leaves. Thus, the glucosyltransferases mainly function in the early stages of seed development.

A cZ-specific glucosyltransferase, cisZOG1, was recently discovered in maize (29); the isozyme cisZOG2 also prefers the cZ isomer to tZ (51). The enzyme utilizes cZ and UDP-glucose but not tZ, DZ, cZR, tZR or UDP-xylose. cisZOG1 predominantly expresses in roots, only weakly in the cob and kernel, and faintly in the leaves and stem (29). The cZ-O-glucoside content is also most abundant in the roots (51).

These O-glucosides are hydrolyzed by β-glucosidase, resulting in the release of the corresponding active bases.

O-Xylosylation
In addition to UDP-glucose, UDP-xylose also can serve as a donor for O-sugar conjugation in some plant species. This reaction is catalyzed by O-xylosyltransferase (EC 2.4.1.204). In *P. vulgaris* the gene for the transferase *ZOX1* is expressed strongly in immature seeds but not in the leaves and roots of older plants (30). Both tZ and DZ, but not tZR or cZ, can be utilized by this xylosyltransferase, which has been purified and characterized. The *Km* values for tZ, DZ and UDP-xylose are 2 μM, 10 μM and 3 μM, respectively at the optimal pH 8.5 (36). The primary structure of ZOX1 is closely related to that of ZOGs (90% identity), indicating that they share a common ancestral gene. So far as is currently known, O-xylosylzeatin is found only in *Phaseolus*.

DEGRADATION

Cytokinin oxidase/dehydrogenase

Cytokinin degradation plays an important homeostatic role in the control of the accumulation of cytokinin metabolites and their distribution in plants. This reaction is catalyzed by cytokinin oxidase/dehydrogenase (CKX, EC 1.5.99.12). CKX irreversibly inactivates cytokinins by cleaving the side chain (Fig. 9). CKX activity was first observed in the early 1970s, and in the following thirty years has been characterized in several plant species. Consequently, it has been shown that CKX is a monomeric enzyme with a molecular mass of about 60 kD; some of the purified enzymes were found to be glycosylated. CKX degrades base and nucleoside forms of isoprenoid cytokinins that have an unsaturated isoprenoid side chain, such as iP, tZ, iPR and tZR. The nucleotide derivatives are not susceptible to CKX. In the case of iP, the degraded products are adenine and 3-methyl-2-butenal. The delta-2 double bond in the isoprenoid side chain of tZ and iP appears to be important for substrate recognition because reduction of the bond confers resistance to CKX. Thus, DZ-type cytokinins are resistant to CKX. Side chain modification such as O-glucosylation also confers CKX resistance.

In terms of the reaction mechanism, CKX had for many years been characterized as a copper-containing amine oxidase, which catalyzes the oxidative conversion of biogenic primary amines to aldehydes, ammonia and hydrogen peroxide via an imine intermediate. However, a recent detailed analysis (57) has raised an important question about the reaction mechanism. CKX purified from wheat was found to be an FAD-containing flavoprotein and did not require oxygen during cytokinin degradation, or produce hydrogen peroxide. Thus, the name of the enzyme is now suggested to be cytokinin dehydrogenase rather than an oxidase (Fig. 9).

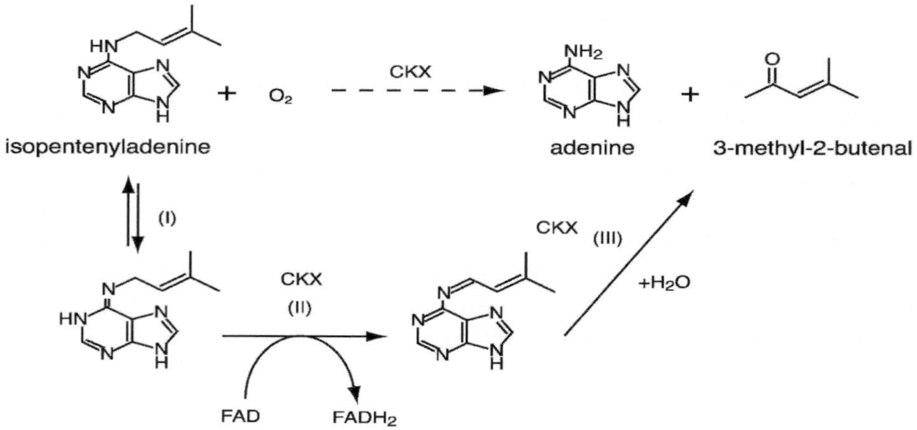

Figure 9. The Reaction mechanism of cytokinin oxidase/dehydrogenase. Interconversion shown in (I) is equilibrated. CKX catalyzes reaction (II) and (III). The formerly accepted reaction is shown as a broken line.

Some phenolic compounds such as acetosyringone (3',5'-dimethoxy-4'-hydroxyacetophenone), 2,6-dimethoxyphenol activate CKX activity *in vitro*. These compounds are reported to serve as a signal compound from host plants and to activate the *vir* genes of the *Agrobacterium* Ti plasmid. Thus the phenolic compounds might control both *vir* gene expression (i.e. cytokinin production) and cytokinin degradation.

Phenylurea-type cytokinins are potent inhibitors of CKX. They work in a non-competitive manner in maize-native CKX, but in a competitive manner in maize-recombinant CKX.

CKX genes in plants

The genes for CKX were first identified in maize (20, 37), and then in *Arabidopsis* (7). In *Arabidopsis*, there are seven genes for CKX whose members show a distinct pattern of expression. The biochemical properties of recombinant CKX have been analyzed, and the K_m values were 1.5 µM for iP, 11 µM for iPR, 14 µM for tZ and 46 µM for cZ (7).

In terms of the regulation of expression, it has been assumed that cytokinin supply positively affects CKX activity. In *Phaseolus vulgaris*, CKX activity was found to be rapidly induced by a transient increase in cytokinin supply (10). Inhibition of induction by cordycepin and cycloheximide suggests that RNA and protein synthesis are required for the response.

CONCLUDING REMARKS

New molecular approaches in recent years have brought about several breakthroughs in the study of cytokinins, including the identification of genes coding for IPT, CKX and cytokinin receptors. This progress has helped to resolve several debated issues, such as the spatial location of cytokinin biosynthesis, degradation and perception. It has also revealed the involvement of two-component systems in the signaling pathway. However, because the molecular analysis is only in its infancy, many other puzzling issues remain.

One of the most important but difficult of these is the elucidation of the physiological significance of the variety of side chain decorations in isoprenoid cytokinins such as iP, tZ, cZ and DZ. A recent study has indicated that cytokinin receptors have a differential preference for the ligand species (54). Thus, the cytokinins might each have their own physiological function. We should soon know the difference of the effect of each cytokinin species on gene expressions that have been reported as cytokinin-regulated genes. A complete understanding of the physiological function of each cytokinin species will provide a full picture of cytokinin metabolism and its action.

The metabolic origin of the isoprenoid side chain has not yet been fully determined. Plants have two distinct biosynthetic routes for the production of DMAPP; the mevalonate pathway and the MEP pathway. The mevalonate pathway is distributed in the cytosol and the MEP pathway is distributed in the plastids. Several studies of cultured cells have implied that the side chain of cytokinins is derived from the mevalonate pathway (25). Elucidation of the subcellular localization of the IPTs will give us useful information on the metabolic origin of the substrate.

References

1. Akiyoshi DE, Klee H, Amasino RM, Nester EW, Gordon MP. (1984) T-DNA of *Agrobacterium tumefaciens* encodes an enzyme of cytokinin biosynthesis. Proc Natl Acad Sci USA 81: 5994-5998
2. Allen M, Qin W, Moreau F, Moffatt B. (2002) Adenine phosphoribosyltransferase isoforms of *Arabidopsis* and their potential contributions to adenine and cytokinin metabolism. Physiol Plant 115: 56-68
3. Åstot C, Dolezal K, Nordström A, Wang Q, Kunkel T, Moritz T, Chua NH, Sandberg G. (2000) An alternative cytokinin biosynthesis pathway. Proc Natl Acad Sci USA 97: 14778-14783
4. Barry GF, Rogers SG, Fraley RT, Brand L. (1984) Identification of a cloned cytokinin biosynthetic gene. Proc Natl Acad Sci USA 81: 4776-4780
5. Bassil NV, Mok D, Mok MC. (1993) Partial purification of a *cis-trans*-isomerase of zeatin from immature seed of *Phaseolus vulgaris* L. Plant Physiol 102: 867-872
6. Bernier G, Havelange A, Houssa C, Petitjean A, Lejeune P. (1993) Physiological signals that induce flowering. Plant Cell 5: 1147-1155
7. Bilyeu KD, Cole JL, Laskey JG, Riekhof WR, Esparza TJ, Kramer MD, Morris RO. (2001) Molecular and biochemical characterization of a cytokinin oxidase from maize. Plant Physiol 125: 378-386
8. Blackwell JR, Horgan R. (1993) Cloned *Agrobacterium tumefaciens ipt1* gene product, DMAPP:AMP isopentenyl transferase. Phytochemistry 34: 1477-1481
9. Brandon DL, Corse J, Higaki PC, Zavala ME. (1992) Monoclonal antibodies for analysis of cytokinin-*O*-glucosides in response to cold stress. *In* Kamínek M, Mok DWS, Zazímalová E, eds, Physiology and Biochemistry of Cytokinins in Plants. SPB Academic Publishing, The Hague, pp447-453
10. Chatfield JM, Armstrong DJ. (1986) Regulation of cytokinin oxidase activity in callus tissues of *Phaseolus vulgaris* L. cv Great Northern. Plant Physiol 80: 493-499
11. Chen C-M, Eckert RL. (1977) Phosphorylation of cytokinin by adenosine kinase from wheat germ. Plant Physiol. 59: 443
12. Chen C-M, Kristopeit SM. (1981) Metabolism of cytokinin: Dephosphorylation of cytokinin ribonucleotide by 5'-nucleotidases from wheat germ cytosol. Plant Physiol 67: 494-498
13. Chen C-M, Leisner SM. (1984) Modification of cytokinins by cauliflower microsomal enzymes. Plant Physiol. 75: 442-446
14. Chen C-M, Petschow B. (1978) Metabolism of cytokinin: Ribosylation of cytokinin bases by adenosine phosphorylase from wheat germ. Plant Physiol 62: 871-874
15. Emery RJN, Leport L, Barton JE, Turner NC, Atkins A. (1998) *cis*-Isomers of cytokinins predominate in chickpea seeds throughout their development. Plant Physiol 117: 1515-1523
16. Entsch B, Parker CW, Letham DS. (1983) An enzyme from lupin seeds forming alanine derivatives of cytokinins. Phytochemistry 22: 375-381
17. Entsch B, Parker CW, Letham DS, Summons RE. (1979) Preparation and characterization, using high-performance liquid chromatography, of an enzyme forming glucosides of cytokinins. Biochim Biophys Acta 570: 124-139
18. Faiss M, Zalubilova J, Strnad M, Schmulling T. (1997) Conditional transgenic expression of the *ipt* gene indicates a function for cytokinins in paracrine signaling in whole tobacco plants. Plant J 12: 401-415
19. Golovko A, Sitbon F, Tillberg E, Nicander B. (2002) Identification of a tRNA isopentenyltransferase gene from *Arabidopsis thaliana*. Plant Mol Biol 49: 161-169
20. Houba-Herin N, Pethe C, d'Alayer J, Laloue M. (1999) Cytokinin oxidase from *Zea mays*: purification, cDNA cloning and expression in moss protoplasts. Plant J 17: 615-626

21. Ihara M, Taya Y, Nishimura S, Tanaka Y. (1984) Purification and some properties of delta 2- isopentenylpyrophosphate:5'AMP delta 2-isopentenyltransferase from the cellular slime mold *Dictyostelium discoideum*. Arch Biochem Biophys 230: 652-660
22. Kakimoto T. (2001) Identification of plant cytokinin biosynthetic enzymes as dimethylallyl diphosphate:ATP/ADP isopentenyltransferases. Plant Cell Physiol 42: 677-685
23. Klämbt D. (1992) The biosynthesis of cytokinins in higher plants: our present knowledge. *In* Kamínek M, Mok DWS, Zazímalová E, eds, Physiology and Biochemistry of Cytokinins in Plants. SPB Academic Publishing, The Hague, pp25-27
24. Krall L, Raschke M, Zenk MH, Baron C. (2002) The Tzs protein from *Agrobacterium tumefaciens* C58 produces zeatin riboside 5'-phosphate from 4-hydroxy-3-methyl-2-(E)-butenyl diphosphate and AMP. FEBS Lett 527: 315-318
25. Laureys F, Dewitte W, Witters E, Van Montagu M, Inze D, Van Onckelen H. (1998) Zeatin is indispensable for the G2-M transition in tobacco BY-2 cells. FEBS Lett 426: 29-32
26. Letham DS. (1963) Zeatin, a factor inducing cell division from *Zea mays*. Life Sci 8: 569-573
27. Letham DS. (1994) Cytokinins as phytohormones-sites of biosynthesis, translocation, and function of translocated cytokinin. *In* Mok DWS, Mok MC, eds, Cytokinins: Chemistry, Activity, and Function. CRC Press, Boca Raton, Florida, pp.57-80
28. Letham DS, Palni LMS, Tao G-Q, Gollnow BI, Bates CM. (1983) Regulators of cell division in plant tissues XXIX. The activities of cytokinin glucosides and alanine conjugates in cytokinin bioassays. J. Plant Growth Regul. 2: 103-115
29. Martin RC, Mok MC, Habben JE, Mok DW. (2001) A maize cytokinin gene encoding an *O*-glucosyltransferase specific to *cis*-zeatin. Proc. Natl. Acad. Sci. USA 98: 5922-5926
30. Martin RC, Mok MC, Mok DW. (1999) A gene encoding the cytokinin enzyme zeatin *O*-xylosyltransferase of *Phaseolus vulgaris*. Plant Physiol. 120: 553-558
31. Martin RC, Mok MC, Mok DW. (1999) Isolation of a cytokinin gene, *ZOG1*, encoding zeatin *O*-glucosyltransferase from *Phaseolus lunatus*. Proc. Natl. Acad. Sci. USA 96: 284-289
32. Martin RC, Mok MC, Shaw G, Mok DWS. (1989) An enzyme mediating the conversion of zeatin to dihydrozeatin in *Phaseolus* embryos. Plant Physiol 90: 1630-1635
33. McGaw BA, Horgan R. (1983) Cytokinin oxidase from *Zea mays* kernels and *Vinca rosea* crown-gall tissue. Planta 159: 30-37
34. Miller CO, Skoog F, Saltza vNH, Strong M. (1955) Kinetin, a cell division factor from deoxyribonucleic acid. J. Am. Chem. Soc. 77: 1329-1334
35. Mok DW, Mok MC. (2001) Cytokinin Metabolism and Action. Annu. Rev. Plant Physiol. Plant Mol. Biol. 52: 89-118
36. Mok DWS, Martin RC. (1994) Cytokinin metabolic enzymes. *In* Mok DWS, Mok MC, eds, Cytokinins: Chemistry, Activity, and Function. CRC Press, Boca Raton, Florida, pp.129-137
37. Morris RO, Bilyeu KD, Laskey JG, Cheikh NN. (1999) Isolation of a gene encoding a glycosylated cytokinin oxidase from maize. Biochem Biophys Res Commun 255: 328-333
38. Murai N. (1994) Cytokinin biosynthesis in tRNA and cytokinin incorporation into plant RNA. *In* Mok DWS, Mok MC, eds, Cytokinins: Chemistry, Activity, and Function. CRC Press, Boca Raton, Florida, pp.87-99
39. Schnorr KM, Gaillard C, Biget E, Nygaard P, Laloue M. (1996) A second form of adenine phosphorybosyltransferase in *Arabidopsis thaliana* with relative specificity towards cytokinins. Plant J. 9: 891-898
40. Shimizu-Sato S, Mori H. (2001) Control of outgrowth and dormancy in axillary buds. Plant Physiol 127: 1405-1413
41. Skoog F, Armstrong DJ. (1970) Cytokinins. Ann. Rev. Plant Physiol. 21: 359-384
42. Spiess LD. (1975) Comparative activity of isomers of zeatin and ribosyl-zeatin on *Funaria hygrometrica*. Plant Physiol 55: 583-585
43. Strnad M. (1997) The aromatic cytokinins. Physiol Plant 101: 674-688

44. Suttle JC, Banowetz GM. (2000) Changes in *cis*-zeatin and *cis*-zeatin riboside levels and biological activity during tuber dormancy. Physiol Plant 101: 68-74
45. Takei K, Dekishima Y, Eguchi T, Yamaya T, Sakakibara H. (2003) A new method for enzymatic preparation of isopentenyladenine-type and *trans*-zeatin-type cytokinins with radioisotope-labeling. J Plant Res. 116: 259-263
46. Takei K, Sakakibara H, Sugiyama T. (2001) Identification of genes encoding adenylate isopentenyltransferase, a cytokinin biosynthesis enzyme, in *Arabidopsis thaliana*. J. Biol. Chem. 276: 26405-26410
47. Takei K, Sakakibara H, Taniguchi M, Sugiyama T. (2001) Nitrogen-dependent accumulation of cytokinins in root and the translocation to leaf: implication of cytokinin species that induces gene expression of maize response regulator. Plant Cell Physiol. 42: 85-93
48. Takei K, Ueda N, Aoki K, Kuromori T, Hirayama T, Shinozaki K, Yamaya T, Sakakibara H. (2004) *AtIPT3*, an *Arabidopsis* isopentenyltransferase gene, is a key determinant of macronutrient-responsive cytokinin biosynthesis. Plant Cell physiol. In press.
49. Tarkowska D, Dolezal K, Tarkowski P, Åstot C, Holub J, Fuksova K, Schmülling T, Sandberg G, Strnad M. (2003) Identification of new aromatic cytokinins in *Arabidopsis thaliana* and *Populus* x *canadensis* leaves by LC-(+)ESI-MS and capillary liquid chromatography/frit-fast atom bombardment mass spectrometry. Physiol Plant 117: 579-590
50. Taya Y, Tanaka Y, Nishimura S. (1978) 5'-AMP is a direct precursor of cytokinin in *Dictyostelium discoideum*. Nature 271: 545-547
51. Veach YK, Martin RC, Mok DW, Malbeck J, Vankova R, Mok MC. (2003) *O*-glucosylation of *cis*-zeatin in maize. Characterization of genes, enzymes, and endogenous cytokinins. Plant Physiol 131: 1374-1380
52. von Schwartzenberg K, Kruse S, Reski R, Moffatt B, Laloue M. (1998) Cloning and characterization of an adenosine kinase from *Physcomitrella* involved in cytokinin metabolism. Plant J 13: 249-257
53. Yamada H, Suzuki T, Terada K, Takei K, Ishikawa K, et al. (2001) The *Arabidopsis* AHK4 histidine kinase is a cytokinin-binding receptor that transduces cytokinin signals across the membrane. Plant Cell Physiol 42: 1017-1023
54. Yonekura-Sakakibara K, Yamaya T, Sakakibara H. (2004) Molecular characterization of cytokinin-responsive histidine kinases in maize: differential ligand preference and response to cis-zeatin. Plant Physiol. 134: (4) 1654-1661
55. Zubko E, Adams CJ, Machaekova I, Malbeck J, Scollan C, Meyer P. (2002) Activation tagging identifies a gene from *Petunia hybrida* responsible for the production of active cytokinins in plants. Plant J 29: 797-808
56. Chen CM, Kristopeit SM. (1981) Metabolism of cytokinins: deribosylation of cytokinin ribonucleoside by adenosine nucleosidase from wheat germ cells. Plant Physiol 68: 1020-1023
57. Galuszka P, Frébort I, Sebela M, Sauer P, Jacobsen S, Pec P. (2001) Cytokinin oxidase or dehydrogenase? Mechanism of cytokinin degradation in cereals. Eur J Biochem 268: 450-461

B4. Ethylene Biosynthesis

Jean-Claude Pech, Alain Latché and Mondher Bouzayen
Ecole Nationale Supérieure Agronomique, UMR INRA-INP/ENSAT 990, Av de l'Agrobiologie, BP 107, F-31326 Castanet-Tolosan Cedex, France
E-mail: pech@ensat.fr

INTRODUCTION

Ethylene, the simplest olefine, is present in nature at trace amounts. It is produced either chemically through the incomplete combustion of hydrocarbons and biologically by almost all living organisms (2). Low levels of ethylene have been found in the expired gases of animals that are generated through a lipid peroxidation pathway. Many microbes among bacteria and fungi produce ethylene from two possible pathways: (i) a methionine and 2-oxo-4-methylthiobutyric acid (KMBA[1]) pathway in which ethylene is formed from KMBA by chemical reaction, and (ii) an α-ketoglutaric acid (KGA) pathway in which KGA is generated from glucose and many other substrates and an ethylene-forming enzyme having very divergent sequence with the enzyme responsible for the last step of ethylene biosynthesis in higher plants (4). However, while animal and microbial organisms are also able to produce ethylene, there is no evidence that ethylene acts as an hormone in these organisms. Primitive plants (liverworts, mosses, ferns, algae) produce ethylene through a still unknown pathway but different from that of higher plants. Some of them, such as semi-aquatic ferns, undergo ethylene-mediated physiological responses suggesting that ethylene acts as a biologically active compound in these organisms.

An ethylene biosynthesis pathway specific to higher plants associated with a perception and transduction pathway (Chapters D3 and D4) has evolved with the most primitive higher plants (32).

Historically, the first treatise reporting indirectly on ethylene synthesis by higher plants was written more than 3 centuries BC (47). A method for

---

[1] Abbreviations: ACC, 1-aminocyclopropane-1-carboxylic acid; ACGT, ACC gamma-glutamyltranspeptidase; ACMT, ACC N-malonyltransferase; ACS, ACC synthase; ACO, ACC oxidase; aminoethoxyvinylglycine (AVG); Br, breaker; CAS, ß-cyanolamine synthase; CH, cycloheximide; IMG, immature green; GACC, 1-(gamma-L-glutamylamino)ACC; KGA, alpha-ketoglutaric acid; KMBA, 2-oxo-4-methylthiobutyric acid; MACC, N-malonylACC; MG, mature green; MTA, 5'methylthioadenosine; MTR, 5-methylthioribose; SAMS, S-adenosylmethionine synthase; SAMase, S-adenosylmethionine hydrolase.

producing wound ethylene without knowing was described in which "scraping with iron claws" of sycamore figs was performed to compensate for the lack of pollination and to stimulate fruit ripening. Ethylene was recognised as affecting plant growth at the beginning of the 20th century (2). Its presence in illuminating gas caused etiolated pea seedling to undergo horizontal growth. Later, this simple observation has served as a major test to screen Arabidopsis mutants altered in ethylene sensitivity or production (Chapters D3 and D4). Ethylene is synthesised naturally by plants and is active at low concentrations ($<1\mu L.L^{-1}$). It is involved in many aspects of plant growth and development and plays a major role in mediating plant responses to biotic and abiotic stresses (2). Work with plants blocked in ethylene biosynthesis has shown that the plants can survive but that soil penetration by seedlings, disease resistance, and fruit ripening are severely impaired.

In the previous edition of this volume all steps of ethylene biosynthesis were described, including the discovery of the key intermediate ACC. The cloning of two major genes involved in ethylene biosynthesis, ACC synthase and ACC oxidase has been performed and subsequently used for the biotechnological control of ethylene synthesis in the tomato. The major progress in recent years is related to the mechanistic aspects of enzyme function, the characterisation of the gene families, and the regulation of gene expression.

THE BASIC ETHYLENE BIOSYNTHETIC PATHWAY IN HIGHER PLANTS

The history of the discovery of the ethylene biosynthetic pathway in higher plants represents a good example of advancement of science through the stepwise utilisation of novel scientific concepts, plant models and methodologies. Ethylene had long been thought to derive by simple chemical reaction from various substrates such as linoleic acid, propanal, β-alanine, and others (28). Because cellular extracts failed to produce ethylene, model systems were used to catalyse the production of hydrocarbon gases, including ethylene. In these experiments, methionine was initially introduced as a possible free radical quenching agent to inhibit ethylene formation derived from lipid peroxidation. Surprisingly, instead of suppression, ethylene production was increased. This observation led to the discovery that methionine was the precursor of ethylene *in vivo* in apple tissues. Tracers studies revealed that C3 and C4 of methionine were converted to ethylene, while C1 and C2 were incorporated into CO_2 and formic acid, respectively (Fig. 1). Upon conversion to ethylene, [^{35}S] methionine released labeled 5'methylthioadenosine (MTA) and 5-methylthioribose (MTR), indicating that S-adenosylmethionine (SAM) was an intermediate in the biosynthesis of ethylene from methionine through an ATP dependent step (Fig. 1). The last step in the pathway between

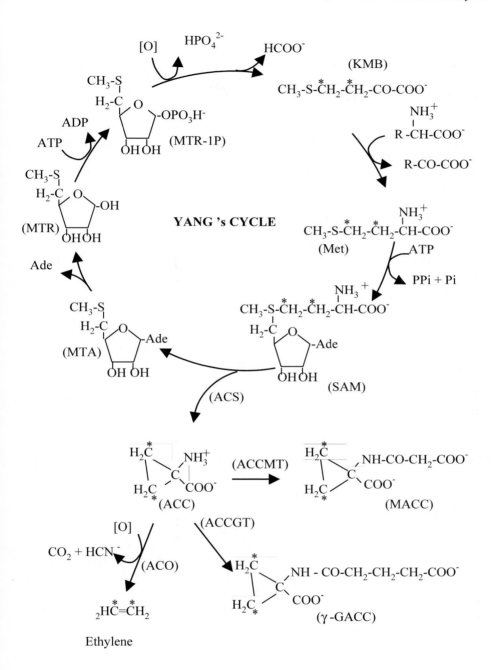

Figure 1. The ethylene biosynthetic pathway

S-adenosyl methionine to ethylene was discovered by incubating apple tissues in the presence of L-[U^{14}]methionine in the absence of oxygen, conditions known to prevent ethylene formation (54).This resulted in the formation of an intermediate compound that was identified as the non-protein amino acid 1-aminocyclopropane-1-carboxylic acid (Fig. 1). Upon return to aerobic conditions, apple tissues release labelled ethylene from ACC. Further, it was demonstrated that (i) the 5'methyl group of MTA was recycled back to methionine (ii) the CH$_3$S group of MTA was converted as a unit to re-form methionine and (iii) the ribose moiety of MTA provides the carbon-skeleton for the 2-aminobutyrate portion of methionine. This recycling process has been referred to as the Yang cycle (Fig. 1).

Side Pathways of ACC Metabolism

ACC can be diverted from its route to ethylene by forming conjugate derivatives. The major ACC conjugate is N-malonylACC (MACC) (54). In some tissues, such as pre-climacteric apples, more than 40% of the ACC synthesized in the skin and 5% in the flesh are diverted to MACC (26). Under normal conditions MACC is not converted back to ACC. However, under high concentrations in the plant tissue MACC can be converted to some extent back to ACC by inducible MACC-hydrolase activity (15,54). Another type of ACC conjugate, 1-(γ-L-glutamylamino)ACC (GACC), can be formed by protein extracts of tomato fruit and seeds though it represents at most a minor form of ACC conjugate in plants (34). The cyanide resulting from C$_1$ of ACC is incorporated into L-cysteine to form β-cyanolalanine which is further metabolized to asparagine or to γ-glutamyl-β-cyanolalanine (54).

Intracellular Distribution and Whole Plant Transport of ACC and MACC

ACC is a neutral amino acid whose uptake by the cell and is dependent only on external concentration. At low concentration ACC uptake is competitively inhibited by other neutral amino acids, including its structural analog α-aminoisobutyric acid (39). The vacuole accumulates the bulk of ACC (around 75%) but the molar concentrations of ACC are similar in the extravacuolar space and in the vacuole (48). ACC can be released freely and efflux is dependent of a concentration gradient. All these data indicate that the intracellular transport of ACC is probably not limiting ethylene production. Contrary to ACC, MACC concentration is higher in the vacuole (48). MACC is synthesized in the cytosol and then transported into the vacuole by an ATP-dependent transtonoplastic carrier. Its sequestration into the vacuole and release from this compartment are dependent on the protonation of this molecule, which is dictated by the vacuolar pH (33).

At the whole plant level, ACC can be translocated both basipetally in the phloem or acropetally through the xylem (2,54).

Enzymes of the Ethylene Biosynthesis Pathway

SAM synthetase
SAM synthetase (SAMS; EC 2.5.1.6) catalyses the biosynthesis of SAM, the immediate precursor of ACC. In addition to its role in ethylene synthesis, SAM is the major methyl group donor in numerous trans-methylation reactions and a precursor for polyamine biosynthesis.

ACC synthase
ACC synthase, (ACS; S-adenosyl-L-methionine methylthioadenosine lyase; EC 4.4.1.14) is capable of generating ACC from SAM. Total purification from wounded zucchini and tomato led to aminopeptide sequencing and subsequent gene isolation (24). The purified ACS proteins have a molecular weight ranging from 48 to 58 KDa depending on the plant species and are active in their monomeric form. However, ACC is present as a homodimer in plant tissues and when expressed in *E. coli* as a recombinant protein. Complementation analysis of ACS mutants indicates that ACS can form heterodimers and that the active site of the enzyme is located at the interface of the protein sub-units (45). The C-terminal end plays an important role in dimerization.

ACS is a pyridoxal phosphate-dependent enzyme that converts SAM into ACC and MTA via α,γ-elimination. It is also capable, by a β-γ elimination process of releasing vinylglycine that binds irreversibly to the active site thus causing suicidal inhibition of the enzyme. Analogs of vinylglycine, rhizobitoxine and amino ethoxy vinylglycine (AVG), as well as hydroxylamine analogs that react with pyridoxal phosphate such as amino-oxyacetic acid (AOA), are therefore used as strong inhibitors of ACS activity (54).

It was initially thought that the deletion of 46 to 52 amino acids from the C-terminal resulted in a hyperactive monomeric enzyme, but this has been attributed to a decreased affinity of the antibody for the truncated form as compared to the wild type enzyme (45). In fact, the C-terminal domain does not have a major effect on the catalytic activity of the enzyme but has rather an effect on the half-life and stability of the protein (9). Deletion of the C-terminal domain of ACS is consistent with the observation that the molecular mass of some ACS isozymes were smaller than the *in vitro* translation products. However, the use of potent proteinase inhibitors during extraction indicated that processing of the C-terminal region does not occur *in vivo*, but is the result of proteolytic cleavage during extraction (46).

The intimate catalytic mechanism of ACC formation is still a matter of discussion. The most probable mechanism involves an inversion process consisting in the nucleophilic displacement of the methylthio-adenosine (MTA) residue (Fig. 2). Displacement of MTA by Tyr^{152} (22) prior to the internal alkylation step is highly improbable as it would result in an overall retention of the C-4 configuration, which is not the case (44). Crystal structure studies of tomato ACS (22) reveal good conservation of the

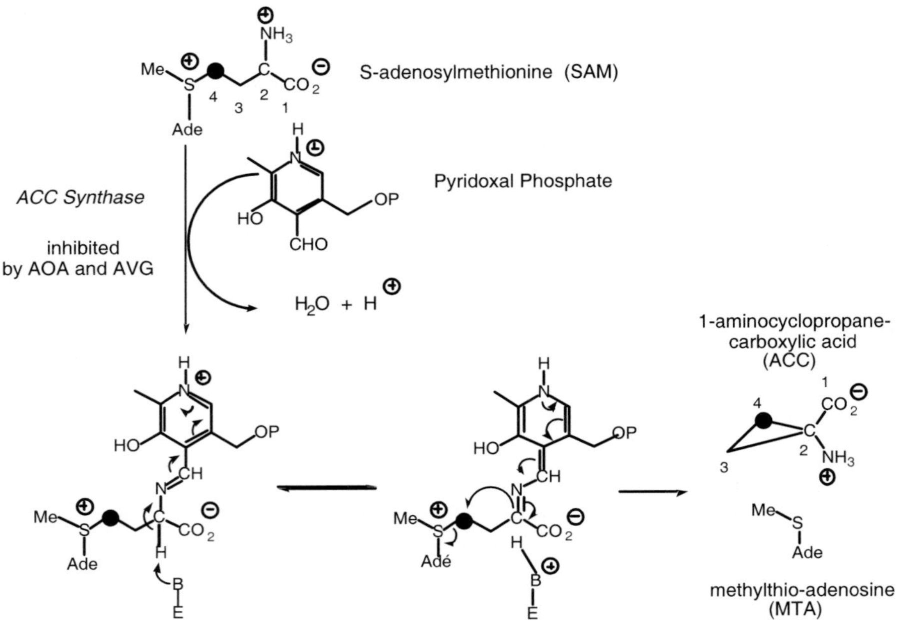

Ethylene biosynthesis

Figure 2. The putative mechanism for the catalytic conversion of SAM to ACC (44).

catalytic residues suggesting a similar catalytic mechanism for ACS and other pyridoxal phosphate-dependent enzymes.

ACC oxidase
Until the early 1990s, all attempts to obtain genuine ACC oxidase (ACO) activity in a cellular fraction failed. Activity was only measured on entire tissues in the presence of saturating concentrations of exogenous ACC and was referred to as the ethylene-forming enzyme (EFE). This feature led to the assumption that ACO is a membrane bound enzyme (28, 54). The mystery of this enzyme was only fathomed once the gene encoding the ACO protein has been isolated, giving a brilliant example of "reverse biochemistry". That is, among the ripening related cDNAs isolated from tomato fruit, the pTOM13 clone was selected as a putative ethylene biosynthetic gene based on its expression during ripening and upon wounding. The expression in tomato plants of an antisense construct of the pTOM13 cDNA resulted in reduced capacity to produce ethylene and significant delay in fruit ripening suggesting that this clone might encode the ACO protein (18). The ultimate identification of the ACO gene was given by functional expression in *Saccharomyces cerevisiae* and in *Xenopus* oocytes (3). Based on the sequence homology between pTOM13 and flavanone-3-hydroxylase genes (18), and the demonstration that iron is an essential co-factor of EFE *in vivo*, a soluble ACO activity could be obtained for the first time from melon fruit and measured in the presence of iron and

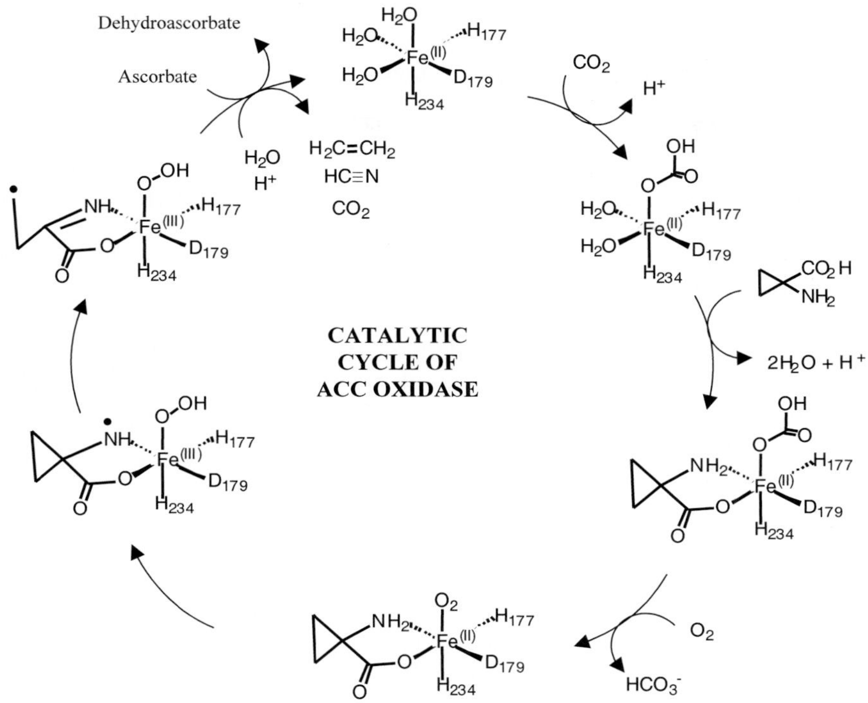

Figure 3. The catalytic cycle of ACC oxidase

ascorbate (36). ACO protein belongs to the family of ferrous-dependent non-heme oxygenases. It requires ferrous iron as a co-factor and oxygen as a co-substrate. Unlike the large majority of the other members of the family, it does not utilize a 2-oxo acid co-substrate, but instead it utilizes ascorbate as co-substrate, exclusive of other reducing molecules. Another specificity is that CO_2 is an essential activator of ACO (36). It has been proposed (55) that CO_2 operates by protecting the active site by complexing with Fe(II).

The conversion of ACC into ethylene by ACO proceeds via the opening of the cyclopropane ring, carbons 2 and 3 giving ethylene and carbon 1 HCN (Fig. 3). The intimate mechanism of the catalytic reaction is still unknown. However it is now established that it does not involve the formation of N-hydroxyl ACC as an intermediate (10). Nevertheless, it is well established that for each mole of ACC consumed, one mole of oxygen is utilized. Also, ascorbate is transformed mole by mole into dehydroascorbate (36). Two histidine and one aspartate serve as Fe(II)-binding sites (20). Recent studies show that ACO operates similarly to other non-heme iron enzymes by opening iron coordination sites at the appropriate time in the reaction cycle, and that CO2 would stabilize the ACO/Fe(II)/ACC complex thus preventing the uncoupled reaction that inactivates the enzyme (55).

Ethylene biosynthesis

The lack of a signal peptide classically required for the protein to cross the plasma membrane suggested a cytosolic targeting for ACO. However, the subcellular localization of the enzyme is still a matter of debate. Radiochemical, cell fractionation and immuno-cytolocalization studies, as well as the use of non-permeant inhibitors, showed that, in tomato and apple fruit, ACO is predominantly located in the apoplast (37) although other immunolocalization studies indicated that ACO was mainly located in the cytosol (11).

Contrary to ACS, ACO is present at high concentration in plant tissues making ethylene. However under catalytic conditions *in vitro*, it displays a very short half-life of less than 20 minutes. One possible explanation is that the enzyme is inactivated though oxidative damage due to an iron(III)-linked peroxide intermediate resulting in partial proteolysis (7). Whether such a process occurs *in vivo* remains to be elucidated

Other Enzymes of ACC Metabolism

ACC N-malonyltransferase and ACC γ-glutamyltranspeptidase
It was suggested that N-malonylation of ACC plays a role in regulating ethylene production by diverting ACC from its route to ethylene (54). Unlike other enzymes of the ethylene biosynthetic pathway, little is known on N-malonyltransferase. Beside ACC this enzyme is also capable of malonylating D-amino acids suggesting a possible role in the cell detoxification process by preventing the incorporation into nascent proteins of abnormal amino acids, often produced under stress conditions. While all attempts to sequence the protein following nearly total purification have failed so far, these studies at least revealed the probable presence of various isoforms (15,24).

Crude protein extracts of tomato exhibit ACC γ-glutamyltranspeptidase activity, but, due to the low levels of GACC in plants, it is probable that the enzyme is far from being crucial for the control of ethylene biosynthesis (34).

β-Cyanolalanine synthase
β-Cyanolalanine synthase, CAS, (EC 4.4.1.9.) converts L-cysteine and HCN into 3-cyano-L-Alanine and H_2S. Interestingly CAS activity is enhanced by ethylene thus preventing HCN, a byproduct of ethylene biosynthesis, from accumulating under conditions where ethylene production is high. It has been hypothesized that cyanide could inhibit cytochrome oxidase and trigger cyanide-resistant respiration, although the level of cyanide in plant tissues is generally considered to be too low to account for an effect on respiration. However, in some conditions, such as the hypersensitive response, HCN concentrations may be locally high enough to account for cell death (29). In addition, physiologically relevant and metabolically safe concentrations of cyanide are capable of stimulating *At-ACS6* gene expression in Arabidopsis (29). In these conditions, the level of HCN derived from ACC conversion to ethylene and activity of CAS could play a role in the auto-stimulation of

ethylene synthesis during stress responses or during normal developmental processes

REGULATION OF ETHYLENE BIOSYNTHESIS GENES

Beside its involvement in a number of developmental processes, ethylene mediates the plant responses to many environmental factors. Under stress conditions, both the conversion of SAM into ACC and the conversion of ACC into ethylene are stimulated. Also, ethylene biosynthesis undergoes feedback regulation through the so-called autocatalytic or autoinhibition process, which has been well described in fruit (26). Other hormones such as auxins and cytokinins are known to stimulate ethylene production in many plant organs (2). These factors participate in regulating the expression of the various members of the gene families involved in ethylene biosynthesis.

The SAM Synthetase Gene Family and its Expression

In the context of increased demand for methylation of lignin monomers, it has been shown that two *SAM synthase* (*SAMS*) genes in Arabidopsis and three in the tomato are highly expressed in vascular or elicited tissues (52). High expression of *AcSAM3* in young developing kiwi fruit has also been reported and correlated with the demand of SAM for the synthesis of polyamines (52). However, SAMS is not considered a limiting factor for ethylene synthesis (54), and in carnation, for example, massive ethylene production is not associated with an increase in SAM mRNA abundance. In contrast, in kiwi fruit there is increasing expression of *AcSAM3* during the peak of ethylene synthesis and *AcSAM1* and *AcSAM* are induced by exogenous ethylene (52). This latter regulation of SAM synthetase gene expression is an indication of a need for the cell to stabilise the intracellular level of SAM. It is not surprising therefore to find situations where steady-state SAM levels need to be maintained during ethylene synthesis such as in ripening fruit (52).

The ACC Synthase Multigene Family and its Transcriptional Regulation

The pioneering work of cloning the first ACS cDNA provided probes and primers for the isolation of homologous genes from a variety of plant species. ACS is encoded by a multigene family of at least nine members in the tomato and 10 members in Arabidopsis (9). They can be classified into 4 classes on the basis of the pI values (15). Phylogenic analysis of deduced amino acid sequences of *ACS* from Arabidopis and tomato (Fig. 4) indicates the presence of at least three branches. These branches correspond to ancestral genes that have evolved before divergence of monocots and dicots (15). Within branch III, a clade comprises 4 Arabidopsis *ACS* (*At-ACS 4, 8, 5* and *9*) possessing a phosphorylation site at the 3' end indicative of post transcriptional regulation. It is possible that these 4 genes have emerged from a common ancestor. However, in the tomato, *Le-ACS2* gene which can undergo phosphorylation is located in branch I in the same clade as *LeACS4*, a closely homolog gene, but lacking the 3' phosphorylation site. In that case, it is probable that phosphorylation capability has emerged after duplication.

In Arabidopsis, a rather extensive pattern of *ACS* gene expression is available for the majority of the gene family members in different plant organs during development as well as under the effect of environmental factors, hormones and chemical effectors (Fig. 5). Only the expression of *At-ACS8* has not yet been studied. The *At-ACS7* which lacks the variable C-terminal region present in other *ACS* genes is expressed in etiolated seedlings only after cycloheximide treatment (53). Expression of *At-ACS4* and *At-ACS6* genes is induced by auxin and four putative several auxin responsive elements have been described in the promoter sequence of the *At-ACS4* gene (1). Interestingly, the translational inhibitor cycloheximide stimulates the expression of *At-ACS1*, *2*, *4*, *5* and *6* (5) indicating (i) possible regulation by some unknown repressors or (ii) retention of mRNA on ribosomes and increased steady state of *ACS* mRNA (5, 50).

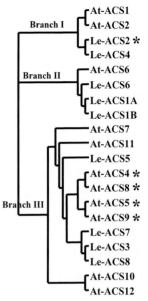

Figure 4. A phylogenetic tree of Arabidopsis and tomato ACC synthase constructed on the basis of the deduced amino acid sequence. Asterisks indicate the presence of phosphorylation site at the C-terminal end.

Many of the genes (*ACS2, ACS4*, and *ACS6*) exhibit stimulation of expression upon wounding, LiCl treatment and anaerobiosis. Of all *At-ACS 1* to *6*, only *At-ACS6* expression was found to be positively stimulated by ethylene (5).

In tomato, data describing the expression pattern in different plant organs of 8 ACS genes are still partial, and so far only concern *Le-ACS1A*, *Le-ACS1B* and *Le-ACS2* to *7* (Fig. 6). The expression of *Le-ACS1B* and *Le-ACS5* has been reported in tomato cell suspension cultures but not in whole plant tissues. In depth studies of ACS gene expression has been performed during fruit ripening and revealed a feed back mechanism leading to either stimulation or inhibition by ethylene (Fig. 7). Before fruit enter the climacteric phase and produce significant amounts of ethylene, the expression of two ACS genes (*Le-ACS6* and Le-ACS1A) is stimulated by developmental factors other than ethylene (3). During the ripening process, fruit undergo a transition from negatively (system 1) to positively regulated (system 2) ethylene production (Fig. 7). This transition correlates with an increase of *LeACS1A* expression, and an induction of *Le-ACS2* and *LeACS4* gene expression. After the initiation of the climacteric, ethylene down-regulates the expression of *Le-ACS1A* and the autocatalytic system is set in

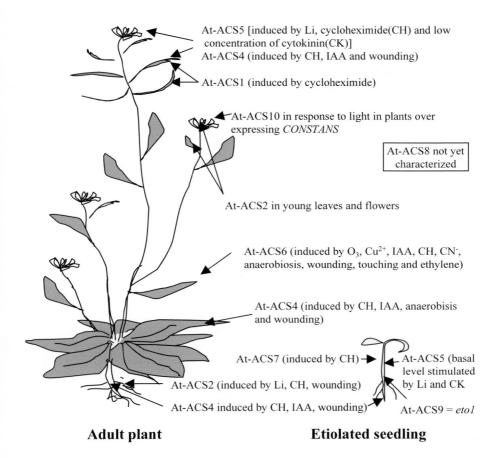

Figure 5. Spatial expression of ACS genes and regulation by chemical and environmental effectors in an adult plant and etiolated seedling of Arabidopsis (from 1, 5, 9, 27, 29, 38, 40, 49, 53)

motion through *Le-ACS2* and *Le-ACS4*. It is worthy of note that neither of these two genes that are up-regulated by ethylene are expressed in the Nr and Rin mutants (Fig. 7). A similar pattern of expression has been described by others (30) except that *Le-ACS1A* and *Le-ACS3* were expressed at a low constitutive level throughout ripening. Nonetheless, transcriptional regulation of ACS cannot account alone for the spatio-temporal regulation of ethylene production. Major post-transcriptional events occur that have been extensively studied in the recent years.

Ethylene biosynthesis

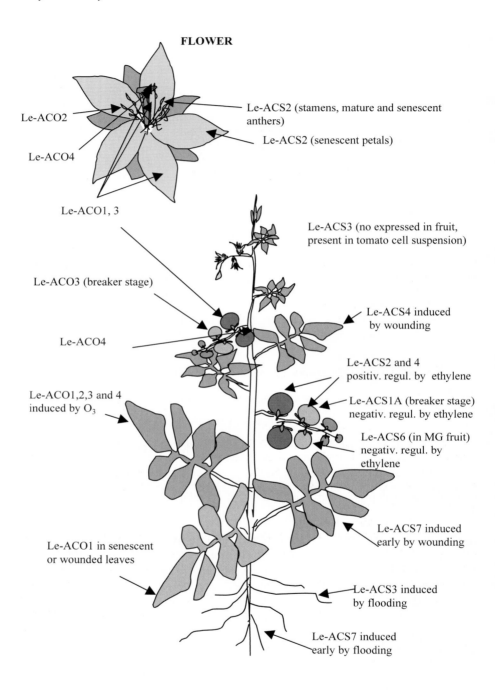

Figure 6: Spatial expression of *ACS* and *ACO* genes and regulation by chemical and environmental effectors in adult plant of tomato.

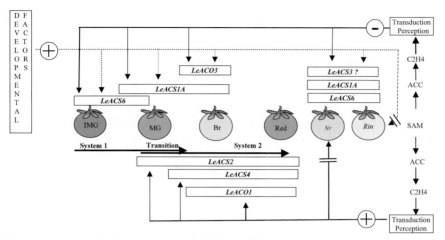

Figure 7. A model for the regulation of *ACS* and *ACO* gene expression in tomato fruit during the transition from system 1 to system 2 and during ripening. The + and – signs indicate positive and negative regulation respectively by ethylene or developmental factors. *Nr* and *Rin*, which fail to ripen completely, represent tomato mutants affected for an ethylene receptor and a *MADS box* gene respectively. IMG = Immature green, MG = Mature green, Br = Breaker

Post-Transcriptional Regulation of ACS Activity

The ethylene over producers (eto) and cytokinin insensitive mutants of Arabidopsis have allowed the elucidation of the role of the C-terminus end in the post-transcriptional regulation of ACS.

Ethylene production of the Arabidopsis mutants *eto* (*eto1, eto2, eto3*) displaying an altered triple response phenotype can be suppressed by AVG, indicating that the mutation probably affected ethylene synthesis (17). In fact, sequencing of *At-ACS5* in the *eto2* mutant revealed a perturbation of its carboxy terminus moiety in which the 12 terminal amino acids were changed with loss of the aginine-valine-serine phosphorylation site corresponding to Ser-400. Furthermore the *eto3* mutation results in a single amino acid mutation in the C-terminal domain of ACS9, which is 91% identical to At-ACS5 at the aminoacid level (9). The locus of the recessive *eto1* mutation has been cloned revealing that, contrary to other *eto* mutants, the *eto3* mutation does not lie within any of the ACS genes (13). The protein encoded by the mutated gene has similarities to proteins with a peptide binding domain and it has been shown to interact with wild-type At-ACS5 in a yeast two-hybrid system. This interaction is disrupted by the *eto2* mutation indicating that the interaction of ETO1 with At-ACS5 occurs at the C-terminal end of the enzyme. Some Arabidopsis mutants insensitive to cytokinins in terms of stimulation of ethylene synthesis are in fact mutated for At-ACS5. Two of these mutants were characterized that had predicted Ser substitution at two very well conserved positions (269 and 201) that probably cause disruption of catalytic activity (49).

Thus the C-terminus of *At-ACS5* and *At-ACS9* appear to play a role in the negative regulation of *ACS*. It is likely to be also the case for *At-ACS4* and *At-ACS8* that share strong similarity with *At-ACS5* and *AT-ACS9* in the C-terminal 18 amino acids. Interestingly, the four genes form a distinct clade within the *ACS* gene family (Fig. 4).

The C-terminal end of ACS controls the stability of the protein.
In many plant tissues, ACS activity displays a short half-life, going from 15 to 25 min in vegetative tissues and from 40 min to 3 hours in green and pink tomato fruit. A possible explanation for the short half-life is a mechanism-based inactivation mentioned earlier in which the vinylglycine residue released by β-γ elimination causes suicidal inhibition of the enzyme (15, 24). It can be speculated that the *eto* mutations somehow affect the rate at which ACS undergoes mechanism-based inactivation *in vivo* with subsequent degradation by protease(s). The fact that the *eto2* mutation resulted in increased stability of the protein indicates that the C-terminal domain indeed impacts the half-life and stability of the protein (9). Since cytokinins have been shown to increase the stability of the protein (Fig. 8) and to stimulate ethylene production in the *eto2* mutant which is affected at the C-terminal region of ACS5, it was concluded that cytokinins act by a mechanism that is partly independent of the C-terminal domain.

The post-transcriptional regulation of ACS involves protein phosphorylation.
Phosphorylation of ACS was first revealed through the use of pharmaceutical drugs. The addition of a Ser/Thr protein kinase inhibitor reduced ethylene production in elicited tomato cells in culture and, conversely, calyculin A, a protein phosphatase inhibitor, induced a rapid increase in ACS activity both in the absence and in the presence of elicitor (42). In *Phalaenopsis* flowers, okadaic acid, a specific inhibitor of type 1 and 2A protein phosphatase, induced *Phal-ACS1* transcript accumulation in the stigma, labelum and ovary, whereas staurosporine, a protein kinase inhibitor, inhibited such accumulation. Phosphorylation is therefore involved in up-regulating the expression of the *Phal-ACS1* gene resulting in increased ethylene production and accelerated senescence of orchid flowers (51).

The first direct evidence for the phosphorylation of an ACS protein was given for the wound-induced *Le-ACS2* gene of tomato (46). It was demonstrated that phosphorylation takes place at residue Ser-460 located in the C-terminal end (Ser-460). ACS proteins lacking this phosphorylation site, such as Le-ACS4, were not phosphorylated. It was also shown that phosphorylation does not affect enzymatic activity *per se*, but rather alters ACS protein turnover. The proposed mechanism (Fig. 8) would imply that the non-phosphorylated C-terminal region binds some interacting proteins, ETO1 being a strong candidate, that induce ACS degradation by proteases and thus reduces the half-life of the protein. Therefore, phosphorylation, by preventing binding would prolong turnover and increase activity. While the phosphorylation-based mechanism may play a major role in regulating the

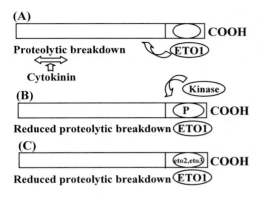

Figure 8: Putative mechanism for the regulation of stability of ACS proteins capable of phosphorylation at the C-terminal end (At-AC5, At-ACS9 and possibly of At-ACS4 and At-ACS8). A: Interaction of the C-terminal end with ETO1 stimulates turnover of the ACS protein while cytokinin partially block the proteolytic breakdown. B: Phosphorylation extends the stability of ACS protein by preventing interaction with ETO1 and proteolytic degradation. C: Alteration of the C-terminal domain by eto2 and eto3 mutations increase stability of the protein (Modified from 9)

activity of some ACS isoforms, the regulation of ethylene production in plant organs may involve post-transcriptional mechanisms other than phosphorylation of the C-terminal domain, and may also occur at the transcription level. An illustration of dual regulatory mechanisms can be found in tomato fruit for Le-ACS2 and Le-ACS4 that play an important role in autocatalytic ethylene production. The two genes likely exhibit two different regulatory controls of expression since Le-ACS2 can be phosphorylated in vivo while Le-ACS4 does not posses the corresponding phosphorylation site and cannot be phosphorylated *in vitro* (46).

ACC Oxidase Gene Family and Expression

ACO is encoded by a small multigene family comprised of three to four members. In tomato, *Le-ACO1, Le-ACO2 and Le-ACO3* encode functional proteins though the ACO isoforms displayed different levels of activity (8). More recently, a cDNA encoding *Le-ACO4* has been isolated from tomato (30) that shares 82.8, 80.0 and 83.6 % amino acid sequence identity with *Le-ACO1, Le-ACO2 and Le-ACO3*, respectively. However, proofs that the encoded protein is functional are lacking. In ripening tomato fruit and senescent leaves both *Le-ACO1* and *Le-ACO3* transcripts display increased accumulation, while *Le-ACO1* is the major gene expressed and *Le-ACO3* message only accumulates transiently at the onset of both leaf senescence and fruit ripening. *Le-ACO1* is wound inducible in leaves, but expression is transient over a few hours (3). The transient accumulation of *Le-ACO3* at the onset of ripening, where ACS genes are already expressed (Fig. 7), strongly suggest that *Le-ACO3* plays a major role in the upsurge of climacteric ethylene production under normal physiological conditions. The *Le-ACO3*-related ethylene would then stimulate *Le-ACO1* gene expression as well as the ethylene-inducible ACC synthases.

In melon, three *ACO* genes have been isolated that show differential expression in different plant organs. *CMe-AC01* is induced in ripe fruit and in response to wounding and ethylene in leaves. *CMe-ACO2* mRNA is detectable at low level in etiolated hypocotyls, while *CMe-ACO3* is expressed in flowers (26). Further information on the expression of ACO gene family members was provided through the characterization of their corresponding promoters. The *CMe-ACO1* promoter was able to drive GUS expression in tobacco leaves in response to wounding and to treatment with ethylene or copper sulfate. It was also rapidly induced in tobacco leaves inoculated with the hypersensitive response (HR)-inducing bacterium *Ralstonia solanacearum*. The *CMe-ACO3* promoter failed to drive GUS expression in response to infection, but stimulated GUS expression during flower development. The *CMe-ACO1*-driven GUS activity increased sharply during leaf senescence, whereas the *CM-ACO3* promoter drove strong GUS expression in green, fully expanded leaves and this declined at the onset of senescence (26).

Pollination in many flowers results in increased synthesis of ethylene, first in stigma and style and then the signal is transmitted to the corolla or petals. In petunia, this corresponds to organ-specific expression of members of the *ACO* gene family comprised of three genes. In tomato only *Le-ACO2* shows some degree of specificity with preferential expression in the anther. *Le-ACO3* mRNA accumulates in the stigma and style after pollination. The late stages of flower senescence are associated with the accumulation of *Le-ACO1* mRNA and further increase in *Le-ACO3* mRNA in petals and stigma+style (15).

Ethylene is involved in the determinism of sex genotypes in cucurbits. In cucumber, this has been associated with differential expression of *ACO* genes, specially in the young leaves near the apex, where sex expression seems to take place (23).

The expression data indicate that different members of the ACO (as well as of ACS) gene family are recruited in different plant organs and at different stages of plant development. The existence of multiple gene families whose members are finely tuned offers multiple ways for the plant to produce ethylene at the right time and right tissue.

Regulation of β-Cyanolalanine Synthase Gene Expression

Several genes of the *ß-substituted Ala synthase* (*Bsas*) family have been recently isolated in Spinach and Arabidopsis that encode cytosolic, chloroplastic and mitochondrial proteins capable of CAS activity (19). Only the mitochondrial Bsas3 isoforms exhibit true CAS activity, while other cytosolic or plastidic proteins had low affinity for cyanide and could rather be considered as cysteine synthases (EC 4.2.99.8). An efficient cyanide detoxification process therefore probably exists in mitochondria where true CAS are located.

CONTROL OF ETHYLENE BIOSYNTHESIS THROUGH BIOTECHNOLOGY

Suppression of ethylene biosynthesis through down-regulation of ethylene biosynthetic genes has enabled a better understanding of the role of ethylene in plant development. Most of the research in this area has been directed towards the control of fruit ripening through genetic manipulation of the two key genes involved in ethylene biosynthesis, *ACS and ACO*. Tomato has long been a model plant but other plant species have been also used in the recent years (43).

Biotechnology of ACC Synthase

Sense or antisense suppression of *ACS* has led to a strong inhibition of ethylene production of around 99% at the peak of ethylene production (31). The development of red color was inhibited, as well as softening and aroma production. Inhibition of ripening could be completely reversed by treating fruit with ethylene ($10\mu L.L^{-1}$). Although residual ethylene production was low (< 0.1 nL g^{-1} h^{-1}), it was sufficient to allow stimulation of the ethylene-responsive *PG* gene expression (26).

Biotechnology of ACC Oxidase

Antisense constructs of the other gene, *ACO*, resulted in a strong reduction of ethylene production during fruit ripening, with profound effects on this process (18). In transgenic lines where ethylene was reduced by 97%, residual ethylene was low enough to slow down the rate of pigment accumulation and the softening of the flesh at the over-ripening stage after harvest. However, it was high enough to allow the onset of color changes and softening of the flesh to occur normally during ripening on the vine.

Cantaloupe melons of the Charentais type have been the second fruit after tomato in which inhibition of ethylene production has been achieved using antisense *ACO* gene (6). One of the transformed lines exhibited more than 99.5% inhibition at the peak so that, in these conditions, ripening of both attached and detached antisense fruit was inhibited, including inhibition of rind yellowing, peduncle detachment (Fig. 9), part of flesh softening, aroma volatiles production and the respiratory climacteric. However, coloration of the flesh and accumulation of sugars and organic acids remained unaffected, indicating that these processes were ethylene-independent. Other studies indicated that antisense ACO ethylene-suppressed melon fruit exhibit better resistance to chilling injury associated with a higher capacity for removing active oxygen species and a lower aptitude to accumulate ethanol and acetaldehyde (26). Inhibition of ethylene synthesis in antisense ACO melon plants resulted in reduced susceptibility to powdery mildew under high disease pressure conditions.

Ethylene suppression through manipulation of ethylene biosynthesis genes has been achieved in other fruit species. Sense or antisense ACC

Figure 9: A photograph of wild type (left) and antisense ACO (right) melon plant 1.5 month after normal harvest and cessation of irrigation. Note that WT fruit have fallen down and are completely rotten while AS fruit are still attached to the plant and have kept a green colour.

synthase and ACC oxidase apples, as well as antisense ACC oxidase plums are under field tests (unpublished). Effects of ethylene suppression in apple are similar to those described in tomato and melon (inhibition of aroma production, softening and coloration).

Strong inhibition of ethylene production (around 90%) has been achieved in carnation by antisense ACO that resulted in a two-fold increase in vase-life (41). However, as in other antisense ACO plants, sensitivity to ethylene remained unaffected. Carnation are sensitive to low levels of exogenous ethylene ($<1\mu L.L^{-1}$) that accelerate wilting and senescence. Such levels of ethylene are readily encountered in the surroundings of flowers during commercial distribution. Transgenic flowers in which ethylene perception has been abolished through either biotechnology or chemical inhibitors are more promising for commercial applications.

A heterologous *ACO* gene of tomato has been transferred in antisense orientation into broccoli via *Agrobacterium rhizogenes*-mediated transformation and regeneration from hairy roots (21). Among selected lines that did not exhibit hairy root-induced morphological changes, two showed less postharvest yellowing associated with the reduction in ethylene production indicating that, similar to ornamental flowers such as carnation, the senescence of broccoli florets can be controlled by an antisense *ACO* gene.

Transgenic tobacco with an antisense *ACO* gene under the control of the pistil specific *Petunia inflata* S3 promoter exhibited reversible ovule development (14), whereas ethylene-deficient ACO antisense tomato plants exhibited delayed leaf senescence and better resistance to flooding (43).

Decreasing the Availability of ACC through Biotechnology

Expressing genes capable of lowering the ethylene precursors is another approach to inhibit ethylene production. The T3 bacteriophage-derived S-adenosylmethionine hydrolase (SAMase) depletes SAM, a precursor of ethylene upstream in the biosynthetic pathway. The *SAMase* gene, has been expressed in tomatoes (16) and melons (12) under the control of the E8 promoter that restricted expression to the ripening fruit. Ethylene synthesis was reduced by at most 80% in either case, which was not enough to profoundly affect ripening.

Limiting the availability of ACC by transforming it into ammonia and β-ketobutyric acid can be achieved by expressing a bacterial *ACC deaminase* gene. Tomato plants have been generated that produced leaves and fruit with up to 90% reduction of ethylene synthesis as compared to the control. Fruit ripened slower when removed from the vine early in ripening. In contrast, fruit that remained attached to the plant ripened much more rapidly, exhibiting little delay relative to control fruit. Fruit detachment is known to induce a strong reduction of internal ethylene due to the diffusion of ethylene out of the stem scar (25). Transgenic tomato plants that expressed *ACC deaminase* were protected significantly from pathogen damage, from flooding, heavy metals and arsenic inhibition (43).

Biotechnology of the E8 Gene

Manipulation of ethylene synthesis can be achieved through genes exerting regulatory functions. An antisense construct of the *E8* Fe(II) dioxygenase gene, related to ACO, stimulated ethylene production during the ripening of tomato fruit detached from the plant at, or well prior to, the onset of ripening (35). This gene is therefore supposed to be involved in the negative feedback regulation of ethylene biosynthesis.

CONCLUDING REMARKS

The ethylene biosynthesis pathway of higher plants is probably the earliest and best established biosynthetic pathway of all the plant hormones. The only gene that has not been isolated so far is related to the conversion of ACC into MACC, a conjugate derivative that in some circumstances may play a regulatory role in ethylene biosynthesis by diverting ACC from its route to ethylene.

In the last decade, a great deal of information has emerged on gene families participating in the pathway and on their pattern of expression. Major progress has been made in demonstrating the presence of post-transcriptional regulation events for at least some of the ACS proteins. The intimate mechanisms of the post-transcriptional regulation and phosphorylation remain, however, to be fully elucidated. Also, their role in the regulation of ethylene synthesis during development and in response to environmental factors is still not well understood. In general, more research is required to discover the developmental factors and regulatory genes that control the expression of ethylene biosynthesis genes.

From the applied side, biotechnological control of ethylene production has long be restricted to a limited number of model plants, mainly tomato. Applications to other species of agricultural interest have increased in the recent years. However the fact that plants remain sensitive to exogenous ethylene will probably be a limitation for commercial applications.

Acknowledgements

The authors wish to thank Islam El Sharkawy (INRA-ENSAT, Toulouse, Fr), Lucien Stella, Marius Reglier and Jalila Simaan (CNRS, Marseille, Fr) and Tony Bleecker (Univ of Wisconsin, USA) for their help in the preparation of the manuscript.

References

1. Abel S, Nguyen MD, Chow W, Theologis A (1995) ACS4, a primary indoleacetic acid responsive gene encoding 1-aminocyclopropane-1-carboxylate synthase in *Arabidopsis thaliana* J Biol Chem 270: 19093-19099
2. Abeles FB, Morgan PW, Saltveit Jr ME (1992) Ethylene in plant biology. Academic Press, New York. 414 p
3. Alexander L, Grierson D (2002) Ethylene biosynthesis and action in tomato: a model for climacteric fruit ripening. J Exp Bot 53: 2039-2055
4. Arshad M, Frakenberger Jr WT (2002) Ethylene: agricultural sources and applications. Kluwer Academic Publishers, New York, 342 p
5. Arteca JM, Arteca RN (1999) A multi-responsive gene encoding 1-amino-cyclopropane-1-carboxylate synthase (ACS6) in mature *Arabidopsis* leaves. Plant Mol Biol 39: 209-219
6. Ayub R, Guis M, Ben Amor M, Gillot L, Roustan JP, Latché A, Bouzayen M, Pech JC (1996) Expression of ACC oxidase antisense gene inhibits ripening of cantaloupe melon fruits. Nature Biotech 14: 862-866
7. Barlow JN, Zhang Z, John P, Baldwin JE, Schofield CJ (1997) Inactivation of 1-aminocyclopropane-1-carboxylate oxidase involves oxidative modifications. Biochem 36:3563-3569
8. Bidonde S, Ferrer MA, Zegzouti H, Ramassamy S, Latché A, Pech JC, Hamilton AJ, Grierson D, Bouzayen M (1998) Expression and characterization of three tomato 1-aminocyclopropane-1carboxylate oxidase cDNAs in yeast. Eur J Biochem 253: 20-26
9. Chae HS, Faure F, Kieber JJ (2003) The eto1, eto2, and eto3 mutations and cytokinin treatment increase ethylene biosynthesis in Arabidopsis by increasing the stability of ACS protein. Plant Cell 15: 545-559
10. Charng YY, Chou SJ, Jiaang WT, Chen ST, Yang SF (2001) The catalytic mechanism of aminocyclopropane-1-carboxylic acid oxidase. Arch Biochem Biophys 385:179-185
11. Chung M-C, Chou S-J, Kuang L-Y, Charng Y, Yang SF (2002) Subcellular localisation of 1-aminocyclopropane-1-carboxylic acid oxidase in apple fruit. Plant Cell Physiol 43: 549-554
12. Clendennen SK, Kellogg JA, Wolff KA, Matsumura W, Peters S, Vanwinkle JE, Copess B, Pieper W, Kramer MG (1999) Genetic engineering of cantaloupe to reduce ethylene biosynthesis and control ripening. In A K Kanellis, et al. (eds), Biology and Biotechnology of the Plant Hormone Ethylene, Kluwer Academic Publishers, Dordrecht, pp. 371-379
13. Cosgrove DJ, Gilroy S, Kao T, Ma H, Schulz C (2000) Plant signaling (2000) Cross talk among geneticists, physiologists, and ecologists. Plant Physiol 124: 499-505

14. De Martinis D, Mariani C (1999) Silencing gene expression of the ethylene-forming enzyme results in a reversible inhibition of ovule development in transgenic tobacco plants. Plant Cell 11: 1061-1071
15. Fluhr R, Mattoo AK (1996) Ethylene biosynthesis and perception. Crit Rev Plant Sciences 15:479-523
16. Good X, Kellogg JA, Wagoner W, Langhoff D, Matsumura W, Bestwick RK (1994) Reduced ethylene synthesis by trangenic tomatoes expressing S-adenosylmethionine hydrolase. Plant Mol Biol 26: 781-790
17. Guzman P, Ecker JR (1990) Exploiting the tripe response of *Arabidopsis* to identify ethylene-related mutants. Plant Cell 2: 513-523
18. Hamilton AJ, Lycett GW, Grierson D (1990) Antisense gene that inhibits synthesis of the hormone ethylene in transgenic plants. Nature 346: 284-287
19. Hatzfeld Y, Maruyama A, Schmidt A, Noji M, Ishizawa K, Saito K (2000) β-cyanoalanine synthase is a mitochondrial cysteine synthase-like protein in spinach and Arabidopsis. Plant Phyiol 123: 1163-1171
20. Hegg EL, Que Jr L (1997) The 2-His-1-carboxylate facial triad, an emerging structural motif in mononuclear non-heme iron(II) enzymes. Eur J Biochem 250: 625-629
21. Henzi MX, McNeil DL, Christey MC, Lill RE (1999) A tomato antisense 1-aminocyclopropane-1-carboxylic acid oxidase gene causes reduced ethylene production in transgenic broccoli. Aust J Plant Physiol 26:179-183
22. Huai Q, Xia Y, Chen Y, Callahan B, Li N, Ke H (2001) Crystal structures of 1-aminocyclopropane-1-carboxylate (ACC) synthase in complex with amino-ethoxyvinylglycine and pyrydoxal-5'-phosphate provide new insight into catalytic mechanism. J Biol Chem 276:38210-38216
23. Kahana A, Silberstein L, Kessler N, Goldstein RS, Perl-Treves R (1999) Expression of ACC oxidase genes differs among sex genotypes and sex phases in cucumber. Plant Mol Biol 41: 517-528
24. Kende H (1993) Ethylene biosynthesis. Annu Rev Plant Physiol Plant Mol Biol 44, 283-307
25. Klee H J (1993) Ripening physiology of fruit from transgenic tomato (*Lycopersicon esculentum*) plants with reduced ethylene synthesis. Plant Physiol 102: 911-916
26. Lelièvre JM, Latché A, Jones B, Bouzayen M, Pech JC (1997) Ethylene and fruit ripening. Physiol Plant 101: 727-739
27. Liang X, Abel S, Keller JA, Shen NF, Theologis A (1992) The 1-amino-cyclopropane-1-carboxylate synthase gene family of *Arabidopsis thaliana*. Proc Natl Acad Sci USA 89: 11046-11050
28. Lieberman M (1979) Biosynthesis and action of ethylene. Annu Rev Plant Physiol 30, 533-591
29. McMahon Smith J, Arteca RN (2000) Molecular control of ethylene production by cyanide in *Arabidopsis thaliana*. Physiol Plant 109: 180-187
30. Nakatsuka A, Murachi S, Okunishi H, Shiomi S, Nakano R, Inaba KY (1998) Differential expression and internal feedback regulation of 1-aminocyclopropane-1-carboxylate synthase, 1-aminocyclopropane-1-carboxylic acid oxidase, and ethylene receptor genes in tomato fruit during development and ripening. Plant Physiol 118: 1295-1305
31. Oeller PW, Min-Wong L, Taylor LP, Pike DA, Theologis A (1991) Reversible inhibition of tomato fruit senescence by antisense RNA. Science 254:437-439
32. Osborne DJ, Walters J, Milborrow BV, Norville A, Stange LM C(1996) Evidence for a non-ACC ethylene biosynthesis pathway in lower plants. Phytochem 42: 51-60
33. Pedreño MA, Bouzayen M, Pech JC, Marigo G, Latché A (1991) Vacuolar release of 1-(malonylamino)cyclopropane-1-carboxylic acid, the conjugated form of the ethylene precursor. Plant Physiol 97:1483-1486
34. Peiser G, Yang SF (1998) Evidence for 1- (malonylamino)cyclopropane-1-carboxylic acid being the major conjugate of aminocyclopropane-1-carboxylic acid in tomato fruit. Plant Physiol 116: 1527-1532

35. Peñarrubia L, Aguilar M, Margossian L, Fischer RL (1992) An antisense gene stimulates ethylene hormone production during tomato fruit ripening. Plant Cell 4: 681-687
36. Prescott AG, John P (1996) Dioxygenases: molecular structure and role in plant metabolism. Annu Rev Plant Physiol Plant Mol Biol 47: 245-271
37. Ramassamy S, Olmos E, Bouzayen M, Pech JC, Latché A (1998) 1-aminocyclopropane-1-carboxylate oxidase of apple fruit is periplasmic. J Exp Bot 49: 1909-1915
38. Rodrigues-Pousada RA, Rycke RD, Dedonder A, Van Caeneghem W, Engler G, Van Montagu M, Van der Straeten D (1993) The Arabidopsis 1-aminocyclopropane-1-carboxylate gene 1 is expressed during early development. Plant Cell 5:897-911
39. Saftner RA, Baker JE (1987) Transport and compartmentation of 1-aminocyclopropane-1-carboxylic acid and its structural analogs, α-aminoisobutyric acid, in tomato pericarp slices.Plant Physiol 84: 311-317
40. Samach A, Onouchi H, Gold SE, Ditta GS, Schwarz-Sommer Z, Yanofsky MF, Coupland G (2000) Distinct roles of CONSTANS target genes in reproductive development of Arabidopsis. Science 288: 1613-1616
41. Savin K W, Baudinette SC, Graham MW, Michael MZ, Nugent GD, Lu CY, Chandler SF, Cornish EC (1995) Antisense ACC oxidase RNA delays carnation petal senescence. HortScience 30:970-972
42. Spanu P, Grosskopf DG, Felix G, Boller T (1994) The apparent turnover of 1-aminocyclopropane-1-carboxylate synthase in tomato cells is regulated by protein phosphorylation and dephosphorylation. Plant Physiol 106: 529-535
43. Stearns JC, Glick BR (2003) Transgenic plants with altered ethylene biosynthesis or perception. Biotech Adv 21:193-210
44. Stella L, Wouters S, Baldellon F (1996) Chemical and biochemical aspects of the biosynthesis of ethylene, a plant hormone. Bull Soc Chim Fr 133: 1141-1145
45. Tarun AS, Theologis A (1998) Complementation analysis of mutants of 1-aminocyclopropane-1-carboxylate reveals the enzyme is a dimer with shared active sites. J Biol Chem 273: 12509-12514
46. Tatsuki M, Mori H (2001) Phosphorylation of tomato 1-aminocyclopropane-1-carboxylic acid synthase, LE-ACS2, at the C-terminal region. J Biol Chem 276: 28051-28057
47. Theophrastus (372-287 BC) Enquiring into plants, (IV II 1-3). Translated to English by A. Hart, 1961, W. Heinemann, Ed. London, vol IV, p293.
48. Tophop S, Martinoia E, Kaiser G, Hartung W, Amrheinn N (1989) Compartmentation and transport of 1-aminocyclopropane-1-carboxylic acid and N-malonyl-1-aminocyclopropane-1-carboxylic acid in barley and wheat mesophyll cells and protoplasts. Physiol Plant 75: 333-339
49. Vogel JP, Woeste KE, Theologis A, Kieber JJ (1998) Recessive and dominant mutations in the ethylene biosynthetic gene ACS5 of Arabidopsis confer cytokinin insensitivity and ethylene overproduction, respectively. Proc Natl Acad Sci USA 95: 4766-4771
50. Wang KL-C, Li H, Ecker JR (2002) Ethylene biosynthesis and signaling networks. The Plant Cell, Sup 2002: S131-S151
51. Wang NN, Yang SF, Charng Y (2001) Differential expression of 1-aminocyclopropane-1-carboxylate synthase genes during orchid flower senescence induced by the protein phosphatase inhibitor okadaic acid. Plant Physiol 126: 253-260
52. Whittaker DJ, Smith GS, Gardner R C (1997) Expression of ethylene biosynthesis genes in *Actinidia chinensis* fruit. Plant Mol Biol 34: 45-55
53. Woeste KE, Ye C, Kieber JJ (1999) Two Arabidopsis mutants that overproduce ethylene are affected in post-transcriptional regulation of 1-aminocyclopropane-1-carboxylic acid synthase. Plant Physiol 119: 521-529
54. Yang SF, Hoffman N E (1984) Ethylene biosynthesis and its regulation in higher plants. Annu Rev Plant Physiol 35: 155-189
55. Zhou J, Rocklin AM, Lipscomb JD, Que Jr L, Solomon EI (2002) Spectroscopic studies of 1-aminocyclopropane-1-carboxylic acid oxidase: molecular mechanism and CO_2 activation in the biosynthesis of ethylene. J Amer Chem Soc 124: 4602-4609

B5. Abscisic Acid Biosynthesis and Metabolism

Steven H. Schwartz and Jan A.D. Zeevaart
MSU-DOE Plant Research Laboratory, Michigan State University, East Lansing, MI 48842, USA E-mail: zeevaart@msu.edu

INTRODUCTION

Abscisic acid (ABA) (Fig. 1) was discovered independently by several groups in the early 1960s. Originally believed to be involved in the abscission of fruit and dormancy of woody plants, the role of ABA in these processes is still not clear. In later work it became evident, however, that ABA is necessary for seed development, adaptation to several abiotic stresses, and sugar sensing. The regulation of these processes is in large part mediated by changes in *de novo* synthesis of ABA. When leaves of mesophytic plants are water-stressed, ABA levels increase strikingly (Fig. 2). The elevated ABA levels lead to stomatal closure, changes in gene expression, and other adaptations that increase the plant's stress tolerance (Chapters D6 & E4). In seeds, elevated ABA levels during mid-embryogenesis are necessary for the expression of genes needed for the accumulation of storage reserves and tolerance to desiccation (Chapter E4).

To understand how the cellular levels of ABA are regulated, it is necessary to elucidate the pathways of ABA synthesis and degradation. Progress in working out the biosynthetic pathway of ABA has been relatively slow, because

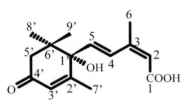

Figure 1. The structure of abscisic acid [(+)-*S*-ABA].

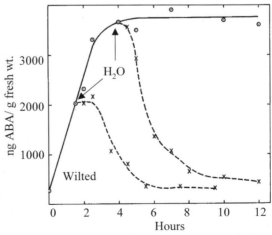

Figure 2. Time course of ABA accumulation in detached and dehydrated leaves of *Xanthium strumarium* and the subsequent decrease in ABA following rehydration (H_2O). Adapted from: Plant Physiol 66: 672-678 (1980).

ABA biosynthesis

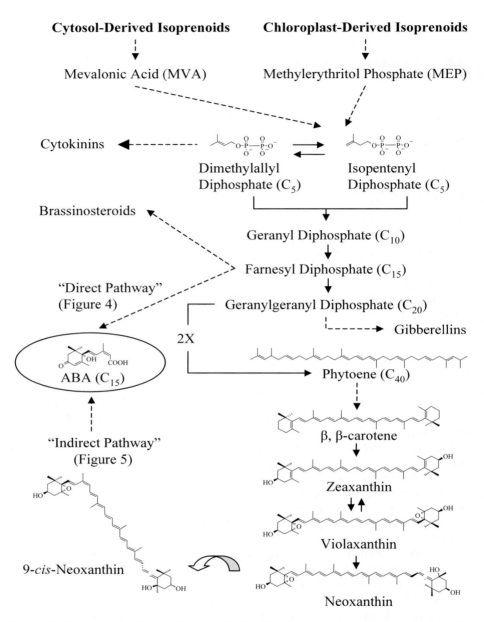

Figure 3. Isopentenyl diphosphate is synthesized in the cytosol via the mevalonic acid (MVA) pathway and in the chloroplast through the methylerythritol phosphate (MEP) pathway. Geranyl diphosphate (C_{10}) is produced by a head-to-tail condensation of isopentenyl diphosphate (C_5) and its isomer dimethylallyl diphosphate (C_5). Addition of another isoprene unit produces farnesyl diphosphate (C_{15}), a precursor of ABA in some fungi. Geranylgeranyl diphosphate (C_{20}) is produced by another condensation reaction. Head-to-head condensation of two geranylgeranyl diphosphate molecules produces phytoene (C_{40}) in the first committed step in carotenoid biosynthesis.

conventional biochemical methods have been of little use. It is only in recent years with the availability of mutants for most steps in the pathway and molecular-genetic approaches that substrates and biosynthetic enzymes have been characterized (32). A number of reviews, some also covering ABA signaling, have appeared in recent years (11, 12, 23, 24, 32, 35, 41, 43, 47).

BIOSYNTHESIS: EARLY STUDIES

The structure of ABA strongly suggests a derivation from isoprenoids. Two distinct pathways have been proposed for the synthesis of ABA (Fig. 3). In the "direct pathway", which occurs in some phytopathogenic fungi, ABA is derived from the C_{15} isoprenoid, farnesyl diphosphate (FPP[1]). When [^{14}C]-mevalonic acid (MVA) was fed to ABA-producing cultures of *Cercospora rosicola*, there was a significant incorporation of label into ABA. Several potential intermediates in the pathway were also labeled and when fed back to cultures these compounds were also converted to ABA (24). Based upon these studies, several likely intermediates for the ABA synthetic pathway in fungi were identified (Fig. 4). More recently, the pathway of ABA biosynthesis has been studied by feeding [1-^{13}C]-*D*-glucose to cultures (15). The ^{13}C-labeling pattern, determined by NMR, was also consistent with a direct pathway from farnesyl diphosphate in fungi.

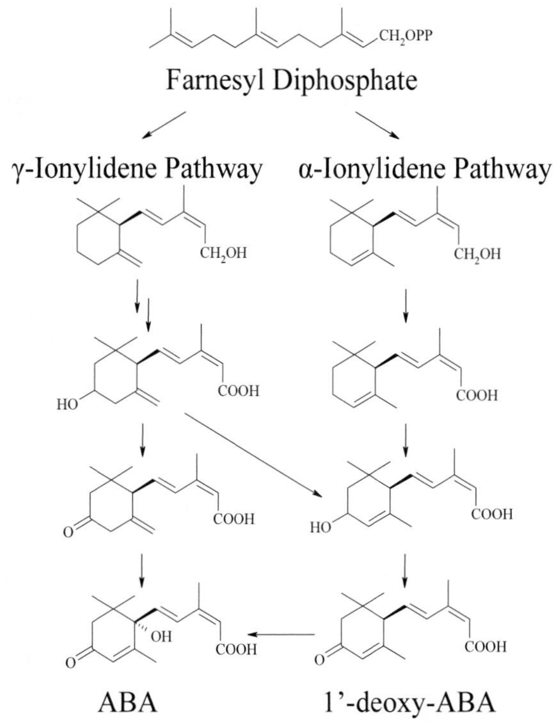

Because of structural similarities, an "indirect pathway" in which ABA is produced from the cleavage of carotenoids also had been proposed. A number of studies indicated that ABA in plants is derived from carotenoids (Fig. 5).

Figure 4. Variations in the "direct pathway" of ABA biosynthesis, which occurs in some phytopathogenic fungi. In this case, 1'-deoxy-ABA is the precursor of ABA.

---

[1] Abbreviations: AAO, abscisic aldehyde oxidase; CCD, carotenoid cleavage dioxygenase; DPA, dihydrophaseic acid; FPP, farnesyl diphosphate; GE, glucosyl ester; IPP, isopentenyl diphosphate; LSD, lignostilbene dioxygenase; MeABA, methyl abscisate; MEP, methylerythritol phosphate; MoCo, molybdenum cofactor; MVA, mevalonic acid; NCED, nine *cis*-epoxy-carotenoid dioxygenase; NCI, negative chemical ionization; PA, phaseic acid; VP, viviparous; ZEP, zeaxanthin epoxidase

ABA biosynthesis

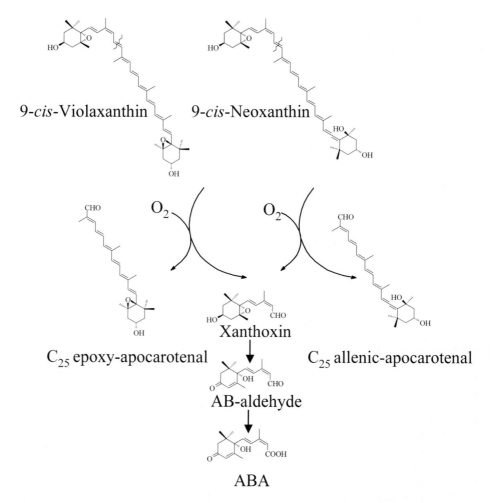

Figure 5. The "indirect pathway" of ABA biosynthesis starting with the first committed step, the oxidative cleavage of a 9-*cis*-epoxy-carotenoid. Here, AB-aldehyde is the immediate precursor of ABA.

Xanthoxin, an oxidation product of violaxanthin, was found to have the same activity as ABA in bioassays for the inhibition of seed germination. It was later shown that xanthoxin is readily converted to ABA by intact leaves and cell-free extracts (discussed later in this chapter).

Initial attempts to study the pathway of ABA biosynthesis in plants by feeding labeled MVA resulted in very low incorporation of label into ABA. This was subsequently explained by the identification of a second pathway for the synthesis of isopentenyl diphosphate (IPP) that occurs in prokaryotes and plants (29). This methylerythritol phosphate (MEP) pathway is the

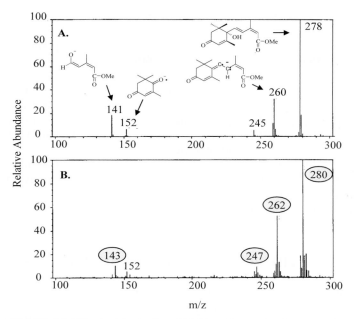

Figure 6. A) Negative chemical ionization (NCI) spectrum of Me-ABA with molecular and fragment ions indicated. B) Spectrum of Me-ABA extracted from wilted *Xanthium strumarium* leaves placed in an atmosphere of $^{18}O_2$. A shift of two mass units corresponds to the presence of one ^{18}O atom in the different ions (shaded). Adapted from: Plant Physiol 91: 1594-1601 (1989).

primary source of IPP that is utilized for the synthesis of carotenoids in plants.

Feeding studies with carotenoids are complicated by their hydrophobic nature and susceptibility to a variety of co-oxidation reactions. As an alternative, labeling studies with $^{18}O_2$ have proved a valuable tool for studying carotenoid metabolism and ABA biosynthesis. In these experiments, intact plant tissue with a high rate of ABA biosynthesis was placed in an atmosphere of $^{18}O_2$ and the incorporation of oxygen at the different positions was analyzed by mass spectrometry. Using negative chemical ionization (NCI), the molecular ion of the methyl ester of ABA (Me-ABA) M$^-$ at m/z 278 is the most abundant peak (Fig. 6A). When leaves are stressed and placed in an $^{18}O_2$ atmosphere, the majority of the ABA is labeled with a single ^{18}O and M$^-$ at m/z 280 (Fig. 6B). This indicates that ABA is derived from a precursor that already contains oxygen at two of the four positions. Based upon the fragmentation pattern, the position of the ^{18}O was found to be primarily in the carboxyl group (Fig. 6B). By $H_2^{18}O$ labeling, it was shown that the oxygen introduced in the final step of the pathway, the oxidation of AB-aldehyde to ABA, is derived from water. These labeling patterns are consistent with a derivation from epoxy-carotenoids. The ^{18}O incorporation would most likely result from an aldehyde introduced by the oxidative cleavage of the carotenoid. The hydroxyl of the epoxy-carotenoid precursor is converted to the 4'-keto and the 1'-hydroxyl of ABA is derived from an epoxy group. Hence, there was no incorporation at these positions. Longer incubations in $^{18}O_2$ do result in a

ABA biosynthesis

gradual increase of ^{18}O label at the other positions as a result of *de novo* synthesis of the epoxy-carotenoids (48).

A similar labeling pattern was observed with several plant species and tissue types studied. There was one exception, apple fruit, in which the majority of the label occurred in the 1'-hydroxyl oxygen. The absence of label in the carboxyl group was explained by the probable exchange of the aldehyde oxygen atom with water (48). The conversion of AB-aldehyde to ABA appears to be relatively slow in apple fruit, allowing the exchange reaction to occur. In most plant tissues the conversion of xanthoxin to ABA appears to occur very rapidly, so that there is no loss of label.

Further support for the indirect pathway of ABA biosynthesis came from experiments with etiolated tissues, which contain relatively low levels of carotenoids. When etiolated bean seedlings are stressed there is a decrease in the level of epoxy-carotenoids and a stoichiometric increase in the levels of ABA and its catabolites (21).

BIOSYNTHESIS: THE ROLE OF ABA-DEFICIENT MUTANTS

The biochemical studies discussed above provided strong circumstantial evidence that ABA is synthesized via an epoxy-carotenoid precursor (Fig. 5). In recent years, this pathway and its regulation have become well established through the identification and characterization of ABA-deficient mutants (Table 1). Several phenotypes associated with an ABA deficiency are illustrated in Figure 7. These mutants have also been the basis for isolating most of the ABA biosynthetic genes (32).

Mutants Impaired in the Early Steps

Isoprenoids are an extremely diverse class of natural products that serve a variety of functions in plants. Gibberellins, brassinosteroids, and cytokinins are members of this family of compounds (Fig. 3). Not surprisingly, mutants that are impaired in the early steps of isoprenoid synthesis have a seedling lethal

Table 1. Abscisic acid-deficient mutants. The ABA-deficient mutants have been identified in

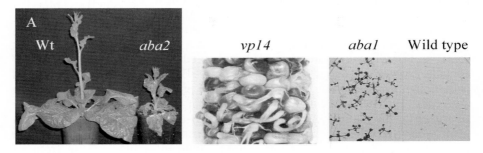

Figure 7 (Color plate page CP4). Phenotypes of ABA-deficient mutants. A) Inability to regulate transpiration results in a wilty phenotype and reduced growth of the *aba2* mutant in *Nicotiana plumbaginifolia*. B) Precocious germination of seeds in the maize mutant, *vp14*. Courtesy of Drs. B.C. Tan and D. R. McCarty. C) A high sugar concentration (10% glucose) inhibits growth of wild-type seedlings of Arabidopsis, whereas *aba1* seedlings are sugar insensitive.

various plants by the following phenotypes: precocious germination, susceptibility to wilting, an increase in stomatal conductance, and an ability to germinate and grow on media containing a high concentration of sucrose or salt. In several cases, allelic mutants that were isolated by different laboratories have been assigned different gene symbols.

Mutant	Species	Enzyme
aba1/npq2/los6	Arabidopsis	Zeaxanthin epoxidase (ZEP)
aba2	Nicotiana plumbaginifolia	ZEP
osaba1	Rice (Oryza sativa)	ZEP
vp14	Maize (Zea mays)	Nine cis-epoxy-carotenoid dioxygenase (NCED)
not (notabilis)	Tomato (Lycopersicon esculentum)	NCED
aba2/gin1/isi4/sis4	Arabidopsis	Short-chain alcohol dehydrogenase
flc (flacca)	Tomato	Molybdenum cofactor (MoCo) sulfurase
aba3/los5/gin5	Arabidopsis	MoCo sulfurase
sit (sitiens)	Tomato	Aldehyde oxidase
aba1	Nicotiana plumbaginifolia	MoCo sulfurase?
nar2A	Barley (Hordeum vulgare)	MoCo synthesis
aao3	Arabidopsis	Abscisic aldehyde oxidase (AAO3)

phenotype. The viviparous mutants in maize display precocious germination of seedlings. In some cases, this is due to reduced ABA levels. Several of these viviparous mutants are impaired in carotenoid synthesis and also have a lethal seedling phenotype.

The *aba1* mutant in Arabidopsis was identified in a screen for suppressor mutations of the gibberellin-deficient mutant, *ga1*. The *aba1* mutant has reduced levels of ABA and in $^{18}O_2$ labeling experiments a large percentage of the 1'-hydroxyl group was labeled, indicating that the pool of ABA precursors was significantly decreased (28). Analysis of the carotenoid composition in *aba1* showed a substantial decrease in epoxy-carotenoids and a corresponding increase in zeaxanthin.

A zeaxanthin epoxidase mutant in *Nicotiana plumbaginifolia*, *aba2*, has been identified and the corresponding gene cloned (22). Based upon sequence similarity, an Arabidopsis epoxidase (*AtZEP*) was cloned and it was shown that this gene was the basis for the *aba1* mutation in Arabidopsis (4). To avoid confusion that may arise from the nomenclature, the gene symbol *ZEP* (zeaxanthin epoxidase) will be used.

The NpZEP protein, which is similar to some flavoprotein monooxygenases from prokaryotes, was able to catalyze the two-step epoxidation of zeaxanthin to violaxanthin *in vitro*. For this activity, it was necessary to add an additional component from chloroplasts. It was later determined that NADPH, ferredoxin, and a ferredoxin reductase are needed to reduce the second atom of oxygen to water.

The Cleavage Reaction of Carotenoids

The first committed step in ABA synthesis is the oxidative cleavage of a 9-*cis*-epoxy-carotenoid. The pathway of ABA synthesis had for many years been a point of contention because of difficulties in demonstrating this activity *in vitro*. This problem was eventually resolved by the identification and characterization of an ABA-deficient mutant in maize, *viviparous 14* (*vp14*). Analysis of *vp14* embryos showed that it had a normal complement of carotenoids, and protein preparations from the mutant were able to convert xanthoxin to ABA (40). These biochemical experiments indicated that *vp14* was most likely impaired in the cleavage reaction.

The *vp14* mutant resulted from a transposon insertion, which allowed the corresponding gene to be cloned (40). At the time the *Vp14* gene was identified, the deduced amino acid sequence was most similar to lignostilbene dioxygenases (LSDs) from *Pseudomonas* (*Sphingomonas*) *paucimobilis*. The LSDs catalyze a double-bond cleavage reaction that is very similar to the cleavage reaction in ABA synthesis. The recombinant VP14 protein was assayed for the ability to cleave 9-*cis*-violaxanthin to xanthoxin and a C_{25} by-product. These assays were analyzed by HPLC and both of the expected cleavage products were identified (33). The characteristics of the cleavage reaction in its substrate specificity and the site of cleavage (11-12 position) were consistent with the proposed pathway. A 9-*cis* double bond in the carotenoid precursor was necessary for activity. The product of this cleavage reaction is *cis*-xanthoxin, which is readily converted to ABA [*cis*-(+)-*S*-ABA] by plants. Cleavage of an all *trans*-isomer would result in *trans*-xanthoxin, which is converted to biologically inactive *trans*-ABA by plants.

Additional ABA synthetic cleavage enzymes have since been identified and characterized in a variety of plant species (10, 19, 25). The recombinant enzymes from these species display the same substrate specificity as VP14. The nomenclature that has been adopted for these enzymes is nine-*cis*-epoxy-carotenoid dioxygenases or NCEDs, which is consistent with either 9-*cis*-violaxanthin or 9'-*cis*-neoxanthin as a substrate.

Although the NCEDs display significant substrate plasticity *in vitro*, circumstantial evidence favors neoxanthin as the primary precursor of ABA. Neoxanthin exists almost entirely as a 9'-*cis*-isomer, whereas only a small proportion of the violaxanthin is present as a 9-*cis*-isomer. The affinity of the NCEDs for various epoxy-carotenoids and the accessibility of these substrates to cleavage are additional factors that may determine the actual precursors of ABA. A large portion of the 9'-*cis*-neoxanthin, for example, is utilized by the light-harvesting complexes and may not be available for cleavage. Definitive evidence of the endogenous substrate would require identification of the C_{25} by-product *in planta*. Previous efforts to identify the C_{25} compounds in vegetative tissue have been unsuccessful. It has been suggested that these compounds are rapidly degraded following the cleavage reaction.

Carotenoids in plants are synthesized within plastids and are associated with the thylakoid and envelope membranes. Therefore, it was expected that the cleavage reaction would also occur in chloroplasts. The PvNCED1 from bean was imported into pea chloroplasts where it was found to associate exclusively with the thylakoid membrane (25). An N-terminal targeting sequence from a cowpea enzyme, VuNCED1, was capable of targeting the green fluorescent protein to chloroplasts (19). Following *in vitro* import assays, the VP14 protein was found in the stroma and on the thylakoid membrane exposed to the stroma (38). In this study, deletion or disruption of a putative amphipathic-helix in the N-terminus of VP14 interfered with the association of VP14 with thylakoids. The binding of VP14 to the thylakoid was saturable, suggesting that it associates with specific components in the thylakoid membrane that have not yet been identified. Of the five NCEDs of Arabidopsis, three (NCED2, NCED3, and NCED6) showed a distribution between the stroma and thylakoid membranes similar to that of VP14. NCED5 was exclusively bound to the thylakoids and NCED9 remained soluble in the stroma (39). Thus, different NCEDs from the same species differ with respect to their suborganellar localization.

The *notabilis* (*not*) mutant in tomato is also impaired in the cleavage step (8). Despite extensive screening for ABA-deficient mutants in Arabidopsis, no mutant impaired in the cleavage reaction has been reported, likely because of redundancy for this step in the pathway. Both the *vp14* and *not* mutants have weak phenotypes relative to other ABA-deficient mutants. The *vp14* null mutant shows only a 35% reduction of ABA levels in stressed leaves and a 70% reduction in developing embryos (40), indicating that there are additional NCEDs involved in ABA synthesis. In avocado, *PaNCED1* and *PaNCED3* were both shown to encode ABA biosynthetic enzymes (10).

In the Arabidopsis genome, there are nine hypothetical proteins that share sequence similarity to NCEDs. In a phylogenetic analysis of the NCEDs and other similar proteins, five of the Arabidopsis proteins are clustered with previously characterized NCEDs. Based upon the degree of sequence similarity and some experimental evidence, five of these genes are thought to encode ABA biosynthetic NCEDs (18). The different NCEDs in Arabidopsis are expressed in different tissues and at different developmental stages (39). The *AtNCED3* gene is an important regulator of ABA levels during water stress in Arabidopsis. The *AtNCED3* transcript is induced by water stress and reduced expression results in a wilty phenotype (18, 39). Expression of *AtNCED9* is also elevated slightly in response to water stress. Expression of other *NCED* genes in Arabidopsis is developmentally regulated. *AtNCED5* and *AtNCED6* remained highly expressed in embryos and endosperm until seed maturity (39). Specific roles for these *NCED*s in pollen and seed development remain to be demonstrated by knockouts of these genes.

A variety of natural products like ABA are derived from the oxidative cleavage of carotenoids. Collectively referred to as apocarotenoids, these compounds serve important functions in a range of organisms (32). Vitamin A, for example, is derived from the central cleavage of β-carotene. Enzymes similar to the NCEDs would presumably catalyze the synthesis of other apocarotenoids. An enzyme from animals that is necessary for the synthesis of vitamin A was identified by sequence similarity to the NCEDs. There are four hypothetical proteins in Arabidopsis that share less sequence similarity with the NCEDs and are not thought to be involved in ABA synthesis. Recent findings demonstrate that at least two of these proteins catalyze a carotenoid cleavage reaction distinct from the one in ABA synthesis (31). To distinguish these enzymes from the NCEDs, the nomenclature of carotenoid cleavage dioxygenases (CCDs) has been adopted. To establish the functions of these compounds in plants a reverse genetics approach will be useful. A number of putative carotenoid cleavage dioxygenases have also been identified in an array of prokaryotes. The function of these proteins in bacteria has not yet been reported.

The Later Steps: Conversion of Xanthoxin to Abscisic Acid

Following the cleavage reaction xanthoxin moves out of the chloroplast and is converted to ABA by cytoplasmic enzymes. In contrast to the cleavage reaction, the later steps in ABA synthesis have been characterized extensively by feeding potential intermediates to intact plants and by cell-free extracts. For the cell-free extracts, the only cofactor required for the conversion of xanthoxin to ABA was NAD or NADP. A variety of potential intermediates were also fed to cell-free extracts to determine the likely sequence of reactions (37). If a substrate was not converted to ABA by the cell-free system it was considered an unlikely intermediate in vivo. These experiments indicated that the first step in the conversion of xanthoxin to ABA is the oxidation of the 4'-hydroxyl to a ketone. The remainder of the ring modifications, i.e. the desaturation of the 2'-3' bond and opening of the epoxide ring, are thought to be non-enzymatic. Indeed, chemical oxidation of the 4'-hydroxyl is sufficient to convert xanthoxin to AB-aldehyde.

The *aba2* and *aba3* mutants in Arabidopsis were first identified by screening for the ability to germinate in the presence of the gibberellin biosynthetic inhibitor, paclobutrazol. By feeding potential intermediates to crude protein preparations from the *aba2* mutant, it was determined that this mutant was impaired in the conversion of xanthoxin to abscisic aldehyde (30). Additional alleles of *aba2* have since been identified in screens for a sugar-insensitive phenotype, altered stomatal conductance, and for growth on a medium containing a high NaCl concentration. The gene corresponding to *aba2* has recently been cloned and the gene product was found to be similar to short chain dehydrogenases/reductases (9, 14). As expected, the ABA2 protein was able to catalyze the conversion of xanthoxin to AB-aldehyde utilizing NAD as a cofactor.

Mutants impaired in the final step of ABA synthesis, the oxidation of AB-aldehyde to ABA, have been identified in a variety of plants. A loss of this AB-aldehyde oxidase activity may result from a mutation in the aldehyde oxidase apoprotein or a lesion in the synthesis of a molybdenum cofactor (MoCo) that the enzyme requires for activity. A lesion in an early step of MoCo synthesis affects a number of enzyme activities. The *nar2a* mutant in barley lacks aldehyde oxidase, xanthine dehydrogenase, and nitrate reductase activities. The

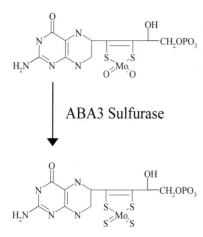

Figure 8. Two forms of the molybdenum cofactor. The *ABA3* gene encodes a sulfur transferase, which introduces a terminal sulfur ligand to the molybdenum cofactor. This desulfo form of the cofactor is necessary for aldehyde oxidase activity.

aba3 mutant in Arabidopsis and the *flacca* mutant in tomato lack aldehyde oxidase and xanthine dehydrogenase activities, but nitrate reductase activity is unaffected. This phenotype results from a defect in the formation of a desulfo moiety of the MoCo that is specifically required by the MoCo hydroxylases. Treatment of a protein preparation from the *aba3* mutant with sodium sulfide resulted in the chemical conversion of the MoCo to the desulfo form and activity of the aldehyde oxidase was restored (Fig. 8) (30). The gene corresponding to *aba3* has been cloned and the N-terminus of the deduced protein is similar to the NifS sulfurase (5, 44). Using cysteine as a sulfur donor, the recombinant protein was able to activate aldehyde oxidase activity (5).

Four aldehyde oxidase genes have been identified in Arabidopsis (*AAO1* through *4*). Of the aldehyde oxidases characterized so far, only *AAO3* uses abscisic aldehyde efficiently as a substrate (34). A wilty, ABA-deficient mutant with a lesion in *AAO3* has been identified (36), demonstrating that AAO3 is responsible for ABA synthesis in vegetative tissues. In contrast to *aba3*, however, the *aao3* mutants are not subject to precocious germination. Another aldehyde oxidase appears, therefore, to be necessary for ABA synthesis in some tissues.

Several variations in the later steps of the pathway may be responsible for generating a small portion of the ABA pool. Both the *flacca* and *sitiens* mutants in tomato are blocked in the final step of the pathway, the oxidation of AB-aldehyde to ABA, and accumulate 2-*trans*-ABA alcohol. These mutants are able to synthesize some ABA by a shunt pathway in which abscisic alcohol is oxidized to ABA (27).

UNRESOLVED QUESTIONS IN ABA BIOSYNTHESIS

There are several steps in ABA biosynthesis preceding the cleavage reaction that are not well characterized. The epoxy-carotenoid precursor must have a 9-*cis* configuration to be cleaved by an NCED and for subsequent conversion to ABA [*cis*-(+)-*S*-ABA]. The formation of these 9-*cis* isomers has not yet been established. An enzyme that catalyzes a similar reaction, the *cis/trans* isomerization of prolycopene to lycopene, has recently been identified (17). This isomerase appears to be necessary only in non-photosynthetic tissue. In light-grown tissue, photo-isomerization of lycopene is sufficient. It has not been established whether the 9-*cis* isomerization of neoxanthin and violaxanthin is an enzymatic reaction.

In most plant tissues, neoxanthin is the predominant carotenoid with a 9-*cis* conformation and is considered the most likely precursor of ABA. Neoxanthin is derived through the opening of an epoxy ring in violaxanthin, followed by an intramolecular rearrangement to form an allenic bond. Allenic carotenoids, such as neoxanthin, are among the most abundant carotenoids in nature. Therefore, an understanding of their synthesis and functions in photosynthetic organisms is of considerable interest. Two genes that encode neoxanthin synthases (NSY) have been identified in potato and tomato (1, 7). The NSY gene products are similar to lycopene cyclases from various plants and a capsanthin-capsorubin synthase from pepper. Transient expression in tobacco and *in vitro* assays both demonstrated that the tomato NSY was capable of converting violaxanthin to neoxanthin (7). However, the *NSY* gene corresponds to the *old-gold* mutant in tomato, which accumulates higher levels of lycopene due to the loss of a fruit-specific lycopene β-cyclase, CYC-B. It has been suggested that NSY is a bifunctional enzyme capable of converting lycopene to β,β-carotene, or violaxanthin to neoxanthin (16). There must be an additional gene responsible for neoxanthin synthesis in plants, because the *old-gold* mutant produces neoxanthin in vegetative tissues and no ortholog of the *NSY* gene is apparent in the Arabidopsis genome. Therefore, the mechanism by which neoxanthin is formed in Arabidopsis and other plants is still unresolved.

REGULATION OF ABA BIOSYNTHESIS

Inhibitors of transcription and translation block stress-induced ABA accumulation, indicating that gene expression is up-regulated for one or more steps in the pathway. Based upon elevated expression, a regulatory role has been proposed for several of the genes.

Because the level of epoxy-carotenoids in green leaves is high relative to the amount of ABA synthesized, it is considered unlikely that the zeaxanthin epoxidase (ZEP) has a regulatory role in these tissues. The expression of *ZEP* transcripts in green tissue does not increase in tobacco, tomato, or cowpea plants that are subjected to osmotic stress. In one study, it was found that the level of *AtZEP* mRNA was induced by drought stress in root tissues, but not in shoots (4). In another study, it was reported that the expression of *AtZEP* mRNA increased in response to osmotic stress or ABA treatment in both roots and shoots (45).

The $^{18}O_2$ labeling experiments, which were instrumental in establishing the indirect pathway of ABA synthesis, also provide some indication of flux through the pathway. The 1'-hydroxyl in ABA, which is derived from the epoxide in the carotenoid precursor, is not labeled in short-term incubations with $^{18}O_2$. Therefore, *de novo* synthesis of epoxy-carotenoids is unnecessary for ABA synthesis in leaves. In etiolated tissues, the concentration of carotenoids is significantly lower and an increase in *ZEP* mRNA expression does correlate with elevated ABA synthesis in roots and seeds (3, 6). The over-expression of *ZEP* in transgenic tobacco resulted in increased seed dormancy (13), providing further evidence that the level of epoxy-carotenoids does limit ABA synthesis in some tissues.

There are several biochemical lines of evidence to indicate that the conversion of xanthoxin to ABA occurs very rapidly and is also independent of stress conditions. If xanthoxin or AB-aldehyde accumulated in plants the observed ^{18}O labeling of the carboxyl group would not occur, due to exchange of the ^{18}O-labeled carbonyl with water. In addition, cell-free extracts prepared from stressed and non-stressed tissue displayed no difference in their ability to convert xanthoxin to ABA (30, 37).

The Arabidopsis *ABA2* transcript level was not affected by stress (14), but was up-regulated by glucose (9). Depending on the process being regulated and the concentrations, both antagonistic and synergistic interactions have been observed for sugars and ABA. For example, high concentrations of sugars inhibit germination and seedling growth (Fig. 7C). Several alleles of ABA-deficient mutants have been identified by their ability to grow on media containing high sugar concentrations. The levels of ABA do increase in response to glucose (2) and it has been suggested that the up-regulation of *ABA2* may be the basis for elevated ABA levels in response to sugars.

Up-regulation of two enzymes necessary for the final step in ABA synthesis may have a regulatory role in the pathway (44, 45). It was found that expression of *ABA3*, which encodes MoCo sulfurase, increased in response to osmotic stress or ABA (5, 44). The *AAO3* mRNA transcript was also elevated in response to stress, but the level of the corresponding protein was unaffected (36). The over-expression of these genes in transgenic plants and any effect that this might have on ABA synthesis has not yet been reported.

Based upon biochemical experiments, expression data, and recent work with transgenic plants it has been shown that the NCEDs play a critical role in regulating ABA biosynthesis in response to stress. In all instances studied to date, stress-induced ABA accumulation correlates well with increased expression of *NCED* mRNA in a variety of plant species and also with NCED protein levels (Fig. 9). Following the alleviation of stress conditions, *PvNCED1* expression and ABA levels decreased rapidly (Fig. 10). The developmental regulation of ABA that occurs in some fruits also appears to be controlled by *NCED* expression. The expression of *PaNCED1* and *PaNCED3* increased prior to the accumulation of high ABA levels during fruit ripening in avocado (10).

The over-expression of *NCEDs* in transgenic plants is sufficient for elevated ABA synthesis. Over-expression of the *LeNCED1* in tomato (42), *PvNCED1* in tobacco (26), and *AtNCED3* in Arabidopsis (18) resulted in increased ABA levels. Tobacco plants over-expressing the *PvNCED1* gene have reduced transpiration rates and are better able to tolerate drought (Fig. 11). For the over-expression of *LeNCED1* (42) and *PvNCED1* (26), increased seed dormancy was also reported.

ABA biosynthesis

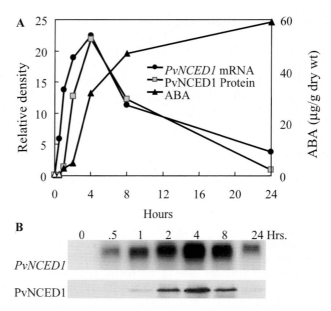

Figure 9. A) Time course of changes in *PvNCED1* mRNA, PvNCED1 protein and ABA levels in bean leaves in response to dehydration. B) Northern analysis of *PvNCED1* expression (upper panel), and western analysis of PvNCED1 protein (lower panel). From (25).

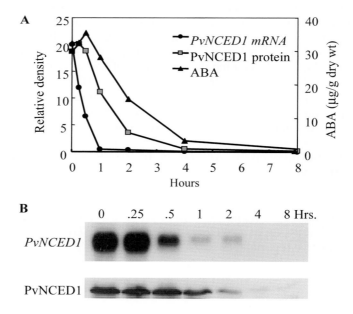

Figure 10. A) Time course of *PvNCED1* mRNA, protein and ABA accumulation in detached bean leaves in response to rehydration. B) Northern analysis of the expression of *PvNCED1* (upper panel), and western analysis of *Pv*NCED1 protein (lower panel). From (25).

CATABOLISM

Control   GVG::PvNCED1

Figure 11 (Color plate page CP4). Over-expression of PvNCED in a transgenic *Nicotiana plumbaginifolia* plant (right) results in elevated ABA levels and increased stress tolerance in comparison with a wild-type plant (left). From (32).

The level of ABA in plants is controlled not only by its synthesis, but also through its catabolism (Fig. 12). There are two major pathways for ABA inactivation. One of the primary catabolites of ABA is phaseic acid (PA), which is biologically inactive. The conversion of ABA to PA begins with the hydroxylation of the 8' position by ABA 8'-hydroxylase. The 8' hydroxyl appears to be an unstable intermediate that spontaneously rearranges to form PA. Phaseic acid is further converted to dihydrophaseic acid (DPA), which can be conjugated to glucose at the 4' position. The 8'-hydroxylase activity has been characterized in a microsomal fraction from maize suspension cultures (20). The requirements for this activity and the sensitivity of the enzyme to inhibition by carbon monoxide and tetcyclacis indicate that the ABA 8'-hydroxylase is a cytochrome P450.

In suspension cultures pretreated with ABA, the rate of conversion of ABA to PA is increased significantly, indicating that the 8'-hydroxylase is itself induced by ABA. This negative feedback regulation is consistent with time course measurements of ABA and PA accumulation in stressed plants. It has also been observed that several ABA-insensitive mutants contain elevated levels of ABA. Once the gene encoding the 8'-hydroxylase has been identified, its manipulation may offer another approach for modifying ABA levels in transgenic plants. Recently, a new catabolite of ABA has been identified. This

Figure 12. Major pathways of ABA catabolism that result in inactivation. The relative importance of the pathways may differ among plant species or even in different tissues of the same plant species.

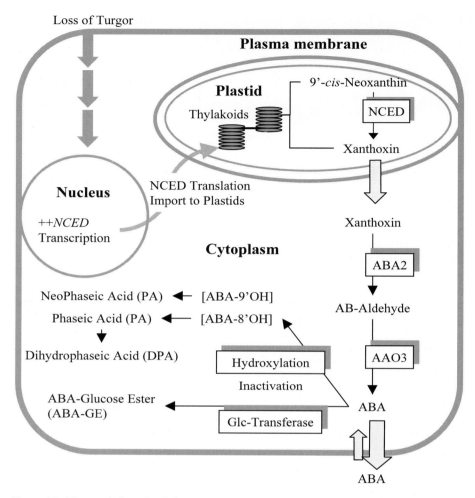

Figure 13. The regulation of cellular ABA levels beginning with a loss of turgor, which leads to elevated *NCED* expression and ABA levels. Inactivation of ABA can occur through conversion to ABA glucose ester or to phaseic and neophaseic acids.

catabolite results from the 9'-hydroxylation of ABA to give 9'-hydroxy ABA. This unstable compound rearranges to form an isomer of PA, called neoPA (Fig. 12) (49). It is not known at present whether the same or different enzymes catalyze hydroxylation at the 8' and 9' positions.

Abscisic acid may also be inactivated by the formation of its glucose ester (ABA-GE), which is stored in the vacuole. The level of ABA-GE continued to increase in plants that were subjected to a series of stress and rewatering cycles. An ABA glucosyltransferase gene from adzuki bean has recently been cloned (46). Interestingly, this gene is also up-regulated by ABA. The physiological significance of ABA-GE formation as well as the potential for engineering ABA levels by decreased catabolism may now be investigated.

FUTURE DIRECTIONS

The biochemical aspects of ABA synthesis, such as the intermediates in the pathway and the sequence of reactions, have become well established. The genes that encode enzymes for many steps in the pathway have now been cloned. The stress and developmental regulation of these ABA synthetic genes has become a primary research focus within the field. It is now clear that osmotic stress results in a loss of turgor that triggers an increase in *NCED* expression and leads to elevated ABA levels (Fig. 13). The perception of turgor loss and the subsequent signal transduction pathway remain to be elucidated. The significance of positive feedback regulation for the latter steps in ABA synthesis and negative feedback regulation leading to the inactivation of ABA is also unclear at this time.

Acknowledgement

Work of the laboratory was supported by grants from the U.S. Department of Energy and the National Science Foundation.

References

1. Al-Babili S, Hugueney P, Schledz M, Welsch R, Frohnmeyer H, Laule O, Beyer P (2000) Identification of a novel gene coding for neoxanthin synthase from *Solanum tuberosum*. FEBS Lett 485: 168-172
2. Arenas-Huertero F, Arroyo A, Zhou L, Sheen J, Léon P (2000) Analysis of *Arabidopsis* glucose insensitive mutants, *gin5* and *gin6*, reveals a central role of the plant hormone ABA in the regulation of plant vegetative development by sugar. Genes Dev 14: 2085-2096
3. Audran C, Borel C, Frey A, Sotta B, Meyer C, Simonneau T, Marion-Poll A (1998) Expression studies of the zeaxanthin epoxidase gene in *Nicotiana plumbaginifolia*. Plant Physiol 118: 1021-1028
4. Audran C, Liotenberg S, Gonneau M, North H, Frey A, Tap-Waksman K, Vartanian N, Marion-Poll A (2001) Localisation and expression of zeaxanthin epoxidase mRNA in *Arabidopsis* in response to drought stress and during seed development. Aust J Plant Physiol 28: 1161-1173
5. Bittner F, Oreb M, Mendel RR (2001) ABA3 is a molybdenum cofactor sulfurase required for activation of aldehyde oxidase and xanthine dehydrogenase in *Arabidopsis thaliana*. J Biol Chem 276: 40381-40384
6. Borel C, Frey A, Marion-Poll A, Tardieu F, Simonneau T (2001) Does engineering abscisic acid biosynthesis in *Nicotiana plumbaginifolia* modify stomatal response to drought? Plant Cell Environ 24: 477-489
7. Bouvier F, D'Harlingue A, Backhaus RA, Kumagai MH, Camara B (2000) Identification of neoxanthin synthase as a carotenoid cyclase paralog. Eur J Biochem 267: 6346-6352
8. Burbidge A, Grieve TM, Jackson A, Thompson A, McCarty DR, Taylor IB (1999) Characterization of the ABA-deficient tomato mutant *notabilis* and its relationship with maize *Vp14*. Plant J 17: 427-431
9. Cheng W-H, Endo A, Zhou L, Penney J, Chen H-C, Arroyo A, Leon P, Nambara E, Asami T, Seo M, Koshiba T, Sheen J (2002) A unique short-chain dehydrogenase/reductase in *Arabidopsis* glucose signaling and abscisic acid biosynthesis and functions. Plant Cell 14: 2723-2743
10. Chernys JT, Zeevaart JAD (2000) Characterization of the 9-cis-epoxycarotenoid dioxygenase gene family and the regulation of abscisic acid biosynthesis in avocado. Plant Physiol 124: 343-353
11. Cutler AJ, Krochko JE (1999) Formation and breakdown of ABA. Trends Plant Sci 4: 472-478
12. Finkelstein RR, Rock CD (2002) Abscisic acid biosynthesis and response. *In* CR Somerville, EM Meyerowitz, *eds*, The Arabidopsis Book. Am Soc Plant Biol, Rockville,

MD, pp doi/10.1199/tab.0058. http://www.aspb.org/publications/arabidopsis/
13. Frey A, Audran C, Marin E, Sotta B, Marion-Poll A (1999) Engineering seed dormancy by the modification of zeaxanthin epoxidase gene expression. Plant Mol Biol 39: 1267-1274
14. González-Guzmán M, Apostolova N, Bellés JM, Barrero JM, Piqueras P, Ponce MR, Micol JL, Serrano R, Rodríguez PL (2002) The short-chain alcohol dehydrogenase ABA2 catalyzes the conversion of xanthoxin to abscisic aldehyde. Plant Cell 14: 1833-1846
15. Hirai N, Yoshida R, Todoroki Y, Ohigashi H (2000) Biosynthesis of abscisic acid by the non-mevalonate pathway in plants, and by the mevalonate pathway in fungi. Biosci Biotechnol Biochem 64: 1448-1458
16. Hirschberg J (2001) Carotenoid biosynthesis in flowering plants. Curr Opin Plant Biol 4: 210-218
17. Isaacson T, Ronen G, Zamir D, Hirschberg J (2002) Cloning of *tangerine* from tomato reveals a carotenoid isomerase essential for the production of ß-carotene and xanthophylls in plants. Plant Cell 14: 333-342
18. Iuchi S, Kobayashi M, Taji T, Naramoto M, Seki M, Kato T, Tabata S, Kakubari Y, Yamaguchi-Shinozaki K, Shinozaki K (2001) Regulation of drought tolerance by gene manipulation of 9-*cis*-epoxycarotenoid dioxygenase, a key enzyme in abscisic acid biosynthesis in *Arabidopsis*. Plant J 27: 325-333
19. Iuchi S, Kobayashi M, Yamaguchi-Shinozaki K, Shinozaki K (2000) A stress-inducible gene for 9-cis-epoxycarotenoid dioxygenase involved in abscisic acid biosynthesis under water stress in drought-tolerant cowpea. Plant Physiol 123: 553-562
20. Krochko JE, Abrams GD, Loewen MK, Abrams SR, Cutler AJ (1998) (+)-Abscisic acid 8'-hydroxylase is a cytochrome P450 monooxygenase. Plant Physiol 118: 849-860
21. Li Y, Walton DC (1990) Violaxanthin is an abscisic-acid precursor in water-stressed dark-grown bean leaves. Plant Physiol 92: 551-559
22. Marin E, Nussaume L, Quesada A, Gonneau M, Sotta B, Hugueney P, Frey A, Marion-Poll A (1996) Molecular identification of zeaxanthin epoxidase of *Nicotiana plumbaginifolia*, a gene involved in abscisic acid biosynthesis and corresponding to the *ABA* locus of *Arabidopsis thaliana*. EMBO J 15: 2331-2342
23. Nambara E, Marion-Poll A (2003) ABA action and interactions in seeds. Trends Plant Sci 8: 213-217
24. Oritani T, Kiyota H (2003) Biosynthesis and metabolism of abscisic acid and related compounds. Nat Prod Rep 20: 414-425
25. Qin X, Zeevaart JAD (1999) The 9-*cis*-epoxycarotenoid cleavage reaction is the key regulatory step of abscisic acid biosynthesis in water-stressed bean. Proc Natl Acad Sci USA 96: 15354-15361
26. Qin X, Zeevaart JAD (2002) Overexpression of a 9-cis-epoxycarotenoid dioxygenase gene in *Nicotiana plumbaginifolia* increases abscisic acid and phaseic acid levels and enhances drought tolerance. Plant Physiol 128: 544-551
27. Rock CD, Heath TG, Gage DA, Zeevaart JAD (1991) Abscisic alcohol is an intermediate in abscisic acid biosynthesis in a shunt pathway from abscisic aldehyde. Plant Physiol 97: 670-676
28. Rock CD, Zeevaart JAD (1991) The *aba* mutant of *Arabidopsis thaliana* is impaired in epoxy-carotenoid biosynthesis. Proc Natl Acad Sci USA 88: 7496-7499
29. Rodríguez-Concepción M, Boronat A (2002) Elucidation of the methylerythritol phosphate pathway for isoprenoid biosynthesis in bacteria and plastids. A metabolic milestone achieved through genomics. Plant Physiol 130: 1079-1089
30. Schwartz SH, Léon-Kloosterziel KM, Koornneef M, Zeevaart JAD (1997) Biochemical characterization of the *aba2* and *aba3* mutants in *Arabidopsis thaliana*. Plant Physiol 114: 161-166
31. Schwartz SH, Qin X, Zeevaart JAD (2001) Characterization of a novel carotenoid cleavage dioxygenase from plants. J Biol Chem 276: 25208-25211
32. Schwartz SH, Qin X, Zeevaart JAD (2003) Elucidation of the indirect pathway of abscisic acid biosynthesis by mutants, genes, and enzymes. Plant Physiol 131: 1591-1601
33. Schwartz SH, Tan BC, Gage DA, Zeevaart JAD, McCarty DR (1997) Specific oxidative cleavage of carotenoids by VP14 of maize. Science 276: 1872-1874
34. Seo M, Koiwai H, Akaba S, Komano T, Oritani T, Kamiya Y, Koshiba T (2000) Abscisic aldehyde oxidase in leaves of *Arabidopsis thaliana*. Plant J 23: 481-488

35. Seo M, Koshiba T (2002) Complex regulation of ABA biosynthesis in plants. Trends Plant Sci 7: 41-48
36. Seo M, Peeters AJM, Koiwai H, Oritani T, Marion-Poll A, Zeevaart JAD, Koornneef M, Kamiya Y, Koshiba T (2000) The *Arabidopsis* aldehyde oxidase 3 (*AAO3*) gene product catalyzes the final step in abscisic acid biosynthesis in leaves. Proc Natl Acad Sci USA 97: 12908-12913
37. Sindhu RK, Walton DC (1987) Conversion of xanthoxin to abscisic acid by cell-free preparations from bean leaves. Plant Physiol 85: 916-921
38. Tan BC, Cline K, McCarty DR (2001) Localization and targeting of the VP14 epoxy-carotenoid dioxygenase to chloroplast membranes. Plant J 27: 373-382
39. Tan B-C, Joseph LM, Deng W-T, Liu L, Cline K, McCarty DR (2003) Molecular characterization of the *Arabidopsis* 9-*cis* epoxycarotenoid dioxygenase gene family. Plant J 35: 44-56
40. Tan BC, Schwartz SH, Zeevaart JAD, McCarty DR (1997) Genetic control of abscisic acid biosynthesis in maize. Proc Natl Acad Sci USA 94: 12235-12240
41. Taylor IB, Burbidge A, Thompson AJ (2000) Control of abscisic acid synthesis. J Exp Bot 51: 1563-1574
42. Thompson AJ, Jackson AC, Symonds RC, Mulholland BJ, Dadswell AR, Blake PS, Burbidge A, Taylor IB (2000) Ectopic expression of a tomato 9-*cis*-epoxycarotenoid dioxygenase gene causes over-production of abscisic acid. Plant J 23: 363-374
43. Xiong L, Zhu J-K (2003) Regulation of abscisic acid biosynthesis. Plant Physiol 133: 29-36
44. Xiong L, Ishitani M, Lee H, Zhu JK (2001) The *Arabidopsis LOS5/ABA3* locus encodes a molybdenum cofactor sulfurase and modulates cold stress- and osmotic stress- responsive gene expression. Plant Cell 13: 2063-2083
45. Xiong L, Lee H, Ishitani M, Zhu JK (2002) Regulation of osmotic stress-responsive gene expression by the *LOS6/ABA1* locus in *Arabidopsis*. J Biol Chem 277: 8588-8596
46. Xu ZJ, Nakajima M, Suzuki Y, Yamaguchi I (2002) Cloning and characterization of the abscisic acid-specific glucosyltransferase gene from adzuki bean seedlings. Plant Physiol 129: 1285-1295
47. Zeevaart JAD (1999) Abscisic acid metabolism and its regulation. *In* PPJ Hooykaas, MA Hall, KR Libbenga, *eds*, Biochemistry and Molecular Biology of Plant Hormones. Elsevier Science, Amsterdam, pp 189-207
48. Zeevaart JAD, Rock CD, Fantauzzo F, Heath TG, Gage DA (1991) Metabolism of abscisic acid and its physiological implications. *In* WJ Davies, HG Jones, eds, Abscisic Acid: Physiology and Biochemistry. BIOS Scientific, Oxford, pp 39-52
49. Zhou R, Cutler AJ, Ambrose SJ, Galka MM, Nelson KM, Squires TM, Loewen MK, Jadhav A, Ross ARS, Taylor DC, Abrams SR (2004) A new abscisic acid catabolic pathway. Plant Physiol 134: 361-369

B6. Brassinosteroid Biosynthesis and Metabolism

Sunghwa Choe
School of Biological Sciences, College of Natural Sciences NS70, Seoul National University, Seoul 151-747, Korea. E-mail: shchoe@snu.ac.kr

INTRODUCTION

The sessile nature of plants requires distinctive regulatory mechanisms to meet the demands of development and environmental challenges. Various plant hormones that act alone or in concert underpin these mechanisms. Brassinosteroids (BRs) collectively refers to naturally-occurring 5α cholestane steroids that elicit growth stimulation in nano- or micromolar concentrations (15). BRs that are biosynthesized using sterols as precursors are structurally similar to the cholesterol-derived, human steroid hormones and insect molting hormones. BRs have been known for decades to be effective in plant growth promotion. However, definitive evidence for their roles in growth and development remained unclear until the recent characterization of BR dwarf mutants isolated from Arabidopsis and other plants. This chapter[1] aims to provide a cohesive summary of information about progress made in the molecular genetic characterization of mutants that are defective in sterol and BR biosynthetic pathways.

HISTORICAL BACKGROUND AND NATURAL OCCURRENCE

A variety of plant growth regulators are involved in the intricate processes of reproduction. Thus, plant scientists recognized that pollen could be a rich source of phytohormones. A search for novel plant hormones from pollen was begun in the 1960s by a United States Department of Agriculture (USDA) group (35). This led to the discovery of a substance, named Brassin, from rape (*Brassica napus*) pollen that stimulated growth in a bean second internode bioassay (35). The first characterized BR, brassinolide (BL), was discovered from bee-collected rape pollen at a concentration of 200 parts per billion (23). The BL structure elucidated by X-ray diffraction technology was (22R,23R,24S)-2α,3α,22,23-tetrahydroxy-24-methyl-B-homo-7-oxa-5α-chol

---

[1]Abbreviations: 22-OHCR, 22-hydroxycampesterol; BL, brassinolide; BRs, brassinosteroids; CN, campestanol; CR, campesterol; CS, castasterone; CT, cathasterone; MVA, mevalonic acid; SR, sitosterol; TE, teasterone; TY, typhasterol.

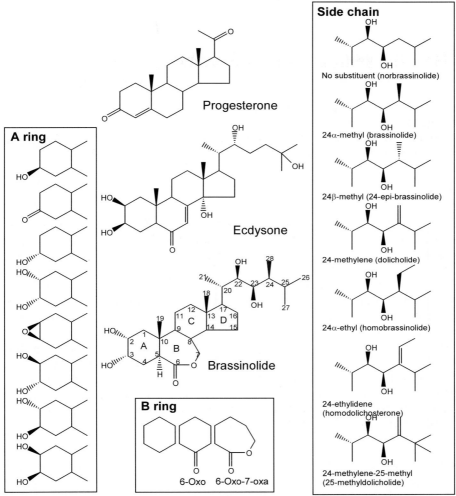

Figure 1. Structural variation in brassinosteroids. Brassinosteroids share structural similarities with the human steroid hormone progesterone and the insect molting hormone ecdysone in that they all have a 4-membered steroid ring backbone and multiple oxidations. Structural variations are primarily based on the different status of oxidation at ring A, B, and the side chain. Different combinations of each element in the three variable regions are reflected in the approximately 50 BRs identified to date. The BR names in the side chain box are applicable when the BL side chain is replaced with one of the structures in the box.

estan-6-one (Fig. 1).

In addition to BL, at least 50 additional BR structures have been identified from various species in the plant kingdom examined to date (20). These include one species of algae (*Hydrodictyon reticulatum*), a pteridophyte (*Equisetum arvense*), 5 species of gymnosperms, and 37 different species of angiosperms (18). Thus, it is conceivable that BRs are ubiquitous in the plant kingdom.

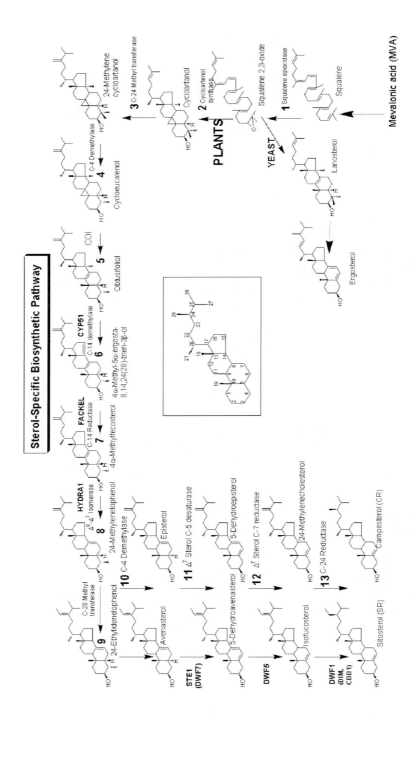

Figure 2. The brassinolide biosynthetic pathway, which starts here and continues on the next page. The biosynthetic pathway is divided into multiple subunits. The sterol-specific pathway refers to the steps from Squalene to CR (or STR).

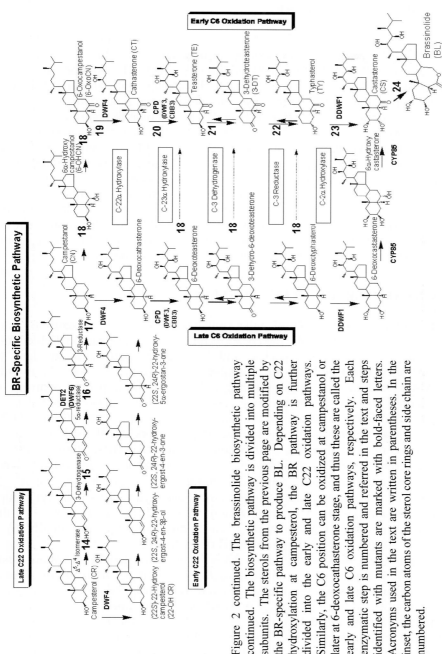

Figure 2 continued. The brassinolide biosynthetic pathway continued. The biosynthetic pathway is divided into multiple subunits. The sterols from the previous page are modified by the BR-specific pathway to produce BL. Depending on C22 hydroxylation at campesterol, the BR pathway is further divided into the early and late C22 oxidation pathways. Similarly, the C6 position can be oxidized at campestanol or later at 6-deoxocathasterone stage, and thus these are called the early and late C6 oxidation pathways, respectively. Each enzymatic step is numbered and referred in the text and steps identified with mutants are marked with bold-faced letters. Acronyms used in the text are written in parentheses. In the inset, the carbon atoms of the sterol core rings and side chain are numbered.

CHEMICAL STRUCTURE

BRs are primarily classified into seven categories depending on the side chain structures that are derived from different parental sterols (Fig. 1). In addition to the side chain variation, two additional structural variations occur: 1) differing numbers of hydroxyl groups with varying configurations at the C-2 and C-3 positions; and 2) either a ketone or a lactone functional group at C-6 and C-7. Among the 50 different BRs, BL was shown to possess the greatest growth-promoting activity (50, 49). As inferred from the chemical structure of BL, it was hypothesized that active BRs should possess the following structural requirements. First, the A and B rings must be in the *trans* configuration, which is determined by an α hydrogen at C-5. Second, the B ring should contain a 6-oxo or a 6-oxo-7-oxa group. Third, the hydroxyl groups at C-2 and C-3 in ring A should be *cis* α-oriented. Fourth, the *cis* α-oriented hydroxyl groups at the C-22, C-23, and the C-24 positions should be occupied either by α-oriented methyl or ethyl groups (36, 1). As a rule, the more these requirements are met, the more active the compound.

BIOSYNTHESIS OF BRASSINOSTEROIDS

BRs belong to the class of molecules known as triterpenoids. Because BRs are a group of modified sterols, the BL biosynthetic pathways can be divided into two major parts: sterol specific and BR-specific pathways, squalene to campesterol (CR) and CR to BL, respectively (Fig. 2). Mevalonic acid (MVA), which serves as the starting molecule of the terpenoid pathway, is condensed and cyclized to produce 2,3-oxidosqualene. 2,3-oxidosqualene is further modified to form the major plant sterols such as sitosterol (SR) and CR. All of these parent sterols then serve as precursors of BL isologs such as homo- or nor-BL (Fig. 1). To become a bioactive BRs, sterols must be processed by the BR specific pathway. In short, sterols are modified to have the following functional groups: 1) saturation of a double bond at Δ^5, 2) formation of a 6-oxo-group, 3) addition of α-oriented vicinal hydroxyl groups at C-22 and C-23, 4) epimerization of a 3β-hydroxyl group to the 3α-configuration, 5) addition of a 2α hydroxyl group, and finally 6) a Baeyer-Villiger type oxidation in the B ring (Fig. 2).

Individual biosynthetic steps have been elucidated by metabolic tests using BR-overproducing cell lines of the periwinkle *Catharanthus roseus* (43). These BL-overproducing lines were developed to overcome low biosynthetic activities in regular plant tissues or cell lines, which technically limits extraction and detection of BRs. BR biosynthesis proceeds through multiple branched pathways. The first branch occurs at CR and the second at campestanol (CN). CR can be either C-22 hydroxylated or C-5 reduced in bifurcated pathways that are termed the early and the late C-22 oxidation pathways, respectively (10) (Fig. 2). In addition, CN proceeds to one of two alternative pathways, the early and the late C-6 oxidation pathways. The early

C-6 oxidation pathway undergoes a two-step oxidation of the C-6 position at the CN stage (20). In the late C-6 oxidation pathway, C-6 is oxidized at the second to last step. The order of chemical substitutions other than the branching steps are conserved between the parallel pathways, such that these reactions are performed by single enzyme acting on both the early and late intermediates (48, 8). The BR biosynthetic pathway that was established in the periwinkle feeding experiments has served as a framework for further validation and modification by results from analyses of dwarf mutants that are defective in the BR biosynthesis and signaling pathways.

GENETICS OF BRASSINOSTEROID BIOSYNTHETIC MUTANTS

BR *dwarf* (*dwf*) mutants are characterized by a suite of characteristic phenotypes including reduced fertility, short but robust stature, small-round and curled leaves, prolonged life span, irregular vascular differentiation, and abnormal skotomorphogenesis (Figs. 3 and 4). Although genetic defects in many other biological processes such as in light signaling pathways and in the synthesis or signaling pathways of other plant hormones result in dwarfism, BR biosynthetic dwarf mutants are, by definition, those that are rescued by exogenous application of BRs. These BR biosynthetic dwarf mutants include Arabidopsis *dwf1*, *cpd/dwf3*, *dwf4*, *dwf5*, *det2/dwf6*, *ste1/dwf7*, pea *lka* and *lkb*, and tomato *dwarf* mutants. The main feature of the characteristic phenotype of a BR dwarf is, by-and-large, a unidirectional reduction in organ

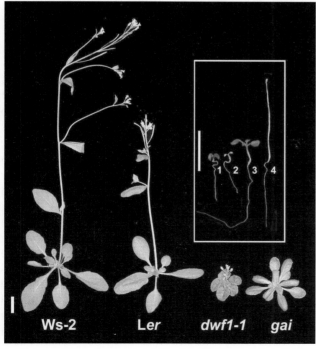

Figure 3. Morphological comparison of a brassinosteroid *dwf1-1* mutant with wild type and the GA dwarf mutant *gai*. As compared to wild type and *gai*, *dwf1-1* displays more round leaves. In the inset, light (1,3) and dark-grown (2,4) seedlings of *dwf4-1* (1,2) and wild type (3,4), respectively, are shown. Dark-grown *dwf4-1* shows an abnormal skotomorphogenesis pattern including short hypocotyls and open cotyledons. Developmental stages of the adult plants and seedlings are 7 (dark) and 30 days old (light). Bar = 1 cm.

Brassinosteroid biosynthesis and metabolism

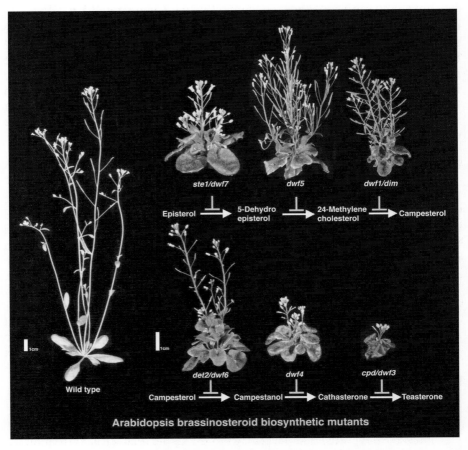

Figure 4 (Color plate page CP4). Arabidopsis *dwarf* mutants and the BR biosynthetic pathways. Select steps identified with Arabidopsis dwarf mutants are shown with mutant phenotypes. Null alleles of each locus display increasingly severe phenotypes as the pathway nears the end product Consistently, the dwarfs display short stature, round and curled leaves as compared to the Ws-2 wild type.

size. The length of pedicels, siliques, roots, and leaf blades are all reduced. The width of leaf blades is, however, not reduced nearly as much as their length. This differential reduction in leaf size gives rise to leaves with a rounder appearance. This is in contrast to the leaf phenotypes of GA biosynthetic and signaling mutants that display shorter and narrower leaves relative to a wild type (Fig. 3). BR *dwf* mutants have a smaller cell volume, but contain the same number of chloroplasts as wild type, which causes the darker green appearance (3).

BR *dwf* mutants display varying degrees of fertility. Generally, the reduced fertility of dwarf plants is ascribed to differential reduction in cell elongation of specific floral organs. Elongation is more severely affected in the filament than the gynoecium, resulting in pollen being shed on the walls of the gynoecium instead of being shed directly on the papillae. When BR dwarf flowers are hand-pollinated, they show increased fertility, suggesting that the sterility is at least partially due to lack of access of pollen. As the siliques are shorter, even when flowers are hand-pollinated, the seeds are smaller than wild type (13).

Mutants defective in light-signal transduction pathways such as *constitutive photomorphogenesis (cop)/de-etiolated (det)/fusca (fus)* mutants display dwarfism, and seedlings of these mutants grown in the dark exhibit differentiated plastids and the gene expression pattern of light-grown wild type plants (38). Dark-grown hypocotyls of these mutants were shown to respond to the exogenous application of BRs (48); however BRs do not rescue their phenotype. Similar to these light signaling mutants, several of the BR dwarfs possess part of the *de-etiolation* (*det*) phenotype in the dark. These include short hypocotyls, expanded cotyledons, absence of an apical hook, and initiation of leaf development in the dark (Fig. 3 inset). Dwarf mutants exhibit shorter hypocotyls in the dark than those of wild type, but the hypocotyls are still capable of responding to darkness with more elongation than light-grown plants (3). This response to darkness suggests that dwarf plants can distinguish light from dark, and that other hormones are also responsible for hypocotyl growth. Presumably, the de-etiolated phenotype of dark-grown dwarfs is primarily due to a generalized defect in cell elongation.

Delayed senescence has been observed in BR dwarf mutants. Dwarf plants stay green and produce flowers for a longer period of time than wild type plants. In the dwarf mutants, the length of time for important developmental events such as bolting, flower opening and senescence are noticeably delayed (9). The prolonged life span is correlated with the degree of reduced fertility in *dwf* mutants. Relatively fertile dwarfs such as *dwf5* have a shorter life span than dwarfs that set few seeds (13).

It has been reported that the primary cause of the reduced stature of dwarf plants is not fewer number of cells but reduced cell elongation (27, 11). Longitudinal sections of *dwf* stems display a dramatic reduction in cell length to approximately 20% of that of wild type, whereas cell width is equal or even greater than wild type. The wider cell files partly contribute to the thicker stems seen in *dwf7* plants (11).

BR dwarfs retain a sensitivity to other plant growth-promoting hormones (27, 48, 3). Although BR dwarfs respond to auxins and GAs, only BRs rescue the dwarf phenotype. Exogenous application of BRs causes elongated internodes, petioles, pedicels, and roots (8). However, the sterility of dwarfs is not completely rescued by exogenous application of BR. Precise spatial and temporal regulation of BR action seems to be important in rescuing all the defects of dwarf mutants.

Mutants that have lesions in the sterol-specific pathways can be divided into two groups based on their response to exogenous application of BRs. One group, which includes *dwf1*, *dwf5*, and *dwf7*, have mutations in the genes regulating the later steps in the sterol-specific pathway; these are successfully rescued by BRs. The members of the other group, including *fackel* and *hydra1*, are defective in the earlier steps in the pathway and display multiple growth defects that are not corrected by exogenously supplied sterols and BRs.

STEPS IN THE BRASSINOSTEROID BIOSYNTHETIC PATHWAYS

Sterols are ubiquitous in eukaryotic organisms. Two major roles played by sterols in biological systems are as precursors of steroid hormones and as a bulk components of biological membranes. Thus, genetic defects in sterol biosynthetic pathways cause a broad spectrum of phenotypes attributable to both deficiency of steroid hormones as well as loss of membrane integrity. A comprehensive description of each step in sterol biosynthesis is well reviewed by Benveniste (4). Each step in the sterol biosynthetic pathway, where a mutant and its gene is known will be highlighted here.

Sterol Specific Biosynthetic Pathways

1. S-adenosylmethionine-sterol-C-methyltransferases1 (SMT1):
 C-24 methylation (Reaction 3)

The presence or absence of an alkyl group at the C-24 position of sterols distinguishes plant and fungal sterols from those of animals. Major vertebrate sterols are devoid of an alkyl group at C-24, whereas fungi have a methyl, and plants possess either a methyl or an ethyl group. In plants, the alkylation at C-24 occurs via two non-consecutive steps. Methylation at C-24 of cycloartanol forms C_{28} sterols such as CR. Additional methylation at C28 of 24-methylenelophenol results in C_{29} sterols such as stigmasterol and SR (Fig. 2). The sterol S-adenosylmethionine-sterol-C-methyltransferases transfer a methyl group from S-adenosylmethionine (SAM) to C-24 of cycloartanol to give rise to 24-methylenecycloartenol.

A mutant for this step (*smt1*) has been recovered from screening for mutants with a developmentally altered root system (15). *smt1* plants are characterized by short malformed roots, greatly reduced fertility, and abnormal embryogenesis. The swelling root phenotype is due to increased sensitivity to exogenous calcium ions in certain types of agar-solidified growth medium. Interestingly, the sterility results from a maternal defect: no seeds are produced when *smt1/smt1* stigma are fertilized with wild type pollen, whereas the reciprocal cross is fertile. In addition, developing *smt1/smt1* embryos display irregular patterns of morphogenesis in that the different stages of embryogenesis are not distinguishable, and the germinating seedlings also show varying cotyledon number and shape. Biochemical analysis of *smt1* plants reveals that the endogenous level of cholesterol (C_{27}) increases while SR (C_{29}) decreases, an indication of suppressed methylation at C-24. Interestingly, the endogenous level of CR (C_{28}), which is used as a precursor of BR biosynthesis, was similar to wild type level, suggesting that a functionally redundant gene may exist in the mutant. Indeed, the Arabidopsis genome contains three *SMT* genes: *SMT1* (At5g13170), *SMT2* (At1g20330), *SMT3* (At1g76090) (15, 4). All three SMT genes can complement the yeast *erg6* mutant that is defective in C-24 methylation, indicating that all three Arabidopsis SMTs are functional homologs of yeast ERG6 (7). Of the three genes, SMT1 is the most similar to ERG6. Decreased C24-methylation activity in *smt1* mutants also suggests that SMT1 is most likely involved in the first methylation reaction in the plant sterol biosynthetic pathway (15).

In relation to the BR dwarf phenotype, the overall growth of *smt1* plants is not significantly reduced as compared to typical BR mutants such as *dwf1*, *dwf5*, or *dwf7*. Thus it is plausible that the unique phenotypes of *smt1* may be due to a deficiency in specific sterols that are required at certain levels in particular cell types and tissues.

SMT2 and SMT3 are likely to mediate the second methylation reaction, which converts 24-methylenelophenol to 24-ethylidenelophenol. The Arabidopsis *cotyledon vascular pattern1* (*cvp1*) mutant is defective in the SMT2 gene. As the mutant name suggests, *cvp1* displays abnormal phenotypes in vascular cell arrangement which results in abnormal leaf vascular patterning (7). Biochemically, the *cvp1* mutant accumulates C_{28} sterols such as 24-methylelenelophenol, whereas C_{29} sterols including 24-methylidenelophenol and its downstream compounds are greatly decreased (7). Consequently, the ratio of CR to SR is shifted from approximately 0.2 in wild type to 1.9 in *cvp1*. These data clearly indicate that SMT2 (CVP1) preferentially mediates the second methylation step in the sterol pathway (Fig. 2). In addition, evidence that Arabidopsis SMT3 is a functional homolog of SMT2 comes from the results of overexpression of *SMT3* using cMV 35S promoter to complement the *cvp1* mutant. The two genes also show partially overlapping gene expression patterns, indicative of the same function in different tissues.

Conversely, overexpression of the SMT2 gene using the 35S promoter in Arabidopsis resulted in increased metabolic flux toward C_{29} sterol pathways and increased SR level at the expense of CR (44). Furthermore, the reduced CR levels led to decreased BL biosynthesis, which resulted in dwarfed plants that could be rescued by exogenous BL treatment (44).

2. *FACKEL/EXTRA LONG LIFESPAN/HYDRA2*: C-14 reduction
(Reaction 7)

The FACKEL-mediated step converts 4α-methyl-5α-ergosta-8,14,24(28)-trien-3β-ol to 4α-methylfecosterol using NADPH as a hydrogenation source (Fig. 2) (4, 20). Three independent groups isolated mutants that are allelic: *fackel* (*fk*) and *hydra2* (*hyd2*) were recovered from screening for mutants possessing abnormal embryonic patterning, and *extra long lifespan* (*ell*) mutants for constitutive cytokinin responses such as longevity. Examination of endogenous sterol and BR content showed that *fk* accumulates $\Delta^{8,14}$ sterols, and thus BR levels are greatly reduced (Fig. 2) (26, 45). Introduction of the Arabidopsis *FK* gene into the yeast *erg24* mutant, defective in the C-14 reduction step, successfully rescued *erg24*, suggesting that *FK* is an ortholog of yeast ERG24. Arabidopsis FK (At3g52940) displays sequence identity to a sterol reductase domain of the human Lamin B receptor and signature sequence of sterol reductases "LLXSGYWGXXRH" (4). Unlike *smt1* mutants, *fk* mutants display a severe growth retardation possibly due to a relatively low level of CR, and accordingly a decreased level of bioactive BRs. However, despite the reduced BR levels in the *fk* mutant, exogenous application of BRs did not rescue the *fk* phenotype, suggesting that the abnormal development in these mutants is partly attributable to unique roles played by sterols. In support of this, *fk* mutants show different gene expression patterns from BR biosynthetic mutants including *dwf4* in that the *TOUCH4* gene, which is normally induced by BL, and repressed in BR mutants is increased (24). This suggests that sterols play an important role as signaling molecules whose molecular mechanism is yet to be discerned.

3. HYDRA1: Δ^8-Δ^7 isomerization (Reaction 8)

A sterol Δ^8-Δ^7 isomerase converts 4α-methylfecosterol to 24-methylenelophenol. This enzymatic step is conserved among vertebrates, fungi, and plants; thus it was possible to isolate an Arabidopsis cDNA clone for this gene by functional complementation of a yeast *erg2* mutant that is defective in this step (21). A loss-of-function mutation for this gene, *hydra1* (*hyd1*), as recovered from screening for mutants that show altered embryonic and seedling cell patterning morphology in Arabidopsis (47). A genomic DNA fragment containing the 2 kb promoter region and the coding sequence of Arabidopsis sterol Δ^8-Δ^7 isomerase (At1g20050) successfully complemented the *hyd1* mutants. Similar to *fk*, *hyd1* displays pleiotropic developmental alterations such as the formation of multiple leaf-like cotyledons. In addition, similar to *cvp1*, *hyd1* shows aberrant vascular patterning. Examination of the sterol profiles in the *hyd1* mutant revealed that CR and SR are merely 12% and 2% that of wild type, respectively, suggesting that molecular lesions exist in the sterol isomerase gene. Although the reduced CR level led to a decreased BL level, exogenous treatment with BL did not rescue the phenotypes of the *hyd1* mutants. However, it is noteworthy that the *dwf7* mutants, which block two steps downstream of this sterol Δ^8-Δ^7 isomerase reaction, display dwarfism and are successfully rescued by exogenous application of bioactive BRs. Thus it is likely that the functional roles of sterols as steroid hormone precursors and as bulk component of membranes could be conferred by the enzymatic modifications catalyzed by sterol Δ^8-Δ^7 isomerase or one step before this.

4. DWARF7/STE1: C-5 desaturation (Reaction 11)

DWARF7 (DWF7) converts episterol (avenasterol) into 5-dehydroepisterol (5-dehydroavenasterol) by a stereo-specific removal of the two protons at C-5 (6). Cytochrome b5, NADH, and a molecular oxygen are involved in this oxidation step (4). A mutant that accumulates sterols without C-5 desaturation was isolated from an Arabidopsis EMS mutant population and named *ste1* (25). The gene for this enzymatic function was identified by isolation of a cDNA clone that functionally complemented the yeast *erg3* mutant. The *erg3*-rescuing clone was shown to encode a protein of 281 amino acids with characteristic His-rich motifs whose molecular role is yet to be discovered (11). When the *35S::STE1* clone was re-introduced into the *ste1* mutant, the genetic defect in C-5 desaturation was completely rescued. Sequencing the C-5 desaturase gene in *ste1* revealed that the threonine at position 114 was replaced by isoleucine, and this resulted in a slight change in the function of the native protein. This can be inferred from the *ste1* mutant phenotype as it has no obvious growth retardation except for the slight reduction in C5 saturation activity. Severe mutants for this gene were independently isolated and characterized by analysis of Arabidopsis as BR dwarf mutants.

Arabidopsis *dwf7* mutants display typical BR dwarf phenotypes, and are successfully rescued by exogenous application of BRs. Examination of the endogenous sterol levels showed that intermediates, such as 24-methylenecholesterol and downstream compounds, are greatly diminished as compared to wild type. In addition, metabolite conversion tests performed by feeding ^{13}C-labeled MVA into *dwf7* seedlings suggested that *dwf7* is not able to produce C-5 desaturated sterols. Sequencing of the C-5 desaturase gene (At3g02580) in *dwf7-1* and *dwf7-2* mutants revealed that mutations caused premature stop codons at position 230 and 60, respectively, and are predicted to be nulls.

The *DWF7* gene (At3g02580) has a contiguous homolog (At3g02590) in the Arabidopsis genome. When the homologous gene to *DWF7* (*HDF7*) was overexpressed in the *dwf7* mutant, the mutant phenotype was complemented (Choe and Tanaka, unpublished data), suggesting that the two genes were duplicated in recent evolutionary time. The Loss-of-function phenotype of *dwf7*, in spite of presence of a functionally equivalent gene suggests that the two genes may be expressed in different cell types in Arabidopsis or different affinities for substrates too. Although a null mutant of *dwf7* exhibits severe dwarfism and reduced fertility, *dwf7* does not display such severe defects in embryo development as are commonly found in upstream sterol mutants such as *smt1*, *fk*, and *hyd1* mutants. This suggests that essential roles played by CR and SR may have been replaced by surrogate sterols in the *dwf7* mutants, accordingly *dwf7* displays more of the phenotype attributable to BR deficiency only.

5. *DWARF5*: C-7 reduction (Reaction 12)

$\Delta^{5,7}$-sterol-Δ^{7}-reductase (S7R) hydrogenates the double bond at the Δ^{7} position. An Arabidopsis S7R gene (At1g50430) has been identified by selecting yeast strains that are resistant to Nystatin after being transformed with an Arabidopsis cDNA expression library (32). Nystatin is toxic to wild-type yeast because it acts on sterols with a double bond at the C-7 position. A yeast strain harboring a functional S7R may have saturated the double bond, and therefore show resistance to this fungicide. Sequencing of the gene selected from the Nystatin-resistant clone revealed that the protein possess the characteristic sterol reductase signature sequence also found in HYDRA2 (LLXSGWWGXXRH). It has long been known that the human sterol metabolic disease Smith-Lemli-Opitz syndrome (SLOS) is due to a genetic defect in the S7R step. Cloning of the Arabidopsis S7R gene accelerated the isolation of a corresponding human gene and subsequent molecular characterization of this devastating genetic disease.

Arabidopsis mutants for the S7R gene were identified from a population of canonical BR dwarf mutants. Unlike the typical sterol mutants *smt1*, *fk*, *cvp1*, and *hyd1*, but similar to *dwf7*, *dwf5* mutants display characteristic BR dwarf phenotype (13). Examination of endogenous sterol and BR levels revealed that intermediates after the S7R step are greatly diminished in *dwf5* mutants. Many of the growth defects in *dwf5* are rescued by exogenous application of BL and its early precursors including 22-hydroxycampesterol (22-OHCR) (Fig. 2). In addition, metabolites from ^{13}C-labeled MVA metabolism tests revealed that C-7 reduced compounds are not detectable in the *dwf5* mutant. Instead, *dwf5* skips the step and forms novel compounds such as 7-dehydroCR and 7-dehydroCN. One characteristic phenotype specific to *dwf5* includes a greatly increased fertility relative to other dwarfs, however, their seed size is small and color is dark-brown compared to wild type (13). Sequencing the S7R gene in *dwf5* showed that mutations are located in splice donor or acceptor sites as well as substitution mutations mostly in the 3' half of the gene. Localization of *dwf5* mutations in the 3' half of the gene suggests that some yet to be identified important domains reside in the C-terminal region of the protein.

6. DWARF1: C-24 reduction (Reaction 13)

A Δ^5-sterol-Δ^{24}-reductase converts the double bond at $\Delta^{24(28)}$ into a saturated single bond. It has been suggested that the reduction step occurs via two consecutive reactions: isomerization of the $\Delta^{24(28)}$ double bond from 24-methylenecholesterol to 24-methyldesmosterol ($\Delta^{24(25)}$) then saturation of the double bond into CR. Arabidopsis *dwf1* is known to be defective in this step. The *DWF1* (At3g19820) gene was cloned long before a precise biochemical role was elucidated by BR intermediate feeding tests and examining the endogenous BR levels because the sequence did not initially indicate the function of the protein (29, 9). Feeding tests showed that altered developmental defects in *dwf1* mutants are rescued by exogenous application of 22-OHCR, which suggests that the biosynthetic defect resides prior to CR. In addition, measurement of endogenous sterol levels in the *dwf1* mutants demonstrated that 24-methylelenecholesterol accumulates to 12 times the level of wild type, whereas the CR level stayed at 0.3% that of wild type (9). These data clearly suggest that *dwf1* is blocked in a step converting 24-methylenecholesterol to CR. The *lkb* mutant of garden pea has a mutation in an orthologous gene of *DWF1* and is defective in the two consecutive steps mediated by the single enzyme DWF1 (41). The amino acid sequence of DWF1 contains a domain identifiable as a flavin adenine dinucleotide (FAD)-binding motif. At least 7 of 10 *dwf1* mutations were mapped to conserved amino acid residues of this domain, a strong indicator of the importance of the FAD-binding domain in proper functioning of this enzyme.

Although fungi have the sterol C-24 reduction step (*erg5*), the protein sequences are divergent from those of the plant. However, human and other eukaryotic organisms besides fungi were found to possess sequences as similar as 40% to DWF1. The human DWF1-homologous gene is called Seladin-1 (GenBank Acc. # Q15392) and is responsible for conferring resistance to Alzheimer's disease-related neurodegeneration as well as oxidative stress (22). Future research will elucidate whether Seladin-1 could mediate a C-24 reduction in the plant sterol biosynthetic pathway.

Brassinosteroid Specific Biosynthetic Pathways

1. DE-ETIOILATED2: C-5 REDUCTION (Reaction 16)

Different sterols such as CR, SR, or cholesterol can be subjected to specific BR biosynthetic pathways resulting in BL, homo-BL, and nor-BL, respectively. These sterol modification steps are collectively referred to as the BR-specific biosynthetic pathway, and consist of reduction, oxidation, and isomerization reactions (Fig. 2). The first enzymatic step that has been identified with mutants in Arabidopsis is mediated by a sterol Δ^5 reductase. In humans, before a reduction occurs, a double bond Δ^5 is isomerized to Δ^4 by the action of a multifunctional enzyme 3β-hydroxysteroid dehydrogenase/$\Delta^{5 \to 4}$-isomerase (34). A reductase then hydrogenates the Δ^4 double bond in the presence of NADPH.

Arabidopsis *de-etiolated2* (*det2*) mutants are defective in this reduction step, and were isolated by screening for mutants that display a light-grown phenotype in the dark (14). *det2* mutants exhibit a typical BR dwarf phenotype including abnormal skotomorphogenic patterns, such as short hypocotyls, open cotyledons, hook opening, and expression of light dependent genes (33). *DET2* (At2g38050) is homologous to human steroid 5α-reductase, and when introduced into human cell lines, it converts progesterone (3-oxo Δ^4) to 4,5-dihydroprogesterone. The reverse is also true; the human steroid 5α-reductase gene functionally complements *det2* mutants (33). The precise biochemical defect in *det2* was resolved by feeding $^2H_6$-labeled CR to *det2* seedlings and subsequently examining the metabolites using gas chromatography-mass spectrometry (GC-MS) (17). Precursors having a 3-oxo Δ^4 structure accumulated 3-fold more in the *det2* mutant than in wild type, whereas CN level is less than 10% that of wild type. Detection of CN in *det2* indicates that either the *det2* mutation may not be null, or another functional homolog may be present in Arabidopsis.

2. DWARF4: C-22 hydroxylation (Reaction 19)

The C-22α hydroxylation step is considered a rate-determining step in the BR biosynthetic pathways based on the findings that the endogenous level of 6-OxoCN was 500 times greater than that of cathasterone (CT). However the bioactivity of CT is as much as 500 times greater than that of its 6-oxoCN in a rice lamina bending assay (16). Recent biochemical evidence has revealed that Arabidopsis C-22α hydroxylase uses various steroids as substrates. These include CR, (24*R*)-ergost-4-en-3-one, (24*R*)-5α-ergostan-3-one, CN, 6-OxoCN, and possibly many other C-24 reduced sterols (Fig. 2) (10, 19).

Arabidopsis *dwf4* mutants have mutations in this enzyme, and display typical BR-deficient dwarf phenotypes. Feeding tests with biosynthetic intermediates showed that only steroids that were C-22 hydroxylated rescued the dwarfism (8). In addition, examination of the endogenous levels of BR biosynthetic intermediates in *dwf4* revealed that substrates such as 6-OxoCN and CN accumulate, whereas the C-22 hydroxylated products are present only in trace amounts (8). Sequence analysis showed that DWF4 (At3g50660) belongs to the cytochrome P450 (CYP90B1) superfamily and shares great similarity with previously identified BR biosynthetic enzymes such as Arabidopsis CONSTITUTIVE PHOTOMOTPHOGENESIS AND DWARFISM (CPD: CYP90A1) and tomato DWARF (CYP85B1) (5, 48). The Arabidopsis genome has 4 members in the CYP90 family: CYP90A1, CYP90B1, CYP90C1, and CYP90D1. Of these, CYP90A1 is another name for CPD that mediates the step after DWF4, but the precise functions of the other two enzymes CYP90C1 and CYP90D1 in BR biosynthesis remain to be elucidated. RNA gel blot analysis revealed that *DWF4* is not highly expressed, suggesting that a lower gene expression level is a part of a mechanism to keep DWF4 enzymatic activity low, resulting in a rate-determining step in the pathway. In accordance with this, overexpression of the *DWF4* gene using the CaMV 35S promoter gave rise to constitutive BL responses including elongated inflorescences, long petioles, elongated leaf blades, an increased number of siliques and consequently an elevated seed production (10).

Completion of the rice genome sequencing revealed that rice has a single *DWF4* homolog. In light of elevated seed production by *DWF4* overexpression in Arabidopsis, characterization and application of the rice *DWF4* gene in rice may result in elevated seed yield from this important crop.

3. CONSTITUTIVE PHOTOMORPHOGENESIS AND DWARFISM:
C-23 hydroxylation (Reaction 20)

Steroid C-23α-hydroxylase is a cytochrome P450 monooxygenase enzyme that adds a hydroxyl group to the C-23 position. The Arabidopsis *cpd* mutant for this enzymatic step has been identified from a T-DNA-tagged mutant population (48). Feeding studies using biosynthetic intermediates indicated that only C-23 hydroxylated steroids such as teasterone (TE) and its downstream compounds could rescue the extreme dwarfism of *cpd*, suggesting that the C-23 hydroxylation step is defective in this mutant. Cross sections of stems from *cpd* mutant plants showed that vascular system differentiation is altered, resulting in extranumerary phloem cells and fewer xylem cells, possibly due to unequal cell division activity in the cambium (48). Examination of gene expression in *cpd*, wild type, and *CPD*-overexpression lines revealed that transcripts of stress-related genes, such as alcohol dehydrogenase, are significantly increased, whereas pathogenesis-related (PR) genes, including *PR1*, *PR2*, and *PR5*, are decreased in *cpd*, suggesting that

BRs are involved in conferring resistance to biotic attacks. The CPD cytochrome P450 protein (At5g05690) is classified as CYP90A1. A genetic defect similar to *cpd* has also been found in the tomato *dumpy* (*dpy*) mutant, where dwarfism is rescued only by C-23 hydroxylated BRs (30).

In addition to CYP90A1 (CPD) and CYP90B1 (DWF4), the Arabidopsis *rotundafolia3* (*rot3*) mutant for CYP90C1 has been isolated. Unlike the severe growth retardation seen in *cpd* and *dwf4*, the *rot3* loss-of-function mutant displays normal stature but the leaf shape is most obviously altered, being more round than the wild type (28). Interestingly, the width of the *rot3* leaves is not affected as much, thus it is thought that CPY90C1 is involved in metabolic pathways that produce molecules which control cell elongation especially in the leaf. Despite significant sequence similarity of ROT3 with CPD and DWF4, the metabolic step that ROT3 mediates has not been clearly defined.

4. Tomato DWARF: C-6 oxidation (Reaction 18)

A C-6 oxidase converts 6-DeoxoBRs to 6-OxoBRs via two consecutive steps: hydroxylation at the C-6 position first, then further dehydrogenation to a ketone group (6). The tomato *Dwarf* (*D*) gene was shown to encode an enzyme catalyzing this step. The loss-of-function mutation (d^x) for the Tomato *D* gene was identified from populations of transposon-tagged mutants (5), and the transposon insertion in the *Dwarf* gene results in extremely small plants. The *D* gene was shown to encode a cytochrome P450 (CYP85B1) protein. This gene was successfully expressed in yeast, and shown to mediate two consecutive steps of C-6 oxidation. Examination of endogenous BR biosynthetic intermediates in wild type, the d^x mutant, and a *D*-overexpression line (*35S::D*) revealed that 6-deoxoCS accumulates about 4-fold in the d^x mutant, whereas the content of 6-DeoxoCS in the *35S::D* line decreased to 1/50 that of wild type (6). This suggests that the tomato *D* gene encodes a C-6 oxidase enzyme in tomato plants.

The Arabidopsis genome has two copies of the tomato D homolog: At5g38970 (CYP85A1) and At3g30180 (CYP85A2). When these two genes were functionally expressed in yeast, both of them converted not only 6-deoxocastasterone (6-deoxoCS), but also 6-deoxoTE, 6-deoxo3DT, and 6-deoxotyphasterol (6-deoxoTY) to their 6-oxidized forms: CS, TE, 3DT, and TY, respectively (46). Real time RT-PCR analysis of the two genes revealed that both of the two genes are highly expressed in apical shoot meristems. In addition, although the CYP85A2 expression level was generally low in the tissues examined, CYP85A1 transcripts are significantly higher in siliques and roots. Thus it is likely that these two functionally redundant genes possess different levels of activity in spatially separate tissues.

REGULATION OF BRASSINOLIDE BIOSYNTHESIS

Studies show that the steady state mRNA levels of the biosynthetic genes are inversely proportional to the endogenous BR levels (37, 40). The expression of the *CPD* gene was shown to be down-regulated by a feed-back mechanism: exogenous application of excess amounts of BRs decreases steady state levels of the *CPD* transcript (37). This suggests that mRNA levels of the biosynthetic genes are a primary target in maintaining homeostasis of endogenous BRs. Although the endogenous BL level in wild-type Arabidopsis plants is a trace amount (approximately 1/10th of a nanogram per gram fresh weight), <u>BR</u>-<u>i</u>nsensitive (*bri1*) mutants and <u>b</u>rassinosteroid <u>in</u>sensitive (*bin2/dwf12*) accumulate as much as 40 times the level of wild type (40, 12). The increased BR levels in these insensitive mutants are accompanied by elevated transcript levels of the biosynthetic genes (40). Thus it is likely that proper BR signaling processes contributed by *BRI1* and *BIN2/DWF12* are essential in maintaining homeostasis of endogenous BRs.

Determining the location of BR biosynthesis is a prerequisite to understanding the BR mode-of-action. Analysis of the expression of the GUS reporter gene driven by the *CPD* promoter revealed that *CPD* is expressed in cotyledons and the uppermost parts of hypocotyls of both dark- and light-grown seedlings. In adult plants, GUS activity was strong in expanding leaf primordia and cauline leaves. When *CPD::GUS* plants were grown in the presence of 1 µM *epi*-BL, GUS activity was rarely detectable, confirming that *CPD* expression is negatively controlled by BR concentration (37).

DWF4 expression is subject to tight regulation in maintaining a proper mRNA level. Different methods including an RNA gel blot analysis, RT-PCR, and *DWF4::GUS* histochemical analysis consistently reveal that the *DWF4* transcript accumulates in actively growing tissues (Fig. 5), such as root tips, the bottom of floral organs (F), axillary buds (H), leaf axils at the bottom of a primary inflorescence (L), and the cotyledons of dark-grown seedlings (D). The endogenous levels of BR biosynthetic intermediates have consistently showed that the metabolic flux after the *DWF4*-mediated step is increased in these tissues (10). Interestingly, although the *35S::DWF4* plants phenocopied the morphology of BR-treated plants, BL was not detectable in this transgenic line. This suggests that increased BR signaling also enhanced BL inactivation.. Histochemical analysis of *DWF4::GUS* transgenic plants showed that exogenous application of BRs and 2,4-D antagonistically regulates *DWF4* expression (Plate 5, page xxxI): DWF4 mRNA levels are decreased by BRs but increased by 2,4-D. Similarly, *DWF4::GUS* expression was up-regulated in the BR-deficient dwarf mutant background, though the change was not noticeable or even weaker in auxin resistant mutants (not shown).

BR biosynthesis is effectively down-regulated by the triazole type inhibitor brassinazole (Brz). Brz has been shown to inhibit DWF4 activity, thus, treatment of Arabidopsis wild-type plants with 5µM Brz phenocopies the *dwf4* morphology (2). The BR signaling pathway was further elucidated by studying Brz-resistant mutants that were isolated for their resistance to Brz-mediated growth retardation. The mutants, including <u>b</u>rassina<u>z</u>ole <u>r</u>esistant1 (*bzr1*) and <u>b</u>ri1 <u>EMS</u> <u>s</u>uppressor1 (*bes1/bzr2*) (51, 52), show

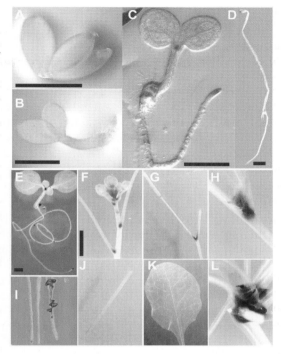

Figure 5 (Color plate page CP5). Temporal and spatial differences in *DWF4* expression. Transgenic plants harboring a *DWF4::GUS* construct were subject to GUS staining. GUS positive tissues look blue. A: an embryo taken from a matured brown siliques; cotyledonary margins and radicle are clearly stained. B: a seedling 1 day after germination (DAG). C: 3 DAG; shoot tip, junction between hypocotyl and root, and root tips are stained. D: dark-grown seedling at 3 DAG. Cotyledons are more widely stained compared to C. E: light grown seedling at 7 DAG, emerging root primordia are also stained. F: shoot tip floral cluster; young flowers are GUS positive. G: adaxial half of the pedicel base is stained. H: emerging axil from the base of a secondary inflorescence is GUS positive. I: from left to right, control, brassinolide treated, 2,4-D treated root tip. Brassinolide abolished GUS stain from the root tip as compared to the control, whereas 2,4-D greatly stimulated secondary root formation and the emerging root primordia are all densely stained. J and K: inflorescence internode and leaf blade are GUS-negative, L: leaf axils at the base of primary inflorescences are rich with leaf and secondary inflorescence primordia, and are heavily stained with GUS. Unit bars in A and B = 0.5 mm, C and D = 1 mm, E = 2 mm, F = 5 mm. Unit bar in F is applicable to F through L.

elongated hypocotyls in the presence of Brz, and suppress *bri1* phenotypes. The two semi-dominant gain-of-function mutations are defective in a novel type of nuclear factor that possesses a nuclear localization signal. In accordance with the resistance to Brz, the steady state level of *CPD* transcripts in the *bzr1* mutants dropped significantly. Furthermore, the endogenous BR levels are greatly decreased in *bzr1* (Table 1), suggesting that BZR1 down-regulates BR biosynthetic genes such as *CPD* and *DWF4*. Consistently, when the *DWF4::GUS* construct was introduced into *bzr1* and *bes1/bzr2*, GUS expression was effectively eliminated, confirming that the BZR proteins act as negative regulators of *DWF4* gene expression (Choe et al., in preparation). Future research should reveal whether BZR1 directly down-regulates *DWF4* transcription or indirectly via another transcriptional regulator.

METABOLISM OF BRASSINOSTEROIDS

A pool of bioactive BRs is modulated not only by influx, such as biosynthesis, but also by efflux, including inactivation and sequestration into subcellular

Table 1. Endogenous BR levels (ng/g fresh weigh) in wild type, *dwf4-1*, *35S::DWF4*, and *bzr1-1D*. The level of total BRs in *35S::DWF4* increased >3-fold compared to Ws-2 wild type control, whereas the level dropped to about 1/100 in *dwf4-1*. *bzr1-1D* exhibits only a quarter of the total BR levels in Columbia wild type.

	Ws-2*	35S::DWF4*	dwf4-1*	Col**	bzr1-1D**
6-DeoxoCT	7.6	28.9	0.05	0.79	0.73
6-DeoxoTE	0.05	0.24	nd	0.23	0.16
6-DeoxoTY	2.33	8.12	0.05	1.59	0.11
6-DeoxoCS	4.01	5.37	0.05	2.31	0.14
CS	0.28	0.23	nd	0.24	0.01
Total BRs	14.27	42.86	0.15	5.16	1.15

nd = not detectable   * From 10   ** From 52

organelles. Bioactive BRs can be inactivated by many different mechanisms, including epimerization, hydroxylation, demethylation, side-chain cleavage, oxidation, sulfonation, acylation, and glycosylation (20) (Fig. 6). Inactivation by epimerization occurs in many plant species including serradella (*Ornithopus sativus*) and tomato, both of which contain epimers such as 3-epiCS, and 3,24-diepiBL. It is likely that these epimers result from CS and 24epiBL being subjected to the reversible 3-epimerization step in the BR biosynthetic pathways (20) (Fig. 6). Interestingly, when the C-3 hydroxyl group is epimerized from α back to the β configuration, the 3β-epimers seem to be preferred substrates for further modification by acylation and glycosylation due to the equatorial configuration in relation to the planar structure of the sterol rings. In addition, it has been found that 24-epimers can be further inactivated by hydroxylation at C-12 in a fungus *Cunninghamella echinulata* (20). In human steroid hormone biosynthesis, the side chain of cholesterol is first cleaved to produce pregnenolone before going through a series of reactions leading to progesterone and testosterone. Similar metabolites are found in serradella. Once C-20 is hydroxylated, it is accompanied by side chain cleavage at the bond between C-20 and C-22. In pregnenolone biosynthesis, this reaction is known to be mediated by a cytochrome P450 enzyme, but the biochemical mechanisms in plants remain to be elucidated.

Active BRs are subjected to change into multiple forms by conjugation and potential degradation. In catabolic pathways, BRs are glycosylated at the C-23, C-25, and C-26 positions (20). Esterification of fatty acids at the 3β hydroxyl group has also been reported (31). In addition a cytochrome P450 dependent hydroxylation at C-25 dramatically reduces the bioactivity of 24-epiBL (31).

A *bas1-D* (*phyB* activation-tagged suppressor 1 – Dominant) mutant that is involved in BR catabolism has been identified by screening for a suppressor mutant of *phyB-4* (39). Overexpression of the *BAS1* gene (CYP72B1, At2g26710) by activation tagging caused the mutant to accumulate C-26 hydroxylated BL, suggesting that CYP72B1 acts as a C-26 hydroxylase. The *phyB-4* mutant possesses a defective phytochrome apoprotein that results in

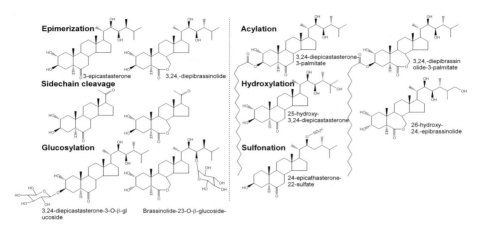

Figure 6. Brassinosteroid metabolism. BRs are subjected to metabolism through mechanisms such as such as inactivation and sequestration. Representative structures of metabolites are shown for the major modification processes.

abnormal light signaling leading to a long hypocotyl phenotype in the light. Shortening of the long hypocotyls of the *phyB* mutant through a reduced level of bioactive BL in the CYP72B1 overexpression line indicates that the light-dependent growth of hypocotyls is regulated in part by the concentration of bioactive BRs.

Sulfonation at the C-22 hydroxyl group by a sulfotransferase (BNST3) is known to be a BL inactivation reaction in *Brassica napus* (42). The BNST3 sulfotransferase takes 24-epiBL as a preferred substrate, thus, it is conceivable that the C-24 of BL is epimerized before being subjected to further reactions of the sulfonated inactivation.

Lessons from studies on other phytohormones suggest that conjugated molecules could serve as pools of inactive BRs that can be converted to active forms by de-conjugation reactions. However similar data are yet to be discovered in BR homeostasis.

Acknowledgements

I thank Drs. Ken Feldmann, Frans Tax, Shozo Fujioka, and Peter Davies for their critical reading and constructive comments on this manuscript. This research was supported by a grant (PF0330201-00) from Plant Diversity Research Center of 21st Century Frontier Research Program funded by Ministry of Science and Technology of Korean government.

References

1. Arteca RN (1996) Plant growth substances: principles and applications. Chapman and Hall, New York
2. Asami T, Mizutani M, Fujioka S, Goda H, Min YK, Shimada Y, Nakano T, Takatsuto S, Matsuyama T, Nagata N, Sakata K, Yoshida S (2001) Selective interaction of triazole derivatives with DWF4, a cytochrome P450 monooxygenase of the brassinosteroid biosynthetic pathway, correlates with brassinosteroid deficiency *in planta*. J Biol Chem 276: 25687-25691
3. Azpiroz R, Wu Y, LoCascio JC, Feldmann KA (1998) An Arabidopsis brassinosteroid-dependent mutant is blocked in cell elongation. Plant Cell 10: 219-230

4. Benveniste P (2002) Sterol metabolism. *In* CR Somerville, EM Meyerowitz, eds, The Arabidopsis Book. American Society of Plant Biologists, Rockville, MD. http://www.aspb.org/publications/arabidopsis/
5. Bishop G, Harrison K, Jones JDG (1996) The Tomato *Dwarf* gene isolated by heterologous transposon tagging encodes the first member of a new cytochrome P450 family. Plant Cell 8: 959-969
6. Bishop GJ, Nomura T, Yokota T, Harrison K, Noguchi T, Fujioka S, Takatsuto S, Jones JD, Kamiya Y (1999) The tomato DWARF enzyme catalyses C-6 oxidation in brassinosteroid biosynthesis. Proc Natl Acad Sci U S A 96: 1761-1766
7. Carland FM, Fujioka S, Takatsuto S, Yoshida S, Nelson T (2002) The identification of *CVP1* reveals a role for sterols in vascular patterning. Plant Cell 14: 2045-2058
8. Choe S, Dilkes BP, Fujioka S, Takatsuto S, Sakurai A, Feldmann KA (1998) The *DWF4* gene of Arabidopsis encodes a cytochrome P450 that mediates multiple 22a-hydroxylation steps in brassinosteroid biosynthesis. Plant Cell 10: 231-243
9. Choe S, Dilkes BP, Gregory BD, Ross AS, Yuan H, Noguchi T, Fujioka S, Takatsuto S, Tanaka A, Yoshida S, Tax FE, Feldmann KA (1999) The Arabidopsis *dwarf1* mutant is defective in the conversion of 24- methylenecholesterol to campesterol in brassinosteroid biosynthesis. Plant Physiol 119: 897-907
10. Choe S, Fujioka S, Noguchi T, Takatsuto S, Yoshida S, Feldmann KA (2001) Overexpression of *DWARF4* in the brassinosteroid biosynthetic pathway results in increased vegetative growth and seed yield in Arabidopsis. Plant J 26: 573-582
11. Choe S, Noguchi T, Fujioka S, Takatsuto S, Tissier CP, Gregory BD, Ross AS, Tanaka A, Yoshida S, Tax FE, Feldmann KA (1999) The Arabidopsis *dwf7/ste1* mutant is defective in the D^7 sterol C-5 desaturation step leading to brassinosteroid biosynthesis. Plant Cell 11: 207-221
12. Choe S, Schmitz RJ, Fujioka S, Takatsuto S, Lee MO, Yoshida S, Feldmann KA, Tax FE (2002) Arabidopsis brassinosteroid-insensitive *dwarf12* mutants are semidominant and defective in a glycogen synthase kinase 3b-like kinase. Plant Physiol 130: 1506-1515
13. Choe S, Tanaka A, Noguchi T, Fujioka S, Takatsuto S, Ross AS, Tax FE, Yoshida S, Feldmann KA (2000) Lesions in the sterol D^7 reductase gene of Arabidopsis cause dwarfism due to a block in brassinosteroid biosynthesis. Plant J 21: 431-443
14. Chory J, Nagpal P, Peto C (1991) Phenotypic and genetic analysis of *det2*, a new mutant that affects light-regulated seedling development in *Arabidopsis*. Plant Cell 3: 445-459
15. Clouse SD, Sasse JM (1998) Brassinosteroids: Essential Regulators of Plant Growth and Development. Annu Rev Plant Physiol Plant Mol Biol 49: 427-451
16. Diener AC, Li H, Zhou W, Whoriskey WJ, Nes WD, Fink GR (2000) Sterol methyltransferase 1 controls the level of cholesterol in plants. Plant Cell 12: 853-870
17. Fujioka S, Inoue T, Takatsuto S, Yanagisawa T, Yokota T, Sakurai A (1995) Biological activities of biosynthetically-related congeners of brassinolide. Biosci Biotech Biochem 59: 1973-1975
18. Fujioka S, Li J, Choi YH, Seto H, Takatsuto S, Noguchi T, Watanabe T, Kuriyama H, Yokota T, Chory J, Sakurai A (1997) The Arabidopsis *deetiolated2* mutant is blocked early in brassinosteroid biosynthesis. Plant Cell 9: 1951-1962
19. Fujioka S, Sakurai A (1997) Brassinosteroids. Nat Prod Rep 14: 1-10
20. Fujioka S, Takatsuto S, Yoshida S (2002) An early C-22 oxidation branch in the brassinosteroid biosynthetic pathway. Plant Physiol 130: 930-939
21. Fujioka S, Yokota T (2003) Biosynthesis and metabolism of brassinosteroids. Annu Rev Plant Biol 54: 137-164
22. Grebenok RJ, Ohnmeiss TE, Yamamoto A, Huntley ED, Galbraith DW, Della Penna D (1998) Isolation and characterization of an Arabidopsis thaliana C-8,7 sterol isomerase: functional and structural similarities to mammalian C-8,7 sterol isomerase/emopamil-binding protein. Plant Mol Biol 38: 807-815
23. Greeve I, Hermans-Borgmeyer I, Brellinger C, Kasper D, Gomez-Isla T, Behl C, Levkau B, Nitsch RM (2000) The human DIMINUTO/DWARF1 homolog seladin-1 confers resistance to Alzheimer's disease-associated neurodegeneration and oxidative stress. J Neurosci 20: 7345-7352

24. Grove MD, Spencer GF, Rohwedder WK, Mandava N, Worley JF, Jr JDW, Steffens GL, Flippen-Anderson JL, Jr JCC (1979) Brassinolide, a plant growth-promoting steroid isolated from *Brassica napus* pollen. Nature 281: 216-217
25. He JX, Fujioka S, Li TC, Kang SG, Seto H, Takatsuto S, Yoshida S, Jang JC (2003) Sterols regulate development and gene expression in Arabidopsis. Plant Physiol 131: 1258-1269
26. Husselstein T, Gachotte D, Desprez T, Bard M, Benveniste P (1996) Transformation of *Saccharomyces cerevisiae* with a cDNA encoding a sterol C-methyltransferase from *Arabidopsis thaliana* results in the synthesis of 24-ethyl sterols. FEBS Let 381: 87-92
27. Jang JC, Fujioka S, Tasaka M, Seto H, Takatsuto S, Ishii A, Aida M, Yoshida S, Sheen J (2000) A critical role of sterols in embryonic patterning and meristem programming revealed by the *fackel* mutants of *Arabidopsis thaliana*. Genes Dev 14: 1485-1497
28. Kauschmann A, Jessop A, Koncz C, Szekeres M, Willmitzer L, Altmann T (1996) Genetic evidence for an essential role of brassinosteroids in plant development. Plant J 9: 701-713
29. Kim GT, Tsukaya H, Uchimiya H (1998) The *ROTUNDIFOLIA3* gene of *Arabidopsis thaliana* encodes a new member of the cytochrome P-450 family that is required for the regulated polar elongation of leaf cells. Genes Dev 12: 2381-2391
30. Klahre U, Noguchi T, Fujioka S, Takatsuto S, Yokota T, Nomura T, Yoshida S, Chua NH (1998) The Arabidopsis *DIMINUTO/DWARF1* gene encodes a protein involved in steroid synthesis [In Process Citation]. Plant Cell 10: 1677-1690
31. Koka CV, Cerny RE, Gardner RG, Noguchi T, Fujioka S, Takatsuto S, Yoshida S, Clouse SD (2000) A putative role for the tomato genes *DUMPY* and *CURL-3* in brassinosteroid biosynthesis and response. Plant Physiol 122: 85-98
32. Kolbe A, Schneider B, Porzel A, Schmidt J, Adam G (1995) Acyl-conjugated metabolites of brassinosteroids in cell suspension cultures of *Ornithopus sativus*. Phytochemistry 38: 633-636
33. Lecain E, Chenivesse X, Spagnoli R, Pompon D (1996) Cloning by metabolic interference in yeast and enzymatic characterization of *Arabidopsis thaliana* sterol delta-7-reductase. J. Biol. Chem. 271: 10866-10873
34. Li J, Nagpal P, Vatart V, McMorris TC, Chory J (1996) A role for brassinosteroids in light-dependent development of *Arabidopsis*. Science 272: 398-401
35. Lorence MC, Murry BA, Trant JM, Mason JI (1990) Human 3B-hydroxysteroid dehydrogenase/D^5 to D^4 isomerase from placenta: expression in nonsteroidogenic cells of a protein that catalyzes the dehydrogenation/isomerization of C21 and C19 steroids. Endocrinology 126: 2493- 2498
36. Mandava NB (1988) Plant growth-promoting brassinosteroids. Ann Rev Plant Physiol Plant Mol Biol 39: 23-52
37. Marquardt V, Adam G (1991) Recent advances in brassinosteroid research. *In* W Ebing, ed, Chemistry of plant protection, Vol 7. Springer Verlag, Berlin, pp 103-139
38. Mathur J, Molnar G, Fujioka S, Takatsuto S, Sakurai A, Yokota T, Adam G, Voigt B, Nagy F, Maas C, Schell J, Koncz C, Szekeres M (1998) Transcription of the Arabidopsis *CPD* gene, encoding a steroidogenic cytochrome P450, is negatively controlled by brassinosteroids. Plant J 14: 593-602
39. McNellis TW, Deng X-W (1995) Light control of seedling morphogenetic pattern. Plant Cell 7: 1749-1761
40. Neff MM, Nguyen SM, Malancharuvil EJ, Fujioka S, Noguchi T, Seto H, Tsubuki M, Honda T, Takatsuto S, Yoshida S, Chory J (1999) *BAS1*: A gene regulating brassinosteroid levels and light responsiveness in *Arabidopsis*. Proc Natl Acad Sci U S A 96: 15316-15323
41. Noguchi T, Fujioka S, Choe S, Takatsuto S, Yoshida S, Yuan H, Feldmann KA, Tax FE (1999) Brassinosteroid-insensitive dwarf mutants of Arabidopsis accumulate brassinosteroids. Plant Physiol. 121: 743-752
42. Nomura T, Kitasaka Y, Takatsuto S, Reid JB, Fukami M, Yokota T (1999) Brassinosteroid/Sterol synthesis and plant growth as affected by *lka* and *lkb* mutations of Pea. Plant Physiol 119: 1517-1526

43. Rouleau M, Marsolais F, Richard M, Nicolle L, Voigt B, Adam G, Varin L (1999) Inactivation of brassinosteroid biological activity by a salicylate- inducible steroid sulfotransferase from Brassica napus. J Biol Chem 274: 20925-20930
44. Sakurai A, Fujioka S (1996) *Catharanthus roseus* (*Vinca rosea*): in vitro production of brassinosteroids. *In* YPS Bajaj, ed, Biotechnology in Agriculture and Forestry, Vol 37. Springer-Verlag Press, Berlin, pp 87-96
45. Schaeffer A, Bronner R, Benveniste P, Schaller H (2001) The ratio of campesterol to sitosterol that modulates growth in Arabidopsis is controlled by STEROL METHYLTRANSFERASE 2;1. Plant J 25: 605-615
46. Schrick K, Mayer U, Horrichs A, Kuhnt C, Bellini C, Dangl J, Schmidt J, Jurgens G (2000) FACKEL is a sterol C-14 reductase required for organized cell division and expansion in Arabidopsis embryogenesis. Genes Dev 14: 1471-1484
47. Shimada Y, Goda H, Nakamura A, Takatsuto S, Fujioka S, Yoshida S (2003) Organ-specific expression of brassinosteroid-biosynthetic genes and distribution of endogenous brassinosteroids in Arabidopsis. Plant Physiol 131: 287-297
48. Souter M, Topping J, Pullen M, Friml J, Palme K, Hackett R, Grierson D, Lindsey K (2002) hydra Mutants of Arabidopsis are defective in sterol profiles and auxin and ethylene signaling. Plant Cell 14: 1017-1031
49. Szekeres M, Nemeth K, Koncz-Kalman Z, Mathur J, Kauschmann A, Altmann T, Redei GP, Nagy F, Schell J, Koncz C (1996) Brassinosteroids rescue the deficiency of CYP90, a cytochrome P450 controlling cell elongation and de-etiolation in Arabidopsis. Cell 85: 171-182
50. Takatsuto S, Yazawa N, Ikegawa N, Takematsu T, Takeuchi Y, Koguchi M (1983) Structure-activity relationship of brassinosteroids. Phytochemistry 22: 2437-2441
51. Thompson MJ, Meudt WJ, Mandava NB, Dutky SR, Lusby WR, Spaulding DW (1982) Synthesis of brassinosteroids and relationship of structure to plant growth- promoting effects. Steroids 39: 89-105
52. Wang ZY, Nakano T, Gendron J, He J, Chen M, Vafeados D, Yang Y, Fujioka S, Yoshida S, Asami T, Chory J (2002) Nuclear-localized BZR1 mediates brassinosteroid-induced growth and feedback suppression of brassinosteroid biosynthesis. Dev Cell 2: 505-513
53. Yin Y, Wang ZY, Mora-Garcia S, Li J, Yoshida S, Asami T, Chory J (2002) BES1 accumulates in the nucleus in response to brassinosteroids to regulate gene expression and promote stem elongation. Cell 109: 181-191

B7   Regulation of Gibberellin and Brassinosteroid Biosynthesis by Genetic, Environmental and Hormonal Factors

James B. Reid, Gregory M. Symons and John J. Ross
School of Plant Science, University of Tasmania, Private Bag 55, Hobart, Tasmania 7001, Australia. E-mail: jim.reid@utas.edu.au

INTRODUCTION

The biosynthesis of plant hormones involves a series of steps that converts intermediates with little or no biological activity into the active form. Usually each step is catalysed by an enzyme that is in turn encoded by a gene, referred to as a hormone "synthesis gene". Mutations in these genes can give rise to "synthesis mutants", which are deficient to varying extents in the hormone in question. The striking phenotypes of some of these mutants provide the most graphic evidence that plant hormones are essential factors for plant growth and development. Striking phenotypes can also be caused by mutations that impair hormone deactivation.

The expression of hormone synthesis genes, or the extent to which they are transcribed, is affected by the environment, as well as by endogenous factors, such as the level of other plant hormones. Thus these factors interact strongly with the hormone biosynthesis pathway to affect the overall phenotype. Indeed, there is an increasing perception among plant biologists that major mechanisms controlling plant development do not operate in isolation, but rather form part of complex regulatory networks.

Understanding these networks is a key challenge for future plant hormone research. In an era when genes, including hormone synthesis genes, are being transferred from one species to another, it is of paramount importance to understand the full range of effects caused by this genetic engineering. For example, it is critical to understand the effects that a synthesis gene for one hormone might have on the levels of other hormones.

In this chapter we illustrate these principles, using examples from the gibberellin (GA[1]) and brassinosteroid (BR) pathways in the garden pea (*Pisum sativum* L.). This species has long been a valuable model for

---

[1] Abbreviations: GA_n, gibberellin A_n; BR, brassinosteroid; GGPP, *trans*-geranylgeranyl-diphosphate; CPP, *ent*-copalyl diphosphate; CS, castasterone; BL, brassinolide; 6-DeoxoCS, 6-deoxocastasterone; Brz, brassinazole; 4-Cl-IAA, 4-chloroindole-3-actetic acid;

hormone studies, because there are numerous pea mutants affected in hormone biosynthesis, and its endogenous hormones are relatively easy to extract and quantify. Accordingly, the GA and BR pathways are well established in *Pisum*. We begin our discussion with genetic regulation, because hormone synthesis mutants provided some of the first evidence for the roles played by these compounds. We then describe how environmental factors, in particular the light regime, can affect hormone levels, and finally turn to the effects that one hormone can have on the biosynthesis and/or deactivation of another.

GENETIC REGULATION OF THE GIBBERELLIN AND BRASSINOSTEROID PATHWAYS

Figure 1. A wild-type pea plant (left) and Mendel's *le-1* mutant (right).

Gibberellin Mutants

The existence of key GA synthesis genes was first inferred from mutant phenotypes. The most famous GA mutant is Mendel's *le-1*, a recessive dwarf that reaches only about 40% of the wild-type height (Fig. 1). The difference between the tall and dwarf characteristics was one of the seven traits utilized by Mendel in his seminal experiments on inheritance. Following the isolation of the gibberellins, GA_3 was applied to *le-1* plants, with spectacular results. This simple procedure restored the height of the mutant to that of the wild type (8).

In the 1980s (2) it was shown that the *le-1* mutation blocks the final, activation step in the biosynthesis of endogenous GAs, namely the 3β-hydroxylation (in broad terms a 3-oxidation) of GA_{20} to GA_1 (Fig. 2). As a result, there is a 10-20-fold drop in GA_1 content in the mutant (50). Importantly, applied GA_{20} stimulates elongation in *LE* plants but is much less effective on *le-1* plants, while GA_1 is highly active on both genotypes (44). The metabolic product of GA_1, GA_8 (Fig. 2) is inactive on both genotypes.

These studies showed the importance of endogenous GA_1 for stem elongation in plants. It was critical to demonstrate that GA_1 can convert *le-1* plants to a phenocopy of the wild-type, since if *le-1* mainly affected plant height via a mechanism unrelated to the GAs (with the effect on GAs a secondary consequence), the phenotype would not be rescued in this way. While GA_{20} was much less active than GA_1 on *le-1* plants, it did cause some growth response. This may have been because GA_{20} possesses some activity (albeit slight), or alternatively because some of the applied GA_{20} was converted to GA_1. Careful measurements of GA levels provided strong

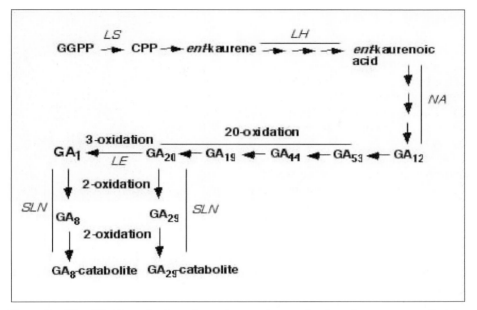

Figure 2. The GA biosynthesis pathway in pea shoots, showing the steps controlled by synthesis and deactivation genes.

evidence for the latter (19).

An even shorter, more GA_1-deficient, mutant at the *LE* locus, *le-2* (previously le^d) was isolated later in the 1980s. Importantly, this discovery confirmed that the *le-1* mutation is "leaky"; in other words, that the *le-1* allele permits some GA_1 production from GA_{20}. In fact, even in *le-2* plants some GA_1 is produced, indicating that *le-2* itself might also be leaky, and/or that other enzymes operate to produce a small amount of GA_1. Differentiating between these possibilities required the molecular cloning of the *LE* gene and the mutant alleles.

The *LE* gene was cloned in two separate laboratories in the 1990s, and encodes a protein predicted to consist of 374 amino acids (25, 29). When the *LE* gene (also known as *PsGA3ox1*) was expressed in the bacterium *Eschericia coli*, an extract from the bacterial culture, containing the enzyme encoded by *LE*, readily converted radioactive GA_{20} to radioactive GA_1 (25). The *le-1* expression product, in contrast, produced 20-fold less GA_1 from GA_{20}. The fact that some GA_1 was produced by the *le-1* expression product again showed that *le-1* is indeed a leaky mutant. The enzyme encoded by *LE* is termed a "GA 3-oxidase". Determining the sequence of bases in the *LE* and *le-1* alleles showed that the *le-1* mutation arose from a G to A substitution at position 685 (25), causing a change from alanine to threonine near the active site of the encoded protein. The *le-2* mutation involves a base deletion of the *le-1* allele at position 376 (the parental line was a *le-1* dwarf), and results in a severely truncated protein with undetectable enzyme activity. Therefore, the *le-2* allele encodes a non-functional protein, is not leaky, and

Regulation of hormone biosynthesis

Figure 3. Phenotypes of (from left to right) wild-type, *lh-2* and *ls-1* seedlings.

Figure 4. Seed abortion in mutant *lh-2*. Left, wild-type pod; right, *lh-2* pod.

is referred to as "null". A third mutant allele, *le-3*, involves a C to T substitution in the cDNA, and dramatically reduces (but does not abolish) 3-oxidase activity (29).

Mutations at other loci block earlier steps in the GA pathway (Fig. 2), reducing the levels of all GAs past the blockage point, and resulting in dwarf phenotypes (44). These additional mutations are all inherited as single-gene recessive mutations. Their effects show the importance of GAs in general for stem elongation, but unlike *le-1* they do not specifically implicate GA_1 as the key bioactive GA. Nevertheless, they have been very instructive for relating the GA economy of the plant to the control of a range of developmental phenomena.

The *ls-1* mutation blocks the first committed step in GA biosynthesis, the conversion of *trans*-geranylgeranyldiphoshate (GGPP) to *ent*-copalyl diphosphate (CPP, Figs. 2 and 3). The *LS* gene was the first pea GA synthesis gene to be cloned, and encodes the enzyme *ent*-copalyl diphosphate synthase (3). The *ls-1* mutation involves a G to A substitution that disrupts the normal RNA splicing of the gene transcript, and consequently the *ls-1* mutant shows reduced enzyme activity (3).

In the *lh-2* mutant (Fig. 3; formerly *lh^i*) GA levels are reduced not only in the shoot but in the young seeds as well (56). Consequently the seeds develop slowly and/or abort completely (Fig 4). These observations were the first to show that GAs are required for normal seed development. The enzyme encoded by the wild-type allele, *LH*, catalyses the three-step oxidation of *ent*-kaurene to *ent*-kaurenoic acid. The *LH* gene has recently been cloned and is also known as *PsKO1*, after *Pisum sativum* kaurene oxidase (12). The *lh-2* mutation, like *ls-1*, disrupts normal RNA splicing. Another mutation at the *LH* locus, *lh-1*, involves a serine to asparagine substitution in the encoded protein and a loss of activity. Both the *lh-1* and *lh-2* mutations are leaky. Interestingly, *lh-2* plants are hypersensitive to the GA biosynthesis inhibitor paclobutrazol. This occurs because only very

Figure 5. Wild-type (left) and *na-1* (right) plants showing reduced root elongation in the mutant

small amounts of normal transcript (and hence enzyme) are produced in *lh-2* plants (12).

One of the shortest pea GA mutants is *na-1* (Fig. 5). Indeed, *na-1* plants are so short that the term "dwarf" is inadequate to describe them, and instead they are referred to as "nana". Importantly, the *na-1* mutation also markedly reduces GA levels in the roots, and root elongation, and *na-1* roots respond to added GA (65). These results indicate that GAs are required for normal root growth.

GA deficiency in *na-1* results from a strong blockage in the three-step conversion of *ent*-kaurenoic acid to GA_{12} (11). In shoots, the *NA* gene encodes the main enzyme for this conversion, termed *PsKAO1* (after *Pisum sativum ent*-kaurenoic acid oxidase). The *NA* gene has been cloned and when expressed in yeast, the expression product converted *ent*-kaurenoic acid to GA_{12} (11). The *na-1* mutation results from a 5-base deletion in the *NA* gene, resulting in a premature stop codon, and the corresponding predicted protein is only 194 amino acids long, as opposed to 485 in the case of *NA*. This change results in dramatically reduced enzyme activity (11).

In peas, the middle section of the GA pathway involves the three-step conversion of GA_{53} to GA_{20}, known as GA 20-oxidation (Chapter B2). Interestingly, there are no known mutants in pea that specifically affect this conversion. In fact, there is only one mutation known to primarily affect 20-oxidation in any species: *ga5*, in Arabidopsis. In pea, two GA 20-oxidase genes have been cloned: *PsGA20ox1* and *PsGA20ox2*, which are expressed mainly in the shoot and developing seeds, respectively (24, 31).

GA biosynthesis is a dynamic process that constantly adds to the pool of GA_1 present in the tissues. In normal circumstance, however, GA_1 does not accumulate beyond a certain level, in part because it is constantly deactivated to GA_8. This process is known as 2-oxidation. GA_{20} can also be 2-oxidized, to the inactive GA_{29}, and GA_{20} is therefore a key branch-point in the GA pathway (Fig 2). So far, the only mutation known to affect GA 2-oxidation in any species is the *sln* mutation of pea (45). In contrast to the dwarf mutants discussed above, *sln* seedlings are very elongated and "slender" (Fig. 6). This trait is especially obvious in the lower internodes.

The cause of the *sln* phenotype was traced to the seeds. Wild-type pea seeds contain relatively high levels of GA_{20} when they are developing in the

Figure 6. Wild-type (left pot) and *sln* plants (right pot).

pod, and are still green, although GA$_1$ is undetectable at this stage. As the seeds mature further, GA$_{20}$ is deactivated first to GA$_{29}$ and then to GA$_{29}$-catabolite. Both these steps are blocked in *sln* seeds and therefore GA$_{20}$ accumulates to very high levels (45). When these seeds germinate, GA$_{20}$ moves into the seedling and is converted to bioactive GA$_1$, which promotes internode elongation.

The *SLN* gene (also known as *PsGA2ox1*) has been cloned and shown to encode an enzyme that converts GA$_{20}$ to GA$_{29}$, GA$_{29}$ to GA$_{29}$-catabolite, and GA$_1$ to GA$_8$ *in vitro* (26, 30). In the *sln* mutant there are only three As rather than the four that are present at bases 744-747 in the wild-type cDNA sequence. This results in a premature stop codon and the predicted protein consequently lacks essential catalytic regions. Consistent with that expectation, the *sln* expression product failed to show 2-oxidase activity *in vitro* (26).

The inheritance of the *sln* slender trait was initially puzzling because the mutant phenotype disappears in the F$_2$ generation of a *sln* x *SLN* cross: all F$_2$ plants are phenotypically normal. This is because all seeds developing on the F$_1$ plants have testae of the maternal genotype *SLN sln*. The presence of one dominant (*SLN*) allele in the testa is sufficient to convert GA$_{20}$ through to GA$_{29}$-catabolite. Thus even in F$_2$ seeds of genotype *sln sln*, GA$_{20}$ does not accumulate, and the resulting seedlings are not slender. However, F$_2$ *sln sln* plants produce slender offspring in the F$_3$ (45).

Further Implications of GA Mutants

Why is it that some mutations only affect the GA pathway in certain organs? This question remained unresolved until the GA genes in pea and other species were cloned. The GA gene cloning revolution of the 1990s revealed that some, and indeed possibly most, GA genes occur in "gene families". In pea, additional gene family members have been cloned in the case of the genes *NA* (11) and *SLN* (26) and, as mentioned above, two genes encoding 20-oxidases have been isolated. Interestingly, all GA synthesis mutations characterised so far are in structural genes that encode enzymes of the pathway, rather than in genes that regulate structural gene expression.

The cloning of a second *NA*-like gene, *PsKAO2*, solved the puzzle of why *na-1* seeds develop normally and synthesise all the GAs normally found

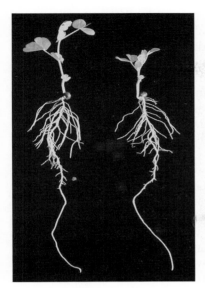

Figure 7. Wild-type (left) and *le-1* (right) plants showing normal root development in the mutant.

in seeds, even though they develop on a grossly GA-deficient shoot. It appears that in seeds, gene *PsKAO2* is the main gene for *ent*-kaurenoic acid oxidation, and therefore GAs can be produced, even in the *na-1* mutant. Consistent with that suggestion, *PsKAO2* is expressed much more strongly in seeds than in other organs, while *NA* (*PsKAO1*) is expressed mainly in stems and only weakly in seeds.

The isolation of a second GA 2-oxidase gene, *PsGA2ox2*, may explain why the *sln* mutation does not completely block GA 2-oxidation in shoots. In fact, in intact *sln* shoots the step GA_1 to GA_8 proceeds more or less normally (51), even though *sln* is a null mutation. It appears that genetic redundancy (expression of an additional 2-oxidase gene or genes) compensates for the lack of functional *SLN* protein in *sln* shoots.

For other genes the existence of additional gene family members can also be inferred from mutant analysis. For example, the roots of *le-2* plants contain normal levels of GA_1 (65), and elongate to the same extent as wild-type plants, despite the fact that *le-2* is a null mutation (29). This indicates that another gene must encode a functional 3-oxidase in pea roots, although this gene has not yet been cloned.

Interestingly, in other cases mutations at the *LE* locus reduce GA_1 content but do not affect the development of the organ concerned. The pods of *le-1* plants, for example, are actually deficient in GA_1 but elongate as well as those of the wild type (52), indicating that GA_1 is not the factor limiting the elongation of pea pods. Because GA_1 appears not to limit the development of some organs, while in other cases additional 3-oxidase gene(s) compensate for the loss of *LE* activity, the effects of the *le-1*, *le-2* and *le-3* alleles are restricted mainly to the internodes.

Studying the effects of individual GA synthesis mutants has yielded valuable information on how GAs affect plant development, but the insight is enhanced still further by comparisons across different GA mutants. For example, the short roots of *na-1* and other mutant plants indicate that GAs are required for normal root development. However, how do we know that these short roots are not simply a consequence of impaired shoot development in the mutants? The answer comes from the dwarf mutants at the *LE* locus (65). The normal roots of these mutants (*le-1*, *le-2* and *le-3*) indicate that shoot dwarfism does not automatically lead to short roots (Fig. 7).

Another valuable comparison can be made between the seed development phenotypes of the *ls-1* and *lh-2* mutants. In both mutants the content of GAs in older, maturing seeds is dramatically reduced. However, only *lh-2* reduces the level of GA_1 in the initial stages of seed development (56), and only *lh-2* causes seed abortion. Young seeds developing on *ls-1* plants contain normal GA_1 levels and therefore do not abort. This indicates that normal GA levels are required for the early stages of seed development, but not for the later stages. Reductions in GA content at the later stages are seemingly without effect, and the function (if any) of GAs at this stage is still not known.

It is instructive also to compare the growth of *na-1* and *le-1* shoots that have been grafted to wild-type stocks (43). Grafting *na-1* shoots to wild-type stocks with expanded leaves leads to a dramatic increase in the elongation of the *na-1* shoot. However, attempts to promote the elongation of *le-1* shoots by grafting have not been successful. These results indicate that an intermediate in the GA pathway past the point blocked by *na-1* (Fig. 2), but before GA_1, can be transported from the wild-type stock across the graft union into the *na-1* scion, where it is converted to GA_1 that in turn stimulates elongation. The inability of the wild-type stocks to promote the elongation of *le-1* scions indicates that endogenous GA_1 itself is not mobile within the pea shoot system.

Finally, mutants in pea and other species have been critical for identifying the negative feedback and feedforward effects that GA_1 and other bioactive GAs exert on the GA pathway (17). In pea, GA_1 down-regulates 20-oxidation and 3-oxidation (31, 46) and up-regulates 2-oxidation (13). These phenomena are described in more detail in Chapter B2.

As can be seen from the foregoing sections, the GA pathway in pea is now extensively characterised. The impetus generated by the careful metabolic studies of the 1970s and 1980s (Chapter B2), which first established the pathway, has been maintained by the isolation of mutants and by the cloning of genes. In pea, only two steps remain for which the genes have not been cloned: 13-hydroxylation and the conversion of CPP to *ent*-kaurene (Fig. 2). Great progress has also been made in Arabidopsis (18), although the small size of that species renders more difficult the physiological analysis of whether GA is required in specific organs.

Brassinosteroid Mutants

During the phase in which several GA-deficient dwarf pea mutants were characterised (1985-1995), it was clear that not all dwarf mutants in this species owed their phenotype to GA deficiency (i.e. while all pea GA-deficient mutants are dwarf, not all pea dwarf mutants are GA-deficient). It was observed that a series of dwarf mutants, *lk, lka, lkb, lkc* and *lkd,* respond poorly to GA_1 application and contain similar GA_1 levels to the wild type (44). These mutants were initially described as "GA response" mutants, although it was suggested that the reduced GA response might not result

Figure 8. Wild-type (left pair), *lkb* (middle pair) and *lka* (right pair) plants.

from a direct interference with the GA signal transduction pathway, but rather from a reduced level of another factor required for normal elongation.

This suggestion is consistent with the fact that these mutants are clearly distinguished from typical GA-deficient mutants in a number of morphological characteristics collectively referred to as an 'erectoides phenotype' (44; Fig. 8). Whilst reduced internode length and a darkened leaf colour are common traits in both GA-deficient and erectoides mutants, the latter generally also exhibit thicker stems, smaller leaves, substantially shorter petioles and peduncles and a corrugated stem surface, referred to as 'banding' (44). The suggestion that the erectoides phenotype may be due to a deficiency in a substance other than GA was subsequently borne out by the discovery that the *lk, lka* and *lkb* mutants are in fact defective in the biosynthesis or perception of brassinosteroids (37, 38).

A crucial role for BRs in plant growth and development is now well established. The events that led to the acceptance of BRs as a new class of plant hormones are outlined in Chapter D7 and our current understanding of BR biosynthesis is discussed in Chapter B6. The committed pathway for BR synthesis starts at the conversion of campesterol to campestanol, which is then converted to castasterone (CS) via one of two alternative routes: the early C-6 oxidation pathway or the late C-6 oxidation pathway (14; Fig. 9). The levels of endogenous BRs indicate that the late C-6 oxidation pathway is more prevalent in a number of species including Arabidopsis and pea, while in tomato it appears to be the only pathway for BR production (39).

The end point of BR biosynthesis, at least in Arabidopsis and *Catharanthus roseus*, is thought to be the conversion of CS to the putative bioactive molecule, brassinolide (BL) (14). The BR pathway has provided an excellent biochemical framework for the characterisation of the pea BR mutants. The characterisation of these mutants has, in turn, helped to refine our understanding of BR biosynthesis and the role of BRs in plant growth and development (4).

The first paper linking BRs with the pea erectoides mutants was published in 1997 and outlined the biochemical characterisation of the *lka* and *lkb* mutants (38). This study showed that internode elongation in *lkb* could be restored to that seen in the wild type by application of bioactive BRs. In addition, *lkb* plants were found to have reduced levels of end-

Regulation of hormone biosynthesis

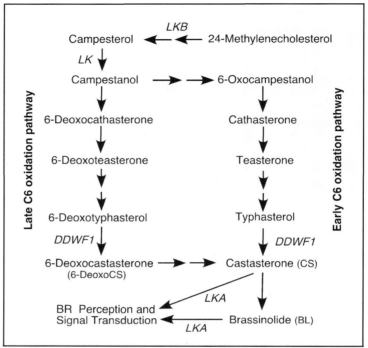

Figure 9. The BR biosynthesis pathway, showing the steps controlled by synthesis genes in pea.

pathway BRs such as CS, 6-deoxocastasterone (6-DeoxoCS) and typhasterol, suggesting a block in an early step in BR biosynthesis (38). The site of this block was pinpointed by demonstrating that *lkb* plants also have reduced levels of the early-pathway BRs campestanol and campesterol but increased levels of their precursor, 24-methylenecholesterol (37). This information suggested that *lkb* is defective in the conversion of 24-methylenecholesterol to campesterol, as had previously been found to be the case in the Arabidopsis *dwarf1* mutant (14). Subsequent feeding experiments using ^{2}H-labelled BRs confirmed that *lkb* plants are unable to isomerise and / or reduce the $\Delta^{24(28)}$ double bond of 24-methylenecholesterol (37).

When the *LKB* gene was cloned, sequence analysis confirmed that it is in fact the pea homologue of the Arabidopsis *DWF1* gene, which encodes a transmembrane protein similar to many oxidoreductases (54). The *lkb* mutation involves a G to A substitution that in turn leads to an arginine to lysine change in the encoded protein. This change, near the transmembrane domain, appears to reduce the efficiency of the resulting enzyme, leading to a reduction in active BR levels and consequently the erectoides phenotype (54). An interesting aspect of the LKB enzyme is that it also catalyses the conversion of isofucosterol to sitosterol in the sterol biosynthesis pathway (37). Thus in addition to a reduction in BR levels, *lkb* plants also exhibit impaired conversion of isofucosterol, and consequently, and have reduced levels of end pathway sterols such as sitosterol and stigmasterol. Interestingly, the reduction in sterol levels in *lkb* plants mainly occurs in the cell membranes (37). The possible contribution that these altered sterol levels play in the development of the *lkb* phenotype is discussed below.

The second of the two erectoides mutants characterised in 1997, *lka*, has an almost identical phenotype to *lkb*. However, BR and sterol levels are not reduced in *lka* plants, and the mutant phenotype could not be rescued by exogenous BRs (37, 38). This led to the suggestion that the *lka* mutant is impaired in its response to the BR signal (38). The increase in the levels of the late pathway BRs 6-DeoxoCS and CS in *lka* plants provided additional evidence for this hypothesis. An increase in late pathway or bioactive hormones is a common characteristic of GA response mutants (17) and is thought to result from the feedback regulation of hormone biosynthesis – in other words it represents an attempt to increase hormone levels in response to the perceived deficiency of the active molecule/s. The initial assessment of the *lka* mutation was subsequently confirmed when the *LKA* gene was shown to encode PsBRI1, an ortholog of BRI1 which is an Arabidopsis leucine-rich-repeat, receptor-like kinase (the BR receptor) (36). Sequence analysis of *PsBRI1* in the *lka* mutant identified a missense mutation that converts the highly conserved aspartic acid residue located in the leucine-rich-repeat, to asparagine (36). The close proximity of the *lka* mutation to the amino acid island in *PsBRI1* would be expected to seriously affect the BR-binding ability of the resultant protein, leading to a phenotype reminiscent of BR-deficiency.

The third pea BR mutant to be characterised, *lk*, is even shorter than *lkb* and *lka*, reducing internode length by up to 85% (Fig. 10; 44). A build-up of campesterol in *lk* plants indicated that the mutation may impair the conversion of campesterol to campestanol, suggesting that the *LK* gene may be the pea homologue of the Arabidopsis *DET2* gene, which encodes the enzyme for this step (14). Molecular studies showed that *LK* and *DET2* are indeed homologous and that the *lk* mutant contains a mutation in the coding region of this gene (39a). This mutation consists of a base change from a G to A, causing an amino acid change from tryptophan to a stop codon, thus resulting in a truncated protein. The loss of function

Figure 10. Wild-type (left pot) and *lk* plants (right pot).

of the LK protein ultimately leads to the inability to produce campestanol, and a deficiency in downstream components of the BR biosynthesis pathway (39a).

Implications of the Pea BR Mutants

Investigating the pea BR mutants has revealed much about the regulation of BR levels and the effect of BR levels on growth and development. In many cases this information has served to reinforce and confirm perspectives gained through other model species. However, several unique aspects of pea as a model species have also provided novel insights into BR biology.

The dwarf phenotypes of the BR mutants clearly show that BRs are essential for normal shoot elongation, but what is the mechanism by which BRs regulate this process? This question can be answered through an examination of the changes in cell length and number in internodes of the BR mutants. As is the case with pea GA-deficient mutants (43), *lk* and *lka* mutants exhibit a reduction in cell length and cell number in both epidermal and outer cortical cells of the stem tissue (44). In contrast, dwarfism in the *lkb* mutant appears to result primarily from a reduction in cell length, as cell numbers are not reduced in epidermal cells and only slightly reduced in cortical cells of *lkb* internodes (44). From these observations we can conclude that reductions in BRs reduce shoot elongation primarily by reducing cell elongation. However, in some circumstances BRs also appear to have an effect on cell division.

The reduction in elongation of the epidermis and outer cortex of the pea BR mutants may also help to explain an interesting feature of these plants, the corrugated appearance of the stem referred to as banding (44; Fig. 11). It is suggested that reduced elongation in the outer stem tissues constrains the growth of inner cortical and pith cells, in which elongation is not reduced (44). As a consequence, these inner cells develop in a zig-zag pattern, which in turn leads to the appearance of banding on the stem surface (Fig. 11). This apparent difference in the rate of elongation of outer (epidermal and cortical) and inner (pith and xylem) cells of BR deficient stems also suggests that BRs may be more important in the elongation of the outer rather than the central portion of pea stems.

Interestingly, analysis of a number of parameters related to stem elongation has revealed important insights into the physical basis for altered cell elongation rates in the *lka* and *lkb* mutants (5). It appears that the *lka* and *lkb* mutations increase the wall-yield threshold, which in

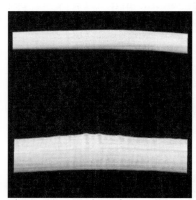

Figure 11. Sections from wild-type (top) and *lkb* (bottom) internodes, showing thickening and banding in the mutant. The sections have been cut to similar lengths.

Figure 12. Leaves (without stipules) from wild-type (left), *lkb* (middle) and *lka* (right) plants

turn reduces cell-wall relaxation and the ability of these cells to elongate. The implication, therefore, is that BRs affect cell elongation by promoting cell-wall loosening (5). This is consistent with the suggested role for BRs in the expression and/or activity of factors involved in cell-wall modification (4).

In addition to the increased wall-yield threshold, both the *lka* and *lkb* mutants exhibit increased osmotic and turgor pressures (5). It is suggested that the increase in turgor pressure is caused by the decrease in cell extension without a concurrent decrease in solute transport into the stem. This combination of factors is thought to result in the increased width and the highly swollen appearance of BR-deficient stems (5; Fig. 11). In contrast, the turgor pressure in the GA-deficient *ls* mutant was similar to the wild type (5), which is consistent with the fact that stem width is not greatly affected in this or in other GA-deficient lines (44). This difference between GA- and BR-deficient mutants suggests a fundamental difference between the mechanisms by which these substances promote shoot elongation.

An additional feature of the pea BR mutants is the reduction in leaf size (44; Fig. 12). Furthermore, BR levels are significantly reduced in the leaves of *lkb* plants, in comparison with the wild type (59). This provides a direct correlation between endogenous BR levels and leaf size, and suggests an important role for endogenous BRs in leaf development. The results from pea are consistent with the finding that exogenous BRs applied to BR-deficient Arabidopsis mutants restores leaf growth to that of the wild type (34).

Feedback regulation of BR biosynthesis
The important role of BRs as regulators of physiological processes suggests that the control of endogenous BR levels may be crucial, and that mechanisms exist for the regulation of BR biosynthesis and/or deactivation. Quantification of intermediates in the late C-6 oxidation pathway in Arabidopsis, pea and tomato revealed a similar profile of BR levels, suggesting that these species share common mechanisms for the regulation of BR biosynthesis (39). For instance, the level of 6-DeoxoCS was an order of magnitude higher than its metabolite, CS, suggesting that the conversion of 6-DeoxoCS to CS is an important rate-limiting step in all three species. This implies that the levels or activity of the enzyme that catalyses this step might be strictly regulated (39). Indeed, there is convincing evidence to suggest that this step is under the control of feedback regulation by BR itself. For instance, in wild-type pea plants the ratio of 6-DeoxoCS to CS is consistently

around 6:1 whilst in the *lka* and *lkb* mutants this ratio is around 2:1 or less (37, 59). This suggests that the conversion of 6-DeoxCS to CS is increased in the *lkb* and *lka* mutants, most likely in response to the real or perceived deficiency of active BRs in these plants, respectively. Indeed these results are entirely consistent with molecular evidence (16, 55) for the feedback regulation of the Arabidopsis *BR6ox1* and *BR6ox2* genes (homologues of the tomato *DWARF* gene which converts 6-DeoxoCS to CS) by bioactive BR levels.

Additional evidence indicates that there may in fact be multiple control points for the regulation of BR biosynthesis (39). For instance, 6-deoxoteasterone levels are two to three orders of magnitude lower than the levels of its precursor, 6-deoxocathasterone, suggesting that this conversion may also be an important control point in BR biosynthesis (39). Indeed, the transcription of the gene involved in this step, *CPD*, appears to be negatively regulated by exogenously applied BRs (14).

Which BR is the biologically active compound?

The accumulation of CS but not BL in *lka* mutant plants (37, 38) raises a number of interesting questions regarding the identity of the bioactive molecule in pea. As discussed previously, it is thought that the accumulation of CS in *lka* plants is due to feedback regulation of the BR pathway (37). A similar accumulation of CS has also been found in BR response mutants in a number of other species, including Arabidopsis, tomato and rice (14). However, if BL is the main bioactive BR in vivo, why are its levels not increased by feedback regulation? This is a fundamental question in BR biology and has implications not only for pea but also for other species.

Several lines of evidence suggest that BL may be the main bioactive BR. For instance, BL is the most biologically active BR in a number of bioassays, and metabolism studies in Arabidopsis and *C. roseus* show that BL may be the endpoint of BR biosynthesis in these species (14). Furthermore, the BR receptor in Arabidopsis, pea and rice, *BRI1*, has a relative binding affinity for BL that is 4-5 times greater than for the direct precursor of BL, CS (62). However, the binding of CS to the receptor suggests that this compound might also have biological activity (62). Furthermore, the conversion of CS to BL has not been demonstrated in cultured cells, explants or intact seedlings of tobacco, rice, or mung bean, suggesting that CS may be the main biologically active compound in these species (66). Similarly, BL has not even been detected in tomato and it has been widely suggested that CS may also be the most active BR in this species (6). Whilst BL has been shown to occur in pea, under many circumstances its levels are below detection limits (approximately 10–fold lower than the endogenous CS levels) (60). It is therefore reasonable to suggest that if CS is only 4-5 times less active than BL, but in many circumstances at least 10–times more abundant than BL, CS should be considered an important, if not the most important, bioactive BR in pea. Clearly, this fundamental aspect of BR biology has yet to be fully resolved.

The role of non-BR sterols in plant growth and development
Since the LKB enzyme plays a role in both BR and sterol biosynthesis, *lkb* plants are deficient not only in BRs but in sterols (such as sitosterol and stigmasterol) as well (37). Whilst BRs are now widely recognised as an important class of plant hormones, other plant sterols are generally regarded as primary components of cellular membranes where they regulate fluidity and permeability (53). The possibility that non-BR sterols could also act as important regulators of plant growth and development has largely been ignored (53). However, studies in Arabidopsis have provided a new insight into this issue. For instance, *dim/dwarf1* are mutations in the Arabidopsis homologue of the pea *LKB* gene (54). As is the case in *lkb*, *dim/dwarf1* plants exhibit a reduction in the levels of both BRs and other sterols. However, the phenotype of this mutant cannot be restored to that of the wild type by application of BRs alone, suggesting that sterols other than the BRs may have a more important role in the regulation of plant growth than first thought (53).

On the other hand, a comparison of the *lka* and *lkb* phenotypes indicates that the contribution of the reduced sterol levels to the formation of the *lkb* phenotype may in fact be minimal (37). The phenotype of *lkb* (which blocks the production of BRs and other sterols) is remarkably similar to the phenotype of *lka* (which blocks the perception of BRs). Indeed, the only morphological difference observed between these two mutants is the substantially greater reduction in peduncle length in *lkb* than in *lka* (44). Similarly, even though the *lk* phenotype (which results from a specific block in BR synthesis) is generally more severe than in *lkb*, both mutants exhibit a very similar range of phenotypic traits (44). Furthermore, internode elongation in *lkb* plants can be completely restored to that of the wild type by application of bioactive BRs alone - despite the fact that this treatment further down-regulates sterol levels (37). Thus it would seem that sterols such as sitosterol and stigmasterol may not play an important role in regulating internode elongation in pea.

ENVIRONMENTAL REGULATION OF THE GIBBERELLIN AND BRASSINOSTEROID PATHWAYS

Gibberellins

Mutants are invaluable for discovering the functions of plant hormones, but do they reflect the variation in hormone level that might normally occur in a natural plant population? Extreme mutants (for example *na-1* in pea) would probably not survive in the wild, although they are arguably "hopeful monsters" that in certain circumstances might be exploited by evolutionary processes to produce a major phenotypic shift. Other mutations might alter hormone levels only slightly, and may therefore persist in the wild.

However, natural variation in hormone levels is perhaps more likely to occur as a result of changes in environmental factors or developmental cues.

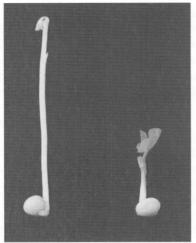

Figure 13. Six-day-old wild-type pea seedlings grown in continuous darkness (left) or continuous white light (right), after emerging from the soil on day 4.

Certainly plant development is highly responsive to changes in the environment, but it is important to realise that only a subset of such responses will be mediated by changes in hormone levels. Nevertheless, research is uncovering clear examples of environmental factors that cause large changes in hormone level and corresponding changes in growth. One such environmental factor is the light regime.

Pea seedlings grown in the dark display an "etiolated" habit (Fig. 13). The stems are longer than those of light-grown plants (although with fewer internodes), and leaf expansion is suppressed. In contrast to the green coloration of light-grown seedlings, the stems of etiolated plants are white and the leaves yellow. The long stems are reminiscent of GA-treated light-grown plants, and early researchers inferred some type of interaction between GAs and the light regime. It was often implied that dark-grown plants might contain higher levels of bioactive GA than plants grown in continuous light.

With the advent of sophisticated physico-chemical methods for hormone analysis, the question of GA levels in dark-grown plants was re-examined. It was shown that pea plants grown in darkness do not contain higher levels of bioactive GA than plants grown in continuous light (63). These results cast doubt on the theory that GAs mediate the effects of light on stem elongation. However, in the late 1990s it was shown that when pea plants are grown in the dark and then transferred to the light (Fig. 14), there is a rapid and substantial drop in their GA_1 content (2, 15, 41). After only 2 hours of exposure to light, the GA_1 content decreases by approximately 6-fold (2, 41). The stem elongation rate is also reduced, and this can be

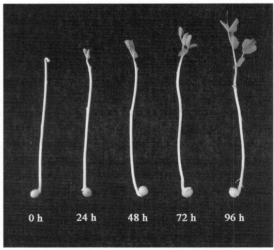

Figure 14. Changes in wild-type (7-day-old) seedlings after transfer from darkness to white light. The time (in hours) since transfer is indicated (after 57).

reversed (at least partially) by applying GA_1.

The drop in GA_1 is quite specific: the levels of other GAs are not substantially altered, and GA_8 levels are actually higher in transferred plants. It fact, it appears that the decrease in GA_1 content in transferred plants is due in part to an up-regulation of the step GA_1 to GA_8. Consistent with this theory is evidence that one of the two known pea GA 2-oxidase genes, *PsGA2ox2*, is rapidly up-regulated after transfer (42). A rapid down-regulation of *LE* expression also appears to contribute to the reduction in GA_1 content on transfer to light (42).

How does the plant perceive the light signal that results in the GA_1 drop? It appears that the photoreceptor phytochrome A plays a key role, since transfer to red or far-red light does not reduce GA_1 levels in the *phyA* mutant, which lacks this form of phytochrome (42).

In summary, after exposure to light, there is an inhibition of growth, caused in part by a rapid drop in GA_1 content. This is a very clear example of how an environmental effect on plant growth can be mediated by a change in hormone content. The initial perception of light by phytochrome A and possibly a blue light photoreceptor results in rapid changes in transcript level of Mendel's *LE* gene and of *PsGA2ox2*, which in turn result in the reduction in GA_1 content. Under natural conditions, the dark to light transition might occur as seedlings emerge from soil or leaf litter under which they initially germinate. Emerging seedlings are known to contain low levels of GA_1 (50).

Three to four days after transfer to light, GA_1 levels increase again, accounting for the observation that long-term exposure to light does not reduce GA_1 content (63). However, despite this increase, elongation remains inhibited compared with dark-grown plants. This long-term growth inhibition is attributable to a reduction in the ability of the tissue to respond to available GA_1 (41). The long-term effects of light on elongation and GA content provide a clear example of how internode length can vary substantially without concomitant changes in GA_1 content.

Brassinosteroids

During the mid to late 1990s evidence emerged suggesting that BRs may also play a role in light-regulated development (9, 27). Two Arabidopsis mutants, *det2* and *cpd*, which were originally isolated on the basis of their partially de-etiolated phenotype (short hypocotyls, expanded cotyledons and developing leaves) when grown in the dark, were subsequently shown to be BR deficient (35). The majority of BR mutants that have since been isolated in Arabidopsis, tomato and rice have also been reported to exhibit a de-etiolated phenotype when grown in the dark (58). In many cases the abnormal dark-grown phenotype of such BR-deficient plants can be, at least partially, restored to a normal wild-type, etiolated phenotype, after application of BRs (35). This evidence has formed the basis of the widely cited suggestion that BRs act as negative regulators of de-etiolation (9, 27). In other words, high BR levels are thought to promote normal etiolated growth in the dark, whilst the development of a de-etiolated phenotype after exposure to light is thought to be mediated (at least in part) by a reduction in BR levels.

Circumstantial evidence supports a role for BRs as negative-regulators of de-etiolation (58). For instance, using the BR biosynthesis inhibitor, brassinazole (Brz), to reduce BR levels in dark-grown Arabidopsis plants induces some morphological characteristics of light grown plants (33). Furthermore, the expression of the light-regulated genes, *RBCS* and *CAB*, are up-regulated in dark-grown, BR-deficient mutants, *det2* and *cpd,* and in dark-grown wild type plants treated with Brz (35). In addition, microarray analysis of gene expression in Arabidopsis plants that were exposed to different light treatments showed that four genes involved in the BR biosynthesis pathway were all down-regulated by light (28). Recent studies also provide an insight into a mechanism by which light could negatively regulate BR levels (21). This work suggested that in pea, a light-repressible protein, Pra2, regulates DDWF1, a cytochrome P450 enzyme involved in brassinosteroid biosynthesis (21; Fig. 9). It has been suggested that this interaction between Pra2 and DDWF1 represents a link between light-signal transduction and endogenous BR levels in pea (10, 21). It is proposed that, on exposure to light, phytochrome and blue light photoreceptors signal the repression of Pra2 (and therefore DDWF1), which leads to a reduction in BR levels, and a slowing of shoot growth (10, 21).

However, the suggestion that BRs negatively regulate de-etiolation is not universally accepted. One reason for this is that a fully de-etiolated phenotype is not a characteristic of all dark-grown BR mutants (58). Indeed, there are several examples of BR mutants in Arabidopsis, tomato and pea, which do not exhibit a clear de-etiolated phenotype in the dark (58). For example, the pea BR-deficient mutants *lk* and *lkb* exhibit neither a de-etiolated phenotype nor altered expression of light-regulated genes when grown in the dark (60). Similarly, dark-grown wild-type pea plants treated with the BR biosynthesis inhibitor, Brz, do not exhibit a de-etiolated phenotype (Fig. 15) (60). Such evidence suggests that BR levels do not play a negative-regulatory role in de-etiolation in pea.

Another problem with the suggestion that BRs mediate the effects of light was the lack of direct BR measurements in light- and dark-grown plants. A reduction in BR levels in light-grown plants is implicit in the argument that BRs play a negative regulatory role in de-etiolation (10). Direct evidence of endogenous BR levels is therefore crucial for testing that

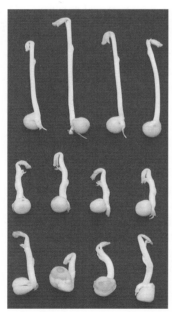

Figure 15. Dark-grown seedlings of the wild-type (top), *lka* (middle) and *lkb* (bottom) genotypes.

theory.

The first measurements of endogenous BR levels in light- and dark-grown pea seedlings were published in 2002. Significantly, BR levels were actually increased, not decreased, in pea seedlings grown in continuous light, compared with seedlings grown in the dark (60). These results are clearly inconsistent with the idea that BRs negatively regulate de-etiolation. This finding was supported by results from a time-course investigation of CS and 6-DeoxoCS levels in etiolated pea seedlings after exposure to light (57). It had previously been suggested that the light-mediated suppression of the pea *Pra2* gene causes a reduction in the levels of the DDWF1 enzyme, which catalyses the formation of 6-DeoxoCS and CS (21). However, no substantial decrease in endogenous 6-DeoxoCS or CS levels was evident in wild-type pea seedlings after exposure to light (57). In fact, 6-DeoxoCS levels were actually increased after prolonged exposure, suggesting an up-regulation of BR biosynthesis, via the late C-6 oxidation pathway, during de-etiolation (57). Together these findings suggest that in pea, BR biosynthesis is not down- regulated by light. Indeed, when we also consider that dark-grown pea BR mutants are not de-etiolated at either the morphological or molecular level (60), it is reasonable to conclude that BRs do not negatively regulate de-etiolation in this species (58).

Comparable outcomes have now emerged from studies of BR levels in other species. For instance, BR levels have been shown to be higher in light than in the dark in Arabidopsis (Y. Shimada and S. Fujioka, unpublished results) and rice (67). These results are in direct contrast to the picture that emerged from the original genetic and molecular studies (9, 27), and strongly suggest that BRs do not play a negative-regulatory role in de-etiolation in these species. In light of these findings we must now ask: exactly what role do BRs play, if any, in light-regulated development?

REGULATION OF HORMONE BIOSYNTHESIS AND DEACTIVATION BY OTHER HORMONES

Hormone biosynthesis can be strongly affected by the levels of other plant hormones. For example, it was recently shown that auxin, the first plant hormone to be discovered, dramatically affects gibberellin biosynthesis and deactivation. Auxin is a growth promoter, with marked effects on stem elongation, particularly in excised segments (Chapter C1). The fact that both auxin and GA promote elongation prompted early researchers to investigate whether or not the two hormones interact, but the question has only recently been resolved.

The first indication that auxin might promote GA biosynthesis came from studies on pea pods. It was suggested that a specific auxin, 4-chloroindole-3-acetic acid, moves from pea seeds into the elongating pods where it promotes the step GA_{19} to GA_{20}. It appears that 4-Cl-IAA exerts this effect by up-regulating GA 20-oxidase gene expression (61).

Next it was reported that application of the classical auxin, indole-3-acetic acid (IAA), strongly promotes GA_1 biosynthesis in decapitated pea stems (48). Decapitation markedly reduces the content of IAA in the stem, by removing the site (the apical bud) in which auxin enters the specialised system by which it is transported towards the base of the plant (Chapter E1). In decapitated stems radioactive $[^3H]GA_{20}$ was converted mainly to the deactivation products $[^3H]GA_{29}$ and $[^3H]GA_{29}$-catabolite. However, applying IAA in lanolin paste to the cut stump enabled the stem to convert $[^3H]GA_{20}$ to $[^3H]GA_1$, and a large $[^3H]GA_8$ peak was also recovered. Thus applying auxin restored the pattern of metabolites to that found in intact stems. The effects of decapitation on the GA pathway were mimicked by applying an auxin transport inhibitor to the stem of intact plants. Importantly, auxin also promoted the conversion of GA_{20} to GA_1 in isolated stem segments (40), showing that the stems themselves are capable of synthesising GA_1.

Auxin promotes the step GA_{20} to GA_1 by up-regulating the expression of Mendel's *LE* gene (*PsGA3ox1*). *LE* transcript (mRNA) levels are dramatically reduced by decapitation, but completely restored by IAA application to the cut stump (48). However, it was discovered, using the stem segment system, that auxin does not up-regulate *LE* expression in the presence of a protein synthesis inhibitor, cycloheximide. This observation indicates that auxin must first up-regulate another gene (a "primary" auxin response gene), which produces a protein that in turn leads to the up-regulation of *LE*. When the synthesis of that protein is inhibited (as by cycloheximide), *LE* is not up-regulated (40). Therefore *LE* appears to be a secondary or late auxin response gene (Chapter D1) (1).

In pea stems other biosynthetic steps leading to GA_1 appear to be less affected by IAA, as shown by quantifying GA_{19} and GA_{20}. However, auxin strongly inhibits the deactivation of GA_1 to GA_8, as indicated by metabolism studies with $[^{14}C]GA_1$. Experiments with the *sln* mutant (40) show that in the wild-type auxin inhibits 2-oxidation by down-regulating the *SLN* gene, a finding consistent with direct measurements of *SLN* mRNA levels (48). It appears that in intact stems, this down-regulation results in another 2-oxidase gene (possibly *PsGA2ox2*) playing a relatively important role in the deactivation of GA_1. This explains why, in intact *sln* shoots, GA_1 deactivation is essentially normal (51).

In summary, auxin from the apical bud maintains the GA_1 content of internodes, by promoting the biosynthesis of GA_1 and by inhibiting its deactivation (Fig 16). These effects are mediated by changes in expression of the key genes *LE* and *SLN*. Auxin up-regulates *LE*, and this well-known gene therefore forms a link between the classical plant hormones auxin and GA. The expression of *SLN*, on the other hand, is down-regulated by IAA, providing an interesting example of opposing effects of auxin on the expression of genes from the same biochemical pathway.

What is the significance of the auxin-GA interaction for growth? That question has been addressed using the isolated stem segment system. Early research had demonstrated that excised pea stem segments, floated on a

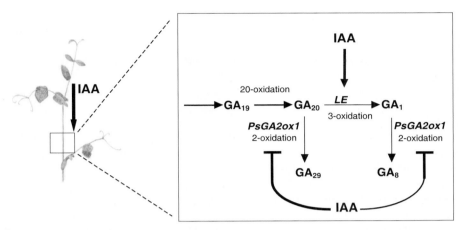

Figure 16. Model showing the effects of IAA on the GA pathway in pea internodes. IAA is transported basipetally into the internodes from the apical bud

liquid medium with or without IAA, can show a strong growth response to the hormone (Chapter C1). It was shown recently that in excised segments IAA maintains the GA_1 content at a relatively high level (47). When IAA was not added to the medium, the endogenous GA_1 content was 10-fold less, after only 6 hours of incubation (47).

It was reasoned that if the auxin-induced GA_1 is important for the growth response to IAA, then IAA would promote elongation to a lesser extent in *le-1* segments than in *LE* segments, because in the mutant, there would be much less auxin-induced GA_1 (the step GA_{20} to GA_1 is blocked in *le-1* plants). This turned out to be the case, showing that auxin acts (at least in part) by increasing GA_1 content (47). The theory that GA_1 is the final hormonal effector of elongation is supported by the observation that decapitated *le-1* internodes respond strongly to GA_1, even though their auxin level is low (49). There seems little doubt, however, that auxin also directly stimulates elongation, because the initial growth response to IAA is very rapid, compared with the response to GA (64).

Although some early hormone researchers suggested that GA might act by affecting auxin levels (that is, the opposite to the recent findings), the effect of GA on auxin content is relatively slight. For example, the large reduction in GA_1 content in Mendel's *le-1* is accompanied by only a 30% reduction in auxin content (22). Similarly, the apical portions of BR-deficient mutants of pea are not deficient in GA_1 (23), although the IAA content of their internodes is somewhat reduced (32). Interestingly, decapitation reduces IAA and GA_1 levels in pea stems, but BR levels are unaffected (59). On these bases, we can construct the three-way scheme shown in Fig. 17, in which the effect of IAA on GA_1 content represents a much stronger interaction than any other relationship shown in the figure. It appears that even large changes in GA_1 or BR levels do not perturb every thread of a web connecting the hormones at the level of hormone content, although effects on hormone responses might be a different matter.

CONCLUDING DISCUSSION

The biosynthesis of plant hormones is critical for the normal growth and development of the pea plant and for plants in general. Disruption of hormone biosynthesis pathways by mutations, as exemplified by the GA and BR synthesis mutants discussed above, leads to dramatic changes in plant growth. The most obvious characteristic affected by GA and BR mutants is shoot height, but GA deficiency also markedly affects seed and root development, and BR deficiency affects leaflet size and shape. Thus mutants have been invaluable for defining the roles played by these hormones. This was graphically reiterated in the case of the BRs during the late 1990s. Before that time the status of BRs as plant hormones was unclear.

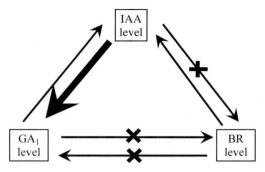

Figure 17. IAA strongly affects GA_1 levels but the other potential interactions shown are either weak (thin arrows) or non-existent (crosses).

Hormone biosynthesis can also be strongly regulated by environmental factors. This provides a means by which factors such as the light regime can affect plant development. An excellent example of this is the rapid, specific, and substantial reduction in GA_1 content after transfer of pea seedlings from darkness to light (2, 15, 41). This decline plays a key role in reducing stem elongation in the transferred plants. However, it appears that the shorter stature of pea plants grown continuously in the light, compared with dark-grown plants, is not due to lower levels of bioactive GA, nor to lower levels of bioactive BRs. In this case, reduced GA responsiveness in the light-grown plants appears to be the major reason for their short stature.

In recent plant hormone research, there has been an increasing emphasis on hormone interactions, and this trend appears set to continue. Hormones will be considered less as individual regulators of plant growth and development, and more as components of complex, web-like signaling systems (7). At the same time, however, it might still be possible to define key pathways within the web that predominate in terms of their effect on the response in question. The sequence auxin → *LE* gene expression → GA_1 → elongation appears to be such a pathway. GA biosynthesis depends on auxin, and at the same time GAs can be viewed as component of the auxin signal transduction pathway. Other key interactions between regulators of plant growth may well await discovery.

Acknowledgements
We thank members of the Hobart developmental genetics group for their support throughout this work, Jennifer Smith for photography, and the Australian Research Council for financial support over many years.

References
1. Abel S, Theologis A (1996) Early genes and auxin action. Plant Physiol 111: 9-17
2. Ait-Ali T, Frances S, Weller JL, Reid JB, Kendrick RE, Kamiya Y (1999) Regulation of gibberellin 20-oxidase and gibberellin 3β-hydroxylase transcript accumulation during de-etiolation of pea seedlings. Plant Physiol 121: 783-91
3. Ait-Ali T, Swain SM, Reid JB, Sun T, Kamiya Y (1997) The *LS* locus of pea encodes the gibberellin biosynthesis enzyme *ent*-kaurene synthase A. Plant J 11: 443-454
4. Altmann T (1999) Molecular physiology of brassinosteroids revealed by the analysis of mutants. Planta 208: 1-11
5. Behringer FJ, Cosgrove DJ, Reid JB, Davies PJ (1990) Physical basis for altered stem elongation rates in internode length mutants of *Pisum*. 94: 166-173
6. Bishop GJ, Nomura T, Yokota T, Harrison K, Noguchi T, Fujioka S, Takatsuto S, Jones JD, Kamiya Y (1999) The tomato DWARF enzyme catalyses C-6 oxidation in brassinosteroid biosynthesis. Proc Natl Acad Sci USA 96: 1761-1766
7. Brady SM, McCourt P (2003) Hormone cross talk in seed dormancy. J Plant Growth Regul 22: 25-31
8. Brian PW, Hemming HG (1955) The effects of gibberellic acid on shoot growth of pea seedlings. Physiol Plant 8: 669-681
9. Chory J, Li J (1997) Gibberellins, brassinosteroids and light-regulated development. Plant Cell Environ 20: 801-806
10. Clouse SD (2001) Integration of light and brassinosteroid signals in etiolated seedling growth. Trends Plant Sci 6: 443-445
11. Davidson SE, Elliott RC, Helliwell CA, Poole AT, Reid JB (2003) The pea gene *NA* encodes *ent*-kaurenoic acid oxidase. Plant Physiol 131: 335-344
12. Davidson SE, Reid JB (2004) The pea gene *LH* encodes *ent*-kaurene oxidase. Plant Physiol 134, 1123-1134
13. Elliott RC, Ross JJ, Smith JJ, Lester DR, Reid JB (2001) Feed-forward regulation of gibberellin deactivation in pea. J Plant Growth Regul 20: 87-94.
14. Fujioka S, Yokota T (2003) Biosynthesis and metabolism of brassinosteroids. Annu Rev Plant Biol 54: 137-164
15. Gil J, García-Martinez JL (2000) Light regulation of gibberellin A_1 content and expression of genes coding for GA 20-oxidase and GA 3β-hydroxylase in etiolated pea seedlings. Physiol Plant 108: 223-228
16. Goda H, Shimada Y, Asami T, Fujioka S, Yoshida S (2002) Microarray analysis of brassinosteroid-regulated genes in Arabidopsis. Plant Physiol 130: 1319-34.
17. Hedden P, Croker SJ (1992) Regulation of gibberellin biosynthesis in maize seedlings. *In* CM Karssen, LC van Loon, D Vreugdenhil, eds, Progress in Plant Growth Regulation. Kluwer Academic Publishers, Dordrecht, pp 534-544
18. Hedden P, Phillips AL (2000) Gibberellin metabolism: new insights revealed by the genes. Trends Plant Sci 5: 523-530
19. Ingram TJ, Reid JB, MacMillan J (1986) The quantitative relationship between gibberellin A_1 and internode growth in *Pisum sativum* L. Planta 168: 414-420
20. Ingram TJ, Reid JB, Murfet IC, Gaskin P, Willis CL, MacMillan J (1984) Internode length in *Pisum*: the *Le* gene controls the 3β-hydroxylation of gibberellin A_{20} to gibberellin A_1. Planta 160: 455-463
21. Kang JG, Yun J, Kim DH, Chung KS, Fujioka S, Kim JI, Dae HW, Yoshida S, Takatsuto S, Song PS, Park CM (2001) Light and brassinosteroid signals are integrated via a dark-induced small G protein in etiolated seedling growth. Cell 105: 625-636
22. Law DM, Davies PJ (1990) Comparative indole-3-acetic acid levels in the slender pea and other pea phenotypes. Plant Physiol 93: 1539-1543
23. Lawrence NL, Ross JJ, Mander LN, Reid JB (1992) Internode length in Pisum. Mutants

lk, *lka* and *lkb* do not accumulate gibberellins. J Plant Growth Regul 11: 35-37
24. Lester DR, Ross JJ, Ait-Ali T, Martin DN, Reid JB (1996) A gibberellin 20-oxidase cDNA (Accession no. U58830) from pea (*Pisum sativum* L.) seed. Plant Physiol 111: 1353
25. Lester DR, Ross JJ, Davies PJ, Reid JB (1997) Mendel's stem length gene (*Le*) encodes a gibberellin 3β-hydroxylase. Plant Cell 9: 1435-1443
26. Lester DR, Ross JJ, Smith JJ, Elliott RC, Reid JB (1999) Gibberellin 2-oxidation and the *SLN* gene of *Pisum sativum*. Plant J 19: 65-73
27. Li J, Nagpal P, Vitart V, McMorris TC, Chory J (1996) A role for brassinosteroids in light-dependent development of *Arabidopsis*. Science 272: 398-401
28. Ma L, Li J, Qu L, Hager J, Chen Z, Zhao H, Deng XW (2001) Light control of Arabidopsis development entails coordinated regulation of genome expression and cellular pathways. Plant Cell 13: 2589-2607
29. Martin DN, Proebsting WM, Hedden P (1997) Mendel's dwarfing gene: cDNAs from the *le* alleles and function of the expressed proteins. Proc Natl Acad Sci USA 94: 8907-8911
30. Martin DN, Proebsting WM, Hedden P (1999) The *SLENDER* gene of pea encodes a gibberellin 2-oxidase. Plant Physiol 121:775-781
31. Martin DN, Proebsting WM, Parks TD, Dougherty WG, Lange T, Lewis MJ, Gaskin P, Hedden P (1996) Feed-back regulation of gibberellin biosynthesis and gene expression in *Pisum sativum* L. Planta 200: 159-166
32. McKay MJ, Ross JJ, Lawrence NL, Cramp RE, Beveridge CA, Reid JB (1994) Control of internode length in *Pisum sativum*. Further evidence for the involvement of indole-3-acetic acid. Plant Physiol 106: 1521-1526
33. Nagata N, Min YK, Nakano T, Asami S, Yoshida S (2000) Treatment of dark-grown *Arabidopsis thaliana* with a brassinosteroid-biosynthesis inhibitor, brassinazole, induces some characteristics of light-grown plants. Planta 211: 781-790
34. Nakaya M, Tsukaya H, Murakami N, Kato M (2002) Brassinosteroids control the proliferation of leaf cells of *Arabidopsis thaliana*. Plant Cell Physiol 43: 239-244
35. Nemhauser J, Chory J (2002) Photomorphogenesis. *In*: CR Somerville, EM Meyerowitz eds. The Arabidopsis Book. American Society of Plant Biologists, Rockville, MD http://www.aspb.org/downloads/arabidopsis/nemhau.p
36. Nomura T, Bishop GJ, Kaneta T, Reid JB, Chory J, Yokota T (2004) The *LKA* gene is a *BRASSINOSTEROID INSENSITIVE1* homolog of pea. Plant J 36: 291-300
37. Nomura T, Kitasaka Y, Takatsuto S, Reid JB, Fukami M, Yokota T (1999) Brassinosteroid/sterol synthesis and plant growth as affected by *lka* and *lkb* mutations of pea. Plant Physiol 119: 1517-1526
38. Nomura T, Nakayama M, Reid JB, Takeuchi Y, Yokota T (1997) Blockage of brassinosteriod biosynthesis and sensitivity causes dwarfism in garden pea. Plant Physiol 113: 31-37
39. Nomura T, Sato T, Bishop GJ, Kamiya Y, Takasuto S, Yokota T (2001) Accumulation of 6-deoxocathasterone and 6-deoxocastasterone in *Arabidopsis*, pea and tomato is suggestive of common rate-limiting steps in brassinosteroid biosynthesis. Phytochem 57: 171-178
39a. Nomura T, Jager C, Kitisaka Y, Takeuchi K, Fukami M, Yoneyama K, Matsushita Y, Nyunoya H, Takasuto S, Fijioka S, Smith J, Kerckhoffs LH, Reid JB, Yokota T (2004) Brassinosteroid deficiency due to truncated 5α-reductase causes dwarfism in the *lk* mutant of pea. Plant Physiol (In press)
40. O'Neill DP, Ross JJ (2002) Auxin regulation of the gibberellin pathway in pea. Plant Physiol 130: 1974-1982
41. O'Neill DP, Ross JJ, Reid JB (2000) Changes in gibberellin A_1 levels and response during de-etiolation of pea seedlings. Plant Physiol 124: 805-812
42. Reid JB, Botwright NA, Smith JJ, O'Neill DP, Kerckhoffs LHJ (2002) Control of gibberellin levels and gene expression during de-etiolation in pea. Plant Physiol 128: 734-741
43. Reid JB, Murfet IC, Potts WC (1983) Internode length in *Pisum*. II. Additional

information on the relationship and action of loci *Le, La, Cry, Na,* and *Lm.* J Exp Bot 34: 349-364.
44. Reid JB, Ross JJ (1993) A mutant-based approach, using *Pisum sativum,* to understanding plant growth. Int J Plant Sci 154: 22-34
45. Reid JB, Ross JJ, Swain SM (1992) Internode length in *Pisum*. A new slender mutant with elevated levels of C19 gibberellins. Planta 188: 462-467
46. Ross JJ, MacKenzie-Hose AK, Davies PJ, Lester DR, Twitchin B, Reid JB (1999) Further evidence for feedback regulation of gibberellin biosynthesis in pea. Physiol Plant 105: 532-538
47. Ross JJ, O'Neill DP, Rathbone DA (2003) Auxin-gibberellin interactions in pea: Integrating the old with the new. J Plant Growth Regul 22: 99-108
48. Ross JJ, O'Neill DP, Smith JJ, Kerckhoffs LHJ, Elliott RC (2000) Evidence that auxin promotes gibberellin A_1 biosynthesis in pea. Plant J 21: 547-552
49. Ross JJ, O'Neill DP, Wolbang CM, Symons GM, Reid JB (2002) Auxin-gibberellin interactions and their role in plant growth. J Plant Growth Regul 20: 346-353
50. Ross JJ, Reid JB, Dungey HS (1992) Ontogenetic variation in levels of gibberellin A_1 in *Pisum*. Implications for the control of stem elongation. Planta 186: 166-171
51. Ross JJ, Reid JB, Swain SM, Hasan O, Poole AT, Hedden P, Willis CL (1995) Genetic regulation of gibberellin deactivation in *Pisum*. Plant J 7: 513-523
52. Santes CM, Hedden P, Sponsel VM, Reid JB, Garcia-Martinez JL (1993) Expression of the *le* mutation in young ovaries of *Pisum sativum* and its effect on fruit development. Plant Physiol 101: 759-764
53. Schaller H (2003) The role of sterols in plant growth and development. Prog Lip Res 42: 163-175
54. Schultz L, Kerckhoffs LHJ, Klahre U, Yokota T, Reid JB (2001) Molecular characterisation of the brassinosteroid-deficient *lkb* mutant in pea. Plant Mol Biol 47: 491-498
55. Shimada Y, Goda H, Nakamura A, Takatsuto S, Fujioka S, Yoshida S (2003) Organ-specific expression of brassinosteroid-biosynthetic genes and distribution of endogenous brassinosteroids in Arabidopsis. Plant Physiol 131:287-297
56. Swain SM, Reid JB, Ross JJ (1993) Seed development in *Pisum*. The lh^i allele reduces gibberellin levels in developing seeds, and increases seed abortion. Planta 191: 482-488
57. Symons GM, Reid JB (2003a) Hormone levels and response during de-etiolation in pea. Planta 216: 422-31
58. Symons GM, Reid JB (2003b) Interactions between light and plant hormones during de-etiolation. J Plant Growth Regul 22: 3-14
59. Symons GM, Reid JB (2004) Brassinosteroids do not undergo long distance transport in pea: Implications for the regulation of endogenous brassinosteroid levels. Plant Physiol (In press)
60. Symons GM, Schultz L, Kerckhoffs LHJ, Davies NW, Gregory D, Reid JB (2002) Uncoupling brassinosteroid levels and de-etiolation in pea. Physiol Plant 115: 311-319
61. van Huizen R, Ozga JA, Reinecke DM (1997) Seed and hormonal regulation of gibberellin 20-oxidase expression in pea pericarp. Plant Physiol 115: 123-128
62. Wang Z-Y, Seto H, Fujioka S, Yoshida S, Chory J (2001) BRI1 is a critical component of a plasma-membrane receptor for plant steroids. Nature 410: 380-383
63. Weller JL, Ross JJ, Reid JB (1994) Gibberellins and phytochrome regulation of stem elongation in pea. Planta 192: 489-496
64. Yang T, Davies PJ, Reid JB (1996) Genetic dissection of the relative roles of auxin and gibberellin in the regulation of stem elongation in intact light-grown peas. Plant Physiol 110: 1029-1034
65. Yaxley JR, Ross JJ, Sherriff LJ, Reid JB (2001) Gibberellin biosynthesis mutations and root development in pea. Plant Physiol 125: 627-633
66. Yokota T (1999) Brassinosteroids. In: PJJ Hooykaas, MA Hall, KR Libbenga (eds*) Biochemistry and molecular biology of plant hormones.* Elsevier Science. pp 277-292
67. Yokota T, Nomura T, Sato T, Tamaki Y (2001) Light regulates brassinosteroid biosynthesis in rice. 17th IPGSA Meeting. Abstract 191.

C. THE FINAL ACTION OF HORMONES

C1. Auxin and Cell Elongation

Robert E. Cleland
Department of Biology, University of Washington, Seattle, Washington 98195-5325, USA. E-mail: cleland@u.washington.edu

INTRODUCTION

One of the most dramatic and rapid hormone responses in plants is the induction by auxin of rapid cell elongation in isolated stem and coleoptile sections. The response begins within 10 minutes after the addition of auxin, results in a 5-10 fold increase in the growth rate, and persists for hours or even days (22). It is hardly surprising that this may be the most studied hormonal response in plants.

Auxin is not the only hormone that plays a major role in controlling cell elongation. Studies with dwarf mutants have shown that in certain cases, gibberellins (Chapter D2) and brassinosteroids (Chapter B6) are essential for maximal cell elongation in stems, and ethylene (Chapter D4) and cytokinin (Chapter D3) can act to decrease the rate of stem cell elongation. The effect of auxin on cell elongation is not always promotive; auxin can also inhibit cell elongation in roots (23). In each of these cases our knowledge about how the hormone exerts its effect is still limited. The promotion of cell elongation in stem and coleoptile cells in *isolated tissues* is still the best understood hormonal growth response, and will be the subject of most of this chapter. At the end, however, the *in vivo* role of auxin in controlling cell elongation will also be discussed[1].

How do auxins produce this growth response? To answer this, the process of cell enlargement must first be considered. Cell enlargement consists of two interrelated processes; osmotic uptake of water, driven by a

---

[1] Abbreviations: CEZ, central elongation zone; DEZ, distal elongation zone; LPL, lysophospholipids; PM, plasma membrane;TIBA, 2,3,5-triiodobenzoic acid; WLF, wall-loosening factors; WLP, wall loosening proteins; XTH, xyloglucan endotransglucosylase/ hydrolase.

water potential gradient across the plasma membrane, and extension of the existing cell wall, driven by the turgor-generated stress within the wall. The process of cell enlargement can be described (31) by two equivalent equations:

$$dV/dt = Lp.\Delta\Psi \qquad \text{Equation 1}$$

$$dV/dt = m(P-Y) \qquad \text{Equation 2}$$

where dV/dt is the rate of increase in cell volume, Lp is the hydraulic conductivity, $\Delta\Psi$ is the water potential gradient across the plasma membrane, m is the wall extensibility, P is the turgor pressure and Y is the wall yield threshold (the turgor that must be exceeded for wall extension to occur). These two equations can be combined into a third equation:

$$dV/dt = m.Lp\,(\Psi_a - \pi - Y)/(m + Lp) \qquad \text{Equation 3}$$

where Ψ_a is the apoplastic water potential and π is the osmotic potential of the cell. In the absence of auxin, the growth rate is low because of a low m, low Lp, low P or high Y (or a combination of these). When auxin initiates rapid cell enlargement, it must do so by increasing m, Lp, Ψ_a or P, or by decreasing π or Y. No matter what the initial effect of auxin is (e.g. gene activation, ATPase activation, or change in membrane permeability), an increase in cell enlargement can be initiated *only* by a change in one or more of these cellular growth parameters.

Characteristics of Auxin-induced Cell Elongation

Auxin-induced cell elongation has a number of distinct characteristics (22). The first is its time course (Fig. 1). Upon addition of auxin there is a lag of at least 8 minutes before the growth rate begins to increase. Then the growth rate increases, and reaches a

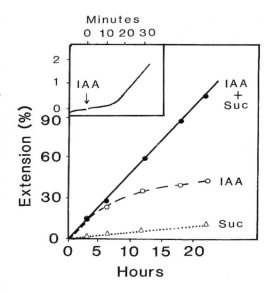

Figure 1. Time course of auxin-induced growth of *Avena* coleoptile sections. Sections were incubated in 10 mM phosphate buffer, pH 6.0, ± 2% sucrose and with 10 μM IAA added at time zero. Rapid elongation is initiated after an 8-10 minute lag. The initial rate is independent of the presence or absence of absorbable solutes (sucrose), but absorbable solutes are required for continual rapid elongation.

205

maximum after 30-60 minutes. The length of the lag cannot be shortened by changing the temperature, by adding excess auxin, or by the addition of any other metabolite or hormone (22). The lag is independent of the rate of protein synthesis; thus it is unlikely to reflect the time needed to synthesize some new protein required for rapid cell elongation.

The rapidity with which the maximum growth rate is achieved, and the magnitude of the growth rate are dependent on the tissue and can be influenced by the past history of the tissue. For example, freshly cut maize coleoptile sections respond slowly and rather poorly to auxin, while sections "aged" in water for 3 hours are far more responsive (22). In contrast, oat coleoptile sections show no change in sensitivity to auxin (8). In oat coleoptile sections, once the maximum growth rate is achieved it can be sustained for up to 18 hours, as long as auxin and an absorbable solute are present (8). In most dicot stem sections, the maximum growth rate is followed by a decline in growth rate, and then at least a partial restoration of the rate thereafter (Fig. 2). Unlike coleoptile sections, dicot stem sections rarely maintain a constant growth rate for more than 2-3 hours after addition of auxin; thereafter the growth rate falls continuously so that by 16-20 hours growth has ceased (22).

In most tissues, the growth rate over the first 2-4 hours is proportional to the log of the external auxin concentration over a range of about 2 decades, but the concentration range varies from tissue to tissue. For example, in Avena coleoptiles the range is from 10^{-8} to 10^{-6} M IAA, while in light-grown pea epicotyl sections it is 3×10^{-7} M to 10^{-4} M (22). This initial rate is independent of the presence of any additional solute (46), although it may require the presence of K^+ in the apoplastic solution (6).

Auxin-induced cell enlargement is an energy-

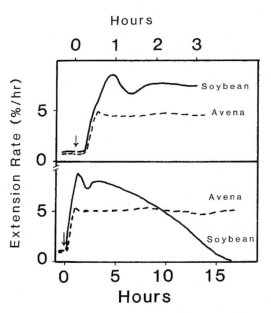

Figure 2. Comparison of the growth kinetics for *Avena* coleoptile and soybean hypocotyl sections, incubated with IAA (10 μM) and sucrose (2%). The growth rate is enhanced after a lag of 10-12 minutes for both. The rate for *Avena* sections reaches a maximum after 30-45 minutes and then remains nearly constant for 18 hours. The rate for soybean hypocotyls increases to a first maximum after 45-60 minutes, then declines before climbing to a second maximum, after which it steadily falls over the next 16 hours as turgor falls due to dilution of the osmotic solutes during growth.

requiring process. All inhibitors of ATP synthesis (e.g. KCN, 2,4-dinitrophenol, azide) or ATPase activity (vanadate, N,N'-dicyclohexyl-carbodiimide [DCCD], diethylstibe-sterol [DES]) block auxin-induced growth within minutes (22). These data indicate that the energy of ATP drives some critical system, and that an ATPase must be involved. Inhibitors of protein synthesis also inhibit auxin-induced growth within minutes after they have inhibited protein synthesis (22).

Another requirement for auxin action is cell turgor in excess of Y. Addition of osmoticum sufficient to reduce $(P-Y)$ to zero inhibits elongation by eliminating the driving force for wall extension, but it also largely blocks the ability of auxin to change m as well. This is shown (Fig. 3) by the fact that when turgor is restored after a period of time with auxin at reduced turgor, there is no burst of growth, as would be expected if auxin had been acting normally on the cellular parameters during the period of low turgor (42).

It must be remembered that auxin-induced growth is a tissue response; because the cells are linked together by their common cell walls, all cells must elongate or not elongate together. Individual cells, however, may differ in their response to auxin. In dicot stem sections, the outer cell layers (epidermis and collenchyma) appear to be the major targets of auxin (33), although the inner cells can respond to auxin under certain conditions. Studies on the cellular nature of auxin-induced cell elongation in dicots should focus on these outer cell layers. In coleoptiles, on the other hand, both epidermal and mesophyll cells are capable of responding to auxin (13).

In conclusion, then, we can say that the initiation of auxin-induced growth requires the continued presence of auxin, a continued supply of ATP and active ATPases, protein synthesis and turgor in excess of Y. Any mechanism that explains how auxin induces cell

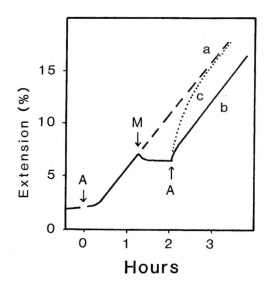

Figure 3. Demonstration that turgor is required for auxin-induced wall loosening in *Avena* coleoptile sections (lack of stored growth). Sections were treated in water with IAA (10 µM) at A, then transferred to IAA + 0.2 M mannitol at M. Curve **a** is a projection of the growth if the section had remained in A. Upon return to A, rapid growth resumes (**b**). If auxin-induced wall loosening had occurred during the period of low turgor, the extension would have followed curve **c**, since the growth potential would have been stored up. Adapted from (42).

elongation must take into account these requirements. In addition, it must explain the 8-10 minute lag that always occurs.

THE INITIATION OF CELL ENLARGEMENT BY AUXIN

The cellular parameters

Auxin must initiate cell elongation by changing one or more of the cellular growth parameters. Which one(s)? Direct measurement of cell turgor, P, using a micro-pressure probe has shown that auxin causes no increase in P in pea stem cells (19). The situation with the wall yield threshold is less clear. With oat coleoptile sections auxin does not cause any change in Y over a period of 90 minutes (10). On the other hand, in *Vigna* hypocotyl sections, P is only slightly greater than Y in the absence of auxin, but when given auxin, there is a decline in Y so that $P-Y$ becomes substantial (36). This decline in Y is mediated by two proteins called *yieldins* (36). Addition of yieldins to *Vigna* epicotyl walls whose endogenous yieldins have been inactivated causes a large decrease in Y, as measured in an *in vitro* extensions assay (Fig. 4). The role of the yieldins in controlling Y in other tissues is not yet known.

Measurement of wall extensibility (m) by any of several techniques including Instron stress-strain analysis, stress-relaxation, creep, or a pressure block technique always has given the same answer; auxin causes a large and rapid increase in wall extensibility (7, 17). Any condition that blocks auxin-induced elongation also blocks the auxin-induced increase in m.

Wall extensibility (m) is not a simple parameter. It consists of two interconnected processes (7, 12, 17). The first is a biochemical cleavage of load-bearing bonds in the wall. Each time such a bond is cleaved, the wall can undergo a limited amount of viscoelastic extension. Wall extensibility is

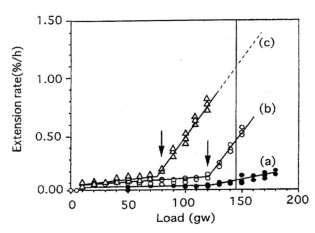

Figure 4. The role of *yieldins* in controlling the wall yield threshold and cell elongation. Glycerinated hollow cowpea hypocotyl sections were heat-treated to inactivate endogenous yieldins, then placed under 10-180 g applied tension and the extension rate was measured. Heat-treated walls incubated at pH 6.2 (**a**) or 4.0 (**b**). Walls reconstituted with recombinant yieldin incubated at pH 4.0 (**c**). Arrows indicate value of Y. Vertical line at 145 g indicates tension in the walls at full turgor. The distance between this line and the arrows indicates $P-Y$. From (36).

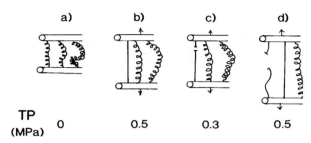

| TP (MPa) | a) 0 | b) 0.5 | c) 0.3 | d) 0.5 |

Figure 5. Model of wall extension. In a cell at incipient plasmolysis (**a**), cellulose microfibrils are crosslinked by polymers which are not under tension or elastically extended. Water uptake results in elastic stretching of the crosslinks until the wall stress causes turgor to rise sufficiently to bring the cell to equilibrium with the apoplastic solution (**b**). The crosslinks are not under equal tension. Cleavage of the crosslink under the most tension (**c**) results in a reduction in stress in the wall and therefore turgor. This also converts the elastic extension into irreversible extension. The cell then takes up water again, resulting in additional elastic extension until water equilibrium is again achieved (**d**).

the product of the rate of bond cleavage times the amount of each viscoelastic extension event. The rate of bond cleavage is dependent on the metabolism of the cell, while the amount of viscoelastic extension will be affected by the amount of P-Y.

The following picture has emerged as to how a cell elongates (Fig. 5). The walls are extended elastically by the force of turgor until the turgor is sufficient to reduce $\Delta \Psi$ to zero. The wall is kept from further extension by load-bearing bonds, namely by crosslinks between wall polymers or by entanglements between the polymers. The first step in cell elongation is breakage of load-bearing bonds, with a consequent rearrangement of wall polymers. This reduces tension in the wall and cell turgor, permitting additional water uptake and thus additional elastic extension. In essence, elastic extension is converted to irreversible extension, and elastic extension is then regenerated; this is a form of viscoelastic extension.

Two major questions arise from this picture. First, how does auxin bring about cleavage of load-bearing bonds? Secondly, what bonds are cleaved? Each of these questions will be discussed in turn.

The Wall-loosening Factor Concept

In order to induce cell elongation, auxin must bind to a receptor, located at the plasma membrane or within the cell (Chapter D1). But extension requires wall loosening, which occurs outside the cell. This means that there must be some communication between the cell and the cell wall, when the cell is stimulated by auxin. This must occur via one or more chemicals, which can be called *wall-loosening factors* (WLF, Fig. 6).

To date, only one WLF has been positively identified; namely protons. In 1970-71, Hager et al. (27) and Rayle and Cleland (41) independently suggested that when coleoptile or stem cells are stimulated by auxin, they excrete protons into the apoplastic solution, where the lowered pH activates wall loosening enzymes (Fig. 6). This "acid-growth theory" can be tested

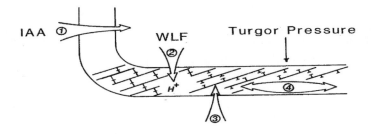

Figure 6. The wall-loosening factor (WLF) concept and the acid-growth theory. Auxin enters the cell (1) and interacts with a receptor. A WLF is then exported to the wall (2), where it induces wall loosening (3). In acid-growth, the WLF is H^+, and the resulting lowered apoplastic pH activates wall polysaccharidases which cleave load-bearing bonds in the wall (3), permitting turgor-driven wall expansion (4).

by means of four predictions. If it is correct, it should be possible to show each of the following. First, auxin causes growing cells to excrete protons. As shown in Fig. 7, peeled *Avena* coleoptile sections excrete protons in response to auxin, with a lag comparable to the lag to induce rapid elongation. As predicted, the rate of elongation in response to auxin is proportional to the rate of proton excretion (Fig. 8). The second prediction is that addition of acid to tissues should substitute for auxin and induce rapid cell enlargement, as long as the acid can penetrate into the walls. With the

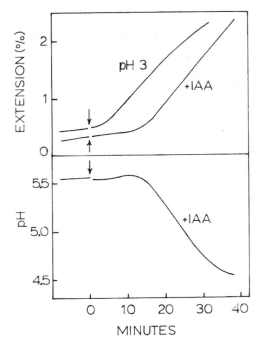

Figure 7. Demonstration that auxin causes H^+-excretion and elongation with similar kinetics, and that exogenous acid can induce elongation comparable to that of auxin. Upper panel: *Avena* coleoptile sections with intact cuticles were incubated in 10 mM phosphate buffer, pH 6.0. At the arrows IAA (10 µM) added, or solution changed to pH 3.0 phosphate buffer. A similar rate of extension would be induced by a pH 5.0 buffer if the cuticle was removed. Lower panel, pH measured on the surface of peeled *Avena* coleoptile sections. IAA (10µM) added at arrow. Note the similar 15 min lag for auxin-induced elongation and H^+-excretion.

cuticle still intact, a pH 3 buffer induces elongation of *Avena* coleoptile sections comparable to that induced by auxin (Fig. 7); if the cuticle is removed, a pH 5.0 buffer will suffice (data not shown). Third, neutral buffers infiltrated into the walls should prevent auxin-induced growth by preventing the decline in wall pH. Finally, any other agent that induces proton excretion, such as the fungal toxin fusicoccin, should also induce rapid cell elongation. This is demonstrated in Fig. 9. All four predictions have been tested for only a few tissues, but in each case, the predictions have been confirmed qualitatively (43), although doubts have persisted about some details (30).

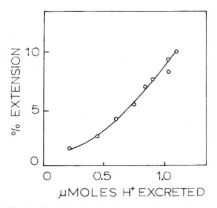

Figure 8. Correlation between growth rate and H^+ excretion for peeled *Avena* coleoptiles given varying amounts of auxin. Sections incubated for 150 min in 5×10^{-9} to 5×10^{-7} M IAA in 1 mM K-phosphate buffer, pH 6.0. (10)

Recently another very different WLF has been suggested (46); namely, hydroxyl radicals (•OH). It is proposed that these radicals are produced in the wall by peroxidase, using a suitable reductant such as NADH. The ability of such radicals to cause cleavage of wall bonds has been demonstrated, but only in the presence of 1 mM Fe^{2+} or Cu^{2+}, conditions which seem unlikely to occur in the apoplast. Nevertheless, the possible role of hydroxyl radicals in cell elongation deserves far more attention.

While acid-induced wall loosening may explain the initiation of growth, it is not sufficient to explain long-term auxin-induced cell elongation (14). There must be other WLFs, but whether they are excreted wall-loosening enzymes, substrates for wall synthesis or some other agent is uncertain.

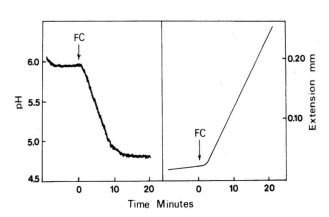

Figure 9. Effect of the fungal toxin fusicoccin (FC) on extension and H^+-excretion of *Avena* coleoptile sections. The pH was measured at the surface of peeled sections. FC (10 µM) was added at the arrow. Extension was measured on unpeeled sections.

The Mechanism of Auxin-induced Proton Excretion

Auxin-induced acidification of the wall is due to excretion of protons, not organic acids. Hager et al. (27) originally suggested that auxin activates a proton ATPase located in the plasma membrane (PM). Such an enzyme exists in the PM of all plant cells, causing an electrogenic export of protons into the apoplastic solution (28). The fact that ATP is required for auxin-induced proton excretion, and that the H^+-excretion is blocked by inhibitors of ATPase activity strongly supports this idea. In addition, auxin can cause the expected hyperpolarization of the membrane potential, with a time-course which matches that of the proton excretion; i.e., both have a lag of about 6-8 minutes (1).

The auxin-induced increase in proton excretion could be due to the synthesis of additional PM ATPase. Continued protein synthesis is required for auxin-induced proton excretion and growth (22), and certain genes are rapidly activated by auxin (Chapter D1). Hager et al. (26) reported that the amount of ATPase protein in the plasma membranes of *Zea mays* coleoptiles, as determined by antibody binding, increased rapidly, starting about 5 minutes after addition of auxin, and nearly doubled by 40 minutes. After addition of cycloheximide, the ATPase protein level in the plasma membrane decreased to the control level within an hour. The PM ATPase, then, might be the labile protein that must be synthesized following addition of auxin. There have been difficulties in repeating these results, but Rober-Kleber et al. (45) have shown that auxin enhances the PM H^+-ATPase in wheat embryos.

A second possibility is that auxin directly enhances the activity of preexisting PM ATPases. For example, both the ATPase activity and the ATP-driven proton transport of plasma membrane vesicles isolated from tobacco leaves was enhanced up to 50% by auxin (28). Similar studies have apparently not been conducted with PM vesicles from stem or coleoptile tissues. In addition, auxin does not bind directly to the PM ATPase (20), making a direct activation of the ATPase unlikely (for a discussion of auxin receptors see Chapter D1)

A third possibility is that auxin indirectly activates proton excretion. After auxin binds to its receptor, a cascade of events can be initiated that ultimately leads to enhanced proton excretion. A number of possible components of the signal cascade have been suggested (Chapter D1). These include IP_3 (L-α-phosphatidylinositol), G-proteins, protein kinases and phospholipases, but as yet none of these factors has been directly connected to the increase in proton excretion. Genetic mutants have identified a number of genes that are involved in auxin action (21), some of which are rapidly induced by auxin and others are essential for at least some auxin responses (Chapter D1). As of yet, none of these genes has been directly linked to auxin-induced proton excretion.

The increase in proton excretion might occur via an activation of the PM ATPase by a decrease in cytoplasmic pH (pH_c). The pH optimum of the PM ATPase is 6.5 (28), but the pH_c of most cells is 7.3-7.6, a pH at which the

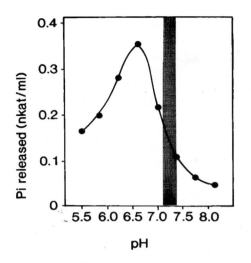

Figure 10. Effect of pH_c on the PM ATPase activity from maize coleoptiles. Plasma membrane vesicles were purified and ATPase activity was assayed in solutions of pH between 5.5 and 8.1. Hatched region denotes the proposed pH_c change that occurs in response to auxin; the right margin is the pH in the absence of auxin, while the left margin is the pH after auxin. A decrease in pH_c of 0.2 units results in a doubling in ATPase activity.

ATPase activity is definitely suboptimal. An auxin-induced decrease in pH_c of 0.2 units has been recorded by two different methods (24, 25); such a decrease in pH should cause a large increase in ATPase activity (Fig. 10). The mechanism of this cytoplasmic acidification is not known; the protons could come from the vacuole via an antiporter, or from the apoplast via a symporter.

Alternatively, auxin might activate phospholipase A_2, resulting in the formation of lysophospholipids (LPL) in the PM. LPLs have been shown to increase the activity of the PM ATPase, and alter its pH profile so that the activity at normal cellular pHs is greatly increased (37).

The activity of the PM ATPase could also be indirectly controlled by the rate of entry of K^+ from the apoplast into the cytoplasm. As the ATPase exports H^+, there must be a comparable uptake of a cation or efflux of an anion to maintain electroneutrality. Phillippar et al. (39) have reported that in maize coleoptiles auxin induces an increase in the density of the K^+ channel ZMK1 in the PM, starting about 10 minutes after addition of auxin. However, in oat coleoptile sections the promotion of K^+ uptake lags behind the enhanced proton excretion rather significantly (9).

Finally, Ray (40) suggested that proton excretion does not directly involve the PM ATPase, but, instead, occurs via secretion of acidic vesicle (the bucket-brigade idea). Recent evidence that the auxin efflux carrier cycles between the PM and some internal region (Chapter E1), raises the possibility that these vesicles might be carriers of protons.

The Mechanism of Auxin-induced Wall Loosening

A result of the export of wall-loosening factors in response to auxin, whether they be H^+ or something else, is the cleavage of load-bearing cell-wall crosslinks. The identity of these crosslinks has been a matter of some controversy, as has the identity of the wall loosening enzymes, the proteins that cause the cleavage of the load-bearing bonds.

The wall is a complex mixture of polymers, with at least three distinct networks coexisting; cellulose crosslinked by hemicelluloses, pectic chains crosslinked by calcium, and structural proteins (4). It is unlikely that calcium crosslinks between pectic chains are major load-bearing crosslinks, as removal of up to 95% of soybean hypocotyl cell wall calcium with Quin-2 (2-[(2-bis-[carboxymethyl]-amino-5-methylphenoxy)methyl]-6-methoxy-8-bis [carboxymethyl]aminoquinoline) resulted in little loosening of these cell walls (48). Likewise, treatment of isolated walls with proteases failed to cause measurable wall loosening (42). Most of the attention has focused on the hemicellulose components of the wall as the site of auxin-induced wall loosening. In dicot walls the principal hemicellulose is xyloglucan (XG), while in monocot walls a (1-3,1-4)-β-glucan and an arabinoxylan are the predominant hemicelluloses (4). The evidence suggests that there are proteins/enzymes, called wall loosening proteins (WLP), that cleave load-bearing bonds in the hemicelluloses (Chapter C4).

One approach to identifying possible wall loosening proteins has been to use an *in vitro* wall extension assay (16, 17). In this assay cell walls are isolated, any existing wall proteins are inactivated or removed, and the walls are placed under applied tension. If the added protein is a WLP analogous to the endogenous WLP it should cause the walls to extend. The WLP involved in the initial phase of auxin-induced cell elongation should only be active at acidic pHs. It should cause no wall loosening in the absence of applied tension, since wall loosening only occurs when P exceeds Y. And the kinetics of extension should mirror the *in vivo* auxin growth response.

A number of enzymes and proteins have been tested in this *in vitro* wall extension assay. One, xyloglucan endotransglucosylase/hydrolase (XTH) has failed to cause wall extension (35). Three others, cellulase, an endo-1.4-β-glucanase (50) and a xyloglucan hydrolase (29) all cause wall extension, *in vitro*, at acidic pHs, but all are active on walls that are not under tension. Only one group of proteins, the expansins (18), has the ability to loosen walls *in vitro* only when the walls are under tension and at an acidic pH. The expansin story will be covered in much greater detail in chapter C4

THE MAINTENANCE OF AUXIN-INDUCED GROWTH

An analysis of long-term auxin-induced growth shows that it is more complex than the initial growth response. While the primary effect of auxin during the initial growth response may be to cause wall acidification, during the prolonged growth other unknown wall-loosening factors appear to be involved (14). For example, while the pH optimum for loosening of oat coleoptile walls is <4.5 in the first hour, it is around 5.5 in the 2-10 hour period. This might indicate a switch from the use of α- to β-expansins for wall loosening (Chapter C4). In addition, at least two additional processes are required for prolonged growth. The first is osmoregulation. As the cells start to enlarge and take up water, the dilution of the osmotic solutes will reduce the effective turgor unless additional solutes are taken up or are manufactured in the cells. When *Avena* coleoptile sections are incubated with auxin but without absorbable solutes, the growth rate declines after the first hour in parallel with the decline in osmotic concentration. In the presence of absorbable solutes such as sucrose or KCl, both the growth rate and the osmotic concentration remain nearly constant for hours (47). The enhanced rate of uptake of solutes into *Avena* coleoptile sections in the presence of auxin is actually in response to the cell enlargement rather than to auxin itself (47). However, in intact stems auxin may play a direct role in facilitating movement of solutes through the phloem to the growing zone (38).

A second important process is the maintenance of the ability of cell walls to undergo auxin-induced wall loosening. When cell walls are isolated by freezing/thawing tissues, then placed under tension and given acidic solutions, they undergo acid-mediated wall extension (41, 42). The capacity for this *in vitro* extension depends on the past history of the tissue (Fig. 11). If coleoptile or hypocotyl sections are incubated in water, this capacity is slowly lost (11). On the other hand, if sections are given auxin, this capacity increases for at least 6-8 hours (11). It would appear that one of the effects of auxin is to regenerate the capacity of walls to undergo this acid-mediated extension. This might involve renewal of the wall-loosening proteins. Alternatively, it might involve synthesis of new cell wall polymers that are susceptible to the wall-loosening enzymes. Synthesis of new wall cannot lead directly to wall loosening, but if wall components are intercalated in after cleavage of load-bearing bonds, it may facilitate further cleavage. The orientation of the wall polysaccharides is another important factor. In order for walls to extend the cellulose must be in a transverse orientation. In maize coleoptiles, auxin causes the newly synthesized cellulose microfibrils in the outer epidermal wall, which are longitudinal in the absence of auxin, to assume a transverse orientation (3).

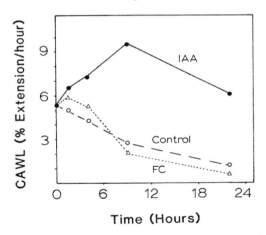

Figure 11. The effect of auxin on the capacity of *Avena* coleoptile cell walls to undergo loosening in response to acid (CAWL) depends on the past history of the tissue. Sections were incubated 0-20 hours in phosphate buffer without (control) or with 10 μM IAA or fusicoccin (FC). The sections were then frozen-thawed, and the rate of extension in response to 20 g tension at pH 3.0 was determined, one hour after addition of acid. Auxin, but not fusicoccin has increased the capacity (CAWL). Adapted from (11).

THE ROLE OF AUXIN IN THE CONTROL OF CELL ELONGATION IN INTACT PLANTS

The Control of Cell Elongation in Stems

Isolated sections taken from the elongation zone of stems can elongate at rates comparable to the elongation rate of intact stems when given auxin, but to a more limited extent in response to either GA or BR alone, and often not at all. Does this mean that auxin is the main hormonal factor controlling cell elongation in stems? Studies with a variety of genetic dwarfs have convincingly shown that both GA and BR play major roles in the control of stem cell elongation; these mutants are discussed in detail in chapter B7. But auxin is also a major factor controlling stem cell elongation. There are three main pieces of evidence to support this claim. First, direct and continued application of auxin to either dwarf or tall pea stems (49) results in a rapid and prolonged promotion of elongation (Fig. 12). Second, a ring of a polar auxin transport inhibitor such as 2,3,5-triiodobenzoic acid (TIBA) placed at the upper end of the elongation region of a pea stem blocks cell elongation below the ring (34). Finally, the rapid asymmetric growth response that occurs during gravitropic curvature of stems has been convincingly shown to be due, primarily, to a change in the auxin concentration on the two sides of the stem (5).

So what is the relationship between auxin, GA and BR in the control of

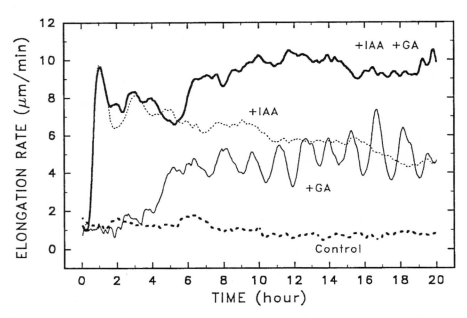

Figure 12. Effects of exogenous IAA and/or GA on stem elongation of dwarf *le* peas (which are deficient in endogenous GAs). IAA (0.2 mM) was added continuously via a wick, while GA_3 (35 µg) was added as a drop to an immature leaf. Note that growth is promoted much more rapidly by IAA than by GA_3. From (49).

stem cell elongation? The rapidity of the altered growth responses that occur during gravitropism or phototropism is consistent with a control of growth by auxin but not GA or BR, since the kinetics of stem growth promoted by either GA or BR is too slow to explain these tropic responses. The fact that in the tropic responses there is an increase in growth rate on one side of the stem and a decreases in the growth rate on the other side suggests that auxin is normally suboptimal, and that increases and decreases in auxin on the two sides are responsible for the change in the growth rate.

One possibility is that BR and GA are permissive factors, rather than regulatory factors in stem cell elongation. This means that they must be present for auxin to promote cell elongation, but that variations in the rate of cell elongation along the stem are not normally due to variations in concentrations of these factors. There is evidence that in BR-deficient stems the yield threshold is increased so that there is insufficient P-Y to allow wall loosening to occur (2). Likewise, GA may be required to maintain the stem cells in a state that allows them to respond to auxin.

Another possibility is that in some stem cells the action of auxin is to promote the synthesis of GA_1, which, in turn, controls the rate of cell elongation. Evidence to support this idea is described in chapter B7.

It is also possible that GA and auxin independently promote stem cell elongation. This possibility is strongly supported by the data of Yang et al. (49), who showed that added auxin and GA promote the growth of pea stems in different areas. As stem cells start to elongate, their rate of elongation may by primarily controlled by GA, and auxin may play a secondary, but essential role of promoting GA_1 biosynthesis. But as the cells enlarge and mature, the control of elongation may shift to a control entirely by auxin. BR may be a permissive factor that is essential for cell elongation in both immature and mature stem cells.

The Control of Cell Elongation in Roots

The role of auxin in cell elongation in roots is far less clear than it is in stems or coleoptiles. Roots contain two distinct cell elongation zones; the distal elongation zone (DEZ), where cells first begin to elongate, followed by the central elongation zone (CEZ), where the majority of the elongation occurs (23). If auxin is added to intact roots, at very low concentrations (e.g. 10^{-10} M) a small promotion of growth can occur, but at higher concentrations (e.g. 10^{-6} M) elongation is strongly inhibited. This inhibition is largely restricted to the CEZ. It occurs with a lag of about 15 minutes, and involves an alkalinization of the apoplast (23, 32). There is conflicting evidence as to whether the growth inhibition is due to this rise in apoplastic pH, or to some other unrelated cause (15, 32).

But does endogenous auxin act to slow down root growth? The fact that a lateral gradient of auxin is rapidly established in gravitroping roots (39) with the higher auxin level on the slower growing lower side, and that the polar auxin transport inhibitor NPA blocks gravitropism suggests that endogenous auxins are acting to inhibit root cell elongation. But the fact that root growth, in the presence of sufficient NPA to inhibit gravitropism is not increased (44), raises doubts about the role of auxin in controlling cell elongation in vertical roots.

References

1. Bates GW, Goldsmith MHM (1983) Rapid response of the plasmamembrane potential in oat coleoptiles to auxin and other weak acids. Planta 159: 231- 237
2. Behringer FJ, Cosgrove DJ, Reid JB, Davies PJ (1990) Physical basis for altered stem elongation rates in internode length mutants of *Pisum*. 94: 166-173
3. Bergfeld R, Speth V, Schopfer P (1988) Reorientation of microfibrils and microtubules at the outer epidermal wall of maize coleoptiles during auxin-mediated growth. Bot Acta 101: 31-41
4. Carpita NC, Gibeaut DM (1993) Structural models of primary cell walls in flowering plants: consistency of molecular structure with the physical properties of walls during growth. Plant J. 3: 1-30
5. Chen R, Rosen E, Masson PH (1999) Gravitropism in higher plants. Plant Physiol 120: 343-350
6. Claussen M, Lüthen H, Blatt M, Böttger M (1997) Auxin-induced growth and its linkage to potassium channels. Planta 201: 227-234
7. Cleland RE (1971) Cell wall extension. Annu Rev Plant Physiol 22: 197- 222
8. Cleland RE (1972) The dosage response curve for auxin-induced cell elongation: a reevaluation. Planta 104: 1-9
9. Cleland RE (1977) Rapid stimulation of K^+-H^+ exchange by a plant growth hormone. Biochem Biophys Res Comm 69: 333-338
10. Cleland RE (1977) The control of cell enlargement. *In* DH Jenning *ed*, Integration of activity in the higher plant, Cambridge Press, Cambridge, pp 101-115
11. Cleland RE (1983) The capacity for acid-induced wall loosening as a factor in the control of *Avena* coleoptile cell elongation. J Exp Bot 34: 676-680
12. Cleland RE (1987) The mechanism of wall loosening and wall extension. *In* DJ Cosgrove, DP Knievel, *eds*, Physiology of Cell Expansion during Plant Growth. Amer Soc Plant Physiol, Rockville, pp 18-27
13. Cleland RE (1991) The outer epidermis of *Avena* and maize coleoptiles is not a unique target for auxin in elongation growth. Planta 186: 75-80
14. Cleland RE (1992) Auxin-induced growth of *Avena* coleoptiles involves two mechanisms with different pH optima. Plant Physiol 99: 1556-1561
15. Cleland RE (2002) The role of the apoplastic pH in cell wall extension and cell enlargement. *In* Z. Rengel, *ed*, Handbook of Plant Growth. pH as the Master Variable. Dekker, New York, pp 131-148
16. Cleland RE, Cosgrove DJ, Tepfer M (1987) Long-term acid-induced wall extension in an *in-vitro* system. Planta 170: 379-385
17. Cosgrove DJ (1993) Wall extensibility: its nature, measurement and relationship to plant cell growth. New Phytol 124: 1-23
18. Cosgrove DJ (1997) Relaxation in a high-stress environment: the molecular basis of extensible cell walls and cell enlargement. Plant Cell 9: 1031-1041
19. Cosgrove DJ, Cleland RE (1983) Osmotic properties of pea internodes in relation to growth and auxin action. Plant Physiol 72: 332-338

20. Cross JW, Briggs WR, Dohrmann UC, Ray PM (1978) Auxin receptors of maize coleoptile membranes do not have ATPase activity. Plant Physiol 61: 581-584
21. Dharmasiri S, Estelle M (2002) The role of regulated protein degredation in auxin responses. Plant Mol Biol 49: 401-409
22. Evans ML (1985) The action of auxin on plant cell elongation. Critical Rev Plant Sci 2: 317-365
23. Evans ML, Ishikawa H, Estelle MA (1994) Responses of Arabidopsis roots to auxin studied with high termporal resolution – comparison of wild-type and auxin-response mutants. Planta 194: 215-222
24. Felle H (1988) Auxin causes oscillations of cytosolic free calcium and pH in *Zea mays* coleoptiles. Planta 174: 495-499
25. Gehring CA, Irving HR, Parish RW (1990) Effects of auxin and abscisic acid on cytosolic calcium and pH in plant cells. Proc Natl Acad Sci US 87: 9645-9649
26. Hager A, Debus G, Edel H-G, Stransky H., Serrano R. (1991) Auxin induces exocytosis and the rapid synthesis of a high-turnover pool of plasma-membrane H^+-ATPase. Planta 185: 527-537
27. Hager A, Menzel H, Krauss A (1971) Versuche und Hypothese zur Primarwirkung des Auxins beim Streckungswachstum. Planta 100: 47-75
28. Jahn T, Palmgren MG (2002) H^+-ATPases in the plasma membrane: physiology and molecular biology. *In* Z Rengel, *ed*, Handbook of Plant Growth. pH as the Master Variable. Dekker, New York, pp 1-22
29. Kaku T, Tabuchi A, Wakabayashi K, Kamisaka S, Hoson T (2002) Action of xyloglucan hydrolase within the native cell wall architecture and its effect on cell wall extensibility in Azuki Bean epicotyls. Plant Cell Physiol 43: 21-26
30. Kutschera U, Schopfer P (1985) Evidence against the acid growth theory of auxin action. Planta 163: 483-493.
31. Lockhart JA (1965) An analysis of irreversible plant cell elongation. J Theor Biol 8: 264-275
32. Lüthen H, Böttger M (1993) The role of protons in the auxin-induced root growth inhibition – a critical reexamination. Bot Acta 106: 58-63
33. Masuda Y, Yamamoto R (1972) Control of auxin-induced stem elongation by the epidermis. Physiol Plant 27: 109-115
34. McKay MJ, Ross JJ, Lawrence NL, Cramp RE, Beveridge CA, Reid JB (1994) Control of internode length in *Pisum sativum*. Further evidence for the involvement of indole-3-acetic acid. Plant Physiol 106: 1521-1526
35. McQueen-Mason SJ, Fry SC, Durachko DM, Cosgrove DJ (1993) The relationship between xyloglucan endotransglycosylase and in-vitro cell wall extension in cucumber hypocotyls. Planta 190: 327-331
36. Okamoto-Nakazato A (2002) A brief note on the study of yieldin, a wall-bound protein that regulates the yield threshold of the cell wall. J Plant Res 115: 309-313
37. Palmgren MG, Sommarin M, Ulvskov P, Jorgensen PL (1988) Modulation of plasma membrane H^+-ATPase from oat roots by lysophosphotidylcholine, free fatty acids and phospholipase A_2. Physiol Plant 74: 11-19
38. Patrick JW (1982) Hormone control of assimilate transport. *In* PF Wareing *ed*, Plant Growth Substances 1982, Academic Press, New York, pp 669-678
39. Philippar K, Fuchs I, Lüthen H, Hoth S, Bauer CS, Haga K, Thiel G, Ljung K, Sandberg G, Böttger M, Becker D, Hedrich R (1999) Auxin-induced K^+ channel expression represents an essential step in coleoptile growth and gravitropism. Proc Nat Acad Sci US 96: 12186-12191

40. Ray PM (1977) Auxin-binding sites of maize coleoptiles are localized on membranes of the endoplasmic reticulum. Plant Physiol 59: 594-599.
41. Rayle DL, Cleland RE (1970) Enhancement of wall loosening and elongation by acid solutions. Plant Physiol 46: 250-253
42. Rayle DL, Cleland RE (1972) The in-vitro acid-growth response: relation to in-vivo growth responses and auxin action. Planta 104: 282-296
43. Rayle DL, Cleland RE (1992) The acid growth theory of auxin-induced cell elongation is alive and well. Plant Physiol 99: 1271-1274
44. Rashotte AM, Brady SR, Reed RC, Ante SJ, Muday JK (2000) Basipetal auxin transport is required for gravitropism in roots of Arabidopsis Plant Physiol 2000 122: 481-490
45. Rober-Kleber N, Albrechtová JTP, Fleig S, Huck N, Michalke W, Wagner E, Speth V, Neuhaus G, Fischer-Iglesias C (2003) Plasma membrane H^+-ATPase is involved in auxin-mediated cell elongation during wheat embryo development. Plant Physiol 131: 1302-1312
46. Schopfer P, Liszkay A, Bechtold M, Frahry G, Wagner A (2000) Evidence that hydroxyl radicals mediate auxin-induced extension growth. Planta 214: 8921-828
47. Stevenson TT, Cleland RE (1981) Osmoregulation in the *Avena* coleoptile in relation to growth. Plant Physiol 67: 749-753
48. Virk SS, Cleland RE (1988) Calcium and the mechanical properties of soybean hypocotyl cell walls: possible role of calcium and protons in wall loosening. Planta 176: 60-67
49. Yang T, Davies PJ, Reid JB (1996) Genetic dissection of the relative roles of auxin and gibberellin in the regulation of stem elongation in intact light-grown peas. Plant Physiol 110: 1029-1034
50. Yuan S, Wu Y, Cosgrove DJ (2001) A fungal endoglucanase with plant cell wall extension activity. Plant Physiol 127: 324-333

C2 Gibberellin Action in Germinated Cereal Grains

Fiona Woodger[1], John V. Jacobsen[2], and Frank Gubler[2]
[1]Australian National University, Canberra, ACT 2601, Australia and [2]CSIRO Plant Industry, PO Box 1600, Canberra, ACT 2601, Australia.
E-mail: Frank.Gubler@csiro.au

INTRODUCTION

The cereal aleurone layer is a secretory tissue that surrounds the starchy endosperm and the embryo. The study of the response of cereal aleurone to gibberellin and abscisic acid (GA and ABA, respectively[1]) has made a significant contribution to our understanding of GA action in plant cells, especially as it relates to the control of gene expression, and also to the function of the endosperm following germination. While much of the work has been carried out using isolated aleurone from barley, it seems that the principles governing this system apply also to other cereal grains.

The effect of incubating isolated aleurone layers in media containing specified concentrations of GA_3 and/or ABA has been extensively studied as a model for hormone action in plants. The main observation is that GA_3 stimulates aleurone cells to secrete a range of hydrolytic enzymes, the major one being α-amylase (Fig. 1). In intact grain these enzymes mobilise stored endosperm reserves that provide the growing seedling with a supply of fixed carbon, reduced nitrogen and other nutrients. The other key regulator of aleurone function, ABA, prevents the action of GA_3 if present in excess and also induces its own set of proteins in isolated aleurone. The aleurone layer is a useful system in which to study GA-response pathways because it is a uniform tissue with well-defined target responses to its two key regulators, GA and ABA. The molecular analysis of the hormonal control of gene expression is also facilitated by the fact that aleurone layers are amenable to manipulations such as protoplast isolation, microinjection and transient expression analyses.

---

[1] Abbreviations: GTP, Guanosine 5'-triphosphate; GDP, Guanosine 5'-diphosphate; AGP, Arabinogalactan protein; CaM, Calmodulin; cGMP, Cyclic guanosine 3'monophosphate; GARE, Gibberellin response element; GARC, Gibberellin response complex.

Gibberellin action in germinated cereal grains

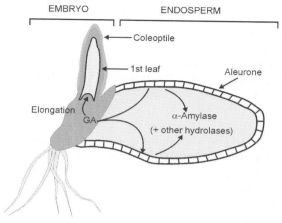

Figure 1. Diagram illustrating some of the principal features associated with reserve mobilization in a wheat or barley grain (and probably other cereals as well) following germination. Gibberellin produced by the embryo stimulates cells of the aleurone layer to synthesise and secrete α-amylase and other hydrolases which degrade starch and other polymeric reserves in the endosperm, providing nutrients for the developing seedling.

This article is a mainly a synopsis of our understanding of GA_3-action and, to a lesser extent, ABA-action in germinating cereal grains, focusing on the aleurone layer. A major part of our attention is given to the recent progress in identifying signal transduction components and transcription factors, which regulate α-amylase gene expression in aleurone cells. Some of these components are common to GA-mediated responses in other tissues, whilst others play a more restricted role. Background information is provided in previous aleurone reviews (2, 11, 24).

EMBRYONIC CONTROL OF ENDOSPERM FUNCTION

Gibberellin Production Following Germination

The high levels of active GAs present in developing grain usually decrease during maturation and dry grain normally contains very low levels. Associated with germination, which is initiated by water uptake, GA levels increase. GA_1 is the predominant species in barley although low amounts of other GAs have been detected. GA_1 starts to accumulate rapidly in wheat and barley embryos after 24 h imbibition (Fig. 2) and this is followed soon after by an increase in GA_1 content in the endosperm of germinated grains (25, 39). However, it is of interest to note that GA_1 does not increase in dormant grain (Fig. 2).

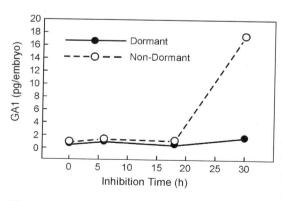

Figure 2. GA_1 content of embryos of dormant and non-dormant barley grain over a 30 h period of imbibition. The timecourse of germination of the non-dormant grains closely paralleled the increase in GA1 in the embryo (25).

222

In barley, the removal of the embryo from the endosperm in the 24 h after germination prevents the increase in GA in the endosperm and the majority of α-amylase production. This indicates that it is GA from the embryo that stimulates the production of α-amylase in the aleurone (Fig. 1). *In situ* localization studies of a key GA biosynthetic gene, GA 3β-hydroxylase, which catalyses the last step in the conversion of inactive GAs to GA_1 indicates that there are two sites of GA biosynthesis in the germinating rice embryos (Fig. 3A-H). Transcripts of two GA_3β-hydroxylase genes *(OsGAox1 and 2)* are localized in the shoot apical region and the scutellar epithelium of germinated rice embryos (27). Expression of one the genes, *OsGA3ox2,* is important for the expression of α-amylase genes in aleurone cells. Genetic studies using rice embryo mutants indicate that the scutellar epithelium cells, and not the shoot, are the major source of GA for α-amylase production in the endosperm of germinating grains. These studies are consistent with earlier studies in wheat showing scutellum is the main site for early steps in GA biosynthesis (39). The pathway by which GA leaves the scutellum and reaches the aleurone is not well understood although there is evidence that there is asymmetric transport of GA towards the apex of the scutellum from which it is released into the aleurone layer (40). GA movement in the endosperm is thought to occur by diffusion (3).

Patterns of Endosperm Breakdown Following Germination

Morphological evidence, based on scanning electron microscopy, shows that initial starch degradation in barley occurs adjacent to the scutellar epithelium (34) and proceeds symmetrically along the grain as a front towards the distal end. The front appears to move faster immediately beneath the aleurone layer on the dorsal surface of the grain. The pattern of endosperm modification broadly agrees with the pattern of expression of hydrolytic enzymes in the germinating grain. Transgenic barley grains transformed with a green fluorescent protein (GFP) reporter gene fused to a high-pI α-amylase gene promoter has allowed *in vivo* monitoring of the temporal and spatial expression of an α-amylase gene in germinating grains (36). In germinating grains transformed with this construct, GFP expression occurs first in scutellar epithelial cells (Fig. 3G) then later in the aleurone layer (Fig. 3H). GFP expression in the scutellum is short-lived, disappearing after 1 day of imbibition. In contrast, aleurone GFP expression began after 1 day in cells adjacent to the scutellum and spread along the grain towards the distal end over the next 3 days (Fig.3C-F). Similar patterns of expression have been detected for other hydrolytic enzymes in barley and rice (7, 32, 47).

It is now clear that the initiation of endosperm hydrolysis early in germination occurs adjacent to the scutellum using enzymes, perhaps specific isoforms, produced and secreted by the scutellar epithelium. Studies of numerous enzymes involved in the hydrolysis of β-glucan, starch, protein and nucleic acids in the starchy endosperm (11) show that a proportion of the

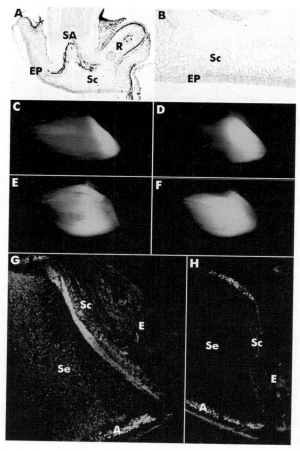

Figure 3 (Colour plate page CP6). **A-B**. Localization of OsGA$_3$Ox2 transcripts in germinating rice embryos. SA, shoot apex; (From 27) **C-H**. GFP fluorescence of grains transformed with Amy:GFP. C-F. Transgenic grains imbibed for 1, 2, 3 and 4 days, respectively. G-H. Confocal images of longitudinal sections of grains incubated for 12h (G) and 2 days (H). (From 36). E, embryo; SE, starchy endosperm; A, aleurone; R, root; Sc, scutellum; Ep, Epithelium.

enzyme activity of most or all of these enzymes originates in the scutellum. In barley, the scutellum accounts for only a minor part (5-10%) of total α-amylase. As germination progresses, enzyme production by the scutellum appears to decline and the aleurone becomes the major source. In sub-tropical cereals such as sorghum, maize, millet and rice, the scutellum might be a more significant source of α-amylase and other enzymes (1).

Although there is abundant evidence that hydrolase synthesis in aleurone is under hormonal control, the question of whether hydrolase synthesis in scutellum is similarly controlled has been more difficult to answer. In the main, the results of experiments involving embryos of barley, wheat and rice as well as the sub-tropical cereals suggest that scutellar enzyme synthesis is largely insensitive to applied GA or GA biosynthetic inhibitors (29, 27). It is possible either that hydrolase genes are not hormonally controlled in embryonic tissue, or that there are already high endogenous levels of GA. However, using a GA-deficient mutant of barley, it has been shown that high-pI α-amylase gene expression is also induced by GA in the scutellum (6). The results of this study are compelling but further work is required to clarify these conflicting results.

RESERVE DEGRADATION

During early seedling growth, the stored reserves (protein, carbohydrate, lipid, nucleic acid and mineral complexes) in the aleurone and starchy endosperm are degraded and mobilised via the scutellum to the growing seedling. The enzymes required for eventual hydrolysis and solubilization of the stored reserves (11) originate mainly in the aleurone layer, but some are also produced in the starchy endosperm and the scutellum.

Morphological changes in aleurone cells and starchy endosperm during endosperm hydrolysis have been examined mainly through light and electron microscopy. Changes in aleurone cells include vacuolation, digestion of the cell walls (Fig. 4) and ultimately cell death (9). Sub-aleurone and starchy endosperm cells are characterised by general disorganisation and cell wall and starch grain hydrolysis. Aleurone cell walls appear to be impermeable to proteins so that hydrolysis of aleurone and probably also endosperm cell walls is likely necessary for efficient release of hydrolytic enzymes from the aleurone layer. In aleurone cells α-amylase is distributed throughout the endoplasmic reticulum and Golgi, and is released initially into the cell wall (Fig. 4A and B). Figure 4C shows the "tunneling" of walls which occurs in the early stages of aleurone cell wall hydrolysis. Digestion is originally localised around plasmodesmata but ultimately involves the whole wall. The gold grains over the digested areas in the micrograph show the localisation of α-amylase in these regions. α-Amylase does not appear to exist in undigested wall in agreement with the apparent impermeability of the wall to proteins. Hydrolysis of aleurone cell walls leaves an innermost layer of wall material (IW) (Fig. 4C) which is apparently resistant to degradation. This layer could be similar to the cell wall connections that persist around plasmodesmata

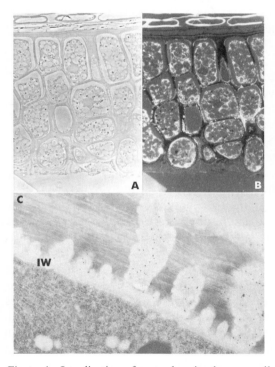

Figure 4. Localization of α-amylase in aleurone cells and in digested pockets within the cell wall of GA-treated tissue as determined by immunofluorescence (A and B) and by immunogold labelling (C). (From 14).

connecting digested aleurone cells, and may contribute to mechanical stability of the aleurone layer following cell wall digestion.

Arabinoxylans (about 80%) and (1-3,1-4)-β-glucan (about 20%) are the major cell wall polysaccharides of barley aleurone. There are five arabinoxylan degrading enzymes which are released from aleurone in response to GA_3, and of these endo-(1-4)-β-xylanase is the one most likely to be involved in the initial cell wall hydrolysis. However there is a discrepancy (11) in timing between the appearance of endoxylanase activity, which is relatively late, and the relatively early commencement of cell wall hydrolysis. This is presently unresolved. The other major cell wall polysaccharide is believed to be degraded by two endo-(1-3,1-4)-β-glucanases which have been extensively characterised.

The starch deposits of wheat and barley exist in two size classes of starch granules embedded in a protein matrix throughout the endosperm. The two glucose polymers forming the starch are amylose, a linear (1-4)-α-glucan and amylopectin, a highly branched form of (1-4)-α-glucan containing (1-6)-α-branches. Following germination of barley the small starch granules are first degraded, primarily from the surface. Large starch granules frequently suffer initial surface pitting, but hydrolysis may then occur inside the granule and proceed through a hollow shell stage.

Starch hydrolysis proceeds through the action of four different enzymes, collectively known as diastase. α-Amylase, an endohydrolase, is probably very important in the initial degradation of amylose and amylopectin within starch grains. There is evidence (33) that different isozyme groups of α-amylase (see below) have different activities on intact starch grains, namely low-pI α-amylase digests starch grains more efficiently than high-pI α-amylase, both in barley and wheat. Although it was once thought that α-amylase was the only enzyme capable of attacking whole starch grains, it is now evident that α-glucosidase, an exohydrolase liberating glucose, has the same ability and that together, the two enzymes are strongly synergistic (48). After the initial attack on starch grains by α-amylase and α-glucosidase, the other diastatic enzymes, β-amylase and limit dextrinase reduce the size of the initial products of hydrolysis. The latter, known also as debranching enzyme, hydrolyses α-1,6 links in amylopectin.

Although relatively little is known about enzymes causing storage protein hydrolysis compared to starch and β-glucan, it is clear that protein is reduced to amino acids and peptides by a number of exo- and endopeptidases (11). Carboxypeptidases appear to be the main exopeptidases involved. Aminopeptidases appear to play no part. Other reserves which survive in mature starchy endosperm in small quantities are also hydrolysed following germination (11). Nucleic acids appear to be hydrolysed by nuclease 1, ribonuclease and nucleosidases and (1-3)-β-glucan, which occurs extracellularly mainly in the sub-aleurone cells, is hydrolysed by GA-induced (1-3)-β-glucanase.

Regulation of reserve degradation

α-Amylase and α-glucosidase are synthesised in active forms in the aleurone layer following germination. β-Amylase accumulates in the starchy endosperm during development and at maturity it exists in the endosperm of the dry grain mostly in an inactive form, disulphide bonded to a component of protein bodies. The increase in β-amylase activity that follows germination results from the activity of GA-induced cysteine endopeptidases that cleave the β-amylase-protein complex, releasing active β-amylase (16). These endopeptidases are synthesised in the aleurone in response to GA_3. The situation with limit dextrinase seems to be complex. Although it is newly synthesised by aleurone cells, it becomes bound in starchy endosperm following germination and, as with β-amylase, proteolysis is required for its activation in the later stages (30).

In the early stages of germination and seedling growth, it would seem that endosperm hydrolysis is initiated in the main by hydrolases which are newly synthesised and secreted by the scutellum and aleurone. This indicates that reserve degradation is controlled by hydrolase availability during this period. However, as seedling growth progresses, the products of reserve hydrolysis accumulate in the starchy endosperm cavity up to 570 milli-osmolar and the concentrations of some of the component products are high enough to inhibit further hydrolase synthesis. Thus several days after germination, osmotic controls may come into play. It is also possible that pH within the starchy endosperm may exercise some limitation on reserve degradation. Many endosperm hydrolases have pH optima in the vicinity of pH 5, and thus would function best in an acidic environment. Such an environment is established in the endosperms of germinating barley and wheat by protons apparently released by the aleurone layer. Barley endosperm is acidified during the latter stages of development as well as following germination (35), but while mature wheat aleurone can acidify (7, 17), developing wheat aleurone does not do so (J.V. Jacobsen, unpublished). Gradients of pH in the endosperm may favour hydrolysis of reserves by providing an optimum environment for the enzymes to function or by increasing accessibility of enzymes to substrates which become more soluble at acid pH (17). However, little is known about endosperm pH and its control except that GA and ABA are involved. Reduced pH in developing barley endosperm is associated with the release of hydrogen ions (as phosphoric and citric acid) from the aleurone cells (8).

Global profiling of gene expression in barley aleurone cells has identified more than 75 genes that are upregulated after 12h treatment with GA (10). While many of these genes encode hydrolytic enzymes which are involved in cell wall and reserve hydrolysis, other genes encoding metabolic enzymes and enzymes involved in programmed cell death are also up-regulated by GA. Therefore, in aleurone cells, GA_3 appears to regulate not only the production of hydrolytic enzymes, but also the metabolic machinery.

On the basis of their response to GA_3, about 18 aleurone hydrolases have been classified into four groups (22). The first group is characterised by a rapid change in activity of the enzyme, with, presumably, little change in amount. Enzymes involved in phospholipid metabolism fall into this group, and at present there is little information concerning the mechanism of their regulation. The second group includes enzymes whose synthesis and secretion are stimulated by GA_3, such as α-amylase and protease. Many aspects of the hormonal control of hydrolase gene expression in aleurone have been studied and this topic will feature in the remainder of this chapter. The third type of hydrolase, exemplified by β-glucanase, acid phosphatase and ribonuclease, increases in activity by virtue of new synthesis in the absence of added GA_3, but also shows an additional increase in activity if GA_3 is added. Finally, there are some hydrolases, including xylopyranosidase and arabinofuranosidase, with an activity which is constant in the absence of GA_3 but which increases in the presence of GA_3. This group of hydrolases also exhibit secreted activity. For many hydrolases, the precise relationship between isozymes formed in the absence or presence of GA_3 is frequently unknown, so it is difficult to assess whether GA_3 is stimulating levels of existing isozymic forms, or causing *de novo* accumulation of new isozymes.

Much research on hormone-induced enzyme changes has been done on a total activity basis, but, as is often the case, examination of enzyme production at the isozyme level has produced interesting results. For example, studies of (1-3, 1-4)-β-glucanase (11, 32, 46) indicate that the early enzyme increase in germinating grain occurs in the scutellum and consists of only isozyme E1. By contrast, most of the later enzyme production occurs in the aleurone and consists of both E1 and E11. The study of α-amylase at the isozyme level has also been informative. Interestingly, the kinetics of production of low- and high-pI α-amylases in aleurone of germinating grains and in isolated aleurone layers are different. In whole grains, high-pI isozymes dominate early in germination while low-pI forms appear later. In isolated layers, α-amylases appear to accumulate coordinately, but whereas high-pI forms decrease, low-pI forms persist (4). Such results highlight the fact that different isozymes can be controlled differentially by GA and reveal the complexity of the aleurone response to this hormone.

Studies of the aleurone response to hormones have been dominated by up-regulation of enzymes, but some enzymes are also down-regulated (26). That GA_3 causes both up- and down-regulation of synthesis of proteins is evident from the very early studies of protein synthesis in aleurone. Studies of the molecular mechanisms of hormone action are also dominated by up-regulation (see later), and little is known about down-regulation.

GIBBERELLIN RESPONSE PATHWAYS LEADING TO HYDROLASE GENE EXPRESSION

In general terms, a cellular response pathway proceeds via a sequence of molecular events that commence with the stimulation of a high-affinity cellular receptor. The primed receptor-complex then activates or represses signaling intermediates, which in turn modulate the activity of various cellular effectors such as transcription factors. It is the activated effector molecules in signaling pathways that generate a specific cellular response, such as a modulation in gene expression. Gibberellin evokes a number of responses in the aleurone cells of germinated cereal grains through a specialised network of signaling pathways. In the following sections, the particular GA-response pathways that elicit increased expression of hydrolase genes will be examined in greater detail. The GA-regulation of genes encoding the α-amylases, which are starch-degrading enzymes, has been most intensively studied and will be featured here. Progress towards understanding these GA-signaling pathways has been made both through biochemical studies in aleurone and genetic studies involving the analysis of dwarf or 'slender' (elongated) GA-response mutants. From this work it appears that GA-signaling is constitutively repressed by negative regulators and the GA-signal operates at least in part through a mechanism of 'derepression'.

Gibberellin Perception

The majority of evidence suggests that GA is perceived at the external face of the aleurone plasma membrane (31). Membrane impermeable forms of GA strongly elicit α-amylase expression while microinjection of GA fails to stimulate such characteristic GA-responses. Also, anti-idiotypic antibodies that contain regions analogous to biologically active domains of GA cause agglutination of aleurone protoplasts, suggesting that GA ordinarily binds to the external surface of aleurone cells.

Two types of signaling pathways have been implicated in transmission of the GA-signal at the plasma membrane – G protein-signaling as well as arabinogalactan proteins are thought to be involved. The G protein heterotrimeric complexes utilise a G protein-coupled plasma membrane receptor that undergoes a conformational change after ligand binding. This stimulates GTP-GDP exchange at the α-subunit of a plasma membrane heterotrimeric G protein complex. The α-subunit then disassociates from the complex and activates downstream targets such as kinases or ion channels and is subsequently deactivated by GTP-hydrolysis. In wild oat aleurone the Mas7 molecule, which artificially stimulates GTP/GDP exchange by heterotrimeric G proteins, elicits α-amylase expression (31). In addition, a hydrolysis resistant analogue of GDP, which cannot be exchanged for GTP, inhibits GA-induction of α-amylase expression. However, the only partially

reduced GA-sensitivity of the rice *Dwarf1* mutant, which is defective in the alpha subunit of a heterotrimeric G protein, indicates this signaling mechanism may operate only indirectly in the GA-response pathway (39).
The arabinogalactan proteins (AGPs) are membrane anchored, hydroxyproline-rich proteoglycans that function as cell fate markers or signaling molecules. Inhibitors of AGPs repress GA-induced α-amylase induction in isolated barley aleurone protoplasts but not in intact aleurone layers (49). This suggests that plasma membrane localised AGPs are involved in transmission of the GA-signal at the plasma membrane. Alternatively, it is possible that AGPs could alter plasma membrane properties that influence GA-perception.

Second messengers

Second messengers are small intracellular signaling molecules such as Ca^{2+} or cAMP which function as signaling intermediates according to their localised concentration within a cell. The concentration of a number of potential second messengers is rapidly altered by the GA-stimulation of aleurone.

Ca^{2+} and calmodulin

One of the earliest events observed in GA-stimulated aleurone is a rapid increase in the concentration of cytosolic Ca^2, $[Ca^{2+}]_i$ (2). In barley aleurone protoplasts, $[Ca^{2+}]_i$ increases approximately 3-fold within 5 h of GA treatment. In wheat aleurone protoplasts, $[Ca^{2+}]_i$ increases approximately 10–fold over 30-90 min and the response is initiated within 2-5 min of GA-stimulation. The calcium driving these changes in $[Ca^{2+}]_i$ is thought to derive from the apoplast because GA does not induce increases in $[Ca^{2+}]_i$ in isolated aleurone cells in the absence of exogenous Ca^{2+}. In addition, the increase in $[Ca^{2+}]_i$ in GA-treated aleurone cells is most concentrated in the peripheral cytoplasm. Accordingly it is possible that a GA-stimulated receptor might activate an aleurone plasma membrane Ca^{2+}-transporter, but a candidate transporter has not yet been identified.

The expression of CaM, a key Ca^{2+}-transduction protein, is also rapidly induced in aleurone in response to GA (2). In barley aleurone, CaM mRNA expression increases five-fold after 1 h and a sustained increase in CaM protein expression is observed for 8 h after GA-stimulation. These changes precede increases in α-amylase mRNA expression, which shows a sustained induction from around 4 h, until 20 h, following GA-treatment. Treatment of barley aleurone protoplasts with ABA prevents the GA-induced increases in $[Ca^{2+}]_i$ and the expression of CaM.

In isolated aleurone layers, the GA-induced synthesis and secretion of a number of α-amylase isoenzymes is dependent on the provision of Ca^{2+} but α-amylase mRNA expression is induced to the same extent by GA irrespective of whether Ca^{2+} is provided in the incubation medium. In addition, the microinjection of caged Ca^{2+}-chelators into barley aleurone blocks the GA-stimulation of the α-amylase secretory pathway but not the GA-induction of α-amylase promoter activity (2). These results suggest that there are Ca^{2+}-independent and Ca^{2+}-dependent GA-signaling pathways in barley aleurone, controlling α-amylase expression, and synthesis and secretion respectively. This interpretation is consistent with the knowledge that α-amylase is a Ca^{2+}-requiring metalloprotein and the observation that secretory vesicle fusion in aleurone is stimulated by Ca^{2+} (53). Although elevated concentrations of Ca^{2+} and expression of CaM protein are not sufficient to induce α-amylase secretion, they can however block the ABA-inhibition of both α-amylase gene expression and α-amylase secretion, suggesting that ABA antagonism of GA-action might work largely through Ca^{2+} and/or CaM levels (2).

cGMP and pH

The cyclic nucleotide monophosphates are known to function as prokaryotic and eukaryotic second messengers, interacting with and activating downstream targets, such as kinases or ion channels. In GA-stimulated barley aleurone cells, the concentration of cGMP increases 3-fold within 3 h, but when cGMP synthesis is prevented by the use of a guanyl cyclase inhibitor GA stimulation fails to upregulate α-amylase synthesis (2). However cGMP analogues do not stimulate α-amylase expression suggesting that cGMP is necessary but not sufficient for the GA-induction of α-amylase expression.

In another early event in GA-stimulated aleurone, cytosolic pH (pH_i) is observed to drop by approximately 0.2 pH units after a 1 h GA treatment and, conversely, is observed to increase by a similar magnitude after a 1 h ABA treatment (18). Artificially altering the pH_i does not affect the activity of an α-amylase promoter in transfected barley aleurone protoplasts (19) indicating that pH_i changes alone are not sufficient to transmit the GA-signal to the α-amylase promoter. However both Ca^{2+} and pH_i are known to act as second messengers in the ABA-regulation of stomatal closure through the control of K^+ channels, an effect thought at least in part to occur through Ca^{2+} and H^+ regulation of protein phosphorylation. It is not yet clear what the role of pH_i might be in GA-signaling and whether there is cross talk with Ca^{2+} signaling.

Signal Transduction proteins

The transient modification of proteins by phosphorylation has a well-established role in the transmission of information in eukaryotic signal transduction pathways. Several lines of evidence implicate this type of modification in GA-signaling in aleurone (31). The treatment of wheat aleurone with the phosphatase inhibitor okadaic acid prevents a number of characteristic GA-responses including the induction in α-amylase gene expression, suggesting that phosphatases are involved in GA-regulated gene expression. The microinjection of Syntide-2, a specific inhibitor of Ca^{2+}- and CaM-dependent protein kinases, into barley aleurone protoplasts reduces the GA-induction of α-amylase mRNA expression, α-amylase secretion and cellular vacuolation, but the GA-mediated increase in $[Ca^{2+}]_i$ is unaffected. Transient expression of the abscisic-acid induced kinase, PKABA1, from wheat suppresses GA-regulated genes including α-amylase in aleurone (31). A salt- and ABA-induced gene from wheat grass resembling a serine-threonine kinase, *Esi47,* has also been found to repress the activity of a low pI α-amylase promoter in barley when transiently expressed, but the barley homologue of this gene is not yet known (45). The placement of these kinases in the GA-signaling pathway awaits identification of their immediate and downstream targets. Currently there exists little information about the involvement of other transduction proteins in the transmission of the GA-signal.

TRANSCRIPTIONAL REGULATION OF HYDROLASE GENES

GA is known to induce expression of numerous genes including those for calmodulin biosynthesis, nucleases, and starch, cell wall and protein degrading enzymes (2). The search for GA-responsive *cis*-elements and their cognate transcription factors has centred on the α-amylase genes. The α-amylase genes form a multi-gene family in cereals and the wheat and barley genes fall into the *Amy1* and *Amy2* subclasses (38). In barley and wheat the α-amylase genes are also classified on the basis of whether the encoded protein possesses a high or low isoelectric point.

cis- elements

Comparisons of the promoter regions from the family of cereal α-amylase genes reveal three highly conserved elements - a $TAACA^A/_GA$ element, a $TATCCA^C/_T$ element and a pyrimidine box (20). Through mutagenesis and the analysis of truncated promoters it has been shown that these elements contribute to GA-response complexes (GARCs) and that the $TACAA^A/_GA$ element is particularly important to the hormonal responsiveness of the α-amylase genes (31). Multimers of a 21 bp region containing the TAACAAA element from the barley *α-amy1/6-4* promoter confer GA-responsiveness to a minimal promoter, and consequently this element has been designated a GARE (gibberellin response element). In the barley and wheat low-pI α-amylase genes, an element corresponding to the Opaque-2 binding sequence (O2S) from maize (known to bind a leucine zipper transcription factor) has also been found to modulate GA-responsiveness (28, 50).

*Trans-*acting factors

The conserved *cis*-elements that form the GARCs in the hydrolase gene

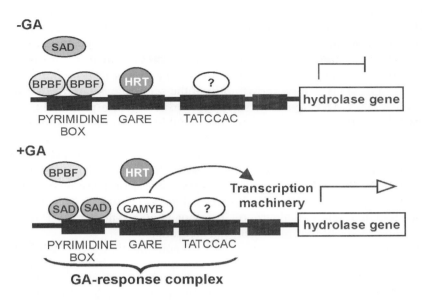

Figure 5. A generalised hydrolase GARC with SAD/BPBF and GAMYB/HRT binding sites. The pyrimidine box is the binding site for SAD/BPBF proteins while GAMYB and HRT proteins bind to the GARE

promoters incorporate binding sites for *trans*-acting factors and numerous interactions between nuclear proteins and these elements have been defined. The best understood events involve antagonistic interactions between a pair of GARE-binding factors and a pair of pyrimidine box-binding factors (Fig. 5).

The GARE in barley α-amylase promoters binds two nuclear proteins - a zinc-finger protein, HRT, and a MYB transcription factor, GAMYB - that exert antagonistic effects on the activity of the promoter. HRT is a nuclear localised zinc-finger protein that binds to the GARE in the barley *Amy1/6-4* promoter. Transiently expressed HRT represses the activity of both *Amy1* and *Amy2* promoters (41). An inverse relationship exists between *HRT* and *Amy1* mRNA expression in barley aleurone tissue, supporting the idea that HRT is a repressor of *Amy1* expression. The regulation of HRT by hormonal signals is not yet understood.

The GARE of the barley high-pI *Amy1* promoter contains a consensus MYB transcription factor binding sequence, TAACAAA. The MYB transcription factors were first identified in animal systems and feature one to three repeats (R1/R2/R3) of a conserved 50 aa DNA binding domain.

Expression of GAMYB mRNA and protein in barley aleurone is GA-inducible (51). In the absence of GA, transient expression of GAMYB is sufficient to activate a high-pI α-amylase promoter while mutation of the GARE in this promoter abolishes this effect. Through transient silencing of the *GAMYB* gene it has been shown that GAMYB is necessary as well as sufficient for induction of α-amylase gene transcription (52). There is a tight correlation between GAMYB binding sequences contained in the GARE of α-amylase promoters, defined *in vitro*, and sequences necessary for transcriptional activity of α-amylase gene promoters in transient expression assays. GAMYB is also known to bind the sequence TAACAGAC *in vitro*, which is the GARE found in barley low-pI α-amylase promoters. Similar motifs, resembling MYB-binding sites, exist in the promoters of other classes of GA-responsive grain hydrolase genes and it has been shown in transient expression assays that GAMYB strongly transactivates the promoters from the barley high-pI α-amylase, (1-3,1-4)-β-glucanase and cysteine protease *EPB-1* genes as well as the wheat α-capthesin-like protease promoter. These results suggest that GAMYB is a broad-acting transcriptional activator of GA-regulated genes in aleurone. It is not known whether HRT is a broad-acting repressor of hydrolase genes in aleurone.

The pyrimidine box, 5'-CTTTT-3', contained in the GARC of many GA-responsive grain hydrolase promoters contains the binding site consensus for the DOF transcription factors – a plant specific family of transcription factors, usually 200-400 aa long and characterised by an N-terminal DNA-binding domain and a C-terminal domain involved in transcriptional regulation. Two DOFs that interact with this domain in hydrolase gene promoters in aleurone have been identified, BPBF and SAD (21, 37). Both of these factors are expressed in aleurone cells of germinated barley grain and bind *in vitro* to the pyrimidine-box motif in the promoter of the cathepsin B-like cysteine protease gene, *Al21*, a GA-inducible gene in aleurone. BPBF also binds to this motif in the *Amy2/32b* gene. The expression of *Pbf* (the gene encoding BPBF) is GA-responsive but ABA-repressible in isolated aleurone. Both BPBF and SAD interact with HvGAMYB in yeast but these factors appear to exert counteractive effects on *AL21* promoter activity. In co-bombardment assays, expression of BPBF inhibits both GA-induction of the *AL21* protease promoter, and the transactivation of this promoter by constitutively expressed HvGAMYB. By contrast transiently expressed SAD and GAMYB cooperatively transactivate the *AL21* gene promoter. It is possible that HvGAMYB function in aleurone might be negatively regulated by BPBF, perhaps as part of programmed cell death. This is consistent with the observation that BPBF is expressed later than HvGAMYB in GA-stimulated aleurone. Since the *Al21* gene promoter contains two DOF-binding sites, its net activity might be the result of a type of combinatorial control exerted by the relative concentrations of SAD and BPBF in an aleurone cell.

The activity of α-amylase promoters in barley aleurone also requires expression of another GAMYB-binding protein, GMPOZ, from the BTB/POZ-domain family (51). The BTB/POZ domain is an NH_3-terminal domain of about 120 amino acids found in a variety of transcriptional regulators and cytoskeletal modifiers and is thought to provide a scaffold for the organisation of higher-order structures such as the cytoskeleton and chromatin. The transient silencing of GMPOZ, a GA-inducible nuclear factor from aleurone, abolishes the activity of a high-pI α-amylase gene promoter. GMPOZ contains no recognisable DNA-binding domain and its exact mode of operation in the GA-response pathway is not yet clear.

REPRESSORS OF GA-RESPONSE PATHWAYS IN ALEURONE

It is thought that GA-response pathways in aleurone are constitutively repressed and are activated mainly by 'de-repression'. Regulated protein degradation and reversible modification of proteins by phosphorylation, or by the addition of glucosamine residues, have been implicated in this process.

SLENDER1

The best-characterised repressors of GA-signaling are the cereal orthologues of the *Arabidopsis GAI/RGA* genes (Chapter D2), including barley and rice *SLENDER1* (SLN1/SLR1), which encode putative plant transcription factors (39). Mutants at this locus exhibit both dwarf and elongated phenotypes representing dominant gain-of-function mutations or loss of function mutations respectively. The dwarfs at this locus contributed to the so-called 'Green Revolution' through the introduction of dwarfing genes into high yielding varieties (42). In the aleurone layers of rice and barley *slender1* (elongated) mutants such as *sln1c*, expression of α-amylase is constitutive rather than GA-dependent (Fig. 6). Conversely, in a dwarf gain-of-function mutant at the barley slender locus (*Sln1d*), the GA-induction of α-amylase expression is substantially repressed.

In aleurone, the barley SLN1 protein is expressed in the nucleus but rapidly degraded in the presence of GA (15) suggesting that wildtype SLN1 acts as a GA-repressible, negative regulator of GA-

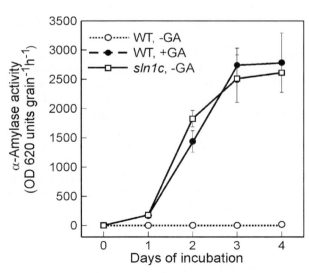

Figure 6. α-Amylase production by half grains of wild type (WT) and *Sln1c* barley grains in response to GA₃ (5).

signaling. Regulated protein degradation is known to be important in plant development and, in the ubiquitin-proteasome pathway, proteins destined for degradation are ubiquitinated and degraded by the proteasome. In aleurone, inhibitors of this pathway prevent GA-mediated degradation of SLN1 and concomitant α-amylase induction (12). Also, no SLN1 degradation occurs in the rice F-box mutant *gid-2*, which is deficient in the component of the ubiquitin-proteasome pathway that recruits substrate to the ubiquitination complex (44). Proteins targeted to the ubiquitin-proteasome pathway are commonly modified by phosphorylation. Since GA induces accumulation of phosphorylated forms of SLN1 in *gid-2* it is thought that GA-dependent phosphorylation ordinarily tags SLN1 for degradation in the wildtype (44). GAMYB has been identified as a target for the repressive action of SLN1. In the barley *Sln1d* dwarf mutant, the GA-induction of GAMYB expression in aleurone is significantly weaker than in wildtype (15). This result suggests that GAMYB expression is controlled by events directly downstream of SLN1 action

Abscisic acid

In seeds, ABA is involved in embryo maturation, the onset and maintenance of dormancy and protection from pathogens (Chapter E4) (11). However, mature aleurone layers retain ABA-sensitivity and ABA generally exerts counteractive effects on GA-regulated processes including hydrolase synthesis and secretion. Presumably this acts to slow down events after the initiation of germination if the environmental conditions are unfavourable. The molecular basis of the ABA-mediated antagonism of GA-signaling in aleurone is incompletely understood. Although exogenous ABA represses the induction of α-amylase gene expression by GA (23), it does not repress GAMYB mediated transactivation of an *a-Amy1* promoter in transient expression assays in barley aleurone (13). Instead, like SLN1, ABA has been observed to inhibit GAMYB transcription rates and protein expression in aleurone (39). However ABA appears to function downstream of SLN1 because the barley *sln1* (elongated) mutant retains ABA-sensitivity and ABA-treatment of wildtype barley aleurone layers has no effect on the GA-induced degradation of SLN1. The transcriptional activity of a 1031 bp GAMYB 5' flanking sequence fused to GUS is induced two-fold by GA treatment in bombarded barley aleurone but is repressed by ABA. However, transient expression of the ABA-induced kinase, PKABA1, also suppresses this activity suggesting that ABA-repression of GAMYB expression is mediated by PKABA1 (13).

SPY and KGM

A second molecular repressor of GA-signaling in aleurone was originally identified through studies of GA-response mutants at the *Arabidopsis SPINDLY* locus (*SPY*) (Chapter D2). SPY is thought to be an *O*-GlcNAc transferase, an enzyme that modifies target proteins through the reversible addition of O-linked N-acetyl glucosamine residues. The barley SPY homologue (HvSPY) represses the GA-induction of α-amylase promoter activity in barley aleurone in transient expression assays (43). However it is not yet clear whether this effect occurs through GAMYB or is mediated by an independent pathway.

A GAMYB-binding protein that inhibits GAMYB activity, rather than GAMYB expression has also been identified as a molecular repressor of aleurone GA-response pathways. KGM is a member of the emerging Mak-subgroup of protein kinases, which are related to the cdc2- and MAP kinases. In transient expression assays KGM specifically inhibits GA-inducible α-amylase promoter activity by repressing the transactivation of this promoter by HvGAMYB (51). This activity is partly dependent on the integrity of a conserved tyrosine residue in the activation loop of KGM, a site that must be phosphorylated in classical MAP kinases to facilitate substrate recognition and catalytic activation. Thus it is possible that phosphorylation and de-phosphorylation are important in the regulation of KGM activity. Despite the functional and physical interaction observed between KGM and HvGAMYB, recombinant GST-KGM fusion protein does not phosphorylate HvGAMYB in *in vitro* kinase assays. It is also possible that KGM modulates HvGAMYB function by a kinase-independent mechanism such as sequestration.

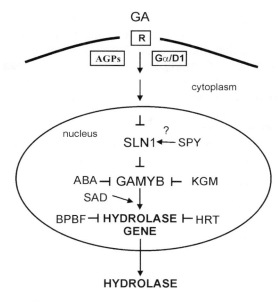

Figure 7. Model of GA signaling in aleurone cells

SUMMARY

The gibberellin response pathways that control hydrolase gene expression in cereal aleurone, while incompletely understood, appear to be activated by regulated degradation of an upstream constitutive repressor, SLN1 (Fig. 7). The identity of the GA-receptor remains elusive but it is apparent that the signaling cascade is initiated at the plasma membrane, possibly in conjunction with a G protein signaling pathway and/or arabinogalactan protein signaling. The relationship between the known GA-signaling intermediates, such as Ca^{2+} and cGMP, and the downstream transcription factors that bind to the GARCs in hydrolase promoters is unclear. The existence of multiple repressors of GA-regulated hydrolase expression, including HvSPY, KGM and ABA is suggestive of the importance of tightly coordinated expression of hydrolases after the onset of germination.

References

1. Aisien AO, Palmer GH, Stark J (1986) The ultrastructure of germinating sorghum and millet grains. J Inst Brew 92: 162-167
2. Bethke PC, Schuurink R, Jones RL (1997) Hormonal signalling in cereal aleurone. J Exp Bot 48: 1337-1356
3. Bruggeman FJ, Libbenga KR, VanDuijn B (2001) The diffusive transport of gibberellins and abscisic acid through the aleurone layer of germinating barley grains. Planta 201: 89-96
4. Chandler PM, Jacobsen JV (1991) Primer extension studies on α-amylase mRNAs in barley aleurone II. Hormonal regulation of expression. Plant Molec Biol 16: 637-645
5. Chandler PM, Marion-Poll A, Ellis M, Gubler F (2002) Mutants at the *Slender1* locus of barley cv Himalaya. Molecular and physiological characterization. Plant Physiol 129: 181-190
6. Chandler PM, Mosleth E (1990) Do gibberellins play an *in vivo* role in controlling α-amylase gene expression? *In*: K Ringlund, E Mosleth, DJ Mares, *eds*, Proceedings of the 5th International Symposium on Pre-Harvest Sprouting in Cereals. Westview Press, Boulder, Colorado p, 100-109
7. Dominguez F, Cejudo FJ (1999) Patterns of starchy endosperm acidification and protease gene expression in wheat grains following germination. Plant Physiol 119: 81-87
8. Drozdowicz YM, Jones RL (1995) Hormonal control of organic and phosphoric acid release by barley aleurone layers and scutella. Plant Physiol 108: 769-776
9. Fath A, Bethke PC, Jones RL (1999) Barley aleurone cell death is not apoptotic: characterization of nuclease activities and DNA degradation. Plant J 20: 305-315
10. Fath A, Hwang YS, Bethke PC, Spiegel YN, Zhu T, Jones RL (2001) Genomic profiling of hormonally regulated events in barley and rice aleurone layers. Proceedings of 17th International Conference on Plant Growth Substances. Brno 2001, pp73
11. Fincher GB (1989) Molecular and cellular biology associated with endosperm mobilisation in germinating cereal grains. Ann Rev Plant Physiol Plant Molec Biol 40: 305-346
12. Fu X, Richards DE, Ait-ali T, Hynes LW, Ougham H, Peng J, Harberd NH (2002) Gibberellin-mediated proteasome-dependent degradation of the barley DELLA protein SLN1 repressor. Plant Cell 14: 3191-3200
13. Gomez-Cadenas A, Zentalla R, Walker-Simmons M, Ho THD (2001) Gibberellin/abscisic acid antagonism in barley aleurone cells: site of action of the protein kinase PKABA1 in relation to gibberellin signaling molecules. Plant Cell 13: 667-679
14. Gubler F, Ashford AE, Jacobsen JV (1987) The release of α-amylase through gibberellin-treated barley aleurone cell walls. Planta 172: 155-161
15. Gubler F, Chandler PM, White R, Llewellyn D (2002) GA signaling in barley aleurone cells: control of SLN1 and GAMYB expression. Plant Physiol 129: 191-200
16. Guerin JR, Lance RCM, Wallace W (1992) Release and activation of barley α-amylase by malt endopeptidases. J. Cer Sci 15: 5-14
17. Hamabata A, Garcia-Maya M, Romero T, Bernal-Lugo I (1988) Kinetics of the acidification capacity of the aleurone layer and its effect upon solubilisation of reserve substances from starchy endosperm of wheat. Plant Physiol 86: 643-644
18. Heimovaara Dijkstra, S., Heistek, J.C., Wang, M (1994) Counteractive effects of ABA and GA3 on extracellular and intracellular pH and malate in barley aleurone. Plant

Physiol 106: 359-365
19. Heimovaara Dijkstra S, Mundy J, Wang, M (1995) The effect of intracellular pH on the regulation of the Rab 16A and the alpha-amylase 1/6-4 promoter by abscisic acid and gibberellin. Plant Mol Biol 27: 815-820
20. Huang N, Sutliff TD, Litts JC, Rodriguez, RL (1990) Classification and characterization of the rice alpha-amylase multigene family. Plant Mol Biol 14: 655-668
21. Isabel-LaMoneda I, Diaz I, Martinez M, Mena M, Carbonero P (2003) SAD: a new DOF protein from barley that activates transcription of a cathepsin B-like thiol protease gene in the aleurone of germinating seeds. Plant J 33: 329-340
22. Jacobsen JV (1983) Regulation of protein synthesis in aleurone cells by gibberellin and abscisic acid. In: A Crozier, ed, The biochemistry and physiology of gibberellins, Vol 2. Praeger, New York, pp 159-187
23. Jacobsen JV, Beach LR (1985) Control of transcription of alpha-amylase and rRNA genes in barley aleurone protoplasts by gibberellin and abscisic acid. Nature 316: 275-277
24. Jacobsen JV, Gubler F, Chandler, P.M. (1995) Gibberellin action in germinating cereal grains. In: PJ Davies, ed, Plant Hormones; physiology, biochemistry and molecular biology. Martinus Nijhoff, Dordrecht, pp 164-193
25. Jacobsen JV, Pearce DW, Poole AT, Pharis RP, Mander LN (2002) Abscisic acid, phaseic acid and gibberellin contents associated with dormancy and germination in barley. Physiol Plant 115:428-441
26. Jones RL, Jacobsen JV (1991) Regulation of synthesis and transport of secreted proteins in cereal aleurone. Int Rev Cytol 126: 49-88
27. Kaneko M, Itoh H, Ueguchi-Tanaka, Ashikari M, Matsuoka M (2002) The α-amylase induction in endosperm during rice seed germination is caused by gibberellin synthesized in epithelium. Plant Physiol 128: 1264-1270
28. Lanahan MB, Ho T-HD, Rogers SW, Rogers JC (1992) A gibberellin response complex in cereal alpha-amylase gene promoters. Plant Cell 4: 203-211
29. Lenton JR, Appleford NEJ, Croker SJ (1994) Gibberellins and α-amylase gene expression in germinating wheat grains. Plant Growth Regul 15: 261-270
30. Longstaff MA, Bryce JH (1993) Development of limit dextrinase in germinating barley (*Hordeum vulgare* L) - Evidence of proteolytic activation. Plant Physiol 101: 881-889
31. Lovegrove A, Hooley R (2000) Gibberellin and abscisic acid signalling in aleurone. Trends Plant Sci 5: 102-110
32. McFadden GI, Ahluwalia B, Clarke AE, Fincher GB (1988) Expression sites and developmental regulation of genes encoding (1-3,1-4)-β-glucanase in germinated barley. Planta 173: 500-508
33. MacGregor AW (1980) Action of malt α-amylases on barley starch granules. MBAA Technical Quarterly 17: 215-221
34. MacGregor AW, Matsuo RR (1982) Starch degradation in endosperms of barley and wheat kernels during initial stages of germination. Cereal Chem 59: 210-216
35. Macnicol PK, Jacobsen JV (1992) Endosperm acidification and related metabolic changes in the developing barley grain. Plant Physiol 98: 1098-1104
36. Matthews PR, Thornton S, Gubler F, White R, Jacobsen JV (2002) Use of the green fluorescent protein to locate α-amylase gene expression in barley grains. Funct Plant Biol 29: 1037-1043
37. Mena M, Cejudo FJ, Isabel-Lamoneda I, Carbonero P (2002) A role for the DOF transcription factor BPBF in the regulation of gibberellin-responsive genes in barley aleurone. Plant Physiol 130: 111-119
38. Mitsui T, Itoh K (1997) The alpha-amylase multigene family. Trends Plant Sci 2: 255-261
39. Olszewski N, Sun T-p, Gubler F (2002) Gibberellin signaling: biosynthesis, catabolism, and response pathways. Plant Cell, Supplement S61-S80.
40. Palmer GH, Shirakashi T, Sanusi LA (1989) Physiology of germination. EBC Congress Proceedings pp. 63-74

41. Raventos D, Skriver K, Schlein M, Karnahl K, Rogers SW, Rogers JC, Mundy, J (1998) HRT, a novel zinc finger, transcriptional repressor from barley. J Biol Chem 273: 23313-23320.
42. Richards DE, King KE, Ait-ali T, Harberd N (2001) How gibberellin regulates plant growth and development: a molecular genetic analysis of gibberellin signaling. Annu Rev Plant Physiol Plant Mol 52:67-88
43. Robertson M, Swain SM, Chandler PM, Olszewski NE (1998) Identification of a negative regulator of gibberellin action, HvSPY in barley. Plant Cell 10: 995-1007
44. Sasaki A, Itoh H, Gomi K, Ueguchi-Tanaka M, Ishiyama K, Kobayashi M, Jeong D-H, An G, Kitano H, Ashikari M, Matsuoka M (2003) Accumulation of phosphorylated repressor for gibberellin signaling in an F-box mutant. Science 299: 1896-1898
45. Shen W, Gomez-Cadenas A, Routly EL, Ho T-HD, Simmonds JA, Gulick PJ (2001) The salt stress-inducible protein kinase gene, *Esi47*, from the salt-tolerant wheatgrasss *Lophopyrum elongatum* is involved in plant hormone signaling. Plant Physiol 125: 1429-1441
46. Slakeski N, Fincher GB (1992) Developmental regulation of (1-3,1-4)-β-glucanase gene expression in barley. Plant Physiol 99: 1226-1231.
47. Sugimoto N, Takeda G, Nagato Y, Yamaguchi J (1998) Temporal and spatial expression of the α-amylase gene during seed development in rice and barley. Plant Cell Physiol 39: 323-333
48. Sun Z, Henson CA (1991) A quantitative assessment of the importance of barley seed α-amylase, β-amylase, debranching enzyme and α-glucosidase in starch degradation. Arch Biochem Biophys 284: 298-305
49. Suzuki Y, Kitagawa M, Know JP, Yamaguchi I (2002) A role for arabinogalactan proteins in gibberellin-induced α-amylase production in barley aleurone cells. Plant J 29: 733-741
50. Tregear JW, Primavesi LF, Huttly AK (1995) Functional analysis of linker insertions and point mutations in the *α-Amy2-54* GA-regulated promoter. Plant Molec Biol 29: 749-758
51. Woodger FJ, Millar A Murray F Jacobsen JV, Gubler F (2003) The role of GAMYB transcription factors in GA-regulated gene expression. J Plant Growth Regulat 22: 176-184
52. Zentella R, Yamauchi D, Ho THD (2002) Molecular dissection of the gibberellin/abscisic acid signaling pathways by transiently expressed RNA interference in barley aleurone cells. Plant Cell 14: 2289-2301
53. Zorec R, Tester M (1992) Cytoplasmic calcium stimulates exocytosis in a plant secretory cell. Biophys J 63: 864-867Zorec R, Tester M (1992) Cytoplasmic calcium stimulates exocytosis in a plant secretory cell. Biophys J 63: 864-867

C3. Cytokinin Regulation of the Cell Division Cycle

Luc Roef and Harry Van Onckelen
Laboratory of Plant Biochemistry and Physiology, University of Antwerp, Universiteitsplein 1, B-2610 Antwerp, Belgium.
E-mail harry.vanonckelen@ua.ac.be

INTRODUCTION

Contrary to the situation in animal systems, plant shape and size are predominantly determined by developmental programs that govern the timely initiation and outgrowth of meristems. In plants, meristematic zones almost exclusively constitute the regions of dividing cells. The role of mitotic cell division in morphogenetic and developmental processes has been the subject of intense debate, but the present view seems to acknowledge the importance of proper cell division regulation for the correct elaboration and execution of developmental programs (29)[1].

Owing to their sessile nature, plants have retained the ability to adapt their developmental programs to changing environmental cues. Plant hormones appear to be major players in the integration of these developmental and environmental stimuli. Together with auxins, cytokinins probably constitute the most conspicuous group of plant hormones regulating cell division, and it has long been known that the balance between these two groups of hormones determines, to a great extent, the developmental fate of a plant (49). The first known "cytokine", kinetin, was isolated some 50 years ago from autoclaved herring sperm DNA as a cell division promoting factor

---

[1] Abbreviations: Cytokinins: BA, N^6-benzyladenine; DZ, dihydrozeatin; DZMP, dihydrozeatin riboside-5'-monophosphate; DZNG, dihydrozeatin-N-glucoside; DZOG, dihydrozeatin-O-glucoside; DZR, dihydrozeatin riboside; DZROG, dihydrozeatin riboside-O-glucoside; iP, isopentenyladenine; iPA, isopentenyladenosine; iPMP, isopentenyladenosine-5'-monophosphate; iPNG, isopentenyladenine-N-glucoside; NAA, 1-naphtalene acetic acid; Z, zeatin; ZNG, zeatin-N-glucoside; ZOG, zeatin-O-glucoside; ZR, zeatin riboside; ZRMP, zeatin riboside 5'-monophosphate; ZROG, zeatin riboside-O-glucoside.
Other abbreviations: CAK, CDK-activating kinase; cAMP, cyclic adenosine 3':5'-monophosphate; CDK, Cyclin-Dependent Kinase (CDKA, CDKB, ...); CKS, CDK Subunit protein; CycB, cyclin B; CycD, cyclin D; DP, Dimerisation Partner of E2F; E2F, adenovirus E2 promoter binding factor; HMG-CoA, 3-hydroxy-3-methylglutaryl coenzyme A; Kip, Kinase Inhibitory Protein; KRP, Kip-related protein; ORC, Origin Recognition Complex; PZ, propyzamide; Rb, Retinoblastoma protein.

(32). About 8 years later the structure of zeatin as a naturally occurring cell division factor from *Zea mays* L. was resolved (25). Since then a number of N^6-substituted adenine derivatives with similar cell division promoting properties were either isolated from natural sources or synthesised, and denominated 'cytokinins'.

This chapter will focus on the role of cytokinins in the regulation of the plant mitotic cell cycle and will try to deal with some questions on the mode of action of some of its representatives.

THE PLANT CELL CYCLE

As the main purpose of the plant mitotic cell cycle is essentially the same as in other eukaryotes (i.e. flawlessly replicate the information contained in the DNA of the cell and have it carefully segregated into the nuclei of two new daughter cells), it is not surprising that the mechanisms involved bear substantial resemblance. As in other eukaryotes the plant cell cycle consists of four phases. DNA is replicated in synthesis phase (S phase) and divided between the daughter cells in mitosis (M phase). Two gap phases separate the S and M phases. They are essential in that they allow for the operation of controls on the accurate and full completion of the previous phase. During the post-mitotic first gap (G_1 phase), the opportunity and necessity to engage into another cell cycle is checked. Parameters such as nutrition status and other external factors are monitored before entrance into the S-phase. In the post-synthetic second gap (G_2 phase) the proper completion of DNA synthesis is verified, while the appropriate mechanical structures that will allow for the separation of the chromosomes are designed. Failure to comply with the set conditions will result in an arrest of the cell cycle mainly at G_1/S or G_2/M boundaries, preventing further erratic cell division. Operation of these checkpoints has been well conserved throughout evolution and so have the basic underlying molecular and biochemical mechanisms.

As in other eukaryotes, the periodic association of regulatory proteins, called cyclins[2], with cyclin-dependent kinases (CDK) drives the plant cell cycle. Their merger in varying combinations of different classes of cyclins and CDK's throughout the cell cycle, results in the formation of active serine/threonine protein kinase complexes that phosphorylate a multiplicity of cell cycle related substrates such as chromatin components, cytoskeleton elements, etc... As such they usher the cell from one phase into the other (Fig. 1). Further analogy with animal systems is found in the sense that activity of cyclin/CDK-complexes is modulated by phosphorylation and

---

[2] Arbitrarily classified as cyclin A, B, C..., (and divided in respective subclasses A1, A2, ...) depending on overall sequence homology, presence of particular conserved motifs and expression patterns. The same applies for CDK's.

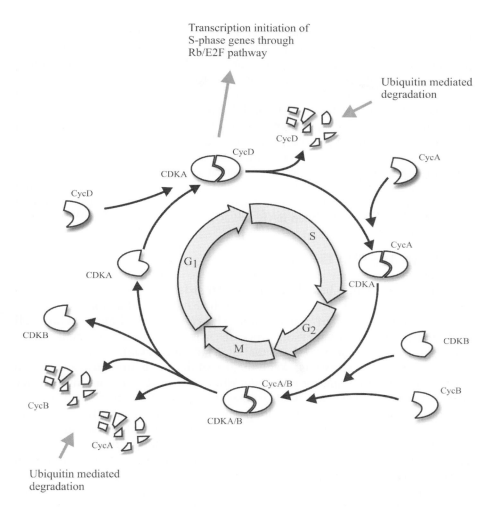

Figure 1. Regulation of the mitotic plant cell cycle. The cycle is driven by the periodic association of cyclin dependent protein kinases (CDK) with varying cyclin partners. CDKA is present throughout the whole cell cycle. CDKB is expressed mainly during G_2 to M. At G_1/S transition, cyclin D appears and interacts with CDKA to initiate transcription of S-phase specific genes through the Rb/E2F pathway. Once in S-phase, cyclin D and E2F are degraded in a ubiquitin dependent manner and cyclin A appears. At the onset of mitosis, cyclin B joins in. CDKA and CDKB associate with the mitotic cyclins A and B to guide the cell through mitosis. Upon completion of M-phase, mitotic cyclins are again removed in a ubiquitin dependent manner. Throughout the cell cycle, the activity of the cyclin/CDK complexes is further regulated by phosphorylation/ dephosphorylation events, and interaction with inhibitory and scaffolding proteins. (Based on 9, 33 and 53).

dephosphorylation events, interaction with Kinase Inhibitory Protein (Kip)-related proteins (KRP), and ubiquitin-mediated degradation of the cyclin partner. Next to their activating or inhibitory action, so called CDK subunit protein (CKS) docking factors can function as scaffold for the CDK complex/substrate interaction. Several other cell cycle related proteins such as retinoblastoma protein (Rb)[3] and heterodimeric E2F/DP transcription factor, which govern the transcription of S-phase specific genes in animal systems, have functional homologues in plants too. On the other hand, the plant cell cycle exhibits features such as the pre-prophase[4] band and the phragmoplast[5] that are not present in other eukaryotes. These differences translate into the existence of cyclins and CDK's not shared with animal systems. The nature of these components and their relation to the plant cell cycle are extensively reviewed in (9; 33; 53).

RESEARCH STRATEGIES

A number of strategies have been adopted in order to detect where cytokinins interface with the cell cycle and to dissect their mode(s) of action.

The correlation of endogenous levels of cytokinins with cell proliferation and differentiation processes has given, and will continue to give, clues to the identity of the actual plants own cytokinin species active in the regulation of these events. With the use of mass spectrometric techniques for the identification and measurement of several different cytokinin species (43; 59) (Chapter G1) one is also able to analyse the metabolic processes underlying the regulation of endogenous levels of bio-active cytokinin species. As these MS techniques will become increasingly sensitive they will allow for smaller and thus more localised sampling (59). For the time being, immunolocalisation remains the method of choice for the elucidation of cytokinin distribution at the cellular and sub-cellular level (10; 21).

Manipulation of endogenous cytokinin levels is another technique that has long been used (6; 32; 49). Numerous examples now exist of the effects of exogenously added cytokinins on the induction of cell division and pattern formation, providing insights into structure/activity relationships, interactions with other plant hormones during these phenomena, and possible modes of action.

---

[3] Tumor suppressor originally identified in retinoblastoma that interacts with transcription factors such as E2F to block transcription of growth regulating genes. The Rb gene plays a role in normal development, not just that of the retina.
[4] Cortical belt-like arrangement of microtubules and actin filaments formed at the onset of mitosis encircling the future division plane. Dissolves as the mitotic spindle is built.
[5] Arrangement of microtubules formed at the end of mitosis. Grows from the inside out at the division plane and organises the synthesis of the new cell wall between the two daughter nuclei.

Table 1. Cytokinin content in leaves of *N. tabacum* cv. Wisconsin 38 plants transformed with the *ipt* gene under the control of a tetracycline inducible promoter. Values obtained after different time periods upon induction with chlortetracycline (1mg.L^{-1}).

Cytokinin type	Cytokinin content (pmol gfw^{-1})					
	control			*ipt*-transformed		
	0 h	8 h	24 h	0 h	8 h	24 h
Z	10	4	6	13	30	120
DZ	2	3	2	1.5	5	4
ZR	10	12	17	17	100	820
DZR	2	4	3.5	4	17	20
ZNG	0.5	2	2	1	<0.3	0.6
DZNG	<0.2	<0.5	<0.8	<0.6	<0.3	<0.3
ZOG	30	20	22	50	140	90
DZOG	10	4	6	13	6	10
ZROG	12	12	15	22	100	830
DZROG	3	2	2	2	7	20
ZMP	3	3	4	30	100	1040
DZMP	7	6	11	5	10	20
iP	7	4	3	8	10	13
iPA	0.7	3	2	0.4	4	6.5
iPNG	0.8	1.1	1.5	1.4	10	1
iPMP	<0.6	<0.8	<1	<1	<1.5	<2

Other approaches exist in altering the cytokinin metabolism by use of transgenic plants. The ectopic expression of the *Agrobacterium* cytokinin biosynthesis *ipt* gene in *Nicotiana tabacum* or Arabidopsis results in an increase of predominantly zeatin-type cytokinins (44) (Table 1). Tobacco somatic mosaics for the *ipt* gene exhibit viviparous leaves with buds appearing at sites of increased cytokinin content. The emanation of both floral and vegetative epiphyllous buds on fully differentiated leaves is also seen in *Nicotiana tabacum* cv. Petit Havana SR1 transformed with the ipt gene under the control of the ribulose biphosphate decarboxylase small subunit promoter (Pssu) (Fig. 2) (Beinsbergen, unpublished) indicating a role for cytokinins in dedifferentiation and the induction of meristematic activity without determining the eventual fate of these meristems (12). In another example, transient expression of the *Zea mays* gene for a β-glucosidase, Zm-p60.1, which stimulates the release of active from inactive endogenous cytokinin species pools, initiated cell division in tobacco protoplasts (3).

Several Arabidopsis mutants exist with deregulated cytokinin responses. Some of these mutants exhibit aberrant hormone content, while in others cytokinin perception or downstream signalling is affected (14). Again, phenotypes are often cell cycle related (hormone autotrophic growth of calli) or involve alterations in meristem initiation. For some, the affected genes

Figure 2. Vegetative (A) and floral (B) epiphyllous buds on Pssu-*ipt* transformed tobacco.

have been isolated and characterised, thus providing clues as to where cytokinins are involved in meristem initiation and induction of cell division (16; 58). In one such instance, a family of histidine kinase-type cytokinin receptors was positively identified together with downstream elements of cytokinin mediated signal transduction (20; 28; 47; 60).

SYNCHRONISED PLANT CELL CULTURE SYSTEMS

Although of extreme value, whole plant approaches *per se* fail to discriminate between events involved in pattern formation on the one hand, and initiation, progression and termination of the plant cell cycle itself on the other. In order to study the plant cell cycle, one needs synchronised systems in which the cell cycle is fully detached from any developmental context. In such systems, the effects of cytokinin addition or deprivation on progress through the different stages of the cell cycle can be studied, or a correlation can be sought between (manipulated) endogenous hormone content and expression and activity of prime cell cycle regulators.

Synchronisation techniques are mainly based on the chemical inhibition of cell cycle specific processes in *in vitro* plant cell cultures, and a subsequent reversal of the inhibition by removal of the chemical. Through the proficient use of combinations of specific inhibitors, coupled with the addition or removal of nutrients, cells can be synchronised at various stages of their cell cycle (41).

Through the years, several plant cell culture systems have been used, that can be synchronised with varying efficiency. These include *Acer pseudoplatanus*, *Glycine max*, *Catharanthus roseus*, *Medicago sativa* and *Daucus carota* (31). Recently, a good synchrony system for the plant genetic model system par excellence, Arabidopsis, was presented (31).

The most effective system to date is the *Nicotiana tabacum* cv. Bright Yellow-2 (BY-2) derived cell culture (37). Addition of aphidicolin, an inhibitor of DNA polymerase α, blocks cells in early S-phase. Upon release from this aphidicolin block BY-2 cells have been reported to reach 50-70% synchrony at mitosis. A double block with aphidicolin and the microtubule destabilising drug, propyzamide, results in a block at mitosis with about 90

% synchrony (36). These synchronisation efficiencies by far outperform those of other systems.

Good synchronicity rates are an important prerequisite to detect short-lived changes in effectors of the cell cycle, and to demonstrate the cell cycle-dependent oscillation of genes, and to characterise their role during the cell cycle. Synchronisation also facilitates the structural and biochemical analysis of cell cycle specific events such as the development of the phragmoplast and the formation of cortical microtubules. Using the systems mentioned above, efforts are now being undertaken to perform genome wide surveys of gene expression during cell cycle progression in both Arabidopsis (30) and tobacco (1).

CYTOKININ LEVELS DURING THE BY-2 CELL CYCLE

The BY-2 cell culture has another edge over many other plant cell culture systems when trying to study the involvement of cytokinins in cell division, i.e. its independence of added cytokinins for cell proliferation. Based on data of a similar plant cell culture (*N. tabacum* cv. Xanthi XD6S) (39), it could be suspected that BY-2 cells provide their own required cytokinins. Studies on the evolution of cytokinin content (and other hormones) during cell cycle progression confirmed that their cytokinin autotroph nature can be attributed to their ability to synthesise cytokinins at the appropriate times in the cell cycle (11; 23; 45). Analysis of cytokinin accumulation in synchronised BY-2 cells indeed revealed the presence of very pronounced short-lived surges in cytokinin levels during cell cycle progression. The most conspicuous behaviour was recorded for zeatin-type cytokinin species. Peak concentrations were observed at early G_1, late G_1 into early S, mid (11) or late S (45) and G_2/M transition (Fig. 3). Less pronounced peaks of dihydrozeatin-type cytokinins were observed at G_1 and late G_1 into early S, and at G_2/M transition. In one study (11), peaks in iP-type cytokinin (at early S-phase) and zeatin-type O-glucoside (at G_2/M) were also recorded. In all studies, the sharp and abundant pre-mitotic peak in zeatin-type cytokinins was most obvious.

Altogether, it is evident that cytokinin levels are highly regulated during cell cycle progression. The data seem to point to three (or maybe four) possible points of interaction of cytokinins with the cell cycle. It is clear that such behaviour could never have been revealed using non-synchronised habituated callus plant material. (This point is further illustrated in table 2, where even more dramatic changes were recorded upon higher synchronicity (see control).)

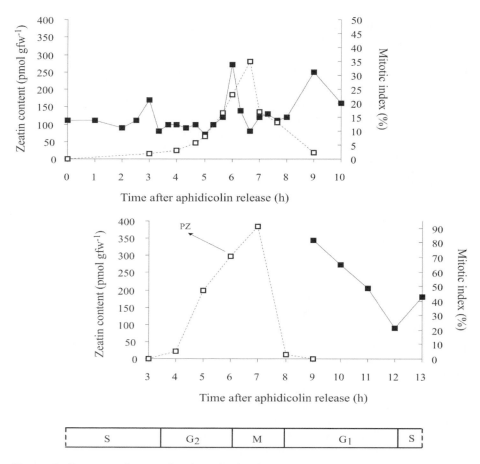

Figure 3. Representative graph of zeatin levels (■) during the cell cycle in BY-2 synchronised by means of an aphidicolin (upper panel) and aphidicolin/propyzamide block (lower panel). Propyzamide (PZ) was added at the time of aphidicolin release and removed at 6h (see lower panel). (□) Mitotic index. Lower bar: estimated stages of the cell cycle (applies to both graphs).

CYTOKININS IN CELL CYCLE INITIATION, PROGRESSION AND TERMINATION?

It has long been known that in the absence of auxins and cytokinins plant cells fail to engage in cell division. Depriving of cells of hormones can help us to identify the points where they are actually needed. Experiments with explanted tissues or protoplasts from differentiated tissue show that cells that have to go through a de-differentiation step, and/or that have to leave a quiescent stage to resume the cell cycle, invariably fail to progress beyond G_1 in the absence of cytokinins (35). This clearly points to the involvement of cytokinins in the initiation of the plant cell cycle.

Lovastatin induced inhibition of cytokinin biosynthesis as a tool to identify points of cytokinin interaction with the plant cell cycle.

The need for cytokinins for proper cell cycle progression in actively dividing cells also becomes obvious when cells are deprived of them. Results from studies in which cytokinins were removed from the medium of dividing plant cell cultures hint at a requirement for cytokinins at G_1/S, G_2/M or S phase (26; 35). The addition at well defined stages of the cell cycle of a specific inhibitor to a highly synchronised cytokinin autotroph plant cell culture with a well documented cytokinin profile would offer an excellent tool to study the effects of cytokinin deprivation at specific points of the cell cycle. Experiments with inhibitors of cytokinin synthesis have been few, mostly because of the seeming lack of specificity (35). However, lovastatin although at first sight a rather aspecific inhibitor, affecting a large number of isoprenoid requiring processes, perfectly suits this purpose. Lovastatin (or mevinolin) is a potent inhibitor of 3-hydroxy-3-methylglutaryl coenzyme A (HMG-CoA) reductase, a component of the pathway leading to the formation of mevalonic acid, precursor of many isoprenoid based biosynthetic pathways (e.g. synthesis of isoprenoid side-chain cytokinins, abscisic acid, gibberellins, ubiquinone, phytosterols, carotenoids, etc.) (Fig. 4).

Exposure of non-synchronously growing BY-2 cells to 1 µM lovastatin severely impairs their ability to grow. Simultaneous addition of 8 µM zeatin, kinetin, benzyladenine or thidiazuron restores cell growth, indicating that at low concentrations lovastatin behaves as a specific inhibitor of cytokinin synthesis (5). When added to synchronised BY-2 cultures, lovastatin dramatically reduces the accumulation of zeatin-type cytokinins (Table 2).

Table 2. Cytokinin content of synchronised BY-2 cells upon addition of 10 µM lovastatin at 4.5 hours after aphidicolin release.

Cytokinin type	Cytokinin content (pmol gfw^{-1})					
	control			lovastatin		
	6 h	7 h	8 h	6 h	7 h	8 h
Z	11	1700	14	1	60	31
DZ	1	7	<1	<1	4	2
iP	<1	15	2	<1	2	3
ZR	1	300	2	<1	8	16
DZR	1	9	1	4	3	5
iPA	<1	4	1	1	1	3

Cytokinins and the plant cell cycle

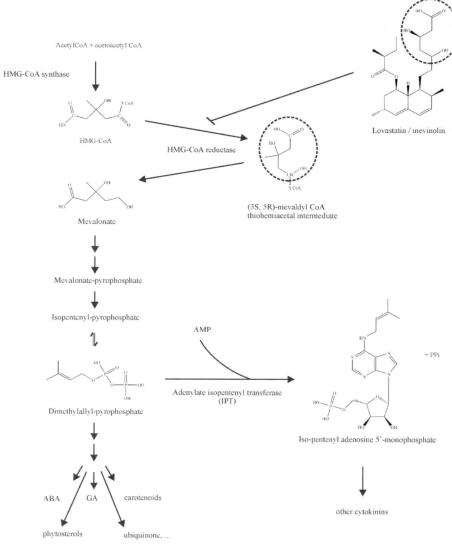

Figure 4. Lovastatin inhibition of the mevalonate pathway. HMG-CoA: 3-hydroxy-3-methylglutaryl coenzyme A.

The effect of lovastatin is almost immediate, proving that the sharp increases in cytokinin levels during cycling are caused by *de novo* biosynthesis rather than by conversion from pre-existing inactive cytokinin pools (23). The rapid decrease is probably executed by the cytokinin-degrading enzyme cytokinin oxidase, which is present throughout the whole cell cycle (11).

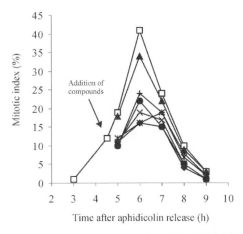

Figure 5. Mitotic activity of synchronised BY-2 cells upon challenge with lovastatin (L) in the presence or absence of cytokinins. Legend: (□) control, (■) 10 µM L, (▲) 10 µM L + 8 µM Z, (×) 10 µM L + 8 µM ZR, (●) 10 µM L + 8 µM DHZ, (✽) 10 µM L + 8 µM iP, (+) 10 µM L + 8 µM kinetin

Two studies have explored the simultaneous effects of lovastatin on both cytokinin accumulation and BY-2 cell cycle progression at respectively G_2/M (23) and at G_1/S (24). The most striking effect of lovastatin was recorded at G_2/M. Challenge of BY-2 cultures with lovastatin at early or late G_2 both dramatically reduced the accumulation of zeatin-type cytokinins at G_2/M, and resulted in a drastic drop in mitotic activity (without harming viability) (23). Attempts to rescue mitotic activity by cytokinin addition revealed a surprisingly specific competence for zeatin at this particular stage of the cell cycle. From a series of different compounds tested, zeatin was the only one that could completely restore mitotic activity (Fig. 5). Other cytokinins, such as zeatin riboside, dihydrozeatin, isopentenyl adenine, isopentenyl adenosine, kinetin, benzyladenine or non-cytokinins, such as NAA and 3', 5'-cAMP failed to do so completely or had only minor effects. This specificity for zeatin is not caused by breakdown or conversion, nor by less efficient uptake of the different compounds. Zeatin therefore seems to be the endogenous cytokinin trigger for G_2/M transition in BY-2.

The observations made upon addition in G_1 are a bit more puzzling. Given in early G_1, lovastatin does result in a significant reduction of accumulation of cytokinins in early G_1 and G_1/S, but according to flow cytometric data it does not induce a block of G_1/S transition (24) (Fig. 6). A similar behaviour was encountered in another study where lovastatin arrested BY-2 cells in late G_1-phase when it was added during mitosis, but where its effect on G_1/S transition gradually decreased upon addition at different time points after M/G_1 (18). Interestingly, application of zeatin, zeatin riboside or isopentenyladenine at early G_1 provoked cell cycle arrest in G_1. These results seem to contradict both the results mentioned above and results from an Arabidopsis cell culture (46). In these cells, strong indications exist that cytokinins promote G_1/S transition through induction of cyclin D3.

One explanation of the observations is that the cytokinin peaks observed in BY-2 at early G_1 and at G_1/S are, respectively, the "on" and "off" switches to G_1. One can then speculate that their transient nature is the key to their signalling function. Prolonged exposure to elevated concentrations of cytokinins (as a result of exogenous addition in early G_1) may cause desensitisation of receptors and prevent cells from detecting the anticipated

Cytokinins and the plant cell cycle

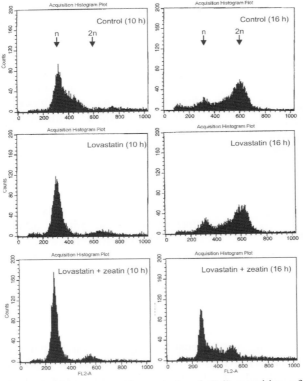

Figure 6. Flow cytometric analysis of G_1/S transition of aphidicolin/propyzamide synchronised BY-2 cells upon challenge with 10 µM lovastatin and 8 µM zeatin. Hours correspond to time after aphidicolin release (see figure 3).

second cytokinin signal at G_1/S, necessary for engaging S-phase. The addition of cytokinins to cytokinin starved Arabidopsis cells, probably arrested in late G_1 (35), may be considered as the equivalent of this second peak and thus a signal to transit the G_1/S border.

An alternative explanation may be that an elevated cytokinin level at early G_1 is a signal to leave the cell cycle, and enter a quiescent state or start to differentiate. The cytokinin peak at early G_1 in BY-2 cells then represents a sub-population of cells leaving the cell cycle. Immunocytochemical data show that high amounts of cytokinins are frequently encountered in differentiating tissues. The presence of high cytokinin concentrations has indeed been successfully correlated to organogenesis in tobacco meristems (8).

The role of increased cytokinin concentrations in S-phase is poorly understood. A diminished mitotic activity upon lovastatin addition at early S-phase was reported by Hemmerlin and Bach (18). Their flow cytometric data do not seem to point at a prolonged stay in S-phase, indicating that loss of the S-phase peak does not impinge on a transition through S. The impaired mitotic activity may just reflect the effect of lovastatin on the pre-mitotic peak of zeatin described earlier. Until combined data on cytokinin accumulation and mitotic activity upon lovastatin challenge are available, it is hard to speculate on cytokinin function in S-phase.

L. Roef and H. Van Onckelen

CYTOKININ TARGETS DURING THE PLANT CELL CYCLE

Since the plant cell cycle is driven by the periodic association of cyclins to cyclin-dependent kinases (CDK), resulting in the formation of active serine/threonine protein kinase complexes (33), it is not surprising that cytokinin signalling in some way interfaces with the cell cycle through these components. Some of the interactions of cytokinins (and other plant hormones) with these complexes have already been extensively reviewed (13; 53). From the data currently available, it appears that not only the final targets for cytokinin action change as one goes along the cell cycle, but that also the primary receptor molecules may differ.

Cytokinin Signalling at G_1/S.

Analysis of GUS expression driven by the promoter of the Arabidopsis CDK *Arath;CDKA;1* in transformed tobacco protoplasts (frequently used as a model for cell cycle initiation) reveals that in the absence of auxins, cytokinin application can induce the expression of this CDK (17). However, it is not until auxin is also present that cells engage in cell division, most likely through the interaction of CDKA with a D-type cyclin partner, whose biosynthesis also turns out to be cytokinin-inducible. Similarly in the conditional presence of auxin (and sucrose), addition of a range of cytokinins to cytokinin starved, G_1-arrested, Arabidopsis cells prompts them to express a cyclin D3. Neither *de novo* protein synthesis nor any cell cycle progression is necessary for this induction of expression. In addition, constitutive expression of this *Arath;CycD3;1* induces cytokinin-independent growth in otherwise cytokinin auxotroph Arabidopsis calli, suggesting that cyclin D3 is one of the earliest targets of cytokinins in the induction of cell division (46). In BY-2, two cyclins with similar behaviour to *Arath;CycD3;1* were isolated. Of these, *Nicta;CycD3;2* (51) is induced in tobacco seedlings upon zeatin addition, indicating that it may fulfil a similar function (7). The other, *Nicta;CycD3;3*, was shown to interact with CDKA to form an active cyclin/CDK complex that can hyperphosphorylate a retinoblastoma-like protein, NtRb1, thus inducing the G_1/S transition (38).

Recent reports implicate the 26S proteasome, a major multisubunit proteolytic complex in eukaryotes, in the cytokinin regulation of cyclin D3. *CycD3;1* transcription is up-regulated in Arabidopsis plants lacking a functional RPN12a, one of the many subunits of the 26S proteasome. These *rpn12a-1* plants also exhibit up-regulation of another cytokinin-induced gene coding for nitrate reductase, NIA1, and in general they display phenotypes typical of cytokinin response deregulation. Interestingly, the ortholog of RPN12 in yeast appears to be involved in G1/S, supposedly through the breakdown of CDK inhibitory protein. How the cytokinin/26S proteasome interaction works is yet unclear, but deserves further study (50).

Immediate targets for cytokinins have long remained elusive, but the recent identification of membrane-bound histidine kinases as receptors for cytokinins is promising. One of these kinases, CRE1 (or AHK4) (20; 54), is allelic to the *wooden leg* (*wol*) mutant (28), a mutant featuring inhibited root growth and lack of lateral roots. Its phenotype seems to be related to a disturbance in the asymmetric cell division during embryonic vascular initial formation. AHK4/CRE1/WOL histidine kinase has been shown to bind different natural and synthetic cytokinins, and AHK4 together with two closely related histidine kinases (AHK2 and AHK3) forms a group of cytokinin receptors (60). A survey of gene expression profiles in synchronised Arabidopsis cell cultures revealed the predominant expression of *AHK4* (and its downstream transcriptional activator targets *ARR4* and *ARR7*) at G_1, making them prime candidates for relaying the cytokinin signal to the cell cycle initiation components (30). These data are corroborated by the observation that inhibitors of bacterial histidine kinase activity severely damage *Catharanthus roseus* cell proliferation (40).

A tentative model for the role of cytokinins at G1/S transition is depicted in figure 7. Initiation of the cell cycle is stimulated by cytokinins in conjunction with auxins through the enhanced expression of CDKA and cyclin D3. The cytokinin response may be mediated by a multi-specific CRE1/AHK4/WOL histidine kinase cytokinin receptor. Auxin and abscisic acid further modulate this response as they are respectively believed to relieve or exacerbate the inhibitory effect of Kip-related CDK inhibitory proteins (KRP) on the cyclin/CDK complex (61). Upon phosphorylation by a so called CDK-activating kinase (CAK) the CDKA/CycD3 complex becomes active and hyperphosphorylates Rb. Hyperphosphorylation causes Rb to release the heterodimeric E2F/DP transcription factor that it normally keeps in an inactive form allowing the latter to transcribe S-phase specific genes (27).

Cytokinin Signalling in S phase.

One of the main targets for cytokinin action in S-phase appears to be the activation and synchronisation of latent origins of DNA replication. In the shoot apical meristems of a number of plants, benzyladenine (BA) addition shortens replicon size and results in the shortening of S-phase (19). In BY-2 cells, cytokinin addition in early S-phase also enhances mitotic activity. Benzyladenine addition (10^{-7} to 10^{-6} M) results in an increase in mitotic index of about 10% but does not cause the maximum in mitotic activity to appear more rapidly. Higher amounts of BA (10^{-5} M) equally result in an increase of about 10%, but even cause a delay of the mitotic maximum (55).

A number of possible interfaces of cytokinins with S-phase phenomena have been extensively reviewed in (2). Putative targets include protein members of the origin recognition complexes (ORC), but conclusive evidence has not yet come forward. Three independent research groups have

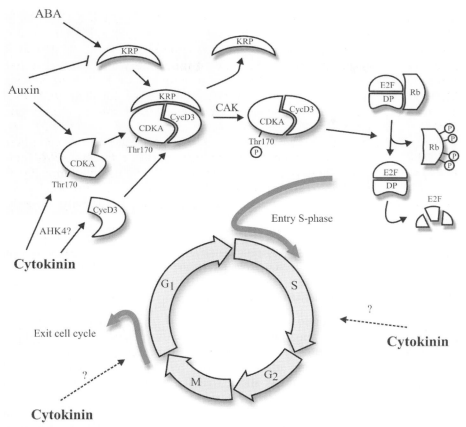

Figure 7. Tentative model for the role of cytokinins in cell cycle initiation and termination in plants. At G_1/S transition, cytokinins enhance expression of both CDKA and CyclinD3. A CycD3/CDK complex forms, is activated by CDK-activating kinase (CAK) through a phosphorylation of the CDK subunit on Thr170 (amino acid numbering according to tobacco), and hyperphosphorylates a retinoblastoma-like protein (Rb). Rb releases E2F bound to its heterodimerisation partner (DP) thus allowing it to start transcription of genes necessary for entry into S-phase. The cytokinin response is modulated by other plant hormones. Abscisic acid inhibits G_1/S transition through an up-regulation of Kip-related protein (KRP). The presence of auxin is not only required for CDKA expression, but has also been shown to counteract expression of certain KRP's. Some reports seem to indicate that a rise in cytokinin concentration at early G_1 may cause cells to leave the cell cycle. The underlying mechanisms are however unknown. The role of cytokinins in S-phase is even less understood.

detected a cytokinin rise in S-phase in cycling tobacco cells. However, the occurrence of this cytokinin peak after DNA synthesis seems to rule out a role in replication origin activation. One can speculate on a possible role in S/G_2 transition, but experiments with lovastatin feeding seem to indicate that the absence of cytokinins is not restrictive for S/G_2 transition.

Cytokinins and the plant cell cycle

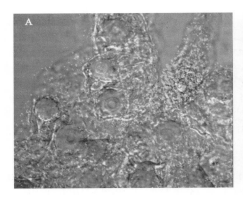 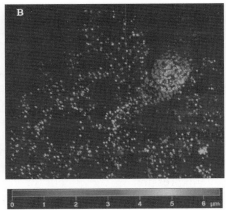

Figure 8 (Color plate page CP7). Confocal laser scanning microscope detection of silver enhanced immunogold labeling with anti-zeatin antibodies in exponentially growing BY-2 cells. (A) superposition of silver induced diffraction (green) on differential interference light image (B) silver induced diffraction at different focal planes (1-6 µm, indicated in colour bar)

Cytokinin Signalling in G_2/M Transition

The nature of the endogenously synthesised cytokinins at G_2/M, in conjunction with the apparent structure-specific rescue upon lovastatin challenge, identifies zeatin as an endogenously produced cytokinin signal essential in G_2/M transition in BY-2 (23). Immunolocalisation studies of three cytokinin types in apical shoot meristems of tobacco during floral transition and flower formation showed a cytoplasmic and perinuclear localisation of all three cytokinins, whereas zeatin also exhibited a clear cut nuclear localisation (Fig. 8) (8). This again indicates a specific physiological role of zeatin in nuclear processes. The patchy aspect of the nuclear zeatin localisation confirms *in vivo* both the ability of single dividing cells to provide their own synthesis of zeatin, and the spatio-temporal isolation of the zeatin signal between neighbouring dividing cells. These findings argue against the role of a membrane bound "cytokinin receptor", such as the AHK cytokinin receptors, at G_2/M. Instead a cytoplasmic or nuclear receptor specific for zeatin is probably at work here.

CDK's themselves have been suggested as immediate targets for cytokinin action. Roscovitine and olomoucine, two cytokinin-like compounds, impose a direct inhibition on plant CDK activity (15; 42). The inhibition of human CDK2 activity by olomoucine and isopentenyl adenine is brought about by a direct contact of these compounds at the ATP-binding pocket, to which they are bound in an orientation opposite to the one of ATP (48). Other cytokinin-like CDK inhibitors have now been reported to induce apoptosis and hence possess anti-proliferative activity in human cells (57). Isopentenyl adenosine and benzyl aminopurine have also been reported to induce programmed cell death in plant cells (4; 34). This effect may also be through direct interaction with CDK's. To date, experimental data for such mechanism of cytokinins with plant CDKs is lacking.

The stimulatory effect of cytokinins on G_2/M transition is through activation of CDKA;1 kinase activity, but data now suggest that this interaction is dependent on a phosphatase activity similar to the cdc25-phosphatase[6] that is responsible for Tyr15 dephosphorylation of the mitotic CDK, cdc2[7], in fission yeast. *Nicotiana plumbaginifolia* cells that are arrested in G_2 upon cytokinin starvation contain increased levels of CDKA carrying an inactivating phosphorylation at Tyr15, probably imposed on it by a Wee1-like kinase[8] (52). Cytokinin addition results in the removal of this phosphate group, thereby permitting subsequent entry into M (62). Ectopic overexpression in these cells of fission yeast cdc25, renders their mitosis cytokinin independent, which can be interpreted as indicating that a plant homologue of CDC25 may be an early target for cytokinin action (22). Cells in cdc25 transgenic tobacco plants divide at smaller cell size, suggesting premature cell division by shortening of G_2 (13). Although an enzymatic phosphatase activity similar to cdc25 was demonstrated (62), the isolation of a plant homologue has proven difficult. Analysis of the complete Arabidopsis genome does not reveal the existence of cdc25 in plants (56).

The above leads to the model shown in figure 9. Transition of BY-2 through G_2/M is specifically dependent on zeatin (under the restrictive condition of auxin presence). Its action probably relies on enhanced CDKA;1 and CycB expression, combined with the regulation of a dual specificity cdc25-like phosphatase removing the inhibitory phosphorylations at Thr14/Tyr15 on CDK. The cytokinin receptor molecule involved in G_2/M transition is not identified yet, but is thought to be specific for zeatin.

GENERAL CONCLUSION

The roles of cytokinin at early G_1 and in S-phase are not completely understood, but may respectively be involved in exit from the cell cycle and S/G_2 transition in ways that are yet unclear. One could imagine a function for the 26S proteasome in the latter.

In general it could be envisaged that cytokinins regulate cell division at three levels. The initiation and termination of the cell cycle are regulated by many cytokinins originating external to the cell cycle. Progression through the cell cycle and more specifically G_2/M transition seems to be under the control of one specific cytokinin (zeatin for BY-2 cells), which is produced by the cycling cell itself.

---

[6] Protein phosphatase allelic to the large phenotype 'cell division cycle' mutant cdc25 in fission yeast
[7] Mitotic cyclin dependent protein kinase allelic to the large phenotype 'cell division cycle' mutant cdc2 in fission yeast
[8] Protein kinase allelic to the 'wee1' (i.e. small) phenotype mutant in fission yeast

Cytokinins and the plant cell cycle

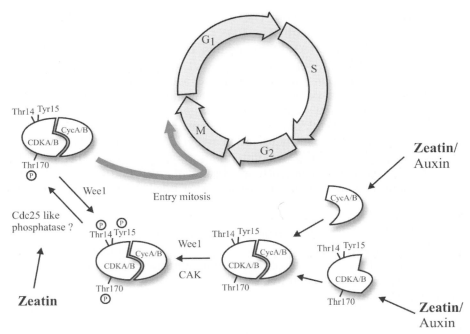

Figure 9. Tentative model for the action of zeatin at G_2/M transtion in the tobacco BY-2 cell culture. Zeatin and auxin cooperate to up-regulate CDKA and cyclin B expression. The activity of the cyclin/CDK complex is enhanced by a stimulatory phosphorylation on the CDK moiety at Thr170 and is inhibited by a phosphorylation at Tyr15 (and possibly also Thr14) by Wee1-like kinase(s). Zeatin counteracts the Wee1 induced inhibition through the action of a cdc25-like phosphatase activity that removes the inhibitory phosphorylation(s) (amino acid numbering according to tobacco).

The future challenge lies in identifying and describing the mode of action of the targets specific for each of these cytokinin actions and its cross-talk with other hormonal or environmental signals.

Acknowledgements

The authors wish to thank Prof. Dr. Russ Newton for critical reading of the manuscript, Dr. Walter Dewitte and Dr. Abdelkrim Azmi for fruitful discussions and Dr. Kris Laukens for his invaluable help with the preparation of figures. Work on the plant cell cycle carried out at the Lab of Plant Biochemistry and Physiology was supported by grants (P4/15 and P5/13) from the "Interuniversity Attraction Poles Programme – Belgian State – Federal Office for Scientific, Technical and Cultural Affairs"

References

1. Breyne P, Dreesen R, Vandepoele K, De Veylder L, Van Breusegem F, Callewaert L, Rombauts S, Raes J, Cannoot B, Engler G, Inze D, Zabeau M (2002) Transcriptome analysis during cell division in plants. Proc Natl Acad Sci U S A 99: 14825-14830
2. Bryant J, Francis D (2001) Plant growth regulators and the control of S-phase. In D. Francis, ed The plant cell cycle and its interfaces. Sheffield Academic Press; CRC Press, Sheffield, UK; Boca Raton, FL, pp 43-55
3. Brzobohaty B, Moore I, Kristoffersen P, Bako L, Campos N, Schell J, Palme K (1993)

Release of active cytokinin by a beta-glucosidase localized to the maize root meristem. Science 262: 1051-4
4. Carimi F, Zottini M, Formentin E, Terzi M, Lo Schiavo F (2003) Cytokinins: new apoptotic inducers in plants. Planta 216: 413-421
5. Crowell DN, Salaz MS (1992) Inhibition of growth of cultured tobacco cells at low concentrations of lovastatin is reversed by cytokinin. Plant Physiol 100: 2090-2095
6. Das NK, Patau K, Skoog F (1956) Initiation and cell division by kinetin and indoleacetic acid in excised tobacco pith tissue. Physiol Plant 9: 640-651
7. de O. Manes C, Van Montagu M, Prinsen E, Goethals K, Holsters M (2001) De novo cortical cell division triggered by the phytopathogen *Rhodococcus fascians* in tobacco. Mol Plant Microbe Interact 14: 189-195
8. Dewitte W, Chiappetta A, Azmi A, Witters E, Strnad M, Rembur J, Noin M, Chriqui D, Van Onckelen H (1999) Dynamics of cytokinins in apical shoot meristems of a day-neutral tobacco during floral transition and flower formation. Plant Physiol 119: 111-122
9. Dewitte W, Murray JAH (2003) The plant cell cycle. Annu Rev Plant Biol 54: 235-264
10. Dewitte W, Van Onckelen H (2001) Probing the distribution of plant hormones by immunocytochemistry. Plant Growth Regul 33: 67-74
11. Dobrev P, Motyka V, Gaudinova A, Malbeck J, Travnickova A, Kaminek M, Vankova R (2002) Transient accumulation of cis- and trans-zeatin type cytokinins and its relation to cytokinin oxidase activity during cell cycle of synchronized tobacco BY-2 cells. Plant Physiol Biochem 40: 333-337
12. Estruch JJ, Granell A, Hansen G, Prinsen E, Redig P, Van Onckelen H, Schwarz-Sommer Z, Sommer H, Spena A (1993) Floral development and expression of floral homeotic genes are influenced by cytokinins. Plant J 4: 379-384
13. Francis D, Sorrell DA (2001) The interface between the cell cycle and plant growth regulators: a mini review. Plant Growth Regul 33: 1-12
14. Frank M, Guivarc'h A, Krupkova E, Lorenz-Meyer I, Chriqui D, Schmulling T (2002) TUMOROUS SHOOT DEVELOPMENT (TSD) genes are required for co-ordinated plant shoot development. Plant J 29: 73-85
15. Glab N, Labidi B, Qin LX, Trehin C, Bergounioux C, Meijer L (1994) Olomoucine, an inhibitor of the cdc2/cdk2 kinases activity, blocks plant cells at the G1 to S and G2 to M cell cycle transitions. FEBS Lett 353: 207-211
16. Helliwell CA, Chin-Atkins AN, Wilson IW, Chapple R, Dennis ES, Chaudhury A (2001) The *Arabidopsis* AMP1 gene encodes a putative glutamate carboxypeptidase. Plant Cell 13: 2115-2125
17. Hemerly AS, Ferreira P, de Almeida Engler J, Van Montagu M, Engler G, Inze D (1993) cdc2a expression in *Arabidopsis* is linked with competence for cell division. Plant Cell 5: 1711-1723
18. Hemmerlin A, Bach TJ (1998) Effects of mevinolin on cell cycle progression and viability of tobacco BY-2 cells. Plant J 14: 65-74
19. Houssa C, Bernier G, Pieltain A, Kinet JM, Jacqmard A (1994) Activation of latent DNA-replication origins - a universal effect of cytokinins. Planta 193: 247-250
20. Inoue T, Higuchi M, Hashimoto Y, Seki M, Kobayashi M, Kato T, Tabata S, Shinozaki K, Kakimoto T (2001) Identification of CRE1 as a cytokinin receptor from *Arabidopsis*. Nature 409: 1060-1063
21. Ivanova M, Rost TL (1997) Cytokinins and the plant cell cycle: Problems and pitfalls of proving their function. In JA Bryant, D Chiatante, eds Plant cell proliferation and its regulation in growth and development. John Wiley & Sons, Chichester, pp 45-57
22. John PCL (1998) Cytokinin stimulation of cell division: essential signal transduction is via Cdc25 phosphatase. J Exp Bot 49 (suppl.): 91
23. Laureys F, Dewitte W, Witters E, Van Montagu M, Inze D, Van Onckelen H (1998) Zeatin is indispensable for the G2-M transition in tobacco BY-2 cells. FEBS Lett 426:

29-32
24. Laureys F, Smets R, Lenjou M, Van Bockstaele D, Inze D, Van Onckelen H (1999) A low content in zeatin type cytokinins is not restrictive for the occurrence of G1/S transition in tobacco BY-2 cells. FEBS Lett 460: 123-128
25. Letham DS (1963) Zeatin, a factor inducing cell division from *Zea mays*. Life Sci 8:
26. Mader JC, Hanke DE (1996) Immunocytochemical study of cell cycle control by cytokinin in cultured soybean cells. J Plant Growth Regul 15: 95-102
27. Magyar Z, Atanassova A, De Veylder L, Rombauts S, Inze D (2000) Characterization of two distinct DP-related genes from *Arabidopsis thaliana*. FEBS Lett 486: 79-87
28. Mähönen AP, Bonke M, Kauppinen L, Riikonen M, Benfey PN, Helariutta Y (2000) A novel two-component hybrid molecule regulates vascular morphogenesis of the *Arabidopsis* root. Genes Dev 14: 2938-2943
29. Meijer M, Murray JA (2001) Cell cycle controls and the development of plant form. Curr Opin Plant Biol 4: 44-49
30. Menges M, Hennig L, Gruissem W, Murray JA (2002) Cell cycle-regulated gene expression in *Arabidopsis*. J Biol Chem 277: 41987-42002
31. Menges M, Murray JA (2002) Synchronous *Arabidopsis* suspension cultures for analysis of cell-cycle gene activity. Plant J 30: 203-212
32. Miller CO, Skoog F, von Saltza MH, Strong FM (1955) Kinetin, a cell division factor from deoxyribonucleic acid. J Amer Chem Soc 77: 1329-1334
33. Mironov V, De Veylder L, Van Montagu M, Inze D (1999) Cyclin-dependent kinases and cell division in plants- the nexus. Plant Cell 11: 509-522
34. Mlejnek P, Procházka S (2002) Activation of caspase-like proteases and induction of apoptosis by isopentenyladenosine in tobacco BY-2 cells. Planta 215: 158-166
35. Murray JAH, Doonan J, Riou-Khamlichi C, Meijer M, Oakenfull EA (2001) G1 cyclins, cytokinins and the regulation of the G1/S transition. In D. Francis, ed The plant cell cycle and its interfaces. Sheffield Academic Press; CRC Press, Sheffield, UK; Boca Raton, FL, pp 19-41
36. Nagata T, Kumagai F (1999) Plant cell biology through the window of the highly synchronized tobacco BY-2 cell line. Methods Cell Sci 21: 123-127
37. Nagata T, Nemoto Y, Hasezawa S (1992) Tobacco BY-2 cell line as the 'HeLa' cell in the cell biology of higher plants. Int Rev Cytol 132: 1-30
38. Nakagami H, Kawamura K, Sugisaka K, Sekine M, Shinmyo A (2002) Phosphorylation of retinoblastoma-related protein by the cyclin D/cyclin-dependent kinase complex is activated at the G1/S-phase transition in tobacco. Plant Cell 14: 1847-1857
39. Nishinari N, Syono K (1986) Induction of cell division synchrony and variation of cytokinin contents through the cell cycle in tobacco cultured cells. Plant Cell Physiol 27: 147-153
40. Papon N, Clastre M, Gantet P, Rideau M, Chenieux JC, Creche J (2003) Inhibition of the plant cytokinin transduction pathway by bacterial histidine kinase inhibitors in *Catharanthus roseus* cell cultures. FEBS Lett 537: 101-105
41. Planchais S, Glab N, Inze D, Bergounioux C (2000) Chemical inhibitors: a tool for plant cell cycle studies. FEBS Lett 476: 78-83
42. Planchais S, Glab N, Trehin C, Perennes C, Bureau JM, Meijer L, Bergounioux C (1997) Roscovitine, a novel cyclin-dependent kinase inhibitor, characterizes restriction point and G2/M transition in tobacco BY-2 cell suspension. Plant J 12: 191-202
43. Prinsen E, Van Dongen W, Esmans EL, Van Onckelen HA (1998) Micro and capillary liquid chromatography tandem mass spectrometry: a new dimension in phytohormone research. J Chromatogr A 826: 25-37
44. Redig P, Schmulling T, Van Onckelen H (1996) Analysis of cytokinin metabolism in ipt transgenic tobacco by Liquid Chromatography-Tandem Mass Spectrometry. Plant Physiol 112: 141-148
45. Redig P, Shaul O, Inze D, Van Montagu M, Van Onckelen H (1996) Levels of

endogenous cytokinins, indole-3-acetic acid and abscisic acid during the cell cycle of synchronized tobacco BY-2 cells. FEBS Lett 391: 175-180
46. Riou-Khamlichi C, Huntley R, Jacqmard A, Murray JAH (1999) Cytokinin activation of *Arabidopsis* cell division through a D-type cyclin. Science 283: 1541-1544
47. Sakai H, Honma T, Aoyama T, Sato S, Kato T, Tabata S, Oka A (2001) ARR1, a transcription factor for genes immediately responsive to cytokinins. Science 294: 1519-1521
48. Schulze-Gahmen U, Brandsen J, Jones HD, Morgan DO, Meijer L, Vesely J, Kim SH (1995) Multiple modes of ligand recognition: crystal structures of cyclin-dependent protein kinase 2 in complex with ATP and two inhibitors, olomoucine and isopentenyladenine. Proteins 22: 378-391
49. Skoog F, Miller CO (1957) Chemical regulation of growth and organ formation in plant tissues cultured *in vitro*. Symp Soc Exp Biol 11: 118-131
50. Smalle J, Kurepa J, Yang P, Babiychuk E, Kushnir S, Durski A, Vierstra RD (2002) Cytokinin growth responses in *Arabidopsis* involve the 26S proteasome subunit RPN12. Plant Cell 14: 17-32
51. Sorrell DA, Combettes B, Chaubet-Gigot N, Gigot C, Murray JA (1999) Distinct cyclin D genes show mitotic accumulation or constant levels of transcripts in tobacco Bright Yellow-2 cells. Plant Physiol 119: 343-352
52. Sorrell DA, Marchbank A, McMahon K, Dickinson JR, Rogers HJ, Francis D (2002) A WEE1 homologue from *Arabidopsis thaliana*. Planta 215: 518-522
53. Stals H, Inze D (2001) When plant cells decide to divide. Trends Plant Sci 6: 359-364
54. Suzuki T, Miwa K, Ishikawa K, Yamada H, Aiba H, Mizuno T (2001) The *Arabidopsis* sensor His-kinase, AHK4, can respond to cytokinins. Plant Cell Physiol 42: 107-113
55. Temmerman W, Ritsema T, Simon-Mateo C, Van Montagu M, Mironov V, Inze D, Goethals K, Holsters M (2001) The fas locus of the phytopathogen *Rhodococcus fascians* affects mitosis of tobacco BY-2 cells. FEBS Lett 492: 127-132
56. The *Arabidopsis* Genome Initiative (2000) Analysis of the genome sequence of the flowering plant Arabidopsis thaliana. Nature 408: 796-815
57. Vermeulen K, Strnad M, Havlicek L, Van Onckelen H, Lenjou M, Nijs G, Van Bockstaele DR, Berneman ZN (2002) Plant cytokinin analogues with inhibitory activity on cyclin-dependent kinases exert their antiproliferative effect through induction of apoptosis initiated by the mitochondrial pathway: determination by a multiparametric flow cytometric analysis. Exp Hematol 30: 1107-1114
58. Vittorioso P, Cowling R, Faure JD, Caboche M, Bellini C (1998) Mutation in the Arabidopsis PASTICCINO1 gene, which encodes a new FK506- binding protein-like protein, has a dramatic effect on plant development. Mol Cell Biol 18: 3034-43
59. Witters E, Vanhoutte K, Dewitte W, Machackova I, Benkova E, Van Dongen W, Esmans EL, Van Onckelen HA (1999) Analysis of cyclic nucleotides and cytokinins in minute plant samples using phase-system switching capillary electrospray-liquid chromatography-tandem mass spectrometry. Phytochem Anal 10: 143-151
60. Yamada H, Suzuki T, Terada K, Takei K, Ishikawa K, Miwa K, Yamashino T, Mizuno T (2001) The *Arabidopsis* AHK4 histidine kinase is a cytokinin-binding receptor that transduces cytokinin signals across the membrane. Plant Cell Physiol 42: 1017-1023
61. Yang SW, Jin E, Chung IK, Kim WT (2002) Cell cycle-dependent regulation of telomerase activity by auxin, abscisic acid and protein phosphorylation in tobacco BY-2 suspension culture cells. Plant J 29: 617-626
62. Zhang K, Letham DS, John PC (1996) Cytokinin controls the cell cycle at mitosis by stimulating the tyrosine dephosphorylation and activation of p34cdc2-like H1 histone kinase. Planta 200: 2-12

C4. Expansins as Agents in Hormone Action

Hyung-Taeg Cho[1] and Daniel J. Cosgrove[2]

[1]Department of Biology, Chungnam National University, Daejeon 305-764, Korea, and [2]Department of Biology, Pennsylvania State University, University Park, PA 16802, USA. E-mail: htcho@cnu.ac.kr

INTRODUCTION

Expansins are wall-loosening proteins implicated in plant responses to most of the major plant hormones. These include cell enlargement, cell proliferation, fruit softening, abscission, senescence, and adaptation to water stress - all hormone-controlled responses in which the cell wall is modified so as to make it more extensible or softer or more easily separated from other walls.

In this chapter [1] we summarize the current evidence implicating expansins in the action of various hormones, beginning with a short summary of the characteristics of this protein and its likely mechanism of action.

EXPANSIN DISCOVERY, STRUCTURE AND MODE OF ACTION

Expansins were originally identified in studies of acid-induced extension of isolated plant cell walls. When cell walls are taken from the growing region of plant tissues and clamped at constant tension in an extensometer, they extend in a pH-dependent manner, with low pH inducing faster extension. This pH-dependent extension is eliminated with a brief heat treatment and can be restored by addition of a crude mixture of wall proteins (Fig. 1). Upon purification, the active proteins in this mixture proved to be ~27 kD proteins, that are now called α-expansins (12, 15).

From the fully-sequenced genomes of Arabidopsis and rice, we now know that the α-expansin (EXP) gene family is relatively large, with 26 genes in Arabidopsis and 32 genes in rice. A second family of expansins,

---

[1] Abbreviations: 1-MCP, 1-methylcyclopropene; ACC, Aminocyclopropane-1-carboxylic acid; BR, Brassinolide; NPA, Naphthylphthalamic acid; RHE, Root hair element, the *cis*-element for root hair specificity; RHF, Root hair factor, the putative transcription factor for root hair specificity; SAM, Shoot apical meristem.

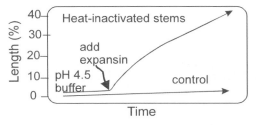

Fig. 1. Scheme for reconstituting acid-induced extension of cell walls. Hypocotyl walls are heated to inactivate the endogenous wall-loosening proteins, then protein extracted from other walls is added back and extension is measured in a constant-load extensometer.

called β-expansins (EXPB), was later identified on the basis of sequence similarity and expansin-like loosening effects on grass cell walls (14). This family consists of 6 genes in Arabidopsis and 19 genes in rice (see web site at http://www.bio.psu.edu/expansins/). The rice and Arabidopsis genomes also contain a third group of related genes called expansin-like (EXPL) and expansin-related (EXPR). This group is only known from its sequence and from expression (microarray) data; the proteins have not yet been isolated nor have their activities been characterized, so it is premature to call them expansins.

Although α- and β-expansins share only ~20% identity in protein sequence, they are structurally homologous to each other, each consisting of two domains (Fig. 2). Domain 1 is distantly related to the glycosyl hydrolase family-45 (GH45, see the CAZY web site at http://afmb.cnrs-mrs.fr/CAZY/). Much, but not all, of the active site for GH45 enzymes is conserved in domain 1 of expansins. Domain 2 may be a polysaccharide binding module, but this notion is still very speculative.

The two bona fide families of expansins (α- and β-) have similar effects on cell walls, namely rapid induction of wall extension and stress relaxation, but they act on different polysaccharides in the cell wall. This conclusion is

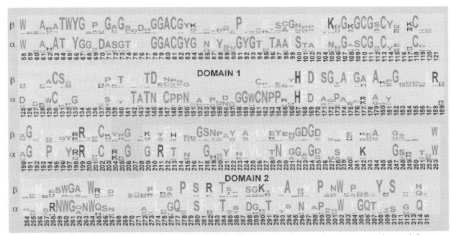

Figure 2 (Color plate page CP7). Alignment of sequence logos for α-expansin and β-expansin. Sequence logos show the magnitude of conservation by the size of the letters (on a log scale) and this alignment shows the conserved residues shared by both families of expansins. Each logo was made from an alignment of 16 sequences (15).

based on the stronger effect of α-expansins on dicot walls in extension assays, as compared with their action on grass walls, whereas the reverse is true for β-expansins, whose action is more marked on grass walls. Grass walls are notable for their relatively high content of arabinoxylan and reduced amounts of xyloglucan and pectin (3), and this difference in wall composition presumably accounts for the different sensitivities of dicot and grass cell walls to α- and β-expansins.

The β-expansins that have been studied in most detail make up a subgroup known in the immunological literature as group-1 allergens from grass pollen. These proteins are very abundant in grass pollen and they probably loosen the cell walls of the grass stigma and style, thereby promoting pollen tube penetration of the stigma and growth towards the ovule. They might also function in the separation of the pollen grains in the anther. The pollen allergen subgroup is distinctive in sequence from the remainder of the β-expansins, which have sometimes been called "vegetative β-expansins", to distinguish them from the pollen allergen class. The β-expansins from maize pollen have a pH optimum of ~5.5, which is notably different from the acidic optimum (pH < 4.5) found for α-expansins. These pollen allergens therefore do not appear to function in acid-induced growth. Whether this is also generally true for the vegetative β-expansins has not yet been tested, but we suspect that they have an optimum more compatible with acid growth, based on the wall extension behavior of oat and maize coleoptiles (29).

The large number of genes in the two families implies redundancy or specialization of the biological function for the different genes. In Arabidopsis most of the α-expansin genes are expressed in specific, distinctive cell types. This observation fits well with the idea that the different genes within a family differ principally in where and when they are expressed (that is, their promoters have differentiated from each other), but not in their loosening effect on the cell wall. However, this hypothesis still needs to be tested directly.

Expansins have a characteristic loosening action on the cell wall. In extension tests under constant load ("creep" tests), expansins rapidly induce wall extension, i.e., the wall begins to extend rapidly within 1-2 min. Expansins also enhance wall stress relaxation over a time range of 100 ms to >100 s. However, α-expansins do not change the plasticity or elasticity of the cell wall (41). These effects on wall mechanics appear to be unique to expansin.

The precise mechanism of wall loosening by expansin is still a bit mysterious. The major matrix polymers of the cell wall are not hydrolyzed or otherwise modified, as far as we can detect. Despite a report that pollen-allergen-type β-expansins have protease activity, this could not be confirmed by further work and seems to be an experimental artifact (22). In our current model, expansin disrupts the non-covalent adhesion of matrix polysaccharides to cellulose or to other scaffolding elements in the cell wall, thereby freeing the wall polymers to move in response to the mechanical forces generated by cell turgor (13).

EXPANSINS IN AUXIN ACTION

The earliest studies of auxin made a connection between this hormone and control of cell enlargement. How auxin induces cell enlargement has long been a vexing problem in the field of plant hormones. From the biophysical perspective, the increase of plant cell volume depends on turgor pressure and cell-wall extensibility. It is now widely accepted that auxin induces cell expansion mainly by increasing the ability of the cell wall to extend, at least for its short-term action.

Acid Growth and Expansins

About 35 years ago, research was focused on an auxin-induced factor hypothesized to enhance cell wall extensibility, a so-called *wall-loosening factor*. The *acid growth hypothesis* proposed protons (H^+) as a mediator between auxin and cell-wall loosening (see Chapter C1). A study of the underlying molecular mechanism of this response led to the discovery of expansins.

Cosgrove (11) characterized acid-induced extension of cucumber hypocotyl walls and concluded that one or more wall proteins were required for acid-induced wall extension. Subsequent investigations showed that a novel class of proteins, eventually named *expansins,* were the principal proteins with this activity (24, 25). In line with the acid growth hypothesis, cell wall extension activity by α-expansins has an acidic optima (pH 3~5.5) in both dicot stems and grass coleoptiles.

A relationship between expansin, acid growth, and auxin has been suggested by several experiments (13). Tobacco suspension cells grew three times faster in response to 0.5 µM fusicoccin, a fungal toxin that stimulates H^+ excretion from the plant cell, thereby inducing "acid growth". Partially purified α-expansins at 1 µg/mL elicited a similar level of growth stimulation as did fusicoccin. Exogenous α-expansins also induced growth of excised Arabidopsis hypocotyls comparable to that induced by optimal auxin concentrations (Fig. 3), and the effects by auxin and expansin were not additive. These results suggest that auxin and expansins stimulate cell expansion through a common pathway (15). Thus, if we accept that auxin stimulates cell wall acidification, at least part of the resulting growth response should be mediated by the expansins residing in the cell wall. This might account for the early stages of auxin-induced growth, but does not exclude the possibility of other responses for later stages.

Auxin-Regulated Expansin Genes during Cell Growth

Studies have shown that auxin regulates expression of some, but not all, expansin genes. For example, expression of an expansin gene (*LeEXP2*) cloned from the hypocotyl of tomato seedlings showed good correlation with auxin-induced growth. The hypocotyl has been a popular model system for study of auxin-induced cell growth. In the hypocotyl, the growth rate is highest at the top and gradually decreases towards the basal region. The expression pattern of *LeEXP2* reflects nicely these spatial differences in growth rate of

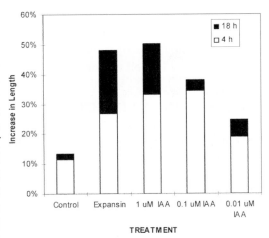

Figure 3. Elongation of Arabidopsis hypocotyl segments by expansin or auxin. The growing regions of etiolated Arabidopsis hypocotyls were excised and floated on a solution supplemented with partially purified cucumber α-expansin (10 µg/mL) or with IAA. Length was measured 4 h and 18 h after treatment (15)

the hypocotyl (Fig. 4A). Auxin treatment markedly increased *LeEXP2* expression in the hypocotyl (Fig. 4B). The *LeEXP2* transcript level increased to 8 fold within 1 h after auxin treatment and reached about 15 fold in 6 h (Fig. 4D). In contrast, gibberellin (GA) had little effect on hypocotyl elongation and the ethylene precursor 1-aminocyclopropane-1-carboxylic acid (ACC) had an inhibitory effect on tomato hypocotyl elongation. Consistent with these growth effects, *LeEXP2* expression level was little changed by GA and decreased by ACC treatment (Fig. 4C). Another growth hormone, brassinolide (BR), showed even stronger growth effect on tomato hypocotyls than did auxin. However, *LeEXP2* gene induction by BR was not significant, suggesting that *LeEXP2* is more intimately regulated by the auxin pathway during cell growth. It is possible that another α-expansin gene is up-regulated by BR, or BR might operate via a separate pathway that does not entail enhancement of expansin gene expression. The *LeEXP2* promoter region contains putative auxin regulatory *cis*-elements (TGTCAC) that exist in *PS-IAA 4 and 5,* the well known auxin-regulated gene.

Gravitropic responses of the stem and the root provide additional evidence for a role of expansin in auxin-regulated cell. In response to gravity, a horizontally laid stem bends upward as a result of differential growth by the top and bottom halves of the horizontal stem. It is generally accepted that this differential growth results from auxin redistribution between the upper and lower parts of the stem (or root). The expression of tomato expansin *LeEXP2* decreased markedly in the top half of the stem within 30 min after

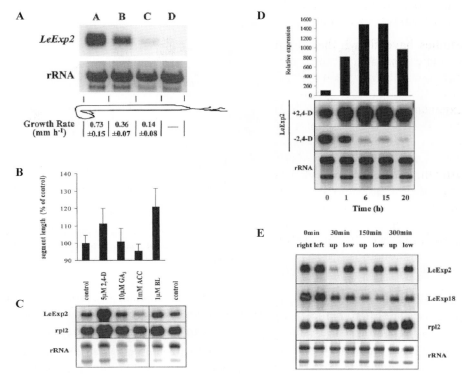

Figure 4. Regulation of expansin genes in the tomato hypocotyl and stem. A, Growth rates and *LeEXP2* mRNA levels in different hypocotyl regions. B, Hypocotyl growth in response to hormones. C, Changes of expansin mRNA levels in hormone treatments. D, Time course of the *LeEXP2* mRNA change after 2,4-D treatment. E, Changes of expansin mRNA levels after gravitropic stimulation of the stem. Total RNA was prepared from the upper or lower halves of the stem after stimulation. A, From (4). B-E, From. (2).

gravity stimulation, while not changing in the bottom half (Fig. 4E) (2). Reduced *LeEXP2* expression in the upper part of the stem could reduce elongation in the upper part, thereby resulting in upward bending. Another study examined gravitropism of the maize root, which bends downward in response to gravitropic stimulus. More expansin protein was detected by immunolocalization in the upper half of the root than in the lower half (Fig. 5). Moreover, naphthylphthalamic acid (NPA), an auxin transport inhibitor, considerably delayed both gravitropism and the asymmetrical distribution of expansin proteins (42). This indicates that auxin is required for differential expansin expression across the gravistimulated root.

With respect to the evolution of cell growth mechanisms, "acid growth" of cell walls appears to be universal in the plant kingdom. Consistent with this, expansin genes have been found in angiosperms, gymnosperms, ferns and bryophytes (23). It seems likely that expansin-mediated wall loosening

Expansins

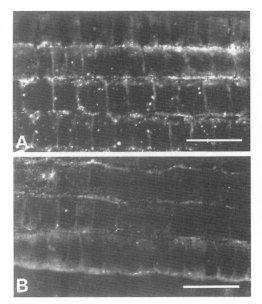

Figure 5. Distribution of expansin proteins during gravitropic stimulation in the maize root. After 30 min of horizontal orientation, the upper part (A) showed stronger expansin signals, remarkably in the periclinal walls, than did the lower part (B). Anti-CsEXP1 (a cucumber expansin) antiserum was used for the primary antibody reaction, and fluorescent secondary antibody was used for visualization. Bars = 50 μm. (42)

and cell expansion evolved early in the origin of land plants and that these processes are often regulated by auxin (and other hormones too; see below).

In yet another study linking expansins to auxin action, expression of a gymnosperm α-expansin (from loblolly pine) increased by as much as 100 fold in response to auxin during adventitious rooting of the hypocotyl, in which active cell growth and differentiation occur (19). Expression of a β-expansin gene from the moss *Physcomitrella patens* also increased within 2 h after auxin treatment of the chloronemal tissue (35). It was proposed that this auxin-regulated β-expansin plays a role in the developmental transition from chloronemata to caulonemata. However, knock-out mutations of the moss expansins did not show marked phenotypes, perhaps due to overlapping expression and a common function of multiple expansin genes expressed in the few cell types in the moss.

Auxin, Lateral Organ Formation, and Expansins

Another auxin-related function of expansins is in leaf primordium initiation on the flanks of the shoot apical meristem (SAM). Initiation of lateral organs such as leaves and flowers on the SAM follows a regular pattern, called *phyllotaxy*. Recent work has implicated auxin as a major regulator of lateral organ formation and patterning in the SAM (31), and also expansin involvement in primordium initiation on the SAM.

Loss-of-function mutation of the Arabidopsis auxin efflux carrier PIN1 or treatment of the tomato shoot apices with NPA (naphthylphthalamic acid, an auxin transporter inhibitor) caused defects in lateral organ formation, but did not affect other SAM processes such as cell division and meristem identity (30). These treatments resulted in a leafless naked stem (resembling a pin, ergo the name of the gene). However, localized application of auxin (IAA) to the naked meristem restored leaf formation (Fig. 6). This experiment supports the idea that local auxin concentrations set up by the auxin transporter modulate the initiation of lateral

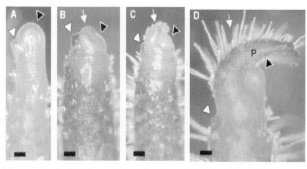

Figure 6. Induction of the leaf by IAA in the NPA-treated tomato apical meristem. A, Four days after treatment with control past (white arrowhead). B-D, Induction of leaf primordium (white arrows) with lanolin paste containing 10 mM IAA (white arrowheads); 1 day (B), 2 days (C), and 4 days (D) after auxin treatment. Black arrowheads point the meristem. Bars=100 μm. (30)

organs and their spatial patterning. Furthermore, it was proposed that auxin affects the spatial organ patterning by regulating the spatial expression pattern of the very early organ identity genes such as *LFY*, *ANT*, and *CUC2*.

Other studies indicate the regulation of lateral organ formation directly by expansin. In-situ hybridization showed that an α-expansin gene was expressed specifically at the location of future leaf primordia initiation on the tomato SAM (32) (Fig. 7). Previous work showed that Sephacryl beads loaded with expansin protein induced leaf-like organs on the tomato SAM and altered leaf phyllotaxis. This result indicates that expansin can induce premature enlargement of incipient leaf primordia, and this then changes the subsequent pattern of phyllotaxy. These conclusions were reinforced by use of tobacco plants transformed with a α-expansin gene whose expression was controlled by the tetracycline-inducible promoter (28). Micro-application of tetracycline to the site of incipient leaf initiation (I2), which normally forms the primordium after I1, induced a leaf primordium, skipping the normal leaf initiation at I1 and reversing the phyllotaxis (Fig. 8A). In contrast to the previous study using exogenous expansin proteins, the leaf primordia stimulated by this technique developed into a normal leaf (Fig. 8C).

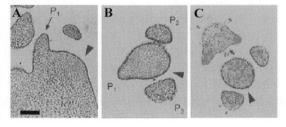

Figure 7 (Color plate page CP7). In situ localization of *LeEXP18* mRNA in the tomato shoot apical meristem. A–B, Localized expression of *LeEXP18* mRNA (in red). A, Longitudinal section. B, Cross section. C, Distribution of histone H4 transcript as a control (in red). Arrowheads indicate the site of incipient leaf initiation. Bar=100μm. (32)

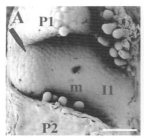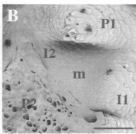

Figure 8. Endogenous expansin-induced leaf morphogenesis in tobacco. A, Scanning electron micrograph (SEM) of a shoot apex where tetracycline-loaded lanolin was placed onto the I2 position of the meristem (m). After 3 days, a leaf primordium bulge (arrow) has formed at this position. B, SEM of an apex treated with buffer at I2 position. No bulge has formed. C, A leaf normally developed from endogenous expansin-induced primordium. D, Altered leaf morphology (arrow) by local induction of endogenous expansin. Tetracycline was applied onto one flank of the leaf primordium (P2 stage) and it was allowed to grow 2-4 weeks. Bars=150 µm (A and B), 0.5 cm (C), and 1 cm (D). (28)

These results show that localized expansin expression can prematurely advance development of leaf primordia. However, they do not demonstrate that expansin expression *alone* is sufficient for leaf initiation (only distal regions containing incipient leaf primordia were shown to development into leaves, other regions of the SAM evidently did not respond in the same way). Interestingly, localized induction of expansin in the young leaf primordium changed the leaf shape, resulting in a lobe (Fig. 8D).

How does expansin modulate organogenesis in the SAM? Expansin itself is not likely to function as a ligand to start a signaling cascade nor is it likely to be a direct regulator for gene expression. Expansin loosens the cell wall and stimulates cell enlargement, which will change the pattern of compressive and tensile forces in the SAM. Paul Green (18) proposed the hypothesis that the pattern of such physical forces within the SAM is part of the mechanism that establishes and maintains the stable pattern of leaf initiation. The reversal of phyllotaxy upon localized expansin expression is consistent with this idea. A stress-sensing mechanism, such as mechano-sensitive ion channels, may be activated by expansin-induced cell growth, or cell wall modification by expansin may stimulate some wall-associated signaling molecules that are sensitive to wall shear. These stress-sensing mechanism could then initiate a signaling cascade regulating cell identity and gene expression.

It is plausible that auxin operates upstream of expansin action in the SAM. As mentioned earlier, local application of auxin induces leaf organogenesis in the SAM as did expansin. Localized auxin molecules may elevate local expression of certain expansin genes. Consistent with this idea, *LeEXP18*, the tomato expansin gene expressed at the site of incipient leaf initiation, is up-regulated by auxin (2). In summary, a tenable, if speculative, model for lateral organ formation in the SAM is as follows: (a) pre-patterning of auxin distribution, which likely is affected by previous local auxin concentration (29) and activity of auxin transporters; (b) local elevation of expansin gene expression via local auxin, (c) changes in the pattern of physical forces between cells as a result of cell enlargement, and consequent mechano-sensitive signaling into the cell from the extracellular matrix, and finally (d) activation of gene expression required for organ formation (division, cell specification, and so on).

Other Wall-Loosening Factors in Auxin-induced Growth

In addition to expansin-mediated acid growth, hydroxyl radicals (•OH) have also been proposed as potential wall-loosening factors (37). Hydroxyl radicals are very aggressive reductants that can react and cleave essentially all biological polymers, including cell wall polysaccharides. Under appropriate conditions they may be generated by reaction of superoxide radicals or other reactive oxygen species with copper or iron ions or by iron-containing peroxidase, which is commonly found in cell walls. Reactive oxygen species, including hydroxyl radicals, are implicated in cell death associated with the hypersensitive response, in cross-linking phenolic polymers in the cell wall, and, more recently, in various signaling functions. The hypothesis advanced by Schopfer and coworkers is that controlled release of •OH in the cell wall may contribute to the cell-wall loosening.

A recent study demonstrated that •OH could induce the irreversible extension (creep) of isolated cell walls, similarly as shown with acid solutions or expansins (36). The creep activity induced by •OH has an acidic optimum and artificially generated •OH also induced elongation of living coleoptile and hypocotyl segments. Relevant to this section, auxin could induce the formation of superoxide radicals ($O_2^{\bullet -}$), a potential precursor of •OH, in the epidermis, and auxin-induced stem growth was inhibited by •OH scavengers (which, it must be said, are relatively nonspecific in action). This work indicates that •OH can be a potent wall-loosening factor in vitro; it remains to be seen whether •OH is generated endogenously in the necessary concentration and in the specific location required for it to play a significant role in auxin-mediate cell growth.

EXPANSINS IN GIBBERELLIN ACTION

The role of expansins in gibberellin (GA) action has been mainly studied in two subjects; cell growth and seed germination, which are also the most actively researched topics in the study of GA. In this section, we describe the regulation of expansin gene expression by GA and their likely role in these two physiological processes.

Expansins in GA-Induced Cell Growth

GA was first discovered by Japanese scientists in 1926 due to its growth-stimulating activity. Many insightful studies of GA-mediated growth have been done using dwarf cultivars defective in GA biosynthesis. The dwarf rice cultivar Tanginbozu shows an extreme sensitivity of growth to GA as low as 3.5 picogram of GA_3. Another model system, deepwater rice, can grow at rates of 25 cm per day upon submergence, mostly through stem internodal growth (20). The marked growth response of the stem takes place via a chain of hormonal interactions upon submergence: (1) increase in ethylene level, (2) lowered abscisic acid (ABA) level, (3) increased ratio of GA to ABA, and (4) elongation of the stem. Thus, direct application of GA to the stem can also induce the growth response. GA-induced and submergence-induced stem growth resulted in more cells and longer cells. In physical terms, GA resulted in greater extensibility of the internodal cell wall. The remarkable growth potential of deepwater rice stems and the GA effect on cell-wall properties led to study of the role of expansins in this growth process (20, 21). Two α-expansin proteins were purified from the growing stems of deepwater rice, and the transcript level of one (*OsEXP4*) of these expansin genes was induced within 30 min after treatment with GA or submergence, preceding the growth response by these stimuli (Fig. 9). Further investigation showed that 5 α-expansin, 4 β-expansin and one expansin-like genes were up-regulated by GA in the deepwater rice stem. Submerged rice stem sections had greater amounts of expansin protein per unit cell-wall material, and the cell wall from the submerged stem showed 2 fold higher acid extension than did that from the unsubmerged stem. These results support the hypothesis that a major part of the GA-induced elongation in deepwater rice is mediated by increased expression and activity of expansins.

The petiole of dicot *Rumex palustris* also elongates rapidly upon submergence, and this growth process follows a similar hormonal cascade as found in deepwater rice, except that auxin is added to GA as the final growth stimulator (38). The mRNA accumulation of a *R. palustris* α-expansin gene (*RpEXP1*) upon submergence was correlated with submergence-induced petiole elongation. In *R. acetosa*, a *Rumex* species

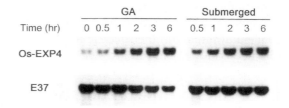

Figure 9. Accumulation of *OsEXP4* transcript in response to GA or submergence. Growing stem sections from deepwater rice, after 2-hr stabilization in water, were incubated for 0 to 6 hr in 50 μM GA_3 or submerged. Total RNA was extracted at the indicated time points from the growing basal 2 cm region. E37 is the loading control. (From 9)

that does not elongate in response to submergence, the transcription of *RpEXP1* did not increase by submergence. These results closely parallel those found in deepwater rice and further support the role of expansins in GA action.

A recent study using transgenic rice plants adds further evidence for the *in-vivo* function of expansin during cell and organ growth (10). Rice plants harboring an antisense *OsEXP4* construct had significantly reduced seedling height, mature plant height, and mesocotyl cell length (Fig. 10). Cell walls were also less extensible at low pH than were controls. Overexpression of *OsEXP4* resulted in the opposite phenotypes, although excessively high expression of *OsEXP4* suppressed plant growth. Moreover, the growth capability of seedlings upon submergence was also considerably reduced in the antisense lines, while the sense lines maintained a similar level to control. This study supports the conclusion that GA induces cell growth at least partially through expansin-mediated loosening of the cell wall.

Expansins during Seed Germination

Both successful germination in favorable conditions and continued dormancy in unfavorable conditions are critical for a plant species to maintain its reproduction by means of seeds. Seed germination is characteristically controlled by an antagonistic rivalry between GA and ABA, where GA promotes germination and ABA inhibits it. For an endospermic seed to germinate, the storage products in the endosperm need to be mobilized. In some cases, the hydrolysis of storage macromolecules includes not only starch, protein, and oil inside the cell, but also polymers outside the cell such as cell wall polysaccharides. For emergence of the radicle, successful growth of the radicle cells and, in cases where the endosperm overlies the radicle, weakening of the endosperm cells on the path of radicle growth are required. Diverse enzymatic activities are thought to be involved in hydrolysis and weakening of the

Figure 10. Transgenic rice plants harboring sense and anti-sense constructs of *OsEXP4*. a, Antisense; b, Control; c, Sense. Expression of the sense and antisense *OsEXP4* constructs were driven by the constitutive ubiquitin 1 (Ubi-1) promoter. (From 10)

Expansins

cell walls during seed germination. A role for expansin in these processes has been hypothesized.

Most studies in this area have focused on the dynamics of gene expression in germinating tomato seeds. Characterization of the effect of expansins on cell walls of germinating seed has not yet been published. Expression of two α-expansin genes was found to correlate with radicle growth and weakening of the endosperm barrier (5, 6). *LeEXP8* is specifically expressed in the elongating region of the radicle, and *LeEXP4* is expressed in the endosperm cap, the very endosperm portion where the radicle must penetrate for emergence. *LeEXP4* expression in the endosperm cap is also accompanied by the expression of other genes encoding wall-hydrolytic enzymes such as xyloglucan endotransglycosylase (*LeXET4*) and endo-β-mannanase (*LeMAN2*) (Fig. 11) (7, 26). Expression of these four genes starts prior to the radicle emergence and is positively regulated by GA. Antagonists of germination, such as ABA and low water potential, partly inhibit α-expansin gene expression in the tomato seed, though this varies between the two genes. These studies suggest that α-expansins, together with wall-hydrolytic enzymes, play a role in growth of the radicle and weakening of the endosperm cap so as to complete the germination process.

EXPANSINS IN ETHYLENE ACTION

Expansins in Ethylene-Mediated Cell Growth

As described previously, ethylene is the first hormone in the hormonal cascade causing stem and leaf growth upon submergence in deepwater rice and *Rumex*. In *Rumex*, the α-expansin gene *RpEXP1* was shown to be induced by ethylene treatment as well. The ethylene-mediated organ growth after submergence is conserved even in certain ferns. When submerged, the rachis of the semiaquatic ferns *Marsilea quadrifolia* and *Regnellidium diphyllum* elongate rapidly. α-Expansin genes, with high sequence similarity to those found in angiosperms, were identified from these ferns, and their

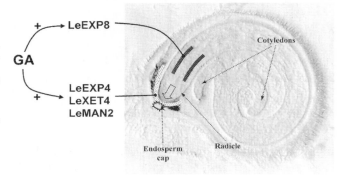

Figure 11. Cross section of a germinating tomato seed showing the expression locations of expansins and other cell wall degrading enzymes under the control of gibberellin

expression in the rachis showed a strong induction by submergence and ethylene. These expression patterns also were closely correlated with changes in rachis elongation and cell-wall extensibility by the two growth stimulators. These results in ferns and in angiosperms suggest that expansins are common targets of growth-inducing stimuli in vascular plants (21).

Expansins during Ethylene-Involved Fruit Ripening

Animals are attracted to well-ripened fruit by its smell, color and the softness. For fruit ripening species are classified as climacteric (bananas, apples, tomatoes, etc) or non-climacteric (citrus species, strawberries, etc), where the former is sensitive, but the latter is minimally- or non- sensitive to ethylene. Fruit softening is accompanied by depolymerization and solubilization of pectins and hemicelluloses. Although cell-wall hydrolases were thought to drive the softening, transgenic experiments suggest that other processes may be important (33).

A tomato α-expansin gene, *LeEXP1*, was first found to be expressed in a ripening-specific manner and up-regulated by endogenous and exogenous ethylene. Subsequent studies with transgenic tomatoes demonstrated the in-vivo role of LeEXP1 in fruit softening. The fruits from antisense lines were slightly firmer throughout fruit development, whereas in sense lines the softening time was advanced, even to the green fruit stages, and softness also increased. These changes of physical characteristics in the transgenic fruits were also accompanied by changes in the status of wall polymers, suggesting that expansins soften tomato fruit primarily through a relaxation of the wall by direct action and through controlling the access of pectinases to the pectins in the wall.

In contrast to expansin genes from climacteric tomatoes, an α-expansin gene (*FaEXP2*) from non-climacteric strawberries was insensitive to ethylene, although its pattern of gene expression was correlated with fruit ripening. Thus, the expression of certain α-expansin genes suggests a role in fruit softening and this is supported by the results with sense and antisense transformed tomato lines (1). The effects noted in the latter experiments were significant, but relatively subtle. Thus, expansins are not the sole catalysts of fruit softening.

Another ethylene-controlled process that shares some similarities to fruit softening is flower petal senescence. Expansin genes are expressed in the flower petals of peas (*Pisum*) and four o'clocks (*Mirabilis*), with a likely, though untested, role in petal growth and opening.

Expansins in Root Hair Development

The root hair is a protrusion from a root epidermal cell. Its growth is a highly polarized cellular process and its initiation by epidermal cells involves and probably requires localized cell-wall acidification and loosening (17).

Root hairs contribute greatly to the surface area of the root, and their initiation and elongation is regulated both by ethylene and auxin, and also by environmental stimuli.

The regulation of two root hair-specific α-expansin genes (*AtEXP7* and *AtEXP18*) from Arabidopsis was studied to dissect the role of ethylene and auxin at the gene regulation level (8). The expression of these genes starts immediately before the root hair bulge appears, indicating a close relationship to hair initiation (Fig. 12A). The Arabidopsis root epidermis consists of alternative longitudinal files of hair and non-hair cells, whose fates are determined by positional information and interactions between transcription factors (34). Exogenous ethylene can induce root hairs from the non-hair cell position accompanied by the induction of the root hair expansin genes. Treatments with ethylene and auxin restored root hairs in the hairless mutant *rhd6*, and similarly restored expression of the hair-specific expansin gene (Fig. 12F-H). Similar results were found with a treatment in which the root was separated from the agar medium (probably causing water stress). Because the ethylene antagonist 1-methylcyclopropene (1-MCP) inhibits the effects of auxin and root separation, ethylene is likely to mediate the effects of these two factors on root hair formation and expansin gene expression. Promoter analyses of *AtEXP7* revealed that signals from all the root hair regulatory factors (developmental, hormonal and environmental cues) converge onto a small promoter region, most likely for binding by a single transcription factor (Fig. 13).

In contrast, treatment of wild type with 1-MCP and mutations of the ethylene signaling components did not affect root hair formation and expansin gene expression. This was interpreted to mean that the ethylene signaling cascade in wild type is not essential to affect these two processes under normal condition. This study proposed that there are two pathways that modulate root hair initiation and expression of related genes: an ethylene-dependent pathway (requiring high ethylene concentration) and a

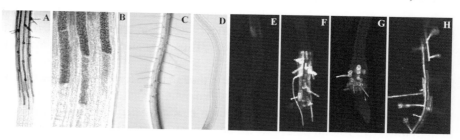

Figure 12 (Color plate page CP8). Expression pattern of *AtEXP7* in the Arabidopsis root. A-B, *AtEXP7* promoter::GUS expression. The reporter gene expressed only in the hair cell files, starting immediately before the bulge emerges. C, Wild-type root. D, Root from hairless *rhd6*. E-H, Confocal images showing the expression of *AtEXP7*::GFP (green fluorescent protein) in *rhd6*. Control (E), 5 μM ACC (the ethylene precursor, F), 30 nM IAA (G), and the root separated from the agar medium (H) (From 8)

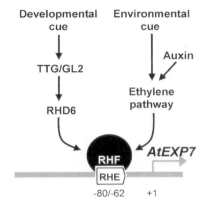

Figure 13. Regulation of root hair-specific α-expansin gene *AtEXP7* by two separate pathways. Defects in TTG or GL2 raise root hairs and *AtEXP7* expression in both hair and non-hair cells, whereas the RHD6 mutation blocks root hair formation and *AtEXP7* expression. RHE, the *cis*-element rendering *AtEXP7* both specified in the hair cell and responsive to ethylene. RHF, a putative transcription factor that binds to RHE. Numbers at the bottom indicate the relative position from the transcription start (From 8)

second internal developmental pathway that is independent of ethylene (Fig. 13). Although several genes have been identified that are required for normal root hair development, none of them have yet encoded an expansin. Thus, while expression of root hair-specific expansin genes is tightly correlated with root hair initiation and elongation, it is not clear whether they are essential for either process.

EXPANSINS IN CYTOKININ ACTION

Cell suspension cultures typically require cytokinins in the medium in order to maintain cell proliferation, which includes both cell division and cell enlargement. When deprived of cytokinin, such cells stop growing and subsequent addition of cytokinin to the medium results in greatly increased transcript levels of a β-expansin, called Cim1 (for cytokinin-induced message) (16). *Cim1* mRNA accumulated 20-60-fold upon cytokinin addition to cytokinin-starved soybean suspension cultures. Cytokinin accomplished this, at least in part, by increasing the stability of *Cim1* mRNA. A further study characterized the Cim1 protein and found that cytokinin acted synergistically with auxin to induce the accumulation and proteolytic processing of Cim1 protein in the culture medium. It is not clear whether such proteolytic processing is simply stepwise degradation of the protein in the medium or has other significance. These studies implicate expansins in cytokinin-induced cell proliferation.

This idea is further strengthened by results of a study of haustorial development in parasitic plants such as *Striga asiatica* and *Triphysaria versicolor* (27, 39). Haustorial formation in *Triphysaria* can be induced in vitro by the cytokinin 6-benzylaminopurine and this response involves localized swelling and proliferation of epidermal hairs near the root tips. The expression of two α-expansin genes was increased upon cytokinin application. Haustorial development could also be induced by exudates from maize roots and by the inducing substance 2,6-dimethoxybenzoquinone. However, these treatments did not increase expression of these α-expansin genes. Thus, these two genes are linked to the cytokinin response, and not specifically to haustorial development.

EXPANSINS IN ABSCISIC ACID ACTION

Three sets of studies have linked expansins to the action of abscisic acid. The first two deal with tomato seed germination and *Rumex* petiole elongation upon submergence; these were described above in the section on GA. The third set involves responses of maize roots to drought.

In many plants, shoot growth is typically inhibited more strongly by low water potential than is root growth. This differential response increases the root:shoot ratio and is considered a favorable adaptation to low water availability. In maize seedlings transplanted to relatively dry rooting medium, the apical part of the root elongation zone continues to grow, whereas the basal half of this zone, as well as the shoots, are strongly inhibited in growth. The continued growth of the root at low water potential occurs because the walls become more extensible, and this involves enhanced levels of expansin protein and transcripts in these root cells (40). Transcript levels for two α-expansins and one β-expansin were rapidly increased upon restriction of water availability. Roots treated with the ABA biosynthesis inhibitor, fluoridone, did not exhibit this growth adaptation to low water potential. This suggests that drought-induced increases in ABA level may act as a signal to activate genes needed for the root adaptation. However, direct application of ABA to well-watered roots did not cause the level of expansin transcripts to increase in the same way that drought did. Thus, while ABA may be required to induce the root response, it does not mimic drought with respect to altered expression of expansin genes. This, in turn, implies that drought modulates expansin gene expression via an ABA-independent pathway.

CONCLUDING REMARKS

The initial discovery of expansins stemmed from studies of cell wall enlargement, but in the intervening decade many unexpected aspects of expansin function have been discovered. We now know that expansins comprise a large multigene family, some of whose members are regulated by one or more of the plant hormones. This regulation is different for different gene, and moreover is specific to certain cell types. For instance, ethylene promotes expression of *EXP7* in Arabidopsis roots, but it does not promote expression of this expansin in other cell types (though it may modulate expression of some expansin genes in other cell types). Many effects of plant hormones involve significant changes in the cell wall, i.e. stimulation of cell enlargement, germination, fruit softening, abscission, etc., and there is abundant evidence that such wall modifications involve the action of expansins. Alteration of expansin gene expression by transgenic methods confirms the role of expansins in wall loosening, especially during cell growth. However, until now, no one has identified an expansin knock-out mutant with a major developmental phenotype. This might be explained by overlapping expression of multiple expansin genes or by compensatory physiological changes by the plant to make up for the genetic defect. A hint of the latter explanation may be seen in transgenic plants overexpressing expansin – some of them have reduced growth despite high levels of expansin activity. Presumably such plants have modulated other properties of the wall in order to limit the effects of high, unregulated expression of expansin. Future work will need to identify the steps between expansin gene expression and the early steps in hormone signal transduction. When that is accomplished, we will have a complete link between initial hormone reception and final physiological effect – something that has eluded plant biologists up to now.

References

1. Brummell DA, Harpster MH, Civello PM, Palys JM, Bennett AB, Dunsmuir P (1999) Modification of expansin protein abundance in tomato fruit alters softening and cell wall polymer metabolism during ripening. Plant Cell 11: 2203-2216
2. Caderas D, Muster M, Vogler H, Mandel T, Rose JK, McQueen-Mason S, Kuhlemeier C (2000) Limited correlation between expansin gene expression and elongation growth rate. Plant Physiol 123: 1399-1414
3. Carpita NC (1996) Structure and biogenesis of the cell walls of grasses. Annu Rev Plant Physiol Plant Mol Biol 47: 445-476
4. Catala C, Rose JK, Bennett AB (2000) Auxin-Regulated Genes Encoding Cell Wall-Modifying Proteins Are Expressed during Early Tomato Fruit Growth. Plant Physiol 122: 527-534
5. Chen F, Bradford KJ (2000) Expression of an Expansin Is Associated with Endosperm Weakening during Tomato Seed Germination. Plant Physiol 124: 1265-1274
6. Chen F, Dahal P, Bradford KJ (2001) Two Tomato Expansin Genes Show Divergent Expression and Localization in Embryos during Seed Development and Germination. Plant Physiol 127: 928-936
7. Chen F, Nonogaki H, Bradford KJ (2002) A gibberellin-regulated xyloglucan endotransglycosylase gene is expressed in the endosperm cap during tomato seed germination. J Exp Bot 53: 215-223
8. Cho HT, Cosgrove DJ (2002) Regulation of root hair initiation and expansin gene expression in Arabidopsis. Plant Cell 14: 3237-3253
9. Cho HT, Kende H (1997) Expression of expansin genes is correlated with growth in deepwater rice. Plant Cell 9: 1661-1671
10. Choi D, Lee Y, Cho HT, Kende H (2003) Regulation of expansin gene expression affects growth and development in transgenic rice plants. Plant Cell 15: 1386-1398
11. Cosgrove DJ (1989) Characterization of long-term extension of isolated cell walls from growing cucumber hypocotyls. Planta 177: 121-130
12. Cosgrove DJ (2000) Expansive growth of plant cell walls. Plant Physiology and Biochemistry 38: 1-16
13. Cosgrove DJ (2000) Loosening of plant cell walls by expansins. Nature 407: 321-326

14. Cosgrove DJ, Bedinger P, Durachko DM (1997) Group I allergens of grass pollen as cell wall-loosening agents. Proc Natl Acad Sci USA 94: 6559-6564
15. Cosgrove DJ, Li LC, Cho HT, Hoffmann-Benning S, Moore RC, Blecker D (2002) The growing world of expansins. Plant Cell Physiol 43: 1436-1444
16. Downes BP, Crowell DN (1998) Cytokinin regulates the expression of a soybean □-expansin gene by a post-transcriptional mechanism. Plant Mol Biol 37: 437-444
17. Gilroy S, Jones D (1999) Through form to function: root hair development and nutrient uptake. Trends Plant Sci 5: 56-60
18. Green PB (1997) Expansin and morphology: a role for biophysics. Trends Plant Sci 2: 365-366
19. Hutchison KW, Singer PB, McInnis S, Diaz-Sala C, Greenwood MS (1999) Expansins are conserved in conifers and expressed in hypocotyls in response to exogenous auxin. Plant Physiol 120: 827-832
20. Kende H, Van der Knaap E, Cho HT (1998) Deepwater rice: A model plant to study stem elongation. Plant Physiol 118: 1105-1110
21. Lee Y, Choi D, Kende H (2001) Expansins: ever-expanding numbers and functions. Curr Opin Plant Biol 4: 527-532
22. Li LC, Cosgrove DJ (2001) Grass group I pollen allergens (beta-expansins) lack proteinase activity and do not cause wall loosening via proteolysis. Eur J Biochem 268: 4217-4226
23. Li Y, Darley CP, Ongaro V, Fleming A, Schipper O, Baldauf SL, McQueen-Mason SJ (2002) Plant expansins are a complex multigene family with an ancient evolutionary origin. Plant Physiol 128: 854-864
24. Li Z-C, Durachko DM, Cosgrove DJ (1993) An oat coleoptile wall protein that induces wall extension in vitro and that is antigenically related to a similar protein from cucumber hypocotyls. Planta 191: 349-356
25. McQueen-Mason S, Durachko DM, Cosgrove DJ (1992) Two endogenous proteins that induce cell wall expansion in plants. Plant Cell 4: 1425-1433
26. Nonogaki H, Gee OH, Bradford KJ (2000) A germination-specific endo-beta-mannanase gene is expressed in the micropylar endosperm cap of tomato seeds. Plant Physiol 123: 1235-1246
27. O'Malley RC, Lynn DG (2000) Expansin message regulation in parasitic angiosperms. Plant Cell 12: 1455-1466
28. Pien S, Wyrzykowska J, McQueen-Mason S, Smart C, Fleming A (2001) Local expression of expansin induces the entire process of leaf development and modifies leaf shape. Proc Natl Acad Sci U S A 98: 11812-11817
29. Rayle DL, Cleland RE (1992) The Acid Growth Theory of auxin-induced cell elongation is alive and well. Plant Physiol 99: 1271-1274
30. Reinhardt D, Mandel T, Kuhlemeier C (2000) Auxin regulates the initiation and radial position of plant lateral organs. Plant Cell 12: 507-518
31. Reinhardt D, Pesce ER, Stieger P, Mandel T, Baltensperger K, Bennett M, Traas J, Friml J, Kuhlemeier C (2003) Regulation of phyllotaxis by polar auxin transport. Nature 426: 255-260
32. Reinhardt D, Wittwer F, Mandel T, Kuhlemeier C (1998) Localized upregulation of a new expansin gene predicts the site of leaf formation in the tomato meristem. Plant Cell 10: 1427-1437
33. Rose JKC and Bennett AB (1999) Cooperative disassembly of the cellulose-xyloglucan network of plant cell walls: parallels between cell expansion and fruit ripening. Trends Plant Sci 4: S-176-183
34. Schiefelbein J (2003) Cell-fate specification in the epidermis: a common patterning mechanism in the root and shoot. Current Opinion in Plant Biology 6: 74-78
35. Schipper O, Schaefer D, Reski R, Fleming A (2002) Expansins in the bryophyte *Physcomitrella patens*. Plant Mol Biol 50: 789-802
36. Schopfer P (2001) Hydroxyl radical-induced cell-wall loosening in vitro and in vivo: implications for the control of elongation growth. Plant J 28: 679-688
37. Schopfer P, Liszkay A, Bechtold M, Frahry G, Wagner A (2002) Evidence that hydroxyl

radicals mediate auxin-induced extension growth. Planta 214: 821-828
38. Voesenek LA, Benschop JJ, Bou J, Cox MC, Groeneveld HW, Millenaar FF, Vreeburg RA, Peeters AJ (2003) Interactions between plant hormones regulate submergence-induced shoot elongation in the flooding-tolerant dicot Rumex palustris. Ann Bot (Lond) 91 Spec No: 205-211
39. Wrobel RL, Yoder JI (2001) Differential RNA expression of a-expansin gene family members in the parasitic angiosperm *Triphysaria versicolor* (Scrophulariaceae). Gene 266: 85-93
40. Wu Y, Thorne ET, Sharp RE, Cosgrove DJ (2001) Modification of expansin transcript levels in the maize primary root at low water potentials. Plant Physiol 126: 1471-1479
41. Yuan S, Wu Y, Cosgrove DJ (2001) A fungal endoglucanase with plant cell wall extension activity. Plant Physiol 127: 324-333
42. Zhang N, Hasenstein KH (2000) Distribution of expansins in graviresponding maize roots. Plant Cell Physiol 41: 1305-1312

D1 Auxin Signal Transduction

Gretchen Hagen[1], Tom J. Guilfoyle[1] and William M. Gray[2]
[1]Biochemistry Department, University of Missouri, Columbia, Missouri 65211 USA, and [2]Department of Plant Biology, University of Minnesota, St. Paul, Minnesota 55108 USA. E-mail: HagenG@missouri.edu

INTRODUCTION

One of the many challenges in plant hormone research is to elucidate the molecular mechanisms of hormone action. The current working model for the auxin signal transduction pathway is based largely on hormone response pathways described in other systems. In this general model, the pathway is activated when receptive cells perceive auxin. Receptors at the cell surface and/or within the cell are thought to recognize and bind auxin with specificity and high affinity. Receptor binding would then trigger a series of biochemical and molecular events that would ultimately lead to observable physiological growth responses, such as cell elongation, division and/or differentiation. Over the past decade, there have been major advances in our understanding of the molecular mechanisms governing auxin action, through the identification of several key regulatory components of the pathway. The results indicate that auxin signaling is complex and may involve multiple pathways. The aim of this chapter is to highlight some of the research findings that have contributed to our current view of the auxin response pathway. A number of reviews on various aspects of auxin action have been published recently, and they are referenced throughout this chapter.

RECEPTORS

Some of the earliest studies on the auxin signaling pathway were aimed at identifying auxin-binding proteins (ABPs[1]), and characterizing their potential role as auxin receptors. The history and progress of research on ABPs have been extensively reviewed (26, 27, 41). The first major auxin binding activity

---

[1]Abbreviations: ABP, auxin binding protein; ARF, auxin response factor; AuxRE, auxin response element; CHX, cycloheximide; CSN, COP9 signalosome complex; DBD, DNA binding domain; DST, downstream element; E1, ubiquitin-activating enzyme; E2, ubiquitin-conjugating enzyme; E3, ubiquitin protein ligase; ER, endoplasmic reticulum; EST, expressed sequence tag; GST, glutathione S-transferase; GUS, bglucuronidase; MAPK, mitogen activated protein kinase; PM, plasma membrane; RUB, related to ubiquitin protein; SCF, multi-subunit ubiquitin-ligase composed of SKP1, Cullin and F-box.

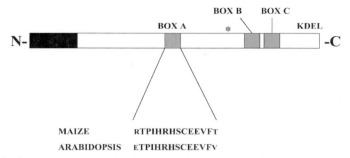

Figure 1. Diagram of the major structural features of ABP1. The protein contains an amino terminal signal peptide (black box), three domains that are highly conserved in amino acid sequence among all ABPs (Box A, B, C; conserved Box A sequences from maize and Arabidopsis are shown below), a conserved N-glycosylation site (*) and a C-terminal ER retention signal KDEL. Residues in Box A are thought to be involved in auxin binding.

to be described in detail came from crude microsomal membranes made from etiolated maize coleoptiles. The protein associated with this activity has been extensively characterized and is referred to as ABP1 (41). Additional proteins with auxin binding activity have been detected using a variety of biochemical techniques including photoaffinity labelling, auxin affinity chromatography and immunoaffinity chromatography. The identity of some of these proteins is known (41), but their role as auxin receptors has not been established. After 30 years of analysis, ABP1 remains the strongest candidate for an auxin "receptor".

Due to its relative abundance in coleoptiles, ABP1 was originally purified from maize (*Zea mays*). The 22kDa ABP1 protein was shown to have a K_D of 5×10^{-8}M for 1-NAA. *ABP1* genes have been characterized from a wide spectrum of plants, and are apparently unique to plants. In general, they appear to be part of a small gene family, although Arabdiposis contains a single *ABP1* gene. A comparison of ABP1 ESTs from different plants reveals a high degree of sequence homology throughout the mature protein (26). They contain an N-terminal signal peptide, which is variable in sequence among all ABPs examined, several conserved domains (termed Box A or D16, B and C; Fig. 1) consisting of 15-20 amino acids and a C-terminal endoplasmic reticulum (ER) lumen retention sequence, KDEL. Polyclonal and monoclonal antibodies raised to the mature protein and peptide fragments have been used to study localization, structure and function of ABP1. A majority (>95%) of the protein is found associated with the ER, which is in agreement with early studies that detected the bulk of auxin-binding activity co-fractionating with ER markers. A small fraction of the activity is associated with the plasma membrane (PM), and it is thought that this ABP1 originates in the ER and escapes ER retention. The conserved Box A (D16) sequence in ABP1 is thought to be the major site for auxin binding. This conclusion comes from data showing that anti-BoxA/D16 antibodies have auxin-like activity. In addition, the protein contains at least one N-glycosylation site. Studies using oligonucleotide and antibody probes have shown that, on the whole plant level, ABP1 is generally expressed at low levels, except in maize coleoptiles. ABP1 is detected in many plant organs (26).

The function of PM-associated ABP1 has been investigated using a variety of assays. To be considered a bona fide auxin receptor, the protein must meet several important criteria: that it specifically binds auxin with high affinity and that the auxin-protein complex is crucial to an auxin response. The demonstration that the purified maize ABP1 does, indeed, bind NAA with high specificity fulfills the first criterion. To define the role of ABP1 in auxin responses, early assays focused on measuring rapid auxin effects at the PM surface. These studies measured hyperpolarization and membrane currents in protoplasts, and ion channel activities in guard cells (22, 41). These responses were affected by auxin agonists and antagonists, and proved to be a convenient way to study auxin action. The results support the hypothesis that ABP1 plays a role in auxin responses at the cell surface. More recently, molecular genetic approaches have been used to further define the role of ABP1 in auxin-regulated growth responses. Tobacco plants induced to over-express ABP1 in leaf strips had larger cells, as compared to wild type leaf cells (2). In the same study, constitutive over-expression of maize ABP1 in a maize endosperm cell line that normally lacks detectable ABP1 resulted in an auxin-dependent increase in cell size that was not observed in non-transformed cells. In another study, antisense suppression of tobacco ABP1 in a tobacco (BY2) cell line was shown to inhibit auxin-induced elongation in these cultured cells (3). Taken together, these studies have led to the conclusion that the primary role of PM-associated ABP1 is as an auxin receptor mediating auxin-regulated cell elongation (2). Interestingly, the loss of the single Arabidopsis *ABP1* gene confers an embryo lethal phenotype, in which cell expansion ceases and the globular embryo arrests (3), suggesting that ABP1 functions in elongation growth responses during embryonic development.

To date, the evidence linking ABP1 to other auxin-regulated growth responses such as cell division and differentiation, or in the well-documented rapid auxin effects at the level of gene expression (11), is lacking. It has been suggested that ABP1 plays no role, or an indirect role, in auxin-regulated cell division (2). As a receptor, ABP1 is an enigma. The protein does not possess characteristics of a classical PM or nuclear receptor. Since it is not an integral PM protein, it is likely that ABP1 associates with a PM protein that is "docked" within the PM (27), although no such association has been detailed. The role of ER-localized ABP1 is also unknown. If ABP1 is, in fact, specific to auxin-regulated cell elongation responses, one might speculate that additional auxin receptors exist. With the diversity of growth responses attributed to auxin, the concept of multiple auxin receptors (and perhaps multiple signal transduction pathways) is plausible. Curiously, extensive genetic screens for auxin insensitive mutants have failed to uncover any candidate receptor proteins. There is some speculation that proteins that are thought to interact with auxin, such as auxin influx and efflux carriers and enzymes involved in auxin metabolism, may be involved in regulating the signal transduction pathway (26, 27), but their role as specific receptors remains uncertain.

DOWNSTREAM COMPONENTS ASSOCIATED WITH THE AUXIN RESPONSE

Research on the mechanisms of auxin action has also focused on the identification of molecular components involved in transducing the auxin signal and regulating the pathway. Several "classical" secondary messengers such as Ca^{2+} and calmodulin (21, 46), G proteins (2, 39) and phospholipases (31) have been implicated in the auxin response. There is good evidence to support the involvement of protein phosphorylation in auxin signaling (5). In Arabidopsis roots, a mitogen activated protein kinase (MAPK) activity is rapidly induced by exogenous auxin, and this activity correlated with the rapid activation of an auxin-responsive reporter gene (24). In a mutant (*axr4*) defective in several auxin responses, this auxin-induced MAPK activation was impaired. To date, the gene encoding this MAPK has not been characterized. In contrast, a MAPK cascade activated by oxidative stress was shown to negatively regulate auxin-inducible transcription (5). Another protein kinase postulated to be involved in auxin responses is the non-receptor serine/threonine protein kinase PINOID (PID), as mutations in this protein result in auxin-related phenotypes (5).

Genetic and molecular approaches have contributed significantly to our current view of the molecular mechanisms of auxin action. Genetic screens designed to recover mutants exhibiting auxin-related phenotypes have identified a number of genes in Arabidopsis that are involved in the auxin response (discussed in detail below) (20). Analysis of these mutants also provides molecular evidence for crosstalk with other signaling pathways and metabolic processes (34, 35). One example of pathway crosstalk involves light and auxin signaling. The connection between these two pathways is well documented from physiological studies, but recent genetic and biochemical studies have provided new molecular evidence. For example, it has been shown that mutations in several Arabidopsis genes belonging to the Aux/IAA family of auxin-regulated transcription factors (e.g. *shy2/iaa3*, *shy1/iaa6*) result in both light and auxin-related phenotypes. Screens for auxin insensitive mutants also identified several members of the *Aux/IAA* gene family (*AXR2/IAA7, AXR3/IAA17*), as well as genes involved in the ubiquitin-proteasome degradation pathway (*AXR1, TIR1*), which revealed a new level of regulation of the auxin response (20; discussed in detail below). Other genetic screens have identified mutations in other auxin-response genes, including mutations in the auxin-inducible *GH3* genes (*FIN219, DFL1*; 18, 25) and several genes in the *A*uxin *R*esponse *F*actor (*ARF*) family of transcription factors (*ARF3/ETT, ARF5/MP and ARF7/NPH4*; 23).

Early observations that auxin could cause rapid and specific changes in gene expression led researchers into taking molecular approaches to identify auxin-responsive genes (11). Advances in this area have led to the discovery of *cis*-acting auxin-responsive elements within the promoters of several genes, and the identification of the proteins that regulate the expression of these genes.

Auxin-Responsive Genes

Over the past 20 years, numerous auxin-responsive sequences have been reported from a wide range of plants (11). The most thoroughly studied gene classes (*Aux/IAAs, SAURs* and *GH3s*) are referred to as early, or primary auxin response genes. These are defined by their rapid induction kinetics (2-15 minutes after auxin treatment), by their specific induction by biologically active auxins and by their induction in the absence of protein synthesis.

Aux/IAA

Aux/IAA sequences were originally described from soybean and pea as mRNAs that were induced within 10-20 minutes after exogenous auxin application (11, 13). Subsequently, *Aux/IAA* genes have been identified in many plants and are encoded by multigene families. Arabidopsis contains 29 *Aux/IAA* genes, most of which are induced by auxin. The proteins encoded by these genes range in size from 20-35 kDa. Several Aux/IAA proteins from pea and Arabidopsis have been shown to be relatively short-lived (ranging from 6 to 80 minutes) and localize to the nucleus (1, 10, 28). Most Aux/IAA proteins contain four conserved motifs, referred to as domains I, II, III & IV, each consisting of 7 to 40 amino acids (Fig. 2). The high degree of amino acid sequence conservation within these domains suggests that each

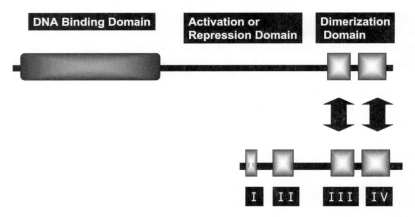

Figure 2. Schematic diagram of an ARF and Aux/IAA protein. ARF proteins (above) contain a conserved N-terminal DNA-binding domain, a nonconserved middle region that functions as an activation or repression domain, and a conserved C-terminal dimerization domain. Aux/IAA proteins (below) contain four conserved motifs or domains, commonly referred to as domains I, II, III, and IV. Domains III and IV are dimerization domains related in amino acid sequence to the C-terminal dimerization domain found in ARF proteins. Domain II is a protein destablization domain that is responsible for the short life times of Aux/IAA proteins. It has been reported that Domain I functions as a homodimerization domain and a repression domain.

domain contributes to a specific function. Aux/IAA proteins can dimerize with one another and with Auxin Response Factors (ARFs) through their C-terminal motifs (i.e., consisting of motifs III and IV). In transfection assays with plant protoplasts, overexpression of Aux/IAA proteins results in repression of auxin-responsive gene expression (12, 13, 36). Dominant or semi-dominant auxin response mutants have been identified in Arabidopsis that result from mutations within domain II of Aux/IAA proteins (see Fig. 7A) (23). These mutant plants, in general, show alterations in auxin sensitivity and responses, including increased resistance to exogenous auxin, defects in lateral root and root hair formation, and impaired root and hypocotyl gravitropism. Domain II mutations result in stabilization of the mutant Aux/IAA proteins, and motif II has been shown to function as a protein degradation signal (see below). Domain I in Aux/IAA proteins has been proposed to function as a dimerization domain (28). The demonstration that Aux/IAA proteins are active repressors of auxin-responsive transcription (36) has led to the recent discovery that domain I is an active repression domain (38). To date, no selective DNA binding activity has been demonstrated for Aux/IAA proteins, leading to the suggestion that these proteins regulate auxin response via their interactions with members of the ARF protein family. Results with Aux/IAA mutant plants support roles for Aux/IAA proteins in a variety of auxin responses. As will be discussed below, there is evidence that auxin plays a role in modulating the lifetime of Aux/IAA proteins (47), and that they are targeted for degradation by the ubiquitin proteasome pathway (9, 20, 30).

SAUR
The *SAUR* (*S*mall *A*uxin *U*p *R*NAs) genes represent some of the most rapidly induced auxin-responsive sequences found in plants. Originally identified in soybean, *SAUR* transcript levels were shown to be transcriptionally induced within 2-5 minutes of exposure to active auxins (11, 13). Auxin-responsive *SAUR* transcripts have been subsequently detected in many plant systems. There are distinct differences in the kinetics and specificity of induction among different plant *SAURs*. For example, Arabidopsis *SAUR-AC1* transcripts are induced by both auxin and cytokinin, as well as by the protein synthesis inhibitor cycloheximide (CHX). While CHX was shown to cause an increase in the levels of soybean *SAUR* mRNA, this was not a transcriptional response to CHX, and is likely the result of stabilization of *SAUR* mRNAs (13). *SAUR* transcripts are highly unstable, and a conserved element (DST) in the 3' untranslated region is thought to play a role in their instability (13). *SAUR* genes encode small proteins ranging from 9-15 kDa. There are over 70 *SAUR* genes in Arabidopsis, many of which are found in clusters on the chromosomes (13). Due to their low abundance in plant organs, the proteins are thought to be short-lived. This is supported by recent data showing that the maize SAUR protein, ZmSAUR2, has a half-life of 7 min (21). The function of SAUR proteins has not been established, but it has been hypothesized that they are nuclear proteins that play a role in the auxin

signal transduction pathway. This is based on studies of the maize SAUR proteins, ZmSAUR1 (46) and ZmSAUR2 (21), which have been shown to bind calmodulin *in vitro* in a calcium dependent manner. In addition, a ZmSAUR2::GUS fusion protein was detected in nuclei of transiently transformed onion epidermal cells (21).

GH3

The original GH3 sequence was isolated as an auxin-inducible cDNA from auxin treated soybean hypocotyls (11). *GH3* mRNA levels are generally low in soybean, but can be induced to high levels in every major organ and tissue after a short exposure to exogenous auxin. The mRNAs are transcriptionally induced within 5 minutes of auxin treatment, and this induction is specific to biologically active auxins. The soybean genes encode proteins of about 70kDa, which are cytoplasmically localized and relatively stable. A number of *GH3* cDNAs and genes have been described from other plants. Arabidopsis has 19 *GH3* genes, each encoding proteins of about 65-70kDa (11, 13). There has been limited characterization of these genes in terms of their specificity and kinetics of auxin induction. Recently, two Arabidopsis mutants with light response phenotypes have been shown to have defects in the expression of two different auxin-inducible *GH3* gene family members, suggesting a role of at least some *GH3* genes in light signaling. The *fin219* mutant was isolated as a suppressor of *cop1* and had a long hypocotyl phenotype in far red light (18). The mutation mapped to the *GH3-11* gene (for nomenclature, see 13), and RNA blot analysis showed decreased expression of *FIN219/GH3-11* in light grown mutant plants. It was reported that the molecular basis of the reduced expression was the result of an epigenetic change involving methylation in the promoter region. Interestingly, the Arabidopsis jasmonic acid (JA) response mutant *jar1* was shown to be defective in the same *GH3-11* gene (33). In contrast to the phenotype of the *fin219* mutant, none of the *jar1* alleles displayed altered hypocotyl elongation in far red light. The basis of this discrepancy has not been determined. The dominant *dfl1-D* mutant was selected from a morphological screen of Arabidopsis activation tagged lines (25). Mutant plants display both light and auxin phenotypes. The *DFL1* gene encodes the GH3-6 protein, and the *dfl1-D* phenotype was shown to be the result of overexpression of the *DFL1/GH3-6* gene. Analysis of the biochemical basis of the Arabidopsis *jar1* mutation led to the discovery that the *GH3* gene family encodes a new class of acyl adenylate-forming enzymes (33). Adenylation activities of JAR1 and 15 other gene family members were tested *in vitro* using plant hormone substrates. JAR1/FIN219/GH3-11 is specifically active on JA, while 6 other family members including DFL1/GH3-6 specifically adenylate IAA. The biological role for adenylation of hormones is not clear, but it has been suggested that it could represent one step in a pathway of hormone metabolism, such as in the formation of hormone conjugates (33). In the case of JA, conjugate formation would positively regulate the signaling pathway, whereas IAA conjugates are generally thought to be inactive. The auxin response phenotypes observed in the *dfl1-D* mutant could be the consequence of elevated levels of IAA conjugates due to increased enzyme activity resulting from overexpression of the *DFL1/GH3-6* gene. These results provide compelling evidence for a link between auxin and JA signaling.

Auxin-Responsive Promoters

The promoters of several auxin-responsive genes have been analyzed in detail. Auxin-responsive elements (AuxREs) have been identified within the promoters of the three major classes of auxin-regulated gene families (11, 13). The core element in the *Aux/IAA* gene from pea (*PS-IAA 4/5*) consists of (G/T)GTCCCAT. Three separate auxin responsive elements each containing a TGTCTC core have been identified in the promoter of the soybean *GH3* gene. An auxin-responsive region of the soybean *SAUR15A* gene was shown to contain both core sequence elements. Each of the 3 AuxREs in the *GH3* promoter was shown to function independently of one another, and found to be part of a composite element containing the TGTCTC core element adjacent to or overlapping with a coupling or constitutive element (Fig. 3). Using a variety of assays, it was shown that the TGTCTC element acts to repress the constitutive element when auxin levels are low. When auxin levels are elevated, the repression is relieved and the composite element is activated (11). Simple AuxREs, consisting of TGTCTC elements that are multimerized and appropriately spaced, have been shown to be highly auxin responsive (DR5, Fig. 3).

Reporter gene constructs with natural composite AuxREs or synthetic, simple AuxREs have provided valuable probes to study gene expression associated with auxin-related growth responses in wild type and mutant plants (13). The synthetic AuxREs-reporter gene constructs have been particularly useful in examining auxin responses at the cell and tissue level (13) (Fig. 4). They have also been used in genetic screens designed to recover auxin response mutants (13).

COMPOSITE AuxREs

D1   CCTCG<u>TGTCTC</u>

D4   CACGCAATCCTT<u>TGTCTC</u>

SIMPLE AuxRE

DR5   <u>TGTCTC</u>CCTTT<u>TGTCTC</u>

Figure 3. Composite and simple AuxREs. The auxin-responsive TGTCTC elements are underlined. The coupling or constitutive elements of the composite AuxREs (D1 and D4) found in the promoter of the *GH3* gene are indicated by arrows. DR5 is a simple AuxRE consisting of direct repeats of TGTCTC separated by 5 base pairs.

ARF Genes and Proteins

The identification of a strong synthetic AuxRE containing palindromic repeats of the TGTCTC element made it possible to isolate a DNA-binding protein, using a yeast one-hybrid screen, that specifically binds to this element (40). This screen yielded a novel protein that was named Auxin Response Factor 1 or ARF1. The C-terminal domain of ARF1 was used as a

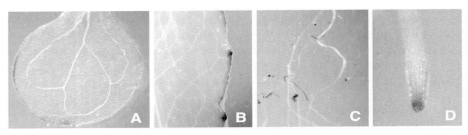

Figure 4 (Color plate page CP8). Histochemical staining for GUS activity in Arabidopsis plants transformed with the auxin-responsive reporter gene *DR5(7X)-GUS*. Plants were grown in soil and stained for GUS activity for 20hr. GUS activity (blue color) was observed in specific regions of cotyledons (A), rosette leaves (B) and roots (C, D). Auxin treatment of these transformed plants induces GUS activity throughout the plant (data not shown; 13).

bait in a yeast two-hybrid screen, and this screen yielded a second ARF, which was named ARF1 binding protein or ARF2 (12). Subsequently, EST and genome sequencing projects, along with mutant screens, have resulted in the identification of 22 *ARF* genes in Arabidopsis (i.e., these are referred to as *ARF1* through *ARF22*; see reference 12 for nomenclature). *ARF* genes are predicted to encode proteins ranging in size from 70-130kDa and contain a conserved N-terminal DNA-binding domain (DBD), a conserved C-terminal dimerization domain (with the exception of ARF3 and ARF17, which lack this domain), and a nonconserved middle region (i.e., just C-terminal to the DBD) which functions as an activation or repression domain (12; Fig. 2).

ARF DBDs are related over a stretch of about 120 amino acids to a conserved B3 domain found in the DBDs of a variety of other plant transcription factors (12). ARF C-terminal dimerization domains are composed of motifs III and IV and are related in amino acid sequence to motifs III and IV in the dimerization domains of Aux/IAA proteins (see above). The motif III and IV domains facilitate the formation of ARF homo- and heterodimers, ARF-Aux/IAA heterodimers, and Aux/IAA homo- and heterodimers. It is also possible that higher order multimers might assemble via motifs III and IV in ARF and Aux/IAA proteins. The non-conserved middle regions of ARFs have biased amino acid sequences (e.g., S-rich, Q-rich), and those with S-, SL-, SG-, and SP-rich middle regions function as repressors (i.e., these are referred to as Class I ARFs and include ARF1, -2, -3, -4, -9, -11, -12, -13, -14, -15, -18, -20, -21, -22) while those with Q-rich middle regions function as activators (i.e., these are referred to as Class II ARFs and include ARF5, -6, -7, -8, -19). A third class of ARFs (Class III), including ARF10, -16, and -17, are more divergent in amino acid sequence in both their DBDs and dimerization domains compared to Class I and II ARFs. It has not been determined whether Class III ARFs function as repressors or activators; however, their middle regions are not Q-rich, but are enriched for S. The middle regions function independently as repression or activation domains when isolated from Class I and Class II ARFs, respectively, when these domains are fused to a heterologous DNA-binding domain and are targeted to promoters that contain binding sites for the heterologous DNA-binding domain (37). Transfection assays using plant protoplasts suggest the C-terminal dimerization domain is required for ARFs to regulate genes in an auxin-dependent manner.

Arabidopsis *ARF* genes, in general, are not regulated by auxin or other plant hormones, although an ARF in rice has been reported to be regulated by auxin (42). *ARF* genes or cDNAs have been found in both monocotyledonous and dicotyledonous plants, gymnosperms, and ferns, but not in fungi, animals, or bacteria. The ARF DBD and the B3 DNA-binding domain, in general, is unique to plants and does not resemble any known DNA-binding domain outside of the plant kingdom.

Mutations in *ARF* genes have provided insight into auxin-dependent processes that ARF transcription factors might regulate (23). *ettin/arf3* mutations affect gynoecium patterning and floral organ number. *monopteros/arf5* mutations affect embryo patterning, vascular tissue formation and root growth. *nph4/msg1/arf7* mutations affect gravitropic and phototropic responses, auxin sensitivity in hypocotyls and leaves, and auxin-regulated gene expression. *In situ* hybridization results with wild-type and ARF mutant plants have shown that selected ARFs display gene expression patterns that are temporally and developmentally regulated in a tissue specific manner (12). Results with ARF mutant plants are consistent with ARFs playing important roles in auxin-regulated transcription and a variety of auxin-regulated growth and developmental responses.

PROTEIN DEGRADATION AND THE AUXIN RESPONSE

Ubiquitin-Mediated Proteolysis

Selective protein degradation has emerged as a common mechanism cells employ to regulate many developmental and metabolic pathways. The ubiquitin-26S proteasome pathway is the major proteolytic pathway in eukaryotic cells (16). Ubiquitin is a 76 amino acid protein that is covalently attached to substrate proteins, marking them for degradation by the 26S proteasome. The conjugation of ubiquitin involves the sequential action of three enzymes (Fig. 5). Ubiquitin is first activated for conjugation by the formation of a thiolester linkage between its COOH-terminus and a conserved cysteine residue of the ubiquitin-activating enzyme (E1). Activated ubiquitin is then trans-esterified to a member of a family of ubiquitin-conjugating enzymes (E2), which then act in concert with a ubiquitin protein ligase (E3), to covalently attach ubiquitin to the e-NH2 group on a lysine residue of specific target proteins. Reiteration of this enzymatic cascade using a specific lysine within the conjugated ubiquitin results in the generation of a polyubiquitin chain. The 26S proteasome recognizes this chain and degrades the tagged protein, releasing free ubiquitin in the process.

The key regulatory step in ubiquitin conjugation lies with the selection of specific substrates by the E3 ubiquitin-ligase. E3 enzymes are

Auxin signal transduction

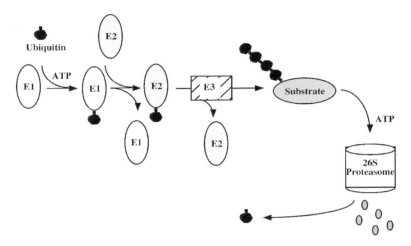

Figure 5. The ubiquitin-26S proteasome protein degradation pathway. Free ubiquitin molecules are conjugated to target proteins by the action of E1, E2, and E3 enzymes. Polyubiquitin chains are recognized by the 26S proteasome, which degrades the target protein and recycles ubiquitin.

functionally diverse, but all facilitate the transfer of ubiquitin from the E2 enzyme to a specific substrate. There are over 1000 predicted E3 enzymes in the completed Arabidopsis genome, the vast majority of which have yet to be assigned a specific function (14). The largest family of ubiquitin-ligases in Arabidopsis is the SCF type E3s (8). These multi-subunit enzymes, named for three of their subunits, SKP1, Cullin, and an F-box protein, employ combinatorial control to mediate ubiquitin-ligase specificity (6). SKP1, Cullin, and the RING-finger protein RBX1 (also called ROC1 and Hrt1) comprise the core of the SCF complex. F-box proteins bind to this core via an interaction between their F-box domain and the SKP1 protein. Each F-box protein acts as an adaptor subunit that identifies specific substrates for ubiquitin conjugation. There are approximately 700 predicted F-box proteins in Arabidopsis (8).

The SCF^{TIR1} ubiquitin-ligase is required for the auxin response

The Arabidopsis *TIR1* gene (auxin Transport Inhibitor Response 1) was originally identified in a screen for mutants resistant to inhibitors of polar auxin transport. Subsequent analysis revealed that this phenotype of *tir1* mutants results from a reduction in auxin response rather than transport. Cloning of the *TIR1* gene revealed that it encodes an F-box protein, and further studies demonstrated that the TIR1 protein assembles into an SCF ubiquitin-ligase with an Arabidopsis SKP1-like protein (either ASK1 or ASK2), the cullin AtCUL1, and the AtRBX1 protein (Fig. 6) (9). The auxin response defects conferred by mutations in *TIR1* are relatively modest, but this is likely due to some degree of functional redundancy with several closely related F-box proteins encoded in the Arabidopsis genome.

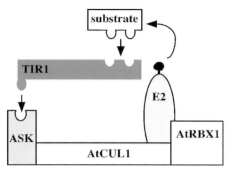

Figure 6. The SCF ubiquitin-ligase model. F-box proteins (e.g. TIR1) act as adaptor subunits that interact with the core SCF subunits ASK* (Arabidopsis SKP1-like), AtCUL1, and AtRBX1. Each F-box protein recruits specific substrates to the SCF complex for ubiquitination. * Arabidopsis contains a large *ASK* gene family. ASK1 and ASK2 have been shown to interact with TIR1.

Genetic studies indicate that the core SCF subunits are also required for auxin response. The *ask1-1* mutant exhibits reduced sensitivity to exogenous auxin and is defective in lateral root development. AtCUL1 is encoded by the *AXR6* (*A*uxin *R*esistant *6*) gene (15). *axr6* mutants display several phenotypes consistent with impaired auxin signaling including agravitropism, defective vascular patterning and reduced apical dominance (17). Studies of transgenic seedlings containing altered AtRBX1 levels indicate that this subunit of the SCF^{TIR1} complex is also required for normal auxin signaling. These findings demonstrate that the SCF^{TIR1} complex is a positive regulator of auxin response and suggest a model involving the ubiquitin-mediated proteolysis of an inhibitor of auxin signaling. Recent evidence indicating that, in response to auxin, the SCF^{TIR1} complex targets members of the Aux/IAA protein family for ubiquitin-mediated proteolysis is strongly supportive of such a model (10).

Mutations in the *AXR2* and *AXR3* genes were isolated in genetic screens for mutants displaying reduced auxin response (29). Due to the dominant gain-of-function nature of the *axr2* and *axr3* mutations, the wild-type gene products have been genetically defined as negative regulators of auxin response. Cloning of the *AXR2* and *AXR3* loci revealed that both genes encode distinct members of the *Aux/IAA* gene family, and that the dominant mutations in the *axr2* and *axr3* alleles all affect a highly conserved region within domain II. Subsequent screens have isolated several additional dominant derivatives of other *Aux/IAA* family members, and all of these mutations similarly map to domain II, strongly suggesting that this region plays a key role in regulating Aux/IAA protein function (Fig. 7A).

Several recent findings suggest that domain II functions as a protein destabilization motif. Early studies of Aux/IAA proteins indicated that at least some members of this family of transcriptional regulators are highly unstable (29). More recently, when expressed as a fusion protein, domain II has been found to confer a dramatic reduction in the stability of unrelated reporter proteins such as luciferase and β-glucuronidase (GUS) (10, 43). In contrast, reporter fusions containing domain II mutations corresponding to the dominant *Aux/IAA* alleles are not destabilized. Consistent with these observations, the axr2-1 and axr3-1 mutant proteins are dramatically more stable than their wild-type counterparts (Fig. 7B).

Auxin signal transduction

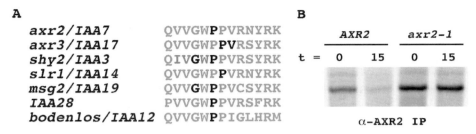

Figure 7. Aux/IAA Domain II. A) The highly conserved 13 amino acid protein degradation signal. Residues affected by dominant mutations are shown in dark type. B) Results of a pulse-chase experiment examining AXR2 stability in *AXR2* and *axr2-1* seedlings. AXR2 protein was immunoprecipitated from extracts prepared from metabolically labeled seedlings immediately after the labeling period (t = 0) or following a 15-minute chase with cold methionine/cysteine.

The molecular basis of Aux/IAA instability appears to be due to domain II-mediated interactions with the SCF^{TIR1} complex, which targets the Aux/IAA proteins for ubiquitin-mediated proteolysis. Domain II is both necessary and sufficient to target Aux/IAA proteins to the SCF^{TIR1} ubiquitin-ligase. This interaction is disrupted by the dominant *axr2* and *axr3* mutations, thus explaining the increased stability displayed by these mutant proteins. Thus, a strong correlation exists between Aux/IAA stability and the ability of these proteins to interact with the SCF^{TIR1} complex. Complementing these findings, analysis of mutants with impaired SCF^{TIR1} function has revealed that Aux/IAA stability is increased relative to that seen in wild-type plants. These data are consistent with the notion that SCF^{TIR1} ubiquitinates the Aux/IAA proteins, thus marking them for degradation by the 26S proteasome. Although SCF^{TIR1}-mediated ubiquitination has yet to be confirmed biochemically, several additional lines of evidence support this possibility. First, Arabidopsis seedlings containing a mutation in the RPN10 subunit of the 26S proteasome exhibit reduced auxin response. And second, the proteasome inhibitor MG132 has been shown to dramatically increase the stability of Aux/IAA-GUS fusion proteins suggesting that Aux/IAA proteins are degraded by the ubiquitin-proteasome pathway.

Regulation of the SCF^{TIR1} complex

Auxin
The findings described above indicate that auxin signaling requires the SCF^{TIR1}-mediated ubiquitination and subsequent proteasome-mediated degradation of Aux/IAA proteins. The role of auxin in this regulation has been addressed in studies of Aux/IAA fusions with GUS and luciferase reporter proteins, and the results suggest that auxin stimulates the proteolysis of Aux/IAA proteins (10, 47). Fusions containing dominant domain II mutations are immune to this hormonal regulation (Fig. 8A). Initial evidence suggests that auxin destabilizes Aux/IAA proteins by facilitating their recognition by the SCF^{TIR1} ubiquitin-ligase. Interactions between Aux/IAA

proteins and the SCF^{TIR1} complex have been detected by both co-immunoprecipitation assays and glutathione S-transferase (GST)-pulldown experiments. In the latter assay, a GST-AXR2 (or AXR3) fusion protein is expressed and purified from *E. coli* and then used as an affinity matrix to purify interacting proteins out of a crude plant protein extract. This assay has been used to demonstrate that auxin promotes the binding of the SCF^{TIR1} complex to the GST-Aux/IAA fusion protein in a dose-dependent fashion (Fig. 8B, C). Thus, a strong correlation exists between auxin-stimulated Aux/IAA degradation *in vivo* and the auxin-induced binding of Aux/IAA proteins to the SCF^{TIR1} complex *in vitro*. Although hormone-treated seedlings were originally used to show auxin-induced binding of Aux/IAA proteins by TIR1, the recent demonstration that auxin promotes this interaction in membrane-free extracts indicates that the auxin receptor is a soluble factor (7).

The possibility that auxin targets Aux/IAA proteins to the SCF^{TIR1}

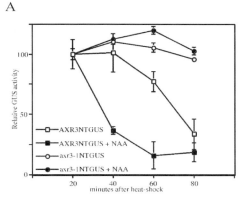

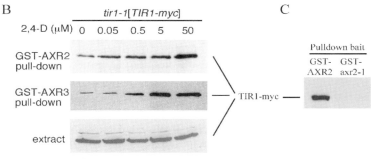

Nature 414:271-276

Figure 8. Auxin regulation of Aux/IAA stability. A) Auxin destabilization of an AXR3-GUS fusion protein in transgenic seedlings. B) Recombinant GST-AXR2 or GST-AXR3 fusion proteins were used as an affinity matrix to purify TIR1 out of crude extracts prepared from control or auxin-treated seedlings expressing a c-myc epitope tagged derivative of TIR1. C) The *axr2-1* mutation abolishes the ability of the GST-AXR2 fusion protein to interact with TIR1.

ubiquitin-ligase is highly attractive, given the results of several studies examining SCF-substrate interactions in other eukaryotes (6). All of the best-characterized substrates for SCF-type E3s cannot bind to their cognate F-box protein unless the substrate is first post-translationally modified. In the vast majority of cases, this modification involves phosphorylation of the substrate by a regulatory kinase (6). In addition, a few alternative regulatory mechanisms have recently been discovered, including glycosylation and proline hydroxylation of the SCF substrate (14). The effect of auxin in the GST-Aux/IAA pulldown assays described above is certainly consistent with the possibility that auxin directs a post-translational modification of Aux/IAA proteins, although the nature of any putative modification is uncertain. Evidence suggesting that Aux/IAA proteins may be modified by phosphorylation comes from *in vitro* studies showing that phytochrome A can bind and phosphorylate the amino terminus of several Aux/IAA proteins (30). However, the finding that domain II is sufficient to confer auxin-induced destabilization of a reporter protein and binding to the SCF^{TIR1} complex, despite the fact that this domain lacks any conserved phosphorylation sites, would argue against auxin-mediated phosphorylation serving as an important triggering mechanism. Further, the addition of protein phosphatase to GST-AXR3 pulldown assays fails to block the auxin-induced interaction with SCF^{TIR1} (W.M.G., unpublished). Regulation by proline hydroxylation, however, is extremely intriguing, given the presence of two adjacent, absolutely conserved proline residues within domain II of the Aux/IAA proteins. The majority of the dominant Aux/IAA mutations that abolish both SCF^{TIR1} binding and auxin-dependent destabilization, affect either of these two prolines (Fig. 7A). Additional biochemical studies will be required, however, to determine whether these prolines might link the auxin signal to the proteolysis of the Aux/IAA proteins.

RUB modification of AtCUL1
An additional layer of regulation on the SCF^{TIR1} complex has been elucidated through studies of the *AXR1* gene product. *axr1* mutants exhibit severe auxin response defects. The AXR1 protein is related to the amino-terminal half of the ubiquitin-activating enzyme (E1), and biochemical studies have revealed that AXR1 functions in the activation of the Related to Ubiquitin (RUB1) protein (also known as NEDD8). Arabidopsis encodes three RUB proteins that share approximately 60% sequence identity with ubiquitin. Like ubiquitin, RUB is conjugated to target proteins in a series of enzymatic reactions employing thiolester intermediates. AXR1 dimerizes with the ECR1 (E1 C-terminal Related) protein to form a RUB-specific activating enzyme. Following transesterification to a RUB Conjugating Enzyme (RCE), RUB is covalently attached to a substrate protein (14). Unlike ubiquitin, poly-RUB conjugates have not been detected, and RUB1 conjugation does not appear to affect the stability of the target protein.

To date, the only known substrates for RUB1 modification are members of the cullin protein family, including the AtCul1 subunit of the SCF^{TIR1}

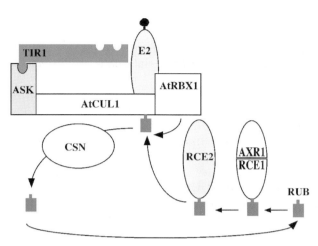

Figure 9. RUB conjugation and de-conjugation to the AtCUL1 protein. The AXR-ECR1 heterodimer activates RUB for conjugation. RUB is transferred to the RUB conjugating enzyme, RCE1, which interacts with AtRBX1 to covalently attach the RUB modifier to a lysine residue near the AtCUL1 COOH-terminus. The CSN contains an isopeptidase activity that cleaves RUB off of AtCUL1. Both conjugation and de-conjugation have been demonstrated to be required for normal SCF^{TIR1} activity.

ubiquitin-ligase (Fig. 9). The SCF subunit, AtRBX1, promotes the conjugation of RUB1 to AtCUL1, perhaps by functioning as a RUB-ligase. The precise function of the RUB1 modification of cullins is unclear, but several lines of evidence from yeast, mammals, and Arabidopsis indicate that it is required for SCF ubiquitin-ligase activity (9). Consistent with this notion, mutations in *axr1* confer a striking increase in Aux/IAA protein stability.

Recent findings suggest that RUB1 removal from the cullin subunit of SCF complexes also plays an important role in SCF function. For example, Arabidopsis seedlings overexpressing AtRBX1 exhibit increased levels of RUB-modified AtCUL1, yet display several phenotypes consistent with reduced auxin response including the stabilization of Aux/IAA proteins. RUB deconjugation appears to involve the COP9 signalosome complex (CSN). The CSN consists of eight subunits, all of which are related to proteins comprising the 19S regulatory particle of the 26S proteasome (32). Several recent findings have implicated the CSN in ubiquitin-mediated proteolysis. The CSN interacts with the SCF subunits AtCUL1 and AtRBX1, and reductions in CSN activity result in several auxin-related phenotypes including reductions in apical dominance, lateral root development and auxin-inducible gene expression. Additionally, Aux/IAA proteins are stabilized, suggesting that the CSN is required for Aux/IAA proteolysis. Recent findings demonstrating that the CSN5 subunit possesses a RUB protease activity suggest that RUB modification of AtCUL1 is a dynamic process, with both conjugation and deconjugation being required for proper SCF function (4).

A Model for Auxin Signaling

Based on existing evidence, the following model has been proposed for the regulation of early auxin response genes that are activated or up-regulated by auxin (Fig. 10). In this model, Class II ARF transcriptional activators reside

Auxin signal transduction

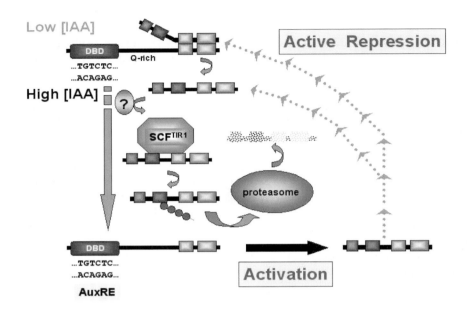

Figure 10. Model for repression and activation of auxin response genes. When auxin concentrations are sub-threshold, Aux/IAA repressors dimerize with ARF transcriptional activators resulting in the active repression of early auxin response genes containing TGTCTC. When auxin levels increase, Aux/IAA proteins are targeted to the SCFTIR1 complex by an undetermined mechanism. SCFTIR1 ubiquitinates the Aux/IAA proteins, marking them for degradation by the 26S proteasome (ubiquitin is shown as dark circles containing "u"). Aux/IAA proteolysis effectively relieves the repression on early auxin response genes, resulting in gene activation. In this model, ARF transcriptional activators are bound to their DNA target sites whether auxin concentrations are low or high, and it is the association/dissociation of Aux/IAA repressors from the ARF activators that regulates the expression of these genes. Curiously, at least some Aux/IAA genes contain TGTCTC AuxREs and are auxin-inducible. At the same time, the Aux/IAA proteins become less stable due to their more rapid degradation via the proteasome pathway when auxin levels are elevated. When auxin concentrations are high, the newly synthesized Aux/IAA proteins would be rapidly degraded and would fail to accumulate to levels required for repression of early response genes. Only when auxin concentrations drop, do the Aux/IAA proteins reach levels sufficient to repress the early genes. This apparent paradox, the auxin-regulated increase in Aux/IAA mRNA abundance and decrease in Aux/IAA protein stability, may represent a negative feedback loop that tightly regulates the early genes in response to changes in auxin concentration.

on early auxin response genes independently of auxin concentration in a cell. When auxin levels are low within a cell, the auxin response genes are either not expressed or expressed at low levels because Aux/IAA repressors are dimerized with ARF activators on the promoters. The dimerization of Aux/IAA repressors to ARF activators results in active repression of the genes. When auxin concentrations are elevated, the Aux/IAA proteins are ubiquitinated by the TIR1 and TIR1-like SCF complexes, and targeted for degradation by the 26S proteasome. SCF activity requires the dynamic conjugation and deconjugation of the RUB modifier on the AtCUL1 subunit. The proteolytic removal of Aux/IAA repressors results in the derepression of early auxin response genes. Gene activation might be further enhanced by Class II ARFs that dimerize with an ARF that is bound to an AuxRE on the promoter. As long as auxin concentrations remain high, Aux/IAA proteins turn over at a high rate, preventing them from functioning as repressors and resulting in continued expression of the auxin response genes. When auxin concentrations are lowered, Aux/IAA proteins accumulate and are able to reassociate with the Class II ARF on the promoter, restoring repression upon the gene.

This model appears to present a paradox, in that auxin promotes both the transcription of *Aux/IAA* genes and the degradation of Aux/IAA proteins. This could be explained by a regulatory mechanism involving a negative feedback loop. *Aux/IAA* genes that contain TGTCTC AuxREs would be activated by auxin, resulting in an increase in mRNA and protein levels. In the continued presence of auxin, the nascent Aux/IAA proteins would be targeted for degradation by the SCF^{TIR1} complex. When auxin levels diminish, however, the newly synthesized Aux/IAA proteins would be stabilized and could shut off the response pathway by forming new dimers with ARFs. A similar type of negative feedback regulation has been observed with the mammalian SCF substrate, IκB (19).

CONCLUSIONS AND PERSPECTIVES

Molecular and genetic approaches have, thus far, identified several major components involved in auxin signaling at the level of gene expression. This information provides the basis for the current model (Fig. 10), but this model does not reflect the complexity that likely exists in the regulation of this pathway. Dissecting this complexity and defining new molecular components in the pathway are challenges for the future.

A central goal in auxin research is to determine how auxin is perceived and how the signal is transmitted to trigger events such as Aux/IAA proteolysis. The failure to identify mutations affecting upstream components of the auxin response pathway may be due to functional redundancy or the fact that such mutations may cause a lethal phenotype that prevents recovery of these mutants. Alternatively, it is also possible that auxin regulates plant growth and development without the complex receptor-activated signaling cascade characteristic of many other hormones and regulatory molecules. Theoretically, auxin could be perceived directly by the Aux/IAA proteins or the SCF^{TIR1} complex, which could serve as the signal that promotes the physical interaction between these factors and leads to Aux/IAA proteolysis. Although there is no direct evidence supporting such a model, the observation that auxin added to crude extracts devoid of membranes and an ATP re-generating system can target Aux/IAA proteins to the SCF^{TIR1} complex, is consistent with the notion that the events upstream of Aux/IAA ubiquitination may be relatively simple (W.M.G., unpublished).

There is still much to be learned about the auxin response genes that are regulated by ARFs and Aux/IAAs. For example, it will be important to identify transcription factors that interact with ARFs on the naturally occurring composite AuxREs, such as those found in the *GH3* promoter. Additionally, the roles of the many Aux/IAA and ARF protein family members need to be clarified. Transient assays using plant protoplasts indicate that many Arabidopsis Aux/IAA proteins function as active repressors (36), but it remains a possibility that some Aux/IAAs might not function as repressors in all cell types. It is also possible that some Aux/IAAs might function as transcriptional activators. In similar assays, it has been shown that some ARF proteins act as activators, while others repress transcription. The role of ARF repressors has yet to be elucidated. The demonstration that Aux/IAAs and ARFs can form homo- and heterodimers suggests that complex combinatorial interactions between these two groups of proteins may be crucial in the regulation of auxin-responsive transcription. It remains to be determined whether there is any specificity in these interactions, and whether temporal and tissue-specific expression patterns of the Aux/IAA and ARF proteins might control these interactions *in vivo*. Understanding these interactions is clearly important, as the myriad of potential Aux/IAA-ARF dimers may be crucial to the control of diverse growth responses to auxin. Another target for discovery will be to identify proteins that function upstream of the Aux/IAA proteins to regulate SCF-substrate interactions.

The identification of other auxin-inducible genes, promoter elements and transcription factors that regulate these genes may give additional insights into auxin signaling. The recently described auxin-inducible *NAC1* gene is of particular interest (44). *NAC1* encodes a member of the NAM/CUC family of transcription factors, and promotes auxin-mediated lateral root development, at least in part by activating the transcription of *AIR3*. Genetic studies indicate that NAC1 functions downstream of the SCF^{TIR1} complex; however, the molecular links connecting these two components of the auxin response pathway remain to be elucidated. Recently, NAC1 was found to be targeted for ubiquitin-mediated proteolysis by SINAT5, a ubiquitin-ligase belonging to a family of E3 enzymes that is distinct from the SCF^{TIR1} complex (45). Microarray analyses may prove to be invaluable in identifying new primary auxin-response genes. There remains the possibility that some auxin-inducible genes will contain auxin-responsive promoter elements other than TGTCTC. The characterization of these elements and the factors that regulate them may define alternative auxin response pathways.

The regulation of complex growth and developmental processes requires the coordination and integration of multiple signaling events. As such, the auxin signal transduction pathway should be viewed as one component of a network of signaling pathways. There is overwhelming evidence that auxin interacts with other pathways, such as other phytohormone and light signaling pathways (34). One such example, the GA biosynthesis gene *LE* in peas, is discussed in chapter E1. The elucidation of how auxin signaling is integrated with these other response pathways provides an important challenge for the future that will ultimately greatly improve our understanding of plant growth and development.

Acknowledgements

The authors wish to acknowledge research support from the National Science Foundation (MCB0080096) and Missouri Food for the 21st Century (to TJG and GH), and the National Institutes of Health (GM067203) (to WMG). The authors (GH and TJG) are grateful to Sean J. Pfaff for assistance in the preparation of the manuscript.

References

1. Abel S, Oeller PW, Theologis A (1994) Early auxin-induced genes encode short-lived nuclear proteins. Proc Natl Acad Sci USA 91: 326-330
2. Chen J-G (2001) Dual auxin signaling pathways control cell elongation and division. J Plant Growth Regul 20:255-264
3. Chen J-G, Ullah, H, Young JC, Sussman MR, Jones AM (2001) ABP1 is required for organized cell elongation and division in Arabidopsis embryogenesis. Genes Dev 15: 902-911
4. Cope GA, Suh GS, Aravind L, Schwarz SE, Zipursky SL, Koonin EV, Deshaies RJ (2002) Role of predicted metalloprotease motif of Jab1/Csn5 in cleavage of Nedd8 from Cul1. Science 298: 608-611
5. DeLong A, Mockaitis K, Christensen S (2002) Protein phosphorylation in the delivery of and response to auxin signals. Plant Mol Biol 49: 285-303
6. Deshaies RJ (1999) SCF and Cullin/Ring H2-based ubiquitin ligases. Annu Rev Cell Dev Biol 15: 435-467
7. Dharmasiri N, Dharmasiri S, Jones AM, Estelle M (2003) Auxin action in a cell-free system. Curr Biol. 13: 1418-1422
8. Gagne JM, Downes BP, Shiu SH, Durski AM, Vierstra RD (2002) The F-box subunit of the SCF E3 complex is encoded by a diverse superfamily of genes in Arabidopsis. Proc Natl Acad Sci U S A 99: 11519-11524
9. Gray WM, Estelle M (2000) Function of the ubiquitin-proteasome pathway in auxin response. Trends Biochem Sci 25: 133-138
10. Gray WM, Kepinski S, Rouse D, Leyser O, Estelle M (2001) Auxin regulates SCFTIR1-dependent degradation of AUX/IAA proteins. Nature 414: 271-276
11. Guilfoyle TJ (1999) Auxin-regulated genes and promoters. *In* KL Libbenga, M Hall, PJJ Hooykaas, eds, Biochemistry and Molecular Biology of Plant Hormones, Elsevier Publishing Co, Leiden, The Netherlands, pp 423-459
12. Guilfoyle TJ, Hagen G (2001) Auxin response factors. J Plant Growth Regul 20: 281-291
13. Hagen G, Guilfoyle T (2002) Auxin-responsive gene expression: genes, promoters and regulatory factors. Plant Mol Biol 49: 373-385
14. Hellmann H, Estelle M (2002) Plant development: regulation by protein degradation. Science 297: 793-797
15. Hellmann H, Hobbie L, Chapman A, Dharmasiri S, Dharmasiri N, del Pozo C, Reinhardt D, Estelle M (2003) Arabidopsis AXR6 encodes CUL1 implicating SCF E3 ligases in auxin regulation of embryogenesis. EMBO J 22: 3314-3325
16. Hershko A, Ciechanover A (1998) The ubiquitin system. Annu Rev Biochem 67: 425-479
17. Hobbie L, McGovern M, Hurwitz LR, Pierro A, Liu NY, et al. (2000) The axr6 mutants of Arabidopsis thaliana define a gene involved in auxin response and early development. Development 127: 23-32
18. Hsieh H-L, Okamoto H, Wang M, Ang L-H, Matsui M, Goodman H, Deng XW (2000) *Fin219*, an auxin-regulated gene, defines a link between phytochrome A and the

downstream regulator COP1 in light control of Arabidopsis development. Genes Dev 14: 1958-1970
19. Karin M, Ben-Neriah Y (2000) Phosphorylation meets ubiquitination: the control of NF-[kappa]B activity. Annu Rev Immunol 18: 621-663
20. Kepinski S, Leyser O (2002) Ubiquitination and auxin signaling: a degrading story. Plant Cell 14: S81-S95
21. Knauss S, Rohrmeier T, Lehle L (2003) The auxin-induced maize gene ZmSAUR2 encodes a short-lived nuclear protein expressed in elongating tissues. J Biol Chem 278: 23936-23943
22. Leyser O (2002) Molecular genetics of auxin signaling. Annu Rev of Plant Biol 53: 377-398
23. Liscum E, Reed JW (2002) Genetics of Aux/IAA and ARF action in plant growth and development. Plant Mol Biol 49: 387-400
24. Mockaitis K, Howell SH (2000) Auxin induces mitogenic activated protein kinase (MAPK) activation in roots of Arabidopsis seedlings. Plant J 24: 785-796
25. Nakazawa M, Yabe N, Ichikawa T, Yamamoto Y, Yoshizumi T, Hasunuma K, Matsui M (2001) *DFL1*, an auxin-responsive GH3 gene homologue, negatively regulates shoot cell elongation and lateral root formation, and positively regulates the light response of hypocotyl length. Plant J 25: 213-221
26. Napier RM (2001) Models of auxin binding. J Plant Growth Regul 20: 244-254
27. Napier RM, David KM, Perrot-Rechenmann C (2002) A short history of auxin-binding proteins. Plant Mol Biol 49: 339-348
28. Ouellet F, Overvoorde PJ, Theologis A (2001) IAA17/AXR3. Biochemical insight into an auxin mutant phenotype. Plant Cell 13: 829-842
29. Reed JW (2001) Roles and activities of Aux/IAA proteins in Arabidopsis. Trends Plant Sci 6: 420-425
30. Rogg LE, Bartel B (2001) Auxin signaling: derepression through regulated proteolysis. Dev Cell 1: 595-604
31. Scherer GFE (2002) Secondary messengers and phospholipase A_2 in auxin signal transduction. Plant Mol Biol 49: 357-372
32. Schwechheimer C, Deng X (2001) COP9 signalosome revisited: a novel mediator of protein degradation. Trends Cell Biol 11: 420-426
33. Staswick PE, Tiryaki I, Rowe RL (2002) Jasmonate response locus *JAR1* and several related Arabidopsis genes encode enzymes of the firefly luciferase superfamily that show activity on jasmonic, salicylic, and indole-3-acetic acids in an assay for adenylation. Plant Cell 14: 1405-1415
34. Swarup R, Parry G, Graham N, Allen T, Bennett M (2002) Auxin cross-talk: integration of signaling pathways to control plant development. Plant Mol Biol 49: 411-426
35. Tian Q, Reed JW (2001) Molecular links between light and auxin signaling pathways. J Plant Growth Regul 20: 274-280
36. Tiwari SB, Wang X-J, Hagen G, Guilfoyle TJ (2001) AUX/IAA proteins are active repressors, and their stability and activity are modulated by auxin. Plant Cell 13: 2809-2822
37. Tiwari SB, Hagen G, Guilfoyle T (2003) The roles of auxin response factor domains in auxin-responsive transcription. Plant Cell 15: 533-543
38. Tiwari SB, Hagen G, Guilfoyle T (2004) Aux/IAA proteins contain a potent transcriptional repression domain. Plant Cell, in press
39. Ullah H, Chen JG, Temple B, Boyes DC, Alonso JM, Davis KR, Ecker JR, Jones AM (2003) The beta-subunit of the Arabidopsis G protein negatively regulates auxin-induced cell division and affects multiple developmental processes. Plant Cell 15: 393-409
40. Ulmasov T, Hagen G, Guilfoyle TJ (1997) ARF1, a transcription factor that binds auxin response elements. Science 276: 1865-1868
41. Venis MA, Napier RM (1995) Auxin receptors and auxin binding proteins. Crit Rev Plant Sci 14: 27-47

42. Waller F, Furuya M, Nick P (2002) OsARF1, an auxin response factor from rice is auxin-regulated and classifies as a primary auxin responsive gene. Plant Mol Biol 50: 415-425
43. Worley CK, Zenser N, Ramos J, Rouse D, Leyser O, Theologis A, Callis J (2000) Degradation of Aux/IAA proteins is essential for normal auxin signalling. Plant J 21: 553-562
44. Xie Q, Frugis G, Colgan D, Chua N-H (2000) Arabidopsis NAC1 transduces auxin signal downstream of TIR1 to promote lateral root development. Genes Dev 14: 3024-3036
45. Xie Q, Guo HS, Dallman G, Fang S, Weissman AM, Chua N-H (2002) SINAT5 promotes ubiquitin-related degradation of NAC1 to attenuate auxin signals. Nature 419: 167-170
46. Yang T, Poovaiah BW (2000) Molecular and biochemical evidence for the involvement of calcium/calmodulin in auxin action. J Biol Chem 275: 3137-3143
47. Zenser N, Ellsmore A, Leasure C, Callis J (2001) Auxin modulates the degradation rate of Aux/IAA proteins. Proc Natl Acad Sci U S A 98: 11795-11800

D2 Gibberellin Signal Transduction in Stem Elongation & Leaf Growth

Tai-ping Sun
Department of Biology, Duke University, Durham, North Carolina 27708, USA. E-mail address: tps@duke.edu

INTRODUCTION

The effect of gibberellin (GA) on promoting stem growth was first discovered in 1930s by studies of the *Bakanae* (foolish seedling) disease in rice (57). *Gibberella fujikuroi*, a pathogenic fungus, produces gibberellic acid (GA_3) that causes the infected rice plants to grow so tall that they fall over. Later studies of dwarf mutants and analysis of their GA contents revealed that bioactive GAs are endogenous hormones that regulate the natural developmental processes including stem growth in plants. An increase in both cell elongation and cell division occurs during stem growth. GA induces transcription of genes involved in these processes. For example, expression of some of the genes encoding xyloglucan endotransglycosylases (XETs[1]) and expansins are upregulated by GA in elongating internodes in rice and in Arabidopsis (7, 62, 66). XET is thought to increase the plasticity of the cell wall because this enzyme is involved in xyloglucan reorganization through cleaving and re-ligating xyloglucan polymers in the cell wall. Expansins are also extracellular proteins that cause plant cell wall loosening, probably by disrupting the polysaccharide adhesion. Transcripts of the genes encoding for cyclin-dependent protein kinases are also elevated in intercalary meristem in rice after GA treatment (12).

The GA perception and signal transduction pathway converts the GA signal into alterations in gene expression and plant morphology. Although the GA receptor has not been identified, significant progress has been made recently in elucidating GA signaling components in higher plants (40, 44). This chapter will focus on GA signaling in stem elongation and leaf growth. Additional developmental stages will be described in other chapters.

---

[1] Abbreviations: GFP, green fluorescent protein; LHR, leucine heptad repeats; NLS, nuclear localization signal; *O*-GlcNAc, *O*-linked N-acetylglucosamine; OGT, *O*-GlcNAc transferase; PAC, paclobutrazol; TPR, tetratricopeptide, XET, xyloglucan endotransglycosylase.

DWARF AND SLENDER MUTANTS

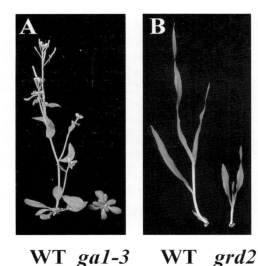

Figure 1. Phenotype of GA biosynthetic mutants in Arabidopsis and in barley. (A) 5-week-old wild-type (WT) Arabidopsis and the GA-deficient *ga1-3* mutant (B) 2-week-old WT Himalaya and a putative GA3ox mutant *grd2* (6)

A genetic approach is a powerful tool in identifying components of complex cellular pathways. One of the most dramatic effects of GA treatment is greatly accelerated stem growth. To study GA signaling, mutants with altered stem heights in a variety of species have been isolated. However, defects in pathways other than GA signaling could also result in a dwarf phenotype. Characterization of GA biosynthetic mutants and GA-treated wild-type plants aided in selecting for mutants that are more likely to affect the GA response specifically. Most of the mutants deficient in GA biosynthesis have a characteristic phenotype, i.e., shorter stature and dark green, compact leaves in comparison to the wild-type isogenic plant (Fig. 1). Additional phenotypes include defects in seed germination and/or flower and fruit development in certain species. Mutants that are impaired in GA signaling resemble the GA biosynthesis mutants, except their phenotype cannot be rescued by GA treatment. On the contrary, mutants with constitutively active GA responses have a very tall (slender) and paler green phenotype, which mimics wild-type plants that are overdosed with GA. Mutant screens based on defects in vegetative growth have the drawback that one might miss mutations in genes that only control specific GA-mediated processes, e.g., seed germination and flower and fruit development. Nevertheless, cloning of genes corresponding to the mutations identified in previous screens has allowed us to identify several positive and negative regulators in GA signaling.

Dwarf mutants are also important in agriculture applications. Reduced-height varieties of wheat and rice were crucial components in improving grain yield during the 'Green Revolution' (20, 52). Application of nitrogen fertilizer did not increase grain yield of the traditional taller wheat and rice cultivars, because the straws grew too tall and fell over. This phenomenon is called lodging. In contrast, the semi-dwarf varieties of wheat and rice responded to the fertilizer nicely with an increased grain production rather than an elevated straw biomass. Their thicker and shorter stalks also allowed

GA signaling

WT   *d1*   *d18*

Figure 2. Phenotype of the *d1* mutant is not as severe as the GA-deficient mutant *d18*. Reproduced from Ueguchi-Tanaka et al. (60).

them to be resistant to lodging. Interestingly, recent molecular studies revealed that the dwarfed rice and wheat are impaired in either the GA biosynthesis (rice) (47) or the GA signaling pathway (wheat) (42). The wheat dwarfing gene will be discussed in more detail later.

POSITIVE REGULATORS OF GA SIGNALING

A positive regulator is one that promotes the effect of the elicitor (in our case GA), whereas a negative regulator blocks the effect of the elicitor. If a positive regulator gene is inactivated the plant cannot respond, whereas if a negative regulator is mutated the phenotype resembles the response in the presence of the elicitor even if it is absent. A number of positive regulators of GA signaling have been identified by characterization of loss-of-function (recessive), GA-unresponsive dwarf mutants. The *dwarf1* (*d1*) (36) and *GA-insensitive dwarf2* (*gid2*) (48) mutants in rice, the *gse* mutants in barley (6) and the *sleepy1* (*sly1*) (54) mutants in Arabidopsis have a semi-dwarf phenotype, but cannot be rescued by GA treatment. The heterotrimeric GTP-binding proteins (G proteins) contain α, β and γ subunits, and they interact with plasma membrane localized receptors (so-called G protein-coupled receptors) and mediate important signaling events in animal cells. Pharmacological studies in cereal aleurones suggested that the heterotrimeric G protein plays a role in GA signaling (28). This hypothesis was strengthened by the finding that *D1* encodes a putative α-subunit of the heterotrimeric G protein (2, 16). However, an alternative GA signaling pathway must exist because the *d1* null mutant is not as short as a severe GA biosynthetic mutant even though the rice genome does not appear to contain any additional gene that is homologous to *D1* (Fig. 2) (60). In Arabidopsis, Gα is also encoded by a single gene *GPA1*. Unlike the rice *d1* mutant, null *gpa1* alleles have normal final plant height, although they are less responsive to GA in seed germination (61). The exact role of the heterotrimeric G proteins in GA signaling requires further investigation.

GID2 and *SLY1* encode homologous proteins with an F-box motif (35, 48). F box motif is a conserved amino acid sequence originally identified in

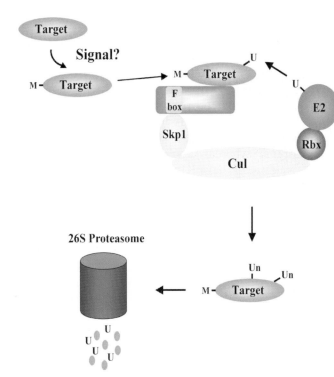

Figure 3. The E3 ubiquitin ligase, SCF complex, mediated ubiquitination and proteolysis of target proteins. A cellular signal induces modification (M) of the target protein, which is recognized by the F-box protein of the SCF complex. Ubiquitin (U) is activated (via E1, not shown), and conjugated via E2 to the target. Once polyubiquitin chain(s) (Un) are formed, the target is degraded by the 26S proteasome and ubiquitins are recycled.

yeast cyclin F. Recent studies showed that many F-box proteins are a subunit of the SCF complex (named after the major components SKP, Cullin and F-box proteins), a class of the ubiquitin protein ligases also called E3 ligases (see below for detail). Ubiquitin is a small protein with only 76 amino acids and is present in all eukaryotes (hence the name). The stability of many cellular proteins is controlled by the so-called ubiquitin-proteasome pathways; i.e., once tagged with ubiquitins, the proteins can be targeted for degradation by the 26S proteasome. In addition to the F-box protein, the SCF complex contains several additional proteins; SKP, Cullin and a zinc-binding RING (for Really Interesting New Protein)-domain protein Rbx (Fig. 3) (9, 22). Through its C-terminal protein-protein interaction domain, the F-box protein in each SCF complex recruits specific target proteins to the SCF complex for ubiquitination, and subsequent degradation by the 26S proteasome. Recently, SCF-mediated proteolysis has been shown to regulate many developmental processes in plants, including floral development, circadian rhythm, light receptor signaling, senescence and hormone signaling (4, 22). GID2 and SLY1 appear to modulate the GA responses by controlling the stability of a class of negative regulators SLR/RGA of GA signaling (see below for details).

PICKLE (*PKL*), another putative positive regulator of GA signaling in Arabidopsis, was identified by the semi-dwarf phenotype of the *pkl* mutants that resembles that of other GA response mutants (38). *PKL* encodes a putative CHD3 chromatin-remodeling factor, which may regulate chromatin architecture and affect transcription of its target genes (39). Interestingly, *pkl* has a unique embryonic root phenotype that is not present in other GA biosynthesis or response mutants (38). This root phenotype has a low penetrance in the *pkl* mutants; i.e. only a small percentage of plants carrying a homozygous *pkl* allele show the embryonic root phenotype. However, treatment with a GA biosynthesis inhibitor paclobutrazol (PAC) increased the penetrance of this phenotype, whereas applications of GA had an opposite effect. These results suggested that PKL may mediate GA-induced root differentiation during germination.

An additional positive regulator of GA signaling, *Photoperiod-Responsive 1* (*PHOR1*), was not found in a mutant screen. *PHOR1* was isolated as a gene whose transcript level increased in potato leaves under short-day conditions (1). Inhibition of *PHOR1* expression in transgenic potato carrying an antisense *PHOR1* construct resulted in a semi-dwarf phenotype. Because this phenotype is similar to GA biosynthetic mutants, GA responsiveness of internode elongation in antisense lines was tested. These lines are, in fact, less sensitive to GA than the wild-type control. Overexpression of PHOR1 confers a longer internode phenotype and an elevated GA response, further supporting the role of PHOR1 as an activator of GA signaling. PHOR1 contains 7 armadillo repeats, which are present in armadillo and β-catenin proteins, components of Wnt signaling in Drosophila and vertebrates and function in controlling polarity during animal development (1). GA appears to promote nuclear localization of transiently expressed PHOR1-GFP in tobacco BY2 cells. These results suggested that PHOR1 may stimulate transcription of GA-induced genes.

NEGATIVE REGULATORS OF GA SIGNALING

Two classes of mutants with opposite phenotypes have contributed to the identification of negative regulators of GA signaling. The first class of mutants is recessive, has a slender phenotype and exhibits a completely or partially constitutive GA response. These phenotypes are due to loss-of-function mutations in the GA signaling repressors. The second class of mutants is semi-dominant and has a GA-unresponsive semi-dwarf phenotype, presumably because the gain-of-function mutations make these repressors constitutively active. Examples of such mutants in barley are shown in Fig. 4.

GA is essential for germination of Arabidopsis seeds. One of the slender mutants, *spindly* (*spy*) in Arabidopsis, was first isolated by screening for mutant seeds that were able to germinate in the presence of the GA biosynthesis inhibitor PAC (27). Additional *spy* alleles were isolated as

sln1c WT *sln1d*

Figure 4. Opposite phenotypes in barley caused by two *sln1* mutations in different parts of the gene. The recessive (loss-of-function) *sln1c* mutant has a slender phenotype, whereas the dominant (gain-of-function) *sln1d* mutant is a dwarf. Reproduced from Chandler *et al.* (5).

suppressors of the GA-deficient dwarf mutant *ga1-3* and a GA-unresponsive dwarf *gai-1* (51, 65). The *GA1* locus encodes an enzyme for the first committed step in GA biosynthesis, namely copalyl pyrophosphate synthase. The null *ga1-3* mutant is a non-germinating, severe dwarf, which only produces male-sterile flowers. Recessive *spy* alleles partially rescued all of the phenotypes in *ga1-3*, indicating that SPY functions to repress an early step in GA signaling. Transient expression of the SPY barley homolog (HvSPY) in aleurone protoplasts inhibited GA-induced α-amylase gene expression (45), further supporting the role of SPY and HvSPY in GA response. However, careful analysis of the *spy* mutant phenotype suggested that SPY may regulate additional cellular pathways, because some of the *spy* phenotypes (e.g., abnormal phyllotaxy of flowers) were not observed in GA-treated wild-type Arabidopsis (55). The SPY sequence is highly similar to that of the Ser/Thr *O*-linked N-acetylglucosamine (*O*-GlcNAc) transferases (OGTs) in animals (26, 46). Animal OGTs modify target proteins by glycosylation of Ser/Thr residues, which in some cases interfere or compete with kinases for phosphorylation sites (8). *O*-GlcNAc modification is a dynamic protein modification, which is implicated in regulating many signaling pathways (63). Purified recombinant SPY protein was shown to have OGT activity *in vitro* (58). Like animal OGTs, the active SPY probably functions as a homotrimer, formed by protein-protein interaction via the tetratricopeptide (34 amino acid) repeats (TPRs) near its N-terminus. Overexpression of the SPY TPRs in transgenic Arabidopsis and petunia conferred a *spy*-like phenotype, suggesting that elevated TPRs alone may block the SPY function by forming poison complexes with SPY and/or by interacting with the target proteins of SPY (25, 59). Because SPY was detected in both cytoplasm and nucleus in plant cells, similar to the localization of animal OGTs, the target proteins of SPY could be present in both cellular compartments (56).

The second type of GA signaling repressors are the RGA/GAI proteins, which are highly conserved in Arabidopsis (RGA, GAI, RGL1, RGL2, RGL3) and several crop plants, including maize (d8), wheat (Rht), rice (SLR1), barley (SLN1) and grape (VvGAI) (3, 40). The RGA/GAI proteins belong to the DELLA subfamily of the plant-specific GRAS protein family

(43). Completion of the Arabidopsis genome sequencing revealed that there are over 30 GRAS family members. All of the GRAS family members contain a conserved C-terminal region, but their N-termini are more divergent, and probably specify their roles in different cellular pathways. A conserved domain (named DELLA, after a conserved amino acid motif) is present near the N-terminus of the RGA/GAI proteins (41, 49). The DELLA domain is unique for RGA/GAI proteins and is absent in other GRAS family members. In addition to the RGA/GAI proteins, the other GRAS proteins with known functions are regulators of diverse developmental pathways, including radial patterning, axillary meristem formation, and phytochrome A signaling (33, 43).

RGA was identified by the ability of loss-of-function *rga* alleles to partially suppress most of the phenotype of the GA-deficient dwarf mutant *ga1-3*, except seed germination and floral development (51). The related gain-of-function (semi-dominant) *gai-1* mutant has a GA-unresponsive dwarf phenotype, indicating that the GA signaling pathway is defective in this mutant (31). Because of the dominant nature of the *gai-1* allele, it was difficult to predict the function of GAI. Subsequently, the loss-of-function *gai-t6* allele was shown to confer resistance to PAC in vegetative growth, suggesting that GAI is a negative regulator of GA response (41). Cloning of *GAI* and *RGA* revealed that the GAI and RGA proteins share 82% identity in their amino acid sequences (41, 49). RGA and GAI have partially redundant functions in maintaining the repressive state of the GA signaling pathway, but RGA plays a more predominant role than GAI. Epistasis analysis showed that *rga* and *gai* null alleles have synergistic effects on processes partially rescued by *rga*, although the *gai* null allele alone has little effect (Fig. 5) (11, 29). Removing both RGA and GAI function allows for a complete de-repression of many aspects of GA signaling, including rosette leave expansion, flowering time and stem elongation. Three additional DELLA subfamily proteins RGL1, RGL2 and RGL3 (for RGA-like) are present in Arabidopsis. Analysis of the mutant alleles at *RGL1* and *RGL2* showed that *RGL2* and perhaps also *RGL1* play a role in inhibiting seed germination (32, 64). Regulation of GA-controlled flower development is more complex, and may require more than one RGA/GAI/RGL protein function.

Studies of RGA/GAI orthologs in various crops indicated that the function of these RGA/GAI proteins in repressing GA signaling is highly conserved between dicots and monocots. Interestingly, only 1 RGA/GAI functional ortholog is present in rice (SLR1) (23) and in barley (SLN1) (5). Unlike in Arabidopsis, GA-independent stem growth in rice and barley was achieved by removing only SLR1 or SLN1, respectively. RGA/GAI proteins have been proposed to function as transcriptional regulators (41, 49). They contain (i) polymeric Ser/Thr motifs (poly S/T) which could be targets of phosphorylation or glycosylation, (ii) Leu heptad repeats (LHR) which may

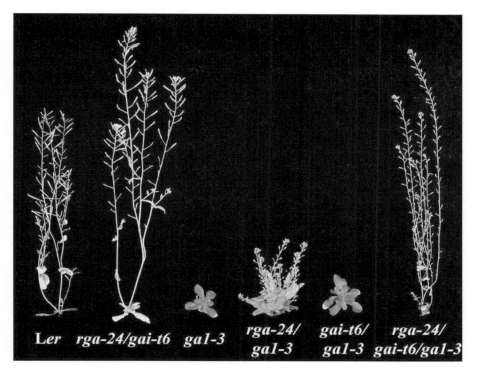

Figure 5 (Color Plate page CP9). Phenotypes of wild-type and homozygous mutant Arabidopsis plants. Each mutant line is homozygous for the allele(s) listed. All plants are 52 days old except for *rga-24/gai-t6,* which is 37 days old. (From 11).

mediate protein-protein interactions, (iii) putative nuclear localization signals (NLS), and (iv) a putative SH2 phosphotyrosine binding domain (42) (Fig. 6). Several RGA/GAI proteins have been demonstrated to direct the GFP fusion protein into plant cell nuclei (40). Although RGA/GAI proteins do not have a clearly identified DNA-binding domain, they may act as co-activators or co-repressors by interacting with other transcription factors that bind directly to DNA sequence of GA-regulated genes.

The unique DELLA domain of the RGA/GAI proteins hinted that this region may specify the role of these proteins in GA response. Indeed, recent studies indicated that the DELLA domain is responsible for modulating the activity of the RGA/GAI proteins in response to the GA signal. The initial evidence came from the discovery that the gain-of-function *gai-1* allele contains a 51-bp in-frame deletion in the *GAI* gene, which results in the loss of 17 amino acids spanning the DELLA motif (41). Peng *et al.* (41) hypothesized that deletions in the gai-1 protein makes it a constitutive repressor of GA response, which can no longer be regulated by the GA signal. Similar internal deletions or N-terminal truncations in other RGA/GAI proteins in different species also resulted in a GA-unresponsive

GA signaling

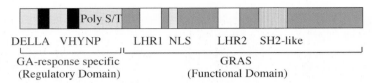

Figure 6. The RGA/GAI proteins. The C-terminal region is conserved among all GRAS family members. The numbers above the diagram correspond to amino acid residues in RGA. Poly S/T, polymeric Ser and Thr. LHR, Leu heptad repeat. NLS, nuclear localization signal. SH2, SH2-like domain.

dwarf phenotype (40). The most notable example is the semi-dwarf wheat cultivars that greatly facilitated an increased grain yield during the Green Revolution in the 1960s and 1970s, which all contain deletions in the DELLA domain of an *Rht* (for reduced height) gene (42). In addition to the deletions in the DELLA domain, studies on *sln1* in barley and *Vvgai* in grape showed that a single amino acid substitution around the DELLA motif could also confer a GA-unresponsive dwarf phenotype (3, 5).

Another negative component of GA signaling, SHORT INTERNODES (SHI) in Arabidopsis was identified by the dwarf phenotype of the dominant *shi* mutant that overexpressed the *SHI* gene (14). SHI contains a zinc binding motif, suggesting its potential role in transcriptional regulation or ubiquitin-mediated proteolysis. Transient expression of SHI in barley aleurone cells inhibited GA-induction of α-amylase expression, further supporting its role in GA signaling (15).

GA DE-REPRESSES ITS SIGNALING PATHWAY BY INDUCING DEGRADATION OF RGA/GAI PROTEINS

Studies on the dwarf mutants containing the DELLA region mutations suggested that GA may activate its signaling pathway by inhibiting the RGA/GAI proteins. To investigate the molecular mechanism by which GA inactivates the RGA protein, we first examined the effect of GA on RGA transcript level. We found that GA treatment resulted in only a slight increase in the amount of the *RGA* mRNA in Arabidopsis seedlings (49). We then tested whether RGA protein localization and/or accumulation are affected by the GA signal. Endogenous RGA levels were monitored by immunoblot analysis using anti-RGA antisera. The subcellular localization of RGA was examined using the GFP-RGA fusion protein (expressed under the control of the *RGA* promoter). This fusion protein is functional in the rescue of the phenotype of the double mutant *rga/ga1-3* (50). The effect of GA treatment on the GFP-RGA fusion protein was analyzed by confocal microscopy (Fig. 7A) and by immunoblotting. It was exciting to see that the levels of both the RGA and GFP-RGA proteins were reduced rapidly by GA

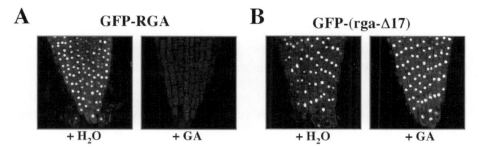

Figure 7. GA induces proteolysis of GFP-RGA, but not GFP-(rga-Δ17). GFP fluorescence in root tips of transgenic Arabidopsis lines expressing the (A) GFP-RGA or (B) GFP-(rga-Δ17) fusion proteins was visualized by confocal laser microscopy. Modified from Dill et al. (10).

treatment (50). GA-induced protein degradation, rather than a decreased translational rate, is likely the cause of the reduction in RGA protein levels (see below). Similar results were also obtained from studies of SLR1 in rice (24) and SLN1 in barley (18). Nevertheless, the GA signal may regulate other DELLA proteins via different mechanisms. For instance, GFP protein fusions with RGL1 or GAI remained stable after GA treatment (13, 64). However, these results need to be verified by analyzing the endogenous RGL1 and GAI proteins.

THE DELLA DOMAIN IS ESSENTIAL FOR THE GA-INDUCED DEGRADATION OF RGA/SLR1/SLN1

As described earlier, deletions within the DELLA domain in RGA/GAI proteins resulted in a GA-unresponsive dwarf phenotype. The sequences missing in gai-1 are identical between GAI and RGA. We found that the same deletion (rga-Δ17) in the RGA gene (with or without the GFP fusion) as in gai-1 conferred a GA-insensitive severe dwarf phenotype in the transgenic Arabidopsis (10). Knowing that GA inhibits RGA protein function by causing a rapid reduction in RGA protein levels, we then tested the effect of GA on the rga-Δ17 mutant protein. Immunoblot analysis and confocal microscopy showed that deletion of the DELLA motif caused the rga-Δ17 and GFP-(rga-Δ17) protein accumulation to be unaffected by GA treatment (Fig. 7B) (10). Thus, GA-inducible RGA protein disappearance depends on the presence of the DELLA motif. These results confirmed that GA-induced protein degradation, rather than reduced translational rate, leads to the rapid disappearance of the RGA protein. We concluded that the GA-unresponsive dwarfing effect is caused by the resistance of the rga-Δ17 and GFP-(rga-Δ17) to GA-induced degradation, and that the DELLA motif is essential for RGA degradation in response to the GA signal. Mutations in the DELLA motif in SLR1 and SLN1 also made the mutant proteins resistant to GA-induced degradation (18, 24).

Additional functional motifs in SLR1 have been identified by overexpressing slr1 mutant proteins with various internal in-frame deletions or a C-terminal truncation in transgenic rice (24). In this study, each *slr* mutant gene was fused to the *GFP* gene and expressed under the control of the constitutive *Act1* promoter. A second highly conserved motif (named VHYNP) is present in the N-terminal DELLA domain (Fig. 6). Deletion of the VHYNP motif or sequences between the DELLA and VHYNP motifs also blocked GA-dependent protein degradation, and conferred a GA-unresponsive dwarf phenotype. The poly S/T/V region present next to the VHYNP sequences may play a regulatory role in SLR1 activity. Expression of poly S/T/V-deleted slr1 conferred a severe dwarf phenotype, but the mutant protein did not accumulate to a higher level than wild-type SLR1 and remains GA-responsive. Overexpression of the LHR1-deleted slr1 did not cause any obvious plant phenotype, even though the mutant protein was not degraded by GA treatment. It was hypothesized that LHR is needed for SLR1 dimerization, which may be required for SLR activity and for GA-induced degradation. The C-terminal truncated slr1 (deletion starts from the VHIID region) was un-responsive to GA, and overexpression of this mutant protein caused a slender phenotype, similar to that of the loss-of-function *slr1* mutant. The dominant-negative phenotype suggested that this truncated protein interferes with the wild-type SLR1, probably by forming non-functional heterodimers.

DEGRADATION OF RGA AND SLR1 MAY BE TARGETED BY SCF^{SLY1} AND SCF^{GID2} THROUGH THE UBIQUITIN-PROTEASOME PATHWAY

As described earlier in this chapter, SLY1 in Arabidopsis and GID2 in rice are positive regulators of GA signaling. Because these proteins contain the F-box domain, their predicted function (as part of an SCF complex) is to modulate the stability of GA-response component(s). The following results support the contention that RGA and SLR1 are the direct targets of SLY1 and GID2, respectively. 1) The RGA and SLR1 proteins accumulated to much higher levels in the *sly1* and *gid2* dwarf mutants, respectively, than in wild type (35, 48). 2) A null *rga* allele partially suppressed the dwarf phenotype of the *sly1* mutant, indicating that the elevated RGA level contributed to the dwarfness of *sly1*, and that RGA is downstream of SLY1 in the GA signaling pathway (35). 3) GID2 was shown to interact with OsSkp2 (one of the Skp homologs in rice) in a yeast two-hybrid assay, showing that GID2 is part of a SCF complex (48).

Post-translational modifications of target proteins, most commonly phosphorylation, are often required for the recognition by the SCF E3 ligases (30). SLR1 protein appears to be present in two forms in the *gid2* mutant, a phosphorylated and an unphosphorylated form, whereas only the unphosphorylated form is detected in wild type (48). GA treatment caused an increase in the phosphorylated SLR1 in *gid2*. Taken together, SLR1 phosphorylation is induced by GA, and GID2 may preferentially target the phosphorylated SLR1 for degradation.

MODEL OF GA SIGNALING PATHWAY

A revised model of the GA signaling pathway in Arabidopsis is shown in Fig. 8. This model only includes SLY1, SPY, RGA and GAI because the genetic interactions among these genes have been examined. RGA and GAI are putative transcriptional regulators, which directly or indirectly inhibit GA-activated genes. Like RGA and GAI, SPY is also a negative regulator of GA response pathway. OGT target sites are often rich in Ser/Thr and near Pro, Val or acidic residues (34). It was hypothesized that SPY may inhibit the GA signaling pathway by activating RGA and GAI because the poly S/T region of the RGA/GAI proteins contains putative OGT modification sites (41, 49). In animal systems, GlcNAc modification could facilitate nuclear localization, increase protein stability and/or alter protein activity (53, 63). The GA signal de-represses its signaling pathway by promoting RGA protein degradation through the SCF^{SLY1}-proteasome pathway. This model also proposed that GAI activity is inhibited by SLY1, but the mechanism involved is not clear because a GFP-GAI fusion protein is unresponsive to GA.

A quantitative response to the amount of GA signal is incorporated into

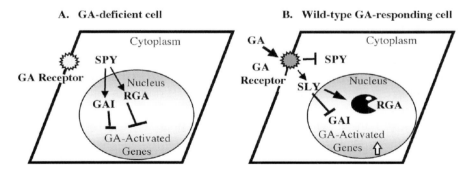

Figure 8. Model of GA signaling in Arabidopsis. RGA, GAI and SPY are repressors, whereas SLY1 is an activator of GA signaling. (A) SPY activates RGA/GAI by GlcNAcylation under GA-deficient conditions. (B) SLY1 mediates GA-induced RGA degradation by the ubiquitin-proteasome pathway.

this model. The concentrations of bioactive GAs determine the degree of activation of the GA response pathway by controlling the extent of RGA/GAI inactivation through SLY1. RGA and GAI are the major repressors that modulate GA-induced leaf and stem growth, and also transition from vegetative to reproductive phase (flowering time). RGL1 and RGL2 have also been shown to play a role in regulating seed germination. It is likely that RGL3 also plays a role in seed germination and/or flower development. RGL1 to RGL3 activities may be also modulated by SPY and SLY1 because *spy* and *sly1* mutations affect all of GA-regulated developmental processes.

Alternative models also exist to explain the results of previous genetic studies. For example, SPY may act downstream from RGA/GAI/RGLs. Thus, biochemical studies are needed to demonstrate that RGA/GAI/RGLs are the direct substrates of SPY. Additional epistasis analysis between *sly1* and *gai/rgl*s will help to determine whether GAI and RGLs are regulated by SLY1. Future biochemical and genetic studies are needed to incorporate the heterotrimeric G protein, SHI and PKL into this Arabidopsis model. In rice, the putative heterotrimeric G protein (D1) functions upstream of SLR1 because the recessive *slr1* is epistatic to *d1* (60). The transcription factor, GAMYB (Chapter C2), is a positive regulator of GA signaling, which acts downstream of SLN1 in barley aleurone cells to induce some of the GA-induced genes during germination (19). GAMYB may also regulate anther development because this gene is strongly expressed in anthers and overexpression of this gene in transgenic barley affected anther development (37). Three GAMYB-like genes have been identified in Arabidopsis (17). In vitro binding assays suggested that AtGAMYB33 may regulate flowering by controlling the expression of the floral meristem-identity gene *LEAFY*. However a specific GAMYB that is involved in GA-regulated stem elongation has not been identified.

FEEDBACK REGULATION OF GA METABOLISM BY THE ACTIVITY IN THE GA RESPONSE PATHWAY.

Accumulating evidence indicated that the activity of GA response pathway is tightly linked to the activities in GA biosynthesis and catabolism by the so-called feedback mechanism (21, 40). GA 20-oxidases (20ox) and GA 3-oxidases (3ox) are enzymes that catalyze the final steps in the synthesis of bioactive GAs, whereas GA 2-oxidases (2ox) are responsible for GA deactivation. In the constitutive GA-response mutants, such as rice *slr1*, barley *sln1*, pea *la cry^s* and Arabidopsis *rga/gai-t6*, the bioactive GAs and/or the transcript levels of the *GA20ox* and/or *GA3ox* genes are present at lower levels than in wild type. In contrast, GA-unresponsive dwarf mutants (e.g., rice *d1*, *gid2*, maize *D8* and Arabidopsis *gai-1*) accumulate higher amounts of bioactive GAs and *GA20ox* and/or *GA3ox* mRNAs. Although *GA2ox* expression has not been examined in these mutants, mRNA levels of some of the *GA2ox* genes were reduced in GA-deficient conditions and elevated by GA treatment. These results indicate that GA homeostasis is achieved by a close interaction between the GA metabolism and GA signaling pathways. Future molecular genetic and biochemical studies will help to elucidate the molecular mechanism involved in the feedback inhibition.

Acknowledgments

I thank Peter Chandler and Makoto Matsuoka for their permissions for using the photos shown in Figures 1B, 2 and 4, and Aron Silverstone for helpful comments on the manuscript. This work was supported by the U.S. Department of Agriculture (01-35304-10892) and the National Science Foundation (IBN-0078003 and IBN-0235656).

References

1. Amador V, Monte E, García-Martínez JL, Prat S (2001) Gibberellins signal nuclear import of PHOR1, a photoperiod-responsive protein with homology to *Drosophila* armadillo. Cell 106: 343-354
2. Ashikari M, Wu J, Yano M, Sasaki T, Yoshimura A (1999) Rice gibberellin-insensitive dwarf mutant gene *Dwarf 1* encodes the α-subunit of GTP-binding protein. Proc Natl Acad Sci USA 96: 10284-10289
3. Boss PK, Thomas MR (2002) Association of dwarfism and floral induction with a grape 'green revolution' mutation. Nature 416: 847-850
4. Callis J, Vierstra RD (2000) Protein degradation in signaling. Curr Opin Plant Biol 3: 381-386
5. Chandler PM, Marion-Poll A, Ellis M, Gubler F (2002) Mutants at the *Slender1* locus of barley cv Himalaya: molecular and physiological characterization. Plant Physiol 129: 181-190
6. Chandler PM, Robertson M (1999) Gibberellin dose-response curves and the characterization of dwarf mutants of barley. Plant Physiol 120: 623-632
7. Cho H-T, Kende H (1997) Expression of expansin genes is correlated with growth in deepwater rice. Plant Cell 9: 1661-1671
8. Comer FI, Hart GW (2000) *O*-glycosylation of nuclear and cytosolic proteins. Dynamic interplay between *O*-GlcNAc and *O*-phosphate. J Biol Chem 275: 29179-29182
9. Conaway RC, Brower CS, Conaway JW (2002) Emerging roles of ubiquitin in transcription regulation. Science 296: 1254-1258
10. Dill A, Jung H-S, Sun T-p (2001) The DELLA motif is essential for gibberellin-induced degradation of RGA. Proc Natl Acad Sci USA 98: 14162-14167
11. Dill A, Sun T-p (2001) Synergistic de-repression of gibberellin signaling by removing RGA and GAI function in *Arabidopsis thaliana*. Genetics 159: 777-785
12. Fabian T, Lorbiecke R, Umeda M, Sauter M (2000) The cell cycle genes *cycA1;1* and *cdc2Os-3* are coordinately regulated by gibberellin in planta. Planta 211: 376-383
13. Fleck B, Harberd NP (2002) Evidence that the Arabidopsis nuclear gibberellin signalling protein GAI is not destabilized by gibberellin. Plant J 32: 935-947
14. Fridborg I, Kuusk S, Moritz T, Sundberg E (1999) The Arabidopsis dwarf mutant *shi* exhibits reduced gibberellin responses conferred by overexpression of a new putative zinc finger protein. Plant Cell 11: 1019-1031
15. Fridborg I, Kuusk S, Moritz T, Sundberg E (2001) The Arabidopsis protein SHI represses gibberellin responses in Arabidopsis and barley. Plant Physiol 127: 937-948

16. Fujisawa Y, Kato T, Ohki S, Ishikawa A, Kitano H, Sasaki T, Asahi T, Iwasaki Y (1999) Suppression of the heterotrimeric G protein causes abnormal morphology, including dwarfism, in rice. Proc Natl Acad Sci USA 96: 7575-7580
17. Gocal GFW, Sheldon CC, Gubler F, Moritz T, Bagnall DB, MacMillan CP, Li SF, Parish RW, Dennis ES, Weigel D, King RW (2000) *GAMYB-like* genes, flowering and gibberellin signaling in Arabidopsis. Plant Physiol 127: 1682-1693
18. Gubler F, Chandler P, White R, Llewellyn D, Jacobsen J (2002) GA signaling in barley aleurone cells: control of SLN1 and GAMYB expression. Plant Physiol 129: 191-200
19. Gubler F, Kalla R, Roberts J, Jacobsen JV (1995) Gibberellin-regulated expression of a *myb* gene in barley aleurone cells: Evidence for Myb transactivation of a high-pI α-amylase gene promoter. Plant Cell 7: 1879-1891
20. Hedden P (2003) The genes of the Green Revolution. Trends Genet 19: 5-9
21. Hedden P, Phillips AL (2000) Gibberellin metabolism: new insights revealed by the genes. Trends Plant Sci 5: 523-530
22. Hellmann H, Estelle M (2002) Plant development: regulation by protein degradation. Science 297: 793-797
23. Ikeda A, Ueguchi-Tanaka M, Sonoda Y, Kitano H, Koshioka M, Futsuhara Y, Matsuoka M, Yamaguchi J (2001) slender rice, a constitutive gibberellin response mutant is caused by a null mutation of the SLR1 gene, an orthologue of the height-regulating gene GAI/RGA/RHT/D8. Plant Cell 13: 999-1010
24. Itoh H, Ueguchi-Tanaka M, Sato Y, Ashikari M, Matsuoka M (2002) The gibberellin signaling pathway is regulated by the appearance and disappearance of SLENDER RICE1 in nuclei. Plant Cell 14: 57-70
25. Izhaki A, Swain SM, Tseng T-s, Borochov A, Olszewski NE, Weiss D (2001) The role of SPY and its TPR domain in the regulation of gibberellin action throughout the life cycle of *Petunia hybrida* plants. Plant J 28: 181-190
26. Jacobsen SE, Binkowski KA, Olszewski NE (1996) SPINDLY, a tetratricopeptide repeat protein involved in gibberellin signal transduction in *Arabidopsis*. Proc Natl Acad Sci USA 93: 9292-9296
27. Jacobsen SE, Olszewski NE (1993) Mutations at the *SPINDLY* locus of Arabidopsis alter gibberellin signal transduction. Plant Cell 5: 887-896
28. Jones HD, Smith SJ, Desikan R, Plakidou-Dymock S, Lovegrove A, Hooley R (1998) Heterotrimeric G proteins are implicated in gibberellin induction of α-amlyase gene expression in wild oat aleurone. Plant Cell 10: 245-253
29. King K, Moritz T, Harberd N (2001) Gibberellins are not required for normal stem growth in Arabidopsis thaliana in the absence of GAI and RGA. Genetics 159: 767-776
30. Kipreos ET, Pagano M (2000) The F-box protein family. Genome Biol 1: 3002.1-3002.7
31. Koornneef M, Elgersma A, Hanhart CJ, van Loenen MEP, van Rijn L, Zeevaart JAD (1985) A gibberellin insensitive mutant of *Arabidopsis thaliana*. Physiol Plant 65: 33-39
32. Lee S, Cheng H, King KE, Wang W, He Y, Hussain A, Lo J, Harberd NP, Peng J (2002) Gibberellin regulates Arabidopsis seed germination via *RGL2*, a *GAI/RGA*-like gene whose expression is up-regulated following imbibition. Genes Dev 16: 646-658
33. Li X, Qian Q, Fu Z, Wang Y, Xiong G, Zeng D, Wang X, Liu X, Teng S, Hiroshi F, Yuan M, Han B, Li J (2003) Control of tillering in rice. Nature 422: 618-621
34. Lubas WA, Frank DW, Krause M, Hanover JA (1997) *O*-linked GlcNAc transferase is a conserved nucleocytoplasmic protein containing tetratricopeptide repeats. J Biol Chem 272: 9316-9324
35. McGinnis KM, Thomas SG, Soule JD, Strader LC, Zale JM, Sun T-p, Steber CM (2003) The Arabidopsis *SLEEPY1* gene encodes a putative F-box subunit of an SCF E3 ubiquitin ligase. Plant Cell 15: 1120-1130
36. Mitsunaga S, Tashiro T, Yamaguchi J (1994) Identification and characterization of gibberellin-insensitive mutants selected from among dwarf mutants of rice. Theor Appl Genet 87: 705-712

37. Murray F, Kalla R, Jacobsen J, Gubler F (2003) A role for HvGAMYB in anther development. Plant J 33: 481-491
38. Ogas J, Cheng J-C, Sung ZR, Somerville C (1997) Cellular differentiation regulated by gibberellin in the *Arabidopsis thaliana pickle* mutant. Science 277: 91-94
39. Ogas J, Kaufmann S, Henderson J, Somerville C (1999) PICKLE is a CHD3 chromatin-remodeling factor that regulates the transition from embryonic to vegetative development in *Arabidopsis*. Proc Natl Acad Sci USA 96: 13839-13844
40. Olszewski N, Sun T-p, Gubler F (2002) Gibberellin signaling:biosynthesis, catabolism, and response pathways. Plant Cell Supplement: S61-S80
41. Peng J, Carol P, Richards DE, King KE, Cowling RJ, Murphy GP, Harberd NP (1997) The Arabidopsis *GAI* gene defines a signalling pathway that negatively regulates gibberellin responses. Genes Dev 11: 3194-3205
42. Peng J, Richards DE, Hartley NM, Murphy GP, Devos KM, Flintham JE, Beales J, Fish LJ, Worland AJ, Pelica F, al. e (1999) 'Green Revolution' genes encode mutant gibberellin response modulators. Nature 400: 256-261
43. Pysh LD, Wysocka-Diller JW, Camilleri C, Bouchez D, Benfey PN (1999) The GRAS gene family in Arabidopsis: sequence characterization and basic expression analysis of the *SCARECROW-LIKE* genes. Plant J 18: 111-119
44. Richards DE, King KE, Ait-ali T, Harberd NP (2001) How gibberellin regulates plant growth and development: a molecular genetic analysis of gibberellin signaling. Annu Rev Plant Physiol Plant Mol Biol 52: 67-88
45. Robertson M, Swain SM, Chandler PM, Olszewski NE (1998) Identification of a negative regulator of gibberellin action, HvSPY, in barley. Plant Cell 10: 995-1007
46. Roos MD, Hanover JA (2000) Structure of *O*-linked GlcNAc transferase: mediator of glycan-dependent signaling. Biochem Biophys Res Comm 271: 275-280
47. Sasaki A, Ashikari M, Ueguchi-Tanaka M, Itoh H, Nishimura A, Swapan D, Ishiyama K, Saito T, Kobayashi M, Khush GS, Kitano H, Matsuoka M (2002) A mutant gibberellin-synthesis gene in rice. Nature 416: 701-702
48. Sasaki A, Itoh H, Gomi K, Ueguchi-Tanaka M, Ishiyama K, Kobayashi M, Jeong D-H, An G, Kitano J, Ashikari M, Matsuoka M (2003) Accumulation of phosphorylated repressor for gibberellin signaling in an F-box mutant. Science 299: 1896-1898
49. Silverstone AL, Ciampaglio CN, Sun T-p (1998) The Arabidopsis *RGA* gene encodes a transcriptional regulator repressing the gibberellin signal transduction pathway. Plant Cell 10: 155-169
50. Silverstone AL, Jung H-S, Dill A, Kawaide H, Kamiya Y, Sun T-p (2001) Repressing a repressor: gibberellin-induced rapid reduction of the RGA protein in Arabidopsis. Plant Cell 13: 1555-1566
51. Silverstone AL, Mak PYA, Casamitjana Martínez E, Sun T-p (1997) The new *RGA* locus encodes a negative regulator of gibberellin response in *Arabidopsis thaliana*. Genetics 146: 1087-1099
52. Silverstone AL, Sun T-p (2000) Gibberellins and the green revolution. Trends Plant Sci 5: 1-2
53. Snow DM, Hart GW (1998) Nuclear and cytoplasmic glycosylation. Int Rev Cytol 181: 43-74
54. Steber CM, Cooney S, McCourt P (1998) Isolation of the GA-response mutant *sly1* as a suppressor of *ABI1-1* in *Arabidopsis thaliana*. Genetics 149: 509-521
55. Swain SM, Tseng T-s, Olszewski NE (2001) Altered expression of SPINDLY affects gibberellin response and plant development. Plant Physiol 126: 1174-1185
56. Swain SM, Tseng T-s, Thornton TM, Gopalraj M, Olszewski N (2002) SPINDLY is a nuclear-localized repressor of gibberellin signal transduction expressed throughout the plant. Plant Physiol 129: 605-615
57. Takahashi N, Phinney BO, MacMillan J (*eds*) (1991) Gibberellins. Springer-Verlag, New York Inc., New York

58. Thornton TM, Swain SM, Olszewski NE (1999) Gibberellin signal transduction presents...the SPY who *O*-GlcNAc'd me. Trends Plant Sci 4: 424-428
59. Tseng T-s, Swain SM, Olszewski NE (2001) Ectopic expression of the tetratricopeptide repeat domain of SPINDLY causes defects in gibberellin response. Plant Physiol 126: 1250-1258
60. Ueguchi-Tanaka M, Fujisawa Y, Kobayashi M, Ashikari M, Iwasaki Y, Kitano H, Matsuoka M (2000) Rice dwarf mutant *d1*, which is defective in the α subunit of the heterotrimeric G protein, affects gibberellin signal transduction. Proc Natl Acad Sci USA 97: 11638-11643
61. Ullah H, Chen J-G, Wang S, Jones AM (2002) Role of a heterotrimeric G protein in regulation of Arabidopsis seed germination. Plant Physiol 129: 897-907
62. Uozu S, Tanaka-Ueguchi M, Kitano H, Hattori K, matsuoka M (2000) Characterization of *XET*-related genes of rice. Plant Physiol 122: 853-859
63. Wells L, Vosseller K, Hart GW (2001) Glycosylation of nucleocytoplasmic proteins: signal transduction and *O*-GlcNAc. Science 291: 2376-2378
64. Wen C-K, Chang C (2002) Arabidopsis *RGL1* encodes a negative regulator of gibberellin responses. Plant Cell 14: 87-100
65. Wilson RN, Somerville CR (1995) Phenotypic suppression of the gibberellin-insensitive mutant (*gai*) of Arabidopsis. Plant Physiol 108: 495-502
66. Xu W, Campbell P, Vargheese AV, Braam J (1996) The Arabidopsis XET-related gene family: environmental and hormonal regulation of expression. Plant J 9: 879-889

D3. Cytokinin Signal Transduction

Bridey B. Maxwell and Joseph J. Kieber
University of North Carolina, Biology Department, CB# 3280, Chapel Hill, NC 27599-3280 USA. Emails: maxwell2@unc.edu; jkieber@unc.edu

INTRODUCTION

Cytokinins influence many aspects of plant growth and development, including apical dominance, leaf expansion and senescence, nutrient mobilization, chloroplast differentiation and activation of shoot meristems (22). This chapter[1] seeks to provide an overview of what is currently known about the signaling mechanisms underlying these cytokinin effects. The biosynthesis, transport, metabolism and specific biological action of cytokinins is covered elsewhere in this book. Over the past ten years many advances have been made toward the understanding of cytokinin signaling including the cloning of several cytokinin receptors and their potential downstream effectors. The primary cytokinin signaling mechanism appears to be analogous to bacterial two-component phosphorelay systems that sense and respond to environmental signals. However, additional pathways may exist and other distinct signaling components have also been implicated in the cytokinin primary response. This chapter will focus on the evidence for early cytokinin signaling events that are similar to the two-component phosphorelay model. In addition, we present summaries of the signaling links between cytokinin and meristem activation and an overview of other genes potentially involved in cytokinin signaling.

---

[1] Abbreviations: AA, amino acid; AHK, Arabidopsis histidine kinase; AHP, Arabidopsis histidine-phosphotransfer protein; APRR, Arabidopsis pseudo-response regulator; ARR, Arabidopsis response regulator; AtDBP, Arabidopsis thaliana DNA-binding protein; BA, N^6 (benzyl) adenine; CCA1, circadian clock associated 1; CHASE, cyclases/histidine kinases-associated sensory extracellular; CheY, chemotaxis Y; CycD3, cyclin D3; EMS, ethyl methanesulfonate; FliM, flagellar motor switch pritein; GARP, GOLDEN/ARR/Psr1; GUS, β-glucoronidase; HOG1, high-osmolarity glycerol 1 gene; Hpt, histidine phosphotransfer protein; iP, $N^6(\Delta^2$-isopentenyl) adenine; KNOX, KNOTTED-like homeobox gene; LHY, late elongated hypocotyl; LUC, luciferase; MAPK, mitogen activated protein kinase; MAPKKK mitogen activated protein kinase kinase kinase; PhyA, phytochrome A; PhyB, photochrome B; PIF3, phytochrome interacting factor 3; SAM, shoot apical meristem; SAUR, small auxins upregulated; TOC1, timing of CAB (chlorophyll la/b binding protein); Zm, Zea maize.

Two-Component Signal Transduction

Two-component signal transduction is an evolutionarily conserved system that enables bacteria to sense and respond to diverse environmental cues. These systems of cellular control are not widely used in eukaryotes, but can be found in yeast, fungi, slime molds, and plants (38). In *Escherichia coli*, over 40 distinct two-component systems have been identified that utilize a total of 62 proteins. The first component is a sensor histidine protein kinase that perceives the external environmental stimulus and the second component is a response regulator that mediates the output, which is often a direct effect on gene transcription. Typically, the sensor is a membrane-bound histidine kinase dimer with an extracellular input domain that controls kinase activity. Upon binding of a stimulus *trans* autophosphorylation occurs, with one kinase domain of the dimer subunit phosphorylating the conserved histidine in the intracellular transmitter domain of the opposing subunit (Fig. 1A). Binding to the input domain can also negatively regulate the histidine kinase domain in some systems. In the active state, the signal is propagated by the subsequent phosphorylation of a response regulator at a conserved aspartate in its receiver domain, causing the activation of an output domain that performs some biological function such as the control of gene transcription. Some response regulators affect the activity of separate proteins via protein-protein interactions, as is the case for CheY in *E.coli*, which directly interacts with the flagellar motor protein FliM to change the direction of flagellar rotation.

In some cases, the basic two-component scheme has been elaborated to become a multi-step system that includes four phosphorylation events that alternate between histidine (His) and aspartate (Asp) residues, often with a third phosphotransmitter protein to make up a phosphorelay (Fig. 1B). For example, budding yeast (*Saccharomyces cerevisiae*) use a multi-step phosphorelay in osmosensing (25). The hybrid histidine kinase, SLN1, is a fusion of an input sensor domain, a histidine transmitter domain and an aspartate receiver domain. Under constant osmotic conditions, the membrane bound SLN1 dimer is active and signaling begins with *trans*-phosphorylation at histidine and the internal relay of a phosphate from histidine to aspartate within SLN1 (Fig. 1C). An intermediary histidine-phosphotransmitter (Hpt) protein, YPD1, then mediates the transfer of a phosphate from the receiver domain of SLN1 to the receiver domain of the response regulator SSK1. Phosphorylated SSK1 inactivates the MAPKKK activity of SSK2 leading to an inactivation of downstream MAPK signaling. When SLN1 detects osmotic stress via its single extracellular domain, its histidine kinase activity is inhibited, the phosphorelay to YPD1 and SSK1 stops and SSK2 and the downstream MAPK pathway are released from inhibition. Ultimately this leads to an increase in the transcription of glycerol-synthesis genes, the production of glycerol and an increase in the osmotic potential of the cytoplasm. This example demonstrates that two-component systems can feed into other signaling pathways such as MAPK

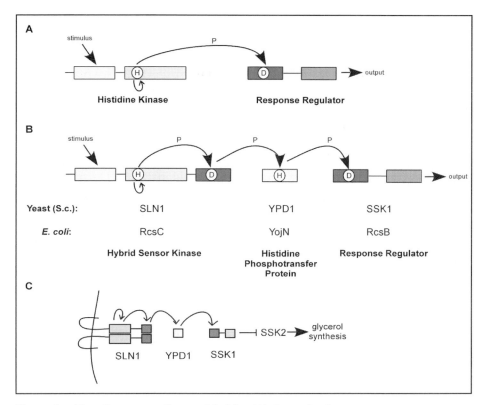

Figure 1. Two-Component System Models. (A) In a simple two-component system, a histidine kinase detects the extracellular stimulus and becomes phosphorylated at the conserved histidine residue. The phosphoryl group is then transferred to the receiver domain of a response regulator that regulates the activity of its output domain. (B) In a multistep phosphorelay, the histidine kinase is fused to a receiver domain. A separate histidine-phosphotransfer protein (Hpt) mediates the transfer of the phosphoryl group to the response regulator. (C) In yeast, a multistep phosphorelay system controls the osmotic potential of the cell in response to external osmotic conditions. The osmosensing hybrid sensor kinase SLN1 phosphorylates the Hpt YPD1 that in turn donates a phosphoryl group to SSK1. Phosphorylated SSK1 represses the MAPKKK activity of SSK2 that ultimately regulates glycerol synthesis. Encircled H and D represent the conserved His and Asp residues that are phosphorylated respectively. P represents the phosphoryl group and arrows denote phosphotransfer events. (For information on Arabidopsis cytokinin signaling genes see http://www.bio.unc.edu/faculty/kieber/lab/)

cascades, and thus the two-component signal does not necessarily end with the response regulators. The existence of over 50 genes with homology to the two-component family of proteins in Arabidopsis suggests that this signaling pathway is likely involved in many aspects of plant cell regulation, one of which is cytokinin action; other stimuli include ethylene, osmotic potential, and possibly light.

Cytokinin signal transduction

HISTIDINE KINASES IN PLANTS

The Arabidopsis genome encodes 16 proteins with ancestral homology to histidine kinases and 32 response regulator related proteins (9, 35). The initial discovery that plants might use two-component signaling systems similar to those found in bacteria was the identification of the ethylene receptor *ETR1* gene (Chapter D4), which encodes a hybrid sensor histidine kinase gene (similar to the SLN1 gene described above). The Arabidopsis histidine kinases are grouped into three main families: the ethylene receptors, the cytokinin receptors (AHKs) and the phytochromes (Fig. 2A). In addition, there are three other histidine kinases in Arabidopsis (CKI1, AHK5, and AHK1) that fall outside these groups. Homologues of these histidine kinase receptors have been found in other plant species including maize, tobacco, rice, Medicago, periwinkle, and the microalga Chlamydomonas. Their overall structure is that of a hybrid sensor histidine kinase containing a variable input domain, several N-terminal transmembrane domains, a transmitter domain with a conserved structure which includes the histidine residue that is the site of autophosphorylation, and a fused receiver domain (Fig. 2B). However, the transmitter domains in the phytochromes are highly divergent and at least one (PhyA) has been shown to have serine/threonine protein kinase activity. The transmitters of several of the ethylene receptors also lack key residues in the highly conserved histidine kinase sequence motifs, and thus are not likely to have His kinase activity. The phytochromes and AHK5 also differ from the typical structure in that they are soluble proteins with no transmembrane domains. Biochemical histidine kinase activity has only been demonstrated for ETR1.

The Cytokinin Receptor Family: AHKs (Arabidopsis Histidine Kinases)

The first indication that two-component-type proteins might be involved in cytokinin signaling came from the identification of the Arabidopsis histidine kinase encoding gene, *CKI1*, as a gene that conferred cytokinin independent (CKI) growth to callus cells when overexpressed (15). A second sensor histidine kinase, *CRE1*, was subsequently isolated in an EMS-mutagenesis screen for callus tissue that displayed reduced shoot formation and greening in response to exogenous cytokinin (Fig. 3A and B) (14). The *cre1* (cytokinin response 1) mutant displays cytokinin insensitivity in root elongation assays and has impaired xylem organization (14). *CRE1* (also called *AHK4* and *WOL*, hereafter *AHK4*), and the highly similar *AHK2* and *AHK3* genes (Fig. 2A), consist of an N-terminal extracellular domain that is flanked by two transmembrane domains, a histidine kinase transmitter domain, and one functional and one degenerate receiver domain (Fig. 2B). These three sensor kinases also share a conserved CHASE (cyclases/histidine kinases-associated sensory extracellular) domain (23) in their predicted extracellular regions, which in the case of AHK4, has been shown to bind cytokinin.

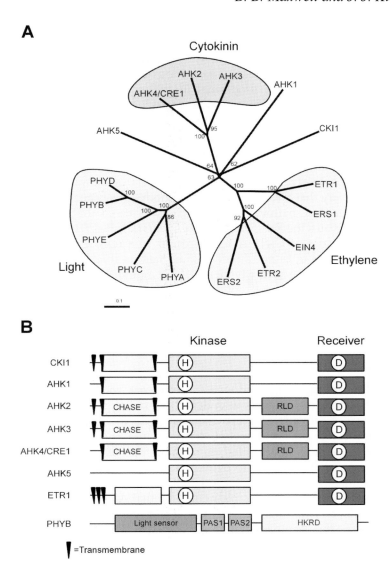

Figure 2. Phylogenetic relationships and domain structures of the 16 histidine kinase related genes from Arabidopsis. (A) Full-length protein sequences were used to generate an unrooted phylogenetic tree with Clustal W[*] that was displayed using TreeView[#]. Numbers at nodes represent bootstrap values expressed as a percentage (1000 trials). The three main receptor classes of histidine kinases, cytokinin, ethylene and light, are labeled. Scale bar denotes number of substitutions per position. (B) Domain structure of the AHKs along with representatives of the ethylene and light receptor classes. Encircled H and D represent the conserved His and Asp residues that are phosphorylated in the kinase and receiver domains respectively. The Cyclase/Histidine kinase-Associated Sensing Extracellular (CHASE) domain is thought to sense cytokinin. RLD = Receiver-like domain, PAS = Per/Arndt/Sim, HKRD = histidine kinase related domain, ED = extracellular domain. The light sensing domain in phytochrome B requires a bound chromophore in order to detect light input. ([*]Clustal W 1.81 by Silicon Graphics at http://www.cmbi.kun.nl, [#]TreeView by R.D. Page at http://taxonomy.zoology.gla.ac.uk/rod/treeview.html)

Cytokinin-Independent: CKI1

CKI1 was uncovered in an activation-tagging screen for hypocotyls that produced callus in the absence of exogenous cytokinin (15). Although CKI1 has been shown to be a functional histidine kinase by complementation experiments in *E. coli*, it fails to bind to cytokinin in yeast assays and does not require cytokinin for activity in the *E. coli* complementation assays (see below). High levels of CKI1 can induce the expression of cytokinin primary response genes (described below) in isolated protoplasts and this induction is dependent on the conserved His and Asp residues in the kinase and receiver domains respectively. However, this activity is cytokinin-independent (10). Thus, it is likely that CKI1 is not a cytokinin receptor, but rather activates the cytokinin signaling pathway when overexpressed in the same promiscuous manner it activates the RcsC pathway in *E. coli* (see Fig.1B for the *E. coli* relay system). The loss of CKI1 function results in a female gametophytic lethal phenotype, which does not directly illuminate its role in cytokinin signaling (26). It remains to be seen whether CKI1 is a modifier of cytokinin signaling under normal physiological conditions or is only able to interact with this pathway when ectopically expressed. It is possible that CKI1 may bind some other signal molecule or a specific type of cytokinin, or may heterodimerize with one of the cytokinin receptors as part of a receptor complex.

CRE1/AHK4/WOL

An *ahk4* mutation, designated *wooden leg* (*wol*), was originally isolated in a screen for short-root mutants with defects in radial patterning of the primary root. Two more alleles were then isolated in a screen for cytokinin-insensitive callus (14) and seven additional alleles were found in a genetic screen for mutants that failed to reduce the expression of phosphate-starvation genes in response to cytokinin. *ahk4* mutants display insensitivity to cytokinin in callus initiation and root elongation assays (Fig. 3). *AHK4* was cloned and shown to complement the lethal $\Delta sln1$ mutation in yeast in a cytokinin-dependent manner (see Fig. 1 for the yeast relay system) (14). Furthermore, this complementation required a functional YPD1 gene (the Hpt gene that acts downstream of SLN1 histidine kinase), indicating that this complementation occurred via the endogenous yeast phosphorelay pathway. These data suggest that AHK4 is a functional histidine kinase that can be activated by cytokinin. Furthermore, mutations in the conserved His or Asp of the kinase and receiver domain, respectively, abolished the ability of AHK4 to complement $\Delta sln1$. Similar results with AHK4 were obtained in complementation experiments using *Schizosaccharomyces pombe* and *E. coli* two-component mutations. The biologically active cytokinins *trans*-zeatin, IP, kinetin, IPA, and BA were shown to activate AHK4 in these heterologous systems, while *cis*-zeatin and other plant hormones had no effect (14). When various AHPs (Arabidopsis Histidine Phosphotransmitters) were co-expressed with AHK4 in the $\Delta RcsC$ histidine-kinase mutant background, they were able to compete for a phosphoryl group with the endogenous Hpt

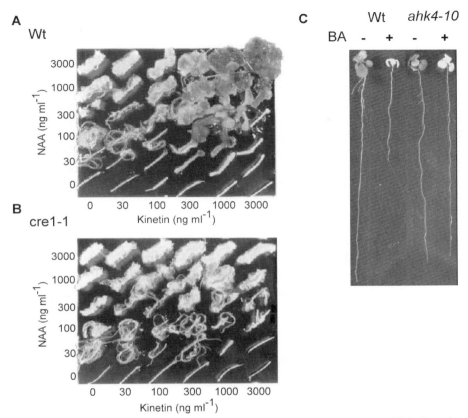

Figure 3. Reduced response of the cytokinin receptor mutant, *cre1-1*, to cytokinin in a shoot initiation assay. (A) Wild-type explants grown on increasing concentrations of auxin and cytokinin generate both root and shoot tissues respectively. (B) The *cre1-1* mutant responds to increasing auxin concentrations with root proliferation but no longer forms shoots in response to cytokinin. *cre1-1* contains a mutation in the histidine kinase domain of AHK4/CRE1. From Inoue et al. (2001). (C) Cytokinin insensitive root phenotype of a T-DNA insertional allele of *cre1*, *ahk4-10*. Seedlings were grown vertically on plates with or without the cytokinin BA for 10 days.

YojN, indicating that AHK4 can phosphorylate Arabidopsis Hpts.

Direct binding with high affinity (Kd = 4.6 nM) by radiolabeled cytokinin (H^3-iP) to AHK4 was shown using AHK4 protein expressed in yeast membrane fractions, confirming the cytokinin binding activity of AHK4 (50). This binding could be competed by other biologically active cytokinins, but not by inactive adenine derivatives, which is similar to the observation that only biologically active cytokinins allowed AHK4 to complement heterologous histidine kinase mutants. The *wol* mutation results in a Thr to Ile substitution in the CHASE extracellular cytokinin-binding domain (18) and disrupts the ability of the protein to bind cytokinin as well as the ability to complement the RcsC mutation in *E. coli* (50). That AHK4 can function as a cytokinin receptor in yeast and bacteria has thus been established, but can AHK4 do the same in plants? Analysis of loss-of-function alleles and expression experiments using Arabidopsis protoplast cells have provided evidence that AHK4 acts as cytokinin receptor in plant cells. Overexpression of AHK4 in protoplast cells can enhance the cytokinin-induced upregulation of the promoter of a cytokinin primary response gene, *ARR6* (10). This activity was dependent on the conserved His and Asp residues in the transmitter and receiver domains of the protein. Mutation of the conserved His also resulted in a dominant negative form of AHK4 that reduced the cytokinin induction of a *pARR6-LUC* promoter-reporter fusion. This data shows that, in plant cells, AHK4 can function as a cytokinin-dependent activator of cytokinin primary response genes and that the conserved His and Asp residues are required for this function. Analysis of *ahk4* mutants has revealed that *AHK4* is required for the cytokinin induction of a subset of type-A genes (Fig. 4A) in roots, although it is not required for the induction of any of the type-As in leaves (16, 27). Recently, it has been demonstrated that AHK4 can act as an osmosensor in yeast in the same manner as SLN1 (51). In this yeast expression study, zeatin-bound AHK4 was rapidly inactivated by changes in osmolarity or turgor pressure. This suggests that AHK4 could function both as a cytokinin receptor and an osmosensor in plants.

AHK2 and *AHK3*

Because AHK2 and AHK3 both contain a CHASE domain that is highly similar to the CHASE domain required for cytokinin binding in AHK4, they are likely to also function as cytokinin receptors. AHK2, but not AHK3, overexpression can enhance the cytokinin-induced expression of *pARR6-LUC* in Arabidopsis protoplasts (10). Evidence for an *in vivo* role for AHK3 in cytokinin signaling is provided by the gain-of-function mutant *ore12*, which is the result of a point mutation in the region encoding the extracellular domain of the AHK3 protein. *ore12* mutants display delayed leaf senescence, consistent with a role for AHK3 in cytokinin signaling.

Studies attempting to show histidine kinase function for AHK2 and AHK3 have produced mixed results. Although one study showed that neither AHK2 nor AHK3 could complement histidine kinase mutations in *E. coli*, in other studies AHK3 displayed His-kinase function in both *E. coli* and yeast complementation assays (8, 15). However, the function of AHK3 was cytokinin-dependent in *E. coli*, but not in yeast, and showed weaker activity relative to AHK4. AHK2 could not complement the yeast mutation that AHK3 could. These results suggest that AHK2 and AHK3 may have diverged to have different specificity for downstream Hpt components from AHK4, from each other and from the heterologous systems so far tested. Direct binding of cytokinins to AHK2 or AHK3 has not yet been tested.

The *WOODEN LEG* Puzzle

wol mutants lack proper cell division patterns in early embryogenesis leading

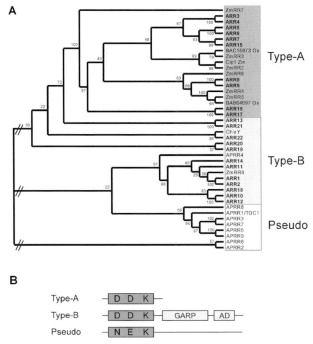

Figure 4. Phylogenetic relationships and domain structures of the Arabidopsis response regulator (ARR) proteins. (A) The receiver domains of the ARRs, along with selected receiver domains from corn (Zm) and rice (Os), and the receiver domain of the bacterial response regulator CheY, were used to generate an unrooted phylogentic tree as described in Figure 2. The Arabidopsis type-A, type-B and pseudo-response regulator classes are labeled. Note that APRR4 has been grouped with the type-Bs because of the tree format and is not considered part of this family. (B) The domain structure of the full-length ARR proteins. Type-As and type-Bs contain a single conserved receiver domain, while pseudo-response regulators lack the conserved D-D-K motif. The type-Bs have a large C-terminal extension that contains a GARP (GOLDEN/ARR/Psr1) DNA-binding domain and a glutamine-rich activation domain (AD). (The C-terminal extension of ARR19 does not appear to contain the GARP domain.) Bold letters represent amino acid residues: D = Asp, K = Lys, N = Asn and E = Glu..

to defects in the vascular system (18). The resulting phenotype is severely reduced root growth likely due to an insufficient number of meristem cells available for recruitment to phloem tissue, resulting in a lack of phloem supply to growing roots. As mentioned above, the *wol* allele is a missense mutation in the CHASE domain of the AHK4 protein that leads to disruption of cytokinin binding (18, 50). Why this should lead to such a dramatic phenotype while the presumed loss of function alleles *ahk4-10 and cre1-2* do not result in a strong root phenotype in the absence of exogenous cytokinin is perplexing. In addition, *ahk4-10* has longer roots than wild-type, the opposite phenotype of the extreme short root in *wol*. One explanation may be that neither the *ahk4-10* nor the *cre1-2* T-DNA insertion allele is a null, but that both result in only a partial loss of function. The *cre1-2* T-DNA insertion occurs in the extreme N-terminus of the encoded protein however, for *ahk4-10* the T-DNA insertion is in the middle of the region encoding the CHASE domain and is likely to be a null allele. These studies indicate complexity in the function of the different cytokinin receptor mutations and further study is needed to clarify the nature of these mutations. A recent study provides one solution this puzzle. García-Ponce de León et al. have found that a second *wol* allele (*wol1-2*), that also has a short root phenotype, can complement the orginal *wol* mutant (a phenomenon called intragenic complemention). *wol1-2* contains a missense mutation in the histidide kinase domain of AHK4 while *wol* contains a missense mutation in the cytokinin binding domain. The short root phenotype is restored to wildtype in the *wol/wol1-2* trans-heterozygotes but these seedlings are still insensitive to exogenous kinetin. This indicates that AHK4 signals involved in root vascular morphogenesis are separable from the signaling involved in cytokinin responsiveness.

AHK1 and *AHK5*

AHK1 is an osmosensing histidine-kinase receptor that can functionally substitute for SLN1 in a cytokinin-independent manner (45). Furthermore, AHK1 allows Δsln1 Δsho1 double mutants to grow in high-osmolarity media and can lead to tyrosine phosphorylation of HOG1 in a NaCl-dependent manner. In plants, *AHK1* is expressed primarily in roots and is up-regulated by changes in external osmolarity. The osmosensing pathway downstream of AHK1 may involve AHP2, ARR4 and ARR5 as AHK1 can interact with AHP2 and high salinity induces the transcription of *ARR4* and *ARR5*. There is no evidence that *AHK1* affects cytokinin signaling, although it is up-regulated in response to cytokinin (27). As previously mentioned, AHK4 may also play a role in osmosensing though there is no evidence for this function in plants as yet.

AHK5 (*CKI2*) was isolated in the same screen as *CKI1* as its overexpression results in callus formation in the absence of cytokinin (15). The potential role of AHK5 in cytokinin signaling has yet to be investigated.

CYTOKININ PRIMARY RESPONSE GENES

The Type-A *ARR*s (Arabidopsis Response Regulators)

A second line of evidence that two-component type pathways are involved in cytokinin signaling comes from the discovery of cytokinin-induced genes that have homology to the response regulator class of proteins. A differential RNA-display screen uncovered two genes, *ARR4* (*IBC7*) and *ARR5* (*IBC6*) that were rapidly (within 10 min) up-regulated by cytokinin (2). These genes belong to a ten member family called the type-A ARRs that contain a single receiver domain with the conserved Asp-Asp-Lys (D-D-K) region, of which the central Asp residue obtains a phophoryl group from the His of a phosphotransmitter (Fig. 4A). There are 32 response regulators encoded by

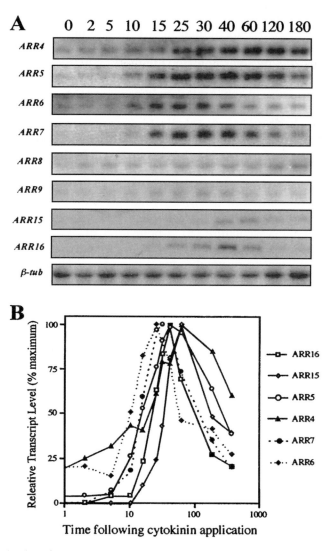

Figure 5. Kinetics of type-A ARR gene induction. (A) Northern-blot analysis of 15 μg of total RNA from 3-d-old etiolated seedlings treated with 5 μM BA for various times (indicated in minutes above each lane) and hybridized with the indicated ARR probe (at left). (B) Quantification of transcript levels from blot depicted in A normalized to the β-tubulin loading control

the Arabidopsis genome, comprised of three main groups: the type-A ARRs, the type-B ARRs and the pseudo-response regulators. The type-B ARRs (discussed below) have a receiver domain fused to a DNA-binding domain and a C-terminal transcriptional activation domain. The pseudo-response regulators have an atypical receiver domain that lacks the conserved D-D-K motif (Fig. 4B) and are likely involved in the control of circadian rhythms.

Only the type-A ARRs are transcriptionally up-regulated by cytokinin.

The ten type-A ARRs are similar in structure to the bacterial flagellar-response regulator gene CheY in that they have only short N and C-terminal extensions beyond the receiver domain (Fig. 4B). The type-A ARR family consists of five highly homologous pairs, and this pairing reflects the most recent duplication event in the Arabidopsis genome (Fig. 4A). Cytokinins can cause the rapid accumulation of the transcripts of at least nine of these type-A genes (Fig. 5) (8, 9, 15). For several type-A ARRs, it has been shown that accumulation of their mRNAs in response to cytokinin does not require *de novo* protein synthesis and is due, at least in part, to increased transcription, indicating that these are primary response genes and that the activation of a transcription factor is involved in this response (2, 27).

The expression pattern of the *ARR5* gene has been examined in depth, and it appears to be expressed primarily in the root and shoot apical meristem and in regions of the vasculature (5). Application of exogenous cytokinin results in expression of *ARR5* throughout the entire seedling, implying that all tissues are competent to respond to cytokinin (5). Preliminary studies of the basal expression of other type-A genes revealed generally overlapping, but not identical patterns of expression (F. Ferreira, J. Deruère, B. Maxwell, J. To, and J. Kieber, unpublished results). In contrast to the *ARR5* gene, the transcriptional induction by cytokinin of some type-As is tissue specific. Several groups have reported the sub-cellular localizations of the type-A proteins using translational fusions to reporters. The majority of the type-A ARRs tested can be found primarily in the nucleus however, several are found in the cytoplasm as well and one, ARR16, appears to be localized exclusively to the cytoplasm (10, 16, 43). The nuclear localization of ARR7 is dependent on a small portion of the C-terminus that contains a putative nuclear localization signal that is not present in the shorter ARR16 protein (16). The localizations of ARR4 and ARR6 are not affected by exogenous cytokinin application (10, 43).

The presence of putative receiver domains in the type-A ARR proteins suggests that, in addition to being induced at the transcriptional level, they may also be phosphorylated in response to cytokinin. Evidence that type-A ARRs can function as phospho-accepting response regulators has been shown in experiments using type-A ARR receiver domains expressed in and purified from *E. coli* in phosphotransfer assays using the *E. coli* Hpt protein ArcB. All three of the ARR proteins tested (ARR3, ARR4 and ARR6) could rapidly (3 min) but transiently acquire a phosphoryl group from ArcB (12). The conserved Asp in the receiver domain of ARR3 was shown to be required for this phosphorylation. Phosphotransfer between AHP1 and AHP2 and several type-A ARRs has also been demonstrated *in vitro* (11). Although it appears that the type-A ARR proteins are functional response regulators that can interact with Arabidopsis Hpts, their phosphorylation in response to cytokinin or other stimuli has not yet been shown. The rapid and transient nature of the Asp phosphorylation event and the intrinsic instability of phospho-aspartate make this a challenging task.

The type-A ARRs have been demonstrated to negatively regulate their own cytokinin-induced expression. In studies using Arabidopsis protoplast cells, the upregulation of an *ARR6* promoter-*LUC* fusion could be induced by exogenous cytokinin. Overexpression of any one of several type-A ARRs repressed the ability of cytokinin to induce *pARR6-LUC* (10). Unexpectedly, mutation of the conserved Asp residue in the receiver domains of ARR4 and ARR6 did not alter their ability to block the expression of *pARR6-LUC* in response to cytokinin. Perhaps the unphosphorylated forms repress signaling while the phosphorylated forms have no effect or act in a positive manner. Quadruple *arr3456* and hextuple *arr345679* T-DNA knockouts of the type-As have demonstrated that these genes act in a partially redundant manner to negatively regulate cytokinin signaling. This can most dramatically be seen as an increase in cytokinin shoot induction response in hypocotyl explants grown on cytokinin containing media (Fig. 6) (54).

Some of the type-As can be induced by other environmental signals such as light, nitrate, high salt and low temperature suggesting that these inputs are somehow integrated with cytokinin, or other two-component type signaling systems. Nitrate has been shown to elevate cytokinin levels in maize, resulting in the expression of maize homologues of response regulator genes (33). Similar to their Arabidopsis counterparts, these maize homologues, *ZmRR1* and *ZmRR2*, are induced by cytokinin in the absence of protein synthesis and their encoded proteins can remove the phosphate from the His residue of a maize Hpt protein, ZmHP2. Interestingly, these genes can be induced in whole plants by both cytokinin and nitrate, but in excised

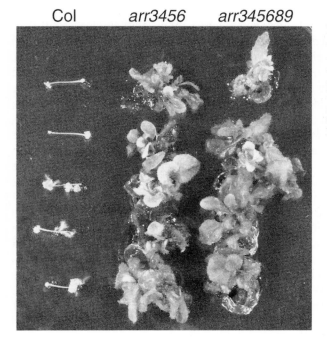

Figure 6. Shoot initiation is enhanced in the type-A ARR mutants *arr3,4,5,6* and *arr3,4,5,6,8,9*. Five excised hypocotyls of the indicated genotypes were grown on the same plate on media containing 100ng/ml NAA and 300ng/ml kinetin.

leaves are only induced by cytokinin (44). These data imply cytokinin may mediate nitrogen signaling from roots to shoots using a two-component style pathway.

The integration of other pathways in addition to nitrate signaling may occur at the level of the type-A genes as well. For instance, light can induce the accumulation of the ARR4 protein, which in turn can stabilize the active form of PhyB resulting in increased sensitivity to red light (43). PhyB has been suggested to associate with gene promoters via an interaction with the DNA binding protein PIF3. Intriguingly, ARR4 has also been shown to interact *in vitro* with two putative DNA-binding proteins, AtDBP1 and AtDBP2 (49). AtDBP1 is expressed at the shoot apex and is induced by auxin also providing a potential link between the cytokinin and auxin response pathways at the transcriptional level.

Other Primary Response Genes

A number of studies have uncovered genes that are cytokinin regulated, but few are cytokinin specific or rapidly responsive (22, 36). In general, cytokinin appears to transcriptionally regulate the expression of a relatively small number of genes. However, tissue type and status as well as interference by endogenous cytokinins are likely to affect the number and type of genes recovered by such studies. Also confounding are the opposing effects cytokinin exerts on the root and shoot meristems, which may cancel each other out in gene-panning experiments using whole seedlings. Finally, negative feedback loops in cytokinin signaling and metabolism may decrease the number of genes that appear to be responsive to exogenous cytokinin. One study using the 8k Affymetrix genechip shows that only a handful of genes (~17) are consistently regulated at the transcriptional level in Arabidopsis seedlings, either positively or negatively, by cytokinin (27). Nearly all of these maintain rapid regulation in the presence of cycloheximide indicating that they are primary response genes. Among the genes newly identified in this class are two AP2-like transcription factors, cytokinin oxidase, auxin related genes (IAA3/SHY2, SAUR-AC1, IAA17/AXR3) and several genes that appear to be related to disease resistance. The promoters of these genes contain at least one, and often multiple copies of the ARR1 and ARR2 DNA-binding site core motif (see below), indicating that the type-Bs are likely to be key players in the induction of most cytokinin primary-response genes.

The small number of genes whose RNA levels change in response to cytokinin suggests that the regulation of protein degradation may play an important role in cytokinin action. This view is supported by analysis of plants containing a mutation in a proteasome subunit, *RPN12*, which results in complex changes in cytokinin-related phenotypes (37). About half of the cytokinin responses observed indicated that *rpn12a-1* mutants are hypersensitive to cytokinin while the other half indicated that *rnp12a-1* plants are less sensitive to cytokinin. *rpn12a-1* roots showed decreased auxin sensitivity but increased cytokinin response. The mRNA of *RPN12* and two other proteasome subunits can be induced by cytokinin (37). Taken together, these data suggest that the proteasome, and thus the degradation of proteins, plays a role in cytokinin responsiveness. The complex effect that mutation of *RPN12* has on cytokinin responses may indicate that cytokinin signaling operates by striking a balance between the stability of positively and negatively acting factors. More direct evidence that cytokinin acts in part via the regulation of protein turnover comes from the finding that exogenous cytokinin has been shown to elevate ethylene biosynthesis in Arabidopsis seedlings through a stabilization of the ACS5 protein, which catalyzes the rate-limiting step in ethylene biosynthesis (4).

The Type-B *ARRs* (Arabidopsis Response Regulators)

The twelve type-B ARR proteins in Arabidopsis have an N-terminal receiver domain followed by a large (~300 AA) C-terminal extension (Fig. 4B). This C-terminus contains a glutamine-rich domain and a motif that is distantly related to the MYB DNA-binding superfamily that has been designated a GARP domain (GOLDEN2/ARR/Psr1). The GARP DNA-binding domain appears to belong to the same general class as those found in the circadian-related MYB transcription factors LHY and CCA1 (7, 31). The type-Bs are linked to cytokinin signaling through overexpression and knockout studies. Overexpression of ARR1, ARR2, ARR10 or ARR11 results in an increase in cytokinin sensitivity and an *arr1* knockout shows reduced cytokinin sensitivity (10, 13, 32). Overexpression of the C-terminal DNA binding domains of ARR14, ARR20 and ARR21 result in different morphological phenotypes generally consistent with their expression profiles (55). In the case of the overexpression of the C-terminus of ARR20, callus tissue forms in young seedlings and a specific subset of type-A mRNAs are upregulated.

The type-B ARRs have been shown to be functional transcriptional activators in both yeast and plant cells. Sequence-specific DNA-binding activity and/or transcriptional activation function has been shown for ARR1, ARR2, ARR10 and ARR11 (13, 17, 31, 32). The glutamine rich C-terminal region of the type-B proteins provides activation function while the GARP domain binds DNA. The receiver domain appears to be a negative regulator of the activation domain although it is not known whether it may also repress the DNA-binding function of the GARP domain.

Oligonucleotide selection experiments along with gel shift studies have determined that the consensus binding site for ARR1, ARR2, ARR10 and ARR11 is (G/A)GAT(T/C) (7, 13, 31). In plant cells, ARR1 and ARR2 can activate transcription from a promoter containing a 6X concatomer of this binding consensus site (31). Overlap in type-B DNA-binding activity is also suggested by the finding that overexpression of the ARR2 DNA-binding domain alone could exert a dominant negative effect, indicating that it competes with other type-Bs for promoter binding sites (10). This binding site consensus can be found in the promoters of nearly all of the cytokinin primary response genes found to date (27). As expected for transcription factors, the type-Bs proteins tested thus far have also been shown to be localized to the nucleus. In the cases of ARR1 and ARR2, it has been shown that addition of cytokinin or removal of the receiver domain has no effect on nuclear localization, which is likely due to the C-terminal nuclear localization signal VRK(R/K)R (10, 31).

In Arabidopsis protoplasts, ARR1, ARR2 or ARR10 overexpression can enhance the cytokinin-induced activation of a type-A promoter (10). Surprisingly, the conserved Asp in the receiver domain of ARR2 does not appear to be required for this activation function. As with parallel results found for the receiver domains of the type-As, this suggests that phosphorylation is not involved in cytokinin activation. Overexpression of ARR2 in whole plants results in delayed senescence and increased shoot and leaf formation in explants in the absence of cytokinin (10). Overexpression of ARR1 in plants results in a hypersensitivity to cytokinin in root elongation, callus growth and type-A induction assays (32). Conversely, *arr1* knockout lines displayed reduced sensitivity in these same assays. Overexpression of ARR1 lacking the receiver domain resulted in a seedling phenotype suggestive of constitutive cytokinin responsiveness, providing further evidence for the negative regulatory function of the receiver domain. These data imply that ARR1 and ARR2 are directly involved in cytokinin signaling and that their levels may be a limiting factor in cytokinin response. *ARR1, ARR10* and *ARR11* mRNA is found in all tissues of the plant. However, the tissues in which *ARR2* is expressed remain unclear, as conflicting data has been gathered for *ARR2* expression patterns (8). One report shows pollen-specific expression and no expression in roots while another reports high levels of *ARR2* mRNA in roots but not in flower. Perhaps the expression of this gene is affected by environmental factors not controlled for in these studies.

If the type-B proteins act as classic two-component response regulators of gene expression, then they must also contain functional receiver domains. Many type-Bs are indeed able to interact with the Arabidopsis Hpt proteins in yeast-two-hybrid assays (8, 15). The receiver domains of ARR1, ARR10 and ARR11 have been shown to be functional in assays using Arabidopsis AHP proteins (13, 41). As discussed above, it is not clear what role phosphorylation plays in the regulation of these proteins.

The Histidine-phosphotransfer Proteins: AHPs

Because the cytokinin receptors have a fused receiver domain with a phospho-accepting Asp and the type-A and B response regulators also have the same Asp phosphorylation residue, an intermediary protein with His phosphotransfer function is required to complete the relay. In Arabidopsis there are five genes, *AHP1-AHP5*, that encode small (~150 AA) proteins with similarity to yeast and prokaryotic His-phosphotransfer domains (Fig. 7). A sixth protein, AHP6/APHP1, also may have His phosphotransfer activity

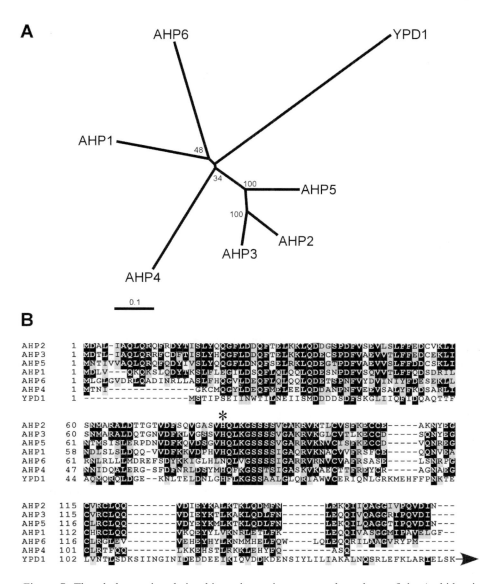

Figure 7. The phylogenetic relationship and protein sequence homology of the Arabidopsis His-phosphotransfer proteins (AHPs). (A) Full length protein sequence of the AHPs and the Hpt protein, YPD1, from yeast was used to generate an unrooted phylogentic tree as described in Figure 2. (B) A protein sequence alignment of the AHPs and YPD1 was created in Clustal W and displayed using BoxShade[‡]. Inverted white residues on a black background are conserved in 40 percent or more of the sequences while gray shading denotes conservative substitutions. The asterisk marks the conserved histidine that is the site of phosphorylation. Full-length protein sequence is shown with the exception of YPD1 which is six amino acids longer as represented by the arrow. ([‡]Boxshade by K. Hofman and M. Baron at http://www.ch.embnet.org).

despite a shift in the His residue from the conserved position (see * in Fig. 7). The steady state level of mRNA for the AHPs are not induced by cytokinins or stress stimuli. AHP1-AHP3 can complement yeast Δypd1 mutants, acquire a phosphoryl group from *E. coli* membrane preparations and compete for phosphoryl groups with the *E. coli* Hpt YojN, showing that these AHPs are functional Hpts (21, 40, 41). In yeast-two-hybrid experiments, AHP1-3 can interact with a subset of sensor histidine kinases. For instance, AHP2 interacts with AHK1, ETR1 and CKI1 while AHP1 only interacts with ETR1 and not the other two. This data suggests that the AHPs may act downstream of several different hybrid-sensor kinases but with some specificity for certain pathways. However, it is not clear whether this assay reflects specificity in interactions between AHPs and AHKs, or is due to differences in protein expression levels in yeast or to other two-hybrid artifacts.

Downstream of the six AHPs are 22 potential response regulator targets belonging to two classes, the type-A and type-B ARRs. Hpts do not have enzymatic activity themselves and transfer of a phosphate must occur by phosphatase activity of the phospho-accepting receiver domain. As discussed above, phosphotransfer experiments have demonstrated that AHPs can donate a phosphoryl group to the receiver domains of several of the type-As and at least one of the type-Bs. These experiments all show a rapid loss of phosphate from the Hpt and a rapid and transient (3 min.) acquisition of phosphate by the response regulator. In yeast-two-hybrid experiments, the AHPs can be shown to interact with several of the type-A and type-B proteins.

A connection to the cytokinin-signaling pathway has been shown for AHP1 and AHP2. In Arabidopsis protoplast cells, AHP1 and AHP2 are transiently translocated from the cytoplasm to the nucleus upon cytokinin treatment, while AHP5 is not (10). This also provides a mechanistic link for phosphotransfer between the plasma membrane-localized AHK receptors and the nuclear-localized members of the ARRs. Furthermore, AHP2 overexpression in whole plants resulted in hypersensitivity to cytokinin in root and hypocotyl elongation assays but no change was detected for responsiveness to ethylene, ABA or auxin (41). When AHP1, AHP2 or AHP5 were overexpressed in Arabidopsis protoplasts, however, no change was observed in responses to cytokinin as assayed by the induction of the promoter of *ARR6* suggesting that, in this system, the AHPs are not a rate limiting step in cytokinin signaling (10). Confirmation of a role for AHPs in cytokinin signaling awaits characterization of their loss-of-function mutant lines. The AHPs are generally expressed in most tissues with higher levels in the roots, although *AHP1* expression may be root-specific. A yeast two-hybrid screen using an AHP2 bait identified the *TCP10* gene, which encodes a potential transcription factor (42). This suggests that AHPs may play roles in signaling other than phosphotransfer to the ARRs and may affect transcription through direct binding to transcription factors, although the interaction with TCP10 should be viewed with caution as it has not yet been confirmed by other means.

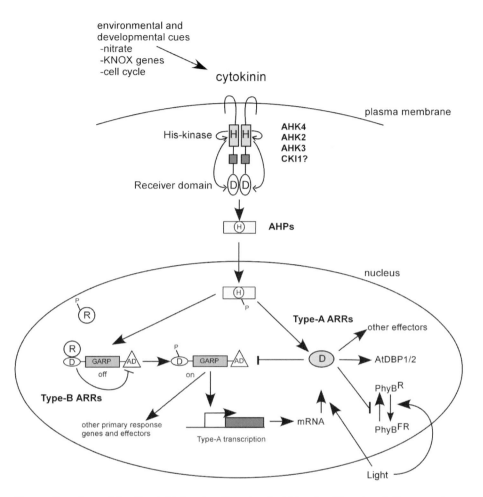

Figure 8. A model of cytokinin signaling in Arabidopsis. Phosphorylation events are presumed based on known prokaryotic two-component systems. Cytokinin is perceived at the plasma membrane by three hybrid sensor histidine kinase receptors. Cytokinin-bound receptors *trans*-phosphorylate at the conserved His residue (H) in the His-kinase domain and the phosphoryl group is then transferred to the conserved Asp residue (D) in the attached receiver domain. Phosphotransfer then occurs between activated receptors and the AHPs. AHPs move to the nucleus where they can transfer a phosphate to the type-A and type-B ARRs. In the absence of phosphorylation the transcriptional activation domains (AD) in the type-B response regulators are functionally repressed by their receiver domains. Upon phosphorylation by the AHPs of either their receiver domain or of a hypothetical repressor protein (R), the type-B ARRs are relieved from repression and activate the transcription of the type-A, and other, primary response genes. Type-A ARR proteins repress their own transcription by negatively regulating the transcriptional activity of the type-B ARRs. The regulation of outputs by the type-A ARR proteins is likely by protein-protein interactions. For instance, the type-A protein, ARR4, has been shown to interact with and stabilize the active form of the red-light receptor PhyB (PhyB^{FR}). Light, in turn, causes an increase the amount of ARR4 protein. GARP = GOLDEN2/ARR/Psr1 DNA-binding domain, AtDBP1/2 = Arabidopsis DNA binding proteins 1 and 2, P = phosphoryl group.

A MODEL OF TWO-COMPONENT CYTOKININ SIGNALING

Cytokinin signaling appears to involve a multistep phosphorelay system that regulates both the expression of genes and the activity of proteins (Fig. 8). The ways in which cytokinin levels are regulated remain largely unknown (Chapter B3). Nitrate, KNOX class homeobox genes and cell cycle phase have all been implicated in playing a role in increasing the synthesis of cytokinins. Extracellular cytokinins are detected at the plasma membrane by the hybrid sensor kinase AHK4 and probably also by AHK2 and AHK3. By analogy to hybrid sensor kinases in other systems, these receptors most likely act as homodimers or heterodimers with each other, and perhaps also with CKI1 and AHK1.

Downstream of the AHKs in the phosphorelay model are the small His-phosphotransfer proteins, the AHPs. The signal relay between the receptors at the cell surface and the nucleus is mediated by the AHPs, some of which have been shown to transiently translocate to the nucleus upon cytokinin treatment. There are two classes of potential response regulator targets of the AHPs: the type-A ARRs, which likely act through protein-protein interactions, and the type-B ARRs, which are transcriptional activators. The type-Bs can activate the transcription of the type-As. The type-As appear to act by negatively regulating the function of the type-Bs, thereby establishing a negative feedback on their own expression. This self-limiting response is apparently reflected in the expression kinetics of the type-A genes (Fig. 5). A rapid (10 min) increase in mRNA level with a peak at about 40 min is seen upon cytokinin treatment, which is followed by a gradual decline over the next several hours before reaching a persistently elevated level. The type-B binding site consensus sequence is found upstream of many of the known primary response genes and thus the negative regulatory loop established by the type-As may also affect the expression of other primary response genes.

The role that phosphorylation plays in the regulation of either the type-As or the type-Bs is not clear. The classical model based on prokaryotic systems suggests that cytokinin binding by the AHK receptors should ultimately result in phosphorylation of the type-B and type-A ARRs leading to the regulation of their function. Transcriptional activators like the type-Bs are most often activated upon phosphorylation of their receiver domains. However, in Arabidopsis protoplasts, ARRs of each type have been shown to have activity despite mutation of the conserved Asp residue in the receiver domain, which is the most likely target of AHP phosphotransfer. In the case of the type-Bs, transcriptional activation of a type-A gene in response to cytokinin was not disrupted by mutation of the phospho-accepting Asp. In the case of the type-As, this same mutation did not alter their ability to inhibit the transcriptional upregulation of a type-A by cytokinin. A simple model that incorporates this data is one in which the type-As act as negative regulators of type-B function in the unphosphorylated state. Upon cytokinin binding to the receptor/s, the type-As become phosphorylated by AHPs and the type-Bs are released from repression. Phosphorylation of the type-B receiver domains may still regulate function, but this may occur in response to signals other than cytokinin. However, no direct interactions between type-A and type-B proteins have been shown that would support this model. Another possibility is that the AHPs control a negative regulator of the type-Bs other than a type-A ("R" in model, Fig. 8). In any case, type-B receiver domains have been shown to negatively regulate the activation domains and thus are likely to be involved in the control of type-B function.

Cytokinin exerts opposite effects on cell division in the shoot and root meristems. Treatment with cytokinin can induce the expression of the *ARR5* gene in all plant tissues indicating that all tissues are capable of responding to cytokinins. How are specific responses to cytokinin generated in the different tissues of the plant? The function of the downstream effectors, type-As, type-Bs or the AHPs may be regulated by cell-type specific co-factors which in turn are regulated by other external cues such as auxin and light to produce a particular cellular response. Alternatively, the expression patterns of the downstream components, type-A, type-B and AHPs, may occur as overlapping sets, and specific combinations of components would produce specific developmental effects. In this scenario, integration with other cues would occur by control over type-A, type-B or AHP levels by cues other than cytokinin such as auxin, stress, and light. Consistent with this, stress and light result in the upregulation of only a subset of the type-A genes. These two models are not mutually exclusive, and a combination may regulate different cellular responses; i.e., control will be exerted at both the level of cell-type specific expression patterns and at the level of functional control by cell-type specific co-factors.

It is not clear whether individual type-B proteins activate specific, or entirely overlapping, sets of promoters. There are slight differences in the *in vitro* binding preferences among the type-B ARRs, suggesting that they may indeed regulate distinct targets. Consistent with this, overexpression of different type-B ARRs results in distinct phenotypes and overexpression of ARR20-C-terminus results in the upregulation of a specific subset of type-A ARRs (55). Overexpression of the type-As results in variable phenotypes such as increased cytokinin sensitivity (ARR4), decreased cytokinin sensitivity (ARR6 and ARR8) or increased light sensitivity (ARR4). The hextuple type-A knockout *arr345689* results in increased cytokinin sensitivity. Thus, the type-A genes are likely to have some specificity of function despite their apparent genetic redundancy. Some of this specificity may be conferred by the short, variable C-terminal extensions of the various type-A proteins.

THE MERISTEM CONNECTION

Cytokinins are primarily defined by their ability to induce cell divisions in cultured tissues that lead to callus formation. One way that cytokinins regulate the cell cycle is through the induction of the *CycD3* gene (Chapter

C3) (28). Constitutive expression of *CycD3* bypasses the requirement for exogenous cytokinin in callus initiation and thus may represent the primary means by which cytokinin affects the cell cycle. D-type cyclins are the main class involved in the G1 to S transition and so play a key role in regulating cell proliferation. The expression of the cytokinin two-component elements *AHK4, ARR4,* and *ARR7* are coordinately regulated during the cell cycle, showing a strong periodic expression peaking at G1 (20). The up-regulation of these genes may reflect an increase in cytokinin synthesis at G1, which in turn could lead to increased *CycD3* expression. The periodic expression of these genes suggests that two-component signaling likely mediates, at least in part, the signaling input of cytokinin into the cell cycle.

Plant shoot apical meristems (SAMs) consist of a small group of pluripotent cells whose proliferation rates are tightly controlled by a group of meristem-specific genes. Mutations in this class of genes generally lead to expansion or depletion of the meristem resulting in the proliferation or absence of lateral organs.

A drop in the overall levels of cytokinin produced by the ectopic overexpression of cytokinin oxidase results in reduced shoot development and enhanced root proliferation while an increase in cytokinin levels leads to a proliferation of leaf primordia (48, 52). Thus cytokinin plays opposite roles in root and shoot meristem development and likely contributes to maintenance of constant cell number in the SAM. The effects of cytokinin on meristems have been shown to be due to changes in the rate of cell proliferation (48), consistent with a role for cytokinin in regulating cell division rates. Interestingly, the level of cytokinins increases in the SAM after plants are induced to flower (53).

An additional group of meristem regulatory genes are the *KNAT* (*KNOTTED*-like in Arabidopsis) family of homeodomain genes, which includes *SHOOT MERISTEM-LESS* (*STM*), *KNOTTED1* (*KNAT1*) and *KNOTTED2* (*KNAT2*). These genes are generally required to maintain meristem cells in an indeterminate state and, more specifically, *STM* is required for maintenance of *WUS* expression in the central zone. *stm* mutants terminate development as seedlings with entirely differentiated cells where meristem cells should be. Cytokinin can elevate the mRNA levels of these three homeobox genes (30), and this transcriptional control of these genes represents an additional input of cytokinin into SAM maintenance/development. Interestingly, the *in vivo* expression of *kn1* (a maize *KNAT1* homologue) under the control of a senescence-induced promoter could delay senescence and was associated with a 15-fold elevation in cytokinin levels over wild-type (24). Furthermore, the overexpression KNOX genes from rice and tobacco can also lead to an increase in cytokinin levels. Taken together, these data indicate that expression of the KNOX gene family members can lead to an increase in cytokinin synthesis and that in turn these genes are up-regulated by cytokinin, suggesting a positive interdependence between cytokinin and the KNOX family of meristem maintenance genes. Because this connection is based primarily on overexpression studies, it remains to be seen whether this is truly the endogenous relationship between these components.

PHOSPHOLIPASE D AND CALCIUM

Inhibitors of phospholipase D (PLD) function can partially inhibit the cytokinin induction of a type-A gene promoter-GUS fusion as well as the endogenous mRNA (29). This implies that PLD, and by association Ca^{2+}, is involved in early cytokinin signaling. Other studies have also suggested a role for calcium in cytokinin signaling (34). In moss cells, cytokinin treatment leads to an increase in intracellular calcium levels that is dependent on the presence of Ca^{2+} in the external medium. Ca^{2+} is taken up by moss protoplast cells in response to cytokinin through voltage-dependent Ca^{2+} channels in the plasma membrane and leads to bud formation. Interestingly, a calcium-dependent protein kinase (CDPK) gene is down-regulated by cytokinin in cucumber and up-regulated in tobacco.

Nitrate signaling in maize offers another indirect connection between cytokinin and Ca^{2+}. Inhibitor studies show that nitrate-dependent (assimilatory related) primary-response gene expression involves calcium ions and protein phosphorylation. The response to nitrate re-supply has been proposed to involve two-component type signaling downstream of cytokinin (33, 44). This correlates cytokinin signaling with a Ca^{2+}-dependent process, further suggesting that calcium may be involved in cytokinin signaling. It remains to be seen whether PLD or calcium play a biological role in the cytokinin signaling pathway in higher plants.

CYTOKININ RESPONSE MUTANTS

Aside from mutations in two-component elements, there have been a number of additional mutants isolated in various screens that show changes in their response to cytokinin. Most are not well characterized and we present here only brief descriptions of these mutants.

Negative Regulators

Disruption of genes that negatively regulate cytokinin responses is generally correlated with ectopic cell proliferation in whole plants or explants. Several genes involved in negative regulation of cytokinin response have been cloned and they encode a range of unrelated proteins. The *POLARIS* gene (*PLS*) encodes a 36 AA peptide that is expressed primarily in the embryonic and seedling root (3). Loss of PLS function results in reduced root growth and vascularization of leaves as well as roots that are hypersensitive to exogenous cytokinins and yet less sensitive to auxin. *pls* mutants are also defective in a subset of both cytokinin and auxin induced gene expression, including *ARR5* but not *ARR6* and *IAA1* but not *IAA2*. Auxin rapidly induces *PLS* while cytokinin represses the expression of this gene and overexpression of *PLS* results in reduced root inhibition response to cytokinin. In addition *pls* can suppress the over-proliferation of roots in the auxin-overproducing mutant rooty/superroot. Thus, auxin may partly block cytokinin response in roots by induction of *PLS*. Overall, the PLS peptide appears to be involved in auxin-cytokinin interactions in root growth control. Because only certain auxin and cytokinin primary response genes are affected by the *pls* mutation, this peptide may be involved in a mechanism that generates specificity in the primary signal transduction of both these hormones.

The *PASTICCINO* (*pas*) mutants were uncovered in a screen for excessive cell proliferation in the presence of cytokinin. Three *pas* mutants, *pas1*, *pas2* and *pas3*, were isolated and two of the corresponding genes have been cloned. The *pas* mutants show ectopic cell proliferation, changes in the SAM and extra cell layers in the hypocotyl. *PAS1* encodes an immunophilin-like protein with unknown biochemical function and *PAS1* mRNA levels are up-regulated by cytokinin (46). *pas1* mutants also have reduced overall root growth and small rosettes. The pleiotropic effects of the *pas1* mutation are correlated with defects in both cell division and elongation. *PEPINO /PAS2* encodes a putative anti-phosphatase that may function by binding, but not catalyzing, phosphorylated residues. *pep* mutants display abnormal shoot growth phenotypes that appear to result from slow tumor-like cell proliferations (6). This phenotype is enhanced by the *amp1* mutation, which results in increased endogenous cytokinin levels, linking PEP to negative regulation of cytokinin signaling. It will be interesting to see how the different PAS proteins interact on a molecular level with cytokinin signaling.

cytokinin hypersensitive 1 and *2* (*chk1* and *chk2*) loss-of-function mutants are short with increased cytokinin sensitivity in tissue culture but have normal levels of cytokinin (15). These genes are proposed to negatively regulate the cell division and greening pathways promoted by cytokinin. *CHK1* and *CHK2* are not yet cloned.

The *zeatin resistant 3* (*zea3*) mutant was isolated from a screen for zeatin resistance and also from a screen for tobacco seedlings unable to grow on low-nitrogen medium. *zea3* mutants grow preferentially on low C/N ratio media and aberrantly repress light-regulated genes involved in nitrate assimilation and photosynthesis in response to sucrose. *zea3* mutants show increased cytokinin sensitivity in root growth assays, by the analysis of expression of a cytokinin inducible gene, msr1, and in cell division and elongation responses (19). The levels of cytokinin and auxin are unaltered in the *zea3* mutant. Thus ZEA3 may be a negative signaling component involved in the cytokinin regulation of sink-source relationships.

Positive Regulators

Several screens have uncovered mutants that display reduced cytokinin responses. The corresponding genes thus may be positive regulators of cytokinin signaling; though none have been cloned.

A subset of cytokinin responses are dependent on ethylene, such as the triple response seen in dark-grown seedlings. Cytokinins can increase the biosynthesis of ethylene in dark-grown seedlings by controlling the activity of the ACS5 protein. The effect of cytokinin on ACS5 activity is via a stabilization of the ACS5 protein (4). Four *cytokinin insensitive* mutants (*cin1* to *cin4*) fail to produce ethylene in response to cytokinin in etiolated seedlings (47). *cin1* is also insensitive to cytokinin in shoot initiation and anthocyanin production assays whereas the others display normal responses in these assays. The *cin4* mutation is a weak allele of the constitutive photomorphogenic mutant *fus9/cop10*, which is an ubiquitin-conjugating enzyme variant that acts together with COP1 and the COP9 signalosome to regulate protein turnover. These mutants may help to uncover the mechanism by which cytokinin interacts with ethylene signaling.

cytokinin resistant 1 (*cyr*)*1* mutants display reduced shoot development, pale leaves, reduced leaf and cotyledon expansion along with reduced cytokinin response in root inhibition, shoot initiation and anthocyanin accumulation assays. However, *cyr1* shows normal induction of the *ARR5* type-A gene and of ethylene production (J. Kieber unpublished). Possibly, *CYR1* affects specific type-A induction as PLS does and will define a branching point in cytokinin regulation.

STUNTED PLANT 1 (*STP1*) is required for rapid root cell expansion and mediates root growth responses to applied cytokinin. *stp1* mutants are dwarves that appear to be generally impaired in cell expansion but are otherwise normal in morphology. *STP1* affects cell division and expansion downstream of cytokinin but not auxin or other hormones (1). However, *stp1* mutants have normal responses to cytokinin in tissue culture callus assay and only root growth shows reduced sensitivity to cytokinin implying that *STP1* is required by cytokinin for negative effects on root cell expansion but not on the expansion of dividing cells in general (1).

CYTOKININ BINDING PROTEINS

There are a number of reports of cytokinin binding proteins (CBPs) isolated by affinity purification to immobilized cytokinin (22). Most of these proteins are relatively small in size (<35kDa) and one is a 59 AA peptide. CBPs have been found in Arabidopsis membrane fractions, mung bean seedlings, wheat, tobacco and other plants. CSBP from mung bean is a member of a pollen antigen family, CBP1 is 90% similar to endochitinase and CBP2 is similar to osmotin-like protein. So far, these proteins have not been connected to cytokinin signaling *in vivo*. It will be of interest to see whether these proteins are involved in signaling or contribute to the storage and/or transport of cytokinins.

CONCLUDING REMARKS

Cytokinins were discovered almost one half a century ago by the pioneering work of Skoog and Miller, but only very recently has an understanding of cytokinin signaling begun to emerge. The cytokinin signal transduction pathway identified is similar to the bacterial two-component phosphorelay paradigm and signaling events have been defined from perception of cytokinin at the cell surface to changes in genes expression in the nucleus. There are still many questions regarding this model for cytokinin signaling, and it is possible that additional pathways may exist for the perception of cytokinin and the transduction of its signal. It remains to be seen if any of the cytokinin response mutants identified affect genes other than two-component elements and/or if these mutations affect the primary cytokinin-signaling pathway. How the activation of the two-component pathway by cytokinin brings about the myriad of growth and developmental changes associated with this hormone is a mystery, though some clues are beginning to emerge. One output appears to be the interaction of type-A ARRs with phytochrome, and another is the regulation of gene expression by the type-B ARRs, but other outputs surely remain to be uncovered. Two important additional potential outputs are the regulation of protein stability and calcium signaling. The analysis of loss-of-function mutations in the various two-component genes should provide definitive evidence for their role in cytokinin signaling and perhaps in other processes as well. The presence of large gene families for the sensor kinase, AHP and the response regulator elements raises the question of how specificity is achieved in these systems. The role of the predicted phosphorylation events in the regulation of this pathway is also unclear. The near future should see the answers to many of these questions as researchers now have many of the tools required for these analyses.

Acknowledgements

Research on cytokinin action in the authors' laboratory is supported by an NIH and an NSF grant to JJK and an NIH fellowship to BBM.

References

1. Beemster, GTS, Baskin, TI (2000) *STUNTED PLANT 1* mediates effects of cytokinin, but not of auxin, on cell division and expansion in the root of Arabidopsis. Plant Physiol. 124: 1718-1727
2. Brandstatter, I, Kieber, JJ (1998) Two genes with similarity to bacterial response regulators are rapidly and specifically induced by cytokinin in Arabidopsis. Plant Cell 10: 1009-1020
3. Casson, SA, Chilley, PM, Topping, JF, Evans, IM, Souter, MA, Lindsey, K (2002) The POLARIS Gene of Arabidopsis encodes a predicted peptide required for correct root growth and leaf vascular patterning. Plant Cell 14: 1705-1721
4. Chae, HS, Faure, F, Kieber, JJ (2003) The *eto1*, *eto2* and *eto3* mutations and cytokinin treatment elevate ethylene biosynthesis in Arabidopsis by increasing the stability of the ACS5 protein. Plant Cell 15: 545-559
5. D'Agostino, I, Deruère, J, Kieber, JJ (2000) Characterization of the response of the Arabidopsis *ARR* gene family to cytokinin. Plant Physiol. 124: 1706-1717

6. Haberer, G, Erschadi, S, Torres-Ruiz, RA (2002) The Arabidopsis gene *PEPINO/PASTICCINO2* is required for proliferation control of meristematic and non-meristematic cells and encodes a putative anti-phosphatase. Dev. Genes Evol. 212: 542-250
7. Hosoda, K, Imamura, A, Katoh, E, Hatta, T, Tachiki, M, Yamada, H, Mizuno, T, Yamazaki, T (2002) Molecular structure of the GARP family of plant Myb-related DNA binding motifs of the Arabidopsis response regulators. Plant Cell 14: 2015-2029
8. Hutchison, CE, Kieber, JJ (2002) Cytokinin signaling in Arabidopsis. Plant Cell 14: S47-59
9. Hwang, I, Chen, H-C, Sheen, J (2002) Two-component signal transduction pathways in Arabidopsis. Plant Physiol. 129: 500-515
10. Hwang, I, Sheen, J (2001) Two-component circuitry in *Arabidopsis* signal transduction. Nature 413: 383-389
11. Imamura, A, Hanaki, N, Nakamura, A, Suzuki, T, Taniguchi, M, Kiba, T, Ueguchi, C, Sugiyama, T, Mizuno, T (1999) Compilation and characterization of *Arabidopsis thaliana* response regulators implicated in His-Asp phosphorelay signal transduction. Plant Cell Physiol. 40: 733-742
12. Imamura, A, Hanaki, N, Umeda, H, Nakamura, A, Suzuki, T, Ueguchi, C, Mizuno, T (1998) Response regulators implicated in His-to-Asp phosphotransfer signaling in *Arabidopsis.* Proc. Natl. Acad. Sci. USA 95: 2691-2696
13. Imamura, A, Kiba, T, Tajima, Y, Yamashino, T, Mizuno, T (2003) In vivo and in vitro characterization of the ARR11 response regulator implicated in the His-to-Asp phosphorelay signal transduction in *Arabidopsis thaliana*. Plant Cell Physiol. 44: 122-131
14. Inoue, T, Higuchi, M, Hashimoto, Y, Seki, M, Kobayashi, M, Kato, T, Tabata, S, Shinozaki, K, Kakimoto, T (2001) Identification of CRE1 as a cytokinin receptor from *Arabidopsis.* Nature 409: 1060-1063
15. Kakimoto, T (2003) Perception and signal transduction of cytokinins. Annu. Rev. Plant Biol. 54: 605-627
16. Kiba, T, Yamada, H, Mizuno, T (2002) Characterization of the ARR15 and ARR16 response regulators with special reference to the cytokinin signaling pathway mediated by the AHK4 histidine kinase in roots of *Arabidopsis thaliana*. Plant Cell Physiol. 43: 1059-1066
17. Lohrmann, J, Buchholz, G, Keitel, C, Sweere, C, Kircher, S, Bäurle, I, Kudla, J, Harter, K (1999) Differentially-expressed and nuclear-localized response regulator-like proteins from *Arabidopsis thaliana* with transcription factor properties. J. Plant Biology 1: 495-506
18. Mähönen, AP, Bonke, M, Kauppinen, L, Riikonon, M, Benfey, P, Helariutta, Y (2000) A novel two-component hybrid molecule regulates vascular morphogenesis of the *Arabidopsis* root. Genes and Dev. 14: 2938-2943
19. Martin, T, Sotta, B, Jullien, M, Caboche, M, Faure, J-D (1997) ZEA3: a negative modulator of cytokinin responses in plant seedlings. Plant Physiol. 114: 1177-1185
20. Menges, M, Hennig, L, Gruissem, W, Murray, JAH (2002) Cell cycle regulated gene expression in Arabidopsis. J. Biol. Chem. 277: 41987-2002
21. Miyata, S-i, Urao, T, Yamaguchi-Shinozaki, K, Shinozaki, K (1998) Characterization of genes for two-component phosphorelay mediators with a single HPt domain in *Arabidopsis thaliana.* FEBS Letters 437: 11-14
22. Mok, DW, Mok, MC (2001) Cytokinin metabolism and action. Annu. Rev. Plant Physiol. Plant Mol. Biol. 89: 89-118
23. Mougel, C, Zhulin, I.B. (2001) CHASE: an extracellular sensing domain common to transmembrane receptors from prokaryotes, lower eukaryotes and plants. Trends Biochem. Sci. 26: 582-584
24. Ori, N, Juarez, MT, Jackson, D, Yamaguchi, J, Banowetz, GM, Hake, S (1999) Leaf senescence is delayed in tobacco plants expressing the maize homeobox gene knotted1 under the control of a senescence-activated promoter. Plant Cell 11

25. O'Rourke, SM, Herskowitz, I, O'Shea, EK (2002) Yeast go the whole HOG for the hyperosmotic response. Trends Genet. 18: 405-412
26. Pischke, MS, Jones, LG, Otsuga, D, Fernandez, DE, Drews, GN, Sussman, MR (2002) An Arabidopsis histidine kinase is essential for megagametogenesis. Proc. Natl. Acad. Sci. USA 99: 15800-15805
27. Rashotte, AM, Carson, SDB, To, JPC, Kieber, JJ (2003) Expression profiling of cytokinin action in Arabidopsis. Plant Physiol. in press
28. Riou-Khamlichi, C, Huntley, R, Jacqmard, A, Murray, JA (1999) Cytokinin activation of Arabidopsis cell division through a D-type cyclin. Science 283: 1541-1544
29. Romanov, GA, Kieber, JJ, Schmulling, T (2002) A rapid cytokinin response assay in Arabidopsis indicates a role for phospholipase D in cytokinin signalling. FEBS Letters 515: 39-43
30. Rupp, H-M, Frank, M, Werner, T, Strnad, M, Schmülling, T (1999) Increased steady state mRNA levels of the STM and KNATI homeobox genes in cytokinin overproducing *Arabidopsis thaliana* indicate a role for cytokinins in the shoot apical meristem. Plant J. 18: 557-563
31. Sakai, H, Aoyama, T, Oka, A (2000) Arabidopsis ARR1 and ARR2 response regulators operate as transcriptional activators. Plant J. 24: 703-711
32. Sakai, H, Honma, T, Aoyama, T, Sato, S, Kato, T, Tabata, S, Oka, A (2001) *Arabidopsis* ARR1 is a transcription factor for genes immediately responsive to cytokinins. Science 294: 1519-1521
33. Sakakibara, H, Hayakawa, A, Deji, A, Gawronski, SW, Sugiyama, T (1999) His-Asp phosphotransfer possibly involved in the nitrogen signal transduction mediated by cytokinin in maize: molecular cloning of cDNAs for two-component regulatory factors and demonstration of phosphotransfer activity in vitro. Plant Mol Biol. 41: 563-573
34. Saunders, MJ (1990) Calcium and plant hormone action. Symp. Soc. Exp. Biol. 44: 271-283
35. Schaller, GE, Mathews, DE, Gribskov, M, Walker, JC (2002) Two-component signaling elements and histidyl-aspartyl phosphorelays. *In* C Somerville and E Meyerowitz, eds The Arabidopsis Book http://www.aspb.org/downloads/arabidopsis/schall1.pdf
36. Schmülling, T, Schäfer, S, Romanov, G (1997) Cytokinins as regulators of gene expression. Physiol. Plant. 100: 505-519
37. Smalle, J, Kurepa, J, Yang, P, Babiychuk, E, Kushnir, S, Durski, A, Vierstra, RD (2002) Cytokinin growth responses in Arabidopsis involve the 26S proteasome subunit RPN12. Plant Cell 14: 17-32
38. Stock, AM, Robinson, VL, Goudreau, PN (2000) Two-component signal transduction. Annu. Rev. Biochem. 69: 183-215
39. Strayer, C, Oyama, T, Schultz, TF, Raman, R, Somers, DE, Mas, P, Panda, S, Kreps, JA, Kay, SA (2000) Cloning of the Arabidopsis clock gene TOC1, an autoregulatory response regulator homolog. Science 289: 768-771
40. Suzuki, T, Imamura, A, Ueguchi, C, Mizuno, T (1998) Histidine-containing phosphotransfer (HPt) signal transducers Implicated in His-to-Asp phosphorelay in Arabidopsis. Plant Cell Physiol. 39: 1258-1268
41. Suzuki, T, Ishikawa, K, Mizuno, T (2002) An Arabidopsis histidine-containing phosphotransfer (Hpt) factor implicated in phosphorelay signal transduction: Overexpression of AHP2 in plants results in hypersensitivity to cytokinin. Plant Cell Physiol. 43: 123-129
42. Suzuki, T, Sakurai, K, Ueguchi, C, Mizuno, T (2001) Two types of putative nuclear factors that physically interact with histidine-containing phosphotransfer (Hpt) domains, signaling mediators in His-to-Asp phosphorelay, in *Arabidopsis thaliana*. Plant Cell Physiol. 42: 37-45
43. Sweere, U, Eichenberg, K, Lohrmann, J, Mira-Rodado, V, Bäurle, I, Kudla, J, Nagy, F, Schäfer, E, Harter, K (2001) Interaction of the response regulator ARR4 with the photoreceptor phytochrome B in modulating red light signaling. Science 294: 1108-1111

44. Takei, K, Sakakibara, H, Taniguchi, M, Sugiyama, T (2001) Nitrogen-dependant accumulation of cytokinins in root and the translocation to leaf: Implication of cytokinin species that induces gene expression of maize response regulator. Plant Cell Physiol. 42: 85-93
45. Urao, T, Yakubov, B, Satoh, R, Yamaguchi-Shinozaki, K, Seki, M, Hirayama, T, Shinozaki, K (1999) A transmembrane hybrid-type histidine kinase in *Arabidopsis* functions as an osmosensor. Plant Cell 11: 1743-1754
46. Vittorioso, P, Cowling, R, Faure, JD, Caboche, M, Bellini, C (1998) Mutation in the Arabidopsis PASTICCINO1 gene, which encodes a new FK506-binding protein-like protein, has a dramatic effect on plant development. Mol. Cell. Biol. 18: 3034-3043
47. Vogel, JP, Schuerman, P, Woeste, KW, Brandstatter, I, Kieber, JJ (1998) Isolation and characterization of Arabidopsis mutants defective in induction of ethylene biosynthesis by cytokinin. Genetics 149: 417-427
48. Werner, T, Motyka, V, Strnad, M, Schmulling, T (2001) Regulation of plant growth by cytokinin. Proc. Natl. Acad. Sci. USA 98: 10487-10492
49. Yamada, H, Hanaki, N, Imamura, A, Ueguchi, C, Mizuno, T (1998) An *Arabidopsis* protein that interacts with the cytokinin-inducible response regulator, ARR4, implicated in the His-Asp phosphorelay signal transduction. FEBS Lett. 436: 76-80
50. Yamada, H, Suzuki, T, Terada, K, Takei, K, Ishikawa, K, Miwa, K, Yamashino, T, Mizuno, T (2001) The Arabidopsis AHK4 histidine kinase is a cytokinin-binding receptor that transduces cytokinin signals across the membrane. Plant Cell Physiol. 41: 1017-1023
51. Reiser, V., Raitt, D.C. and Saito, H. (2003) Yeast osmosensor Sln1 and plant cytokinin receptor Cre1 respond to changes in turgor pressure. J. Cell Biol. 161: 1035-1040
52. Werner, T., Motyka, V., Laucou, V., Smets, R., Van Onckelen, H. (2003) Cytokinin-deficient transgenic *Arabidopsis* plants show multiple developmental alterations indicating opposite functions of cytokinins in the regulation of shoot and root meristem activity. Plant Cell 15: 2532-2550
53. Corbesier, L., Prinsen, E., Jacqmard, A., Lejeune, P., Van Onckelen, H., Périlleux, C., Bernier, G. (2003) Cytokinin levels in leaves, leaf exudate and shoot apical meristem of *Arabidopsis thaliana* during floral transition. J. Exp. Bot. 54:2511-2517
54. To, J.P.C., Haberer, G., Ferreira, F.J., Deruère, J., Mason, M.G., Schaller, G.E., Alonso, J.M., Ecker, J.R., Kieber, J.J. (2004) Type-A *Arabidopsis* response regulators are partially redundant negative regulators of cytokinin signaling. Plant Cell online preview at www.plantcell.org/cgi/doi/10.1105/tpc.018978
55. Tajima, Y., Imamura, A., Kiba, T., Amano, Y., Yamashino, T., Mizuno, T. (2004) Comparative studies on the type-B response regulators revealing their distinctive properties in the His-to-Asp phosphorelay signal transduction of *Arabidopsis thaliana*. Plant Cell Physiol. 45:28-39

D4. Ethylene Responses in Seedling Growth and Development

Ramlah B. Nehring and Joseph R. Ecker
Plant Biology Laboratory, The Salk Institute for Biological Studies, La Jolla, California 92037 USA. E-mail: rnehring@biomail.ucsd.edu

INTRODUCTION

The gaseous plant hormone ethylene is an olefin hydrocarbon produced by all plants. Despite its simple chemical structure it orchestrates a myriad of complex functions. Ethylene controls processes as diverse as germination, root hair development, root nodulation, senescence of organs (including fruit ripening), differential cell growth, abscission, stress responses and resistance to necrotrophic pathogens (1). It is known that ethylene is effective to induce biological responses at nanomolar concentrations and that its response only takes minutes to be induced. Due to the important nature of its signaling the production of this hormone is a tightly regulated process, controlled by both developmental signals and response to environmental stimuli. To further the understanding of ethylene signaling in plants we need to fully comprehend how the hormone is synthesized, perceived and signal is transduced[1].

ETHYLENE IN SEEDLINGS: THE TRIPLE RESPONSE

Exposure to ethylene induces dramatic morphological changes in the growth of seedlings. Normally an etiolated (dark-grown) seedling displays a closed apical hook, a long slender hypocotyl and an elongated root. When exposed to ethylene seedlings exhibit exaggerated curvature of the apical hook, radial swelling of the hypocotyl, and inhibition of hypocotyl and root elongation (20). This change associated with exogenously applied ethylene was first

---

[1]Abbreviations: ACC, 1-Aminocyclopropane-1-Carboxylic Acid; AVG, Aminoethoxyvinylglycine; C-END, Carboxy-terminal end; CTR, Constitutive Triple Response; EIN, Ethylene Insensitive; ER, Endoplasmic Reticulum; EREBP, Ethylene Response Element Binding Protein; ETO, Ethylene Overproducers; His, Histidine; JA, Jasmonic Acid; LNM, Low Nutrient Medium; MAP, Mitogen-activated Protein; MAPK, Mitogen-activated Protein Kinase; MS, Mechanical Stimuli; PERE, Primary Ethylene Response Element; PP2A, Protein Phosphatase 2A; PR, Pathogenesis-Related; SA, Salicylic Acid; SE, Strong EIN; WE, Weak EIN.

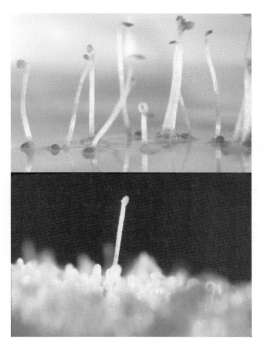

Figure 1 (Color plate page CP9). The ethylene triple response. Seedlings were grown in the dark and treated with ethylene gas three days after germination. (Top, center) Eto⁻ or Ctr⁻ mutants that are identified as mutants that display the seedling triple response grown on agar media in the dark without the addition of ethylene or its precursors. (Bottom, center) Ein mutants are identified because they fully lack the ability to respond to ethylene when grown in the dark in the presence of ethylene. Tissue specific mutants are selected by identifying the organs that do not respond to ethylene treatment (not shown).

discovered over 100 years ago in pea seedlings and was termed the "triple response" (33). In this context, ethylene is a stress-induced hormone with the triple response mimicking the natural response that is produced when seedlings encounter a physical barrier as they attempt to penetrate the soil. The exaggeration of the apical hook, due to the production of ethylene, protects the delicate apical meristem from the physical damage that may occur during seedling emergence (15).

Plant biologists have exploited the highly reproducible triple response phenotype in Arabidopsis to discover mutants that affect components in the ethylene signal transduction pathway (Fig. 1). Three main classes of mutants defective in the ethylene response have been identified (16). These classes are: the ethylene insensitive (*Ein*) mutants, tissue specific mutants (such as a non-responsive apical hook or root) and "constitutive" mutants that demonstrate a triple response without the addition of exogenous ethylene. The last category can be further dissected into two groups: ethylene overproducers (*Eto*⁻) and constitutive signaling pathway mutants (*Ctr*⁻).

THE ETHYLENE SIGNAL TRANSDUCTION PATHWAY

Ethylene Perception

Receptors
A family of five membrane-localized receptors perceives ethylene in

Ethylene signal transduction in stem elongation

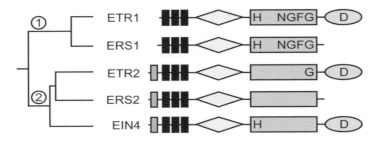

Figure 2. The Ethylene Receptor Family in Arabidopsis. Predicted primary protein structures of the five-member family are indicated. Black bars represent transmembrane segments. Gray bars represent putative signal sequences. Diamonds indicate GAF domains. Rectangles indicate histidine kinase domains. Ovals indicate receiver domains. The conserved phosphorylation sites upon histidine (H) and aspartate (D) are indicated if present. Conserved motifs (NGFG) within the histidine kinase domain are indicated if present. There are two subfamilies of ethylene receptors (sub-family 1 and 2) based on sequence and phylogenetic analysis (From 41).

Arabidopsis (Fig. 2). The receptors share homology with two-component histidine (His) kinases that are involved with sensing environmental changes, originally discovered in bacteria. Two component systems normally consist of a membrane localized "sensor" protein kinase which detects the input signal and a "response regulator" that mediates the output (25). In plants this family consists of mostly hybrid kinases which contain both the kinase and receptor domains and can be involved in phosphorelay activities, passing along a single phosphate from the receptor through an intermediate phosphorelay protein and finally onto the response regulator (21). The ethylene receptors are one of four families of His protein kinases found in plants (25). They are further subdivided into two subfamilies (41). All of the ethylene receptors contain predicted amino terminal transmembrane domains, and for several of the receptors this domain has been found to contain the site for ethylene binding (40). ETR1 and ERS1 make up the first subfamily; these two proteins have complete conservation of the His kinase protein motifs and each has three predicted amino terminal membrane spanning domains (Fig. 2). The other three receptors (ERS2, EIN4 and ETR2) are part of the second subfamily, characterized by a signal sequence whose putative function may be to target proteins to the secretory pathway. This subfamily lacks some of the amino acids previously shown to be essential for His kinase activity, casting some doubt on the designation that these are functional His kinases. All members of this subfamily contain four predicted amino terminal transmembrane domains.

The receptors have been hypothesized to require a transition metal in order to bind the olefin ethylene (9). Based upon the in vitro biochemical association of copper ions and the binding domain of ETR1 (38), and the in

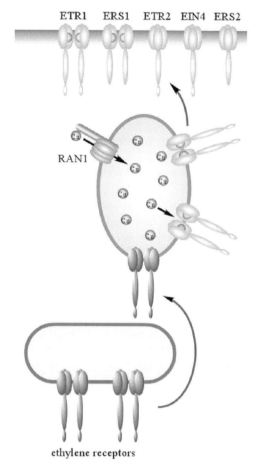

Figure 3 (Color plate page CP9). Assembly of ethylene receptors. Five transmembrane proteins, ETR1, ETR2, ERS1, ERS2, and EIN4, some localized to the endoplasmic reticulum, make up the family of ethylene receptors in *Arabidopsis*. ETR1 and ERS1 contain three transmembrane domains and a conserved histidine kinase domain, and have been shown to function as homodimers. Ethylene binding occurs at the N-terminal transmembrane domain of the receptors, and a copper co-factor is required for the binding. RAN1, a copper transporter, is involved in the delivery of copper to the ethylene receptor.

planta identification of RAN1, a copper-transporter, the metal in the ethylene receptors has been revealed to be copper (Fig. 3). Defects in RAN1 cause altered ligand specificity of the ethylene receptors and reduced function of the entire ethylene pathway (23). Silver ions, similar to copper in their capability to interact with ethylene, cause an inhibitory effect on all plant responses to ethylene, phenocopying the ethylene insensitive mutants (7).

Ethylene is readily diffusible through both aqueous and lipid environments allowing for one of the receptors, ETR1 to be localized to the endoplasmic reticulum of *Arabidopsis* cells (13). The receptors are expressed throughout plant tissues and have overlapping domains of expression. *ETR1* mRNA is down-regulated in the apical hook of etiolated seedlings, while the other receptors are shown to be up-regulated in adult leaves (3). Mutant analysis revealed dominant mutations in the receptors result in ethylene insensitivity (12). Interestingly, single null alleles of the receptors do not show phenotypes, but when combining *etr1* with other null receptor alleles, these plants display a constitutive activation of the ethylene pathway. The genetic prediction based on these two opposite phenotypes suggests that when ethylene is unable to bind to the receptors that they remain active, and that the receptors normally function as negative regulators of the downstream ethylene signal transduction pathway (21). Further insight has been made into the differences between the type 1 and type 2 subfamilies of receptors using biochemical approaches and newly discovered null mutations in one of the receptors. The protein level of ETR1 has been found to be up-regulated in the *etr1* ethylene insensitive mutants (*etr1-1*, *etr1-2*, *etr1-3* and *etr1-4*) (52). The ethylene insensitivity of these mutants was equivalent to treating the seedlings with silver, demonstrating that silver mimics an ethylene insensitive mutation. It had been previously postulated that loss of one receptor might be compensated for by up-regulation of another receptor. This was not shown to be the case with ETR1, because in the mutant backgrounds, *etr2-3*, *ein4-4*, and *ers2-3* (including combinations of single, double and triple mutants) the level of ETR1 protein was similar to wild type levels. Null alleles for four of the receptor genes have been previously described. Another null allele of *ERS1* has recently been discovered and characterized (49). Like all of the other null alleles of the individual receptors, the null allele in *ers1-2* did not have any discernible phenotype as a single mutant. The lack of a phenotype was not due to the functional compensation by ETR1, its subfamily member, as neither the protein or mRNA expression level of ETR1 were affected in *ers1* plants. However, when combined in the same plant, these two null mutants for the type 1 subfamily (*ers1-2* and *etr1-7*) caused a dramatic constitutive ethylene signaling phenotypes both at the seedling and adult stages (49). The adult phenotypes observed in the *etr1 ers1* double mutant including small rosette size, delayed flowering, and defects in fertility and flower morphology. These phenotypes are even more severe than the adult phenotypes of the constitutive ethylene signaling mutant, *ctr1* (22). Whether the pleiotropic defects seen in the double mutant plants were ethylene related or ethylene independent was determined by generating a triple mutant of *ers1-2*, *etr1-7* and *ein2*. These mutants showed an *ein2*-like ethylene insensitive phenotype and therefore demonstrated that the receptor defects are dependant on EIN2 (22). The double mutant phenotypes in both light and dark grown seedlings could be rescued by either of the wild type cDNAs for *ETR1* or *ERS1*, but not by the cDNAs for the type 2 subfamily of receptors, *EIN4*, *ERS2* or *ETR2* (49). The main difference between the type 1 and type 2 receptors is the conserved histidine kinase domain. To test whether the lack of this domain was responsible for the double mutant phenotype, double mutants were transformed with an ETR1 fragment containing a kinase-inactivated domain (49). Interestingly, this kinase-inactive domain was also able to rescue the mutant phenotype, indicating that the histidine kinase phosphotransfer activity is not essential for signaling from the ethylene receptors to their downstream effectors.

Downstream Signaling Pathway Components

The central signaling pathway component CTR1 and a MAPK Pathway
The first ethylene signaling pathway gene that was cloned in plants was

identified as a T-DNA tagged allele of the constitutive ethylene signaling mutant, *ctr1* (27). Plants containing loss of function mutations in *CONSTITUTIVE TRIPLE RESPONSE1* look as if they are constantly exposed to ethylene in both seedlings and adults, indicating that this protein acts as a negative regulator of ethylene signaling. The protein encoded by the CTR1 gene shares homology with the mammalian Raf serine/threonine protein kinase family. The role of these proteins in mammals is acting as a mitogen-activated protein kinase kinase kinase (MAPKKK). The similarity of CTR1 to MAPKKKs suggested a downstream MAP kinase cascade in the ethylene pathway (Fig. 4).

The CTR1 protein is a very likely candidate for the next step in the putative phosphorelay cascade proposed for the ethylene receptors. It has been demonstrated to posses in vitro Ser/Thr kinase activity. Biochemically the amino terminus of CTR1 has also been shown to interact directly with the ethylene receptor ETR1, via a putative CTR specific protein-protein interaction domain called the CN box (24). These studies suggest that in the absence of ethylene, the ethylene receptors are active and stimulate the kinase activity of CTR1, allowing it to phosphorylate downstream targets and causing the ethylene pathway to be repressed. Weak associations between CTR1 and other ethylene receptors, such as ERS1 and ETR2 have also been recognized (10), implicating both subfamily 1 and 2 of the receptors in the binding/localization of CTR1. Interestingly, the histidine kinase activity of ETR1, when expressed in yeast, is not required for the binding of CTR1 to the receptor (18). Domains within CTR1, the amino terminal regulatory domain and the kinase domain have also been found to associate with themselves (29). This might be a mechanism of negative regulation causing an inhibition of CTR1 kinase activity. CTR1 is postulated to bind to the receptors by being associated within the same location in the cell (18). CTR1 has been isolated as part of an ethylene receptor signaling complex, including the receptor ETR1. Using sucrose density centrifugation CTR1 was localized to the ER of *Arabidopsis* microsomes (18). The ethylene receptors are required for the membrane localization of CTR1, because when multiple receptors were mutated CTR1 was relocalized to the soluble fraction of the cell (18).

Ethylene signaling does not appear to be completely dependant upon CTR1 signaling. Plants containing strong alleles of *ctr1*, which do not have any kinase activity, still respond weakly to ethylene treatment (27). It has also been shown that quadruple ethylene receptor mutants demonstrate a constitutive phenotype that is more severe than the strongest *ctr1* alleles. Additionally the strong *ran1-3* allele, of the copper transporter, also has a more severe ethylene phenotype (50). The "residual" ethylene response observed in *ctr1* plants may be a result of the activity of other CTR1-like proteins that are as of yet undetermined.

Ethylene signal transduction in stem elongation

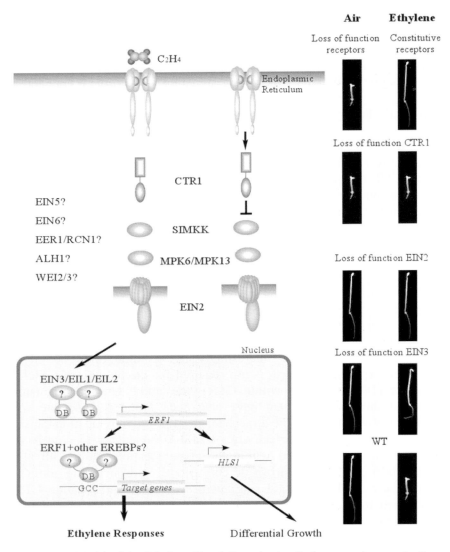

Figure 4. A Model of the Ethylene Signal Transduction Pathway. As seen in figure 3, ethylene binds to the family of transmembrane receptors in a copper dependant manner. In the absence of an ethylene signal, ethylene receptors activate a Raf-like kinase, CTR1, and CTR1 in turn negatively regulates the downstream ethylene response pathway, possibly through the MAP-kinase cascade including SIMKK, MAPK6 and MAPK13. Binding of ethylene inactivates the receptors, resulting in the deactivation of CTR1, which allows EIN2 to function as a positive regulator of the ethylene pathway. EIN2 contains an N-terminal hydrophobic domain similar to the Nramp metal transporter proteins and a novel hydrophilic C terminus. EIN2 positively signals downstream to EIN3 and its family of related transcription factors, located in the nucleus. EIN3 binds to the promoter of ERF1 and activates its transcription in and ethylene-dependent manner. The transcription factors ERF1 and the other EREBPs can interact with the GCC box in the promoter of target genes to activate the downstream effects of the ethylene pathway (modified from Wang et al 2002).

Recent studies using *Medicago* and *Arabidopsis* now suggest a direct association between a MAP kinase pathway and ethylene signaling (34). This study showed the association of components of ethylene with a mitogen activated protein kinase cascade. In brief, *Arabidopsis* plants overexpressing *Medicago* SIMKK, a MAPKK, showed a constitutive triple response phenotype in the absence of exogenous ethylene, and this phenotype was not inhibited by blocking of ethylene biosynthesis using an inhibitor of ACC synthase, aminoethoxyvinylglycine (AVG). In addition, ethylene treatment of plants was found to result in the activation of an *Arabidopsis* MAPK (MPK6). The order of action of the pathway components was determined using either *in vitro* kinase assays, and/or the ability to the SIMKK transgene to induce ethylene target genes in various mutant backgrounds. Both SIMKK transgenic phenotypes and MPK6 activity were positioned downstream of ETR1 but upstream of EIN2, placing the corresponding genes in the predicted location for the proposed MAPK cascade. Many interesting questions remain regarding the proposed MAPK pathway. For example, what effects are observed when mutations are introduced into these genes, how is the phosphorylation signal transmitted to the downstream components in the pathway and how are the downstream ethylene target genes activated.

The Central Signaling Component EIN2
Loss of function mutations in *EIN2* confer complete plant ethylene insensitivity, indicating that EIN2 is an essential positive regulator of the ethylene signal transduction pathway (2). Over 25 *EIN2* alleles have been discovered, more than of any other ethylene signaling pathway component. In part, this is because *ein2* plants, relative to all other mutants, display the least response to ethylene; they display the greatest degree of ethylene insensitivity). In addition, many alleles of ein2 have been uncovered in mutant screens for plants with defects in the other signaling pathways including responses to cytokinins, abscisic acid, sugars and in screens for delayed senescence (48). The appearance of *EIN2* occurs so recurrently in such a divergent set of mutant screens indicates that this protein plays a essential role in ethylene signaling and that ethylene plays a role in modulating responses to a diverse set of signals, each of which also shows unique plant responses that are unrelated to ethylene.

EIN2 encodes an integral membrane protein with 12 predicted transmembrane domains at its amino terminus (2). This domain shares homology with the Nramp family of metal-ion transporters. This family is made up of members such as yeast SMf1p, *Drosophila* Malvolio and mammalian DCT1 (3). No studies have been able to identify a metal transporting activity for EIN2, nor is EIN2 able to complement metal uptake deficient yeast strains as shown for authentic *Arabidopsis* Nramps genes (41). The carboxy-terminal region of EIN2 protein shares little homology with other proteins of known function. This hydrophilic region contains domains that may be associated with protein-protein interactions. Interestingly, it is this unknown part of the EIN2 protein that is sufficient to elicit the ethylene response in Arabidopsis plants. Plants overexpressing the unique carboxy-terminal end (C-END) of the protein show a constitutive ethylene response in both light grown seedlings and in adult plants. These plants also constitutively express (at the mRNA level) all ethylene-regulated genes. The overexpression of EIN2 C-END is also sufficient to activate the downstream nuclear pathway through EIN3 (2). However, overexpression of the C-END was not able to induce the triple response in etiolated seedlings, indicating that the Nramp domain must be necessary for ethylene mediated effects in dark grown seedlings. These results suggest that the Nramp domain of EIN2 is required for sensing the upstream ethylene signal while the carboxy-terminal domain of EIN2 is necessary for transducing the signal to downstream ethylene pathway components.

Nuclear Events: The EIN3 Family and its Target Genes
The plants' response to ethylene gas is known to involve changes in gene expression (45). However, until identification of the *ein3* mutant and cloning of the *EIN3* gene there was no direct evidence of nuclear regulation in the ethylene signaling pathway (11). Mutations in *EIN3* in Arabidopsis cause reduced response to ethylene, although these plants are more sensitive to ethylene than *ein2* mutants. This reduced response is seen in an ethylene insensitive dark grown seedling, reduced ethylene-dependent expression of genes and reduced ethylene-mediated leaf senescence (11). This insensitivity phenotype displayed in all loss of function mutations indicates that *EIN3* is a positive regulator of ethylene signal transduction.

The protein encoded by the *EIN3* gene is a member of a family of six related proteins in Arabidopsis. These nuclear-localized transcriptional activators contain an acidic domain at their amino terminus along with a proline-rich region, a coil structure and several regions that contain highly basic amino acids. The amino terminus of these proteins is involved in DNA binding (44). Members of this family are closely related in sequence and function; in fact, two of the *EIN3*-like genes (*EILs*), *EIL1* and *EIL2* can rescue the mutant phenotype of *ein3*, indicating that these related proteins must also be involved in ethylene signal transduction. Mutants for *ein3* show only a reduction of function for ethylene signaling, unlike the strong loss of function mutants such as *ein2* or *etr1*. Possible functional overlap between EIN3 family members might account for this reduced phenotype (11). Additionally, overexpression of either *EIN3* or the *EIL1* induces a constitutive ethylene response similar to the phenotype seen in the *ctr1* mutant, demonstrating the sufficiency of these proteins to activate the ethylene response pathway in both seedlings and adult plants (11). The *EIN3* overexpression phenotype is independent of the presence of a functional EIN2 protein indicating that the EIN3/EIL family functions downstream of

EIN2 in the ethylene pathway. Interestingly, *EIN3* gene expression is not affected by ethylene treatment(48). Recent studies have revealed the nature of the post-translational regulation of EIN3 (19, 35). EIN3 protein can be stabilized when proteasome function is inhibited, mimicking the effect of exogenous ethylene treatment on the regulation of the protein (19). Two F box proteins, EBF1 and EBF2 have been found to directly interact with EIN3 and EIL1, leading to EIN3 proteolysis (35) (and probably that of EIL1 as well). Mutants in the genes encoding *EBF1* and *EBF2* result in hypersensitivity to ethylene and an increase in EIN3 protein accumulation. The double mutant *ebf1,ebf2* results in *ctr1* like phenotypes, and suppresses *ein2*, indicating that these genes function downstream of, or parallel to, EIN2 in the ethylene signaling pathway (35). The transcription of the *EBF1* and *EBF2* genes is disrupted in the *ein3* mutant. These studies indicate the presence of a possible negative-feedback mechanism, where on one side the production of ethylene acts to stabilize and prevent degradation of the EIN3 protein, whereas, on the other side, EBF1 and EBF2 are induced by the production of ethylene, which stabilizes EIN3; EBF1 and EBF2 then target EIN3 for proteasome-mediated degradation. The ubiquitin/proteasome pathways have been demonstrated or implicated in most of the other plant hormone signaling pathways, leaving us to wonder whether this type of protein regulation contributes to the well-documented cross-talk between various phytohormones.

One of the targets of the EIN3/EIL proteins was discovered when an ethylene response element binding protein (EREBP), *ERF1*, was found to require *EIN3* function for its expression (44). *ERF1* mRNA was not detectable in *ein3* mutants, and its steady-state levels were dramatically increased in EIN3 overexpressing plants. The amino-terminal end of EIN3 was found to specifically bind to the promoter of the *ERF1* gene. In addition the related proteins EIL1/2 were also able to bind to the EIN3 binding site in the *ERF1* promoter (44). The EIN3/EIL proteins were also able to bind to the gene promoter of another unrelated transcription factor EDF1 (48). A short palindromic fragment of DNA was identified as the EIN3 binding site. This binding site has been classified as a primary ethylene response element (PERE), and is similar to sequences in promoter regions required for ethylene regulated gene expression in other species. EIN3 binds to this site as a homodimer, and, from in vitro DNA binding studies, it does not form heterodimers with its related family members (44).

Another known target of EIN3, *ERF1*, is a transcription factor itself. This suggests a transcriptional cascade where EIN3 binds to the promoter of *ERF1*, and ERF1 protein in turn regulates other EREBPs (44). Using in vitro binding studies, ERF1 has been shown to bind specifically to the promoters of ethylene-regulated genes containing the GCC box. ERF1 is also a positive regulator of ethylene signaling similar to EIN3, and yields a constitutive ethylene phenotype similar to EIN3 when overexpressed in plants. The

EREBPs were originally identified because they contained a highly conserved DNA binding domain (the ERF domain) which is able to bind to the GCC box, a *cis*-element found in the promoters of many ethylene responsive pathogen-related genes. Many EREBP genes exist in Arabidopsis and other monocots and dicots (37), but only a few of these related genes are directly regulated by ethylene. These genes are, however, also regulated by various abiotic stress responses such as those induced by cold, drought, salt, wounding and pathogens. The EREBPs can function as either transcriptional activators or transcriptional repressors (17). As transcriptional repressors, they are active repressors suppressing the transactivation of other transcription factors without competing for the same DNA binding site.

Other Ethylene Response Mutants

Numerous other ethylene mutants have been identified, both through triple response screens as well as other mutant screening approaches. Three other alleles of ethylene insensitive mutants have yet to be characterized (EIN5, EIN6 and EIN7) (39). All of these alleles show a reduced response to ethylene, but do not display as severe ethylene insensitive when compared with *ein2*. Using double mutant analysis, these ethylene insensitive mutants are found to be epistatic to CTR1. Two recessive alleles of *ein5*, along with one semi-dominant allele of *ein7* were originally discovered. The *ein6* mutant was originally characterized as a recessive mutation with reduced gametophytic transmission. The ethylene insensitive phenotype of this mutant has now been attributed to not one but two mutations in the background (32). The double mutant resulting in the *ein6* phenotype has a number of other interesting phenotypes such as taxol hypersensitivity (39) and alteration in the response of touch sensitive genes (51). The taxol result is interesting because cortical microtubule reorientation, from transverse to longitudinal, has been implicated as a downstream effect of the ethylene pathway. Since only single alleles were generated for these mutants perhaps screening for triple response mutants has not been yet saturated.

Altering the triple response assay has found a number of other new ethylene mutants. One of these mutants was found in a screen for plants that had enhanced ethylene responsiveness (EER). Compared to wild type, Eer⁻ plants displayed greater sensitivity and increased amplitude of the response to ethylene. One mutant isolated from this screen, *eer1*, has a short thick hypocotyl in response to treatment with ACC, ethylene gas and also in response to ethylene antagonists (28). The *eer1* hypocotyls are only about 50% the length of wild-type seedlings in the presence of exogenous ethylene. This phenotype was not caused by a constitutive signaling defect, due to the fact that it can be reversed by treatment with an inhibitor of ethylene biosynthesis (AVG). In combination with other known ethylene mutants *eer1* showed a number of interesting phenotypes. Many of these were suppressed by the ethylene insensitive mutants *ein2* and *etr1*, but several

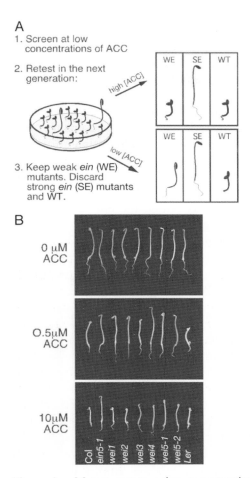

Figure 5. Mutant screen using nonsaturating levels of ethylene. (A) Schematic representation of the screening strategy. Mutagenized *Arabidopsis* plants were screened at a low concentration of the ethylene precursor ACC. Plants that showed hormone insensitivity were selected and retested in the next generation. Only putative mutants that showed weak hormone insensitivity were characterized further. (B) Comparison of the phenotypes of *wei* mutants in different concentrations of ACC. Seedlings were grown in the dark for 3 days in the presence of the indicated concentrations of ACC (from 2).

phenotypes showed an additive effect when combined with *ctr1*. The mutant also showed an increase in expression of a known ethylene response reporter gene, *BASIC-CHITINASE*. Taken together, these studies suggest that the phenotype of this mutant is not simply the result of ethylene overproduction, but these plants may be highly sensitive to endogenous ethylene. Recent cloning of *EER1* revealed that it encodes a previously described Arabidopsis PP2A regulatory subunit, also known as *RCN1* (29). Placing EER1 in the ethylene pathway is difficult due to the complex phenotypes. It has been proposed to be partially responsible for the regulation of CTR1, or another related MAPKKK, either through the known ethylene pathway or through an alternative ethylene-signaling pathway (28, 29).

Another variation on the standard triple response assay involves looking for ethylene mutants that are only partially lack the normal response (Fig. 5). Using low doses of ethylene, five weakly ethylene-insensitive (Wei⁻) mutants were identified (4). Three of the five mutants, *wei1*, *wei4* and *wei5*, were found to correspond to previously known genes, *TIR1*, *ERS1* and *EIL1* respectively. The other two mutants *wei2* and *wei3* are thought to be previously unidentified members of the ethylene response pathway. TIR1/WEI1 is thought to participate as an ubiquitin ligase in the ubiquitin-mediated degradation of auxin response proteins. Uncovering *ers1/wei4* in a mutant screen is the first time that this gene has ever been found in a mutant screen, validating that this assay is uncovering new mutants. Identification of alleles in the ethylene receptors similar to this one may give us insights into the normal developmental context of gene function. The EIL genes had previously been suggested to be involved in the ethylene response (11) and the isolation of *eil1/wei5* confirms the involvement of *EIL1* in the ethylene-signaling pathway. The remaining two WEI mutants, *wei2* and *wei3*, show mainly root-specific phenotypes. In the past other ethylene insensitive root mutants have been discovered, but these genes are actually related to auxin responses (see section on ethylene and other growth regulators). Interestingly the Wei⁻ mutants show normal sensitivity to exogenous auxin and a normal growth response to gravity.

The number of newly uncovered mutants within the ethylene pathway continues to grow, demonstrating that genetic screens have not been saturated, although new twists on the old standby (triple response) might yield the best results.

OTHER ETHYLENE RESPONSES IN THE SEEDLING

The Response of the Seedling to Light and Ethylene

Ethylene has been shown to dramatically inhibit cell expansion. In the earliest studies on the hormone it was shown that exposing pea seedlings to ethylene caused an inhibition of epicotyl elongation (33). Similarly, prominent effects of ethylene in Arabidopsis include short roots and inflorescences, and an overall stunted appearance (27). One more recently described phenotype induced by the hormone is an *increase* in the length of the hypocotyl of Arabidopsis seedlings grown in the light in the presence of ACC or ethylene (Fig. 6) (43). Most dramatically the length was increased 2-fold when the seedlings were grown on a nutrient-deficient growth medium. This increase in length is not a result of increased cell division, but instead of an increase in cell elongation and overall cell volume. The

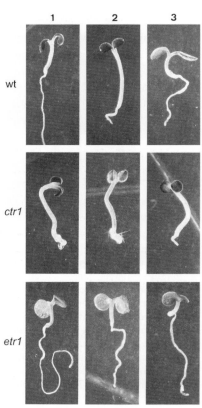

Figure 6. Effect of ACC treatment on seedlings of Col-0 (wild type,wt), *etr1-3*, and *ctr1-1* grown on LNM. Column 1, LNM; column 2,LNM with 50 μMACC; column 3, LNM with 50 μMACC and 100 μM AgNO$_3$ (from 43)

increase in hypocotyl length was confirmed to be caused by ethylene treatment because it could not be induced in the ethylene insensitive mutant *etr1,* and was reversed by treatment with the ethylene antagonist, silver. This effect of ethylene might be employed as a screening method to uncover additional ethylene signaling mutants.

Ethylene and Mechanical Stimuli

Changes in the growth of plants in response to physical environmental forces with which they interact are known as thigmomorphogenesis. When plants physically interact with their environment through processes such as rubbing, touching, rain, wind or due to growth against an object, they alter their growth habit to offset these mechanical stimuli (MS). This process is known to be calcium dependent and causes a rapid elevation in the amount of cytosolic free-calcium. Plants up-regulate a specific set of genes known as the touch genes (*TCH*) within a few minutes of receiving a stimulus (8). While plant hormones/growth regulators may somehow mediate this process, the signal transduction pathway from perception to response is still unknown. In this regard, MS can induce ethylene gas production and ethylene can induce the expression of the gene *TCH3* in the absence of a mechanical stimulus (42). More recently, another link between MS and ethylene was found in the requirement of the genes encoded by the double mutant *ein6/een* for *TCH3* induction (51). This result seems contradictory to the finding other ethylene insensitive mutants demonstrate a wild type response to MS, as measured both by gene expression and physiological response (26). However, the observed change in *TCH3* expression is independent of ethylene and therefore, the effect of EIN6/EEN seems to act downstream of calcium to control MS gene expression.

A second connection between ethylene and mechanical stimuli can been seen in seed germination. Ethylene is known to promote seed germination and certain ethylene insensitive mutants (*etr1*) are known to have lower germination rates then wild type (6). Vibration, a type of MS, is known to promote seed germination as well. Vibration-induced promotion of germination is not seen in the seeds of the Arabidopsis ethylene insensitive mutant *etr1* (46). This response was also lacking when wild-type seedlings were treated with the inhibitor of ethylene production, AVG. This finding indicates that alternate signal transduction pathways in response to MS may be present in the seed and germinated-seedling stages of development.

ETHYENE AND OTHER GROWTH REGULATORS

Ethylene has been found to play a central role in global hormone responses. It controls not only its own response pathway, but has also been found to regulate the synthesis of and response to other hormones and growth regulators.

Ethylene and Auxin

A number of genetic screens for mutants with aberrant responses to ethylene application have yielded mutants in genes required for auxin signaling. One of the earliest discovered was the mutant *HOOKLESS1* (*HLS1*)(20). *hls1* mutants lack a normal differential growth response to ethylene in the apical region of the hypocotyl; these plants do not form an apical hook in ethylene. These mutants also show abnormally enlarged cells in the hypocotyl and cotyledon. This defect can be phenocopied when seedlings are treated with either auxin or auxin-transport inhibitors, suggesting a link between auxin and ethylene signaling. The promoter region of HLS1 contains an ethylene inducible GCC-box, making it a possible target of the EREBPs (45). The HLS1 protein encodes a putative N-acetyltransferase, which may function in acetylating a protein involved in auxin transport or signaling (30). There are a number of other mutants with tissue specific defects in response to ethylene, which are now implicated in auxin signaling. Three mutants, *axr1*, *aux1* and *eir1* all display ethylene insensitivity in the root (3). AUX1 and EIR1 were found to be proteins which regulate polar auxin distribution. AUX1 is an auxin influx carrier, and plants containing mutations in this gene are also resistant to auxin. EIR1 is an auxin efflux carrier, and plants containing mutations in this gene are not resistant to high levels of auxin. Cloning of *AXR1* revealed that it might be involved in the ubiquitin-mediated degradation of a part of the auxin-signaling cascade. The mutant *nph4* shows a specific defect in its auxin-mediated response: the hypocotyl of this mutant is unable to bend in response to laterally applied blue-light or after auxin application. The auxin-related phenotype observed in *nph4* is complemented by exogenous application of ethylene. Interestingly, this ethylene effect is blocked in the *hls1* mutant, demonstrating a synthetic requirement of HLS1 for blue-light-mediated (auxin-dependent) differential growth. Because *NPH4* encodes a member of the ARF family of auxin response (transcription) factors, *ARF7*, these results suggest that ethylene compensation of bending in *nph4* mutants may occur by activation of other ARF family members (3).

More recently, there have been two additional ethylene-related mutants identified that show "cross-talk" between ethylene and auxin responses. The mutant *eer1* (see the section on Downstream Signaling Pathway Components) was discovered due to its enhanced response to ethylene. Upon cloning the gene it was discovered to be a mutation in the previously characterized gene *RCN1*. Seedlings mutant for *rcn1* show altered auxin transport, specifically increased basipetal auxin transport, due to their reduced phosphatase activity (36). The effect of ACC or ethylene on light grown seedlings has been exploited to find new mutants with altered responses to ethylene (43). Recently a mutant, *ACC-related long hypocotyl 1* (*alh1*) was isolated in a screen for mutants, which displayed elongated hypocotyls when grown in the light on nutrient-deficient growth medium without the presence of hormone (47). These mutants were found to overproduce ethylene, but were also altered in their response to auxin. It is thought that this mutant may affect ethylene-auxin cross talk, possibly by regulating the transport of auxin in the hypocotyl.

Ethylene and the Stress Hormones

Responses to ethylene are known to overlap with two other stress signals: jasmonic acid (JA) and salicylic acid (SA). The combinatorial action of these signals regulates plant defense responses to a variety of pathogens.

Ethylene and JA signaling can be interdependent, functioning as both positive and negative regulators of each other to provide the correct pattern of expression of the systemically induced defense-related genes. This interaction is mediated, at least partially, through the gene *ERF1*. Recent findings have indicated that *ERF1* is a common downstream target of both ethylene and JA in response to pathogen attack. This key gene can be induced by exogenous treatment by either hormone. However, the induction of *ERF1*, and its target genes, requires both JA and ethylene pathways to be intact, as seen by lack of *ERF1* gene expression in the ethylene mutant, *ein2*, and the JA mutant, *coi1* (31). The induction of downstream defense genes such as P*DF1.2* and *BASIC CHITINASE* (*b-CHI*) also requires that upstream elements of both these pathways are functional (44). The induction of downstream target genes can be fully rescued by introduction of *ERF1* expression in the ethylene/JA mutant backgrounds. *ERF1* is also sufficient to confer resistance to necrotrophic fungi such as *B. cinerea* (5). Finally, expression of the *ERF1* gene requires both ethylene and JA, demonstrating the necessity of both pathways in plant resistance to necrotrophic pathogens.

While ethylene and JA act in a synergistic manner to provide plants with resistance to necrotrophic pathogens, SA has been shown (in some instances) to be an antagonist. Overexpression of ERF1 or activation of the ethylene pathway has a detrimental effect on plant resistance to *Pseudomonas syringae tomato* DC3000 (5). There are, however, considerable interactions between these two pathways, or at least the outputs of these two pathways can overlap (or have common targets). With the advent of expression arrays, it has been shown that many genes respond to two or more defense signals when treated with SA, JA, ethylene or pathogen infection (48). Positive interactions between ethylene-dependant and SA-dependant pathways mediating disease responses have also been found, such as the necessity for both pathways to be active in the response to the pathogen *Plectosphaerella cucumerina* (5). Interestingly an SA dependant, *NPR1* (non-expresser of *PR-1*) independent pathway has been discovered which requires a loss of function of the ethylene-signaling pathway (for example, a mutation in *ein2*) to fully abolish *PR-1* gene expression (14). This response involves two mutants that have been found to constitutively express *PR* genes, *cpr5* and *cpr6*. The mutant, *ein2*, also controls the level of SA in these mutants, causing accumulation in *cpr5* and decreasing accumulation in *cpr6*. Yet another player has been discovered between SA and JA/ethylene dependent pathways in the suppressor of *npr1*, *ssi1*. This mutant constitutively expresses *PDF1.2* in an SA dependent manner, linking it to both pathways along with CPR5 and CPR6.

CONCLUSIONS

Understanding the myriad roles of ethylene in plant development and disease has greatly advanced. In particular, studies of the effects of ethylene on the development of dark grown seedlings of the reference plant *Arabidopsis thaliana*, have allowed the identification of many mutants in this signaling pathway. In turn, these mutants have lead to identification of many of the components of the ethylene signaling pathway and provided insight into the mechanisms of ethylene regulation of plant growth/development and response to pathogens. However, there are still many more pieces of the ethylene puzzle to uncover. New methods, such as additional novel mutant screens, genome-wide expression studies, the identification of transcription factor binding (EIN3/EIL/ERF1/EREBP) sites in combination with chemical and reverse genetic approaches, will need to be employed to further our understanding of mechanisms of action of this critical plant hormone.

References

1. Abeles, FB, Morgan, PW, and Saltveit, Jr., ME (1992) Ethylene in Plant Biology, Ed 2 Academic Press, San Diego
2. Alonso JM, Hirayama T, Roman G, Nourizadeh S, and Ecker JR (1999) EIN2, a bifunctional transducer of ethylene and stress responses in Arabidopsis. Science 284: 2148-2152
3. Alonso JM and Ecker JR (2001) The ethylene pathway: a paradigm for plant hormone signaling and interaction. Sci STKE 70: RE1
4. Alonso JM, Stepanova AN, Solano R, Wisman E, Ferrari S, Ausubel FM, Ecker JR (2003) Five components of the ethylene-response pathway identified in a screen for weak ethylene-insensitive mutants in Arabidopsis. Proc Natl Acad Sci 100: 2992-2997
5. Berrocal-Lobo M, Molina A, Solano R (2002) Constitutive expression of ETHYLENE-RESPONSE-FACTOR1 in Arabidopsis confers resistance to several necrotrophic fungi. Plant J 29: 23-32
6. Bleecker A, Estelle M, Somerville C and Kende H (1988) Insensitivity to ethylene conferred by a dominant mutation in *Arabidopsis thaliana*. Science 241: 1086-1089
7. Bleecker AB and Kende H (2000) Ethylene: a gaseous signal molecule in plants. Annu Rev Cell Dev Biol. 16: 1-18
8. Braam J and Davis RW (1990) Rain-, wind-, and touch-induced expression of calmodulin and calmodulin-related genes in Arabidopsis. Cell. 1990 60: 357-64
9. Burg, SP and Burg EA (1967) Inhibition of polar auxin transport by ethylene Plant Physiology 42: 1224-8
10. Cancel JD, Larsen PB (2002) Loss-of-function mutations in the ethylene receptor ETR1

cause enhanced sensitivity and exaggerated response to ethylene in Arabidopsis. Plant Physiol. 129: 1557-67
11. Chao Q, Rothenberg M, Solano R, Roman G, Terzaghi W, and Ecker JR (1997) Activation of the ethylene gas response pathway in Arabidopsis by the nuclear protein ETHYLENE-INSENSITIVE3 and related proteins. Cell. 89: 1133-44
12. Chang C and Stadler R (2001) Ethylene hormone receptor action in Arabidopsis. Bioessays. 23: 619-27
13. Chen YF, Randlett MD, Findell JL, and Schaller GE (2002) Localization of the ethylene receptor ETR1 to the endoplasmic reticulum of Arabidopsis. J Biol Chem. 277: 19861-6
14. Clarke JD, Volko SM, Ledford H, Ausubel FM, and Dong X (2000) Roles of salicylic acid, jasmonic acid, and ethylene in cpr-induced resistance in arabidopsis. Plant Cell. 12: 2175-90
15. Darwin, C and Darwin, F (1881) Darwins Gesammelte Werke Bd. 13. Schweizerbart'sche Verlagsbuchhandlung, Stuttgart
16. Ecker JR (1995) The ethylene signal transduction pathway in plants. Science 268: 667-75
17. Fujimoto SY, Ohta M, Usui A, Shinshi H, and Ohme-Takagi M (2000) Arabidopsis ethylene-responsive element binding factors act as transcriptional activators or repressors of GCC box-mediated gene expression. Plant Cell. 12: 393-404
18. Gao Z, Chen YF, Randlett MD, Zhao XC, Findell JL, Kieber JJ, Schaller GE (2003) Localization of the Raf-like kinase CTR1 to the endoplasmic reticulum of Arabidopsis through participation in ethylene receptor signaling complexes. J Biol Chem. 278: 34725-32
19. Guo H and Ecker JR (2003) Plant Responses to Ethylene Gas are Mediated by $SCF^{EBF1/EBF2}$-Dependent Proteolysis of EIN3 Transcription Factor. Cell 115: 667-677
20. Guzman P and Ecker JR (1990) Exploiting the triple response of Arabidopsis to identify ethylene-related mutants. Plant Cell 2: 513-23
21. Hall, AE, JL Findell, GE Schaller, and Bleecker AB (2000) Ethylene perception by the ERS1 protein of *Arabidopsis*. Plant Physiology 123:1449-1457
22. Hall, AE and Bleeker AB (2003) Analysis of combinatorial loss-of-function mutants in the Arabidopsis ethylene receptors reveals that the *ers1 etr1* double mutant has severe developmental defects that are EIN2 dependent. Plant Cell. 15: 2032-41.
23. Hirayama T, Kieber JJ, Hirayama N, Kogan M, Guzman P, Nourizadeh S, Alonso JM, Dailey WP, Dancis A, and Ecker JR (1999) RESPONSIVE-TO-ANTAGONIST1, a Menkes/Wilson disease-related copper transporter, is required for ethylene signaling in Arabidopsis. Cell. 97: 383-93
24. Huang Y, Li H, Hutchison CE, Laskey J, and Kieber JJ (2003) Biochemical and functional analysis of CTR1, a protein kinase that negatively regulates ethylene signaling in Arabidopsis. Plant J. 33: 221-33
25. Hwang I, Chen HC, and Sheen J (2002) Two-component signal transduction pathways in Arabidopsis. Plant Physiol. 129: 500-15
26. Johnson KA, Sistrunk ML, Polisensky DH, and Braam J (1998) Arabidopsis thaliana responses to mechanical stimulation do not require ETR1 or EIN2. Plant Physiol. 116: 643-9
27. Kieber JJ, Rothenberg M, Roman G, Feldmann KA, and Ecker JR (1993) CTR1, a negative regulator of the ethylene response pathway in Arabidopsis, encodes a member of the raf family of protein kinases. Cell. 72: 427-41
28. Larsen PB, and Chang C (2001) The Arabidopsis eer1 Mutant Has Enhanced Ethylene Responses in the Hypocotyl and Stem. Plant Physiol. 125: 1061-73
29. Larsen PB, and Cancel JD (2003) Enhanced ethylene responsiveness in the Arabidopsis eer1 mutant results from a loss-of-function mutation in the protein phosphatase 2A A regulatory subunit, RCN1. Plant J. 34: 709-18
30. Lehman A, Black R, and Ecker JR (1996) HOOKLESS1, an ethylene response gene, is required for differential cell elongation in the Arabidopsis hypocotyl. Cell. 85: 183-94
31. Lorenzo O, Piqueras R, Sanchez-Serrano JJ, and Solano R (2003) ETHYLENE RESPONSE FACTOR1 integrates signals from ethylene and jasmonate pathways in

plant defense. Plant Cell. 15: 165-78
32. Nehring RB and Ecker JR, personal communication
33. Neljubow D (1901) Ueber die horizontale Nutation der Stengel von *Pisum sativum* und einiger Anderer. Pflanzen Beih. Bot. Zentralbl. 10: 128-39
34. Ouaked F, Rozhon W, Lecourieux D, and Hirt HA (2003) MAPK pathway mediates ethylene signaling in plants. EMBO J. 22: 1282-8
35. Potuschak T, Lechner E, Parmentier Y, Yanagisawa S, Grava S, Koncz C and Genschik P (2003) EIN3-dependent regulation of plant ethylene hormone signaling by two Arabidopsis F box proteins: EBF1 and EBF2. Cell 115: 679–689
36. Rashotte AM, DeLong A, and Muday GK (2001) Genetic and chemical reductions in protein phosphatase activity alter auxin transport, gravity response, and lateral root growth. Plant Cell. 13: 1683-97
37. Riechmann, J.L. and Meyerowitz EM (1998) The AP2/EREBP family of plant transcription factors. *Biol. Chem.* 379: 633-646
38. Rodriguez FI, Esch JJ, Hall AE, Binder BM, Schaller GE, Bleecker AB (1999) A copper cofactor for the ethylene receptor ETR1 from Arabidopsis. Science. 283: 996-8
39. Roman, G., B. Lubarsky, J.J. Kieber, M. Rothenberg, and Ecker JR (1995) Genetic analysis of ethylene signal transduction in *Arabidopsis thaliana*: Five novel mutant loci integrated into a stress response pathway. *Genetics* 139: 1393-1409
40. Schaller, GE and Bleecker AB (1995) Ethylene-binding sites generated in yeast expressing the Arabidopsis ETR1 gene. Science. 270: 1809-11
41. Schaller, GE and Kieber JJ (Sept 30, 2002) Ethylene, The Arabidopsis Book, eds. C.R. Somerville and E.M. Meyerowitz, American Society of Plant Biologists, Rockville, MD, doi/10.1199/tab.0009, http://www.aspb.org/publications/arabidopsis/
42. Sistrunk ML, Antosiewicz DM, Purugganan MM, and Braam J (1994) Arabidopsis TCH3 encodes a novel Ca2+ binding protein and shows environmentally induced and tissue-specific regulation. Plant Cell. 6: 1553-65
43. Smalle J, Haegman M, Kurepa J, Van Montagu M, and Straeten DV (1997) Ethylene can stimulate Arabidopsis hypocotyl elongation in the light. Proc Natl Acad Sci U S A. 94: 2756-2761
44. Solano R, Stepanova A, Chao Q, and Ecker JR (1998). Nuclear events in ethylene signaling: A transcriptional cascade mediated by ETHYLENE-INSENSITIVE3 and ETHYLENE-RESPONSE-FACTOR1. Genes Dev. 12: 3703–3714
45. Stepanova AN, and Ecker, JR (2000) Ethylene signaling: From mutants to molecules. Curr. Opin. Plant Biol. 3, 353–360
46. Uchida A, and Yamamoto KT (2002) Effects of mechanical vibration on seed germination of Arabidopsis thaliana (L.) Heynh. Plant Cell Physiol. 43: 647-51
47. Vandenbussche F, Smalle J, Le J, Saibo NJ, De Paepe A, Chaerle L, Tietz O, Smets R, Laarhoven LJ, Harren FJ, Van Onckelen H, Palme K, Verbelen JP, and Van Der Straeten D (2003) The Arabidopsis mutant alh1 illustrates a cross talk between ethylene and auxin. Plant Physiol. 131: 1228-38
48. Wang KL, Li H and Ecker JR (2002) Ethylene biosynthesis and signaling networks.Plant Cell. 14 Suppl:S131-51
49. Wang W, Hall AE, O'Malley R, Bleecker AB (2003) Canonical histidine kinase activity of the transmitter domain of the ETR1 ethylene receptor from Arabidopsis is not required for signal transmission. Proc Natl Acad Sci U S A. 100: 352-7
50. Woeste KE and Keiber JJ (2000) A Strong Loss-of-Function Mutation in *RAN1* Results in Constitutive Activation of the Ethylene Response Pathway as Well as a Rosette-Lethal Phenotype. Plant Cell. 12: 443-455
51. Wright AJ, Knight H, and Knight MR (2002) Mechanically stimulated TCH3 gene expression in Arabidopsis involves protein phosphorylation and EIN6 downstream of calcium. Plant Physiol. 128: 1402-9
52. Zhao XC, Qu X, Mathews DE, and Schaller GE (2002) Effect of ethylene pathway mutations upon expression of the ethylene receptor ETR1 from Arabidopsis. Plant Physiol. 130: 1983-91

D5. Ethylene Signal Transduction in Fruits and Flowers

Harry J. Klee and David G. Clark
Plant Molecular and Cellular Biology Program, University of Florida, Gainesville, FL 32611-0690, USA. E-mail: hjklee@ifas.ufl.edu

INTRODUCTION

Persons reading this book will appreciate that phytohormones have vital roles in assimilating many aspects of plant growth and development. While it is well established that hormonal action is regulated at the level of synthesis, many developmental processes are also regulated at the level of hormone perception (8). Sensitivity to hormones is regulated both spatially and temporally during growth and development. For example, adjacent cell layers respond differentially to hormones during organ abscission. Likewise, sensitivity of an organ can change over time as occurs during fruit ripening. How such alterations in hormone responses are regulated is not well understood. However, as the individual components of signaling pathways are identified, a basic understanding of hormone responsiveness is emerging. In the context of this article, the term "sensitivity" refers to the response of a tissue or organ to the hormone. A change in sensitivity indicates that the concentration of hormone that is able to initiate a set response is altered.

ETHYLENE

Ethylene is a simple gaseous hormone with profound effects on many aspects of plant growth and development. In terms of development, ethylene has an essential role in the control of fruit ripening as well as flower fertilization, senescence and organ abscission. These topics will be addressed in detail within this chapter[1]. Some of the other well defined roles for ethylene in such processes as stem elongation and vegetative senescence are discussed in other chapters (D4 and E6 respectively). In addition to its developmental roles, ethylene is an essential component of wide range of responses to biotic and abiotic environmental stresses. Further, many of these stress responses

---

[1]Abbreviations: ACC, 1-aminocyclopropane-1-carboxylic acid; HK, Histidine protein kinase; STK, Serine/threonine protein kinase; AOA, Aminoethoxyacetic acid;
AVG, Aminoethoxyvinylglycine; STS, Silver thiosulfate; 1-MCP. 1-aminocyclopropene.

involve integration of ethylene signaling into more complex circuitry involving, among others, salicylate and jasmonate signaling. It is in this context that we consider ethylene action. Understanding how a plant can use this simple hydrocarbon to control such a diverse set of developmental and environmental responses is particularly challenging. This integration process must involve tight control of both synthesis and perception of the hormone. The focus of this chapter will be principally on the perception of ethylene and evidence supporting the hypothesis that regulation of ethylene signaling is a critical control point.

ETHYLENE AND FRUITS

Climacteric fruits are defined by the concomitant increase in respiration and ethylene synthesis during ripening. Ethylene is an essential component of climacteric ripening and blocking its synthesis or action prevents ripening. Whereas ethylene treatment of immature climacteric fruits hastens the onset of ripening, it has no effect on non-climacteric fruits. Many commercially important fruits including tomato, apple, peach, avocado and papaya are classified as having climacteric ripening. A number of ripening-associated phenomena are intimately tied to ethylene synthesis. Many ripening-associated genes are considered to be ethylene-regulated, encoding enzymes involved in color change, fruit softening, cell wall breakdown, pathogen defense and nutrient composition (1).

Particularly relevant to discussion of developmental control of ripening is the concept of System 1 and System 2 ethylene. This terminology refers to the pattern of ethylene synthesis in immature vs. mature fruits. Immature fruits produce low levels of ethylene and exogenous ethylene treatment does not stimulate further synthesis (System 1). In contrast, System 2 ethylene synthesized by ripening fruits is autocatalytic (55). The difference between the two systems can be explained at a molecular level by the lack of induction of ACC synthase, the rate-limiting enzyme in ethylene synthesis, during System 1, and induction during System 2. There are several interesting aspects of this phenomenon related to ethylene perception. First, immature fruits do recognize and respond to ethylene, as measured by increased expression of certain ethylene-inducible genes. But mature fruits respond in a different way with ethylene induction of a large number of genes. This differential expression is exemplified by the E8 gene (28). E8 is highly ethylene inducible in a dose-dependent manner only in ripening fruits, indicating a clear developmental component to ethylene regulation. All fruits perceive and respond to ethylene but ripening fruits respond in a different manner. This suggests that combinations of transcription factors, both developmental and ethylene-regulated, control expression of some genes.

In addition to temporal control of ripening, there is also spatial control. Fruits do not ripen uniformly. Ripening begins in internal tissue. It then proceeds toward external tissue progressing from the blossom end toward the calyx. Ethylene is a readily diffusible gas within the confines of a fruit. In fact, the skin of a tomato fruit is relatively impermeable to ethylene diffusion and the gas builds up to quite high internal levels throughout the fruit. Thus, the differential spatial ripening within the fruit can only be explained by differential signal transduction.

Another level of complexity in the developmental regulation of ethylene responses relates to the "memory" of immature fruits with respect to ethylene exposure. Although an immature fruit will not initiate ripening upon exposure to exogenous ethylene, that exposure will hasten the onset of ripening (55). Thus tomato, as well as apple and avocado fruits, possess a capacity to measure cumulative ethylene through development. Conversely, inhibition of the low level of ethylene synthesized by pre-climacteric fruits delays the onset of ripening. The mechanisms that mediate this measuring capacity are completely unknown.

Outline of the Signaling Pathway

A critical breakthrough in our understanding of ethylene signal transduction was the recognition that germination of seeds in the presence of ethylene could be used as a screen for insensitive mutants (4). This assay is elegant in its simplicity. Seedlings of many dicotyledonous plant species germinated in the dark grow tall and spindly. In the presence of ethylene they undergo the so-called triple response; relative to air-grown controls, hypocotyls and roots are shortened and thickened and the apical tip exhibits an exaggerated hook. Mutants in ethylene responses are significantly taller than the wild type and elongation of a mutant is directly proportional to the loss of ethylene response. The beauty of the assay is that thousands of seeds can be rapidly assayed for defects in ethylene signaling. Using this assay, several groups have extensively screened Arabidopsis for mutants. These mutants, in turn, have defined many of the elements composing the ethylene signaling pathway (3, 5).

Epistatic analysis of ethylene-related mutants has permitted researchers to place the genetic elements in an order. In some cases, physical interactions have been measured *in vitro*, but the proposed pathway is still very much a model and there are likely to be modifications as the model becomes further refined. Nonetheless, the proposed model does provide a framework for experimentation. Briefly, the signaling pathway begins with a family of receptors. In Arabidopsis, there are five distinct proteins, presumed to be functional receptors (described below and in chapter D4). Downstream from the receptors is the Raf-like protein kinase, CTR1. Genetic evidence indicates that CTR1 is a negative regulator, as loss-of-function mutants exhibit constitutive activation of ethylene signaling. Since CTR1 is homologous to a MAPKKK (mitogen-activated protein kinase), there may be roles in ethylene signaling for additional proteins homologous to MAPKK and MAP kinases. The next identified component is EIN2. It has homology to Nramp metal transporters and is likely to be an integral membrane protein. Its mode of action in ethylene signaling has yet to be determined but loss-of-function mutants are completely ethylene insensitive. At the end of the signaling pathway are EIN3 and ERF1. Both are transcription factors and it has been shown that EIN3 binds to the promoter of ERF1. Loss of function mutations in EIN3 are partially ethylene insensitive, very likely because the gene is part of a family of at least three members. ERF1 is a member of a large family of transcription factors. Overexpression of ERF1 in transgenic plants recapitulates a partial constitutive ethylene response, indicating that ERF1 is a positive transcription factor but there must be one or more additional transcription factors involved in the overall ethylene response (29). With this as a background, we will proceed to a more detailed analysis of individual signaling elements as they relate to regulation of ethylene perception. While we will describe the systems as they relate to fruits and flowers, the Arabidopsis paradigm is critical to proper understanding of the regulatory mechanisms in place.

Receptors

The first step in hormone signaling must necessarily start with the receptor. Intuitively, it is logical that a hormone receptor would be a key point of regulation. Thus, we begin discussion of ethylene signaling with receptors. Much of the critical work to define ethylene receptor structure and action has been performed with Arabidopsis. The ethylene receptor, ETR1, was the first protein to be definitively assigned the role of a hormone receptor in plants. It was also the first protein with homology to histidine kinases to be identified in a higher eukaryote (12). ETR1 is homologous to the prokaryotic family of signal transducers known as two-component regulators. In bacteria, the two components, the sensor and the response regulator, modulate responses to a wide range of developmental and environmental stimuli (42). Multiple mutant alleles of *ETR1* have been identified and all of these confer dominant ethylene insensitivity. Both genetic and biochemical data confirm that *ETR1* encodes an ethylene receptor. It functions as a dimer and exhibits copper-mediated high affinity ethylene binding (34, 35). ETR1 is a member of a family consisting of five proteins (*ETR1, ETR2, EIN4, ERS1* and *ERS2*) (5). Ethylene receptor proteins can be structurally separated into three domains. The numbering of the amino acids refers to the ETR1 protein.

> The *sensor* domain (amino acids 1–313) contains three hydrophobic, putative transmembrane stretches. Ethylene binding occurs within this amino terminal hydrophobic region and all of the known ETR1 mutations are located within the hydrophobic stretches. Three of the receptors, EIN4, ETR2 and ERS2 are predicted to have a fourth membrane-spanning domain. Ethylene binding is abolished in several of the ethylene-insensitive mutant proteins (such as *etr1-1*). The amino

terminal domain also contains the amino acids necessary for dimerization and copper binding. Although homodimerization has been demonstrated (35), proof of heterodimer formation is lacking. The first domain also contains a region referred to as GAF, whose function has yet to be established. GAF is defined in terms of conserved secondary, rather than primary structure and is present in a widely disparate set of proteins.

The *kinase* domain (amino acids 314-581) has extensive sequence homology to histidine kinases (HK). There are five sub-domains that define the catalytic core of histidine kinases (H, N, G1, F, G2). While ETR1 and ERS1 contain all of these sub-domains, the other three receptors lack one or more of them. Notably, ETR2 and ERS2 lack the histidine that is autophosphorylated. This histidine is not essential for the dominant ethylene insensitivity conferred by *etr1-1* (49). ETR1 has been shown to have *in vitro* HK activity (18). This is the only ethylene receptor with demonstrated HK activity. We have shown that the other four Arabidopsis receptors actually have serine/threonine kinase (STK) activity (Moussatche et al., manuscript in preparation). This phosphorylation is actually more consistent with a report of STK activity in a tobacco ethylene receptor (54). It is also consistent with the phosphorylation observed in phytochrome, a protein with homology to the ETR1 kinase domain (30). The STK activity would also explain the lack of conservation of the histidine kinase catalytic domains. The ethylene receptor family thus presents an interesting snapshot into an active process of protein evolution. However, linkage of any protein kinase activity to ethylene signal transduction has yet to be demonstrated.

The *receiver* (or *response regulator*) domain (amino acids 582-738). This region has sequence identity to the output portion of bacterial two-component systems and contains an aspartate that is active in phosphorelay in bacterial proteins. In prokaryotes, there are two classes of first-component proteins, those that contain a receiver domain and those that do not. However, there is always a separate protein that is the ultimate receiver in signal transduction. In the yeast SLN1 and some bacterial two-component systems, the receiver domain is integral to transfer of the phosphate to the response regulator. As in bacteria, some members of the plant ethylene receptor family are missing the receiver domain; ERS1 and ERS2 lack it while the other three contain it. That some of these proteins maintain the receiver domain with a high degree of conservation while others completely lack it, suggests an important but undetermined function for this domain.

Based on structural and DNA sequence comparisons, the receptors have been classified as Type 1 or Type 2. The Type 1 receptors, ETR1 and ERS1, have the highest conservation of the histidine kinase elements. Overall sequence comparisons, including intron positioning, support this classification. Despite these structural and functional differences, most of the genetic evidence is consistent with redundant receptor function.

All of the receptor mutants isolated in screens display semidominant ethylene-insensitive phenotypes. Single gene knockouts, in contrast, have no obvious phenotype. Experiments using combinations of receptor knockouts indicate that the receptors act as negative regulators of ethylene responses (22). Single and double loss-of-function mutants do not show an obvious ethylene-related phenotype; the only exception is an *etr1/ers1* double mutant that knocks out the two most highly expressed receptors (49). This mutant and the triple mutants exhibit constitutive ethylene hypersensitivity. A quadruple mutant is more severe yet and does not reach maturity. The model predicts a default state in which the receptor is actively suppressing expression of ethylene-inducible genes. Ethylene binding inactivates that suppression. The most logical explanation is that kinase activity acts to suppress ethylene responses and a receptor incapable of binding ethylene cannot be inactivated. Based on this model, less receptor increases sensitivity to ethylene while more receptor reduces sensitivity. The triple and quadruple mutants respond to basal levels of ethylene constitutively made by plants because it takes less ethylene to inactivate the remaining receptors. An important aspect of the model is that plants should be able to modulate ethylene responses by altering the expression of receptor genes. This is indeed what is observed.

Tomato Ethylene Receptors.

In tomato, six genes encoding ethylene receptors (*LeETR1-6*) have been characterized (27). The structures of the six receptor proteins are presented schematically in Figure 1. The first of these to be cloned encodes the well characterized ripening mutant, *Never ripe* (*Nr*) (51). Like the many Arabidopsis receptor mutants, *Nr* confers dominant ethylene insensitivity. Mutant flowers do not wilt or abscise nor do the fruits ripen (Fig. 2C), demonstrating the absolute requirement for ethylene in these processes. The predicted structures of the tomato receptor family are very similar to those of the Arabidopsis receptors. The tomato receptors are quite divergent, exhibiting less than 50% identity in primary sequence at the extremes. Three receptors have a potential extra amino terminal membrane-spanning domain. Only one receptor, NR, lacks the receiver domain. Three (LeETR4-6) are missing one or more conserved HK domains, thus resembling the Type 2 Arabidopsis receptors. Despite the extensive structural differences between them, all are receptors, as defined by their ability to bind ethylene (F. Rodriguez, A. Bleecker and H. Klee, unpublished). Further, the Type 1 NR can function *in vivo* to complement loss of the Type 2 LeETR4, despite their being less than 50% identical (45). Thus, available evidence is consistent with a structurally divergent but functionally redundant tomato receptor

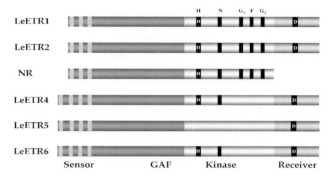

Figure 1. Schematic representation of the six members of the tomato ethylene receptor family. The three or four bars at the amino terminus indicate the locations of potential membrane-spanning domains. The black bars within the kinase domain indicate conserved elements of histidine protein kinases (H, N, G_1, F, G_2). (H), the potentially phosphorylated histidine; (D), potentially phosphorylated aspartate within the receiver domain.

family.

Each tomato receptor gene has a distinct pattern of expression throughout development and in response to external stimuli (27). *LeETR1* and *LeETR2* are expressed at constant levels in all tissues throughout development with *LeETR1* expressed at about five-fold higher level than *LeETR2*. In contrast, expression patterns of the other four genes are highly regulated. For example, during fruit development, ovaries express high levels of *NR* mRNA at anthesis. The level then drops ~10-fold until the onset of ripening whereupon it rises ~20-fold. The ripening-associated rise is an example of developmentally dependent ethylene inducibility, i.e., the gene is ethylene inducible during ripening but not in immature fruit (51). The *LeETR4*, *LeETR5* and *LeETR6* genes are expressed abundantly in reproductive tissues (flowers and fruits) and less so in vegetative tissues. The levels of *LeETR4* and *LeETR5* also increase significantly as fruits mature and ripen. *NR* and *LeETR4*, but not the other genes, are induced by pathogen infection. The pathogen inducibility of *LeETR4* is associated with increased ethylene synthesis in the case of infection with an avirulent pathogen. This induction is an important component of the defense response, functioning to reduce ethylene sensitivity of the infected tissue, thus limiting tissue damage (13) (described below).

Is Receptor Gene Expression Related to Function?

All of the genetic evidence supports a model in which ethylene receptors act as negative regulators. It is in this context that we consider the importance of receptor expression for regulating overall ethylene response. A model for ethylene signaling is outlined in Figure 3. A receptor in the absence of ethylene actively suppresses ethylene-inducible gene expression. For clarity, the receptor is shown as a histidine kinase and the kinase activity is necessary

Ethylene signal transduction in fruit and flowers

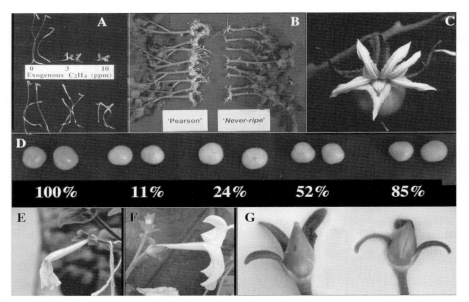

Figure 2 (Color plate page CP10). Top: Ethylene responses in wild type (Pearson) and ethylene-insensitive (*Nr*) tomato plants. A. Seedling triple response of Pearson (top) and *Nr* (bottom) seedlings to increasing levels of exogenous ethylene. B. adventitious root formation in Pearson and *Nr*. C. Failure of *Nr* flowers to senesce and abscise following fertilization. Middle: The effects of reduced *LeEIN3* expression on tomato fruit ripening. Each pair of fruits in the series are from different antisense *LeEIN3* lines. The numbers under each pair of fruits refer to the percent of EIN3 expression. Thus, the pair on th eleft express wild type levels of the three LeEIN3 genes (100%) while the pair to the immediate right express only 11% of wild type levels. Bottom: Long-lived flowers in ethylene-insensitive transgenic petunias. E. Non-transgenic control three days after pollination. F. Transgenic 35S-*etr1-1* flower expressing the dominant ethylene-insensitive Arabidopsis receptor pollinated at anthesis and photographed eight days later. Note the expanding carpel at the base of the corolla. G. Delayed carpel senescence in 35S-*etr1-1* (right) vs. wild type (left) petunia.

established that suppression of ethylene responses is mediated by kinase activity *per se* and it is likely that activity is not mediated by histidine kinase activity alone. This model predicts that there should be an inverse correlation between receptor levels and ethylene sensitivity of a tissue. More ethylene will be required to inactivate high receptor levels than to inactivate low receptor levels. This is why the multiple receptor knockouts in Arabidopsis exhibit constitutive ethylene responsiveness despite unaltered levels of the hormone; basal levels of ethylene synthesis are sufficient to inactivate the full complement of receptors. A further factor to consider is that receptors apparently have a very long half-life for ethylene dissociation. The measured K_D for yeast-expressed ETR1 was approximately 12 hours (34). This K_D is most likely an underestimate since it does not account for protein turnover. Dissociation could be substantially longer than 12 hours and may even be irreversible. Once a receptor has bound ethylene, it cannot repress ethylene-

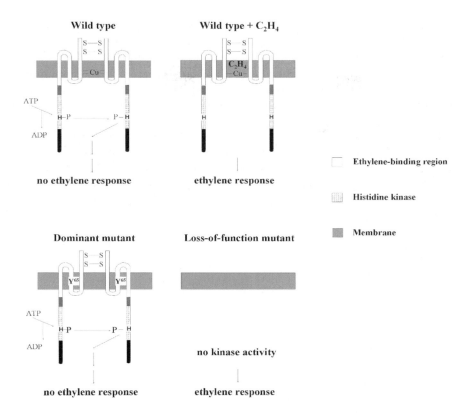

Figure 3. A model for ethylene receptor action. The model is based on the Arabidopsis ETR1, a known histidine kinase. In the absence of ethylene, the receptor actively suppresses downstream ethylene responses. Upon ethylene binding, that suppression is abolished, presumably following an ethylene-mediated conformational change. Mutants are dominant because they do not bind or recognize ethylene and therefore do not undergo the conformational change. Loss-of-function mutants show constitutive ethylene responses because they do not have functional receptors actively suppressing responses.

inducible genes for a long time, if ever. Therefore, the only way that a tissue can rapidly turn off an ethylene response is via synthesis of new receptor. For terminal, irreversible responses such as abscission, flower senescence or fruit ripening, there is no need to reverse the process. But ethylene also mediates a wide range of environmental responses. Many of these responses, such as water stress or wounding, can be transitory and must be terminated after some period of time. A plant must be capable of shutting down the ethylene response, when necessary. A rapid return to normal growth would likely require synthesis of new receptors. It is noteworthy that turnover has not been measured *in planta* and ethylene dissociation kinetics may be different *in vivo*. An accurate measure of binding kinetics and protein turnover in plants is certainly warranted. Although the model is consistent with most available data, it will require further refinement. Several receptors are clearly not histidine kinases and the dominant *etr1-1* mutant effectively suppresses ethylene responses even when the phosphorylated histidine is changed to another amino acid. Thus, its kinase activity may not be essential for suppression of ethylene responses.

With this model in mind, we consider receptor gene regulation in tomato. We have used transgenic plants to test the model developed in Arabidopsis. While most of the data measure only RNA accumulation, we have measured the levels of NR protein with antibodies in both over- and under-expressing lines and there is a good correlation between RNA and protein levels. Plants constitutively over-expressing the wild type *NR* cDNA accumulate more protein and are less sensitive to ethylene. Thus, more receptor leads to reduced ethylene response. Conversely, antisense reduction in expression of most receptors does not affect ethylene sensitivity. This lack of phenotype indicates that there is a degree of redundancy built into the system. There is, however, a notable exception; plants with reduced *LeETR4* expression are severely affected, exhibiting a constitutive ethylene response. The effects, including epinasty, loss of flowers and substantially earlier fruit ripening, occur without any increase in ethylene synthesis. These plants are more sensitive to ethylene, exhibiting a significant reduction in the dose of ethylene required to initiate a biological response. This phenotype is due to functional compensation (45); expression of *LeETR4* increases to compensate for reductions of other receptors. Thus, the overall receptor content of *NR* antisense lines is not substantially affected whereas the receptor content in *LeETR4* antisense lines is substantially reduced.

Increased expression of *LeETR4* in response to pathogen infection is an important aspect of disease response. This gene is induced during the hypersensitive response triggered by infection with *Xanthomonas campestris* pv. *vesicatoria* (13). In antisense lines with greatly reduced *LeETR4* expression, an accelerated hypersensitive response occurs upon infection with the pathogen. Increases in ethylene synthesis and pathogenesis-related gene expression are greater and more rapid in the infected antisense line, indicating a hastened defense response. However, this response has negative consequences since damage to the plant is not restricted to the immediate infection site. If the receptor level does not increase in response to infection, the tissue is overly sensitive to ethylene. Thus, the increase in receptor has the effect of dampening the subsequent ethylene response to limit overall damage to the plant.

There are significant alterations in expression of multiple receptors in tomato, either during development or in response to external stimuli. In every known ethylene response, expression of receptors always increases. Even though reduced expression can increase ethylene sensitivity, there are no examples where this response has been documented to occur. Rather, tomato initiates an ethylene response through the finely tuned system of ethylene synthesis. When an ethylene response is initiated, increased ethylene synthesis is frequently followed by increased receptor synthesis. While it may seem counterproductive to reduce hormone sensitivity shortly after synthesis, this is a typical phytohormone response. A rapid increase in a hormone is followed by induction of mechanisms to inactivate the response. Usually this involves synthesis of enzymes that inactivate the hormone directly. For ethylene, there are no known inactivating enzymes. Indeed, because of its rapid diffusion, there is probably no need for inactivation. In summary, it is entirely normal for a plant to act to reduce a hormone response shortly after it is initiated. Increased ethylene receptor synthesis likely serves this purpose.

Can the patterns of receptor gene expression be reconciled with the known patterns of fruit ethylene perception? Receptor levels are generally high in ovaries at anthesis and decline until the onset of ripening, when there is a large increase that coincides with the climacteric burst. Thus, at the time when ethylene exerts its greatest effect on fruit development, receptor gene expression is at its highest level. This higher rate of receptor synthesis must reduce ethylene responsiveness of the tissue. In the context of receptors as negative regulators of ethylene responses, this ripening-associated increase is paradoxical. However, during fruit ripening ethylene is synthesized far in excess of what is needed to drive the process forward. Any reduction in ethylene sensitivity caused by higher receptor gene expression would be more than offset by climacteric ethylene synthesis.

It is well established that ethylene acts as a clock to regulate the initiation of ripening in tomato and other climacteric fruits. Is there a molecular explanation for this phenomenon? Possibly receptor gene expression and steady state levels of NR protein are relatively low in immature fruits. The limited available data suggest that ethylene binding inactivates receptors for an extended time. If the rate of System 1 ethylene synthesis exceeds the rate of receptor synthesis, there would be an effective depletion of receptors, leading to increased ethylene sensitivity of fruit tissue. Assuming a constant rate of replenishment of receptors in immature fruits, receptors could act as ethylene "clocks". At some point in development, sensitivity to ethylene would rise to a level where ripening could be initiated. Fruits with reduced rates of receptor synthesis should initiate ripening earlier than normal. This is indeed the case. Antisense *LeETR4* fruits, depleted in the most highly expressed receptor RNA, do ripen significantly earlier than controls. While this model remains hypothetical, it is consistent with the available information. How the tomato fruit uses ethylene as a molecular clock is an interesting and important unanswered question that can be addressed with transgenic plants.

The increased expression of ethylene receptors in association with fruit ripening is not limited to tomato. In muskmelon, expression of *CmETR1* increased in parallel with climacteric ethylene synthesis, similar to the tomato *NR* receptor (32). Ripening-associated increases in receptor gene expression are common in climacteric fruits, having been observed in peach, mango, pear and passion fruit.

Downstream Signaling Elements and Fruits

There is less information available about the genes downstream of the receptors in fruits. The available information argues against ripening control at the level of transcription of the downstream components. The first known signaling component downstream of the receptor in Arabidopsis is CTR1. The Arabidopsis gene is constitutively expressed. In contrast, there are at least two genes encoding proteins with significant homology to CTR1 in tomato. One of these, *LeCTR1*, has been shown to functionally complement the Arabidopsis *ctr1* mutation. Since CTR1 is a negative regulator of ethylene responses, its expression would decrease during ripening if it was a key regulatory element in ethylene signaling. In fact, the opposite is true; *LeCTR1* is more highly expressed in ripening fruits than in unripe fruits. Thus, *LeCTR1* behaves much like the ethylene receptors in that its expression goes up in response to ethylene. Transcriptional regulation of this gene cannot explain the observed differential ethylene responsiveness of fruits.

In contrast to the receptors and *LeCTR1*, genes encoding the downstream components of ethylene signaling that have been examined do not show any degree of regulation at the level of transcription. *LeEIN2* appears to be encoded by a single gene, as it is in Arabidopsis. The gene expression is unaltered during fruit development and it is not ethylene inducible. Antisense reduction of expression delays ripening, as would be predicted upon loss of function (Tieman, Ciardi and Klee, unpublished). In the case of EIN3, there is a family of three genes, similar to Arabidopsis (44). Each of these genes was shown to functionally complement the Arabidopsis *ein3* mutation. The tomato genes are functionally redundant and expression of all three must be reduced before ethylene signaling is measurably affected. There is little alteration in gene expression throughout growth and development and none of the genes is ethylene inducible. Thus, the evidence does not support any regulatory roles for *CTR1*, *EIN2* or *EIN3* at the transcriptional level. Presently there is no evidence for post-transcriptional regulation of any component of ethylene signaling, but it certainly cannot be ruled out.

Genetic Engineering

Because of the economic importance and short shelf lives of climacteric fruits, many groups have attempted to extend postharvest life of these fruits by manipulating ethylene synthesis or perception. Since the fruits must ultimately ripen, most efforts have emphasized control of synthesis (26). By reducing ethylene production, ripening is slowed or stopped. When ripe fruits are needed, the ethylene can be added back. Large-scale facilities for gassing fruits such as tomatoes and bananas are widespread and the industry is comfortable manipulating ripening with appropriate postharvest handling techniques. Given the existing handling systems, it makes more sense to manipulate ethylene synthesis rather than perception. However, technically, it is entirely possible to produce fruits with extended shelf life by reducing ethylene sensitivity. The key to such a strategy is to reduce, but not eliminate, ethylene responses. This can easily be accomplished with either dominant receptor mutants (50) or partially reduced expression of downstream genes such as *EIN2* or *EIN3* (44) (Fig. 2D). Although less sensitive to ethylene than wild type, these fruits can be fully ripened by ethylene treatments.

Although tomato has been the model for much delayed ripening work, there is currently no genetically engineered, extended shelf life tomato in commercial production. However, the technology clearly works well. This paradox is explained because of the widespread commercial adoption of *rin* (*r*ipening *in*hibited) hybrids. These non-engineered mutants have comparable shelf life to the engineered varieties. Thus, producers have no incentive to use genetically engineered tomatoes. However, it is clear that ethylene technology is directly applicable to any climacteric fruit. Much effort is being placed into transformation of highly perishable tropical fruits in particular. It is only a matter of time before we see commercial introduction of engineered extended shelf life fruits.

FLOWERS

Flowers – An introduction

The main goal of all higher plants is to successfully complete development, dispersing offspring in the form of seeds. Angiosperms have evolved a diversity of flowering mechanism to ensure reproductive success. There are many forms of floral structures and a wide variety of means by which pollination is facilitated. Flowers serve as a food source by providing nectar to pollinators, attracting pollinators by both visual (color and form) and olfactory (scent) stimuli. Since ethylene is known to be involved in so many other plant processes, it is not surprising that it is involved in multiple aspects of floral development from flower initiation through senescence. Most of the research with respect to ethylene has focused on induction or acceleration of premature flower or petal senescence and abscission.

Flower petals have evolved to attract pollinators; after pollination, floral attraction is no longer needed. Once pollinated, flowers of many different plant species change in many different ways, often mediated by ethylene (41, 46). After pollination or ethylene treatment, some flowers close their corollas, while others change colors. In some species such as carnation, petunia and orchids, petals wilt in response to pollination or ethylene. In other species such as *Pelargonium* and *Digitalis*, petals abscise after pollination or ethylene treatment (17). Surveys of large numbers of plant genera have shown that, in general, species displaying flower and/or petal abscission are usually highly sensitive to ethylene (48). Flower abscission occurs in both monocots and dicots, while petal abscission is rare in monocots and common in dicots. Petal wilting or senescence can be either ethylene-sensitive or ethylene-insensitive, and both types can be found in both monocots and dicots.

In higher plants, flowers that synthesize ethylene are usually responsive to it. As in fruits, many flowers synthesize basal levels of ethylene (System 1), then shift to an autocatalytic mode of ethylene synthesis (System 2). In both fruits and flowers, this shift requires induction of genes encoding ethylene biosynthetic enzymes (25, 53). The shift to autocatalytic ethylene synthesis also requires that tissues become increasingly sensitive to ethylene, and many fruits and flowers become more sensitive to ethylene as they mature.

Although there are some exceptions, most plant species in which pollination accelerates petal senescence or abscission tend to be ethylene-sensitive. With such a wide range of possible changes during development and after pollination, advancements in our understanding of the role of ethylene in floral processes has depended on research conducted on a wide range of plant species. Many gains are being made in the area of flower abscission in *Arabidopsis* because its complete genome has been unraveled, and there are a plethora of mutants and transgenic plants with altered flower abscission available (31, 43). However, *Arabidopsis* is not ideal both because of its small size and because it abscises fully turgid petals in response to ethylene and pollination. Species such as carnation, orchid and petunia have proven to be better experimental models for evaluation of floral development and senescence programs. These species have visually pronounced post-pollination responses that clearly involve ethylene, their physiology is well studied, and their flowers are large enough to provide reasonable amounts of experimental tissue.

Roles of Ethylene in Pollination and Senescence

Pollination initiates the transition of floral function from one of facilitating pollination to supporting fruit and seed development. Successful pollination signals the end of petal function usually leading to their senescence or abscission. Petals on all flowers eventually senesce or abscise, whether pollination occurs or not. In some species, pollination accelerates petal senescence, while in others it does not. In species where pollination affects flower color or form, treatment with exogenous ethylene usually produces the same visible symptoms (47). Since ethylene can substitute for pollination in initiating and accelerating changes in flowers, and since pollination induces ethylene synthesis, ethylene has long been thought to participate in the inter-organ communication between pollinated stigmas and senescing corollas. For ethylene sensitive flowers, changes in synthesis and sensitivity to the hormone are equally important in the regulation of senescence and abscission. In plants such as *Pelargonium*, petals on freshly opened flowers do not abscise after exposure to ethylene at high concentrations (100 μL L^{-1}), while petals 2-3 days post-anthesis abscise after exposure to ethylene concentrations as low as 0.5 μL L^{-1} (17). In these plants, a short burst of ethylene within minutes after pollination is sufficient to cause complete petal abscission within two hours (14, 17). Other plants such as petunia, carnation, and orchid, become more sensitive to ethylene as they age; they become competent to respond to ethylene at the stage of development that is optimal for pollination and continue to gain sensitivity as they progress towards natural senescence. Concomitant with this increase in ethylene sensitivity, they gain the ability to synthesize ethylene, whether they are senescing naturally or as a result of pollination. In these plants, two or three phases of pollination-induced ethylene synthesis from the gynoecium are required to induce visible changes in flowers. In all of these species, the first peak of ethylene synthesis is short-lived, and is usually detectable within minutes after pollination. In *Petunia inflata*, this initial burst of ethylene has been shown to promote early pollen tube growth in the pistil (21). Later, higher levels of ethylene are produced in response to fertilization of the ovaries, resulting in autocatalytic ethylene production in ovaries and/or petals. It is this sustained production of ethylene by the gynoecium, that is critical to pollination-induced petal senescence.

Despite much research on ethylene's role in post-pollination phenomena, the precise mechanisms by which pollen-borne signals initiate this process are unknown. Several candidates for pollen-borne signal that leads to the early burst of ethylene in styles of pollinated flowers have been proposed. Initially, ACC found in tobacco and petunia pollen was proposed to be a primary pollen signal feeding the early burst of ethylene synthesis in the style. However, this idea has been dismissed (52). Pollen-borne auxin or physical stress generated in the transmitting tract by the growing pollen tube have also been suggested as candidates for initiating post-pollination changes in flowers (39, 56). It is likely that the nature of the agents that initiate signals from the stigma to the petals are not universal among plant species. Regardless of the identity of the pollen signals, one of the main consequences is induction of endogenous ethylene synthesis, which apparently coordinates the development of multiple floral organs.

The experimental systems receiving the most attention with regard to the role of ethylene in petal senescence and abscission have been orchid, carnation, and to a lesser extent, Arabidopsis and petunia. Arabidopsis displays programmed floral organ abscission in a way that is similar to other model abscission systems. Addition of exogenous ethylene up-regulates cell wall hydrolytic enzymes in petal abscission zones, and accelerates abscission. Addition of auxin suppresses these activities as well as abscission. Knockout mutants and transgenic plants with altered floral abscission have been generated and they are proving valuable in advancing the field of abscission. Antisense ACS plants by (16) were observed to have reduced ethylene synthesis, which led to the delay of abscission ripening and senescence. Delayed floral abscission has been observed in several different genetic mutants. These plants have been shown to have mutations in hormone responsive genes, pathogen response genes, cell wall associated genes, MADS-box transcription factor genes, and several genes with unknown function (31, 43). Ethylene response mutants such as *etr1*, *ein2*, *ein3* and *ers2* show changes in the progression of floral organ abscission (6). In most of these mutants, floral organ abscission is delayed, but once initiated, proceeds normally.

Although much progress has been made on understanding auxin signaling in Arabidopsis, studies pertaining to the involvement of auxin in floral organ abscission are lacking. Nonetheless, ethylene and auxin have long been known to have an antagonistic relationship as regulators of abscission (43). Factors that affect supply of auxin to abscission zones can affect sensitivity of the abscission zone to ethylene. Conversely, ethylene can inhibit auxin transport while increasing sensitivity of abscission zone cells to itself. Auxin can also stimulate ethylene synthesis and subsequently accelerate flower and petal abscission. Since floral organ diversity is the norm among plant species, it is no wonder that the hormonal factors affecting abscission are multifactorial, and encompass both increases and declines in synthesis of and sensitivity to auxin and ethylene. Interestingly, the Arabidopsis loss of function mutant *nph4*, with reduced auxin-dependent changes in gene expression, has a normal phenotype restored by the application of ethylene (20). This would suggest that ethylene enhances the sensitivity or activity of at least part of the auxin signaling system. It is likely that future experiments directed toward the elucidation of auxin and ethylene intercommunication will require both biochemical and genomic approaches, simply due to the number of possible factors involved in any given response scenario.

Elegant models for ethylene biosynthesis after pollination have been described for orchid and carnation, but information on expression of ethylene receptors and signal transduction genes is just beginning. Thus the precise role of ethylene in post-pollination events of these plants is still unknown. The models for ethylene biosynthesis in orchid and carnation both predict that pollen-borne compounds initiate ethylene synthesis in the stigma, and that *de novo* autocatalytic ethylene synthesis coordinates post-pollination events in the flower. The most complete model for ethylene biosynthesis in orchids was proposed by Bui and O'Neill (9). This model, shown in Figure 4, predicts that pollen-borne signals (including auxin) induce *Phal-ACS2* expression in the stigma, leading to increased ACS activity and ACC synthesis. This ACC is then oxidized by constitutive ACO, leading to the production of ethylene, which induces the *Phal-ACS1* and *Phal-ACO1* genes, leading to autocatalytic synthesis. ACC and/or ethylene are then translocated from the stigma to the labellum and perianth. In the labellum, *Phal-ACS1* and *Phal-ACO1* expression are induced, leading to autocatalytic ethylene production and subsequent senescence of the labellum. In the perianth, ACC from the stigma, ovary and labellum are oxidized by ACO to ethylene, leading to senescence of the perianth. In the ovary, pollen borne auxin and other unknown factors induce *Phal-ACS3* expression, leading to increased ACS activity and ACC accumulation. Due to a low level of ACO, and subsequent low levels of ethylene production, it has been suggested that the ovary may serve as a potential ACC source for the perianth. Unfortunately, little is known about regulation of ethylene and auxin signaling genes in orchids. Once this information is accumulated, a complete model for the role of ethylene in orchid flower senescence should be possible.

In carnation, the pollination signal produced by the stigma/style is translocated through the ovary (gynoecium) to the petals, where it induces autocatalytic ethylene synthesis responsible for petal senescence (23, 24, 37). Along with transport of the signal, sequential expression of ACS and ACO

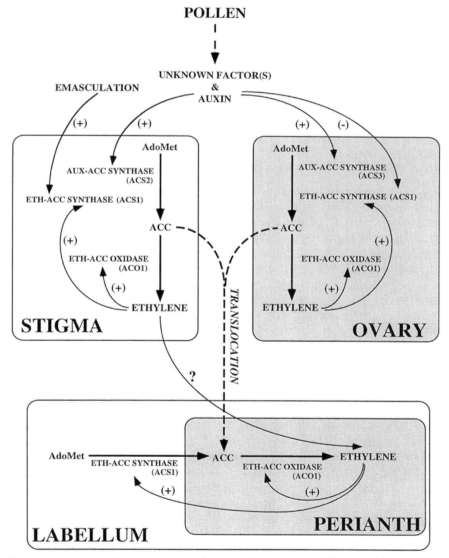

Figure 4. A model of interorgan regulation of ACC synthase (ACS) and oxidase (ACO) gene expression in pollinated orchid flowers. ETH, ethylene-induced; AUX, auxin-induced. Reprinted from (10) with permission.

genes occurs in the respective floral organs (23). The primary signal for petal senescence is ethylene, and ACC may act as a secondary stimulus (24). The ethylene produced in the gynoecium acts as a diffusible signal and is perceived by the petals, resulting in ACS and ACO expression, and subsequent autocatalytic ethylene production and rapid petal wilting. Recently, several members of the carnation ethylene receptor gene family were cloned and characterized (38). Two of these receptors appear to be active in floral senescence, both are spatially regulated, and both show decreases in mRNA accumulation after exogenous ethylene treatment. It is not known whether the down regulation of these genes is a result of a decrease in transcription or whether it is due to stimulation of mRNA degradation due to programmed cell death factors, but reduced expression could potentially lead to increased sensitivity. Unfortunately, the role of ethylene perception and signaling in pollination-induced senescence is still very much unknown. In recent years, several groups have isolated ethylene receptor and ethylene signaling genes from several flowering species, but they have yet to be placed in the most advanced models.

Control of Ethylene Biosynthesis and Sensitivity in Flowers

Much effort has been focused on ways to diminish the detrimental effects of ethylene on the display life of cut flower and potted plants used in the ornamental industry. Abscised or wilted flowers reduce the visual impact of these plants, and can also increase the incidence of problems associated with saprophytic pathogens. Reduced vase life can be caused by endogenous ethylene that is produced and perceived in the flowers, or ethylene generated from other sources contained in the same transit or storage containers. Regardless of the source, ethylene is detrimental to product quality, and significant means are undertaken by the industry to avoid its effects.

Several methods of chemical control have been utilized to keep flowers from making or perceiving ethylene. Chemicals such as aminoethoxyacetic acid (AOA) and aminoethoxyvinylglycine (AVG) significantly reduce ethylene biosynthesis and delay floral senescence and abscission. However, these chemicals have not received much commercial attention because they are expensive and do not alter ethylene sensitivity, and thus are ineffective in preventing the effects of exogenous ethylene during storage and transport. Cameron and Reid (10) first described the use of silver as an effective means to render flowers insensitive to ethylene. Applied as a spray, or pulsed into stems through vase solutions, silver thiosulfate (STS) effectively binds to ethylene receptors preventing floral senescence and abscission. Unfortunately, STS is a heavy metal pollutant, and in recent years there have been rising concerns about its potential as a groundwater contaminant. Additionally, STS is classified by the EPA as a growth regulator and is not registered for legal use in the US ornamental plant industry. Even with these limitations, STS has been the industry standard for blocking the effects of ethylene on flowers for almost two decades, and is widely used in several parts of the world today.

In the mid 1990's, a new chemical, 1-methylcyclopropene (1-MCP) was developed as a gaseous, non-toxic alternative to STS (36). It works by irreversibly binding to ethylene receptors, thus preventing ethylene binding (40). Despite the increased effort required to commercially apply a gas (1-MCP) instead of a spray (STS) to flowers, it has been widely adopted for use in the ornamental plant industry. 1-MCP effectively reduces abscission or senescence in many plant species as well as climacteric fruit ripening. However, 1-MCP apparently does not control senescence for as long as STS does. Following application, STS stays within the tissues and continues to bind to newly synthesized receptors. Being a gas, 1-MCP blocks the receptors that are present in tissues at the time of treatment, but once removed, newly synthesized receptors are not protected from ethylene. Thus, for example, 1-MCP application to ripening tomato fruits will completely stop the ripening process for 4-5 days, whereupon it will resume. The transient nature of the benefits achieved with 1-MCP may limit its commercial utility in some plants, especially ones that rapidly synthesize new receptors. Begonias treated with 1-MCP were initially protected against ethylene treatment, but showed no real increases in display life in an interior environment (36). Similar observations have been reported for geraniums (11). In this crop, multiple applications of 1-MCP are required to significantly maintain ethylene insensitivity at a level that allows for reduced petal abscission over an extended period of time.

The most obvious way to circumvent the issues associated with chemical control of ethylene action in flowers is through genetic control. Much research has been conducted to produce transgenic floriculture crops that produce less ethylene, or are less sensitive to it. Transgenic antisense ACO carnation (33) and torenia plants (2) produce significantly less ethylene, and have delayed flower senescence. As was the case with chemical approaches to control the effects of ethylene in ornamental crops, the method of control receiving the most research attention in terms of genetic engineering has been at the level of ethylene sensitivity. Wilkinson et al. (50) transformed petunia with a dominant mutant *Arabidopsis* ethylene receptor, *etr1-1*, under the control of a constitutive Cauliflower Mosaic Virus 35S promoter to produce ethylene-insensitive plants. Flowers on these plants had delayed natural and pollination-induced flower senescence compared to wild-type plants (Fig. 2E and F), but had physiological side effects that would ultimately limit their commercial use (19, 50). These plants showed a significant reduction in adventitious root formation that could not be reversed by exogenous treatments with auxin (15). Clearly the key to manipulation of ethylene sensitivity lies in the promoters used to drive their transcription. Effective temporal and spatial control of transgene expression will be required to produce plants that have longer lasting flowers with no negative side effects that would limit commercial production. For example, Bovy et al. (7) produced transgenic ethylene-insensitive carnations by driving the *etr1-1* transgene with the flower specific transcriptional promoter fbp1 from petunia. These plants have a delayed flower senescence phenotype, but an extensive investigation of horticultural performance characteristics was not reported. Future experiments directed toward discovering the appropriate promoter to drive transgenes conferring ethylene insensitivity will likely be the key to success for the development of long lasting flowers in the future. It will be imperative to conduct field and greenhouse trials of these plants to eliminate the possibility of negative side effects.

CONCLUSIONS

Ethylene clearly has a major role in controlling ripening in fruits and senescence in flowers. The effects of ethylene on these developmental processes are regulated both at the level of synthesis and perception of the hormone. While the mechanisms regulating ethylene synthesis are well established, the mechanisms regulating differential perception during development are not well understood. Receptor synthesis is highly regulated and alterations in receptor levels do affect the overall sensitivity of tissues to ethylene. But the mechanisms regulating the altered perception are far from understood.

References

1. Abeles FB, Morgan PW, Saltveit ME (1992) Ethylene in plant biology. 2 ed. San Diego: Academic Press.
2. Aida R, Yoshida T, Ichimura K, Goto R, Shibata M (1998) Extension of flower longevity in transgenic torenia plants incorporating ACC oxidase transgene. Plant Sci. 138:91-101.
3. Alonso JM, Ecker JR (2001) The ethylene pathway: a paradigm for plant hormone signaling and interaction. Science's stke www.stke.org/content/full/OC-sigtrans;2001/70/re1
4. Bleecker A, Estelle M, Somerville C, Kende H (1988) Insensitivity to ethylene conferred by a dominant mutation in *Arabidopsis thaliana*. Science 241:1086-9.
5. Bleecker A, Kende H (2000) Ethylene: a gaseous signal molecule in plants. Ann. Rev. Cell Dev. Biol 16:1-18.
6. Bleecker AB, Patterson SE (1997) Last exit: senescence, abscission, and meristem arrest in Arabidopsis. Plant Cell 9:1169-1179.
7. Bovy AG, Angenent GC, Dons HJM, van Altvorst AC (1999). Heterologous expression of the Arabidopsis etr1-1 allele inhibits the senescence of carnation flowers. Molecular Breeding 5:301-308.
8. Bradford K, Trewavas A (1994) Sensitivity thresholds and variable time scales in plant hormone action. Plant Physiol. 105:1029-36.
9. Bui AQ, O'Neill SD (1998) Three 1-aminocyclopropane-1-carboxylate synthase genes regulated by primary and secondary pollination signals in orchid flowers. Plant Physiol. 116:419-428.
10. Cameron AC, Reid MS (1982) The use of silver thiosulfate as a foliar spray to prevent flower abscission from potted plants. Sci. Hort. 19:373-378.
11. Cameron AC, Reid MS (2001) 1-MCP blocks ethylene-induced petal abscission of Pelargonium peltatum but the effect is transient. Postharvest Biol. Technol. 22:169-177.
12. Chang C, Kwok SF, Bleecker AB, Meyerowitz EM (1993) *Arabidopsis* ethylene-response gene *ETR1*: similarity of products to two-component regulators. Science 262:539-44.
13. Ciardi JA, Tieman DM, Jones JB, Klee HJ (2001) Reduced Expression of the Tomato

Ethylene Receptor Gene LeETR4 Enhances the Hypersensitive Response to *Xanthomonas campestris* pv. *vesicatoria*. Molecular Plant-Microbe Interactions 14:487-95.
14. Clark DG, Richards C, Hilioti Z, Lind-Iversen S, Brown K (1997) Effect of pollination on accumulation of ACC synthase and ACC oxidase transcripts, ethylene production and flower petal abscission in geranium (*Pelargonium Xhortorum* L.H. Bailey). Plant Mol. Biol. 34:855-865.
15. Clark DG, Gubrium EK, Klee HJ, Barrett JE, Nell TA (1999). Root formation in ethylene insensitive plants. Plant Physiol. 121:53-59.
16. Ecker JR, Theologis A (1994) Ethylene: a unique plant signaling molecule. In E.M. Meyerowitz, C.R. Somervillle, eds., Arabidopsis. Cold Spring Harbor Press, Plainview, NY, pp 485-521.
17. Evensen K (1991) Ethylene responsiveness changes in Pelargonium x domesticum florets. Physiol. Plant. 82:409-412.
18. Gamble R, Coonfield M, Schaller GE (1998) Histidine kinase activity of the ETR1 ethylene receptor from Arabidopsis. Proc.Nat.Acad.Sci.USA 95:7825-9.
19. Gubrium EK, Clark DG, Barrett JE, Nell TA (2000) Reproduction and horticultural performance of transgenic ethylene insensitive petunias. J. Amer. Soc. Hort. Sci. 125:277-281.
20. Harper RM, Stowe-Evans EL, Luess DR, Muto H, Tatematsu K, Watahiki MK, Yamamoto K, Liscum E (2000) The NHP4 locus encodes the auxin response factor ARF7, a conditional regulator of differential growth in aerial Arabidopsis tissue. Plant Cell 12:757-770.
21. Holden MJ, Marty JA, Singh-Cundy A (2003) Pollination-induced ethylene promotes the early phase of pollen tube growth in *Petunia inflata*. J. Plant Physiol. 160:261-269.
22. Hua J, Meyerowitz EM (1998) Ethylene responses are negatively regulated by a receptor gene family in *Arabidopsis thaliana*. Cell 94:261-71.
23. Jones ML, Woodson WR (1997) Pollination-induced ethylene in carnation: role of stylar ethylene in corolla senescence. Plant Physiol. 115:205-212.
24. Jones M, Woodson WR (1999) Inter-organ signaling following pollination in carnations. J. Amer. Soc. Hort. Sci. 124:598-604.
25. Kende H (1993) Ethylene biosynthesis. Ann. Rev. Plant Physiol. 44:283-307.
26. Klee HJ, Clark DG (2002) Manipulation of ethylene synthesis and perception in plants: The ins and the outs. Hortscience 37:450-452.
27. Klee HJ, Tieman DM (2002) The tomato ethylene receptor gene family: form and function. Physiol. Plant. 115:336-341.
28. Lincoln J, Fischer R (1988) Diverse mechanisms for regulation of ethylene-inducibel gene expression. Molec. Gen. Genet. 212:71-75.
29. Lorenzo O, Piqueras R, Sanchez-Serrano J, Solano R (2003) ETHYLENE RESPONSE FACTOR 1integrates signals from ethylene and jasmonate pathways in plant defense. Plant Cell 15:165-178.
30. Montgomery BL, Lagarias JC (2002) Phytochrome ancestry: sensors of bilins and light. Trends Plant Sci. 7: 357-366.
31. Patterson SE (2001) Cutting loose. Abscission and dehiscence in Arabidopsis. Plant Physiol. 126:494-500.
32. Sato-Nara K, Yuhashi K, Higashi K, Hosoya K, Kubota M, Ezura H (1999) Stage- and tissue-specific expression of ethylene receptor homolog genes during fruit development in muskmelon. Plant Physiol. 120:321-329.
33. Savin KW, Baudinette SC, Graham MW, Michael MZ, Nugent GD, Lu CY, Chandler SF, Cornish EC (1995) Antisense ACC oxidase RNA delays carnation petal senescence. HortScience 30:970-972.
34. Schaller GE, Bleecker AB (1995) Ethylene binding sites generated in yeast expressing the *Arabidopsis ETR1* gene. Science 270:1809-11.
35. Schaller GE, Ladd AN, Lanahan MB, Spanbauer JM, Bleecker AB (1995) The ethylene response mediator ETR1 from *Arabidopsis* forms a disulfide-linked dimer. J. Biol. Chem. 270:12526-30.

36. Serek M, Sisler EC, Reid MS (1994) Novel gaseous ethylene binding prevents ethylene effects in potted flowering plants. J. Amer. Soc. Hort. Sci. 119:1230-1233.
37. Shibuya K, Yoshioka T, Hashiba T, Satoh S (2000) Role of the gynoecium in natural senescence of carnation (*Dianthus caryophyllus* L.) flowers. J. Exp. Bot. 51:2067-2073.
38. Shibuya K, Nagata M, Tanikawa N, Yoshioka T, Hashiba T, Satoh S (2002) Comparison of mRNA levels of three ethylene receptors in senescing flowers of carnation (*Dianthus caryophyllus* L.). J. Exp. Bot. 53:399-406.
39. Singh A, Evensen KB, Kao T-H (1992) Ethylene synthesis and floral senescence following compatible and incompatible pollinations in Petunia inflata. Plant Physiol. 99:38-45.
40. Sisler EC, Dupille E, Serek M (1996) Effect of 1-methylcyclopropene and methylene-cyclopropane on ethylene binding and ethylene action on cut carnations. Plant Growth Regul. 18:79-86.
41. Stead AD (1992) Pollination-induced flower senescence: a review. Plant Growth Regulation 11:13-20.
42. Stock AM, Robinson VL, Goudreau PN (2000) Two-component signal transduction. Ann. Rev. Biochem. 69:183-215.
43. Taylor JE, Whitelaw CA (2001) Signals in abscission. New Phytol. 151:323-339.
44. Tieman DM, Ciardi JA, Taylor MG, Klee HJ (2001) Members of the tomato *LeEIL* (*EIN3-like*) gene family are functionally redundant and regulate ethylene responses throughout plant development. Plant J. 26:47-58.
45. Tieman DM, Taylor MG, Ciardi JA, Klee HJ (2000) The tomato ethylene receptors NR and LeETR4 are negative regulators of ethylene response and exhibit functional compensation within a multigene family. Proc. Nat. Acad. Sci. USA 97:5663-8.
46. van Doorn W (1997) Effects of pollination on floral attraction and longevity. J. Exp. Bot. 314:1615-1622.
47. van Doorn WG (2002a) Does ethylene treatment mimic the effects of pollination on floral lifespan and attractiveness? Ann. Bot. 89:375-383.
48. van Doorn WG (2002b) Effect of ethylene on flower abscission: a survey. Ann. Bot. 89:689-693.
49. Wang WY, Hall AE, O'Malley R, Bleecker, A. (2003) Canonical histidine kinase activity of the transmitter domain of the ETR1 ethylene receptor from Arabidopsis is not required for signal transmission Proc. Natl. Acad. Sci. USA 100:352-357.
50. Wilkinson JQ, Lanahan MB, Clark DG, Bleecker AB, Chang C, Meyerowitz EM, Klee HJ (1997) A dominant mutant receptor from *Arabidopsis* confers ethylene insensitivity in heterologous plants. Nature Biotechnol. 15:444-447.
51. Wilkinson JQ, Lanahan MB, Yen H-C, Giovannoni JJ, Klee HJ (1995) An ethylene-inducible component of signal transduction encoded by *Never-ripe*. Science 270:1807-9.
52. Woltering EJ, Somhorst D, van der Veer P (1995) The role of ethylene in interorgan signaling during flower senescence. Plant Growth Regul. 12:1-10.
53. Woodson WR (1994) Molecular biology of flower senescence in carnation. In: Scott, R.J., Stead, A.D. (eds.) Molecular and cellular aspects of plant reproduction (Society for experimental biology seminar series, vol 55). Cambridge University Press, pp. 255-267.
54. Xie C, Zhang JS, Zhou HL, Li J, Zhang ZG, Wang DW, Chen SY (2003) Serine/threonine kinase activity in the putative histidine kinase-like ethylene receptor NTHK1 from tobacco. Plant J. 33:385-393.
55. Yang SF (1987) The role of ethylene and ethylene synthesis in fruit ripening. In: Thompson W, Nothnagel E, Huffaker R, eds. Plant Senescence: Its Biochemistry and Physiology. Rockville, MD: The American Society of Plant Physiologists; pp 156-165.
56. Zhang XS, O'Neill SD (1993) Ovary and gametophyte development are coordinately regulated by auxin and ethylene following pollination. Plant Cell 5:403-418.

D6. Abscisic Acid Signal Transduction in Stomatal Responses

Sarah M. Assmann
Biology Department, Penn State University, 208 Mueller Laboratory, University Park, PA, 16802-5301 USA. E-mail: sma3@psu.edu

INTRODUCTION

ABA plays major roles in the regulation of several important plant processes. ABA is important for plant acclimation to drought, cold, and salinity, for development of seed dormancy, and for inhibition of seed germination. ABA at higher concentrations inhibits root growth, but detailed studies have also shown that some ABA is required to maintain root elongation during stress (54). ABA also affects the transcription of a large number of genes (48).

The focus of the present chapter[1] is on ABA regulation of a specific cell type, the stomatal guard cell (2, 4, 17, 45, 52, 53). Pairs of guard cells in the aerial epidermes of terrestrial plants define and regulate the width of microscopic pores through which plants take up carbon dioxide for photosynthesis, but also, inevitably, lose water through evapotranspiration (Fig. 1). In

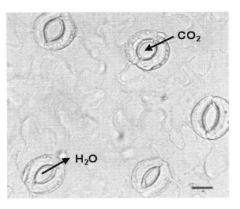

Figure 1. Epidermis of *V. faba*, illustrating pavement epidermal cells and pairs of guard cells surrounding stomatal pores. Loss of water vapor and uptake of CO_2 occurs through the stomata.

---

[1] Abbreviations: ABC, ATP-binding-cassette transmembrane regulator; cADPR, Cyclic ADP ribose; CDPK, calcium-dependent protein kinase; CPTIO, 2-carboxyphenyl-4,4,5,5-tetramethylimidazoline -1-oxyl-3-oxide; DAG, diacylglycerol; DMS, N,N-dimethylsphingosine; GSNO, S-nitrosoglutathione; GPCR, G-protein coupled receptors; $InsP_3$, inositol 1,4,5 trisphosphate; $InsP_6$, myo-inositol hexakisphosphate; L-NAME, NG-nitro-L-arginine methyl ester; PIP_2, phosphatidylinositol bisphosphate; PI3P, phosphatidylinositol 3-phosphate; PI4P, phosphatidylinositol 4-phosphate; PLC, phospholipase C; PLD, phospholipase D; PP2A, protein phosphatase 2A; PP2C, protein phosphatase 2C; ROS, reactive oxygen species; SIP, sphingosine-1-phosphate; SNP, sodium nitroprusside; snRNP, small nuclear ribonucleoprotein; SV, slow vacuolar.

response to ABA, guard cells respond so as to narrow stomatal apertures, thereby reducing transpirational water loss (Fig. 2). These responses can occur quite rapidly, probably without changes in transcription, while longer-term drought episodes also alter gene expression.

In this chapter, ABA delivery to and perception by the guard cell is first briefly described. The mechanisms by which ABA evokes changes in guard cell volume are summarized, followed by an overview of the multitude of guard cell signal transduction pathways implicated in transduction of the ABA signal. Emerging evidence for a role of ABA in RNA metabolism is then discussed. The chapter ends with a discussion of the biotechnological potential of manipulating guard cell ABA responses, and a list of unanswered questions and directions for future research.

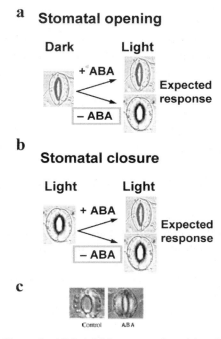

Figure 2. ABA inhibits stomatal opening (a) and promotes stomatal closure (b). (c): The ABA response of *V. faba* stomata (c from 23).

CELL-SPECIFIC APPROACHES IN GUARD CELL RESEARCH.

A strength of whole-plant research on ABA effects on stomatal conductance is the variety of plant species that have been assayed, including both the model plant species, Arabidopsis, and crop species such as cotton, maize, and sunflower. Many of these species are not, however, amenable to methods used in cell-specific molecular and biochemical analyses of guard cell function, which require isolation of large amounts of pure guard cells. Until recently, *Vicia faba* and tree tobacco were the only species for which such large scale methods had been developed; such protocols are now also available for Arabidopsis (46) (Fig. 3). Given the favorable characteristics of Arabidopsis with regard to genetic manipulation, it is rapidly becoming a system of choice for guard cell physiologists.

The starting point for the isolation of pure guard cell material is epidermal peels, i.e. strips of the epidermis isolated from the rest of the plant. In addition to *V. faba*, tree tobacco, and Arabidopsis, only a few species have been found in which the epidermis readily detaches from the interior mesophyll. Thus, most of the cell biological research on guard cell responses to date has been conducted on the species named abov Commelina communis, and the *argenteum* mutant of pea. C. comm. *V. faba* have been favored by guard cell researchers because the larg the guard cells in these species renders them particular amei measurements of stomatal aperture in isolated epidermal peels, and to microinjection techniques. *V. faba* has proven to be the easiest species on which to obtain electrophysiological data, followed by Arabidopsis. Electrophysiological techniques include patch clamp measurements on isolated guard cell protoplasts (Fig. 3h), recording from guard cells in epidermal peels impaled with double-barreled electrodes, and impalement studies on guard cells *in situ* in the intact leaf (49).

ABA DELIVERY TO AND PERCEPTION BY GUARD CELLS

Reductions in atmospheric humidity result in an increased driving force for plant water loss. Guard cells respond so as to close stomata, thereby limiting evapotranspiration. Transduction of the humidity signal does not require ABA sensitivity, as stomatal closure in response to decreases in humidity occurs normally in the Arabidopsis ABA insensitive mutants, *abi1-1* and *abi2-1*, and in the ABA biosynthetic mutant, *aba1* (reviewed in 45, 57). However, if atmospheric drying is sufficiently severe so as to decrease leaf water potential, ABA signaling comes into play. Cellular dehydration increases bulk leaf ABA concentrations, and acidifies cytosolic pH, thereby favoring exit to the apoplast of increased quantities of ABA in the lipid-permeant, ABAH form (57). In addition, guard cells themselves are a site of ABA synthesis (45).

Upon soil drying, either root-sourced chemical signals or altered shoot water potential may serve as signals that restrict stomatal opening. The extent to which each of these processes function may be species specific (57), and ABA is not the only chemical messenger signaling soil dehydration. However, there is little doubt that ABA plays a central role in

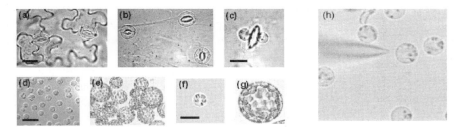

Figure 3. Preparation of guard cell protoplasts. Panels (a) – (d) Successive stages in isolation of guard cell protoplasts from Arabidopsis. e) Arabidopsis mesophyll cell protoplasts. (f) Isolated Arabidopsis guard cell protoplast. g) Isolated Arabidopsis mesophyll cell protoplast (h) Guard cell protoplasts of *V. faba* with approaching patch pipette. ((a) - (g) from 46; (h) modified from Plant Physiol. 125: 139-141).

root to shoot signaling in the majority of species studied to date. ABA synthesis by the root is stimulated by cellular dehydration, and this ABA is then carried in the xylem sap to the terminus of the transpiration stream, the stomata. In addition, xylem sap pH increases during drought, a phenomenon that will tend to favor ABA retention in the apoplast, by promoting formation of the ABA^- form, which is impermeant through the lipid bilayer by virtue of its charge.

Once ABA arrives at the guard cell, it must be perceived. It is ironic that despite intensive research into ABA responses of guard cells, the guard cell ABA receptor(s) remains unidentified. In general, stomatal sensitivity to externally applied ABA increases as external pH is lowered. This result indicates that either guard cells perceive symplastic ABA, and/or that there is a plasma-membrane sensor of apoplastic ABA that binds ABAH preferentially over ABA^-. Stomatal closure, production of reactive oxygen species (ROS) and cytosolic calcium elevation is induced by ABA injected directly into the guard cell symplast (1, 60). Inhibition of K^+ channels mediating K^+ uptake is also observed in guard cells in which the cytosol or cytosolic side of the membrane is perfused with ABA (1, 60). In addition, plasma membrane calcium channels of guard cells are activated by ABA supplied on the symplastic face of the membrane (15). These results suggest that guard cells can perceive internal ABA. In apparent contradiction to these results, guard cells microinjected with ABA did not show inhibition of stomatal opening when assayed 1-2 hours later (1). Calculations of apoplastic ABA concentrations and guard-cell internal ABA concentrations, as compared with stomatal apertures, suggest that stomatal apertures are better correlated with the former than the latter under mild to moderate water stress, whereas the converse is true under more severe water deficits (45, 57). Thus, both externally-facing and internal ABA receptors may operate in guard cells, though in neither case has their identity been determined. One 42 kDa ABA-binding protein has been purified from guard cells of *V. faba* (59) and a key experiment will be to determine whether this protein is resident in the plasma membrane, and, if so, whether it binds symplastic or apoplastic ABA.

MECHANISMS OF ABA-INDUCED VOLUME CHANGE IN GUARD CELLS.

Stomatal opening occurs when osmotically-driven influx of water causes the two guard cells to swell and bow apart, thus widening the stomatal pore. Conversely, stomatal closure results from osmotically-driven reductions in guard cell volume that cause the two guard cells to deflate against each other, thereby narrowing the pore. ABA both inhibits stomatal opening and promotes stomatal closure, and targets a variety of ion transport mechanisms to do so (2, 4, 17, 45, 52, 53) (Fig. 4).

Signals that stimulate stomatal opening include blue and red light, the fungal toxin fusicoccin, low concentrations of intracellular CO_2, and high

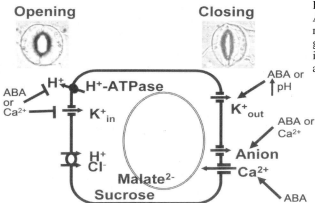

Figure 4. Known sites of ABA action on plasma membrane transporters of guard cells. Pointed arrows indicate activation; blunted arrows indicate inhibition.

ambient humidity (2, 13, 45, 52). For light and fusicoccin at least, it is known that these stimuli activate plasma-membrane localized H^+ ATPases that extrude protons, thereby hyperpolarizing the membrane potential, i.e. shifting the transmembrane voltage to a more negative value. Membrane hyperpolarization both creates a driving force for K^+ uptake and activates voltage-regulated K^+ channels that mediate this uptake. In Arabidopsis, KAT1, KAT2, AKT1 and AKT2/3 genes encoding inward K^+ channels are expressed in guard cells (55).

Guard cells also take up a moderate amount of Cl^-, presumably via a H^+/Cl^- symporter or Cl^-/OH^- antiporter, although the mechanism has yet to be identified at the molecular level. In addition, malate^{2-} synthesis from starch breakdown provides a counterion for K^+. Depending on environmental conditions, sugars, particularly sucrose imported from the apoplast, may contribute significantly to osmotic build-up (45).

Nitrate (NO_3^-) concentrations can also increase in guard cells up to 3-fold during stomatal opening (14). Mutants (*chl1*) in the dual-affinity nitrate transporter gene, *AtNRT1*, show reduced stomatal opening and transpiration. Nitrate is more permeant through S-type anion efflux channels (see below) than are either Cl^- or malate^{2-} (51), suggesting that NO_3 efflux can contribute to stomatal closure. However, the *chl1* mutant does not exhibit alterations in either ABA-inhibition of stomatal opening or ABA-promotion of stomatal closure, suggesting that this anion is not obligately tied to the ABA response. Instead, the extent to which NO_3^- plays a significant role in guard-cell turgor changes appears to correlate with NO_3^- availability in the soil (14).

Experimental evidence from *V. faba* indicates that ABA both inhibits H^+ ATPase activity (13) and decreases the number of inward K^+ channels available for activation; the latter phenomenon has also been documented in Arabidopsis (4, 52). The combined effect of these processes is to limit the K^+ uptake required to drive stomatal opening. In addition, ABA activates ion channels in the guard cell membrane that mediate Ca^{2+} uptake (15, 47). Uptake of positively charged Ca^{2+} depolarizes the membrane potential, contributing, along with H$^+$ ATPase inhibition, to production of a free energy gradient for K$^+$ that opposes passive K$^+$ uptake. In addition, as discussed in the next section, cytosolic Ca^{2+} plays a significant role in ABA signal transduction pathways.

During ABA promotion of stomatal closure, the above processes presumably function and are joined by anion efflux through two types of anion channels, called R-type (rapid) and S-type (slow) by virtue of their kinetic profile (4, 16, 52, 53). Anion efflux both increases cellular water potential, thereby driving water efflux, and contributes to membrane depolarization. The combined promotion of membrane depolarization by H$^+$ ATPase inhibition, Ca^{2+} channel activation, and anion channel activation, provides a driving force for K$^+$ efflux via outward K$^+$ channels that are voltage-activated in response to depolarizing potentials. In Arabidopsis, use of genetic knockouts has implicated the product of one K$^+$ channel gene in particular, *GORK1*, in sustained K$^+$ efflux (18). GORK-type currents are enhanced by ABA in both Arabidopsis and *V. faba*, and biophysical analysis of these currents indicates that ABA increases the number of channels available for voltage-dependent activation. Another outwardly-rectifying K$^+$ current, I_{AP} is transiently activated during membrane depolarization; the channel responsible for mediating the I_{AP} current has not yet been identified at the molecular level, and its modulation by ABA has not been reported (52). Intriguingly, some research indicates that an ATP-binding-cassette transmembrane regulator (ABC protein) may mediate ABA effects on both slow anion channels and outward K$^+$ channels, as inhibitors of sulphonylurea receptor-type ABC proteins inhibit slow anion channels, outward K$^+$ channels, and ABA-induced stomatal closure (32).

Guard cells undergo changes in ion concentration of several tenths molar during stomatal opening and closure, and much of this ion content is stored within the vacuole (37, 45). Direct responsiveness of tonoplast channels to ABA has not been demonstrated, thus they are not addressed further in this chapter.

TRANSDUCING THE ABA SIGNAL

Vesicle Trafficking, the Actin Cytoskeleton, and Small GTPases

Guard cells undergo significant volume changes during stomatal opening and closure. These changes may be accompanied by increases and decreases in plasma-membrane surface area. Thus, vesicle trafficking would appear to be a likely target of ABA action. Syntaxins are important for vesicle trafficking and fusion with target membranes in mammalian systems, and a tobacco syntaxin, Nt-Syr1, also appears to play this role (12). A truncated version of Nt-Syr1 interferes with ABA regulation of inward and outward K$^+$ channels and S-type anion channels. The lethal neurotoxin, botulinum Cl, which inhibits syntaxin-dependent vesicle trafficking, also inhibits trafficking when microinjected into guard cells (33).

Vesicle trafficking occurs along the microtubule and actin cytoskeletons. Based on experiments utilizing either immunolocalization or tubulin visualization by labeling with the microtubule-associated protein MAP4-GFP, the guard cell radial microtubule array does not reorganize in response to ABA (8, 30). By contrast, reorganization of the actin cytoskeleton appears consistently in response to ABA (Fig. 5). ABA induces a disruption of radial actin filaments in *C. communis, V. faba*, and Arabidopsis that is complete within 60 min. (8, 22, 30). Because actin antagonists disrupt both stomatal opening and stomatal closure, it has been proposed that depolymerization of actin is required for changes in stomatal aperture but does not determine the direction of the change (8, 22). According to this hypothesis additional proteins would need to be involved in determining the specificity of this event.

In mammals and yeast, small GTPases of the Rho class regulate the actin cytoskeleton. Plant Rho GTPases or "Rops" comprise an 11 member family of small GTPases in plants, and several family members have been implicated in cytoskeleton reorganization. Two members of this family, ROP6 (also called AtRAC1) and ROP10, have been implicated in the guard cell ABA response.

In transgenic Arabidopsis lines allowing actin visualization via a GFP-tagged actin-binding protein, talin, ABA effects on the cytoskeleton have been evaluated (30) upon inducible expression of either a ROP6 mutant with 10x reduced GTPase activity, i.e. 'constitutively active', or 'dominant-positive', or a ROP6 mutant with highly reduced affinity to GTP and GDP, i.e., 'constitutively inactive' or 'dominant-negative'. While the latter mimicked ABA effects on actin disruption and stomatal closure, the former blocked ABA-induced stomatal closure. These results implicate ROP6 as a negative regulator of the ABA response.

Null mutants of a second small G protein, ROP10, show hypersensitivity to ABA in stomatal closure, seed germination, and root elongation (61). The role, if any, of ROP10 in actin organization in guard cells, remains to be described. ROP10 localizes to the plasma membrane and such localization is typically important for small GTPase function. ROP10 contains a putative farnesylation motif that could mediate this process. Therefore, it was of interest to ascertain whether ROP10 localization was disrupted in the *era1-2* genetic background. ERA1 encodes the β-subunit of a protein farnesyltransferase,

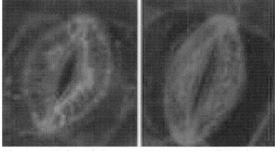

Figure 5 (Color Plate page CP10). ABA-disruption of the radial actin cytoskeleton in *C. communis* guard cells. Left panel: control. Right panel: after 30 min. of ABA. Actin was visualized by staining fixed cells with rhodamine-phalloidin. (Figure courtesy of Dr. Youngsook Lee and Yunyung Choi).

and *era1-2* mutants also show an ABA-hypersensitive guard cell phenotype, including enhanced anion channel activation, cytosolic Ca^{2+} elevation and stomatal closure (52). In the *era1-2* background, however, plasma membrane localization of ROP10 was only slightly reduced, indicating that the *era1-2* phenotype probably cannot be attributed to mislocalization of ROP10.

Ca^{2+} and pH

Upon the advent of techniques to measure cytosolic Ca^{2+}, it was discovered that both external Ca^{2+} and externally applied ABA stimulate oscillations in cytosolic Ca^{2+} that correlate with stomatal closure (5, 9, 17, 50, 53). These oscillations exhibit different waveforms, and information contained in their frequency, number, duration, amplitude, and probably spatial characteristics as well, determines the extent of stomatal closure (3, 17, 50) (Fig. 6). In wild-type guard cells, experimental abrogation of Ca^{2+} oscillations impedes ABA-induced stomatal closure (52). In several ABA-insensitive mutants (discussed in more detail in the section on kinases and phosphatases), including *abi1-1*, *abi2-1*, *rcn1*, and *gca2*, aberrant Ca^{2+} oscillations in response to ABA are observed. In *gca2* plants, artificial manipulation of cytosolic Ca^{2+} to mimic wild-type oscillations induces stomatal closure (50, 52, 53). The *gca2* gene has not yet been identified, but such experiments verify the importance of the Ca^{2+} signal.

However, it also must be emphasized that in virtually every study of ABA-induced Ca^{2+} transients (24), not every cell examined exhibits such transients. These results demonstrate that at least under some conditions, backup or redundant mechanisms for ABA responses occur in cells where the Ca^{2+} signal does not operate. In addition, Ca^{2+}-oscillations may be more important for maintaining the ABA-induced state, since they often proceed after aperture

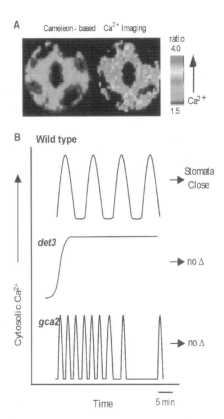

Figure 6 (Color plate page CP10). Ca^{2+}-signaling in guard cells. (A) Cytosolic calcium concentrations can be quantified using fluorescent Ca^{2+} indicators. (B) As shown schematically, Ca^{2+}-oscillations are altered in Arabidopsis mutants *det3* and *gca2*, which exhibit altered sensitivity to ABA (from 50).

responses have reached steady-state levels (9).

Both Ca^{2+} uptake and Ca^{2+} release from internal stores contribute to ABA-induced Ca^{2+} oscillations. ABA-stimulated Ca^{2+} uptake currents have been recorded from guard cells of Arabidopsis and *V. faba* (15, 47). Release of Ca^{2+} from internal stores, stimulated by inositol phosphates and sphingosine-1-phosphate (S1P) contribute to transducing the ABA signal (5, 9, 44). These lipid-based signaling pathways are described in more detail in a subsequent section. In addition, cADPR serves as a Ca^{2+}-mobilizing agent in guard cells. In mammalian cells, the enzyme ADP-ribosyl cyclase produces cyclic ADP ribose from NAD; cADPR then interacts with the ryanodine receptor channel to release Ca^{2+} from the endomembrane system. In plants, the target membrane is probably the tonoplast, as cADPR activates a Ca^{2+}-permeable channel in this organelle (5, 17, 50, 52).

Elevated Ca^{2+} contributes to actin depolymerization (22) and many of the ion transport pathways that are ABA-modulated are also sensitive to cytosolic Ca^{2+} concentrations. Plasma membrane H^+ ATPase activity is inhibited by both ABA and Ca^{2+} (13). At the plasma membrane, Ca^{2+} influx channels exhibit feedback inhibition by elevated cytosolic Ca^{2+} (15, 53); this may be vital so that Ca^{2+} elevation does not propagate to such a high level that insoluble $CaPO_4$ precipitates are formed. Both ABA and elevated intracellular Ca^{2+} inhibit the plasma membrane H^+ ATPase and inward K^+ channels, and slow anion channels are activated by both ABA and Ca^{2+} (4, 50, 52, 53). Ca^{2+} activates rapid anion channels as well (16). At the tonoplast, Ca^{2+} also may activate slow vacuolar (SV) Ca^{2+}-permeable channels, leading to the phenomenon of Ca^{2+}-induced Ca^{2+} release (52). Cytosolic Ca^{2+} also activates vacuolar K^+ (VK) channels, leading to K^+ efflux from vacuolar stores to the cytosol (37, 45, 53). All of these effects are consistent with the reductions in guard cell osmotica known to be engendered by ABA.

One transport mechanism which appears to be independent of the Ca^{2+} signal are the outward K^+ channels, which are Ca^{2+} insensitive. However, these channels are activated by elevations in cytosolic pH, which also occur in response to ABA (4). The pH effect on the outward K^+ channels occurs in isolated membrane patches, suggesting either that pH directly affects the channel protein, or that it activates signaling metabolites that are membrane-derived or associated (2).

A major question facing guard cell physiologists today is how the ABA signal is transduced into these ionic changes, and how changes in cytosolic Ca^{2+} and H^+ concentration are transduced into signals that ultimately regulate the ion transport molecules. The sections below discuss secondary messengers implicated in ABA action, and, where data are available, describe whether these messengers operate upstream, downstream, or independently of the Ca^{2+} signal.

H_2O_2 and Reactive Oxygen Species (ROS)

Use of a fluorescent reporter dye indicates that ABA stimulates ROS generation in guard cells, and DPI, an inhibitor of NADPH oxidases, partially interferes with ABA-induced stomatal closure (41, 47, 60). Like ABA, application of H_2O_2 initiates Ca^{2+}-oscillations and stimulates stomatal closure in Arabidopsis, *V. faba*, and *C. communis*. H_2O_2 activates plasma membrane Ca^{2+}-permeable channels in Arabidopsis and *V. faba* guard cells by increasing channel open probability without altering single channel amplitude (15, 26, 47). H_2O_2 inhibits inward K^+ channel activity but does not promote the activity of outward K^+ channels (26), perhaps because these channels are regulated by pH rather than Ca^{2+}. Both ABA and H_2O_2 fail to induce Ca^{2+} channel activity in the recessive mutant *gca2* (47), although ROS production still occurs in this mutant in response to ABA.

The above results implicate ROS as secondary messengers for ABA action, but do not identify the specific intracellular ROS species generated, which could be H_2O_2, O_2^-, $^1O_2$, and or OH^-. In addition, it is important to note that NADPH oxidase inhibitors such as DPI may also interfere with generation of another ABA-related secondary messenger, nitric oxide (NO), described in the following section.

Nitrate Reductase, and Nitric Oxide

In plants, nitric oxide (NO) was first implicated in stress responses and more recently it has been assigned a specific role in guard cell responses to ABA (7, 11). A suite of pharmacological tools has been developed that assists in characterization of NO action. SNP and GSNO are NO donors, while CPTIO is a NO scavenger and L-NAME is an inhibitor of the enzyme NO synthase. NO can also be synthesized from nitrite in an NAD(P)H dependent reaction catalyzed by the enzyme nitrate reductase (NR). NO production can be visualized as an increase in fluorescence emission from the cell-permeant probe, DAF-2 DA.

In guard cells of pea and *V. faba*, ABA application leads to an increase in DAF-2 DA fluorescence, suggesting ABA-dependent production of NO (Fig. 7). The NO donors SNP and GSNO induce stomatal closure while the NO scavenger CPTIO inhibits ABA-induced stomatal closure. While L-NAME blocks ABA-induced NO production in pea, it does not do so in Arabidopsis, where the NR-dependent pathway of NO production apparently operates (7). Thus, in the NR double mutant, *nia1 nia2*, ABA production of NO in nitrite-supplied guard cells is blocked (7). Stomatal closure in response to ABA or nitrite also

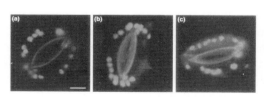

Figure 7 (Color plate page CP10). ABA stimulates production of NO in guard cells as visualized by DAF-2 DA fluorescence. (a) Control. (b) treatment with 150 µM SNP (an NO donor) (c) Treatment with 10 µM ABA. (From 11).

no longer occurs in this mutant. However, it is of interest that ABA inhibition of stomatal opening is retained at wild-type levels in the mutant, despite the fact that application of NO releasers inhibits stomatal opening (in *V. faba* (11)). Taken together, these results imply that ABA-induced stomatal closure relies on NO, while ABA-inhibition of stomatal opening can bypass the NO mechanism.

Heterotrimeric G Proteins and Lipid Signaling

In mammalian systems, many hormones are perceived by G-protein coupled receptors (GPCRs), which are membrane proteins with 7 transmembrane domains. Upon ligand binding, conformational changes in the GPCR activate a heterotrimeric G protein comprised of α, β, and γ subunits, and induce GTP for GDP exchange on the α subunit. The α subunit separates from the $\beta\gamma$ subunit pair and both components may initiate downstream signaling events, including the regulation of many types of ion channels. Signaling is terminated when the intrinsic GTPase activity of the α subunit results in GTP hydrolysis. The GDP-bound form of the Gα subunit then reassociates with the $\beta\gamma$ subunit pair (8. 9).

The Arabidopsis genome encodes one canonical Gα subunit, one Gβ subunit, and two Gγ subunits. Based on the mammalian paradigm, a potential role of G-protein signaling in ABA regulation of guard-cell ion-channels was evaluated, by taking advantage of the availability of T-DNA insertional null mutants in the Gα gene. These plants show altered sensitivity to ABA, characterized by an increased rate of water loss, and insensitivity to ABA-inhibition of stomatal opening and inward K^+ channel inhibition (Fig. 8). However, wild-type responses to ABA-promoted closure of open stomata are observed in the *gpa1* knockouts (56). In patch-clamped *gpa1* guard cells, ABA regulation of slow anion channels is abrogated if cytosolic pH is clamped, but occurs normally if cytosolic pH is allowed to vary, suggesting that GPA1 functions in a pH-independent pathway of ABA-based anion channel regulation that is not absolutely required for normal levels of stomatal closure to occur (56).

A link has been established between G-proteins and sphingolipid signaling in mammals, and this link operates in guard cells as well (6). Sphingolipids are long-chain amine alcohols. The phosphorylated sphingolipid, sphingosine-1-phosphate (S1P) is an interesting molecule because, in mammalian cells, it is generated internally and activates Ca^{2+} release from endomembrane sites, but is sufficiently hydrophobic that it also crosses the cell membrane. Once outside, S1P interacts with a family of G-protein coupled receptors (GPCRs) originally called the EDG receptors but now renamed the S1P receptors (5).

Sphingolipids constitute only minor components of the lipid bilayer, yet S1P is emerging as a major player in guard cell signaling (5, 6, 44). S1P triggers stomatal closure and Ca^{2+} elevation in *C. communis* guard cells (5, 44; Fig. 9). ABA-induced stomatal closure and ABA inhibition of stomatal

Abscisic acid signal transduction in stomata

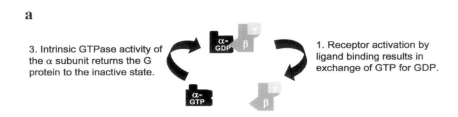

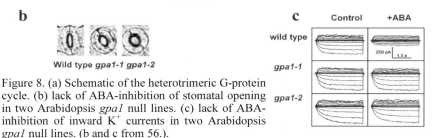

Figure 8. (a) Schematic of the heterotrimeric G-protein cycle. (b) lack of ABA-inhibition of stomatal opening in two Arabidopsis *gpa1* null lines. (c) lack of ABA-inhibition of inward K^+ currents in two Arabidopsis *gpa1* null lines. (b and c from 56.).

opening are opposed by S1P antagonists (6, 44). Detailed biochemical analyses have shown that ABA stimulates sphingosine kinase activity and S1P production in Arabidopsis guard cells (6, Fig. 9). In further analysis, ABA effects on inward K^+ currents and slow anion currents are opposed by the S1P antagonist, DMS, and are mimicked by S1P administration (6).

To test the hypothesis that S1P signals via heterotrimeric G proteins in plants as in mammals, S1P effects were evaluated in *gpa1* null lines. Just as for the ABA response of these mutants, S1P inhibition of stomatal opening and inward K^+ channels were eliminated in the *gpa1* null lines (6). ABA-activation of sphingosine kinase still occurs in these mutants, suggesting that S1P can still be produced in response to this hormone, but can no longer be sensed in the absence of a functional G-protein heterotrimer.

Interestingly, S1P promotion of stomatal closure was also eliminated in these *gpa1* lines, even though ABA-promotion of stomatal closure is retained

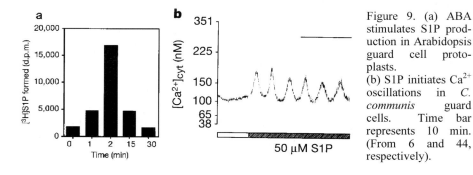

Figure 9. (a) ABA stimulates S1P production in Arabidopsis guard cell protoplasts. (b) S1P initiates Ca^{2+} oscillations in *C. communis* guard cells. Time bar represents 10 min. (From 6 and 44, respectively).

402

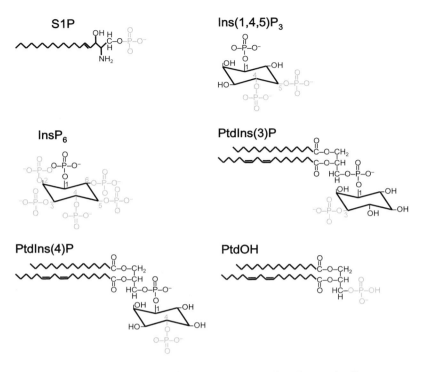

Figure 10. Lipids and lipid metabolites implicated in ABA signaling in guard cells.

(6, 56). These results suggest a bifurcation of the pathway for ABA-induced stomatal closure: one branch requires the ability to perceive S1P (via GPA1), while the other branch does not. It will be interesting to determine whether or not Ca^{2+}-oscillations in response to ABA and S1P are altered in the *gpa1* null mutants.

Several other lipids have also been implicated in ABA signaling in guard cells (17; Fig. 10). The first lipid metabolite implicated in the guard-cell ABA response was inositol 1,4,5 trisphosphate, or $InsP_3$. $InsP_3$ and diacylglycerol (DAG) are produced from phosphatidylinositol bisphosphate (PIP_2) by activation of phospholipase Cs (PLCs). Biochemical experiments have shown ABA-stimulation of IP_3 production in *V. faba* guard cells, and release of caged IP_3 within the guard cell cytosol results in inhibition of inward K^+ channels, Ca^{2+} elevation, and stomatal closure (2, 4, 17). The importance of PLC in guard-cell ABA signaling has been confirmed by genetic suppression of PLC synthesis (21). By analogy to mammalian systems, IP_3 is thought to activate IP_3-binding Ca^{2+}-release channels located in endomembranes (17, 50). Another inositol-phosphate, myo-inositol-hexakisphosphate (InsP6) is also produced in guard cells in response to ABA, and inhibits inward K^+ channels, in a Ca^{2+}-dependent manner (31, 52).

Phosphatidylinositol 3-phosphate (PI3P) and phosphatidylinositol 4-phosphate (PI4P) have also been implicated in the ABA response. PI 3-

kinase (PI3K) and PI 4-kinase (PI4K) catalyze synthesis of PI3P and PI4P, respectively. Wortmannin and Ly294002, pharmacological inhibitors of these enzymes, reduce ABA-stimulated stomatal closure in *V. faba* and *C. communis*, and decrease the percentage of Arabidopsis guard cells exhibiting Ca^{2+} transients in response to ABA (24).

In addition to PLC, a second phospholipase, phospholipase D (PLD), has also been implicated in ABA action (23). PLD hydrolyzes phospholipids, generating phosphatidic acid and the lipid head group. ABA stimulation of PLD activity has been documented in *V. faba* guard cells. Phosphatidic acid inhibits stomatal opening, promotes stomatal closure, and inhibits inward K^+ channels. 1-butanol, an inhibitor of phosphatidic acid production, partially opposes both ABA-inhibition of stomatal opening and ABA-promotion of stomatal closure. These partial effects suggest that PLD acts in parallel with other ABA secondary messengers. Dual administration of 1-butanol and the cADPR inhibitor, nicotinamide, increases the net inhibition of the ABA response, but dual administration of 1-butanol and the PLC inhibitor U73122 does not. These results suggest that PLC and PLD may be acting in the same pathway, while cADPR may be acting in a parallel pathway. Because administration of phosphatidic acid does not elicit a rise in cytosolic Ca^{2+}, PLC and Ca^{2+}-release may function upstream of PLD.

Protein Kinases and Phosphatases

K252A, a broad-spectrum inhibitor of serine/threonine protein kinases, opposes ABA-induced stomatal closure in a variety of species, and reduces slow anion channel activity (52). Given the important role of Ca^{2+} in the ABA response, K252A may be targeting calcium-dependent protein kinases or CDPKs, which comprise a 20+ member gene family in Arabidopsis. By biochemical approaches it has been demonstrated that guard cell CDPK phosphorylates the KAT1 inward K^+ channel, and, as assayed in a heterologous system, decreases K^+ channel activity (45, 52). In addition, there is evidence to suggest that in pea guard cells, ABA activates a MAP kinase (52). However, the strongest evidence for involvement of a protein kinase in the ABA response comes from a serine/threonine protein kinase that is not a MAP kinase and is Ca^{2+}-independent. This protein kinase, ABA-activated protein kinase (AAPK) or ABRK (ABA-related protein kinase) was first characterized biochemically from *V. faba* guard cells (35, 40). AAPK is activated within minutes by physiological concentrations of ABA but is not responsive to other plant hormones. *De novo* sequencing by tandem mass spectrometry of AAPK peptides provided sequence information that allowed the eventual cloning of the *AAPK* cDNA (35). Biolisitic expression in *V. faba* guard cells (Fig. 11) of a dominant negative form of AAPK results in guard cells that are insensitive to ABA-induced stomatal closure and ABA activation of slow anion channels, confirming a central role of this kinase in ABA response (35). AAPK also phosphorylates the KAT1 inward K^+ channel in *in vitro* assays (40). The observations that phosphatase inhibitors enhance Ca^{2+} channel activity and that ATP hydrolysis is required for Ca^{2+}

channel activation suggests that kinases promote activation of the Ca^{2+} channel as well (25). It will be of interest to determine the activity of AAPK against guard-cell Ca^{2+} channels.

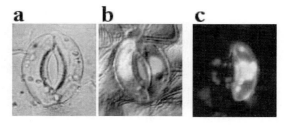

Figure 11 (Color plate page CP10). Biolistic-mediated expression in (a) *V. faba* guard cells of (b) GFP-tagged *V. faba* AKIP1; and (c) GFP-tagged *V. faba* AAPK.

The Arabidopsis genome encodes multiple kinases with significant homology to AAPK and genetic approaches have identified one of them, *OST1*, as a functional homolog of AAPK (42, 58). There are also 9 other OST1-like (OSKL) kinases encoded in the Arabidopsis genome (42). *ost1* mutants were identified in a genetic screen for mutants that exhibited cooler leaf temperatures following water stress, indicating ABA-insensitivity of stomatal closure (Fig. 12). The *ost1* mutants exhibit ABA insensitivity of both stomatal opening and stomatal closure but show wild-type responses to Ca^{2+} and H_2O_2, suggesting that these secondary messengers operate downstream of OST1 (42). Insertional mutation of *OST1* (also named *SRK2E*) also results in a weakening of the stomatal closure response to decreased humidity, and reduced ABA-induction of gene expression (58).

Kinases and phosphatases are partners in the regulation of protein phosphorylation states. Inhibitors of serine/threonine protein phosphatases 1A and 2A can either promote or interfere with guard cell ABA signaling, depending on the particular response being assayed, the physiological state of the guard cells (open or closed at the initiation of the assay), and the plant species (4, 28). MacRobbie has presented

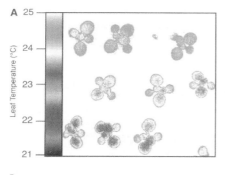

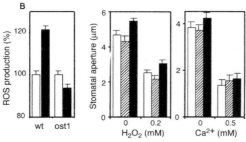

Figure 12 (Color Plate page CP10). (a) Compared to L*er* wild-type (top row) Arabidopsis *ost1* mutants (middle and bottom rows) exhibit increased transpirational leaf cooling; (b) Left panel: ROS production in response to 50 μM ABA (black bars) is impaired in the *ost1-2* mutant. Middle, Right panels: Stomata in *ost1-1* (hatched bars) and *ost1-2* (black bars) mutants respond to exogenous Ca^{2+} and H_2O_2 similarly to wild-type (white bars). (From 42).

pharmacological evidence that tyrosine phosphatases may be involved in regulation of ABA-stimulated K^+ release from the vacuole (38). The *abi1-1* and *abi2-1* mutations are dominant mutations in protein phosphatase 2Cs (PP2Cs) that confer strong ABA insensitivity and a wilty phenotype. These mutations disrupt ABA-induced stomatal closure, slow anion channel activation (41, 52, 53), Ca^{2+}-channel activation (41), and Ca^{2+}-increases (52, 53). In addition, ABA-induced actin depolymerization is blocked in *abi1-1* (8). However, because these mutations are dominant, they may have neomorphic effects and indeed, study of recessive mutants of *abi1* showed that recessive mutation of this gene results in a phenotype of ABA hypersensitivity, suggesting that the intact, wild-type gene product is a negative regulator of ABA response (39).

Another recessive phosphatase mutant is *rcn1*, a mutant in a protein phosphatase 2A (PP2A) regulatory A subunit. The *rcn1* mutation has pleiotropic effects, including altered auxin transport and gravitropism, and root curling in the presence of the auxin transport inhibitor, NPA. Guard cells of *rcn1* plants show impaired cytosolic Ca^{2+}-response to ABA and little ABA activation of anion channels. However, stomatal closure occurs normally in response to exogenous application of Ca^{2+} or H_2O_2, suggesting that pathway components downstream of Ca^{2+} retain their wild-type character (28).

ABA AND RNA IN GUARD CELLS

ABA triggers responses that lead within minutes to changes in guard cell volume and stomatal aperture in intact plants, and the above discussion has highlighted secondary messengers that are involved in transducing such short term responses. However, upon longer drought episodes, ABA-induced changes in gene expression are evoked.

Surprisingly, given the number of genes whose transcript levels are known to change in response to ABA (48), there are relatively few studies of ABA-regulation of transcription in guard cells. One study using potato epidermal RNA, enriched in guard cell transcripts, assessed expression of specific genes following 2-4 days of water stress (27). Genes encoding components involved in stomatal opening, namely an inward K^+ channel and a H^+ ATPase, showed decreased expression. Genes encoding enzymes of carbon metabolism also showed altered expression, and some of these responses were observed in guard cells but not in whole-leaf tissue. In such gene-by-gene approaches, dehydrins are another class of proteins whose steady-state transcript levels have been shown to be up-regulated in response to ABA in legume guard cells; these proteins are hypothesized to assist in the maintenance of membrane integrity during stress. As summarized by Kopka et al. (27), an Arabidopsis aquaporin gene shows a similar response. The dearth of more comprehensive reports probably results from the difficulty in collecting enough pure Arabidopsis guard cells to obtain sufficient RNA for microarray analysis. However, this obstacle has now been overcome (19, 46), and one expects such publications to surface shortly.

Recent genetic and biochemical approaches have implicated ABA in the regulation not only of RNA transcription but also of RNA processing. Multiple RNA binding proteins of diverse types assist in the synthesis, processing, export and translation of RNA molecules (10). By phenotypic analysis of Arabidopsis null mutants, two RNA-binding proteins, ABA-hypersensitive 1 (ABH1) and SAD1, have been implicated in guard cell responses to ABA (19, 20). ABH1 is an mRNA cap-binding protein; such proteins interact with the capped 5' end of the RNA and are implicated in splicing. The cap-binding complex also forms part of the ribonucloprotein complex that exits the nucleus. Consistent with its assignation as a cap-binding protein, GFP-ABH1 expression is observed primarily in the nucleus, but also to a limited extent in the cytoplasm. *abh1* null mutants show hypersensitivity to ABA in assays of stomatal closure and cytosolic Ca^{2+} elevation, and also exhibit drought resistance (19). SAD1 is an Sm-like small nuclear ribonucleoprotein (snRNP); proteins of this class are involved in mRNA splicing, export and stability. Like *abh1* mutants, *sad1* mutants show hypersensitivity of ABA-induced stomatal closure (20). Null mutation of a third protein, the double-stranded RNA binding protein HYL1, results in altered phenotypes in whole-plant response to ABA, auxin, and cytokinin (10). However, HYL1 may not be involved specifically in guard cell responses, as wild-type levels of ABA-induced stomatal closure are observed in *hyl1* mutants (20).

While the hypersensitive nature of the *abh1* and *sad1* phenotypes indicate that these RNA binding proteins are negative regulators of the ABA response in wild type plants, biochemical and molecular analyses implicate another RNA binding protein as a possible positive effector of ABA signaling. The ABA-activated protein kinase, AAPK (35), phosphorylates an heterogeneous ribonucleoprotein-like (hnRNP-like) *V. faba* RNA binding protein dubbed AKIP1 ("AAPK interacting protein 1") (Fig. 11). Phosphorylation confers upon AKIP1 the ability to bind dehydrin mRNA (34). hnRNPs are diverse single-stranded RNA binding proteins with numerous functions, including transcriptional and translational regulation, telomere maintenance, splicing, and RNA export. An Arabidopsis protein related to AKIP1 by sequence homology, UBA2a, has been implicated in transcript stabilization (29). If AKIP1 functions in a similar manner, it will be of interest to ascertain whether this single-stranded RNA binding protein preferentially stabilizes transcripts encoding proteins known to be important for plant acclimation to drought, salt or cold stress. One interesting characteristic of ABH1, SAD1, and AKIP1 is that their expression is not limited to guard cells. *abh1* and *sad1* null lines also show ABA-related phenotypes in other aspects of plant function.

BIOTECHNOLOGY AND THE GUARD CELL ABA RESPONSE

It is clear that mutants are a powerful tool for dissecting the guard cell response to ABA. Recessive mutation of the *GCA2* gene, of a PP2A regulatory subunit (*rcn1*), of the G-protein α subunit (*gpa1*), of an ABA-activated protein kinase (*ost1*), and dominant mutations in two PP2C genes (*abi1-1* and *abi2*), and in a small GTPase (*rop6*) confer ABA insensitivity. Conversely, recessive mutations in *ABI1*, *ROP10*, *ERA1*, and *ABH1* confer ABA hypersensitivity. Recessive mutation of *ABI2* results in a wild-type sensitivity to ABA; however double recessive mutants in both *ABI1* and *ABI2* result in greater ABA sensitivity than either of the parental mutants alone (39). It has been proposed that mutations conferring guard-cell hypersensitivity to ABA could have biotechnological value in improving drought resistance. One caveat is that, at least for *era1-2*, *abh1*, and recessive mutants of *ABI1*, these mutations also confer ABA hypersensitivity in the seed, likely to be an undesirable trait. Therefore, expression driven by guard-cell specific promoters of dominant mutations conferring hypersensitivity, or use of genes which evince guard-cell-specific expression would be desirable.

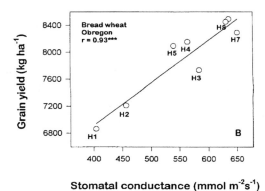

Figure 13. Increased yield correlates with increased stomatal conductance in cultivars of breadwheat grown at high temperatures. (From 36).

At first glance it might seem that ABA-insensitive mutants, would not be agronomically useful, but this may not be true. Such mutants could confer more rapid rates of drying on crops that are dried in the field, such as feed corn and soybean. In addition, for crops such as cotton and breadwheat, it has been observed that cultivars which show the highest yields when these crops are grown at supraoptimal temperatures also exhibit the highest transpiration rates (Fig. 13). Experimental analysis indicates that the higher productivity is not because of a decreased limitation on CO_2 uptake, but rather because of increased transpirational cooling (36). Therefore, controlled manipulation of high transpiration rates may also be agronomically useful.

FUTURE QUESTIONS: RECEPTORS, REDUNDANCY, SENSITIVITY AND INTERPLAY

While our knowledge of ABA signal transduction is increasing rapidly, major questions remain unanswered, foremost among them the identity of the ABA receptor. As is known to be the case for ethylene, there may be multiple ABA receptors. Potentially there may exist receptors for both apoplastic and symplastic ABA, and receptors that are specific for guard cells or other particular cell and tissue types. Identification of the ABA receptor(s) remains a paramount goal of research in this field.

As is obvious from the discussion above concerning ABA signaling mechanisms, there is great potential for both redundancy and cross-talk in the secondary messenger pathways of ABA response. Do all of these pathways function simultaneously or are some of them evoked only when primary pathways are blocked genetically or pharmacologically? Which of these pathways operate only during ABA-inhibition of stomatal opening or during ABA-promotion of stomatal closure? With regard to these questions, we have some clues, but few definitive answers. Finally, for most of the internally-generated signals, their epistatic relationships remain to be established. For example, in mammalian systems, certain PLC isoforms function downstream of G proteins in mediating cellular responses; however we do not know as yet whether or not ABA stimulation of these phospholipases is blocked in the *gpa1* mutants. Thus, another major goal for the immediate future is further determination of the connectivity between identified components of the ABA response.

A final, but very important area for future research is determining the extent to which the pathways uncovered in model systems will also be found to function in crop species. Such knowledge is essential if we are to propose biotechnological solutions to improve crop productivity in marginal habitats and under stressful climatic conditions.

Acknowledgments

Research on ABA signaling in the author's laboratory is supported by grants from the National Science Foundation and the U.S. Department of Agriculture. The author thanks Drs. Sylvie Coursol and Carl Ng for assistance in preparation of Figures 2 and 10, and Figure 11, respectively. All figures are reproduced by permission of the respective journals cited below.

References

1. Assmann, SM (1994) Ins and outs of guard cell ABA receptors. Plant Cell 6: 1187-1190
2. Assmann SM, Wang X-Q (2001) From milliseconds to millions of years: guard cells and environmental responses. Curr Opin Pl Biol 4: 421-428
3. Allen GJ, Chu SP, Harrington CL, Schumacher K, Hoffmann T, Tang YY, Grill E, Schroeder JI (2001) A defined range of guard cell calcium oscillation parameters encodes stomatal movements. Nature 411: 1053-1057
4. Blatt MR (2000) Cellular signaling and volume control in stomatal movements in plants. Annu Rev Cell Dev Biol 16: 221-41
5. Brownlee C (2001) Intracellular signalling: sphingosine-1-phosphate branches out. Curr Biol. 11: R535-R538
6. Coursol S, Fan L, Le Stunff H, Spiegel S, Gilroy S, Assmann SM (2003) Sphingolipid signalling in *Arabidopsis* guard cells involves heterotrimeric G proteins. Nature 423: 651-654
7. Desikan R, Griffiths R, Hancock J, Neill S (2002) A new role for an old enzyme: nitrate reductase-mediated nitric oxide generation is required for abscisic acid-induced stomatal closure in *Arabidopsis thaliana*. Proc Natl Acad Sci USA 99: 16314-16318
8. Eun S, Bae, S, Lee Y (2001) Cortical actin filaments in guard cells respond differently to abscisic acid in wild-type and *abi1-1* mutants. Planta 212: 466-469
9. Evans NH, McAinsh MR, Hetherington AM (2001) Calcium oscillations in higher plants. Curr Opin Plant Biol 4: 415-420
10. Fedoroff NV (2002) RNA-binding proteins in plants: the tip of an iceberg? Curr Opin Plant Biol. 5: 452-459

11. Garcia-Mata C, Lamattina L. (2003) Abscisic acid, nitric oxide and stomatal closure - is nitrate reductase one of the missing links? Tr Pl Sci 8: 20-26
12. Geelen D, Leyman B, Batoko H, Di Sansebastiano GP, Moore I, Blatt MR, Di Sansabastiano GP (2002) The abscisic acid-related SNARE homolog NtSyr1 contributes to secretion and growth: evidence from competition with its cytosolic domain. Plant Cell 14: 387-406
13. Goh CH, Kinoshita T, Oku T, Shimazaki K. (1996) Inhibition of blue light-dependent H^+ pumping by abscisic acid in Vicia guard-cell protoplasts. Plant Physiol 111: 433-440
14. Guo F, Young J, Crawford NM (2003) The nitrate transporter AtNRT1.1 (CHL1) functions in stomatal opening and contributes to drought susceptibility in Arabidopsis. Plant Cell 15: 107-117
15. Hamilton DW, Hills A, Kohler B, Blatt MR. (2000) Ca^{2+} channels at the plasma membrane of stomatal guard cells are activated by hyperpolarization and abscisic acid. Proc Natl Acad Sci USA 97: 4967-72
16. Hedrich R, Busch H, Raschke K (1990) Ca^{2+} and nucleotide dependent regulation of voltage dependent anion channels in the plasma membrane of guard cells. EMBO J 9: 3889-3892
17. Hetherington AM (2001) Guard cell signaling. Cell 107: 711-714
18. Hosy E, Vavasseur A, Mouline K, Dreyer I, Gaymard F, Poree F, Boucheez J, Lebaudy A, Bouchez D, Very A, Simonneau T, Thibaud J, Sentenac H (2003) The *Arabidopsis* outward K^+ channel *GORK* is involved in regulation of stomatal movements and plant transpiration. Proc Natl Acad Sci USA 100: 5549-5554
19. Hugouvieux V, Kwak JM, Schroeder JI (2001) An mRNA cap binding protein, ABH1, modulates early abscisic acid signal transduction in Arabidopsis. Cell 106: 477-487
20. Hugouvieux V, Murata Y, Young JJ, Kwak JM, Mackesy DZ, Schroeder JI (2002) Localization, ion channel regulation, and genetic interactions during abscisic acid signaling of the nuclear mRNA cap-binding protein, ABH1. Pl Physiol 130: 1276-1287
21. Hunt L, Mills, LN, Pical C, Leckie CP, Aitken FL, Kopka J, Mueller-Roeber B, McAinsh MR, Hetherington AM, Gray JE (2003) Phospholipase C is required for the control of stomatal aperture by ABA. Plant J 34: 47-55
22. Hwang JU, Lee Y (2001) Abscisic acid-induced actin reorganization in guard cells of dayflower is mediated by cytosolic calcium levels and by protein kinase and protein phosphatase activities. Plant Physiol 125: 2120-2128
23. Jacob T, Ritchie S, Assmann SM, Gilroy S (1999) Abscisic acid signal transduction in guard cells is mediated by phospholipase D activity. Proc Natl Acad Sci USA 96: 12192-12197
24. Jung J, Kim Y, Kwak JM, Hwang J, Young J, Schroeder JI, Hwang I, Lee Y (2002) Phosphatidylinositol 3- and 4-phosphate are required for normal stomatal movements. Plant Cell 14: 2399-2412
25. Köhler B, Blatt MR (2002) Protein phosphorylation activates the guard cell Ca^{2+} channel and is a prerequisite for gating by abscisic acid. Plant J 32: 185-194
26. Köhler B, Hills A, Blatt MR (2003) Control of guard cell ion channels by hydrogen peroxide and abscisic acid indicates their action through alternate signaling pathways. Plant Physiol 131: 385-388
27. Kopka J, Provart NJ, Muller-Rober B (1997) Potato guard cells respond to drying soil by a complex change in the expression of genes related to carbon metabolism and turgor regulation. Plant J 11: 871-882
28. Kwak JM, Moon J, Murata Y, Kuchitsu K, Leonhardt N, DeLong A, Schroeder JI (2002) Disruption of a guard cell-expressed protein phosphatase 2A regulatory subunit, *RCN1*, confers abscisic acid insensitivity in Arabidopsis. Plant Cell 14: 2849-2861
29. Lambermon MH, Fu Y, Kirk DA, Dupasquier M, Filipowicz W, Lorkovic ZJ. (2002) UBA1 and UBA2, two proteins that interact with UBP1, a multifunctional effector of pre-mRNA maturation in plants. Mol Cell Biol 2: 4346-4357
30. Lemichez E, Wu Y, Sanchez J, Mettouchi A, Mathur J, Chua N (2001) Inactivation of AtRac1 by abscisic acid is essential for stomatal closure. Genes & Dev 15: 1808-1816
31. Lemtiri-Chlieh F, MacRobbie EAC, Brearley CA (2000) Inositol hexakisphosphate is a

physiological signal regulating the K$^+$-inward rectifying conductance in guard cells. Proc Natl Acad Sci USA 97: 8687-8692
32. Leonhardt N, Vavasseur A, Forestier C (1999) ATP binding cassette modulators control abscisic acid-regulated slow anion channels in guard cells. Plant Cell 11: 1141-1151
33. Leyman B, Geelen D, Quintero F, Blatt MR (1999) A tobacco syntaxin with a role in hormonal control of guard cell ion channels. Science 283: 537-540
34. Li J, Kinoshita K, Pandey S, Ng C K-Y, Gygi SP, Shimazaki K-I, Assmann SM (2002) Modulation of an RNA-binding protein by abscisic-acid-activated kinase. Nature 418: 793-797
35. Li J, Wang X-Q, Watson MB, Assmann SM (2000) Regulation of abscisic acid-induced stomatal closure and anion channels by guard cell AAPK kinase. Science 287: 300-303
36. Lu Z, Percy RG, Qualset CO, Zeiger E (1998) Stomatal conductance predicts yields in irrigated Pima cotton and bread wheat grown at high temperatures. J Exp Bot 49: 453-460
37. MacRobbie EAC (2000) ABA activates multiple Ca^{2+} fluxes in stomatal guard cells, triggering vacuolar K$^+$ (Rb$^+$) release. Proc Natl Acad Sci USA 97: 12361-12368
38. MacRobbie EAC (2002) Evidence for a role for protein tyrosine phosphatase in the control of ion release from the guard cell vacuole in stomatal closure. Proc. Natl. Acad. Sci. USA 99: 11963-11968
39. Merlot S, Gosti F, Guerrier D, Vavasseur A, Giraudat J (2001) The ABI1 and ABI2 protein phosphatases 2C act in a negative feedback regulatory loop of the abscisic acid signalling pathway. Plant J 25: 295-303
40. Mori IC, Uozumi N, Muto S (2000) Phosphorylation of the inward-rectifying potassium channel KAT1 by ABR kinase in Vicia guard cells. Plant Cell Physiol 41: 850-856
41. Murata Y, Pei Z, Mori IC, Schroeder J (2001) Abscisic acid activation of plasma membrane Ca^{2+} channels in guard cells requires cytosolic NAD(P)H and is differentially disrupted upstream and downstream of reactive oxygen species production in *abi1-1* and *abi2-1* protein phosphatase 2C mutants. Plant Cell 13: 2513-2523
42. Mustilli AC, Merlot S, Vavasseur A, Fenzi F, Giraudat J (2002) Arabidopsis OST1 protein kinase mediates the regulation of stomatal aperture by abscisic acid and acts upstream of reactive oxygen species production. Plant Cell 14: 3089-3099
43. Neill SJ, Desikan R, Clarke A, Hancock JT (2002) Nitric oxide is a novel component of abscisic acid signaling in stomatal guard cells. Plant Physiol 128: 13-16
44. Ng CK, Carr K, McAinsh MR, Powell B, Hetherington AM (2001) Drought-induced guard cell signal transduction involves sphingosine-1-phosphate. Nature. 410: 596-599
45. Outlaw WH Jr. (2003) Integration of cellular and physiological functions of guard cells. Crit Rev Pl Sci, in press
46. Pandey S, Wang X-Q, Coursol SA, Assmann SM (2002) Preparation and applications of *Arabidopsis thaliana* guard cell protoplasts. New Phytol 153: 517-526
47. Pei Z, Murata Y, Benning G, Thomine S, Klüsener B, Allen GJ, Grill E, Schroeder JI (2000) Calcium channels activated by hydrogen peroxide mediate abscisic acid signaling in guard cells. Nature 406: 731-734
48. Rock CD (2000) Tansley Review No. 120. Pathways to abscisic acid-regulated gene expression. New Phytol 148: 357 - 396
49. Roelfsema MR, Steinmeyer R, Staal M, Hedrich R (2001) Single guard cell recordings in intact plants: light-induced hyperpolarization of the plasma membrane. Plant J 26: 1-13
50. Sanders D, Pelloux J, Brownlee C, Harper JF (2002) Calcium at the crossroads of signaling. Plant Cell 14: S401-417
51. Schmidt C, Schroeder JI (1994) Anion selectivity of slow anion channels in the plasma membrane of guard cells (large nitrate permeability). Plant Physiol 106: 383-391
52. Schroeder JI, Allen GJ, Hugouvieux V, Kwak JM, Waner D (2001) Guard cell signal transduction. Annu Rev Plant Physiol Plant Mol Biol 52: 627-658
53. Schroeder JI, Kwak JM, Allen GJ (2001) Guard cell abscisic acid signaling and engineering drought hardiness in plants. Nature 410: 327-330
54. Sharp RE, LeNoble ME (2002) ABA, ethylene and the control of shoot and root growth

under water stress. J Ex Bot 53: 33-37
55. Szyroki A, Ivashikina N, Dietrich P, Roelfsema MR, Ache P, Reintanz B, Deeken R, Godde M, Felle H, Steinmeyer R, Palme K, Hedrich R (2001) KAT1 is not essential for stomatal opening. Proc Natl Acad Sci U S A 98: 2917-2921
56. Wang X-Q, Ullah H, Jones AM, Assmann SM (2001) G protein regulation of ion channels and abscisic acid signaling in *Arabidopsis* guard cells. Science 292: 2070-2072
57. Wilkinson S, Davies WJ (2002) ABA-based chemical signaling: the co-ordination of responses to stress in plants. Plant Cell Environ 25: 195-210
58. Yoshida R, Hobo T, Ichimura K, Mizoguchi T, Takahashi F, Aronso J, Ecker JR, Shinozaki K (2002) ABA-activated SnRK2 protein kinase is required for dehydration stress signaling in Arabidopsis. Plant Cell Physiol 43: 1473-1483
59. Zhang DP, Wu ZY, Li XY, Zhao ZX (2002) Purification and identification of a 42-kilodalton abscisic acid-specific-binding protein from epidermis of broad bean leaves. Plant Physiol 128: 714-725
60. Zhang X, Zhang L, Dong F, Gao J, Galbraith DW, Song C (2001) Hydrogen peroxide is involved in abscisic acid-induced stomatal closure in *Vicia faba*. Plant Physiol 126: 1438-1448
61. Zheng A, Nafisi M, Tam A, Li H, Crowell DN, Chary SN, Schroeder JI, Shen J, Yang Z (2002) Plasma membrane-associated ROP10 small GTPase is a specific negative regulator of abscisic acid responses in Arabidopsis. Plant Cell 14: 2787-2797

D7. Brassinosteroid Signal Transduction and Action

Steven D. Clouse
Department of Horticultural Science, North Carolina State University, Raleigh, North Carolina, 27695 USA. Email: steve_clouse@ncsu.edu

INTRODUCTION

Brassinosteroids (BRs[1]) are endogenous plant growth-promoting hormones that are found throughout the plant kingdom in seeds, pollen and young vegetative tissue, functioning at nanomolar concentrations to influence cellular expansion and proliferation (13, 34, 43). BRs also interact with other plant hormones and environmental factors to regulate the overall form and function of the plant, and examination of the phenotype of mutants affected in BR biosynthesis or signaling provides genetic evidence that BRs are essential for normal organ elongation, vascular differentiation, male fertility, timing of senescence, leaf development and responses to light (1). The biosynthesis and metabolism of BRs are covered in detail in chapter B6. Here, the rapid acceleration in our understanding of BR signal transduction is discussed along with studies on BR-regulated gene expression. The impact of these studies on clarifying the physiological mode of action of BRs will also be considered.

ACTIONS OF BRASSINOSTEROIDS

Biochemical and genetic analysis of BR signaling and biosynthetic mutants has increased our understanding of the physiological mode of action of BRs, with mechanistic studies on cell elongation, differentiation and division reported. Besides direct effects on growth, BRs mediate abiotic and biotic stresses, such as salt and drought stress, temperature extremes and pathogen attack (43).

---

[1]Abbreviations: BAK1, BRI1-Associated Kinase 1; *bes1-D, bri1-EMS-suppressor 1-Dominant*; *bin2, brassinosteroid insensitive 2*; BRI1, Brassinosteroid Insensitive 1; BRS, BRI1 Suppressor; *bzr1-D, brassinazole resistant 1-Dominant*; *cbb, cabbage*; *dim, dimunito*; CPD, Constitutive Photomorphogenesis and Dwarfism; *cu3, curl3*; *det2, deetiolated 2*; *dwf, dwarf*; GFP, Green Fluorescent Protein; GSK-3, Glycogen Synthase Kinase-3; KD, kinase domain; LRR, Leucine-Rich Repeat; RLK, Receptor Like Kinase; TGF, Transforming Growth Factor; *TCH4, Touch4*; TRIP, TGF-β Receptor Interacting Protein; *ucu1, ultracurvata1*; XTH, Xyloglucan Endotransglucosylase/Hydrolase.

Cell Elongation

The primary cell walls in dicotyledonous and non-Poaceae monocotyledonous plants are thought to consist of cellulose microfibrils tethered into a network by non-covalent attachment to hemicelluloses (primarily xyloglucans) which, in turn, are embedded in a pectic gel matrix (7). Regulation of the synthesis and activity of wall-modifying enzymes and other proteins such as XTHs, glucanases, expansins, sucrose synthase and cellulose synthase, is an obvious target for hormones involved in cell elongation. BR regulation of genes encoding XTHs and expansins has been demonstrated in soybean, tomato, rice and Arabidopsis. Consistent with a wall-modifying affect, BRs have been shown by biophysical measurements to promote wall loosening in soybean epicotyls and hypocotyls of *Brassica chinensis* and *Cucurbita maxima* (4, 13, 43).

The dwarf stature of BR-deficient mutants (Chapter B6) and their rescue to wild-type phenotype specifically by BR treatment, points to a critical role for BRs in normal plant growth. Examination of cell files by light and electron microscopy in wild-type Arabidopsis and *cbb*, *dwf*4, *cpd* and *dim* mutant plants, provides direct physical evidence that longitudinal cell expansion is dramatically reduced in BR mutants (1, 43). The overexpression of DWF4, which encodes the enzyme controlling the putative rate-limiting step in BR biosynthesis, resulted in increased hypocotyl length in Arabidopsis, suggesting that increasing endogenous BR levels leads directly to increased cell expansion (8). The arrangement of cortical microtubules is also known to be an important factor in regulating cell elongation, and BRs have been shown by physiological and genetic experiments to affect re-configuration of microtubules to the transverse orientation which allows longitudinal growth (13). In addition to alterations in cell wall properties, BRs may also influence cell elongation by affecting transport of water via aquaporins and by regulating the activity of a vacuolar H^+-ATPase subunit (18, 37).

Cell Division

BRs promote cell division as well as elongation in bean second internodes, and have been shown to stimulate cell proliferation in the presence of auxin and cytokinin in cultured parenchyma cells of *Helianthus tuberosus*, and in protoplasts of Chinese cabbage and petunia (13, 43). A role for BRs in Arabidopsis cell division has been suggested by the observation that 24-epibrassinolide treatment of *det*2 cell suspension cultures increases transcript levels of the gene encoding cyclinD3, a protein involved in the regulation of G1/S transition in the cell cycle. CyclinD3 is also regulated by cytokinins and it may be significant that 24-epibrassinolide effectively substitutes for zeatin in the growth of Arabidosis callus and cell suspension cultures (25).

Cell Differentiation

A variety of experiments in several different species provide evidence that BRs play an important role in vascular differentiation. Low concentrations of BR stimulate tracheid formation in *H. tuberosus* explants and isolated mesophyll cells of *Zinnia elegans,* and in the *Zinnia* system BRs also regulate the expression of several genes associated with xylem formation (20). BRs also regulate the expression of an XTH in soybean epicotyls that is expressed at high levels in paratracheary parenchyma cells surrounding vessel elements, suggesting a role for BRs and XTHs in xylem formation (13). Microscopic analysis of BR mutants has also shown a role for endogenous BRs in vascular differentiation. For example, the BR-deficient mutant *cpd* exhibits unequal division of the cambium, producing extranumerary phloem cell files at the expense of xylem cells (45). Moreover, the sterol and BR-deficient mutant *dwf7* also exhibits an increase in phloem vs. xylem cells and the number of vascular bundles is reduced from eight in the wildtype to six in the mutant, with irregular spacing between vascular bundles (9).

Reproductive Biology and Senescence

Reduced fertility or male sterility is commonly observed in BR-deficient and insensitive mutants. Pollen and immature seeds are generally the richest sources of endogenous BRs, and *in vitro* studies in *Prunus avium* suggest that pollen tube elongation may be partly BR-dependent (13). The male sterility of the BR-deficient *cpd* mutant has been explained by the inability of pollen to elongate during germination (45). However, in the *dwf4* BR biosynthetic mutant, affected in the step immediately preceding CPD, the pollen appears to be viable and sterility is due to the reduced length of stamen filaments which results in the deposition of pollen on the ovary wall rather than the stigmatic surface. The sterol and BR-deficient mutant *dwf5-1* has wild-type fertility and interestingly, is also the only BR mutant with stamens longer than the gynoecium (11). Despite its increased fertility, *dwf5-1* produces seeds that do not develop normally and that require exogenous BR application for full germination. A possible endogenous role for BRs in Arabidopsis seed germination has been proposed. ABA and GA play well-documented antagonistic roles in establishing and breaking dormancy during seed development and germination. BRs can rescue the germination defect in GA biosynthetic and insensitive mutants, and the BR mutants *det2* and *bri*1 are more susceptible to inhibition of germination by ABA than wildtype. Therefore, BR signaling may be required to reverse ABA-induced dormancy and to stimulate germination (4).

Most of the BR mutants also exhibit an extended life span and delayed senescence when compared to wild-type plants grown under the same conditions. The extent of the delayed senescence is correlated with reduced fertility. Sterile mutants such as *bri*1 have the most delayed development, and it has been proposed that the infertile mutants are unable to produce signals for the onset of senescence which leads to the observed extended life span. Consistent with this hypothesis, the fertile *dwf*5-1 mutant does not show delayed senescence (11). Delayed senescence in *Arabidopsis* BR mutants suggests a role for BRs in accelerating senescence in normal plants, however it is not yet clear whether BRs play a critical function in the intrinsic program of senescence in vegetative tissue.

STEROIDS AS SIGNALLING MOLECULES

Steroids are essential signaling molecules in vertebrates and invertebrates with multiple roles in regulating development and adult homeostasis. The binding of a specific steroid to its cognate receptor initiates a cascade of regulatory events and because of the immense practical importance of steroids to human and animal health, mammalian receptors have been studied in detail for several decades. The classical animal intracellular steroid receptor is composed of modular domains including a variable N-terminal region often associated with transcriptional activation, a highly conserved DNA-binding domain with two zinc fingers, and a multifunctional domain mediating ligand-binding, receptor dimerization and ligand-dependent transcriptional activation (2). The steroid-receptor complex is nuclear localized and recognizes specific motifs in the promoters of steroid-responsive genes, resulting in altered gene expression with consequent changes in cellular biochemistry and physiology. A large superfamily of nuclear steroid receptors exists in vertebrates and invertebrates, which recognize steroids such as androgens, estrogens, progesterone, ecdysones, glucocorticoids and mineralocorticoids.

Plants contain many of the same steroids as animals (21), but it was not until reports on the chemical structure and biological activities of BRs began to appear that it was suggested that steroids could play a major role in regulating plant growth and development (34). Because of the close structural similarity of BRs to animal steroid hormones, it was postulated that plants might have similar nuclear steroid receptors and initial studies of BR signal transduction focused on screening plant sequences with probes or primers based on conserved regions of the DNA-binding domains of animal steroid receptors. Such experiments failed to reveal any plant homologs of animal nuclear steroid receptors and examination of the completed genome sequence of Arabidopsis further indicated that plants may have conserved the steroid signal but not the classical steroid receptors found in animal systems. As is the case with several plant hormone analyses, genetics screens of Arabidopsis for hormone insensitivity served as the foundation on which our growing understanding of BR signal transduction has been assembled. The components of BR signaling so far identified consist of two cell-surface localized receptor-like kinases, a cytoplasmic kinase that serves in a negative regulatory role, and two closely related positive activators that are nuclear localized in response to BR treatment. The emerging picture of BR signal transduction is one that is distinctly different from the classical model of animal steroid signaling, but one that does share certain similarities to the more recently defined "non-genomic" steroid pathways in animals which involve perception of steroids by cell surface receptors, and to non-steroid eukaryotic signaling pathways such as TGF-β and the Wingless/Wnt pathway of *Drosophila* and mammals (14).

BR SIGNAL TRANSDUCTION

The initial phase of work on BR signal transduction focused on mutant screens for BR insensitivity, with the assumption that such mutants would likely reflect defects in the BR receptor or other components of the signaling pathway. It is a common observation that many plant hormones, when applied at appropriate concentrations, can inhibit primary root elongation in Arabidopsis. BRs also demonstrate this effect and the first BR-insensitive mutant, *bri*1, was identified by screening 70,000 EMS-mutagenized seedlings for elongated roots in the presence of 0.1 μM 24-epibrassinolide (15). The *bri*1 mutant elongates roots over a wide range of BR concentrations, but retains normal sensitivity to auxins, cytokinins, GA, and ethylene, and shows hypersensitivity to ABA. Besides the root phenotype, *bri*1 exhibits extreme dwarfism, altered leaf morphology, male sterility, delayed development, and reduced apical dominance, particularly in older plants. More than two dozen *bri*1 alleles have been identified in a variety of independent screens, which has greatly facilitated precise chromosomal mapping and positional cloning of the *BRI1* locus (27, 30, 40). Table 1 shows the known alleles of *bri*1 and the location of the mutations within the *BRI1* gene. BR-insensitive mutants have also been identified in pea, rice and tomato (36, 49). The BR-insensitive tomato mutant *cu3* is strikingly similar to Arabidopsis *bri1* in appearance and root response to various hormones (29).

BRI1 is an LRR-Receptor Kinase Required for BR Perception

The structure of the *BRI1* gene was first determined in Arabidopsis by positional cloning and mutations in this gene were shown to be responsible for the *bri1* phenotype by sequencing numerous mutant alleles (30). The predicted protein is an 1196 amino acid leucine-rich repeat (LRR) receptor kinase that contains all three major domains common to animal and plant receptor kinases; including the extracellular domain, putatively involved in ligand binding; the single pass transmembrane domain which anchors the protein in the plasma membrane; and the cytoplasmic kinase domain which propagates the signal downstream by phosphorylating protein substrates. Searching a rice expressed sequence tag database identified a clone with sequence similarity to BRI1. The full-length cDNA corresponding to this clone rescued the rice BR-insensitive dwarf mutant *d61*, confirming this gene as the rice homolog (OsBRI1) of Arabidopsis BRI1 (49). Alignment of the

Table 1. Mutant alleles of BRI

Allele	Alias	Predicted Mutation	Location of Mutation
Arabidopsis severe/null *bri*1 alleles			
bri 1-1		Ala-909-Thr	Kinase subdomain II
*bri*1-2	*cbb*2	Premature Stop	Truncated after LRR18
*bri*1-3	*dwf*2-32	Premature Stop	Truncated after LRR4
*bri*1-4	*dwf*2-2074	Premature Stop	Truncated after kinase subdomain II
*bri*1-101	*bin*1-1	Glu-1078-Lys	Kinase subdomain IX
*bri*1-102	*bin*1-2	Thr-750-Ile	Between LRR25 and second paired Cys
*bri*1-103, 104	*bin*1-3, 4	Ala-1031-Thr	Kinase subdomain VII
*bri*1-105-107	*bin*1-5-7	Gln-1059-Stop	Truncated after subdomain VIII
*bri*1-108-112	*bin*1-8-12	Arg-983-Gln	Kinase subdomain VIa
*bri*1-113	*bin*1-13	Gly-611-Glu	Island domain between LRR21 and LRR22
*bri*1-114, 116	*bin*1-14, 16	Gln-583-Stop	Truncated in Island domain before LRR22
*bri*1-115	*bin*1-15	Gly-1048-Asp	Kinase subdomain VIII
*bri*1-117, 118	*bin*1-17, 18	Asp-1139-Asn	Kinase subdomain XI
*bri*1-201		Premature Stop	Truncated after 44 amino acids
Arabidopsis weak and intermediate *bri*1 alleles			
*bri*1-5	*dwf*2-w41	Cys-69-Tyr	First paired Cys in extracellular domain
*bri*1-6, 119	*dwf*2-399	Gly-644-Asp	Island domain between LRR21 and LRR22
*bri*1-7	*dwf*2-WM3-2	Gly-613-Ser	Island domain between LRR21 and LRR22
*bri*1-8	*dwf*2-WM6-2	Arg-983-Asn	Kinase subdomain VIa
*bri*1-9	*dwf*2-WMB19	Ser-662-Phe	LRR22
*bri*1-301		Glu-989-Ile	Kinase subdomain VIa
Rice *bri*1 alleles			
d61-1		Thr-989-Ile	Kinase subdomain IX (weak allele)
d61-2		Val-491-Met	LRR17 (more severe allele)
Tomato *bri*1 alleles			
cu3		Glu-749-Stop	Truncated after LRR25 (severe/null allele)
cu3^{-abs}		His-1012-Tyr	Kinase subdomain VIb (weak allele)

Arabidopsis and rice BRI1 sequences revealed areas of high homology which guided the synthesis of degenerate primers for isolation of the tomato BRI1 homolog, subsequently named tBRI1. Sequence analysis of tBRI1 in the *cu3* mutant background identified a nonsense mutation in the extracellular domain that would likely result in a null mutant, accounting for the severe phenotype of *cu3*. This mutation also generated a TspRI restriction site polymorphism which co-segregated with the mutant phenotype, providing further evidence that the developmental defects in *cu3* are caused by a mutation in the *tBRI1* gene (36).

Figure 1 shows the various functional domains of Arabidopsis BRI1 and the percent sequence identity of tBRI1 and OsBRI1 to each of these domains. In all three species the extracellular domain begins with an N-terminal signal peptide, putatively involved in delivering the protein to the cell surface, and in fact an Arabidopsis BRI-GFP fusion has been shown to localize to the plasma membrane. In Arabidopsis and tomato BRI the signal peptide is

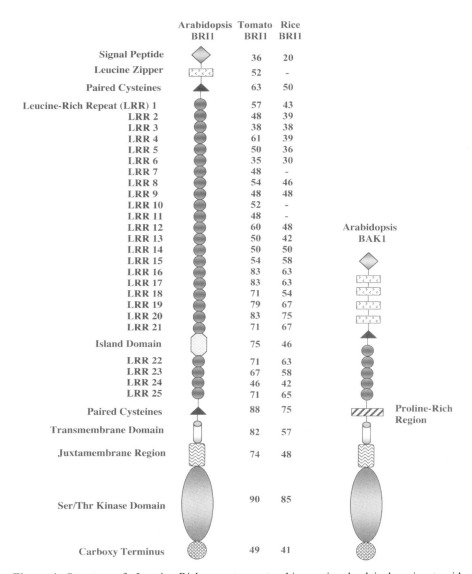

Figure 1. Structure of Leucine-Rich repeat receptor kinases involved in brassinosteroid signal transduction. BRI1 and BAK1 may form a heterodimer on the cell surface and are both involved in BR signaling. BRI1 has been shown to be at least one component of the BR receptor. Percent sequence identity of tomato and rice BRI1 to the Arabidopis protein is shown to the right of each domain. All segments of the protein N-terminal to the transmembrane domain are extracellular, while those portions C-terminal of the transmembrane domain lie in the cytoplasm.

followed by a leucine zipper motif, which is not as conserved in OsBRI1. The major portion of the extracellular domain consists of 22 to 25 tandem copies of a 24-amino acid LRR flanked on each end by conservatively spaced paired cysteines. Both leucine zippers and LRRs are known to be involved in protein-protein interactions and it has been observed that many receptor kinases dimerize in response to ligand binding (24). The molecular structure of the BRI1 extracellular domain suggests possible formation of homo or heterodimers, as discussed below. All known BRI proteins contain a 68-70 amino acid island inserted four LRRs upstream of the membrane spanning region, and mutational analysis has assigned an important biological function to this region (18). Following the extracellular region there is a predicted hydrophobic transmembrane domain spanning amino acids 793-814 of Arabidopsis BRI1. The cytoplasmic domain consists of a juxtamembrane region (amino acids 815-882), followed by a Ser/Thr kinase domain (amino acids 883-1155) showing all invariant amino acid residues and conserved sub-domains of eukaryotic kinases. Finally, a sequence of approximately 40 amino acids represents the carboxy-terminal segment of the cytoplasmic domain.

BRI1 function
A discussion of the functional analysis of BRI1 will be enhanced by a summary of how receptor-like kinases (RLKs) in general participate in signaling pathways. The Arabidopsis genome encodes 417 RLKs with an organization of functional domains similar to that of animal receptor kinases (3). Several of these plant RLKs have proven functional roles in the regulation of plant growth, morphogenesis, disease resistance, and responses to the environment, but the functions of most members of this large and important family of signal transduction molecules remain unknown. The 417 Arabidopsis RLKs can be grouped into 21 families based on the structure of the extracellular domain, with LRRs predominating (216 out of 417). The LRR-RLK family can be further classified into 13 sub-families based on similar structural organization and sequence alignment. Besides BRI1, several individual LRR subfamily members have been extensively characterized, including BAK1, discussed in detail below (30, 33, 39, 47); CLAVATA1, controlling meristematic cell fate; ERECTA, specifying organ shape; HAESA, which plays a role in organ abscission; and AtSERK1, involved in early embryogenesis (3).

The mechanism of action of numerous animal receptor kinases has been studied in detail and generally involves ligand-mediated homo- or heterodimerization of the receptor followed by autophosphorylation and activation of the intracellular kinase domain. Kinase activation leads to recognition and phosphorylation of downstream components of the signaling pathway, resulting ultimately in altered expression of specific genes (44). Based on molecular structure and several preliminary studies, plant RLKs are likely to follow the same general mechanism of action (3). Differential function of an RLK could arise at several levels, including the specific ligand recognized by the extracellular domain, heterodimerization with different partners, variation in sites of autophosphorylation leading to generation of specific docking sites for downstream signaling components, and tissue-specific expression patterns of the RLK during development. A thorough characterization of plant RLK function requires understanding the mechanism of receptor dimerization and its affect on kinase domain autophosphorylation, including identification of specific autophosphorylation sites whenever possible. Identification of RLK kinase domain substrates and their downstream interacting partners, along with specific genes that are ultimately regulated in response to binding of the ligand to the receptor kinase, complete the picture of RLK signaling. Successful approaches to understanding RLK function include mutational analysis, functional genomics, bioinformatics, kinase biochemistry and phosphoprotein analysis using mass spectrometry (14).

All of the above approaches are being applied to the functional analysis of BRI1. The large number of mutant alleles available (Table 1) clearly show the biological significance of both the extracellular and kinase domains. Point mutations in critical residues of various kinase subdomains result in complete loss of Arabidopsis BRI1 activity with a resultant severe mutant phenotype, indicating the importance of a fully functional kinase domain for normal BRI1 action. In the extracellular domain, several mutations cluster in the island between LRR21 and LRR22 with the severity of the mutation depending on the specific residue altered. Similarly, mutations in the LRR and paired cysteine regions have a variable affect from severe to quite weak mutant phenotypes. Mutations in the extracellular domain would be expected to affect BR binding and/or interactions with other proteins. The functional significance of specific BRI1 domains can also be assessed by comparing sequence conservation in known BRI1 proteins. The kinase domain shows the highest sequence conservation followed by the second set of paired cysteines and the membrane spanning domain. The island domain and the juxtamembrane region both show similar levels of identity between Arabidopsis and tomato (75%) and between Arabidopsis and rice (46%). Interestingly, the LRRs flanking the island domain are more highly conserved than those closer to the N-terminus.

BRI1 is involved in binding BR
Direct biochemical approaches have demonstrated the role of BRI1 as at least one component of the BR receptor. Radiolabeled brassinolide binds to microsomal fractions from wildtype Arabidopsis plants and to an even greater extent from transgenic plants overexpressing BRI1-GFP fusions (47). The BR-binding activity could be precipitated by antibodies to GFP, was competitively inhibited by active BRs, and was of high affinity (K_d = 7.4 +/- 0.9 nM), consistent with concentrations of BR known to regulate physiological responses *in planta*. Moreover, binding was eliminated by mutations in the extracellular domain but not the kinase domain of BRI1. Binding of labeled BR to purified recombinant BRI1 has not been shown, so it is currently not possible to distinguish mechanisms of direct BR binding to the extracellular domain of BRI1 vs. binding of BR to an unidentified steroid-binding protein which then complexes with BRI1.

LRRs are generally not known for binding small molecules such as steroids, although this possibility cannot be ruled out. However, if BR does bind BRI1 directly, it might be more likely through the island domain embedded in the LRR region. Alternatively, the Arabidopsis genome contains sequences with similarity to animal high affinity binding proteins for sex steroids that are not in the classical superfamily of steroid receptors. These include putative membrane progesterone receptors and the soluble sex hormone binding globulins (17). These Arabidopsis proteins represent candidates for possible accessory BR-binding proteins and it would be of great interest to determine if recombinant proteins derived from these sequences can bind labeled BR directly. It is conceivable that such a BR-protein complex would then bind BRI1 via the LRRs and/or island domain. A dominant gain-of-function mutant, *brs1-D*, that suppresses the weak *bri1-5* extracellular domain mutant was identified by activation tagging (31). The BRS1 protein is a putative secreted serine carboxypeptidase that may act to process a protein involved in early events of BR signaling, hypothetically the putative steroid binding proteins mentioned above, some of which contain sequences for carboxypeptidase cleavage. Interestingly, tBRI1 is identical to SR160, which has been shown by biochemical methods to bind the tomato peptide hormone systemin, involved in defense responses, strengthening the argument that BRI1 may be able to bind both peptides and steroids (36).

BRI1 is an active Ser/Thr kinase

Biochemical approaches have also been applied to study of the kinase domain (KD) and show that BRI1 is an active Ser/Thr kinase (19, 41). Using recombinant BRI1-KD and Matrix-Assisted Laser Desorption/Ionization Mass Spectrometry, at least twelve sites of *in vitro* autophosphorylation were identified, five uniquely and seven with some remaining ambiguity because of multiple Thr or Ser residues within the peptide (41). Some of the BRI-KD autophosphorylated Ser and Thr residues are conserved at corresponding positions in related plant Ser/Thr kinases, with the region of greatest conservation occurring in the peptide 1038-DTHLSVSTLAGTPGY-VPPEYYQSFR-1062, in the activation loop of kinase subdomain VIII. As mentioned above (Table 1) several *bri1* mutant alleles are located in this region. Autophosphorylation in subdomain VIII likely leads to general kinase activation, while multiple autophosphorylation sites in the juxtamembrane and carboxy terminal regions, which are not conserved in other kinases, might indicate specific docking sites for interacting cytoplasmic partners of BRI1.

BRI1 phosphorylation has also been shown to occur *in planta* in a BR-dependent manner. When BR biosynthetic mutants or wild-type plants treated with a BR biosynthesis inhibitor are incubated with BL, BRI1 mobility shifts to a slower-moving form. Treatment with alkaline phosphatase indicates that the slower mobility form is phosphorylated BRI1. This shift in mobility does not occur in a BRI1 kinase mutant, suggesting that autophosphorylation is responsible for the mobility shift in wild-type plants

(47). In summary, molecular studies have shown that BRI1 is part of the BR receptor complex and binding of the ligand activates the kinase domain by autophosphorylation. Thus BRI1 appears to follow the general mechanistic paradigm of animal receptor kinases. It is well known that animal receptor kinases form homo- or heterodimers and, as discussed below, recent genetics screens have shown that BRI1 is associated with a second LRR receptor-like kinase *in vivo*.

BAK1 is an LRR-Receptor-Like Kinase Associated with BRI1

BAK1 was identified independently by activation tagging for *bri1-5* suppressors and by a yeast two-hybrid screen for BRI1 interacting proteins. BAK1 is a member of the LRRII-RLK subfamily that includes AtSERK1, expressed in closed flower buds before fertilization and during stages 1-7 of Arabidopsis embryo development (3). BAK1 and AtSERK1 exhibit 80% amino acid identity, yet show markedly different patterns of gene expression. The 12 other members of the LRRII subfamily share 44 to 91% identity with BAK1 or AtSERK1. In contrast to BRI1, BAK1 has only five leucine-rich repeats in its extracellular domain, lacks the 70 amino acid island of BRI1 and has a proline-rich region between the terminal LRR and the membrane-spanning domain (Fig. 1). BAK1 is expressed in all tissues of the plant, similar to the global expression pattern of BRI1 and confocal laser microscopy of transgenic plants expressing a BAK1-GFP fusion, shows plasma membrane localization in a pattern similar to BRI1-GFP. Direct physical interaction between BRI1 and BAK1 was demonstrated both in yeast cells and Arabidopsis plants by co-immunoprecipitation experiments with tagged proteins (33, 39).

Genetic analysis demonstrated a clear role for BAK1 in BR signaling. The dominant mutant *bak1-1D* has *BAK1* mRNA levels that are 30-fold higher than wildtype or *bri1-5*, suggesting that overexpression of *BAK1* leads to the suppressed phenotype of *bri1-5* in the activation tagging lines. This was further supported by phenotypic suppression of *bri1-301*, another weak allele of *bri1*, in transgenic mutant plants overexpressing *BAK1* (33, 39). However, phenotypic suppression was not observed in double mutants of *bak1-1D* and null mutants of *bri1*, or when *bri1-5 bak1-1D* plants were crossed into a BR-deficient mutant background, indicating that both partial BRI1 activity and the presence of BR are required for suppression. Knockout mutants of *BAK1* exhibit a weak *bri1*-like phenotype including decreased sensitivity to BR in the root inhibition assay (33, 39). The weaker phenotype of *bak1* null mutants compared to severe *bri1* mutants might result from functional redundancy, since *BAK1* is a member of a multigene family. The genetic evidence also supports a direct *in vivo* interaction between BRI1 and BAK1 in plant membranes. Overexpression of a kinase-deficient mutant form of *BAK1* in *bri1-5* led to a severe dwarf phenotype, suggesting a dominant-negative effect, most likely arising by the poisoning of a possible heteromeric complex between BRI1 and BAK1 (33).

Biochemical studies with radiolabeled BL showing BAK1 is an essential component of the BR receptor complex have not yet been reported, but the BR-binding results with BRI1 mentioned above do not rule out the inclusion of BAK1 in such a complex, since BAK1 was not discovered at the time of the earlier publication. For example, immunoprecipitation of BRI1-GFP BR binding activity by GFP antibody might have also co-immunoprecipitated BAK1 as part of a complex. Further evidence of a BRI1/BAK1 interaction comes from expression studies of full-length BRI1 and BAK1 in yeast cells. When expressed alone, neither BRI1 nor BAK1 showed appreciable autophosphorylation. However when co-expressed, transphosphorylation between BRI1 and BAK1 predominated and mutant analysis showed that the kinase activities of both BRI1 and BAK1 were required for phosphorylation, even though physical association in a heterodimer still occurred in mutant kinases. Interestingly, application of BR to yeast cells did not enhance BRI1/BAK1 phosphorylation, suggesting the possibility that yeast cells lack a required accessory protein for BR binding, perhaps one of the plant steroid binding proteins described above. Recombinant cytoplasmic domains of BRI1 and BAK1 purified from bacterial cells can both autophosphorylate independently, but also showed trans-phosphorylation of each other *in vitro*, with BAK1 showing a noticeable increase in phosphorylation when incubated with BRI1 (33).

Possible Mechanisms of BRI1/BAK1 Action

If heterodimerization of BRI1 and BAK1 is dependent on BR binding, it would suggest BR signal transduction shares mechanistic similarities to animal receptor tyrosine kinases and/or the TGF-β Ser/Thr receptor kinases. The TGF-β family of polypeptide growth factors regulate cell proliferation, lineage determination, differentiation and death, and play prominent roles in the development and homeostasis of multicellular organisms from fruit fly to human (35). The TGF-β receptor complex is a heterotetramer of receptor pairs called type I (RI) and type II (RII). RII receptors homodimerize in the absence of TGF-β ligand and exhibit constitutive kinase activity. TGF-β binding by RII induces formation of the heterotetramer with RI and results in phosphorylation of RI by RII on specific Thr and Ser resides in a region immediately preceding the RI kinase domain. Once activated by phosphorylation, RI propagates the signal by phosphorylating proteins, termed Smads, which translocate to the nucleus and form complexes with transcription factors to regulate the expression of TGF-β-responsive genes (Fig.2).

The RII receptor also phosphorylates specific cytoplasmic substrates, including TRIP-1, a unique WD-40 domain protein that functions both as a modulator of TGF-β-regulated gene expression and as an essential subunit of the eukaryotic translation initiation factor, eIF3 (12). eIF3 is the largest eukaryotic translation initiation factor, consisting of 10 to 11 subunits, and is required for binding of mRNA to the 40S ribosomal subunit. While the role

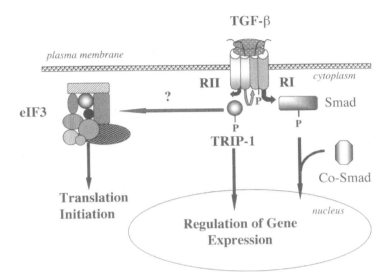

Figure 2. TGF-β signaling pathway. Binding of the TGF-β polypeptide ligand by the Ser/Thr receptor kinase RII leads to assembly of the heterotetramer and unidirectional phosphorylation of RI by RII. RI phosphorylates signaling proteins termed Smads which move to the nucleus where they associate with transcription factors and other regulatory proteins to modulate gene expression. RII phosphorylates TRIP-1 directly which is also involved in regulation of gene expression. TRIP-1 has a dual role as an essential component of the eIF3 translation initiation factor. It is currently not known whether phosphorylation of TRIP-1 by RII plays a role in the assembly or activity of eIF3.

of TRIP-1 in TGF-β signaling has been established, it is currently not known whether TRIP-1 phosphorylation by RII is essential for assembly or activity of eIF3. If such were the case, it would represent a non-genomic regulation of cellular activity by TGF-β, i.e. control of translation initiation. The Arabidopsis homolog of TRIP-1 has been shown to be a functional component of the plant eIF3 complex (6), and its possible role in BR signaling will be discussed further below.

Several of the known facts about BRI1 and BAK1 are consistent with a TGF-β type model (Fig. 3). First, BRI1 has a much larger extracellular domain than BAK1, including the island domain important for BRI1 function. If BR (or a BR-sterol binding protein complex) binds BRI1 directly, it might initiate heterodimerization with BAK1, much like RII binding of TGF-β followed by oligomerization with RI. Second, while transphorylation in both directions between BRI1 and BAK1 was shown in yeast cells, *in vitro* experiments with recombinant kinase domains showed that BRI1 stimulated BAK1 phosphorylation (33, 39). Moreover, earlier experiments revealed that *in vitro* autophosphorylation of BRI1-KD most likely occurred by an intramolecular mechanism (41). The critical *in planta* experiments regarding BRI1 phosphorylation of BAK1 have not been

Brassinosteroid signaling and action

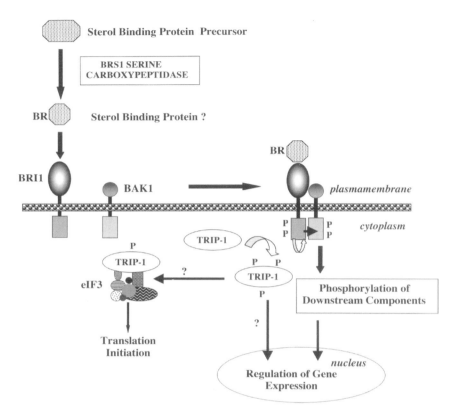

Figure 3. TGF-β model for interaction of BRI1 and BAK1. In this model, BRI1 binds BR (either directly or in association with a sterol-binding protein that has been processed by the BRS1 carboxypeptidase), which leads to association with and phosphorylation of BAK1. BAK1 phosphorylates unknown downstream components which eventually leads to regulation of gene expression. In a manner analogous to TGF- β RII, BRI1 phosphorylates TRIP-1. While *in vitro* phosphorylation of TRIP-1 by BRI1 has been shown, as well as *in vivo* association of the two proteins, it is not yet clear what functional role TRIP-1 plays in BR signaling.

undertaken, but a useful working hypothesis is one in which BR binds BRI1, leading to autophosphorylation, followed by dimerization with BAK1 and predominant phosphorylation of BAK1 by BRI1 in a manner similar to RII phosphorylation of RI (Fig. 3). Finally, in keeping with the analogy of BRI1 with RII, it is quite interesting that the Arabidopsis TRIP-1 homolog is phosphorylated by BRI1 *in vitro*, as discussed in more detail below. It should be noted, however, that although the general mechanism may have similarities, there is not extensive sequence identity between BRI and RII, or BAK1 and RI.

While the TGF-β analogy is intriguing, current data cannot rule out a mechanism closer to animal receptor tyrosine kinases (Fig. 4), which

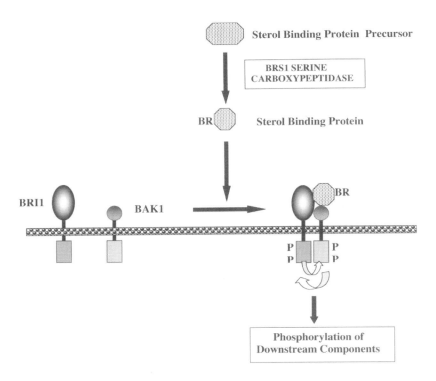

Figure 4. Animal receptor tyrosine kinase model for interaction of BRI1 and BAK1. In this model, BR binding (either directly or in the presence of a sterol binding protein) promotes or stabilizes heterodimerization of BRI1 and BAK1 leading to reciprocal trans-phosphorylation. The activated kinases phosphorylate unknown downstream intermediates, leading eventually to gene regulation.

generally exist in momomeric form in the absence of ligand, with some notable exceptions including the insulin receptor family that exist as covalent heterotetramers (24). Ligand binding either promotes or stabilizes oligomerization, which can be either homomeric or heteromeric, depending on the receptor family. Ligand-induced oligomerization then results in trans-phosphorylation of tyrosine residues in the activation loops of the kinase domains, followed by further phosphorylation of docking sites within the cytoplasmic domains, and subsequent phosphorylation of downstream signaling molecules. A model for BR signaling has been proposed in which BRI1 and BAK1 occur primarily in monomeric form with a small proportion of heterodimers (39). Binding of BR shifts the equilibrium towards the heterodimer and leads to trans-phosphorylation of the BRI1 and BAK1 kinase domains.

Putative Substrates of the BRI Receptor Kinase

Identification of cytoplasmic binding partners and kinase domain substrates for BRI and BAK1 is critical for understanding downstream signaling events. Both molecular genetic and biochemical approaches can be employed to identify putative *in vivo* binding partners including yeast two-hybrid analysis, interaction cloning, immunoprecipitation followed by purification of receptor-protein complexes, and use of synthetic peptides to identify putative binding motifs and substrate recognition consensus sequences (14). A study of BRI1-KD phosphorylation recognition sequences in synthetic peptides showed that positive residues at P-3, P-4, P+5, and P+6 (relative to the phosphorylated Ser) were important for phosphorylation with a preference for Lys over Arg (41). A preliminary consensus sequence [RK]-[RK]-X(2)-S-X(2)-[LMVIFY]-X-[RK]-[RK] was identified, and one interesting protein that contains elements of this recognition motif is the plant homolog of mammalian TRIP-1, mentioned above as a substrate of the TGF-β Type II receptor kinase. We have recently found that recombinant Arabidopsis TRIP-1 is strongly phosphorylated by BRI-KD *in vitro*, predominately on three specific Thr residues that are all in the context of a positive residue at P-3 or P-4. Moreover, BRI1 and TRIP-1 co-immunoprecipitate from Arabidopsis plant extracts using native antibodies or various combinations of tagged proteins, strongly suggesting an *in vivo* interaction between the two proteins. Thus, TRIP-1 may be the best candidate to date for a BRI1 substrate (16).

The morphological characteristics of transgenic lines expressing antisense TRIP-1 RNA are also consistent with a possible role for TRIP-1 in BR signaling (26). Many of the antisense lines were lethal, but those that survived beyond the seedling stage showed dramatic alterations in several developmental pathways that shared several characteristics with BR-deficient and –insensitive mutants, including dwarfism, altered leaf morphology, delayed development, reduced apical dominance, reduced fertility, and an abnormally long life span with a very bushy habit in older plants. Several of the lines, however, exhibited unique developmental defects not seen in BR mutants, particularly in floral development, suggesting TRIP-1 is also involved in other signaling pathways. As mentioned above, TRIP-1 is a component of the eIF3 translation initiation complex and it is conceivable that BRI1, or one of the many other plant receptor kinases involved in developmental control might phosphorylate TRIP-1 *in vivo*, enhancing its competence to initiate assembly of the eIF3 complex. Such phosphorylation of eIF3 subunits might present an avenue by which signal transduction pathways controlling development could impinge directly on translation initiation. While the exact role of TRIP-1 and its BR dependence remain unknown, the preliminary data and parallels with TGF-β signaling make this protein an interesting target for further study.

Downstream Components of BR Signaling

The perception of BR by LRR receptor kinases has been discussed at length because of the relatively extensive literature on this segment of BR signaling.

However, several recent papers have revealed novel downstream components of the BR signal transduction pathway. A careful re-examination of BR-insensitive Arabidopsis dwarfs obtained in one of the genetic screens yielding several new *bri1* alleles, also uncovered *bin2*, a dwarf mutant that in the homozygous state closely resembles the *bri1* phenotype (32). Unlike all known *bri1* alleles, which are recessive loss-of-function mutations, *bin2* is a semi-dominant gain-of-function mutant. BIN2 encodes a cytoplasmic Ser/Thr kinase with a variable N-terminal domain, a kinase domain and a C-terminal domain, whose overall structure most closely resembles the Drosophila shaggy kinase and mammalian GSK-3 (32). For example, alignment of BIN2 with human GSK-3 reveals 44% identity in the N-terminal domain, 71% identity in the kinase domain, and 49% identity in the C-terminal region. Two independent genetic screens (10, 42) identified additional alleles of *bin2*, which were named *dwf12* and *ucu1*. Interestingly, all known *bin2/dwf12/ucu1* mutations are clustered in a four amino acid sequence termed the TREE domain because of its sequence of threonine, arginine and two glutamic acids.

In both vertebrates and invertebrates GSK-3/shaggy-like kinases often function as negative regulators of signaling pathways controlling metabolism, cell fate determination and pattern formation (28). Specifically, in humans GSK-3 is a negative regulator of the insulin pathway where it phosphorylates glycogen synthase, inhibiting its activity. When insulin binds to a cell surface receptor tyrosine kinase, a signaling cascade is initiated that results in the activation of Akt, a kinase that phosphorylates and inactivates GSK-3. With GSK-3 inactivated, the non-phosphorylated form of glycogen synthase accumulates and converts glucose to glycogen. GSK-3/shaggy kinases also play an important negative regulatory role in the Wingless/wnt signaling pathway by phosphorylating β-catenin, promoting its proteasome-dependent degradation. Ligand binding to a Frizzled family seven-pass transmembrane receptor leads to GSK-3/shaggy kinase inactivation resulting in an accumulation of unphosphorylated β-catenin, which escapes degradation by the proteasome and moves to the nucleus where it interacts with transcription factors to regulate the expression of genes essential for developmental pattern formation (28).

Following the pattern described for animal GSK-3/shaggy kinases, BIN2 has been shown by biochemical and genetic analysis to be a negative regulator of BR signal transduction. The severity of the mutant phenotype in transgenic plants overexpressing *BIN2* is correlated with the level of *BIN2* transcripts, and recombinant *bin2* mutant kinase has greater activity in a peptide kinase assay than wildtype BIN2 recombinant protein (32). Thus *bin2* has characteristics of a hypermorphic mutant whose increased GSK-3/shaggy kinase activity negatively regulates BR signaling. By further analogy with GSK-3/shaggy kinase action, it is likely that BIN2 phosphorylates and inactives an important positive regulator of BR signal transduction and that inactivation of BIN2, perhaps by phosphorylation, releases its negative control of BR signaling. Such phosphorylation may occur on the Thr residue in the TREE domain and is likely to be initiated by BR binding to BRI1 or a putative BRI1/BAK1 complex. Genetic and biochemical experiments suggest that BRI1 is unlikely to phosphorylate BIN2 directly, but BAK1 has not been tested. It is also possible that intermediate steps, as yet undiscovered, are required between receptor activation and BIN2 inactivation (32).

Downstream interacting partners of the BIN2 kinase have been identified in genetic screens for suppressors of *bri1* mutants and for plants resistant to the BR biosynthesis inhibitor, brassinazole (46, 50). The semidominant or dominant mutants, *bes1-D* and *bzr1-D*, show constitutive brassinosteroid responses including long petioles, curled leaves, accelerated senescence and constitutive expression of BR-regulated genes. BES1 and BZR1 are closely related, sharing 88% amino acid identity, and both contain bipartite nuclear localization signals and multiple consensus sites (S/TXXXS/T) for phosphorylation by GSK-3 kinases such as BIN2. Indeed, recombinant BIN2 strongly phosphorylates BES1 and BZR1 *in vitro*, and experiments with transgenic plants showed that BIN2 activity negatively affected the level of tagged BES1 and BZR1 proteins *in vivo* (23, 50). BR treatment affects the phosphorylation state and cellular localization of BES1 and BZR1. Unphosphorylated forms of both proteins accumulate upon BR treatment resulting in increased nuclear localization which is associated with the overexpression of several genes encoding wall-modifying enzymes involved in cell expansion (46, 50). In a further demonstration of mechanistic similarity to the Wingless/wnt signaling pathway, it was found that a specific inhibitor of proteasome activity led to the accumulation of the phosphophorylated form of BZR1 in the cytoplasm, suggesting that phosphorylation by BIN2 targets BZR1 for proteasome-mediated degradation (23). While this mechanism is similar to the behavior of β-catenin in the Wingless/wnt pathway (Fig. 5), it should be noted that neither BES1 nor BZR1 has any sequence homology with β-catenin.

BR-REGULATED GENE EXPRESSION

The terminal end of BR signal transduction consists of the specific genes that are regulated by interaction with nuclear localized BR signaling components. Classical methods of studying differential gene expression such as subtractive hybridization, and more recently global analysis of gene expression using DNA microarrays, have both been employed to identify numerous genes which appear to be regulated by BR signaling. Some of these genes are specifically regulated by BR, while others show regulation by additional hormones such as auxin, gibberellins, ethylene and jasmonic acid; and by environmental stimuli including light, temperature and mechanical perturbation. Several genes are consistently reported as BR-regulated while some are specific to particular growth conditions, plant genotypes and BR application methods. Genes regulated by BR either transcriptional or

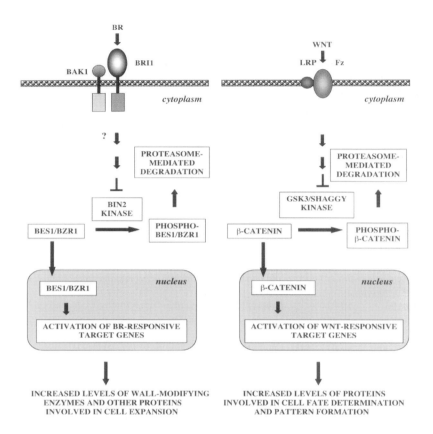

Figure 5. Mechanistic parallels between BR and Wingless/wnt signal transduction pathways. The BIN2 kinase of Arabidopsis shares 70% amino acid identity in the catalytic domain with GSK3/shaggy kinases of vertebrates and invertebrates. These kinases are cytoplasmic and act to negatively regulate signaling by phosphorylating positive regulators of the pathway, leading to their proteasome-mediated degradation. Ligand binding to cell surface receptors initiates a signaling cascade that inactivates the BIN2/GSK3 kinases, allowing accumulation and nuclear localization of the unphosphorylated form of the positive regulators.

posttranscriptional mechanisms have been identified, although a specific BR response element has yet to be reported (4, 14, 18). A cautionary note should be raised here about how one defines a gene as truly BR-regulated. Application of exogenous BR is quite common but may not reflect endogenous regulation because of problems with uptake and distribution of the applied compound. Moreover, wild-type plants may already have optimal levels of endogenous BRs and might not respond to exogenous treatment. One can use BR-deficient mutants as the treated plants, but the severe phenotype of these plants might also yield secondary effects that are not directly related to BR regulation. Table 2 summarizes several genes proposed to be regulated by BR under a variety of experimental conditions.

Brassinosteroid signaling and action

Table 2. Examples of Arabidopsis BR-Regulated Genes Determined by Microarray Analysis[a]

Accession No.	Gene	Gene Function
Genes up-regulated by BR treatment and/or with lower expression in BR-deficient mutant		
AF051338	TCH4	Xyloglucan endotransglucosylase/hydrolase
AF060874	AGP4	Arabinogalactan protein
U30478	EXP5	Expansin
U18406	IAA3	Auxin-inducible gene
U49075	IAA19	Auxin-inducible gene
S70188	SAUR-AC1	Small auxin up RNA
L39650	ZFP7	Zinc finger protein
U09335	HAT2	Homeobox protein
Y08892	Hsc70-GB	Heat shock protein
AB00487	COR47	Dehydration and cold-regulated gene
AF053345	KCS1	Fatty acid elongase 3-ketoacyl-CoA synthase 1
U92460	OPR1	12-Oxophytodienoate reductase (jasmonate biosynthesis)
AF016100	ACO2	1-Aminocyclopropane-1-carboxylic acid oxidase
AL021960	AAT1	Amino acid transport protein
Genes down-regulated by BR treatment and/or with higher expression in BR-deficient mutant		
AF100166	PIF3	Phytochrome-interacting factor 3
X13434	NIA1	Nitrate reductase NR1
U40154	AKT2	Potassium channel
U43489	XTR7	Putative xyloglucan endotransglycosylase/hydrolase
AF044216	DWF4	Steroid 22-hydroxylase (BR biosynthesis)
X87367	CPD	Steroid 23-hydroxylase (BR biosynthesis)
AB008097	ROT3	Cytochrome P450 (steroid biosynthesis)
AF176000	MYB55	MYB-like gene
Z95741	MYB14	R2R3-MYB transcription factor

[a]Data from references 22, 38, 50.

Given the dwarf stature of BR-deficient and insensitive mutants, it is not surprising that BRs are intimately involved in the cell expansion process. The plant cell wall forms a highly cross-linked, rigid matrix that opposes cell expansion and differentiation. In order for elongation and other morphogenetic processes to occur, the cell wall must be modified in multiple ways including changes in mechanical properties, orientation of microtubules and incorporation of new polymers into the extending wall to maintain wall integrity. BRs have been shown to increase transcript levels of many genes encoding wall modifying proteins including XTHs and expansins (43). Besides cell expansion-related genes, BRs have also been shown to increase mRNA levels of heat shock protein 90 in bromegrass cell suspension cultures and to affect the expression of genes controlling late stages of tracheary element differentiation of *Zinnia elegans* mesophyll cells, including those encoding phenylalanine ammonia lyase, cinnamic acid 4-hydroxylase and a cysteine protease (13, 20). Genes associated with translation initiation and cell division, such as cyclins and histones, have been shown to be BR regulated, as has an extracellular invertase (43). Down-regulation of genes encoding sterol and BR biosynthetic enzymes (1), as well as certain GA-regulated genes has also been reported (5).

Examination of BR-regulated gene expression in the *bes1-1D*, *bri1*, *det2* and *dwf1* mutants using Arabidopsis oligonucleotide microarrays (Table 2), verified the BR regulation of numerous genes encoding wall-modifying proteins and showed that several genes associated with auxin signal transduction are also BR-regulated (22, 38, 50). Moreover, a genetic analysis of the Arabidopsis *ucu1* mutants (allelic to *bin2*) suggests that the BIN2/UCU1 kinase may also be involved in auxin signal transduction, indicating a point of possible cross-talk between BR and auxin signaling (42). While both BR and auxin promote elongation, their kinetics are quite different. Auxin generally shows a very short lag time of 10 to 15 min between application and the onset of elongation, with maximum rates of elongation reached within 30 to 45 min. In contrast, BR has lag times of at least 45 min with elongation rates continuing to increase for several hours. This difference in kinetics is also seen at the level of gene expression in Arabidopsis, where auxin induces members of the IAA, SAUR and GH gene families generally much more rapidly than BR. Other differences in the effect of auxin and BR on elongation have been observed in physiological and molecular studies. However, synergisms between BR and auxins occur in many systems, and this potential interaction deserves much more detailed mechanistic studies (43).

While a number of genes have been identified as BR-regulated, only one report on the extensive analysis of a BR responsive promoter has appeared so far (14). The Arabidopsis *TCH4* gene encodes an XTH whose activity is associated with cell wall modification during elongation and morphogenesis. *TCH4* gene expression is transcriptionally regulated by both BR and auxin, and by environmental stimuli such as touch, darkness and temperature (48). *TCH4* expression in response to IAA is maintained in the BR-insensitive mutant *bri1*, although the magnitude of the response is reduced compared to wild type. *TCH4* mRNAs also accumulate in *bri1* in response to temperature extremes, mechanical perturbation and darkness. Therefore, the signal transduction pathway used for BR activation of *TCH* expression must be distinct from the signaling pathways employed by other stimuli, or alternatively, the signaling pathways activated by these diverse inducing stimuli converge at some point downstream of BR perception. Induction of expression by these diverse stimuli is conferred to reporter genes by the same 102-bp *TCH4* promoter region. However, upstream regions influence the magnitude and kinetics of expression and likely harbor regulatory elements that are redundant with those located in the 102-bp region. Substitution of the proximal regulatory region sequences in the context of distal elements does not disrupt inducible expression, making the identification of the specific BR response element difficult (14).

CONCLUDING REMARKS

In this chapter we have seen that plant steroids regulate a wide range of developmental programs, but by signal transduction mechanisms that are distinct from the classical pathway of animal steroid hormone action. Still, BR signaling logic has parallels with several animal pathways including cell surface perception of steroids in the "non-genomic", rapid action pathway; growth regulator pathways involving receptor tyrosine kinases; the TGF-β pathway; and downstream elements of the Wingless/Wnt signal transduction pathway. The pace of recent discoveries in BR signaling has been rapid but several critical questions remain, such as; 1) Do BRI and BAK1 form true co-receptors *in vivo*? 2) Are accessory steroid-binding proteins required for BR binding to BRI1 and/or BAK1? 3) What are the *in vivo* autophosphorylation sites of BRI1 and BAK1 and what are their BR dependence? 4) What are the number and nature of cytoplasmic binding partners of the BRI1 and BAK1 kinase domains? 5) What is the mechanism of BR-dependent inactivation of BIN2 kinase activity? And 6) How do BES and BZR1 participate in the regulation of specific genes?

With respect to BR-regulated gene expression, the recent availability of nearly full-genome Arabidopsis microarrays provides a unique opportunity to characterize the complete spectrum of BR-regulated genes. As with other plant hormones, the effect of BRs on growth are pleiotropic and the cataloging and functional analysis of newly discovered gene sets may increase our understanding of the range of physiological events influenced by BRs. Comparative analysis of upstream sequences from these large gene sets may uncover motifs that can be tested by a variety of methods for their potential function as BR response elements, thus clarifying the terminal end of BR signal transduction. In summary, a greater understanding of the molecular mechanisms of BR signal transduction will deepen our knowledge of how these unique steroids regulate multiple events critical for normal plant growth and development.

References

1. Altmann T (1999) Molecular physiology of brassinosteroids revealed by the analysis of mutants. Planta 208:1-11
2. Beato M, Herrlich P, Schutz G (1995) Steroid hormone receptors: many actors in search of a plot. Cell 83:851-857
3. Becraft PW (2002) Receptor kinase signaling in plant development. Annu Rev Cell Dev Biol 18:163-192
4. Bishop GJ, Koncz C (2002) Brassinosteroids and plant steroid hormone signaling. Plant Cell 14:S97-110
5. Bouquin T, Meier C, Foster R, Nielsen ME, Mundy J (2001) Control of specific gene expression by gibberellin and brassinosteroid. Plant Physiol. 127:450-458
6. Burks EA, Bezerra PP, Le H, Gallie DR, Browning KS (2001) Plant initiation factor 3 subunit composition resembles mammalian initiation factor 3 and has a novel subunit. J Biol Chem 276:2122-2131
7. Carpita N, Gibeaut D (1993) Structural models of the primary cell walls in flowering plants: consistency of molecular structure with the physical properties of the walls during growth. Plant J. 3:1-30
8. Choe S, Fujioka S, Noguchi T, Takatsuto S, Yoshida S, Feldmann KA (2001) Overexpression of DWARF4 in the brassinosteroid biosynthetic paththway results in increased vegetative growth and seed yield in Arabidopsis. Plant J 26:573-582

9. Choe S, Noguchi T, Fujioka S, Takatsuto S, Tissier CP, Gregory BD, Ross AS, Tanaka A, Yoshida S, Tax FE, Feldmann KA (1999) The Arabidopsis *dwf7/ste1* mutant is defective in the delta7 sterol C-5 desaturation step leading to brassinosteroid biosynthesis. Plant Cell 11:207-221
10. Choe S, Schmitz RJ, Fujioka S, Takatsuto S, Lee MO, Yoshida S, Feldmann KA, Tax FE (2002) Arabidopsis brassinosteroid-insensitive *dwarf12* mutants are semidominant and defective in a glycogen synthase kinase 3beta-like kinase. Plant Physiol 130:1506-1515
11. Choe S, Tanaka A, Noguchi T, Fujioka S, Takatsuto S, Ross AS, Tax FE, Yoshida S, Feldmann KA (2000) Lesions in the sterol delta reductase gene of Arabidopsis cause dwarfism due to a block in brassinosteroid biosynthesis. Plant J 21:431-443
12. Choy L, Derynck R (1998) The type II transforming growth factor (TGF)-beta receptor-interacting protein TRIP-1 acts as a modulator of the TGF-beta response. J Biol Chem 273:31455-31462
13. Clouse S, Sasse J (1998) Brassinosteroids: Essential regulators of plant growth and development. Annu Rev Plant Physiol Plant Mol Biol 49:427-451
14. Clouse SD (2002) Brassinosteroid signal transduction. Clarifying the pathway from ligand perception to gene expression. Mol Cell 10:973-982
15. Clouse SD, Langford M, McMorris TC (1996) A brassinosteroid-insensitive mutant in *Arabidopsis thaliana* exhibits multiple defects in growth and development. Plant Physiol 111:671-678
16. Ehsan H, Ray WK, Huber SC, Clouse SD (2003) An Arabidopsis protein with sequence similarity to animal TGF-beta Receptor Interacting Protein is phosphorylated by the BRI1 receptor kinase *in vitro*. American Society of Plant Biologists Abstracts http://abstracts.aspb.org/pb2003/public/M12/5068.html
17. Falkenstein E, Tillmann H-C, Christ M, Feuring M, Wehling M (2000) Multiple actions of steroid hormones---A focus on rapid, nongenomic effects. Pharmacol Rev 52:513-556
18. Friedrichsen D, Chory J (2001) Steroid signaling in plants: from the cell surface to the nucleus. Bioessays 23:1028-1036
19. Friedrichsen DM, Joazeiro CA, Li J, Hunter T, Chory J (2000) Brassinosteroid-insensitive-1 is a ubiquitously expressed leucine-rich repeat receptor serine/threonine kinase. Plant Physiol 123:1247-1256
20. Fukuda H (1997) Tracheary element differentiation. Plant Cell 9:1147-1156
21. Geuns JMC (1978) Steroid hormones and plant growth and development. Phytochemistry 17:1-14
22. Goda H, Shimada Y, Asami T, Fujioka S, Yoshida S (2002) Microarray analysis of brassinosteroid-regulated genes in Arabidopsis. Plant Physiol 130:1319-1334
23. He JX, Gendron JM, Yang Y, Li J, Wang ZY (2002) The GSK3-like kinase BIN2 phosphorylates and destabilizes BZR1, a positive regulator of the brassinosteroid signaling pathway in Arabidopsis. Proc Natl Acad Sci U S A 99:10185-10190
24. Heldin C (1995) Dimerization of cell surface receptors in signal transduction. Cell 80:213-224
25. Hu Y, Bao F, Li J (2000) Promotive effect of brassinosteroids on cell division involves a distinct CycD3-induction pathway in Arabidopsis. Plant J 24:693-701
26. Jiang J, Clouse SD (2001) Expression of a plant gene with sequence similarity to animal TGF-beta receptor interacting protein is regulated by brassinosteroids and required for normal plant development. Plant J 26:35-45
27. Kauschmann A, Jessop A, Koncz C, Szekeres M, Willmitzer L, Altmann T (1996) Genetic evidence for an essential role of brassinosteroids in plant development. Plant J 9:701-713
28. Kim L, Kimmel AR (2000) GSK3, a master switch regulating cell-fate specification and tumorigenesis. Curr Opin Genet Dev 10:508-514
29. Koka C, Cerny R, Gardner R, Noguchi T, Fujioka S, Takatsuto S, Yoshida S, Clouse S (2000) A putative role for the tomato genes *DUMPY* and *CURL-3* in brassinosteroid biosynthesis and response. Plant Physiol 122:85-98

30. Li J, Chory J (1997) A putative leucine-rich repeat receptor kinase involved in brassinsteroid signal transduction. Cell 90:929-938
31. Li J, Lease KA, Tax FE, Walker JC (2001) BRS1, a serine carboxypeptidase, regulates BRI1 signaling in *Arabidopsis thaliana*. Proc Natl Acad Sci U S A 98:5916-5921
32. Li J, Nam KH (2002) Regulation of brassinosteroid signaling by a GSK3/SHAGGY-like kinase. Science 295:1299-1301
33. Li J, Wen J, Lease KA, Doke JT, Tax FE, Walker JC (2002) BAK1, an Arabidopsis LRR receptor-like protein kinase, interacts with BRI1 and modulates brassinosteroid signaling. Cell 110:213-222
34. Mandava NB (1988) Plant growth-promoting brassinosteroids. Annu Rev Plant Physiol Plant Mol Biol 39:23-52
35. Massague J (1998) TGF-beta signal transduction. Annu Rev Biochem 67:753-791
36. Montoya T, Nomura T, Farrar K, Kaneta T, Yokota T, Bishop GJ (2002) Cloning the tomato *curl3* gene highlights the putative dual role of the leucine-rich repeat receptor kinase tBRI1/SR160 in plant steroid hormone and peptide hormone signaling. Plant Cell 14:3163-3176
37. Morillon R, Catterou M, Sangwan RS, Sangwan BS, Lassalles JP (2001) Brassinolide may control aquaporin activities in *Arabidopsis thaliana*. Planta 212:199-204
38. Mussig C, Fischer S, Altmann T (2002) Brassinosteroid-regulated gene expression. Plant Physiol 129:1241-1251
39. Nam KH, Li J (2002) BRI1/BAK1, a receptor kinase pair mediating brassinosteroid signaling. Cell 110:203-212
40. Noguchi T, Fujioka S, Choe S, Takatsuto S, Yoshida S, Yuan H, Feldmann K, Tax F (1999) Brassinosteroid-Insensitive dwarf mutants of Arabidopsis accumulate brassinosteroids. Plant Physiology 121:743-752
41. Oh MH, Ray WK, Huber SC, Asara JM, Gage DA, Clouse SD (2000) Recombinant brassinosteroid insensitive 1 receptor-like kinase autophosphorylates on serine and threonine residues and phosphorylates a conserved peptide motif *in vitro*. Plant Physiol 124:751-766
42. Perez-Perez JM, Ponce MR, Micol JL (2002) The *UCU1* Arabidopsis gene encodes a SHAGGY/GSK3-like kinase required for cell expansion along the proximodistal axis. Dev Biol 242:161-173
43. Sakurai A, Yokota T, Clouse SD (*eds*) (1999): Brassinosteroids: Steroidal Plant Hormones Tokyo: Springer, pp 1-253
44. Schlessinger J (2000) Cell signaling by receptor tyrosine kinases. Cell 103:211-25
45. Szekeres M, Nemeth K, Koncz-Kalman Z, Mathur J, Kauschmann A, Altmann T, Redei GP, Nagy F, Schell J, Koncz C (1996) Brassinosteroids rescue the deficiency of CYP90, a cytochrome P450, controlling cell elongation and de-etiolation in Arabidopsis. Cell 85:171-182
46. Wang ZY, Nakano T, Gendron J, He J, Chen M, Vafeados D, Yang Y, Fujioka S, Yoshida S, Asami T, Chory J (2002) Nuclear-localized BZR1 mediates brassinosteroid-induced growth and feedback suppression of brassinosteroid biosynthesis. Dev Cell 2:505-513
47. Wang ZY, Seto H, Fujioka S, Yoshida S, Chory J (2001) BRI1 is a critical component of a plasma-membrane receptor for plant steroids. Nature 410:380-3
48. Xu W, Purugganan MM, Polisenksy DH, Antosiewicz DM, Fry SC, Braam J (1995) Arabidopsis *TCH4*, regulated by hormones and the environment, encodes a xyloglucan endotransglycosylase. Plant Cell 7:1555-1567
49. Yamamuro C, Ihara Y, Wu X, Noguchi T, Fujioka S, Takatsuto S, Ashikari M, Kitano H, Matsuoka M (2000) Loss of function of a rice brassinosteroid insensitive1 homolog prevents internode elongation and bending of the lamina joint. Plant Cell 12:1591-1606
50. Yin Y, Wang ZY, Mora-Garcia S, Li J, Yoshida S, Asami T, Chory J (2002) BES1 accumulates in the nucleus in response to brassinosteroids to regulate gene expression and promote stem elongation. Cell 109:181-191

E. THE FUNCTIONING OF HORMONES IN PLANT GROWTH AND DEVELOPMENT

E1. The Transport of Auxins

David A. Morris[1], Jiří Friml[2], and Eva Zažímalová[3]
[1]School of Biological Sciences, University of Southampton, Bassett Crescent East, Southampton, SO16 7PX, UK. [2]Centre for Plant Molecular Biology, University of Tübingen, Auf der Morgenstelle 3, 72076 Tübingen, Germany. [3]Institute of Experimental Botany, Academy of Sciences of the Czech Republic, Rozvojová 135, 16502 Prague 6, Czech Republic.
E-mail: dam3@soton.ac.uk

INTRODUCTION

Auxins play a crucial role in the regulation of spatial and temporal aspects of plant growth and development[1]. As well as being required for the division, enlargement and differentiation of individual plant cells, auxins also function as signals between cells, tissues and organs. In this way they contribute to the coordination and integration of growth and development in the whole plant and to physiological responses of plants to environmental cues (63). At the individual cell level, fast changes or pulses in hormone concentration may function to initiate or to terminate a developmental process. In contrast, the maintenance of a stable concentration of the hormone (homeostasis) may be necessary to maintain the progress of a developmental event that has already been initiated. It should be stressed that both transmembrane transport and metabolic processes such as biosynthesis, degradation and

---

[1] Abbreviations: ARF, auxin response factor; ABC, ATP-binding cassette; BFA, brefeldin A; CHX, cycloheximide; 2,4-D, 2,4-dichlorophenoxyacetic acid; GEF, guanine nucleotide exchange factor; MNK, Menkes copper-transporting ATPase; NAA, 1-naphthaleneactic acid; NPA, 1-N-naphthylphthalamic acid; NBP, NPA-binding protein; PM, plasma membrane; TIBA, 2,3,5-triiodobenzoic acid.

Auxin transport

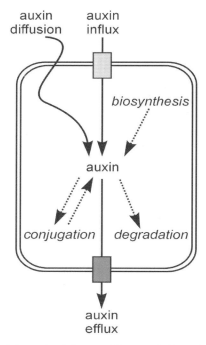

Figure 1. Scheme of the metabolic and transport processes involved in regulating the momentary level of auxin in a plant cell or its compartments. Full arrows: transport processes; dotted arrows: metabolic processes.

conjugation, operate together to regulate the momentary concentration of the active hormone molecule in target cells and cell compartments (97; Fig. 1).

Auxins such as indole-3-acetic acid (IAA) mediate interactions between cells, tissues and organs over both short distances (for example, between adjacent cells) and over very long ones (for example between the shoot apex and sites of lateral root initiation). Characteristically, auxin transport through cells and tissues other than mature vascular elements is strongly polar. This feature endows an auxin signal with directional properties and, *inter alia*, the vectorial transport of IAA is important in the regulation of spatial aspects of plant development. Consequently IAA transport may play a crucial role in the initiation and/or maintenance of cell and tissue polarity and axiality, upon which pattern formation depends (31).

New approaches to the physiology of auxin transport, and rapid advances in our knowledge of the molecular and genetic mechanisms that underlie it, have dramatically increased our understanding of the process in the nine years that have elapsed since auxin transport was discussed in the last edition of this book (48). It has become increasingly clear that polar auxin transport is a very dynamic and complex process and one that is regulated at many different levels.

This chapter describes and analyses these recent developments. The earlier work on which they were based has been reviewed elsewhere (33, 42); several comprehensive reviews analyse the recent discoveries relating to the mechanism, regulation, molecular genetics and significance in development of polar auxin transport (4, 19, 22, 59, 63, 64, 97). The reader is referred to these for further background information.

It is impossible to discuss polar auxin transport without reference to the direction of transport in the plant and the orientation of auxin-transporting cells with respect to plant poles and axes. The terminology in general use has the potential to confuse, not least because in plant developmental biology the base of the developing embryo equates to the apex (i.e. the youngest, meristematic, end) of the growing root. Thus the basal end of a cell refers to the end nearest the root pole and the apical end to that nearest the shoot pole,

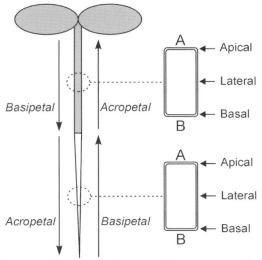

Figure 2. The nomenclature used in the text to define the directions of auxin transport in a plant or plant organ and apical and basal ends of root and shoot cells. The apical end of a cell is defined here as the end nearest the shoot tip, regardless of whether the cell is located in the root or the shoot. A = Apical end of cell; B = basal end of cell.

regardless of whether the cell is in the root or shoot. In roots, this contrasts with the terms acropetal (towards the root tip) and basipetal (towards the base of the root), generally used to denote direction. To avoid possible confusion, the terminology that has been adopted throughout this chapter is defined in Fig. 2.

LONG-DISTANCE AUXIN MOVEMENT - PATHWAYS AND MECHANISMS

Two physiologically distinct and spatially separated pathways function to transport auxin over long distances through plants. Firstly, auxin is translocated rapidly by mass flow with other metabolites in the mature phloem. Secondly, auxin is transported downwards towards the root tips from immature tissues close to the shoot apex by a much slower (ca. 7-15 mm h^{-1}), carrier-dependent, cell-to-cell polar transport (33). Early experiments (14) suggested that some of the auxin reaching the root tip might be re-exported basipetally through the root, possibly in the cortex.

Non-Polar Translocation of Auxin in Vascular Tissues

IAA is a natural constituent of phloem sap and is exported in physiologically significant quantities by source leaves (1). Furthermore, labeled auxins applied directly to source leaves, and IAA synthesized by source leaves from applied tryptophan, are also rapidly loaded into the phloem and exported (1, and references therein). Only traces of endogenous IAA are normally found in xylem sap (1) and it is unlikely that the mature xylem functions as a major pathway for long-distance auxin movement.

The IAA that is loaded into mature phloem in source leaves is translocated passively in solution in the phloem sap to sink organs and tissues where unloading of phloem-mobile assimilates occurs. Thus the direction and speed of auxin translocation in this pathway will be subject to all the factors that influence phloem translocation (58). This makes the phloem an unlikely route for the transmission of a signal molecule involved in the fine-tuning of growth and development. Nevertheless, significant quantities of IAA occur in phloem sap and it is clear that the total amount of auxin that is delivered by the phloem to sink tissues is considerable. Consequently, the high concentration of IAA detected in young (sink) leaves at the shoot apex (often uncritically assumed to indicate high rates of local biosynthesis of IAA) results from both local synthesis (47) and IAA accumulation following unloading of phloem-mobile auxin synthesized in source leaves (1).

In addition to having roles in the loading and unloading of phloem mobile assimilates in source and sink tissues (58), the IAA translocated in the phloem also is a major source of the IAA transported in the long distance polar auxin transport pathway (see below).

The Long Distance Polar Auxin Transport Pathway

Auxin exported by the apical tissues of intact dicotyledonous plants moves downwards through the stem and root by a mechanism that has all the characteristics of the slow, polar transport first described from coleoptile and stem segments (33, 59). In stems, this transport occurs basipetally, but not acropetally (Fig. 2), at velocities around 10 mm h^{-1}; is blocked by inhibitors of polar auxin transport (see below); and is competed by other growth-active auxins such as 1-naphthaleneactic (NAA) and 2,4-dichlorophenoxyacetic acid (2,4-D). In both herbaceous and woody dicotyledonous species the pathway for the root-directed long distance polar transport of auxin includes the vascular cambium and its partially differentiated derivatives, including differentiating xylem vessels and xylem parenchyma (44, 62). Environmental treatments and experimental conditions that reduce or inhibit cambial activity correspondingly reduce long distance polar auxin transport.

Communication Between Polar and Non-Polar Pathways

Of particular interest and importance is the question of whether the two long-distance auxin transport pathways – the polar pathway in immature cells and the translocation pathway in mature vascular elements – are isolated from each other, or whether auxin moving in one pathway can enter the other. The answer to this question has considerable bearing on the sources of auxin present in the polar transport pathway.

There is little evidence to indicate that auxin enters the phloem from the polar transport pathway; on the contrary, a number of reports suggest that it does not (10). In contrast, there is both physiological and genetic evidence that phloem-translocated auxin is transferred to the long distance polar auxin transport system (10). For example, when [1-^{14}C]IAA was applied to a mature leaf of pea (*Pisum sativum* L.) it was exported in the phloem (indicated by recovery of label in aphid colonies on the internodes above and below the fed leaf). However, efflux of [^{14}C] from 30 mm stem segments excised 4h after labelling occurred in the basipetal, but not in the acropetal, direction and was prevented by inclusion of 1-N-naphthylphthalamic acid

(NPA), an inhibitor of polar auxin transport (see later). This polar, NPA-sensitive efflux clearly suggests that auxin had entered the polar transport system from the phloem. The quantity of auxin that effluxed basipetally from the stem segments was dramatically reduced when very young tissues at the shoot apex were excised, suggesting that the transfer of auxin from the phloem to the polar transport pathway occurred mainly in the younger sink tissues of the shoot apex. A possible explanation for this is that the transfer of IAA from the phloem to the polar transport system may require carrier-mediated radial transfer through undifferentiated cells.

PHYSIOLOGICAL ASPECTS OF POLAR TRANSPORT

Carrier-Mediated Auxin Transport: The Role of Efflux Carriers

An early indication that polar auxin movement involved carrier-mediated cell-to-cell transport was the observation that 2,3,5-triiodobenzoic acid (TIBA), a known inhibitor of polar auxin transport, stimulated the net uptake of labelled IAA in *Zea mays* L. coleoptile segments (37). This indicated that efflux rather than influx of IAA was inhibited by TIBA and therefore that efflux was more important than uptake in determining polar transport. Together with the already known facts that polar transport could take place against an overall concentration gradient, was sensitive to inhibitors of energy metabolism, and could continue in the absence of cytoplasmic continuity between adjacent cells (33, 42), these observations led to suggestions that auxin uptake and efflux by cells were mediated by energy-dependent influx and efflux carriers (37). Polarity of transport was believed to result from small differences in net auxin efflux between the two ends of a cell, and polar transport was envisaged as involving active (polar) "secretion" of auxin by carriers from the end of one cell, diffusion of auxin through the intervening wall space, and active uptake of auxin by an adjacent cell.

An important milestone in the efforts to explain the mechanism of polar auxin transport was the proposal that net polar transport of auxin through cells could be explained by differences between their two ends in their relative permeabilities to dissociated and undissociated auxin molecules (76, 79; Fig. 3). Being relatively lipophilic, undissociated molecules of some auxin species (e.g. IAA and NAA, but not 2-4,D) readily enter cells by diffusion across the plasma membrane (PM) from the surrounding wall space. Because the cytoplasm is normally far less acid than the wall space (typically around pH 7.0 and pH 5.5 respectively), a high proportion of the auxin molecules which enter the cells by diffusion dissociate after crossing the PM, the extent to which they do so depending on the dissociation constant of the auxin species concerned. Because auxin anions cannot readily penetrate the PM, they become "trapped" in the cytoplasm. Undissociated auxin will continue to move into the cell as long as an inwardly directed concentration gradient in undissociated auxin persists. As concentration equilibrium is approached, the total concentration of auxin in the cell (dissociated plus undissociated) will exceed the total concentration in the more acidic wall space, where the concentration of auxin anions will be considerably lower. Thus, when a pH gradient is maintained across the PM, diffusion alone can appear to drive accumulation of auxin against a gradient in its concentration! Unlike the polar secretion model (see above), which assumes primary activation of the transport catalysts, the "chemiosmotic polar diffusion model" requires only secondary energy expenditure in order to maintain the transmembrane pH gradient (Fig. 3).

As in the polar secretion model, an asymmetric distribution of auxin anion efflux carriers in the PM will lead to a polarized leakage of auxin anions from a cell (79), and if the asymmetry of carrier distribution is repeated in each cell in a file of contiguous cells, it would amplify the polarized leakage of auxin and lead to a net polar movement of auxin through tissue.

Auxin Uptake Carriers

Despite the prominence given to the efflux carrier in polar transport, it quickly became clear that a functionally distinct class of auxin anion carriers is also involved in auxin uptake. This was first suggested by observations that auxin uptake by suspension cultured cells (79) and tissue segments (15), possessed a saturable component. Elegant experiments demonstrated that auxin uptake into sealed outside-out zucchini (*Cucurbita pepo* L.) hypocotyl PM vesicles was a saturable, carrier-mediated process specific for growth active auxins. Subsequently, it was found that carrier-mediated uptake was electrogenic and probably involved a proton symport in which two protons were transported for each IAA$^-$ anion (48).

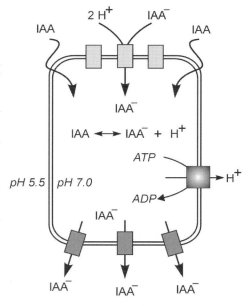

Figure 3. Components of transmembrane auxin transport according to the chemiosmotic polar diffusion model (33). A membrane pH gradient (maintained by plasma membrane H$^+$-ATPases) drives diffusive accumulation of undissociated auxin molecules. At the higher pH of the cytoplasm, some of the auxin molecules which enter the cell dissociate. The plasma membrane is relatively impermeable to auxin anions (IAA$^-$), which are "trapped" in the cytoplasm and can only exit or enter the cell through the action of specific influx (upper; light shading) and efflux (lower; heavy shading) carrier systems. Asymmetry in the distribution of the two carrier systems, more especially the efflux carrier, results in a net polar transport of auxin through the cell.

Although undissociated IAA can readily enter cells by diffusion across the PM, at the low concentrations of endogenous IAA likely to be present in the extracellular wall space in intact plants, it is probable that high affinity uptake carriers provide a more efficient means of effecting auxin uptake than diffusion. A key role for auxin influx carriers in polar auxin transport and in the loading of auxin into and unloading from the long distance transport pathway in the phloem is now supported both by physiological and biochemical observations and by evidence from recent molecular genetics studies (discussed in detail later).

Auxin Efflux Carriers Are Multi-Component Systems

It is probable that auxin efflux carriers are multi-component systems consisting of at least two, but possibly more individual protein components, each with a different function (61). The synthetic auxin transport inhibitor NPA strongly inhibits auxin efflux and consequently stimulates auxin accumulation by cells (59, 78). The mechanism by which NPA inhibits polar auxin transport remains a matter of debate and will be discussed later in this chapter, but available evidence suggests that it is mediated by a specific, high affinity, NPA-binding protein (NBP; 78). Protein synthesis inhibitors such as cycloheximide (CHX) rapidly uncouple carrier-mediated auxin efflux from the inhibition of efflux by NPA, but in the short-term do not affect auxin efflux itself or the saturable binding of NPA to microsomal membranes (61). These observations suggest that the NBP and the efflux catalyst are separate proteins that may interact through a third, rapidly turned over transducing protein (16, 49, 59, 63).

Polar Auxin Transport Inhibitors and NPA-Binding Proteins

Inhibitors of polar auxin transport (Fig. 4) have been very useful tools with which to investigate both the mechanism and regulation of polar transport and the role auxin transport plays in the control of plant development. The auxin transport inhibitors NPA and TIBA have been mentioned already. NPA, the most widely used of these, belongs to a class of

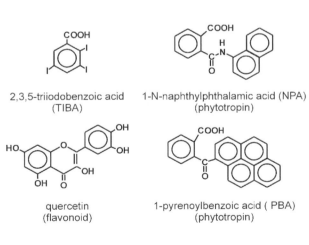

Figure 4. Examples of inhibitors of polar auxin transport. Of these, only quercetin occurs naturally.

structurally related synthetic compounds – phytotropins – characterized by their inhibition of auxin-dependent tropic responses, (78). Most reports indicate that the phytotropins interfere with polar auxin transport by strongly inhibiting auxin efflux from cells and a characteristic feature of their action is that auxin accumulates in treated cells. When stems of intact plants are ringed with a preparation containing NPA, root-directed basipetal transport of auxin is blocked and auxin accumulates in tissues above the NPA-treated region (e.g. 10). Despite the fact that most responses to NPA are best interpreted in terms of inhibition of efflux, a few reports suggest that under some circumstances influx carrier activity may also be sensitive to NPA (17).

Phytotropin action is mediated through a putative phytotropin-binding (NPA-binding) receptor protein, the NBP. For reasons already discussed, it is likely that the NBP is a separately synthesised and targeted regulatory protein that is functionally associated with the efflux catalyst. Interestingly, an NBP detected by an indirect immunofluorescence technique, using monoclonal antibodies raised against a membrane associated NPA-binding protein from zucchini hypocotyl tissue, exhibited a predominantly polar (basal) localization in cells associated with the vascular tissue in pea stem sections (40). This was the first direct evidence for the polar localization in cells of a protein believed to be associated with polar auxin transport.

Despite one report that the NBP is an integral membrane protein (8), the weight of evidence indicates that the NBP is probably a peripheral membrane protein that is located on the cytoplasmic face of the PM and is associated with the actin cytoskeleton (63, 64). Several studies have demonstrated that TIBA does not compete with NPA for high affinity binding sites on the NBP, implying the probable existence of different binding sites for different classes of polar auxin transport inhibitors.

All the first-discovered polar auxin transport inhibitors were synthetic compounds. Nevertheless, NBPs are almost universally distributed in the plant kingdom, suggesting that the NBP performs an essential function in plant cells. Although this function may not necessarily involve the regulation of polar auxin transport, the possibility remains that natural equivalents of the phytotropins exist and that through association with the NPA-binding site, they regulate auxin carriers. A screen for phenolic compounds which both promoted net accumulation of labelled IAA by zucchini hypocotyls segments and competed with labelled NPA for binding sites on membrane fractions isolated from zucchini hypocotyls, led to the discovery of a group of naturally-occurring flavonoids which were both ligands for the NPA binding site and inhibitors of IAA efflux (41). When several of these compounds were compared, their ability to inhibit NPA binding was directly correlated with their ability to stimulate IAA accumulation (41). Flavonoids, therefore, might be natural regulators of auxin efflux carriers. In support of this, the flavonoid-deficient Arabidopsis mutant *tt4* (*transparent testa 4*) was found to have altered patterns of auxin distribution resulting from higher than normal auxin efflux (67). The lesion in auxin transport could be corrected by application of the missing intermediate in the flavonoid biosynthetic pathway, naringenin.

Rapid Turnover and Cycling of the Auxin Efflux Carrier

A significant recent discovery was that auxin efflux catalysts turnover very rapidly in the PM. This was first revealed by investigations of the effect of the inhibitors of vesicle traffic to the PM, monensin and brefeldin A (BFA), on the accumulation of IAA by zucchini (*Cucurbita pepo* L.) hypocotyl segments (59, 77; Fig. 5), and of the effect of BFA on NAA accumulation by suspension-cultured tobacco cells (*Nicotiana tabacum* L.; 16). Both monensin and BFA very rapidly stimulated IAA or NAA accumulation (depending on system), but in neither system did they affect the accumulation of 2,4-D (16, 60). Because 2,4-D is a readily transported substrate for auxin uptake carriers but not, in most species, for auxin efflux carriers (17), these observations indicated that the target for BFA action was the efflux carrier system. This is supported by an observed reduction in the efflux of NAA or IAA (but not of 2,4-D) from pre-loaded cells and tissue segments in the presence of BFA (16, 60). Also consistent with this conclusion is the ability of BFA to abolish polar (basipetal) auxin transport in long (30 mm) pea and zucchini stem segments (77). Significantly, BFA does not affect saturable NPA binding to microsomal preparations from zucchini providing additional evidence that the NBP and the auxin efflux catalyst are different proteins (77). Compared with responses to CHX, the lag time for which may be up to 2.0 h in zucchini, the lag for BFA action is very short (minutes or less; 16, 77). Furthermore, the response to BFA is unaffected by CHX (Fig. 5).

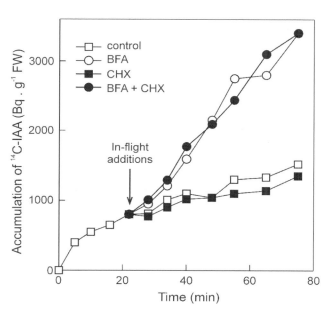

Figure 5. The effect of inflight additions (arrow; 22 min from start) of brefeldin A (BFA; 30 µM)) and cycloheximide (CHX; 10 µM) on the time-course of accumulation of [^{14}C]IAA by 2 mm zucchini hypocotyl segments. (Redrawn from 77).

Taken together, these observations provided strong indirect evidence that an essential component of the auxin efflux carrier system (possibly the efflux catalyst itself) is targeted to the PM through the BFA-sensitive secretory system, and that this component turns over very rapidly at the PM without the need for concurrent protein synthesis; this implies the existence of internal pools of the carrier protein (16, 60, 77). It has been argued that because such pools would be of finite capacity, the high rates of efflux carrier turnover revealed by these experiments could only be sustained if a proportion of the carriers cycled between the PM and the proposed pools (77).

Rapid cycling of PM-located carriers and receptors in animal cell systems is well documented. Examples of particular interest in relation to the auxin efflux carriers include the Menkes copper-transporting ATPase (MNK) and the insulin-dependent GLUT4 glucose transporter. Similarities exist between the cycling of GLUT4 in insulin-sensitive mammalian cells and the putative cycling of auxin efflux catalysts in plant cells (59, 64). Cycling of GLUT4 involves an endosomal pathway, whilst MNK cycles between the PM and the *trans* Golgi network. The steady-state distribution of MNK protein between the PM and the *trans* Golgi network depends on substrate (Cu ions) levels in the cell (59). This property is of particular interest because there is indirect evidence to show that the targeting of auxin carriers to the PM also may be stimulated by the carrier substrate itself, namely IAA (59).

The Role of the Cytoskeleton in Auxin Carrier Traffic and Cycling

The actin cytoskeleton plays an important role in directing vesicular traffic to delivery sites at the PM and to other membrane-enclosed compartments in the cell. It has been known for several years that the NBP is closely associated with actin filaments (49, 63, 64).

The link between the high affinity NBP and the actin cytoskeleton suggests that NBPs may play a role in directing the polar distribution of auxin efflux carriers, which have been shown to cycle along actin filaments (30). However, the link remains to be proven experimentally.

Irrespective of whether NPA acts through the NBP, NPA action does not involve changes in the structure of the actin cytoskeleton; NPA has been shown recently to have no effect on the arrangement of either microtubules or actin filaments in BY-2 tobacco cells, even at concentrations well above that required to saturate the inhibition of auxin efflux (72).

THE GENETICS OF AUXIN TRANSPORT

Although physiological and biochemical approaches have provided much information about the process of polar auxin transport and its role in plant development, they have told us little about the molecular identity and molecular regulation of components of the polar auxin transport system. Recent developments in plant molecular genetics, particularly the analysis of Arabidopsis mutants affected in polar transport, or in responses to auxins or auxin transport inhibitors, have enabled major advances to be made in our understanding of the process. Some of these developments are now described.

AUX1 Proteins – Components of Auxin Influx

A mutant referred to as *auxin 1* (*aux1*) has been pivotal in attempts to identify genes that encode auxin influx carrier catalysts (3). The root agravitropic and auxin-resistant phenotype of *aux1* (Fig. 6B) is consistent with a defect in auxin influx, but similar phenotypes have been observed also in mutants defective in auxin response (46). The *AUX1* gene encodes a protein that shares significant similarity with plant amino acid permeases – this favours a role for AUX1 in the uptake of the tryptophan-like IAA (3). Despite the fact that final biochemical proof of AUX1 function as an auxin uptake carrier is still lacking, several lines of evidence strongly support its involvement in auxin influx. The strongest support comes from a detailed analysis of the *aux1* phenotype. It has been demonstrated that the membrane permeable NAA rescues the *aux1* root agravitropic phenotype much more efficiently than the less membrane permeable IAA or 2,4-D and that this rescue coincides with restoration of polar auxin transport in this mutant (53, 96). Moreover, the *aux1* phenotype can be mimicked by growing seedlings on media containing the recently identified (39) inhibitors of auxin influx, 1-naphthoxyacetic and 3-chloro-4-hydroxyphenylacetic acids (71). The most direct support for AUX1 as an influx component comes from auxin uptake assays in *aux1* and wild type roots. Roots of *aux1* accumulated significantly less auxin influx substrate 2,4-D than do wild-type roots but no such difference was found for the membrane-permeable NAA or the IAA-like tryptophan (53).

Recently the AUX1 protein was localized within Arabidopsis root tissue using an epitope tagging approach (86). It was detected in a remarkable pattern in a subset of protophloem, columella, lateral root cap and epidermal cells, exclusively in the root tips. Most interestingly, the polar localization of AUX1 has been detected in the protophloem cells at the apical side opposite to that of PIN1 protein (Fig. 6C). However, the root tips of *aux1* contain less free auxin than wild type, probably due to defects in long distance auxin distribution from apex to the root. This paradox between local expression and long-distance effect of AUX1, taken together with its localization at the upper side of protophloem cells, suggests a role of AUX1 protein in the protophloem in unloading auxin from the mature phloem into the polar auxin transport system in the root tip. *AUX1* is a member of the small gene family in Arabidopsis. However the characterization of the other members of this family has not yet been reported (87).

Auxin transport

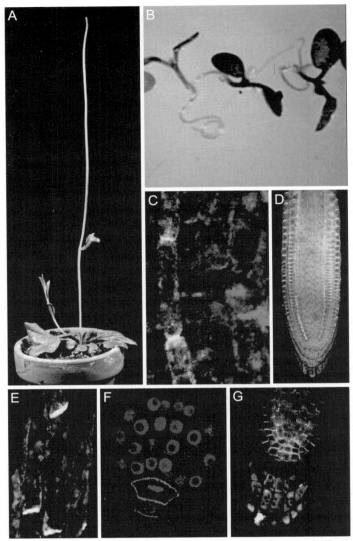

Figure 6 (Color plate page CP11). A. The "knitting needle" like phenotype of the Arabidopsis *pin* mutant. B. Root agravitropic phenotype of the Arabidopsis *aux1* mutant. C. Polar localization of the putative auxin influx carrier AUX1 (green) at the apical side and the putative auxin efflux carrier PIN1 (red) at the basal side of protophloem cells in an Arabidopsis root. D. Polar localization of PIN2 (green) at the apical side of lateral root cap and epidermis cells in an Arabidopsis root. The less polar localization of PIN2 in cells of the cortex is also visible. E. Basal localization of PIN1 (yellow) in xylem parenchyma cells in a longitudinal section of an Arabidopsis inflorescence. F. PIN4 localization (red) around the hypophysis and at the basal side of the subtending suspensor cell, in a globular stage Arabidopsis embryo. DAPI stained nuclei are depicted in blue. G. PIN4 localization (green) in the central root meristem. Basal localization in vascular and endodermis/cortex initials and their daughter cells; less pronounced basal localization in the quiescent centre; and non-polar localization in columella cells. (A, E from 70; B, C from 86; D from 21; F, G from 20).

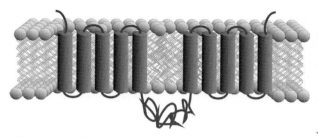

Figure 7. Predicted topology of the AtPIN1 protein showing conserved transmembrane domains at the N- and C-termini and the large, central, hydrophilic loop.

PIN Proteins – Components of the Auxin Efflux Machinery

In the early nineties, the knitting needle-like *pin-formed 1* (*pin1*) mutant phenotype of Arabidopsis was described (Fig. 6A). This strongly resembled plants treated with inhibitors of auxin efflux. Furthermore, basipetal auxin transport in this mutant is strongly reduced (69). The *AtPIN1* gene has now been cloned by transposon tagging and has been found to encode a transmembrane protein with similarity to a group of transporters from bacteria (Fig. 7). It has been suggested that AtPIN1 represents an important component of auxin efflux carrier (26). Alternatively, and equally likely on the basis of currently available genetic evidence, AtPIN1 could act as a regulator of auxin efflux. The Arabidopsis *PIN* gene family consists of eight members, four of which have been characterized in detail in relation to their expression, localization and role in plant development (22). Homologous genes have been found in all other plant species examined, including maize, rice and soybean. To date the proposed auxin efflux catalyst function of AtPIN proteins has not been verified biochemically. Nevertheless, there are several lines of evidence which strongly support such a role for the AtPIN proteins:

1. *The AtPIN protein primary sequences and predicted topology suggest a transport function.*

The AtPIN proteins share more than 70% similarity and have a closely similar topology - two highly hydrophobic domains with five to six transmembrane segments linked by a hydrophilic region (Fig. 7). Transporters of the major facilitator class display similar topology. Moreover AtPIN proteins were demonstrated to share limited sequence similarity with known prokaryotic and eukaryotic transporters (12, 51, 65, 70, 90).

2. *Yeast cells expressing AtPIN2 accumulate less auxin and related compounds.*

The only heterologous system used to address PIN transport activity has been yeast cells with an altered ion homeostasis (51). When AtPIN2 is overexpressed in such yeast cells they exhibit enhanced resistance to the toxin fluoroindole, a substance with a limited structural similarity to auxin (51). Other experiments demonstrated that these cells also retain less radioactively labeled auxin (12). The decreased accumulation of labeled auxin or its analogs support an auxin efflux function for AtPIN2 in yeast.

Nevertheless, attempts to demonstrate changes in auxin transport rate in such cells have so far failed, leaving this issue unresolved.

3. AtPIN proteins have a polar distribution in auxin transport-competent cells.

The chemiosmotic hypothesis (see earlier) predicts that auxin efflux carriers are asymmetrically localized in cells and that this polar localization determines the direction of the net auxin flux (76, 79). Coincidence between the polarity of distribution and the known direction of the auxin flux has been demonstrated for several AtPIN proteins. AtPIN1 protein is localized at the basal side of elongated parenchymatous xylem and cambial cells of Arabidopsis inflorescence axes where polar auxin transport occurs in the basipetal direction (26, 70). In contrast, the AtPIN2 protein was polarly localized at the upper side of the lateral root cap and epidermis cells (Fig. 6D; 21, 65) again in accordance with a basipetal auxin stream from the root tip. The AtPIN3 protein localizes predominantly to the lateral side of shoot endodermal cells (Fig. 8B; 23) and the polar localization of AtPIN4 in root tip directs towards columella initials, the site of auxin accumulation (Fig. 6G; 20).

4. Atpin mutants are defective in polar auxin transport.

One of the strongest arguments for the involvement of PIN proteins in auxin transport is a reduction of polar auxin transport in *Atpin* mutants, which directly correlates with loss of *AtPIN* expression in the corresponding tissue. This was demonstrated for basipetal auxin transport in stem of the *Atpin1* mutant and in the root of the *Atpin2* mutant (69, 74).

5. Disruption of AtPIN function cause changes in cell specific auxin accumulation.

Auxin accumulation has been indirectly monitored in Arabidopsis by determining the activity of auxin-responsive constructs such as *DR5::GUS* (81). Where compared, this seems to correlate well with direct IAA measurements (11, 20). Using this approach, it has been found that changes in cell-specific auxin accumulation in several *Atpin* mutants are correlated with a loss of AtPIN expression in the corresponding cells. Such correlations have been demonstrated for *Atpin2* (*eir1*) mutant roots after gravistimulation (51) and for *Atpin4* mutant roots and embryos (20). In addition, *Atpin2* (*agr1*) root tips preloaded with radioactively labeled IAA retain more radioactivity than similarly treated wild type roots (12).

6. Phenotypes of Atpin mutants can be phenocopied by auxin efflux inhibitors.

Defects observed in *Atpin* mutants are in processes that are known to be regulated by polar auxin transport and they can be phenocopied by treatment of wild type plants with auxin efflux inhibitors. Examples include the *Atpin1* embryonic and aerial phenotype (69); the defect in root and shoot

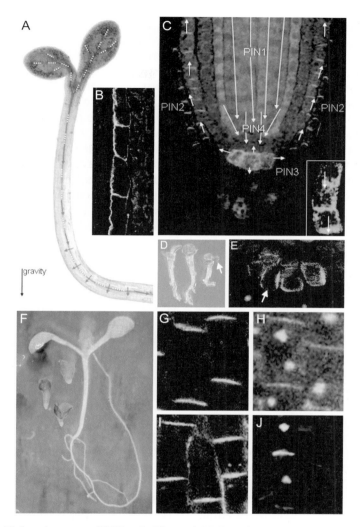

Figure 8 (Color plate page CP12). A. Elevated DR5 auxin reporter response (blue) in the outer layer of a gravistimulated Arabidopsis hypocotyl. The direction of basipetal (red arrows) and lateral (orange arrows) polar auxin flows is depicted, as well as the non-polar phloem transport of auxin (dashed line). B. Lateral localization of PIN3 in endodermal cells seen in longitudinal section of an Arabidopsis inflorescence. C. Directions of auxin flow derived from the polar localization of various PIN proteins in the Arabidopsis root. PIN3 localization (green) symmetrically around the columella cells is visible. Relocation of PIN3 to the basal side of columella cells after gravity stimulation (inset). D. Seedling phenotypes of *smt1* mutant. Arrows shows triple cotyledon seedling. E. Changes in PIN3 polar localization in the columella of the *smt1* mutant. F. Comparison of wild type Arabidopsis and *gnom* mutant seedlings. G. Polar localization of PIN1 in Arabidopsis root cells. H. Internalization of PIN1 after the BFA treatment. I. Reconstitution of PIN1 polar localization after the BFA removal. J. No internalization of polarly localized PIN1 (red) and internalisation of KNOLLE (green) after BFA treatment in the BFA-resistant GNOM transgenic plant. (A, C from 19; B from 23; D, E from 95; F and J reproduced by courtesy of Niko Geldner).

gravitropism in *Atpin2* and *Atpin3* mutants, respectively (23, 52); the defect in hypocotyl and root elongation in light, in apical hook opening and in lateral root initiation, which have been reported for *Atpin3* mutants (23); and the *Atpin4* root meristem pattern aberrations, which can be also found in seedlings germinated on low concentrations of auxin efflux inhibitors (20).

The data accumulated so far provides an extensive body of evidence to argue that AtPIN proteins are crucially involved in auxin efflux. Nevertheless, the central question of whether PIN proteins represent transport or regulatory components of auxin efflux still remains unresolved. To answer this question, auxin transport assays will have to be developed to establish directly the carrier functions of different the PIN proteins and to determine their substrate specificities, affinities and kinetic properties.

ABC Transporters – More Auxin Transport Proteins?

Recently, a combination of genetic and biochemical approaches has implicated another protein family in auxin transport, namely the so-called multidrug resistance (MDR) proteins, a sub-family of the ATP-binding cassette (ABC) transporters (54). In mammalian system these transmembrane proteins enhance the export of chemotherapeutic substances. Two of them, AtMDR1 and AtPGP1, were originally identified in Arabidopsis as having anion channel-related functions. Nevertheless, the corresponding mutants and double mutants also exhibited phenotypic aberrations consistent with defects in polar auxin transport, such as reduced apical dominance, a lower rate of basipetal auxin transport and also reduced NPA binding. The AtMDR1 and AtPGP1 proteins may transport auxins across both the PM and across intracellular membranes (68; 50). The proteins were isolated by affinity chromatography and were also able to bind NPA *in vitro* or when expressed in yeast cells. In spite of this, membranes prepared from *Atmdr1* mutants still exhibited up to 60% of the NPA binding found in wild type and NPA was able to reduce auxin transport to background levels in the mutant (68). This implies that other NPA-binding protein(s), in addition to MDRs, must be present in Arabidopsis cells (64).

Another ABC transporter, AtMRP5, has also been implicated in auxin biology (25). The corresponding mutant *mrp5-1* displays reduced root growth, increased lateral root formation and higher than normal auxin levels in roots. Thus although much detailed information for ABC transporters is still lacking, especially on their expression and subcellular localization, their direct or indirect connection to the polar auxin transport, and especially to NPA-mediated processes, seems to be established.

Mutations Indirectly Affecting Polar Auxin Transport

In a mutant screen designed to identify components of polar auxin transport based on the fact that auxin efflux inhibitors reduce root elongation, several mutants were isolated in which roots were able to elongate in the presence of such inhibitors. These mutants were designated *tir* (*transport inhibitor response*; 80), and seven *tir* loci (*tir1* - *tir7*) have now been identified. Several of these mutants have been molecularly characterized and this has revealed that the primary defects are in auxin signaling rather than in polar auxin transport. The most relevant for polar auxin transport research is the *tir3* mutant. This displays a variety of morphological defects including reduced elongation of root and inflorescence stalks, decreased apical dominance and reduced lateral root formation. Both auxin transport and NPA binding activity are reduced in the *tir3* mutant (80). It has been suggested that the *TIR3* gene may encode the NBP (see above) or some closely related protein (38). The corresponding gene has been characterized and renamed *BIG* to reflect the unusually large size of the protein it encodes. This has been identified as a protein with several putative Zn-finger domains homologous to the *Drosophila* CALOSSIN/PUSHOVER (CAL/O) protein (32). A defect in CAL/O interferes with neurotransmitter release in *Drosophila*. In Arabidopsis the *big/tir3/doc1* mutations interfere with an effect of auxin efflux inhibitors on subcellular movement of a putative auxin efflux carrier AtPIN1 (32), supporting a role for BIG in vesicle trafficking, although the mechanism of this action is not clarified yet (22, 49, 64).

Another mutant, *pis1* (*polar auxin transport inhibitor-sensitive 1*), isolated in a similar screen, displays a phenotype in many respects opposite to that of the *tir* mutants – it is hypersensitive to some auxin efflux inhibitors (24). This has led to speculation that *PIS1* may encode a negative regulator of the auxin efflux inhibitor pathway. However, lack of molecular data leaves this an open question.

Other genetic work has pointed to a role for phosphorylation in polar auxin transport. A mutant called *rcn1* (*roots curl in NPA*) has been isolated, the roots of which curl in the presence of NPA, in contrast to straight root growth in wild type (27). This mutant has reduced root and hypocotyl elongation and is defective for apical hook formation. The *RCN1* gene was found to encode a regulatory subunit of protein phosphatase 2A. RCN1 may control the level of phosphorylation and thereby the activity of a component involved in polar auxin transport (27; and see below). This possibility has been strengthened by the recent finding that the *rcn* mutant displays enhanced basipetal auxin transport, a phenotype feature which has also been observed in plants treated with a phosphatase inhibitor cantharidin (75). In addition, the *pinoid* (*pid*) mutants in the gene coding for a serine-threonine protein kinase (13) display defects in the formation of flowers and cotyledons, thus resembling plants treated with auxin efflux inhibitors. Moreover basipetal auxin transport in *pid* inflorescences is reduced. The PID protein has been proposed to be involved either in polar auxin transport or auxin signaling. Detailed studies correlating PID expression data with knock out and tissue specific overexpression phenotypes demonstrate that PID action is not cell or organ autonomous and it is sensitive to auxin efflux inhibitors (2). These data favor the hypothesis that PID is involved in long distance signaling and functions as a positive regulator of auxin transport (2).

Several other Arabidopsis mutants with possible roles in polar auxin transport have been identified in a screen for plants with early developmental aberrations (57). Seedlings of the mutant *monopteros* (*mp*) lack roots and display defects in cotyledon establishment (5). The adult plants show measurable reduction in basipetal auxin transport which, however, might be the result of a reduced vasculature. The *MP* gene codes for auxin response factor 5 (ARF5), a transcription factor, which mediates auxin dependent activation of gene expression (55). It is possible that MP regulates expression of polar auxin transport components. Another mutant with strong seedling phenotype *cephalopod/orc* has been shown to display defects in auxin transport and cell polarity (95). *CPH* encodes sterol methyl transferase 1 (SMT1) – a protein involved in biosynthesis of plant sterols (83). The polar localization of the PIN efflux components is defective in the *cph/orc/smt1* mutant (Fig. 8E), which can account for the defect in polar auxin transport. However, cause and effect in the relationship between PIN localization and cell polarity, and how the defect in sterol membrane composition interferes with this remains unclear.

The connection to polar auxin transport is more clearly characterized in the case of another early development mutant, *gnom* (*gn*), which was isolated from the same screen as *mp* (57). Embryos of *gn* mutants display a variety of aberrations, including defects in apical-basal patterning, and fused or improperly placed cotyledons (Fig. 8F). Most of these defects are reminiscent to defects observed when embryos are cultivated in the presence of polar auxin transport inhibitors (35). The *GN* gene was cloned and the GN protein demonstrated to have guanine nucleotide exchange factor (GEF) activity for small ARF-type GTPases (85). These small GTPases are known to play a role in the control of intracellular vesicle trafficking. It has been shown that *gn* mutant embryos have defects in the correct localization of the polar auxin transport component PIN1 (85). Moreover, recent elegant experiments with GN engineered to be resistant to the inhibitor of vesicle trafficking, BFA (29), show that GN is directly involved in the trafficking of PIN proteins from the endosomes to their polarly localized domain in the PM (Fig. 8J). These observations complete the connection between the polar auxin transport related *gn* phenotype and GN function in vesicle trafficking.

SUBCELLULAR DYNAMICS OF POLAR AUXIN TRANSPORT COMPONENTS

Physiological evidence, which indicates that efflux carrier proteins have a very short half-life in the PM and probably cycle between the PM and an unidentified intracellular compartment, was described above. Subsequent work at the genetic and molecular levels has confirmed the dynamic nature of the asymmetrically localized PIN proteins. It has been discovered that mutations in the Arabidopsis intracellular vesicle trafficking regulator GNOM, an ARF GEF (see above), interfere with the correct localization of the AtPIN1 protein at the PM during embryogenesis (85). Similarly, chemical inhibition of ARF GEFs by BFA causes the disappearance of PIN1 label from the PM (Fig. 8G) and its intracellular accumulation (Fig. 8H; 30). This reversible BFA effect (Fig. 8I) also occurs in the presence of the protein synthesis inhibitor CHX, thus demonstrating that the internalized AtPIN1 originated from the PM and not as a result of *de novo* protein synthesis (Fig.5).

Together, therefore, physiological and molecular studies have provided good evidence that AtPIN1 cycles rapidly between the PM and a so-called "BFA" compartment, recently characterized as an accumulation of endosomes (29). This model has been corroborated by electron microscopy studies, which detected the homologous PIN3 protein not only at the PM but frequently also in intracellular vesicles (20). The action of drugs that disrupt the structure of the cytoskeleton indicated that AtPIN1-containing vesicles were transported predominantly along the actin cytoskeleton. However in dividing cells, tubulin was also required for correct PIN1 traffic (30).

Recently, experiments using BFA have also demonstrated turnover in the PM of the auxin influx component – the AUX1 protein (34). It seems that the cycling of polar auxin transport components is an essential part of auxin transport since interfering with the cycling process by treatment with BFA, auxin efflux inhibitors, or drugs that depolymerize actin, as well as genetically through mutation of GN, also interfere with auxin efflux and polar auxin transport regulated plant development (16, 30, 60).

Thus previous models of polar auxin transport which envisaged long-lived carriers located asymmetrically in the PM, must now be substantially modified to take account of the new information which demonstrates that auxin carriers (or, perhaps, carrier complexes) are much more dynamic structures than they were originally believed to be. An important unresolved question is the role of AtPIN cycling in polar auxin transport. Several possible scenarios can be conceived:

Firstly, a high turnover of polar auxin transport components would provide the flexibility to allow rapid changes to be made in carrier distribution in the PM and provide a mechanism for the rapid redirection of auxin fluxes in response to environmental or developmental cues (19, 59). This, in turn, would contribute to developmental plasticity underlying adaptive processes such as tropisms, initiation of lateral organs or regulation of meristem activity (see below).

A second possibility is that components of polar auxin transport may have a dual receptor/transporter function (36). In this case cycling might be part of a mechanism for signal transduction and receptor regeneration, as is known for some other kinds of receptors. The issue of dual sensor and transport functions has been extensively discussed in relation to sugar carriers in yeast cells and in plants (45).

Perhaps the most exciting possibility is that vesicle trafficking itself is a part of the auxin transport machinery and that, in a manner analogous to the mechanism of neurotransmitter release in animals, auxin is a vesicle cargo, released from cells by polar exocytosis (22). In this model PIN localization in endosomes and recycling vesicles would have an entirely new significance. Instead of being PM localized "auxin channels", PIN proteins would mediate the accumulation or retention of auxin in the vesicles in which auxin would be translocated to the corresponding cell pole. Some support of this scenario comes from experiments using anti-IAA antibodies and electron microscopy, in which auxin was found in small vesicles near the PM (84). Moreover, the BIG protein, which is involved in polar auxin transport and PIN1 subcellular trafficking (see above), is a homologue of calossin (32), a protein that mediates vesicle recycling in *Drosophila* during synaptic transmission.

Regardless of how well any of these scenarios (or combinations of them) eventually turn out to fit the true picture, recent advances have made it abundantly clear that gaining an understanding of the cellular mechanisms controlling the subcellular dynamics of the auxin carriers will be crucial if we are to fully understand polar auxin transport.

REGULATION OF POLAR AUXIN TRANSPORT

It might be expected that a system as complex as the one that mediates polar auxin transport will be subject to regulation at a variety of different levels. Polar transport requires the expression of many genes and the synthesis of the corresponding proteins; it requires the targeting of these proteins to defined locations in the cell and at the PM; it involves their metabolic turnover and their cycling; and, in the case of both the transport catalysts themselves and probably also the NBPs, the direct or indirect regulation of their activity. A brief description of some aspects of the regulation of these processes is given here.

The Role of Phosphorylation

The results of physiological experiments indicate that at least some of the processes involved in polar auxin transport are energy-dependent and that phosphorylation/dephosphorylation processes are probably involved in the regulation of the activity of some of the carrier systems. The role of reversible phosphorylation in the regulation of auxin transport has been comprehensively reviewed elsewhere (63) and will be discussed only briefly here. The likely involvement in this regulation of a protein kinase encoded by the Arabidopsis *PINOID* gene (*AtPID*), and of protein phosphatase 2A (PP2A), the regulatory A subunit of which is coded by the *AtRCN1* gene, was described above. Studies of auxin transport in seedling roots of *rcn1* mutants (in which PP2A activity is reduced) and in cultured tobacco cells treated with various kinase and phosphatase inhibitors, suggest that reversible phosphorylation acts at several different loci in the regulation of polar auxin transport (16, 63). Furthermore, it is likely that reversible phosphorylation and phytotropin action (see below) interact in this regulation. In tobacco cells, kinase inhibitors very rapidly and strongly inhibit auxin efflux without affecting auxin influx (16). In roots of *rcn1* seedlings the sensitivity of acropetal transport to NPA is drastically reduced, suggesting a role for PP2A in NPA function; in contrast, the basipetal return flow of auxin from the root tip is increased substantially and is unaffected by NPA. In this case, PP2A seems to act as a negative regulator of the polar transport system (63, 75).

Nevertheless, it is still not clear how directly, or at what levels the products of *RCN1* and *PID* intervene in the polar auxin transport process and whether they influence the expression and localization, or the activity of the proteins involved. It is also not known how protein phosphorylation-dephosphorylation affects the interaction between the NBP and the auxin efflux carrier system; one recent suggestion is that the reversible phosphorylation loop may mimic or control the activity of the putative, and apparently dynamic, transducing protein proposed to couple the NBP and efflux catalyst action (61, 97). Gaining a clear understanding of the putative role of reversible phosphorylation in the regulation of auxin transport remains a challenge for the future.

The Role of Phytotropins

Phytotropins (such as NPA) have contributed significantly to our understanding of the auxin transport machinery (59, 78). The point has also been made that phytotropins are synthetic compounds with, as yet, no unequivocally identified endogenous equivalent. Some flavonoids have been shown to be competitors of saturable NPA binding to membrane preparations and to inhibit polar auxin transport (9, 41, 78). There is some evidence to show that these endogenous compounds, perhaps together with aminopeptidases, are involved in regulation of auxin transport (66).

How do phytotropins act? At a functional level they simply inhibit auxin efflux, resulting in auxin accumulation by cells. This action appears to require interaction with the NBP, but little is known about the molecular mechanisms involved. Besides, phytotropins may participate in two other cellular processes related to polar transport of auxin: Firstly, some experimental data (73) suggest that NBP is involved in the establishment and the maintenance of polarity of cell division (see below). Secondly, phytotropins have been shown to play a role in auxin efflux carrier cycling through a general inhibitory action on vesicle-mediated traffic to the PM (30, but see72). Studies of the trafficking of PIN1 and other unrelated proteins in Arabidopsis roots have revealed that high concentrations of auxin transport inhibitors interfere with vesicle traffic to and from the PM (30). This has led to the suggestion that the action of phytotropins on the inhibition of auxin efflux is in fact the result of a non-specific inhibition of protein trafficking, including that of auxin efflux carriers. However, recent observations on BY-2 tobacco cells have revealed that the inhibition of auxin efflux by NPA is much more efficient than the inhibition caused by the well-established inhibitor of protein traffic, BFA, and that BFA, but not NPA, affects some of the intracellular structures related to the trafficking machinery, including the actin cytoskeleton (72). These findings argue against a causal link between a general role of NPA and other phytotropins in vesicle-trafficking and auxin efflux inhibition. Rather, they suggest that a population of auxin efflux catalysts exists which is NPA-sensitive but insensitive to inhibitors of vesicle-mediated traffic.

Generally, the mechanism of regulation of auxin afflux by phytotropins and other inhibitors of polar auxin transport still remains unclear and it is apparent that more work on this topic is needed.

Synthetic Inhibitors of Auxin Influx

In contrast to auxin efflux carriers, work on the physiology of auxin influx carriers has suffered from lack of specific inhibitors. Like efflux carriers, influx carriers may be inhibited by phytotropins such as NPA, but to a very much smaller extent (17). Recently, a promising new group of auxin influx inhibitors of the aryl and aryloxyalkylcarboxylic acid type has been identified (39). Of these, 1-naphthoxyacetic and 3-chloro-4-hydroxyphenylacetic acids have been reported to inhibit auxin influx carrier activity substantially. Both compounds also disrupt root gravitropic responses and in wild type plants mimic the auxin influx carrier mutation *aux1* (71; see above). However, further work in this area is required and we still know nothing about the molecular mechanism by which these compounds inhibit auxin influx carrier function.

Regulation by Auxin and Other Hormones

Auxin itself is needed for the induction of new polar auxin transport pathways, for example during induction and axial development of new vascular tissues (7). A continuous supply of auxin is also necessary to maintain the auxin polar transport system itself. The ability of tissue to transport auxin in a polar manner rapidly decreases following loss of the auxin source, although NPA-sensitive accumulation of auxin is unaffected (33). In both these cases a role for auxin in the regulation of auxin carrier distribution is indicated (59). It remains largely unknown how auxin gradients themselves might establish (or re-establish) new polar transport pathways or change the direction of existing polar transport pathways in response to environmental or internal cues.

Data about the action of other plant hormones on auxin transport are scarce. Much of the currently available information relates to the effects of ethylene on auxin transport in excised segments of various plant organs, such as maize coleoptiles, bean petioles, cotton stems and petioles (33). In contrast to ethylene, cytokinins have been reported to prevent the decrease in auxin transport in excised tissues in the absence of an auxin source (33). Recently an Arabidopsis mutant was characterised (91) showing alteration in the collaboration between ethylene and auxin, possibly at the level of auxin transport. Nevertheless, the mechanism of the ethylene response is still unclear.

There are some indications that other hormones, including abscisic acid, brassinosteroids and gibberellins, may be involved in the control of auxin transport (88). The physiological importance of mutual control of transport processes for individual phytohormones, including IAA, is obvious; gaining an understanding of the mechanisms by which individual hormones interact at the level of transport remains a major challenge for future research.

AUXIN TRANSPORT AND PLANT DEVELOPMENT

The developmental responses of plants to modifications to the normal patterns of auxin flow make it abundantly clear that auxin transport plays a crucial role in the regulation of development. A bewildering variety of processes may be affected, ranging from cell division, through establishment of polar axes, cell differentiation, pattern formation, histogenesis and organogenesis, to the coordination and integration of biochemical and physiological activities in different regions of the plant body. In some of these processes, auxin transport clearly delivers a signal that acts as a switch, or regulates the progress and rate of a process. In other cases the direction of auxin flow, or, perhaps, tissue gradients in auxin concentration that result from transport, endow auxin transport with both directional qualities and some features characteristic of a morphogen. Here, a few selected examples of the relationship between auxin transport and development will be discussed in order to explore the diversity of this relationship and of the mechanisms probably involved (93).

Cell Division, Initiation of Polarity, and Cell Elongation

Cell division and cell differentiation underpin plant development and in most plant cells differentiation is associated with expansion. The majority of cells in the plant body enlarge by diffuse growth, which is frequently polar in character; i.e. cells tend to elongate preferentially along one axis. Therefore, a mechanism is required to ensure that traffic of secretory vesicles containing material destined for the cell wall and PM is targeted in such a way that it is distributed towards the proper regions of the cell. Auxin seems to be part of the mechanism involved, although to date experimental data relating to this mechanism is fragmentary and incomplete.

Attempts to explore the role of polar auxin transport in the initiation of cell division and in the establishment of polarity at the single cell level have been hampered by lack of a suitable model system. The wide diversity of cell types within even a relatively simple structure, such as the seedling root of Arabidopsis, makes it almost impossible to investigate either biochemical or some cytological aspects of auxin transport at the cell level in a meaningful way. Because of this, attention is increasingly focussing on the development of cell suspension culture systems in which individual cells

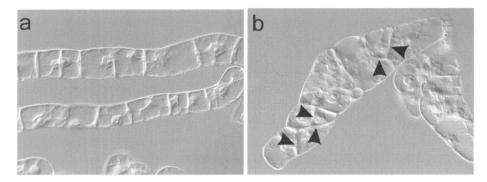

Figure 9. Effect of 1-N-naphthylphthalamic acid (NPA) on the polarity of cell division of suspension-cultured cells of *Nicotiana tabacum* L. line VBI-0. **a**. Cells grown in control medium, day 9. **b**. Cells grown in control medium supplemented with NPA (final concentration 10 µM), day 9. Note abnormal cell division planes (black arrows). (Modified from 73).

separate easily and in which the stages of cell development can be readily synchronised. Although so far no such system is available for Arabidopsis, recent studies suggest that some well-defined tobacco cell lines (28) may provide a suitable alternative model system in which to conduct parallel studies of the genetic, cellular and biochemical aspects of polar auxin transport.

Immobilised cultured cells of *Nicotiana tabacum* L., derived from leaf mesophyll protoplasts, have been used to study cell elongation in initially spherical cells after the addition of NAA (92). No polar ion fluxes, typical of those found in tip growth, were detected, indicating that the positional information required to establish a polar axis was delivered by a different signal, possibly auxin itself. A model was proposed whereby an auxin flux initiated a signal cascade that resulted in a reorientation of microtubules which guided deposition of cellulose in the cell wall. The resulting modification of the cell wall mechanics promoted elongation growth.

Changes in auxin (NAA) accumulation and auxin efflux carrier activity during the growth cycle of another model tobacco cell line have also been studied in detail (73). Although the level of NAA accumulation remained relatively stable over a subculture period, the sensitivity of cells to NPA changed markedly and reached a maximum at the onset of division. Treatment of cells with NPA delayed the onset of cell division but did not prevent it. However, when cell division commenced a significant proportion of the NPA-treated cells exhibited loss of cell polarity and abnormal orientation of cell division (73; Fig. 9). These observations, together with data suggesting that the NBP may affect the localization of auxin efflux carriers (32, 49), indicate that in tobacco cells the NPA-sensitive directed traffic of auxin efflux carriers to the specific regions at PM may regulate the orientation of cell division (73).

An interesting link between cell polarity and the sterol composition of cell membranes has been described recently (95) from *cephalopod/orc* mutants of Arabidopsis (see above). In the mutants the PIN proteins, but not AUX1, are mislocalized and the mutants show a wide range of polarity defects. As neither membrane fluidity nor vesicle trafficking seems to be impaired, mutation of the *CPH/ORC/SMT1* gene may disrupt the docking of proteins to specific membrane microdomains or lipid rafts (95). Together the data reveal the importance of a balanced membrane sterol composition for auxin efflux and for promoting cell polarity in Arabidopsis.

Auxin Transport and Vascular Development

The role of auxin in the induction of vascular tissues is covered in depth in chapter E2 and treatment of the topic here will be confined to a brief examination of the possible part played by polar auxin transport itself in determining the axes and pathways of vascular development.

Vascular differentiation normally takes place along well defined pathways initially laid down in the embryo as provascular strands of narrow, elongated cells early in the development of an organ (6). Although usually predictable, the ease with which new axes of vascular development may arise following wounding and/or hormone application demonstrate considerable flexibility in vascular patterning (6, 82). It is now clear that vascular differentiation occurs along axialized (canalized) auxin flows. These may be generated naturally by auxin originating in the apical regions of developing organs, including the developing embryo itself, or experimentally by organ removal and replacement with exogenous auxin (6, 82). Canalized auxin flows may themselves by mediated by PIN proteins (see earlier). In auxin transport mutants or in plants treated with polar auxin transport inhibitors, abnormal patterns of vascular development frequently occur. For example, following application of increasingly high concentrations of NPA to Arabidopsis seedlings, the normal pattern of rosette leaf vein formation becomes increasingly disrupted (56). This observation is consistent with the canalization hypothesis (82,), which postulates that through a feedback mechanism, an initially weak polar auxin flow through a file of cells renders them more conductive and increases polar auxin flow. This will tend to "drain" surrounding tissues until a point is reached at which the concentration of auxin in the conducting cells exceeds some critical threshold required for vascular differentiation. The postulated feedback mechanism is unknown, but the ability of auxin to stimulate the accumulation of H^+-ATPase protein at the PM and for BFA to inhibit auxin-mediated growth and secretion of cell wall proteins leads to speculation that auxin may promote Golgi-dependent traffic of components of the polar auxin transport system itself (59). Precisely how the canalized auxin flow regulates the differentiation of vascular elements remains uncertain, but the transported auxin probably up-regulates the activity of ARF proteins such as MP by removing their inhibition by AUX/IAA proteins such as BODENLOS

(BDL). This enables the ARFs to interact with appropriate auxin responsive elements and activate genes involved in early axialization and development of provascular strands (93).

Tropisms

Tropisms are permanent changes in the direction of growth of an organ caused by differential growth rates on either side of the organ in response to a directional environmental stimulus such as light (phototropism) or gravity (gravitropism). The Cholodny-Went hypothesis (94) proposed that the different growth rates on either side of an organ resulted from an unequal distribution of auxin between its two sides in response to the stimulus. To explain the different directions of root and shoot growth in response to a stimulus (i.e. roots are positively gravitropic, shoots negatively gravitropic) the hypothesis proposed that in shoots a high auxin concentration promoted, and in roots inhibited, cell elongation resulting in bending responses in opposite directions. Many experiments have demonstrated a differential distribution of auxin or of auxin response following stimulation, and these have been correlated with the change in growth rate and the bending response (Fig. 8A). Because auxin efflux inhibitors interfere with the asymmetric distribution of auxin and tropisms, polar auxin transport has been implicated as the process by which asymmetric auxin distribution is achieved (19). Radial polar transport of auxin has been proposed to facilitate the exchange of auxin between the main basipetal stream in shoot vasculature and peripheral regions where control of elongation occurs. Molecular support for these proposals was obtained following the identification of auxin efflux component PIN3. The PIN3 protein is involved in hypocotyl and root tropisms and is localized in shoot at the lateral side of endodermal cells (Fig. 8B), where it is perfectly positioned to regulate radial auxin flow (23). Since the lesions in tropic responses in *pin3* mutants are rather subtle, other PIN proteins may functionally replace PIN3 (23).

In roots gravitational stimuli are perceived in the root cap but the growth response occurs in the elongation zone where elevated auxin levels on the lower side inhibit growth, causing downward bending. Results have shown that the initial response is a lateral redistribution of auxin towards the lower side in the root cap itself, from where it is transported basipetally to the elongation zone (Fig. 8C; 74, 81). Both efflux and influx components of polar auxin transport are involved in these processes. Localization and mutant studies suggest that the putative influx carrier AUX1 facilitates auxin uptake into the lateral root cap and epidermis region and the efflux regulator PIN2 mediates directional translocation towards the elongation zone (Fig. 8C; 19)

How is polar auxin transport linked to the perception of a stimulus, such as gravity? Gravity is perceived by sedimentation of starch containing organelles (statoliths) in the columella region of the root cap and in shoot endodermis ("starch sheath"). The presence of PIN3 in these cells suggested the possibility that gravity perception and auxin redistribution are coupled *via* PIN3. This possibility has now been tested in gravistimulated Arabidopsis roots. During normal downward growth of the root the majority of PIN3 is located symmetrically at the columella cell boundaries. However, within two minutes of gravistimulation, the position of PIN3 changes and it becomes relocated towards the new lower end of the cells (Fig. 8C, inset). Thus PIN3 is ideally placed to mediate an auxin flow towards the lower side of root (Fig. 10). Interestingly, AUX1 is also localized in columella cells, both at the PM and internalized. One possibility is that AUX1 mediates auxin influx into the columella after gravistimulation, thereby creating a temporary pool of auxin needed for asymmetric redistribution in a PIN3-dependent pathway.

But how is PIN3 so rapidly relocated after the perception of a stimulus? The answer may lie in the rapid cycling of PIN proteins along the actin cytoskeleton, as already discussed above. Such a mechanism would provide the necessary rapidity and flexibility of response, which is unlikely to be achieved through *de novo* protein synthesis and targeting.

An important question remains for future investigation in relation to gravitropism: How is the sedimentation of statoliths connected to PIN3 relocation? Several observations suggest that the actin cytoskeleton reorganizes during statolith sedimentation. Thus, the actin dependent intracellular traffic of PIN3 could be redirected along the sedimentation routes and PIN3 would preferentially accumulate at the lower side of the cell.

However, so far there is only indirect evidence that it is indeed PIN3

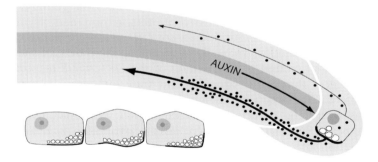

Figure 10. Model of root gravitropism. Auxin is provided to the root tip through the stele, is laterally distributed symmetrically from the columella and is basipetally transported through the lateral root cap and epidermis to the elongation zone of the root. After reorientation of the root, statoliths in the columella sediment to the lower side of the cells (inset), PIN3 auxin transport proteins are relocated and facilitate auxin transport to the lower side of root. From there auxin is transported to the elongation zone, where it inhibits elongation resulting in downward bending.

relocation that mediates auxin redistribution. The presence of both statoliths and PIN3 in the shoot endodermis suggests that a similar mechanism involving relocation of PIN3 and/or other PIN proteins also operates during shoot tropisms - but this too remains to be demonstrated unequivocally.

Another important issue concerns phototropism. Although it seems to be well established that PIN-dependent asymmetric auxin distribution also underlies this process, the mechanism by which light direction acts to influence lateral redistribution of auxin remains unknown and is a topic for future investigations (63).

Patterning - Polar Auxin Transport and the Maintenance of a Morphogen Gradient

The role of auxin and polar auxin transport in roots is not restricted to growth responses. Exogenous manipulation of auxin levels as well as analysis of mutants impaired in auxin signaling, have demonstrated a role for auxin in the regulation of patterns of cell division and differentiation (Fig. 11; 43, 80) and have renewed the debate about a possible role for auxin as a morphogen (81). From an animal standpoint, rigorous definition of a morphogen requires: (i) the formation of a stable concentration gradient of the compound in question; (ii) that the compound itself should directly "instruct" the responding cells (and not work through another signalling pathway); and (iii) that the magnitude of the response (of a cell) should depend on morphogen concentration. Evidence is accumulating, albeit mainly indirect, to demonstrate that auxin shares some of these properties. Chemical analysis has revealed a graded distribution of free auxin in Scots pine stem (89) and along the *Arabidopsis* root (11). These findings have been corroborated indirectly by the use of auxin response reporters (*DR5::GUS*) to indirectly visualize auxin gradients within the root meristem (Fig. 11), forming a maximum in the columella initial cells (20, 81). Intriguingly, the auxin efflux component PIN4 is expressed in this region and asymmetrically positioned towards cells with increased *DR5::GUS* response (Fig. 6G). Moreover *pin4* mutations, as well as chemical inhibition of auxin efflux, disrupt the spatial pattern of distribution of the *DR5::GUS* response. These findings suggest that efflux-driven auxin transport actively maintains the auxin gradient. The observed changes in the auxin gradient are accompanied by various patterning defects and correlate well with changes in cell fate. Thus it appears that efflux-dependent auxin distribution is linked with cell fate acquisition and thus correct meristem patterning. However, as yet we know too little about auxin gradient perception and downstream signalling to be able to pinpoint the direct effect on instructed cells. The emerging model of active, regulated, polar transport-driven auxin gradients differs from the classical view of animal morphogens, which are supposed to freely diffuse from a localized source. Interestingly, in the animal field too, the morphogen concept is being revised, since the gradients of classical morphogens such as Decapentaplegic or Wingless appear to be actively maintained by vesicular trafficking (18), thus showing similarities to GNOM-regulated polar auxin transport (see earlier).

It seems that auxin and its graded distribution also plays a role in patterning of embryos. Exogenous manipulation of auxin homeostasis in *in vitro* cultured *Brassica* embryos demonstrated a role for polar auxin transport in the establishment of the apical-basal axis and in cotyledon separation (35). Genetic studies with mutants impaired in auxin signalling, such as *mp*, *bdl* and *auxin resistant 6* (*axr6*), as well as with the *gn* mutant impaired in proper localization of PIN proteins, further strengthen this hypothesis. Possible auxin distribution and polar auxin transport routes in embryos can be indirectly inferred from the polar localization of the PIN1 and PIN4 proteins, which are expressed in embryos (Fig. 6F). In such a model, auxin would be transported through the outer layers towards the future cotyledon tips and then through the centre towards the embryo base (corresponding to the location of the *DR5::GUS* response maximum) and further exported through the suspensor (Fig. 12). Nonetheless there is still a substantial amount of experimental work needed to verify this model and to clearly pinpoint the role of auxin in embryo development. So, despite the fact that we can not with our present knowledge decide whether auxin rigorously fulfils the definition of a morphogen, it certainly meets the most important criteria of it – it exhibits a graded distribution and it is involved in patterning processes. However the important issue of the interpretation of auxin gradients remains a challenge for future research.

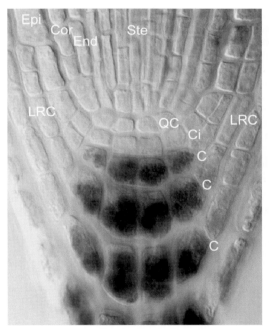

Figure 11. Arabidopsis root meristem pattern. The quiescent centre (QC) in the middle is surrounded by undifferentiated initials, including the columella initials (Ci), which give rise to columella (C), where starch containing statoliths occur (stained here by lugol). Other differentiated cell types such as epidermis (Epi), cortex (Cor), endodermis (End), stele (Ste) and lateral root cap (LRC) are indicated. This regular and invariant pattern correlates with the auxin gradient, which displays its maximum in the columella initials. (Reproduced with permission from 19).

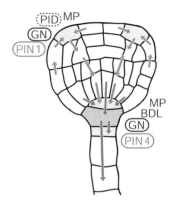

Figure 12. Model of auxin transport and distribution during Arabidopsis embryogenesis. Most auxin is transported in a PIN1- and PIN4-dependent manner to the basal part of the embryo, where the root meristem is specified. *mp*, *bdl*, *gn* and *pin4* show defects in root pole establishment. Additional sites of auxin accumulation appear at the tips of forming cotyledons. *mp*, *gn*, *pid* and *pin1* are defective in cotyledon establishment. Presumed sites of auxin accumulation are indicated by shading (grey). Arrows indicate routes of auxin efflux. Proteins involved in embryo patterning and related to auxin transport (encircled) or responses are depicted.

Acknowledgements

The authors acknowledge the support for their work from the UK Royal Society and the Academy of Sciences of the Czech Republic under the European Science Exchange Scheme (DAM), VolkswagenStiftung (JF) and the Grant Agency of the Academy of Sciences of the Czech Republic (EZ, project No. A6038303).

References

1. Baker DA (2000) Long-distance vascular transport of endogenous hormones in plants and their role in source:sink regulation. Israel J. Plant Sci 48: 199-203
2. Benjamins R, Quint A, Weijers D, Hooykaas P, Offringa R (2001) The PINOID protein kinase regulates organ development in *Arabidopsis* by enhancing polar auxin transport. Development 128: 4057-4067
3. Bennett MJ, Marchant A, Green HG, May ST, Ward SP, Millner PA, Walker AR, Schulz B, Feldmann KA (1996) *Arabidopsis AUX1* gene: a permease-like regulator of root gravitropism. Science 273: 948-950
4. Bennett MJ, Marchant A, May ST, Swarup R (1998) Going the distance with auxin: unravelling the molecular basis of auxin transport. Philos Trans R Soc Lond B Biol Sci 353: 1511-1515
5. Berleth T, Jürgens G (1993) The role of the *monopteros* gene in organising the basal body region of the *Arabidopsis* embryo. Development 118: 575–587
6. Berleth T, Mattsson J, Hardtke CS (2000) Vascular continuity, cell axialisation and auxin. Plant Growth Regul 32: 173-185
7. Berleth T, Sachs T (2001) Plant morphogenesis: long-distance coordination and local patterning. Curr Opin Plant Biol 4: 57-62
8. Bernasconi P, Patel BC, Reagan JD, Subramanian MV (1996) The N-1-naphthylphthalamic acid-binding protein is an integral membrane protein. Plant Physiol 111: 427-432
9. Brown DE, Rashotte AM, Murphy AS, Normanly J, Tague BW, Peer WA, Taiz L, Muday GK (2001) Flavonoids act as negative regulators of auxin transport in vivo in Arabidopsis. Plant Physiol 126: 524–535
10. Cambridge AP, Morris DA (1996) Transfer of exogenous auxin from the phloem to the polar auxin transport pathway in pea (*Pisum sativum* L). Planta 199: 583-588
11. Casimiro I, Marchant A, Bhalerao RP, Beeckman T, Dhooge S, Swarup R, Graham N, Inze D, Sandberg G, Casero PJ, Bennett M (2001) Auxin transport promotes Arabidopsis lateral root initiation. Plant Cell 13: 843-852
12. Chen R, Hilson P, Sedbrook J, Rosen E, Caspar T, Masson PH (1998) The *Arabidopsis thaliana AGRAVITROPIC 1* gene encodes a component of the polar-auxin-transport efflux carrier. Proc Natl Acad Sci USA 95: 15112-15117

13. Christensen SK, Dagenais N, Chory J, Weigel D (2000) Regulation of auxin response by the protein kinase PINOID. Cell 100: 469–478
14. Davies PJ, Mitchell EK (1972) Transport of indoleacetic acid in intact roots of *Phaseolus coccineus*. Planta 105: 139-154
15. Davies PJ, Rubery PH (1978) Components of auxin transport in stem segments of *Pisum sativum* L. Planta 142: 211-219
16. Delbarre A, Muller P, Guern J (1998) Short-lived and phosphorylated proteins contribute to carrier-mediated efflux, but not to influx, of auxin in suspension-cultured tobacco cells. Plant Physiol 116: 833-844
17. Delbarre A, Muller P, Imhoff V, Guern J (1996) Comparison of mechanisms controlling uptake and accumulation of 2,4-dichlorophenoxy acetic acid, naphthalene-1-acetic acid, and indole-3-acetic acid in suspension-cultured tobacco cells. Planta 198: 532-541
18. Entchev EV, González-Gaitán MA (2002) Morphogen gradient formation and vesicular trafficking. Traffic 3: 98-109
19. Friml J (2003) Auxin transport – shaping the plant. Curr Opin Plant Biol 6: 7-12
20. Friml J, Benková E, Blilou I, Wisniewska J, Hamann T, Ljung K, Woody S, Sandberg G, Scheres B, Jürgens G, Palme K (2002) AtPIN4 mediates sink-driven auxin gradients and root patterning in *Arabidopsis*. Cell 108: 661-673
21. Friml J, Benková E, Mayer U, Palme K, Muster G (2003) Automated whole mount localisation techniques for plant seedlings. Plant J 34: 115-124
22. Friml J, Palme K (2002) Polar auxin transport - old questions and new concepts? Plant Mol Biol 49: 273-284
23. Friml J, Wisniewska J, Benková E, Mendgen K, Palme K (2002) Lateral relocation of auxin efflux regulator PIN3 mediates tropism in *Arabidopsis*. Nature 415: 806-809
24. Fujita H, Syono K (1997) PIS1, a negative regulator of the action of auxin transport inhibitors in *Arabidopsis thaliana*. Plant J 12: 583-595
25. Gaedeke N, Klein M, Kolukisaoglu U, Forestier C, Müller A, Ansorge M, Becker D, Mamnun Y, Kuchler K, Schulz B, Mueller-Roeber B, Martinoia E (2001) The *Arabidopsis thaliana* ABC transporter *At*MRP5 controls root development and stomata movement. EMBO J 20: 1875-1887
26. Gälweiler L, Guan C, Müller A, Wisman E, Mendgen K, Yephremov A, Palme K (1998) Regulation of polar auxin transport by AtPIN1 in *Arabidopsis* vascular tissue. Science 282: 2226-2230
27. Garbers C, DeLong A, Deruère J, Bernasconi P, Söll D (1996) A mutation in protein phosphatase 2A regulatory subunit affects auxin transport in Arabidopsis. EMBO J 15: 2115-2124
28. Geelen DNV, Inzé D (2001) A bright future for the Bright Yellow-2 cell culture. Plant Physiol 127: 1375-1379
29. Geldner N, Anders N, Wolters H, Keicher J, Kornberger W, Muller P, Delbarre A, Ueda T, Nakano A, Jürgens G (2003) The *Arabidopsis* GNOM ARF-GEF mediates endosomal recycling, auxin transport, and auxin-dependent plant growth. Cell 112: 219–230
30. Geldner N, Friml J, Stierhof Y-D, Jürgens G, Palme K (2001) Auxin transport inhibitors block PIN1 cycling and vesicle trafficking. Nature 413: 425-428
31. Geldner N, Hamann T, Jürgens G (2000) Is there a role for auxin in early embryogensis? Plant Growth Regul 32: 187-191
32. Gil P, Dewey E, Friml J, Zhao Y, Snowden KC, Putterill J, Palme K, Estelle M, Chory J (2001) BIG: a calossin-like protein required for polar auxin transport in *Arabidopsis*. Genes Dev 15: 1985-1997
33. Goldsmith MHM (1977) The polar transport of auxin. Annu Rev Plant Physiol 28: 439-478
34. Grebe M, Friml J, Swarup R, Ljung K, Sandberg G, Terlou M, Palme K, Bennett MJ, Scheres B (2002) Cell polarity signaling in *Arabidopsis* involves a BFA-sensitive auxin influx pathway. Curr Biol 12: 329-334
35. Hadfi K, Speth V, Neuhaus G (1998) Auxin-induced developmental patterns in *Brassica juncea* embryos. Development 125: 879-887

36. Hertel R (1983) The mechanism of auxin transport as a model for auxin action. Z Pflanzenphysiol 112: 53-67
37. Hertel R, Leopold AC (1963) Versuche zur Analyses des Auxintransports in der Koleoptile von *Zea mays* L. Planta 59: 535-562
38. Hobbie LJ (1998) Auxin: Molecular genetic approaches in *Arabidopsis*. Plant Physiol Biochem 36: 91-102
39. Imhoff V, Muller P, Guern J, Delbarre A (2000) Inhibitors of the carrier-mediated influx of auxin in suspension-cultured tobacco cells. Planta 210: 580-588
40. Jacobs M, Gilbert SF (1983) Basal localization of the presumptive auxin transport carrier in pea stem cells. Science 220: 1297-1300
41. Jacobs M, Rubery PH (1988) Naturally occurring auxin transport regulators. Science 241: 346-349
42. Kaldewey H (1984) Transport and other modes of movement of hormones (mainly auxins). *In* TK Scott, *ed*, Encyclopedia of Plant Physiology, New Series, Vol 10, Hormonal Regulation of Development II. Springer-Verlag, Berlin, Heidelberg, pp 80-148
43. Kerk N, Feldman L (1994) The quiescent centre in roots of maize - initiation, maintenance, and role in organization of the root apical meristem. Protoplasma 183: 100-106
44. Lachaud S, Bonnemain JL (1982) Xylogénèse chez les dicotylédones arborescentes. III. Transport de l'auxine et activité cambiale dans les jeunes tiges de Hêtre. Can J Bot 60: 869-876
45. Lalonde S, Boles E, Hellmann H, Barker L, Patrick JW, Frommer WB, Ward JM (1999) The dual function of sugar carriers: transport and sugar sensing. Plant Cell 11: 707-726
46. Lincoln C, Britton JH, Estelle M (1990) Growth and development of the *axr1* mutants of Arabidopsis. Plant Cell 2: 1071–1080
47. Ljung K, Bhalerao RP, Sandberg G (2001) Sites and homeostatic control of auxin biosynthesis in *Arabidopsis* during vegetative growth. Plant J 28: 465-474
48. Lomax TL, Muday GK, Rubery PH (1995) Auxin transport. *In* PJ Davies *ed*, Plant Hormones: Physiology, Biochemistry and Molecular Biology, Ed 2. Kluwer Academic Publishers, Dordrecht, Boston, London, pp 509-530
49. Luschnig C (2001) Auxin transport: Why plants like to think BIG. Curr Biol 11: R831-R833
50. Luschnig C (2002) Auxin transport: ABC proteins join the club. Trends Plant Sci 7: 329-332
51. Luschnig C, Gaxiola RA, Grisafi P, Fink GR (1998) EIR1, a root-specific protein involved in auxin transport, is required for gravitropism in *Arabidopsis thaliana*. Genes Dev 12: 2175-2187
52. Maher EP, Martindale SJB (1980) Mutants of *Arabidopsis thaliana* with altered responses to auxins and gravity. Biochem Genet 18: 1041–1053
53. Marchant A, Kargul J, May ST, Muller P, Delbarre A, Perrot-Rechenmann C, Bennett MJ (1999) AUX1 regulates root gravitropism in *Arabidopsis* by facilitating auxin uptake within root apical tissues. EMBO J 18: 2066-2073
54. Martinoia E, Klein M, Geisler M, Bovet L, Forestier C, Kolukisaoglu Ü, Müller-Röber B, Schulz B (2002) Multifunctionality of plant ABC transporters – more than just detoxifiers. Planta 214: 345-355
55. Mattsson J, Ckurshumova W, Berleth T (2003) Auxin signaling in Arabidopsis leaf vascular development. Plant Physiol 131: 1327-1339
56. Mattsson J, Sung ZR, Berleth T (1999) Responses of plant vascular systems to auxin transport inhibition. Development 126: 2979-2991
57. Mayer U, Ruiz RAT, Berleth T, Misera S, Jürgens G (1991) Mutations affecting body organization in the *Arabidopsis* embryo. Nature 353: 402-407
58. Morris DA (1996) Hormonal regulation of source-sink relationships: an overview of potential control mechanisms. *In* E Zamski, AA Schaffer *eds*, Photoassimilate distribution in plants and crops. Marcel Dekker Inc, New York, Basel, Hong Kong, pp 441-465

59. Morris DA (2000) Transmembrane auxin carrier systems - dynamic regulators of polar auxin transport. Plant Growth Regul 32: 161-172
60. Morris DA, Robinson JS (1998) Targeting of auxin carriers to the plasma membrane: differential effects of brefeldin A on the traffic of auxin uptake and efflux carriers. Planta 205: 606-612
61. Morris DA, Rubery PH, Jarman J, Sabater M (1991) Effects of inhibitors of protein synthesis on transmembrane auxin transport in *Cucurbita pepo* L. hypocotyl segments. J Exp Bot 42: 773-783
62. Morris DA, Thomas AG (1978) A microautoradiographic study of auxin transport in the stem of intact pea seedlings (*Pisum sativum* L.). J Exp Bot 29: 147-157
63. Muday GK, DeLong A (2001) Polar auxin transport: controlling where and how much. Trends Plant Sci 6: 535-542
64. Muday GK, Murphy AS (2002) An emerging model of auxin transport regulation. Plant Cell 14: 293-299
65. Müller A, Guan C, Gälweiler L, Tänzler P, Huijser P, Marchant A, Parry G, Bennett M, Wisman E, Palme K (1998) *AtPIN2* defines a locus of *Arabidopsis* for root gravitropism control. EMBO J 17: 6903-6911
66. Murphy AS, Hoogner KR, Peer WA, Taiz L (2002) Identification, purification and molecular cloning of N-1-naphthylphthalamic acid-binding plasma membrane-associated aminopeptidases from Arabidopsis. Plant Physiol 128: 935-950
67. Murphy A, Peer WA, Taiz L (2000) Regulation of auxin transport by aminopeptidases and endogenous flavonoids. Planta 211: 315-324
68. Noh B, Murphy AS, Spalding EP (2001) *Multidrug resistance*-like genes of Arabidopsis required for auxin transport and auxin-mediated development. Plant Cell 13: 2441-2454
69. Okada K, Ueda J, Komaki MK, Bell CJ, Shimura Y (1991) Requirement of the auxin polar transport system in the early stages of *Arabidopsis* floral bud formation. Plant Cell 3: 677-684
70. Palme K, Gälweiler L (1999) PIN-pointing the molecular basis of auxin transport. Curr Opin Plant Biol 2: 375–381
71. Parry G, Delbarre A, Marchant A, Swarup R, Napier R, Perrot-Rechenmann C, Bennett MJ (2001) Novel auxin transport inhibitors phenocopy the auxin influx carrier mutation *aux1*. Plant J 25: 399-406
72. Petrášek J, Černá A, Schwarzerová K, Elčkner M, Morris DA, Zažímalová E (2003) Do phytotropins inhibit auxin efflux by impairing vesicle traffic? Plant Physiol 131: 254-263
73. Petrášek J, Elčkner M, Morris DA, Zažímalová E (2002) Auxin efflux carrier activity and auxin accumulation regulate cell division and polarity in tobacco cells. Planta 216: 302-308
74. Rashotte AM, Brady SR, Reed RC, Ante SJ, Muday GK (2000) Basipetal auxin transport is required for gravitropism in roots of Arabidopsis. Plant Physiol 122: 481-490
75. Rashotte AM, DeLong A, Muday GK (2001) Genetic and chemical reductions in protein phosphatase activity alter auxin transport, gravity response, and lateral root growth. Plant Cell 13: 1683-1697
76. Raven JA (1975) Transport of indoleacetic acid in plant cells in relation to pH and electrical potential gradients, and its significance for polar IAA transport. New Phytol 74: 163-172
77. Robinson JS, Albert AC, Morris DA (1999) Differential effects of brefeldin A and cycloheximide on the activity of auxin efflux carriers in *Cucurbita pepo* L. J Plant Physiol 155: 678-684
78. Rubery PH (1990) Phytotropins: receptors and endogenous ligands. Symp Soc Exp Biol 44: 119-146
79. Rubery PH, Sheldrake AR (1974) Carrier-mediated auxin transport. Planta 188: 101-121
80. Ruegger M, Dewey E, Hobbie L, Brown D, Bernasconi P, Turner J, Muday G, Estelle M (1997) Reduced naphthylphthalmic acid binding in the *tir3* mutant of Arabidopsis is associated with a reduction in polar auxin transport and diverse morphological defects. Plant Cell 9: 745-757

81. Sabatini S, Beis D, Wolkenfelt H, Murfett J, Guilfoyle T, Malamy J, Benfey P, Leyser O, Bechtold N, Weisbeek P, Scheres B (1999) An auxin-dependent distal organizer of pattern and polarity in the *Arabidopsis* root. Cell 99: 463-472
82. Sachs T (2000) Integrating cellular and organismic aspects of vascular differentiation. Plant Cell Physiol 41: 649-656
83. Schrick K, Mayer U, Martin G, Bellini C, Kuhnt C, Schmidt J, Jürgens G (2002) Interactions between sterol biosynthesis genes in embryonic development of *Arabidopsis*. Plant J 31: 61-73
84. Shi L, Miller I, Moore R (1993) Immunocytochemical localization of indole-3-acetic acid in primary roots of *Zea mays*. Plant Cell Environ. 16: 967-973
85. Steinmann T, Geldner N, Grebe M, Mangold S, Jackson CL, Paris S, Gälweiler L, Palme K, Jürgens G (1999) Coordinated polar localization of auxin efflux carrier PIN1 by GNOM ARF GEF. Science 286: 316-318
86. Swarup R, Friml J, Marchant A, Ljung K, Sandberg G, Palme K, Bennett M (2001) Localization of the auxin permease AUX1 suggests two functionally distinct hormone transport pathways operate in the *Arabidopsis* root apex. Genes Dev 15: 2648-2653
87. Swarup R, Marchant A, Bennett MJ (2000) Auxin transport: providing a sense of direction during plant development. Biochem Soc T 28: 481-485
88. Swarup R, Parry G, Graham N, Allen T, Bennett M (2002) Auxin cross-talk: integration of signalling pathways to control plant development. Plant Mol Biol 49: 411–426
89. Uggla C, Mellerowicz EJ, Sundberg B (1998) Indole-3-acetic acid controls cambial growth in Scots pine by positional signaling. Plant Physiol 117: 113-121
90. Utsuno K, Shikanai T, Yamada Y, Hashimoto T (1998) *AGR*, an *Agravitropic* locus of *Arabidopsis thaliana*, encodes a novel membrane-protein family member. Plant Cell Physiol 39: 1111–1118
91. Vandenbussche F, Smalle J, Le J, Saibo NJM, De Paepe A, Chaerle L, Tietz O, Smets R, Laarhoven LJJ, Harren FJM, Van Onckelen H, Palme K, Verbelen J-P, Van Der Straeten D (2003) The Arabidopsis mutant *alh1* illustrates a cross talk between ethylene and auxin. Plant Physiol 131: 1228-1238
92. Vissenberg K, Feijó JA, Weisenseel MH, Verbelen J-P (2001) Ion fluxes, auxin and the induction of elongation growth in *Nicotiana tabacum* cells. J Exp Bot 52: 2161-2167
93. Vogler H, Kuhlemeier C (2003) Simple hormones but complex signalling. Curr Opin Plant Biol 6: 51-56
94. Went FW (1974) Reflections and speculations. Annu Rev Plant Physiol 25: 1-26
95. Willemsen V, Friml J, Grebe M, van den Toorn A, Palme K, Scheres B (2003) Cell polarity and PIN protein positioning in Arabidopsis require *STEROL METHYLTRANSFERASE1* function. Plant Cell 15: 612-625
96. Yamamoto M, Yamamoto KT (1998) Differential effects of 1-naphthaleneacetic acid, indole-3-acetic acid and 2,4-dichlorophenoxyacetic acid on the gravitropic response of roots in an auxin-resistant mutant of Arabidopsis, aux1. Plant Cell Physiol 39: 660-664
97. Zazimalova E, Napier RM (2003) Points of regulation for auxin action. Plant Cell Rep 21: 625-634

E2. The Induction of Vascular Tissues by Auxin

Roni Aloni
Department of Plant Sciences, Tel Aviv University, Tel Aviv 69978, Israel
E-mail: alonir@post.tau.ac.il

INTRODUCTION

The vascular system connects the shoot organs with the roots and enables efficient long-distance transport between them. In higher plants it is composed of two kinds of conducting tissues: the *phloem*, through which organic materials are transported and the *xylem*, which is the pathway for water and soil nutrients. In angiosperms, the functional conduits of the *phloem* are the *sieve tubes*, and the most specialized conduits of the *xylem* are the *vessels* (2). Vascular development in a plant is an open type of differentiation, continuing as long as the plant grows from apical and lateral meristems. The continuous development of new vascular tissues enables regeneration of the plant and its adaptation to changes in the environment. The differentiation of vascular tissues along the plant is induced and controlled by longitudinal streams of inductive signals (4, 52). In spite of the structural and developmental complexity of vascular tissues (47, 51), there is evidence that the differentiation of both the vessels and the sieve tubes is induced by one major hormonal signal, namely, the auxin indole-3-acetic acid (IAA), produced mainly by young leaves (11, 33, 39). Such evidence raises endless questions as to how this hormonal signal induces and controls complex patterns of xylem and phloem, and emphasizes the need to understand where and how the IAA signal is produced and transported, what are the mechanisms that control the formation of various cell types, their relationships, dimensions, differentiation and maturation patterns in the vascular system. Nevertheless, it should be emphasized that additional growth regulators may influence vascular differentiation, such as cytokinins which increase the sensitivity of the vascular cambium to auxin (3, 4), gibberellin that induces fibers (2, 47), and ethylene which reduces vessel diameter and retards fiber formation (12). These hormonal signals are beyond the scope of this article and the reader is directed to reviews on these topics (2, 36, 51, 58).

Molecular genetic approaches, using Arabidopsis as a major model system, have substantially increased interest in the subject and have boosted our understanding in the processes of vascular differentiation and vein pattern formation. However, these analytical tools, which have already yielded impressive contributions (17, 56, 61), are in an early phase of uncovering new genes and identifying their role in vascular differentiation. They promise to provide new insights on the mechanisms that control organized vascular development. The models by which we can currently explain vascular pattern formation are limited (4, 11, 17, 23, 43, 52, 56). Hence, there is need for additional analysis at the molecular, cellular and organismic levels, which should be integrated for better understanding (4, 43, 52).

Recently, the sites of auxin production in shoot organs were discovered and their role in vein pattern formation has been demonstrated (4, 11). Additional basic unexplained vascular patterns (e.g., diameter changes in the primary vessels of roots and internal phloem formation in bicollateral bundles) require better understanding of the role of auxin transport pathways in organized vascular differentiation. Likewise, also, are possible changes which may have occurred during plant evolution in the pathways of auxin in the vascular tissues: from transport through the active cambium of trees to transport through xylem parenchyma in the advanced herbaceous species in which cambium activity is minimized, or absent (in most of the monocots). Another key issue in vascular differentiation that requires clarification is the process of xylem maturation during bundle formation, which should not be confused with the induction of vascular bundles. Xylem maturation, which is characterized by the initiation and progression of discontinuous basipetal and acropetal patterns, will also be analyzed here because of its importance for correct interpretations.

The aim of this chapter is to present general concepts on the induction and control of vascular tissues by IAA. It provides a summary of major topics in organized vascular differentiation and the recent advances made in each. I start with recent findings on the sites of auxin production and accumulation, then proceed to propose a new hypothesis for elucidating IAA transport pathways, and finally focus on the role of auxin in controlling basic patterns of xylem and phloem formation in plants.

SITES OF AUXIN PRODUCTION

IAA Production and Vein Pattern Formation in Leaves

The pioneering study of Jacobs (33) demonstrated that auxin produced in young leaves is a limiting and controlling factor in xylem regeneration around a wound. Although fifty years have elapsed since Jacobs' (33) discovery, until recently there was no physiological or molecular understanding of how and where auxin is synthesized in a young leaf and how this regulates vascular differentiation and venation pattern formation (4, 56). In an attempt to confront this challenge, auxin transport inhibitors were applied to Arabidopsis plants (40, 53) revealing that the inhibitors restricted the vascular tissues to the leaf margin, indicating that vascular tissues in a leaf depend on an inductive signal, likely IAA, from the leaf margin.

To explain how auxin induces vascular patterns in leaves, I proposed (4) that during primordium development there are orderly shifts in the sites and concentrations of auxin production, and it is these shifts that control vein pattern formation in a leaf. Supporting evidence shows that an exogenous IAA source applied along a cut made in a very young cucumber cotyledon induced tracheary element patterns as in the hydathodes (4), thus mimicking the role of differentiating hydathodes as the primary sites of auxin production in leaves.

RT-PCR analysis of the mRNA expression of enzymes involved in auxin metabolism and transport (indole-3-glycerol synthase, nitrilase, IAA-amino acid hydrolase, chalcone synthase and PIN1) has revealed a succession of auxin production events during leaf-primordium development, starting with *de novo* build-up of a massive bound-IAA pool in the youngest primordia before vascularization (11), and from this bound-IAA reservoir the free IAA is later released by hydrolysis in a gradual basipetal pattern (Fig. 1).

Visualization of total auxin distribution (both free and conjugated auxin) in Arabidopsis leaf primordia, by immunolocalization with specific IAA antibodies and fluorescent secondary antibody conjugates, demonstrate high auxin concentrations in the chloroplasts of the mesophyll and the guard cells (11), suggesting that chloroplasts play a major role in auxin biosynthesis and metabolism.

Localization of the bioactive auxin, namely, free IAA, can be visualized in leaf primordia by auxin-response element-*GUS* expression in the DR5 Arabidopsis transformant (59). Analysis of the youngest shoot region of the DR5 Arabidopsis reveals that the promeristem and the youngest meristematic

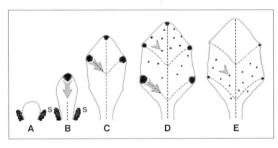

Figure 1. Schematic diagrams showing the gradual changes in sites (black spot locations) and concentrations (black symbol sizes) of free-IAA production during leaf primordium development in Arabidopsis. The arrows show the experimentally confirmed direction of the basipetal polar IAA movement, descending from the differentiating hydathodes (B-D). Recent incision experiments using the DR5 Arabidopsis transformant (R. Aloni, unpublished data), revealed that in a young leaf primordium, although the midvein matures acropetally (B), it is induced by the basipetal polar IAA flow (arrow) originating in the primordium tip. Arrowheads show the location of low free-auxin production sites in the lamina (D-E). The ontogeny of the midvein and secondary vascular bundles is illustrated by broken lines (marginal and minor veins are not shown). A. Early high auxin production occurs only in the stipules (s) of a very young leaf primordium, before auxin is detectable in the tip and prior to midvein differentiation. B. Primary auxin production in the tip of a fast growing primordium induces acropetal midvein differentiation, and illustrating leaf apical dominance. C. High primary auxin production at the differentiating hydathodes of the upper lobes inducing differentiation of the upper secondary vascular bundles. D. High primary auxin production at the differentiating hydathodes of lower lobes inducing the lower secondary bundles, and randomly distributed sites of secondary auxin production in the lamina. E. Maintenance low auxin production in hydathodes and secondary auxin production in the lamina during a late phase of leaf development.

leaf primordia do not show free-IAA production (do not show GUS activity) (11, 13), although these primordia are loaded with bound IAA (11). Stipules, which are outgrowths of a leaf primordium, are the earliest sites of intense free-IAA production (Fig. 1A,B). In a more developed leaf blade, hydathodes, the water secreting glands, which develop in the tip and later also in the lobes (Figs. 1B-E, 2A), are the primary sites of free-auxin production, while trichomes (Fig. 2B) and mesophyll cells (11) are the secondary sites. During early stages of primordium development, an apical dominance within the leaf is evidenced by strong *DR5::GUS* expression in the elongating tip (Fig. 1B), possibly suppressing the production of IAA in the leaf tissues below it (11). During leaf-primordium development there are gradual shifts in the sites and concentrations of free-auxin production, progressing from the elongating tip (Fig. 1B), continuing downward along the expanding blade margins (Fig. 1C-E) and ending in the central regions of the lamina (Fig. 1D,E) (11, 39). This IAA pattern was confirmed by auxin analysis by GC-MS/MS techniques (42). These successive IAA shifts are presumed to control the basipetal development of the primordia and vein pattern formation (see below; Fig. 7) in Arabidopsis leaves (11). The intense production of IAA in the differentiating hydathodes (Fig. 2A) induces the midvein and the secondary bundles (Fig. 1B-D), while low auxin levels produced by developing trichomes (Fig. 2B) and mesophyll cells induce the minor tertiary and quaternary veins as well as the freely ending veinlets (11).

IAA Production and Vascular Differentiation in Flowers

I suggest that, as in foliage leaves (4, 11, 39), free-auxin production starts at the tip of each floral organ – so that vascular tissues develop basipetally, from the site of IAA production downward. Experimental evidence obtained with the DR5 Arabidopsis transformant supports this suggestion (R. Aloni et al., unpublished data; (R.A.)). The stigma is a site of high IAA production (Fig. 3B,C), and illustrates how a major source of IAA induces immediately below it a wide fan pattern of the stylar xylem. This fanning xylem is characterized by a decreasing density of vessel elements from the auxin-producing stigma downward, as predicted from the decreasing auxin gradient down the style (see below; 9). This vascular tissue below the stigma supplies the water required for pollen hydration (Fig. 3C).

IAA Production in Decapitated Organs

Experimental analysis of *DR5::GUS* expression patterns in decapitated shoot organs (e.g., Arabidopsis stamens, gynoecia, and inflorescence internodes) shows that the distal parenchyma cells, located immediately below the cut, start to produce free auxin (R.A.). This demonstrates that the upper cells of decapitated shoot organs become active sites of free-IAA production.

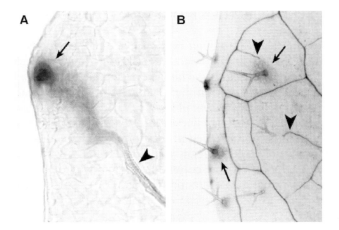

Figure 2 (Color plate page CP13). Free-IAA production detected by *DR5::GUS* gene expression (the blue GUS staining) in transformed Arabidopsis, showing histochemical localization of GUS activity during leaf morphogenesis. A. Center of strong *GUS* expression (arrow) in a developing hydathode, from which a decreasing pattern of the blue staining marks the auxin pathway towards the differentiating secondary bundle (arrowhead), x150 (x200 in color plate). B. *GUS* expression at the base of a few trichomes (arrows). Two of the trichomes are associated with freely ending veinlets (arrowheads), x30 (x40 in color plate).

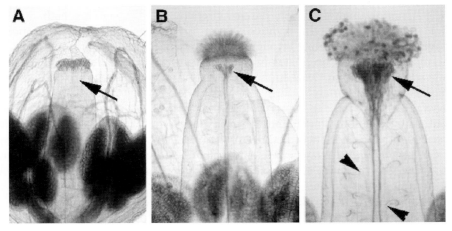

Figure 3 (Color plate page CP13). Free-IAA production detected by *DR5::GUS* expression in transformed Arabidopsis, showing localization of GUS activity during flower morphogenesis. All photographs x 65. A. Young flower bud, prior to detectable IAA production and vascular development (arrow) in the gynoecium. At this developmental stage, the stamens produce high auxin concentrations (stained dark blue). B. In a young gynoecium, free-IAA production (detected by blue *GUS* expression) in the stigma induces the central bundle and an early stage of xylem fan formation (arrow) below the stigma. C. Mature gynoecium with germinating pollen grains (stained dark blue). A typical well developed wide fan xylem pattern (arrow) descending into two central bundles was induced by the IAA-producing stigma (stained blue). The discontinuous short veinlets (arrowheads) induced by the ovules do not connect with the gynoecium's central bundles, which are well supplied with auxin produced by the stigma.

SITES OF AUXIN ACCUMULATION

When either a young leaf or a young inflorescence of Arabidopsis is removed from the plant and kept in humid conditions, free IAA starts to accumulate at the petiole or the basal internode. This auxin accumulation was detected by *DR5::GUS* expression (R.A.). Similarly, after removing a distal part of the main root from a young Arabidopsis plant, free IAA starts to accumulate at the root base, above the cut (R.A.). These observations are in accordance with normal auxin accumulation observed at the tip of the main root, or the tips of emerging lateral roots (11, 49). Recently, a model involving an active AtPIN4-dependent sink for auxin in the columella region was suggested, possibly implying that auxin gradients are sink driven (26, 27).

THREE PATHWAYS FOR AUXIN TRANSPORT ALONG THE PLANT

The molecular mechanisms that control IAA transport are discussed in chapter E1. Here the focus is on the routes of IAA transport which are complex and poorly understood. A detailed analysis of IAA routes is required for understanding the mechanisms controlling basic patterns in vascular systems, which will be discussed below. Auxin produced in young leaves moves polarly in the vascular tissues (4, 47, 51), specifically along their cambium (39, 54, 57), and through the xylem parenchyma (18, 28, 44), starch sheath and the root pericycle (27). In addition, there is evidence for rapid non-polar movement of the hormone through sieve tubes. This non-polar auxin flow in the phloem conduits originates in mature leaves (41). Furthermore, there are indications that auxin flows in the epidermis (16, 27, 55). Surprisingly, these fragments of important information have never been incorporated into a general concept for explaining where IAA moves in the plant body. This confusing situation hinders proper analysis of new findings, which could consequently result in incorrect interpretations. I propose that free auxin flows along three main pathways in the entire plant. These three pathways, and how they lead to differences in the control of development, differentiation and IAA activity along these routes, will be described bellow.

All living cells in the plant body are capable of transporting auxin, but only those through which IAA is canalized become specialized to transport the hormone rapidly, resulting in canalized files of cells (51). During plant development, initial auxin flows are canalized into three main routes of IAA transport. This canalization starts during embryogenesis. The first two pathways induce and control the vascular and protective tissues, while the third pathway controls the activity of the phloem conduits.

(**1**) The *internal route* – which is the most complex pathway, and can be subdivided into the following longitudinal components (Fig. 4): primary shoot (I), primary root (III), and secondary body (II) which might be produced between the primary parts in woody dicotyledonous and gymnosperm species. Each of these components has its unique anatomy and physiology as follows:

In the primary shoot (I), IAA moves polarly through the vascular bundles in three distinct streams: (i) via the vascular meristem (M), namely, the procambium or early stages of cambium, (ii) in the surrounding (B) bundle sheath at the phloem pole, and (iii) through parenchyma cells at the (X) xylem pole.

In the primary root (III), IAA moves polarly in the vascular cylinder through: (i) the vascular meristem (M), and (ii) the pericycle (Pe).

In the secondary body (II), the polar IAA streams descending from the primary shoot (I) merge into one pathway, which occurs in the cambium (C) and differentiating vascular elements.

(2) The *peripheral route* – which courses through the protective and adjunct tissues. The auxin in this pathway originates in the epidermal cells, like the auxin-producing trichomes (Fig. 2B) and stomata, and moves polarly towards the root tips through the epidermis (E) and subepidermal cell layers, and the phellogen (Ph) that produces the cork.

(3) The *non-polar route* – which courses in the phloem conduits, where IAA moves rapidly up and down via the sieve tubes (S). This fast auxin flow should be considered as a house-keeping signal that controls callose levels in the sieve tubes (3, 10).

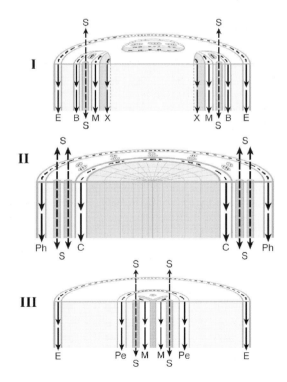

Figure 4. Schematic diagrams showing free-IAA transport pathways in the primary shoot (I), the secondary body (II), and the primary root (III). In the **internal route**, IAA moves polarly through the (M) vascular meristem (in the primary body), which is the (C) cambium in secondary body, the (X) xylem parenchyma cells, the (B) bundle sheath (in primary shoot) and the (Pe) pericycle (in primary root). In the **peripheral route**, IAA moves polarly through the (E) epidermis and the (Ph) phellogen. In the **non-polar route**, IAA moves up and down in the (S) sieve tubes.

The induction of vascular tissues by auxin

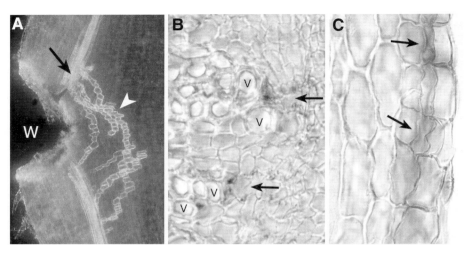

Figure 5 (Color plate page CP14). Pathways of auxin transport detected by the pattern of xylem regeneration (white arrowhead) around a wound (W) in a decapitated young internode of *Cucumis sativus* treated with auxin (0.1% IAA in lanolin for 7 days), which was applied to the upper side of the internode immediately after wounding and removing the leaves and buds above it (A), and by *DR5::GUS* expression (blue GUS staining) in an intact hypocotyl of a 3-month-old transformed Arabidopsis (grown under short day conditions), showing localization of GUS activity in the vascular cambium (B) and the phellogen (C). A. Longitudinal view of a polar pattern of xylem regeneration induced by the basipetal polar flow of auxin, observed after clearing with lactic acid and staining with phloroglucinol, x 25. The photograph shows a typical polar regeneration characterized by a dense appearance of regenerative vessel elements (arrow) immediately above the wound that differentiated close to the wound surface. Below the wound there are a few defined files that connect to the damaged bundle at greater distances from the wound. The pattern of xylem regeneration follows the path of auxin movement around the wound. As vascular bundles are preferable pathways for auxin flow, the applied IAA moved basipetally in the damaged bundle to where it was interrupted by the wound and was forced (arrow) to find new pathways around the obstacle. B and C are transverse sections in a well developed Arabidopsis hypocotyl, both have the same orientation; their right side is located towards the stem periphery. B. The internal secondary tissues, showing that auxin (detected by *GUS* expression) moves through the cambium and differentiating vascular elements, (all show blue spots of GUS activity, located in front and beside the arrows), forming radii patterns adjacent to the latest formed secondary vessels (v), likely indicating the sites where new vessels will differentiate, x 600. C. The peripheral secondary tissue, showing the path of free-IAA transport (detected by blue GUS staining) through the phellogen (arrows), x 400.

Regular passages (especially within the auxin streams of the internal route) and environmentally-induced exchanges of IAA occur between the auxin routes following light or gravity stimulation, in response to wounding, at merging points, and via the vascular rays in the secondary body.

The auxin which induces the vascular tissues flows polarly in the internal route, mainly through the vascular meristem (procambium or cambium) and the associated parenchyma sheathes. The constant polar IAA movement through the vascular meristem maintains vascular continuity. When vascular tissues are wounded, the interrupted IAA flow could form bypassing pathways via parenchyma or cambial cells resulting in the regeneration of vascular tissues around a wound (Fig. 5A). Free IAA (detected by *DR5::GUS* expression) moves via intact vascular cambium in the hypocotyls of well developed Arabidopsis plants, preferably between the differentiating vessels and sieve tubes (Fig. 5B), thus explaining the formation of radial patterns of secondary vessels and sieve tubes.

When a cambium shows low activity, or occurs mainly inside the vascular bundles (fascicular cambium), as in organs of many herbaceous plants, it is likely that the polar auxin descending from leaves and flowers would move preferably through xylem parenchyma cells, as has already been found in the inflorescence of Arabidopsis (18, 28, 44). However, in growing trees possessing a wide vigorous cambium, undergoing active cell divisions, the polar auxin is transported through the cambium (54, 57). This is especially true for conifers, where most of the axial cells in their xylem are dead tracheids. It should be noted, that trees have developed earlier than did the specialized herbaceous species, and that during plant evolution, the cambium activity of trees has been minimized in herbaceous plants. Since the extremely advanced species of the monocots do not produce cambium, the polar IAA transport in their internal route is expected to occur via their xylem parenchyma cells.

The associated sheaths that surround the vascular tissues in the primary body, e.g. the pericycle (in roots) and bundle sheaths (like the starch sheath in shoots), are specialized envelopes of cells through which auxin moves polarly. Their auxin flow could boost auxin concentration locally above sites where the polar IAA transport is interrupted (e.g., by wounding, or local ethylene production), resulting in apical meristem formation (e.g., lateral root initiation from the pericycle, or adventitious root formation associated with vascular bundles in the stem). The auxin flowing in the sheaths as well as in the sieve tubes could promote vascular regeneration in cases of wounding.

Auxin flow in the enveloping sheaths could influence normal vascular differentiation within shoot vascular bundles and the root vascular cylinder. Thus, in primary roots, although the metaxylem and protoxylem vessels start differentiation almost simultaneously, the auxin flow in the pericycle enhances the differentiation of the neighboring protoxylem vessels, which consequently lay down their secondary walls earlier, resulting in narrow vessels. Contrariwise, the metaxylem vessels (differentiating away from the pericycle) have more time to expand before secondary wall deposition and therefore become wide vessels.

In shoots, the polar IAA flow via the bundle associated cells at the xylem pole may induce an internal phloem. While the most common bundle type in gymnosperms and angiosperms is the *collateral* bundle, in which the phloem occurs on one side (phloem toward the outside, xylem toward the inside), there are stem bundles in some angiosperm families (e.g., Cucurbitaceae, Solanaceae) where an internal low-level auxin stream at the xylem pole induces an internal phloem, thus forming a *bicollateral* bundle, which has phloem on both sides of the xylem.

In the peripheral route, the polar IAA flow controls the following major activities: (a) promotes tropic growth of young organs following modifications in auxin concentrations induced by environmental stimulations; (b) maintains the continuity of the protective tissues by promoting suberin lamellae and cork formation after wounding; and (c) induces and controls phellogen activity.

We have recently found that IAA moves along the epidermis and phellogen (Fig. 5C), forming longitudinal strips, or a continuous sheath pattern (all around the stem and root circumference). This was observed in the epidermis and phellogen of stems and roots of *Trifolium* (detected by high *GH3::GUS* expression), and the phellogen of well developed hypocotyls of Arabidopsis (detected by strong *DR5::GUS* expression) (R.A.). Recent findings in Arabidopsis indicate polar auxin flow in the outermost tunica layer of the shoot vegetative meristem, the epidermis of young leaf primordia (45) and epidermis near the root tip (27, 55), but these studies have not yet clarified experimentally the direction of the auxin flow in the outermost layer of the shoot (where both AUX1 and PIN1 partially overlapped; 45) and near the root tip (55), which need to be determined by physiological means. From the published photographs (55) it looks as if the AUX1 permease in the epidermis is enhanced near the root tip, where the highest *DR5::GUS* expression is detected (11, 49), possibly indicating that the high IAA concentrations accumulated at the root tip intensify the expression of the specialized auxin transport protein at this region. As mentioned above, developing hydathodes in leaves (11) and tips of floral organs (R.A.) are major sites of auxin production, whereas the root tips are the sites where auxin is accumulated (11, 49). Therefore, both extremities are sites of high auxin concentrations resulting from either free-IAA production (by hydrolysis) in the shoot organs, or auxin accumulation in the root tips. At these shoot and root tips, the local high IAA concentrations could enhance expression of genes associated with the metabolism and transport of auxin.

ORGANIZED DIFFERENTIATION OF VASCULAR TISSUES

Control of Vascular Bundle Formation by a Signal Flow

In an elegant series of experiments using pea seedlings, Sachs (51) provided evidence supporting his hypothesis that canalization of the auxin flux determines the orderly pattern of vascular tissues from leaves to roots. He proposed (51) that auxin flow, which starts by diffusion, induces a polar auxin transport system which promotes auxin movement and leads to canalization of auxin flow along a narrow file of cells. The continuous polar transport of auxin through these cells induces a further complex sequence of events which terminates in the formation of a vascular bundle. Once developed, a vascular bundle remains the preferable pathway of IAA transport. Consequently, new streams of auxin produced by young leaves are directed towards the developed vascular bundles. A preexisting vascular bundle that is not supplied with auxin (e.g., one descending from an old leaf) acts as a sink for a new stream of auxin (50) and therefore, a new bundle will differentiate towards the preexisting bundle that has a low supply of IAA. Thus, the canalization of auxin flow through meristematic, or parenchyma cells, induces the orderly pattern of continuous vascular bundles from leaves to roots.

It has also been shown that IAA could have an inhibitory effect on vascular tissue formation, where by a newly induced vascular bundle will not interlink with a preexisting vascular bundle that is well supplied with high auxin concentrations (50). A naturally-occurring instance of such an inhibition was recently encountered in gynoecia of Arabidopsis flowers, where the short veinlets (arrowheads in Fig.. 3C; Fig. 8C) induced by the ovules did not form connections with the longitudinal bundles descending from the IAA-producing stigma (Fig. 3B,C) because these longitudinal bundles were well supplied with auxin.

Application of auxin transport inhibitors increased the number of vessels in the midvein and secondary vascular bundles of Arabidopsis leaves (40, 53) and in the petals (Fig. 6B). However, when an exogenous auxin (NAA) was applied (on and off with TIBA, but not on the same day), drainage of the extra auxin did occur, consequently resulting in narrow secondary bundles with a limited number of vessels (Fig. 6C). Furthermore, when only auxin (40μM NAA) was applied daily, it did not exert any effect on vascular differentiation (R.A.). These findings demonstrate that both movement and drainage of auxin are required for normal bundle formation, and that neither presence of the hormone, nor its concentration is sufficient *per se*.

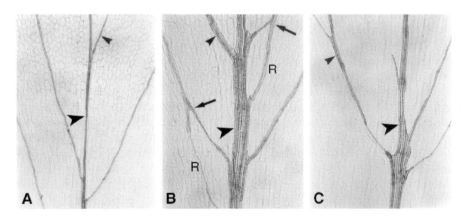

Figure 6. Effects of auxin transport inhibitor (2,3,5-triiodobenzoic acid, TIBA) and auxin (NAA) spray applications (over a 14-d period) on vascular differentiation in wild-type Arabidopsis petals, all x 85. A. Typical delicate vein system in the middle of an intact petal, consisting of two to three vessels in the midvein (large arrowhead) and one vessel in each secondary vein (small arrowhead). B. The auxin transport inhibitor (20μM TIBA) induced a very wide midvein (comprised of 10 to 12 vessels), and wide secondary veins (2 to 4 vessels). In addition, random sites of auxin retardation (arrows) in the secondary bundles, promoted new bypassing pathways resulting in additional regenerative bundles (R). C. Repeated alternate-day treatments of 20μM TIBA (on one day) and 40μM NAA (on the next), caused an increase in vessel number at the base of the midvein and a moderate effect in the secondary veins.

Control of Vessel Size and Density Along the Plant Axis

Auxin acts as a morphogenetic signal forming polar concentration gradients along the plant, from the shoot organs to the root tips. These gradients provide directional and locational information to differentiating cells along the morphogenetic fields. A decreasing gradient of IAA concentrations from leaves to roots may result in polar patterns of growth and differentiation along the plant axis (4, 9, 49).

The control of vessel diameter is an important parameter for assessing the ascent of water and minerals from roots to leaves and the adaptation of plants to their environment (2). The specific purpose of research on this topic is to develop a reliable concept for understanding and explaining the mechanisms that control vessel size and density (number of vessels per transverse-sectional area) along the plant. Such a concept may also serve as a tool for analyzing developmental patterns of vessel size and density in specific sites (e.g. below the stigma, see Fig. 3C) in the vascular system.

The longest known gradient in cell size along plants is encountered in their vascular tissues (as much as 100 m long in very tall trees), where vessel diameter (or tracheid size in conifers) along the plant axis increases gradually with increasing distance from the leaves (9, 37). The narrow vessels differentiate near the leaves, where the highest auxin concentrations are expected, while the widest vessels are formed in the roots, at the greatest distance from the auxin sources. The gradual increase in vessel diameter from leaves to roots is associated with a gradual decrease in vessel density. Hence, vessel density is generally greater in branches, where the vessels are narrow, than in roots, where they are wide (9, 37).

We proposed that the general increase in vessel size and decrease in vessel density along the plant axis is due to a gradient of decreasing auxin concentrations from leaves to roots (9). This is based upon the assumption that the steady polar flow of IAA from leaves to roots controls these polar changes in the vascular system. High auxin concentrations near the young leaves induce narrow vessels because of their rapid differentiation, allowing only limited time for cell growth. Contrariwise, low IAA concentrations further down result in slow differentiation, which permits more cell expansion before secondary wall deposition, and thereby results in wide vessels. Vessel density is controlled by auxin concentration, to wit: high concentrations (near the sites of IAA production) induce greater density, while low concentrations (further down, towards the roots) diminish density. Consequently, vessel density decreases from leaves to roots (4, 9). This hypothesis was experimentally confirmed by showing that various auxin concentrations applied to decapitated bean stems induce substantial gradients of increasing vessel diameter and decreasing vessel density from the auxin source towards the roots (9). High auxin concentration yielded numerous vessels that remained small because of their rapid differentiation. Low auxin concentration resulted in slow differentiation and therefore in fewer and larger vessels. Auxin concentration also influenced the patterns of vessels in the secondary xylem of bean.

Immediately below an auxin source the vessels were arranged in layers. Further down along the stem, where lower levels of auxin were expected, the vessels grouped into bundles (9).

Studies on transgenic plants with altered levels of IAA confirmed the general relations between IAA concentration and vessel size and density. Auxin-overproducing plants (i.e., ones overexpressing the *iaaM* gene) contain many more vessel elements than do control plants, and their vessels are narrow (34). Conversely, plants with lowered IAA levels (i.e., expressing the *iaaL* gene as an anti-auxin gene) contain fewer vessels of generally larger size (48).

In spring, the first wide (up to 500 μm) earlywood vessels in temperate deciduous ring-porous trees initiate a few weeks (up to 6 weeks) before the onset of leaf expansion. These wide vessels are induced by low-level streams of auxin produced by dormant looking buds (before swelling), only because the cambium of these trees requires extremely low auxin levels for reactivation (4).

Control of Phloem and Xylem Relationships in Axial and Foliar Organs

A major problem in studying vascular differentiation is the difficulty of observing the phloem (2, 3, 47). Not surprising, then, that information on phloem differentiation or on gene expression in the phloem is limited, with most of the studies focusing on xylem differentiation in organized vascular bundles (2, 17, 51, 56, 61) and in isolated *Zinnia* tracheary elements (36).

The first organized vascular system encountered in lower members of the plant kingdom consists of phloem with no xylem (3). Thus, we find the first developed sieve tubes in members of the brown algae. Much later in the evolution of plants, during their transition from aquatic to terrestrial habitats, the water conducting system developed.

In tissue cultures, low IAA concentrations induced sieve elements but not tracheary elements, while high auxin concentrations resulted in the differentiation of both phloem and xylem (1, 2), but even in these cultures, at the surface farther away from the auxin-containing medium, only phloem with no xylem developed (1). In the course of vascular development along the plant axis, phloem differentiation precedes that of xylem (25) and therefore vascular bundles consisting of phloem with no xylem are common in young internodes. In very young internodes of *Coleus*, there are more phloem-only bundles than collateral bundles (47). During internode development, vessels will differentiate in some of these phloem bundles.

Plant vascular systems are usually composed of phloem and xylem, and insofar as the relative proportions of phloem and xylem are concerned, there is a major difference between foliar and axial organs. Thus, along the plant axis, xylem does not differentiate in the absence of phloem, though bundles of phloem (with no xylem) and phloem anastomoses are common in stems of many plant species (5, 6, 7, 8, 47). To emphasize the abundance of these phloem ramifications we pointed out that in a mature internode of *Cucurbita maxima* (150 mm long) there are about 10,000 phloem anastomoses between the vascular bundles (5), which means that the phloem may form dense reticulated patterns. The anastomoses are variable in size, consisting of one or more sieve tubes that are difficult to visualize in conventional light-microscope sections because of their sinuous nature. It has been suggested that the differentiation of phloem bundles and phloem anastomoses between the bundles is induced by streams of low-IAA levels (2, 3). High auxin concentration applied to decapitated *Luffa* stem induced xylem differentiation in its phloem anastomoses (3) indicating the need for high hormonal stimulation for xylem differentiation. These networks of phloem anastomoses operate as an emergency system which enables the plant to respond to damage by providing alternative pathways for assimilates around the stem (7). They also enable xylem regeneration within phloem anastomoses in mature internodes (after the parenchyma cells between the bundles had lost the ability to redifferentiate to vessel elements) before they produce interfascicular cambium (5).

Conversely, in leaves, the differentiation of xylem in the absence of phloem is a common feature and occurs in freely ending veinlets (32, 38) and hydathodes (4). In *Oxalis stricta* there are virtually no sieve tubes in any terminal vein, while *Polygonum convolvulus*, at the other extreme, has sieve tubes extending to the tips of most terminal veins (32). In most of the studied species freely ending veinlets may display disparate relations between phloem and xylem in the same plant, e.g., vein endings lacking sieve tubes or having them up to the tip, and sieve tubes that end at some intermediate point (32, 38). In leaves, the proximity between the sites of IAA production and the sites of differentiating vascular cells probably results in relatively high local auxin concentrations at the differentiating sites (4), which may explain why xylem can differentiate in the absence of phloem at the freely ending veinlets (32, 38) and hydathodes. However, when Arabidopsis plants are subjected to limiting conditions, like very 'short' days (4h light / 20h darkness), some of their tertiary veins, which had developed near the leaf base, produce phloem-only bundles (R.A.). This finding demonstrates that limited hormonal stimulation at the base of these small leaves grown under very short days may not be enough to produce xylem in the phloem-only bundles.

Control of Discontinuous Xylem and Phloem Patterns during Early Stages of Bundle Maturation in Wild-Type Plants

A major issue in vascular differentiation deserving of clarification is the process of xylem maturation during bundle formation. It should be emphasized that during early stages of bundle formation in wild-type plants, normal vessel differentiation initiates and progresses in discontinuous basipetal and acropetal patterns (Fig. 7C-E). Although the incipience of xylem and phloem discontinuities during early stages of leaf and shoot morphogenesis is well known (25), there is no understanding of how these discontinuities are controlled. Understanding these basic vascular patterns is important for correct interpretations and they therefore need detailed explanation. The direction of xylem maturation by itself does not provide enough information to determine the source of IAA production, or IAA transport direction. Although xylem and phloem regeneration is induced by IAA originating in the young leaves above a wound (4, 33, 51), the regeneration of xylem and phloem around the injury is characterized by basipetal, acropetal and discontinuous developmental patterns (5,6), emphasizing that the basipetal polar auxin flow can induce vascular maturation in opposite directions and may also result in discontinuous patterns.

Generally in shoots, a basipetal maturation pattern starts from the free-IAA production site, whereas an acropetal maturation pattern results from auxin accumulation above a specific location. At the end of the maturation process, these acropetal and basipetal vascular patterns will join into a functional bundle (R.A.).

In leaves, the naturally-occurring discontinuous xylem patterns follow the configuration of the vascular meristem. Loops of procambium develop basipetally (Fig. 7A,B) in very young leaf primordia of Arabidopsis. The first procambium loop commences at the tip (Fig. 7A) and additional loops will gradually differentiate basipetally towards the developing leaf base (Fig. 7B). This very early basipetal pattern of provascular development is likely induced by incipient low auxin stimulation produced by the tip hydathode (11, 39, 60). The late formed basal loops may also be supplied with additional auxin, which is produced by the developing hydathodes in the lobes (11, 39). This basic procambium framework (midvein and loops) determines the patterns of xylem and phloem differentiation, which will follow the procambium pattern during leaf morphogenesis (4).

Naturally-occurring discontinuous basipetal patterns of early-differentiating vessel fragments can be detected at the developing hydathode of the tip (Fig. 7C) and at the differentiating hydathodes of the lobes (Fig. 7D-E). The initiation of fast differentiating isolated vessel elements in the lobes of wild-type plants is promoted by the high local IAA concentrations produced by the differentiating hydathodes (Figs. 1B-D, 2A). Analogously, normal discontinuous basipetal patterns also characterize the early stages of vessel differentiation in the floral organs, where the initiation of vessel elements starts from the sites of high auxin production at the anthers (Fig. 8A) and immediately below the auxin-producing stigma (Fig. 8B). Hence, it is crucial to realize that normal vascular differentiation in shoot-organ primordia of wild-type plants starts from the sites of IAA production and progresses basipetally in discontinuous patterns, which should never be confused with possible vascular discontinuities found in mature organs of defective mutants.

On the other end, acropetal vascular development may progress from the base of shoot organs upward. Such acropetal progress of xylem and phloem differentiation (which characterizes the midvein and the base of secondary bundles) is suggested to result from a local build-up of IAA concentration above a local interruption to the basipetal auxin flow, which slows IAA movement locally. Thus, above locations which slow auxin movement (e.g., at the base of the petioles of leaves, where the abscission zone will develop) a local IAA accumulation (11) will induce an acropetal pattern of xylem

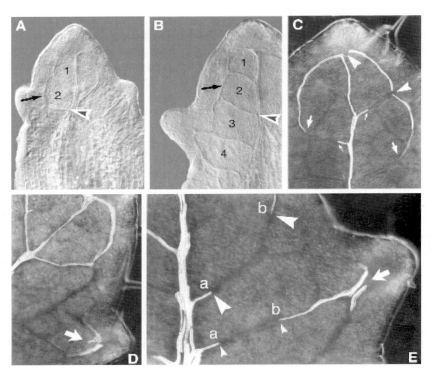

Figure 7. Procambium development (A,B) and naturally-occurring discontinuous vein patterns (C-E) in cleared wild-type Arabidopsis leaf primordia, photographed under Nomarski illumination. Under this illumination the vessels have a white appearance. Magnifications: A-D x 35; E x 50. A. A very young primordium with two procambial loops (loop 1 differentiated before loop 2). B. A more developed primordium with four procambial loops numbered according to their basipetal development. The black arrowhead marks the analogous site in A and B, emphasizing the basipetal development of the procambium (arrow). C. Discontinuous xylem (arrowheads) at the tip, showing basipetal (arrows), and acropetal differentiation (small arrows). D. Typical discontinuous xylem initiation (arrow) in a lobe, where the high auxin concentration is produced by the differentiating hydathode. E. Typical early stages of secondary bundle maturation, showing two discontinuous vessels (each vessel is marked by the same size arrowheads) progressing acropetally (a) and basipetally (b).

development. This is true also in junctions between the midvein and the secondary bundles (Fig. 7C,E), where the auxin streams originating in the marginal hydathodes merge with that of the midvein. Only because the midvein is well supplied with the IAA descending from the tip hydathode, does the junction become a site of elevated auxin concentration which promotes fast local acropetal vessel development at the base of the secondary bundle. Such typical acropetal patterns of xylem development (Fig. 7C,E) are normally detected during early stages of leaf-primordium ontogeny.

At the end of the bundle-maturation process, the vessel fragments which develop in acropetal and basipetal directions (Fig. 7) will join into one functional vessel. This gradual and fragmented bundle maturation, progressing

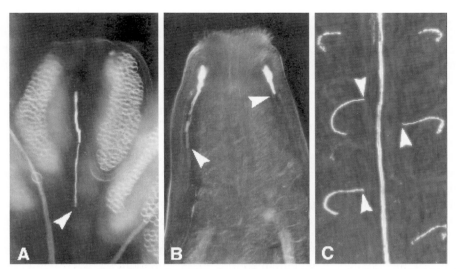

Figure 8. Naturally-occurring discontinuous vein patterns in cleared wild-type Arabidopsis flowers photographed under Nomarski illumination, all x 70. Xylem maturation starts immediately beneath the sites of free-IAA production at the tips of very young floral-organ primordia (A,B), whereas in mature gynoecium (C) the typical short ovule veinlets never connect with the central vascular bundle. A. In a young stamen primordium, a discontinuous vessel (arrowhead) commences maturation from the anthers, and progresses basipetally. B. Two sites of xylem initiation in a very young gynoecium, starting beneath the stigma and progressing downward (arrowheads). C. In a mature gynoecium, the short veinlets (arrowheads) originating from the ovules do not connect with the central bundle of the gynoecium.

in opposite directions, demonstrates that a bundle is composed of a population of differentiating cells that may have different rates of maturation. Those procambial cells exposed to high IAA concentrations (either at the hydathodes, or at the midvein junctions) will differentiate faster than the intervening procambial cells.

To avoid any possible confusion between naturally-occurring and defective-mutant discontinuities, the latter should be analyzed in mature organs (at a time when mature wild-type organs do not show vascular discontinuities). Xylem discontinuities in mature organs of defective mutants (19, 20, 24, 35, 46) can be regarded as early developmental stages that have become fixed during differentiation owing to a failure to complete the maturation process, or due to other possible interruptions to vascular differentiation. Similar discontinuities in mature phloem have been observed also in defective mutants of Arabidopsis (19, 46). Such discontinuous phloem patterns have been correlated with analogous discontinuities in xylem patterns (19).

Genetic Approaches Provide New Insights on the Mechanisms that Control Patterned Vascular Differentiation

Molecular biology techniques are powerful tools for studying vascular differentiation and vein pattern formation (17, 56, 61). In order to identify genes that determine venation pattern formation and control vascular differentiation, mutants of Arabidopsis and other model plants are being screened for altered vascular patterns. Such genetic screening may uncover mutations in genes that specifically disrupt the normal pattern of vascular differentiation (20, 24, 31, 35, 46).

Molecular biology tools are very useful for studying early events in vascular differentiation at the stages where it is difficult to observe procambium formation and initiation of vascular cells. Handy markers for this purpose are the *AtHB8, TED3* and *VH1* genes which are expressed during procambium and early stages of tracheary element differentiation (14, 22, 39). When these reporter genes are fused with the *GUS* gene, their activity can be visualized before clear anatomical features can be detected. Likewise, uncovering genes which are specifically expressed during phloem differentiation would be an important contribution for analyzing sieve-tube formation.

Vascular differentiation is induced and controlled by the basipetal polar flow of the auxin signal and, therefore, it is important to understand the molecular basis of auxin influx and efflux and the genes involved in this process. Polar auxin flow is attributed to an asymmetric distribution of specific transport proteins involved in the efflux of auxin (15). These and genes regulation their biosynthesis, location and recycling are described in chapter E1.

Recent findings indicate that sterols may play a crucial role in vascular differentiation; there is now evidence that sterols may be involved in the establishment of plant cell polarity, by regulating positioning of proteins in the plasma membrane. Sterols which regulate fluidity and interact with lipids and proteins within the plasma membrane can modulate the activity of membrane-bound proteins required for correct auxin signaling (20, 60).

Mutants with defective vascular patterns in their mature cotyledons and leaves may fail to establish uniformly aligned vascular cells and xylem discontinuities may occur in their mature leaves (19, 24, 35, 46). The genes detected in these mutants may be involved in auxin signal transduction during organized bundle formation. The most studied mutant with discontinuous patterns is *monopteros (mp)*, which is characterized by discontinuous bundles and improperly aligned vessel elements especially in the leaves (17, 46). The *MP* gene, which has been cloned, encodes a transcriptional factor capable of binding to auxin response elements in the control regions of auxin regulated genes (30).

Additional to the above, novel patterns of vascular networks may emerge in defective mutants (24, 35), indicating modifications in the auxin pathways and transport during the process of vascular differentiation. Mutants defective in their ability to form continuous vascular networks may result from mutations in genes encoding components of the polar auxin transport machinery, e.g., genes causing defected basipetal transport of IAA (21), reduced capacity for polar transport of auxin (46), or modified sensitivity to IAA (24). Genes affecting early stages of vascular patterning, prior to provascular network formation, may promote differentiation along wide pathways rather than narrow canals, owing to failure to establish efficient canals of IAA flow. An ineffective broad pathway resulting from a defect in the polar auxin transport machinery may fail to establish continuous wide xylem strands and can therefore result in fragmented patterns of vascular islands (24, 35).

SUMMARY AND CONCLUSIONS

Differentiating hydathodes (water secreting glands), that develop in the tip and later also along the leaf margins, are the primary sites of free-IAA production. In a leaf primordium, the basipetal polar auxin flow descending from the tip hydathode induces the midvein, and IAA streams from the marginal hydathodes induce the secondary vascular bundles. Trichomes and mesophyll cells, which produce low free-IAA concentrations, induce the minor tertiary and quaternary veins, as well as the freely ending veinlets. Similarly, the stigma of a young gynoecium is a major site of free-auxin production which induces a wide xylem fan pattern immediately below it, whereas the ovules with their low auxin content induce short veinlets. The polar auxin flow from the shoot organs to root tips induces the entire plant vascular system.

From the sites of free-IAA production in shoot organs, the auxin moves along the plant through three main transport pathways: (1) the vascular tissues, (2) protective tissues, and (3) sieve tubes. The continuous polar auxin flow through the vascular tissues controls their differentiation and regeneration. In vascular tissues of trees, the IAA flows mainly via the cambium. However, during plant evolution, because the cambium activity of trees has been minimized in herbaceous plants, I suggest that in advanced herbaceous species, with minimized, or absence (in monocots) of cambium activity, the polar IAA transport in the vascular tissues occurs mainly through their xylem parenchyma cells. The polar transport of auxin through parenchyma sheaths around the vascular tissues affects vascular pattern formation (e.g., by controlling the diameter of primary vessels in roots and internal phloem formation in bicollateral bundles) and could promote vascular regeneration and root initiation.

Polar IAA transport that induces vascular differentiation is attributed to asymmetric distribution of specific transport proteins regulated by vesicle trafficking, which are usually confined to the apical/basal end-poles of auxin-transporting cells. The fast non-polar IAA flow inside the sieve tubes is likely a house-keeping signal, which controls callose levels and possibly polarizes the actin cytoskeleton along the apical-basal axis in these phloem conduits.

Auxin acts as a morphogenetic signal, forming polar concentration gradients along the plant from the free-IAA producing hydathodes to the root tips and inducing polar patterns of increasing vessel diameter and decreasing vessel density form leaves to roots.

Phloem differentiation is induced by low-auxin streams, while xylem differentiation requires higher IAA concentrations. Consequently, near sites of high-auxin concentrations in a leaf primordium, veins of xylem (with no phloem) differentiate frequently. Conversely, away from the sites of IAA production, phloem-only bundles and phloem anastomoses between bundles are common along the stem in many plant species.

Early stages of vein maturation in wild-type plants are characterized by initiation and development of discontinuous basipetal and acropetal patterns. Generally, in shoots the basipetal maturation patterns start at the free-IAA production sites, whereas acropetal maturation patterns result from auxin accumulation above sites where IAA movement is locally retarded. At the end of this gradual process, the discontinuities will join into one functional vein. In order to avoid possible confusion between the naturally-occurring and defective-mutant discontinuities, the latter should be analyzed in mature organs.

Molecular approaches are expected to provide further new insights into the mechanisms that control vascular differentiation and vein pattern formation, and they should be integrated with physiological analyses at the cellular and organismic levels for correct interpretations.

References

1. Aloni R (1980) Role of auxin and sucrose in the differentiation of sieve and tracheary elements in plant tissue cultures. Planta 150: 255-263
2. Aloni R (1987) Differentiation of vascular tissues. Annu Rev Plant Physiol 38: 179-204
3. Aloni R (1995) The induction of vascular tissues by auxin and cytokinin. *In* Plant Hormones: Physiology, Biochemistry and Molecular Biology, PJ Davies, *ed*, Kluwer, Dordrecht, pp 531-546
4. Aloni R (2001) Foliar and axial aspects of vascular differentiation - hypotheses and evidence. J Plant Growth Regul 20: 22-34
5. Aloni R, Barnett JR (1996) The development of phloem anastomoses between vascular bundles and their role in xylem regeneration after wounding in *Cucurbita* and *Dahlia*. Planta 198: 595-603
6. Aloni R, Jacobs WP (1977) The time course of sieve tube and vessel regeneration and their relation to phloem anastomoses in mature internodes of *Coleus*. Amer J Bot 64: 615-621
7. Aloni R, Peterson CA (1990) The functional significance of phloem anastomoses in stems of *Dahlia pinnata* Cav. Planta 182: 583-590
8. Aloni R, Sachs T (1973) The three-dimensional structure of primary phloem systems. Planta 113: 345-353
9. Aloni R, Zimmermann MH (1983) The control of vessel size and density along the plant axis - a new hypothesis. Differentiation 24: 203-208
10. Aloni R, Raviv A, Peterson CA (1991) The role of auxin in the removal of dormancy callose and resumption of phloem activity in *Vitis vinifera*. Can J Bot 69: 1825-1832
11. Aloni R, Schwalm K, Langhans M, Ullrich CI (2003) Gradual shifts in sites of free-auxin production during leaf-primordium development and their role in vascular differentiation and leaf morphogenesis in *Arabidopsis*. Planta 216: 841-853
12. Aloni R, Wolf A, Feigenbaum P, Avni A, Klee HJ (1998) The *Never ripe* mutant provides evidence that tumor-induced ethylene controls the morphogenesis of *Agrobacterium tumefaciens*-induced crown galls on tomato stems. Plant Physiol 117: 841-847
13. Avsian-Kretchmer O, Cheng J-C, Chen L, Moctezuma E, Sung ZR (2002) IAA distribution coincides with vascular differentiation pattern during *Arabidopsis* leaf ontogeny. Plant Physiol 130: 199-209
14. Baima S, Nobili F, Sessa G, Lucchetti S, Ruberti I, Morelli G (1995) The expression of the *Athb-8* homeobox gene is restricted to provascular cells in *Arabidopsis thaliana*.

Development 121: 4171-4182
15. Baluška F, Šamaj J, Menzel D (2003) Polar transport of auxin: carrier-mediated flux across the plasma membrane or neurotransmitter-like secretion? Trends Cell Biol 13: 282-285
16. Barker-Bridgers M, Ribnicky DM, Cohen JD, Jones AM (1998) Red-light regulated growth. II. Changes in the abundance of indoleacetic acid in the maize mesocotyl. Planta 204: 207-211
17. Berleth T, Mattsson J, Hardtke CS (2000) Vascular continuity and auxin signals. Trends Plant Sci 5: 387-393
18. Booker J, Chatfield S, Leyser O. (2003) Auxin acts in xylem-associated or medullary cells to mediate apical dominance. Plant Cell 15: 495-507
19. Carland FM, Berg BL, FitzGerald JN, Jinamornphongs S, Nelson T, Keith B (1999) Genetic regulation of vascular tissue patterning in *Arabidopsis*. Plant Cell 11: 2123-2137
20. Carland FM, Fuioka S, Takatsuto A, Yoshida S, Nelson T (2002) The identification of CVP1 reveals a role for sterols in vascular patterning. Plant Cell: 14: 2045-2058.
21. Carland FM, McHale NA (1996) *LOP1*: a gene involved in auxin transport and vascular patterning in *Arabidopsis*. Development 122:1811-1819
22. Clay NK, Nelson T (2002) VH1, a provascular cell-specific receptor kinase that influence leaf cell patterns in *Arabidopsis*. Plant Cell 14: 2707-2722
23. Dengler N, Kang J (2001) Vascular patterning and leaf shape. Curr Opin Plant Biol 4: 50-56
24. Deyholos MK, Cordner G, Beebe D, Sieburth LE (2000) The *SCARFACE* gene is required for cotyledon and leaf vein patterning. Development 127: 3205-3213
25. Esau K (1965) Vascular Differentiation in Plants, Holt, Rinehart and Winston, New York
26. Friml J, Benkova Elilou I, Wisniewska J, Hamann T, Ljung K, Woody S, Sandberg G, Scheres B, Jürgens G, Palme K (2002) AtPIN4 mediates sink-driven auxin gradients and root patterning in *Arabidopsis*. Cell 108: 661-673
27. Friml J, Plame K (2002) Polar auxin transport - old questions and new concepts? Plant Mol Biol 49: 273-284
28. Gälweiler L, Guan C, Müller A, Wisman E, Mendgen K, Yephremov A, Palme K (1998) Regulation of polar auxin transport by AtPIN1 in *Arabidopsis* vascular tissues. Science 282: 2226-2230
29. Geldner N, Friml J, Stierhof Y-D, Jürgens G, Palme K (2001) Auxin transport inhibitors block PIN1 cycling and vesicle trafficking. Nature 413: 425-428
30. Hardtke CS, Berleth T (1998) The *Arabidopsis* gene *MONOPTEROS* encodes a transcription factor mediating embryo axis formation and vascular development. EMBO J 17: 1405-1411
31. Hobbie L, McGovern M, Hurwitz LR, Pierro A, Yang Liu N, Bandyopadhyay A, Estelle M (2000) The axr6 mutants of *Arabidopsis thaliana* define a gene involved in auxin response and early development. Development 127: 23-32
32. Horner HT, Lersten NR, Wirth CL (1994) Quantitative survey of sieve tube distribution in foliar terminal veins of ten dicot species. Amer J Bot 81: 1267-1274
33. Jacobs WP (1952) The role of auxin in differentiation of xylem around a wound. Amer J Bot 39: 301-309
34. Klee HJ, Horsch RB, Hinchee MA, Hein MB, Hoffmann MB (1987) The effects of overproduction of two *Agrobacterium tumefaciens* T-DNA auxin biosynthetic gene products in transgenic petunia plants. Gene Dev 1: 86-96
35. Koizumi K, Sugiyama M, Fukuda H (2000) A series of novel mutants of *Arabidopsis thaliana* that are defective in the formation of continuous vascular network: calling the auxin signal flow canalization hypothesis into question. Development 127: 3197-3204
36. Kuriyama H, Fukuda H (2001) Regulation of tracheary element differentiation. J Plant Growth Regul 20: 35-51
37. Leitch MA (2001) Vessel-element dimensions and frequency within the most current growth increment along the length of *Eucalyptus globulus* stems. Trees 15: 353-357
38. Lersten NR (1990) Sieve tubes in foliar vein endings: review and quantitative survey of *Rudbeckia laciniata* (Asteraceae). Amer J Bot 77: 1132-1141

39. Mattsson J, Ckurshumova W, Berleth T (2003) Auxin signaling in *Arabidopsis* leaf vascular development. Plant Physiol 131: 1327-1339
40. Mattsson J, Sung ZR, Berleth T (1999) Responses of plant vascular systems to auxin transport inhibition. Development 126: 2979-2991
41. Morris DA, Kadir GO, Barry AJ (1973) Auxin transport in intact pea seedlings (*Pisum sativum* L.): the inhibition of transport by 2,3,5-triiodobenzoic acid. Planta 110: 173-182
42. Müller A, Düchting P, Weiler EW (2002) A multiplex GC-MS/MS technique for the sensitivity and qualitative single-run analysis of acidic phytohormones and related compounds, and its application to *Arabidopsis thaliana*. Planta 216: 44-56
43. Nelson T, Dengler N (1997) Leaf vascular pattern formation. Plant Cell 9: 1121-1135
44. Palme P, Gälweiler L (1999) PIN-pointing the molecular basis of auxin transport. Curr Poin Plant Biol 2: 375-381
45. Reinhardt D, Pesce E-R, Stieger P, Mandel T, Baltensperger K, Bennett M, Traas J, Friml J, Kuhlemeier C (2003) Regulation of phyllotaxis by polar auxin transport. Nature 426:255-260
46. Przemeck GKH, Mattsson J, Hardtke CS, Sung ZR, Berleth T (1996) Studies on the role of the *Arabidopsis* gene *MONOPTEROS* in vascular development and plant cell axialization. Planta 200: 229-237
47. Roberts LW, Gahan PB, Aloni R (1988) Vascular Differentiation and Plant Growth Regulators. Springer-Verlag, Berlin
48. Romano CP, Hein MB, Klee HJ (1991) Inactivation of auxin in tobacco transformed with the indoleacetic acid-lysine synthetase gene of *Pseudomonas savastanoi*. Genes Dev 5: 438-446
49. Sabatini S, Beis D, Wolkenfelt H, Murfett J, Guilfoyle T, Malamy J, Benfey P, Leyser O, Bechtold N, Weisbeek P, Scheres B (1999) An auxin-dependent distal organizer of pattern and polarity in the *Arabidopsis* root. Cell 99: 463-472
50. Sachs T (1969) Polarity and the induction of organized vascular tissues. Ann Bot 33: 263-275
51. Sachs T (1981) The control of patterned differentiation of vascular tissues. Adv Bot Res 9: 151-262
52. Sachs T (2000) Integrating cellular and organismal aspects of vascular differentiation. Plant Cell Physiol 41: 649-656
53. Sieburth LE (1999) Auxin is required for leaf vein pattern in *Arabidopsis*. Plant Physiol 121: 1179-1190
54. Sundberg B, Uggla C, Tuominen H (2000) Cambial growth and auxin gradients. *In* RA Savidge, JR Barnett, R Napier, eds, Cell and Molecular Biology of Wood Formation, BIOS Scientific Publishers, Oxford, pp 169-188
55. Swarup R, Friml J, Marchant A, Ljung K, Sandberg G, Palme K, Bennett M (2001) Localization of the auxin permease AUX1 suggests two functionally distinct hormone transport pathways operate in the *Arabidopsis* root apex. Genes Dev 15: 2648-2653
56. Turner S, Sieburth LE (2002) Vascular patterning. The Arabidopsis Book, CR Somerville, EM Meyerowitz, eds, American Society of Plant Biologists. Rockville, MD, http://www.aspb.org/publications/arabidopsis/toc.cfm
57. Uggla C, Moritz T, Sandberg G, Sundberg B (1996) Auxin as a positional signal in pattern formation in plants. Proc Nat Acad Sci USA 93: 9282-9286
58. Ullrich CI, Aloni R (2000) Vascularization is a general requirement for growth of plant and animal tumours. J Exp Bot 51: 1951-1960
59. Ulmasov T, Murfett J, Hagen G, Guilfoyle TJ (1997) Aux/IAA proteins repress expression of reporter genes containing natural and highly active synthetic auxin response elements. Plant Cell 9: 1963-1971
60. Willemsen V, Friml J, Grebe M, van den Toorn A, Palme K, Scheres B (2003) Cell polarity and PIN protein positioning in *Arabidopsis* require *STEROL METHYLTRRANFERASE1* function. Plant Cell 15: 612-625
61. Ye Z-H (2002) Vascular tissue differentiation and pattern formation in plants. Annu Rev Plant Biol 53: 183-202

E3. Hormones and the Regulation of Water Balance

Ian C. Dodd and William J. Davies
The Lancaster Environment Centre, Lancaster University, LA1 4YQ, UK
E-mail: W.Davies@lancaster.ac.uk

INTRODUCTION

Soil water status limits both species distribution and crop yield and a long-held view is that plant water status is the key variable that mediates the influence of soil drying. Plant water status can be a highly dynamic variable that fluctuates over varying time scales. For example, over the course of a day, plant water status of even a well-watered plant declines as solar noon approaches, as transpiration increases with increasing evaporative demand. The plant will hydrate again as temperatures (and transpiration) decrease towards the end of the day. In between rainfall (or irrigation) events, plant water status will generally decrease along with soil water status. In many circumstances, we are content with the suggestion that as soil dries, reduced uptake of water from drying soil results in shoot water deficit which closes stomata to restrict water loss and further shoot dehydration.

However, this is not always the case. Under some situations, some plants show so-called 'iso-hydric' behaviour, when water status is strongly regulated, usually by partial stomatal closure (Fig. 1). Here, it is necessary to suggest that the stomata are either very finely tuned to reductions in water supply to shoots or that they are responding to some other measure of soil water availability. In this chapter[1], we advance the hypothesis that chemical signalling between roots and shoots can tune stomatal (and other physiological and developmental) responses to soil drying such that plant water status can be effectively regulated.

Total plant water loss is the sum of water loss of each individual leaf and therefore limiting leaf expansion provides another means of limiting

---

[1] Abbreviations: A, CO_2 assimilation; ABP1, auxin-binding protein 1; ACC, aminocyclopropane-1-carboxylic acid; AVG, aminoethoxyvinylglycine; g_s, stomatal conductance; ipt, isopentenyl transferase; Lp, root hydraulic conductivity; Ψ_{leaf}, leaf water potential; Ψ_{soil}, soil water potential; PRD, partial rootzone drying; θ, soil water content; WT, Wild-type; WUE, water use efficiency; [X-ABA], Xylem sap ABA concentration.

Hormones and water balance

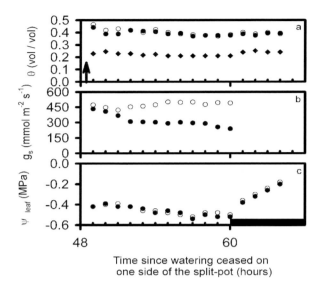

Figure 1. Soil moisture content of pots watered daily (indicated by arrow) on both sides of a split-pot (O), and of the watered (●) and drying (♦) sides of split-pots watered only on one side (a). Stomatal conductance (b) and leaf water potential (c) of plants watered daily on one (●) or both (O) sides of the split-pot (b-c). In (b), points are means of 5 leaflets per leaf. Dark shading on the time axis indicates the night period. (W.Y Sobeih, I.C. Dodd, M.A. Bacon & W.J. Davies, unpublished data).

plant water loss over periods of several days. Root growth and root activity can also exert considerable influence over plant water uptake. Integration of these processes (stomatal regulation, leaf expansion and root growth) will contribute to effective maintenance of plant water balance, and hormone-based signalling may be a means of achieving this. Such a central role for signalling in the plant's drought responses may provide targets for crop improvement.

STOMATAL REGULATION

Leaf conductance depends on the distribution of stomatal pore size across the leaf, and stomatal frequency. While leaf hormonal status can affect stomatal differentiation and frequency, we emphasise here the important role of chemical signalling in the dynamic regulation of stomatal aperture and water balance. Without regulation of this type, plants as we know them could not survive in the rapidly changing, challenging terrestrial environment into which they have evolved. Although exogenous hormone applications have demonstrated that stomata can respond to all 5 classical hormone classes (10) and several other chemical species, the role of ABA in eliciting stomatal closure is highlighted here. This is largely because of the now-recognised potency of this hormone in the regulation of stomatal behaviour (across several orders of magnitude of concentration) and its stress-induced accumulation across a similar concentration range.

Abscisic Acid

Abscisic acid – receptors and signal transduction

Although the plant hormone concept suggests that a hormone molecule must interact with a receptor protein in order to effect a physiological response (Chapter A2), identification of an ABA receptor has proved elusive. Physiological experiments have sought to determine where in the stomatal guard cells such receptors might reside. ABA is not readily able to cross the cell membrane at pH 8, yet incubation of epidermal strips in solutions of ABA at this pH can elicit stomatal closure (14), which has been taken as evidence of extracellular ABA receptors. The presence of intracellular ABA receptors has also been suggested, as injection of ABA into individual guard cells causes stomatal closure, indicating that stomata can perceive symplastic hormone concentrations (1).

In terms of long-distance signalling between roots and shoots, the extracellular receptors must encounter hormones arriving in the apoplast of the leaf and therefore offer a sensitive link between hormone delivery and response. Irrespective of the location of ABA receptors, much effort has been expended in elucidating the intracellular signal transduction pathway resulting in stomatal closure. Observations that extracellular calcium and ABA acted synergistically in causing stomatal closure (Fig. 2) provided early clues that calcium acted as an important second messenger of ABA action. Perfusion of isolated guard cells with ABA increased the free calcium concentration in the cytosol (23). However, exposing plants to ABA at low temperature for 48 hours closed stomata in the absence of increases in concentration of free calcium in the cytosol (1), indicating that ABA-induced stomatal closure can use both calcium-dependent and calcium-independent pathways. Much research has concentrated on defining the intracellular source and dynamics of this elevated calcium, and its interaction with downstream parts of the ABA signal transduction pathway (Chapter D5).

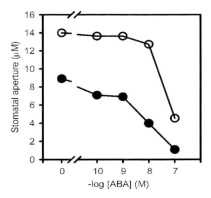

Figure 2. Stomatal aperture of *Commelina communis* as a function of ABA concentration when incubated with (●) and without (○) 10^{-4} M $CaCl_2$. Data re-drawn from (9).

Synthesis and distribution of abscisic acid

Much progress has been made in elucidating the pathway of ABA biosynthesis (Chapter B5). Although it is important to note that even well-watered plants contain some ABA, bulk leaf ABA concentration increases in detached leaves in response to cellular dehydration (53). The capacity of leaf tissue to synthesise additional ABA varies between different genotypes (33), can be enhanced by nutrient deficiency and decreased by high (40°C) temperature (35). Although root tissue generally contains lower ABA concentrations than leaves, dehydration of detached roots from various species, and of various ages and branching orders also stimulated ABA synthesis (8).

Synthesis and transport of ABA conjugates provides another mechanism of increasing ABA concentrations in various compartments. Apoplastic glucosidases can liberate free ABA from the ABA glucose ester, which can be found in relatively high concentrations in the xylem sap (39). Xylem sap is also reported to contain an ABA precursor or adduct, which is hydrolysed at alkaline pH (28). Although such compounds may be important in long-distance transport, their contributions to the ABA budgets of specific organs have yet to be evaluated.

Since ABA is a weak acid (pKa = 4.75), its distribution in plant tissues will be governed by the Henderson-Hasselbach equation. At a cytosolic pH commonly found in many well watered plants (pH 7.3), ABA present in the protonated form moves into the alkaline chloroplast stroma (pH 7.9) where it dissociates to form an anion which is not as readily permeable (14). For this reason, most of the ABA in unstressed leaves is assumed to be in the chloroplasts (which act as an anion trap for ABA). Leaf dehydration alkalises the apoplast, increasing apoplastic ABA concentrations (14). Redistribution of ABA between different compartments of the leaf provides an attractive possibility for stomatal regulation in response to drought.

Prior to the quantification of the effects of drought on ABA accumulation and the demonstration of the effects of this hormone on stomata, it was widely assumed that stomatal response to soil drying was effectively a direct response to a change in leaf turgor. Stomatal closure was then attributed to the accumulation of ABA in leaves but careful work showed that the hormone accumulated only after leaf conductance declined (8, 54). Results such as these are not altogether surprising, given that most of the ABA in a leaf will be sequestered in the mesophyll chloroplasts and unavailable to active sites at or within the guard cell (14). It was then proposed that drought-induced changes in leaf water potential liberated ABA from the mesophyll chloroplasts, and that this ABA would move to the guard cells to initiate stomatal closure (22). However, this hypothesis does not explain stomatal closure without leaf dehydration (Fig. 1) and it is therefore necessary to argue for signalling from elsewhere in the plant to explain responses of this type.

ABA as a long distance signal of soil drying
For ABA to act as a root signal molecule that regulates stomatal behaviour, it was necessary to show increased delivery of the hormone into the leaf apoplast adjacent to putative extracellular binding sites on the guard cells. After overnight immersion of roots in an ABA solution (which was removed prior to the lights coming on), transpiring leaves showed stomatal closure during the following morning with the degree of closure being proportional to the accumulation of ABA in the leaf epidermis (8). More recently, using a combination of microdissection and a sensitive immunoassay, the accumulation of ABA has been measured in individual guard cells (57). These kinds of measurements cannot be made in any great numbers in crop level studies (as sample collection is labour intensive), but there is evidence that guard cell ABA accumulation can correlate with whole leaf and xylem ABA concentration, [X-ABA] (57).

Evidence that [X-ABA] was dynamically linked to changes in stomatal conductance (g_s) was first provided by Loveys (21). Subsequently, [X-ABA] was shown to increase much earlier and to a greater extent than bulk leaf ABA concentration during a soil drying cycle, and that this increase correlated with decreased g_s (54). Several comprehensive data sets from field and glasshouse studies indicate an excellent correlation between [X-ABA] and g_s (when xylem sap was collected from the same leaves in which g_s was measured) in a diverse range of species (46, 6, 4). Importantly, the [X-ABA] / g_s relationship for a given species is commonly unified across different growing conditions, from leaf to leaf on individual plants, and from day to day as the plant develops, providing good evidence of a causal relationship. There is considerable species to species variation in the [X-ABA] at which stomatal closure occurs (Fig. 3), suggesting it is not the absolute ABA concentration, but the relative increase in ABA concentration that is physio-logically important. A similar analysis can be applied to some ABA-deficient genotypes, which show a decreased capacity to synthesise ABA. Although such genotypes have a lower absolute ABA concentration (and thus higher g_s), they show a similar sensitivity of g_s to [X-ABA] (4).

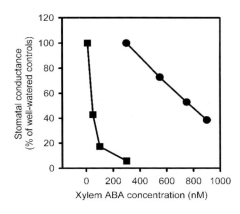

Figure 3. Relative stomatal conductance as a function of xylem ABA concentration for field-grown maize (■) and grape (●). Data re-drawn from (46) (■) and (6) (●).

Irrespective of the site of sap collection, as the soil dries, the concentration of all xylem sap constituents should passively increase as transpirational fluxes decrease. Thus any solute, irrespective of any effect on stomata, could show a negative relationship with flux, similar to that shown between [X-ABA] and g_s (Fig. 3). Consequently, more rigorous tests of the physiological significance of ABA have been formulated. *Correlation* and *duplication* experiments have shown that soil application or stem injection of synthetic ABA to well-watered plants generates a similar relationship between [X-ABA] and g_s to that found in droughted plants. *Deletion* and *re-instatement* experiments test the specificity of hormone action by manipulating endogenous hormone levels. Removal of ABA from maize xylem sap using an immunoaffinity column eliminated its antitranspirant activity, as assessed using a detached leaf transpiration assay (8).

Although xylem sap is assumed to be in direct contact with the leaf apoplast (and thus available to extracellular guard cell receptors), the site and nature of xylem sap collection is important to accurately assess [X-ABA] (19). Xylem sap collected from leaves will more closely reflect apoplastic sap adjacent to the guard cells than sap collected from the root system, due to gains or losses in xylem solutes during long-distance transport through the stem. Leaf xylem sap can be obtained by growing plants in a whole plant pressure chamber or by pressurising entire detached leaves in a Scholander-type pressure chamber. If xylem sap must be collected from the root system, sap should flow at rates equivalent to whole plant transpiration rate, since the concentrations of many xylem constituents increase exponentially with decreasing sap flow rate (40). Sap collection from de-topped root systems under root pressure alone over-estimates true xylem sap concentration (40).

Even when xylem sap is collected from the same leaf in which g_s is measured, xylem sap can be modified en route to the leaf apoplast, by exchange with xylem parenchyma within the petiole, or by mesophyll catabolism and sequestration. The importance of the mesophyll tissue in modifying stomatal behaviour is revealed by experiments comparing dose-response relationships of stomata to apoplastic or xylem-supplied ABA (Fig. 4). In isolated epidermal strips, where hormone concentrations are controlled at the guard cell apoplast, stomatal closure occurred at ABA concentrations as low as 0.1 µM. In contrast, 1 µM ABA was needed to elicit substantial stomatal closure in detached leaves supplied with ABA via the xylem. Inhibition of ABA catabolism with tetcyclacis enhanced stomatal closure in leaf pieces (47). The implication is that xylem ABA concentrations are sufficient to keep stomata closed most of the time were it not for the mesophyll sequestering and breaking down ABA.

Although variation in [X-ABA] often accounts for variation in g_s, the origin of this ABA continues to attract debate. During some drying cycles, root ABA concen-tration and xylem sap concentration increase in parallel, prior to any increase in leaf ABA concentration (8), suggesting that xylem ABA is root-derived. However, considerable re-circulation of ABA between

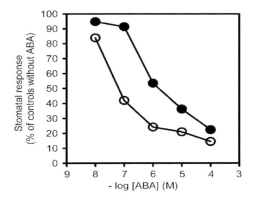

Figure 4. Relative stomatal response of *Commelina communis* as a function of incubation solution ABA concentration for epidermal strips (O) and ABA concentration supplied via the xylem to detached leaves (●). Data re-drawn from (47).

xylem and phloem can occur (52) thus not all ABA in a xylem sap sample is likely to be root-derived. Droughted plants can first lose turgor in older leaves (54), which might enhance any root-sourced ABA signal to younger leaves. In some species, ABA found in the rhizosphere can be efficiently transferred across the root tissues into the xylem (11). The importance of root-sourced ABA in mediating drought-induced stomatal closure has been addressed by reciprocal grafting of wild-type (WT) and ABA-deficient genotypes and comparing the stomatal responses and ABA concentrations of the graft combinations (15). Irrespective of whether WT shoots were grafted on WT or ABA-deficient roots, stomatal closure occurred in both graft combinations, despite a 4-fold difference in [X-ABA]. Such data suggests that other chemical species, in addition to ABA, can act as signals of the degree of soil drying and can be important in regulating stomatal behaviour.

Does apoplastic pH determine the ABA response ?
Several studies indicate that variation in [X-ABA] alone cannot always explain the extent of drought-induced stomatal closure. Studies with detached leaves have suggested the presence of other antitranspirant compounds in wheat and barley xylem sap (27). Also, drought-induced stomatal closure can precede increases in [X-ABA].

Alkalisation of xylem sap is a common response to various edaphic stresses (50) and supplying detached *Commelina* and tomato leaves with neutral or alkaline buffers (pH ≥ 7) via the transpiration stream can close stomata (Fig. 5). These alkaline buffers increased apoplastic pH, thus decreasing sequestration of ABA by mesophyll cells, causing increased apoplastic ABA concentrations, which closed stomata (51). Stomatal closure in response to xylem-supplied alkaline buffers was ABA-dependent, as leaves detached from an ABA-deficient mutant (*flacca - flc*) did not show stomatal closure when fed pH 7 buffers (Fig. 5), and in some cases transpiration actually increased (50). This pH response of detached *flc* leaves is consistent with g_s increasing when *flc* plants dried the soil within a certain range of soil water contents (15).

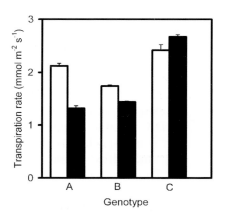

Figure 5. Transpiration rate of detached leaves fed phosphate buffers at pH 6-6.25 (hollow bars) and 7-7.75 (filled bars). Genotypes were *Commelina communis* (A), *Lycopersicon esculentum* cv. Ailsa Craig (B) and the ABA deficient tomato mutant *flacca* (C). Data re-drawn from (51) (*C. communis*) and (50) (*L. esculentum*).

Stomatal closure in response to sap alkalisation may help explain observations where stomatal closure could not be readily explained in terms of an increased [X-ABA]. During soil drying, changes in xylem sap pH may initiate stomatal closure as these changes may precede changes in [X-ABA]. Further, alkalisation of xylem sap may be responsible for the "unexplained" antitranspirant activity of some saps. However, in some species, xylem sap pH can acidify (*Ricinus communis;* 40) or show no change (*Nicotinia plumbaginifolia;* 4) in response to drought.

Variations in the aerial environment can also alkalise xylem sap. In well-watered *Forsythia x intermedia* plants grown in a greenhouse, more alkaline xylem sap pH was correlated with an increased light intensity and stomatal closure (7) although it not yet known which component of the aerial environment (e.g., light intensity, temperature, vapour pressure deficit) is responsible for the pH variation. Such phenomena allow integration of signals from both aerial and root environments, and may also explain observations showing that the effectiveness of ABA in closing stomata depends on prevailing environmental variables.

Environmental modification of ABA response

Stomatal responses to ABA vary considerably according to nutritional factors (14, 35) and environmental factors such as temperature and leaf water potential (Ψ_{leaf}). Such interactions may be important in the minute-by-minute control of stomatal aperture in a fluctuating environment, and allow drought-stressed plants to open their stomata to maximise photosynthesis under conditions (lower temperature and vapour pressure deficit) where transpirational losses can be minimised.

Since temperature and vapour pressure deficit and thus whole plant transpiration rate co-vary, an increase in temperature is likely to increase ABA delivery to the shoot. For this reason, demonstrations that the effects of ABA are truly temperature-dependent have often used epidermal strips to control apoplastic ABA concentrations. In maize epidermal strips, application of ABA stimulated stomatal opening below a threshold

I. C. Dodd and W. J. Davies

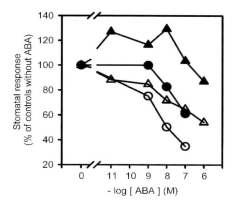

Figure 6. Relative stomatal response as a function of ABA concentration for epidermal strips of *Bellis perennis* incubated at 10°C (▲) and 30°C (△) and *Commelina communis* incubated at 7°C (●) and 25°C (○). Data re-drawn from (16) (*B. perennis*) and (49) (*C. communis*).

temperature, yet caused stomatal closure as temperature increased (36). Similarly, stomata of several species (*Bellis perennis, Cardamine pratensis, Commelina communis*) were relatively insensitive to ABA when incubated at 10°C (Fig. 6) and in some cases showed stomatal opening, but showed normal ABA-induced stomatal closure at 20°C or 30°C (16). The pretreatment history of the plants can also be important, as a 48 hour exposure to cold temperatures (< 15°C) decreased the rate of stomatal closure of plants subsequently challenged with ABA (1). The decreased stomatal response to ABA at chilling temperatures was not related to thermotolerance in a comparison of two species (49): cold-tolerant *Commelina communis* showed decreased stomatal sensitivity to ABA at low temperature while cold-intolerant *Nicotinia rustica* showed a similar stomatal sensitivity at both 7°C and 27°C. The physiological bases for temperature-dependent ABA-induced stomatal closure, and for such genotypic differences, remain poorly defined.

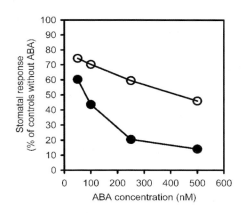

Figure 7. Relative stomatal response as a function of ABA concentration for *Commelina communis* epidermal strips incubated on solutions with an osmotic potential of –0.3 MPa (○) and –0.5 MPa (●). Modified from (46).

In field-grown maize, the relationship between [X-ABA] and g_s varied diurnally, with the most sensitive stomatal closure occurring at lower Ψ_{leaf} (46). Since an increased Ψ_{leaf} increases the rate of catabolism of xylem-supplied ABA (20), differences in the amounts of ABA reaching the guard cells might occur at different Ψ_{leaf}. However, increased stomatal sensitivity to ABA was seen when *Commelina* epidermes were incubated on ABA solutions of decreasing medium osmotic potential (Fig. 7). This interaction may be thought of as a sensitive dynamic feedback control mechanism to ensure homeostasis of Ψ_{leaf}. Any decrease in Ψ_{leaf} (e.g., caused by the sun appearing from behind a cloud) will enhance stomatal response to ABA thus decreasing transpiration and returning Ψ_{leaf} to its original value. Interactions between ABA concentration and the environmental and plant variables discussed above enable the plant to integrate stomatal responses to a wide range of factors, all of which have the potential to modify plant water balance.

Auxins

Although attention has been given to stomatal responses to auxin, there is still far too little information on the effects of water stress on auxin delivery to the shoot, and leaf auxin concentrations. Abaxial stomata of *Commelina communis* can be relatively insensitive to IAA concentration in CO_2-free air, yet increasing IAA concentrations abolishes CO_2-induced stomatal closure (Fig. 8). IAA-induced stomatal opening can be an indirect effect of auxin-induced ethylene production. An inhibitor of ACC synthase, aminoethoxyvinylglycine (AVG), prevented auxin-induced stomatal opening in *Vicia faba* epidermes (24). Addition of ACC (the immediate precursor of ethylene) restored auxin-induced opening in a concentration-dependent manner when AVG was also present in the incubation solution (24).

Evidence that IAA can affect stomatal behaviour via ethylene-independent pathways is provided by the observation that antibodies to auxin-binding protein 1 (ABP1), a putative auxin receptor, induced stomatal opening (12). Overexpression of ABP1 in transgenic plants enhanced the sensitivity of guard cells to auxin, as measured by the changes in the auxin dose-response of inward and outwardly rectifying K+ currents (3).

While stomatal responses to auxin are often studied in isolation, it seems likely that the interaction of auxin with other hormones will be important *in vivo*. Incubation of *Commelina* epidermes in solutions containing 10 or 100 µM IAA antagonised ABA-induced stomatal closure (10). IAA overproducing and underproducing transgenics may allow IAA antagonism of ABA-induced stomatal closure to be evaluated *in planta*, and it is

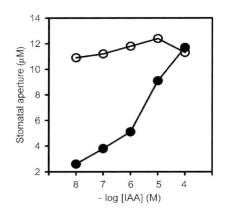

Figure 8. Stomatal aperture of *Commelina communis* as a function of IAA concentration at 700 µL L^{-1} CO_2 (●) or without CO_2 (○). Modified from (44).

important to give attention to potential hormonal interactions.

Cytokinins (CKs)

Stomatal responses to CK application are rather variable (Fig. 9) according to the species and the particular CK applied (10, 18). In isolated epidermes, micromolar CK concentrations are usually required to stimulate stomatal opening, but stomata can respond to CK concentrations as low as 10 nM (10). Xylem CK concentrations typically range between 1-50 nM according to the species, growth conditions and perhaps most importantly, sap flow rate during collection. When supplied via the xylem to detached leaves, nanomolar CK concentrations stimulate transpiration of some monocotyledonous species (18), yet much higher CK concentrations apparently have no effect on dicotyledonous species (Fig. 9). The apparent insensitivity of stomata to CKs in some studies may occur because endogenous CK concentrations are already optimal for stomatal opening. Another explanation is that CKs most effectively promote stomatal opening in ageing leaves and most studies use the youngest fully expanded leaf (18), suggesting that stomata might become sensitive to CKs only once endogenous concentrations decrease.

Perhaps the most convincing demonstration that endogenous CKs increase g_s and plant transpiration is provided by plant transformation with bacterial *ipt* (isopentenyl transferase, which catalyses *de novo* CK biosynthesis) or *zmp* (which encodes a protein capable of cleaving CK glucosides into active forms) genes (32). However, such transformations can also alter the concentrations of other hormones, which might affect g_s. Tobacco *ipt* and *zmp* transformants showed decreased and increased leaf IAA levels respectively, yet both showed increased g_s and increased leaf CK concentrations (32), suggesting that CKs and not auxin caused the observed stomatal phenotype.

Although CKs can promote stomatal opening when applied in isolation, they can also antagonise ABA-induced stomatal closure.

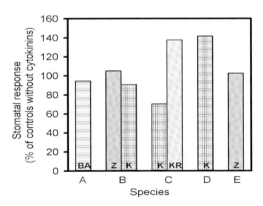

Figure 9. Relative stomatal responses to 1 μM solutions of various CKs in different species. Epidermal strips of *Commelina benghalensis* (A), *C. communis* (B) and *Vicia faba* (C) were incubated in the light in CK solutions; while detached barley (D) and lupin (E) leaves were supplied with CKs via the xylem. CKs applied were BA, benzyladenine; K, kinetin; KR, kinetin riboside and Z, zeatin. Data re-drawn from references in (10).

Incubation of *Commelina* epidermes in solutions containing 50 µM benzyladenine and incubation of maize leaf pieces in solutions containing 10 or 100 µM zeatin or kinetin antagonised ABA-induced stomatal closure (10). However, similar experiments supplying CKs via the xylem to detached leaves have not always shown similar results, suggesting that xylem CK concentrations can be less important to stomatal responses than leaf (or apoplastic) CK concentrations. Supply of 0.1-10 µM kinetin to detached barley leaves and supply of 0.1-10 µM zeatin to detached lupin (*Lupinus cosentinii*) leaves did not reverse ABA-induced stomatal closure (10), perhaps due to rapid CK metabolism by the mesophyll. More recently, ABA-induced stomatal closure in detached sunflower leaves was partially reversed by nanomolar concentrations of zeatin riboside, which are typically found in sunflower xylem sap (13). The importance of CK antagonism of ABA-induced stomatal closure to *in vivo* responses might be tested by submitting CK overproducing transgenics along with WT plants to gradual soil drying and analysing stomatal sensitivity to [X-ABA].

ROOT GROWTH AND WATER UPTAKE

While stomata can regulate water loss from the shoot to slow the development of harmful water deficits, such regulation relies on the root system to sustain an adequate water supply. Mild soil drying commonly first depletes the water content of the soil immediately adjacent to the root and plant water uptake can therefore be restricted, even at quite high bulk soil water content. Under these circumstances, variation in root hydraulic conductivity (Lp) will only have a small impact on plant water uptake but it may be important in certain circumstances that root Lp does increase as soil dries (perhaps as a result of changed aquaporin activity) and that this response can be mimicked by ABA application to roots (56). However, prolonged and more severe soil drying decreases root Lp due to cell death and structural modifications such as suberisation to prevent water loss from dehydrated roots.

Although drought usually decreases overall plant (and root system) size, the proportional allocation of dry matter to the roots is usually increased (31) and root elongation can be increased (41). While drying of surface soil layers may cause local root mortality, root exploration of wetter and sometimes deeper parts of the soil profile may act to minimise the development of high soil/root interface resistances for water uptake and as such can help to maintain plant water status.

An experimentally convenient system to analyse root growth responses to soil water deficit, that circumvents the influence of the shoot controlling carbon supply to the roots, is the growth of primary roots in the dark at fixed soil water potentials (Ψ_{soil}). ABA seems to have an important role in root growth maintenance at low Ψ_{soil} since genetic or chemical reductions in ABA accumulation inhibited primary root elongation of dark-grown maize seedlings (Fig. 10). In contrast, at high Ψ_{soil}, ABA application to WT plants inhibited primary root elongation (42). At low Ψ_{soil}, ABA-deficient roots were not only shorter but thicker, suggesting a drought-induced increase in ethylene synthesis, which has been confirmed (43). Further work suggests that ABA accumulation in roots at low Ψ_{soil} limits their ethylene synthesis, and that this allows some root extension even at very low soil/tissue water potentials.

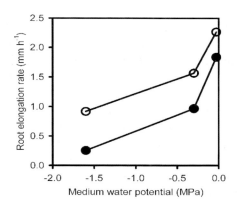

Figure 10. Primary root elongation rate as a function of medium water potential for the ABA deficient mutant *vp5* (●) and WT maize (○). Data re-drawn from (37, 42).

In light-grown plants, root growth is influenced by shoot control of assimilate availability. Growth analysis of ABA-deficient mutants provides some support for another role for ABA in maintaining root growth under various soil stresses, as the mutants often partition a decreased proportion of assimilates to the roots compared to WT plants (26), perhaps because the mutants are experiencing water deficit as indicated by a decreased Ψ_{leaf}. Reciprocal grafting experiments may avoid confounding effects of shoot water and ABA status. When WT scions were double-grafted onto both ABA-deficient (*sitiens*) and WT rootstocks and exposed to several cycles of soil drying, the *sitiens* root system accumulated 1.45-fold more biomass than the WT root system (15), suggesting that ABA deficient roots had a greater sink strength independent of shoot water status. Thus the role of root ABA status in whole plant resource partitioning requires further evaluation.

LEAF GROWTH

While stomatal closure limits water loss from existing leaf area, another water-saving measure is to decrease the extent of leaf area expansion. Several research groups have now demonstrated that leaf expansion of plants grown in drying soil can be decreased even when plant water status is maintained (8). In such situations, chemical signals generated as a result of the interaction between root systems and drying soil can inhibit leaf growth. Although many studies suggest a dominant role for the plant hormone ABA in the control of stomatal behaviour under drought (4), the role of ABA in the control of leaf growth is still debated.

Early experiments showed that incubation of leaf discs on ABA solutions decreased leaf expansion by decreasing cell wall extensibility (48). The relationship between leaf growth inhibition and [X-ABA] was similar for maize plants fed ABA hydroponically in a nutrient solution and those from which water was withheld (55), indicating a possible regulatory role for ABA in controlling leaf growth of droughted plants. Detached maize shoots supplied with ABA via the xylem showed a similar relationship to that generated in intact plants between [X-ABA] and leaf growth (Fig. 11). However, in some cases leaf growth is apparently restricted prior to an increased [X-ABA] (26).

Recently, it was demonstrated that increased xylem sap pH correlated with drought-induced leaf growth inhibition, and that feeding leaves more alkaline buffers via the xylem inhibited leaf growth (2). This response is directly analogous to the effect of alkaline buffers on detached leaf transpiration (discussed earlier). Feeding more alkaline buffers to an ABA-deficient barley mutant (*Az34*) did not inhibit leaf growth unless an ABA concentration typical of well-watered plants was also present in the buffer. It was suggested that alkaline pH allowed ABA access to sites of action within the leaf elongation zone, inhibiting growth. At a well-watered, more acidic apoplastic pH, ABA is presumably partitioned into alkaline compartments in the symplast and away from sites of action regulating leaf growth. Apoplastic pH variation proves to be extremely sensitive to reduction in soil water availability and allows redistribution of ABA into different compartments to modify plant behaviour without the necessity for substantial *de novo* ABA synthesis.

Initial work growing ABA-deficient mutants at low Ψ_{soil} supported the contention that drought-induced ABA accumulation inhibits shoot growth. Shoots of the *vp5* mutant grew more than WT shoots when transplanted to vermiculite at −0.3 MPa (37). However, extending the experiment beyond 60 hours after transplanting resulted in the WT shoots growing faster than *vp5*, indicating that the effects of

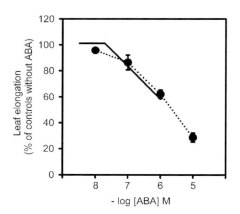

Figure 11. Response of leaf elongation to [X-ABA] in intact maize plants (solid line) and detached shoots (●). Solid line, redrawn from (55), was for whole plants grown in drying soil or fed a nutrient solution containing ABA. Symbols are means ± S.E. of n experiments for 10^{-8}M (n=3), 10^{-7}M (n=9), 10^{-6}M (n=11) and 10^{-5}M (n=8) ABA (I.C. Dodd, unpublished data).

ABA varied with the stage of plant development (43).

Further work with ABA-deficient mutants has reinforced the view that an important function of ABA is to restrict ethylene synthesis (43). Several changes in the soil environment such as soil compaction (17) can increase ethylene evolution, inhibiting shoot growth in most circumstances. In one report, intact droughted plants did not show increased ethylene evolution (25). However, soil drying can also increase soil strength, and it therefore seems possible that many soil drying episodes can perturb ethylene evolution, thus modifying leaf growth.

HORMONAL TARGETS TO INFLUENCE CROP WATER BALANCE AND YIELD ?

A framework (Fig. 12) to analyse the effects of manipulating hormone status or response on crop yield has been formulated (29), stating that:

Yield = Water transpired x Efficiency of water use x Harvest Index.

Increasing the amount of water transpired (moving along the slope – arrow A in Fig. 12) can increase yield, as can increasing the carbon gain per unit of water transpired (increasing the water use efficiency - WUE) (arrow B in Fig. 12). An increasingly important agricultural objective in many water-limited areas is to decrease the amount of water used (by decreasing water supplied to the crop) with minimal yield penalty (arrow C in Fig. 12). Can manipulating plant hormone status or response deliver any of these outcomes?

Manipulating leaf ABA accumulation has been the target of several crop breeding programs (34). ABA-induced stomatal closure in flag leaves could result in stomatal limitation of photosynthesis and decrease assimilate supply to the developing grains, since yield of cereal crops is often proportional to flag leaf photosynthesis. Although some field studies indicated that decreased ABA accumulation was associated with higher yields, others showed that enhancing ABA accumulation was beneficial to crop yield (34). The discussion above shows that there are many possible explanations for both types of response and that temperature, evaporative demand, nutrient supply and many other variables might determine the impact of a genetic modification in hormone accumulation. It seems likely that to generate a predictable impact on

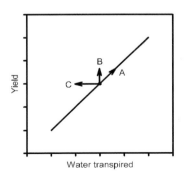

Figure 12. Relationship between crop yield and transpiration (from 29). For description of arrows A, B, C see text.

plant water balance and yield, any change in hormone production and/or distribution will have to be specifically targeted to a specific plant part, stage of development or even a specific climatic regime.

One problem with much breeding for variation in ABA accumulation is that variation is often defined by a standard laboratory test for ABA accumulation following a given degree of leaf dehydration (33). Lines may behave differently in the field where leaf ABA accumulation will depend on leaf water status and the contribution of xylem-supplied ABA from the roots. For this reason, some breeding programs have used field ABA accumulation as a selection criterion. In some cases, lines with low leaf ABA accumulation have a higher leaf water status (presumably due in part to an increased ability to extract soil water) which is correlated with an increased force required to uproot the plant (38). In this way, leaf ABA accumulation indirectly (and fortuitously) acts as a selection criterion for a potentially desirable root system property that might increase yield in dry soil. Although ethylene synthesis can limit elongation of roots in drying soil (43), roots of ethylene insensitive or deficient lines are less able to penetrate compacted soil (5, 17) and therefore manipulation of ethylene synthesis may not always confer agronomic advantages in the field.

Desirable root system properties are likely to depend on the environment, specifically soil water availability throughout the cropping period. Increased partitioning of resources to roots may restrict shoot growth and yield and may be of limited benefit to an individual growing in a monoculture in environments where water supply is assured. In water-limited environments, root systems that are able to explore deeper horizons of the soil profile (as a means of sustaining crop water use throughout the entire season) might be useful. It is argued (30) that in such climates where the soil is wet at the beginning of the season and dries thereafter (Mediterranean environments), a desirable plant profile (crop ideotype) would:

a) cover the ground rapidly to avoid soil water loss when water is available. This saves water for later flowering and fruiting, which is vitally important to yield.
b) have roots that will continue to slowly explore the soil and have a high resistance to water flow so that soil water is not used too fast.
c) show some turgor maintenance but also limitation of leaf growth later so the plant has resources for reproductive development.
d) show high WUE. However, since high resource use efficiency is usually linked with slow growth, this link must be broken early in the season.

It should be clear that while one of these characters can be changed via manipulation of plant hormone status or response, achieving all desirable traits would require manipulation of many specific genes at specific times during the crop life cycle. Given physiological uncertainties in the value of a crop ideotype approach, and the biotechnological complexity required to

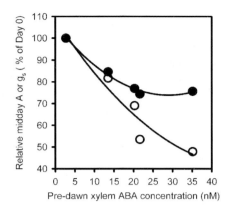

Figure 13. Relative responses of stomatal conductance (○) and CO_2 assimilation (●) to changes in pre-dawn [X-ABA] in *Trifolium subterraneum*. Data re-drawn from (45).

achieve it, perhaps agronomic manipulation provides better opportunities to apply our knowledge of hormone responses to crop improvement.

Decreasing the amount of irrigation water applied (deficit irrigation) is a legitimate irrigation strategy to decrease plant water use, but can restrict yield or at least result in some lesion in plant growth. A recent development of deficit irrigation is partial root drying (PRD). The technique works in drip or furrow irrigated crops where each side of the row can be watered independently. When the crop is irrigated, only one half of the plant root system receives water and the other is allowed to dry the soil. Water use is commonly halved under PRD irrigation. The wet side of the root system is able to supply enough water to the shoot to prevent harmful shoot water deficits. In response to drying soil, the dry part of the plant root system produces chemical signals, which are transmitted to the shoots to close the stomata (decreasing water loss). Given the relationship between yield and water use (Fig. 12), the projected decrease in water use (caused by stomatal closure) by PRD crops should decrease crop yield. However, stomatal conductance is more responsive to increased xylem ABA concentration than are changes in photosynthesis (Fig. 13), allowing considerable decreases in stomatal conductance to occur with minimal impacts on CO_2 assimilation. Also, plants grown with PRD may allocate proportionally more dry matter to the harvested portion of the crop (increasing the harvest index). Much remains to be learnt about whether (and how) drought-induced changes in plant hormone status alter biomass allocation within the plant. However, early indications from PRD trials in different parts of the world (7) suggest that agronomic exploitation of plant long-distance signalling may allow more efficient water use in agriculture.

Ackowledgements

We thank BBSRC for a ROPA award for continued support of research on soil drying.

References

1. Allan AC, Fricker MD, Ward JL, Beale MH, Trewavas AJ (1994) Two transduction pathways mediate rapid effects of abscisic acid in *Commelina* guard cells. Plant Cell 6: 1319-1328

2. Bacon MA, Wilkinson S, Davies WJ (1998) pH-regulated leaf cell expansion in droughted plants is abscisic acid dependent. Plant Physiol 118: 1507-1515
3. Bauly JM, Sealy IM, Macdonald H, Brearley J, Droge S, Hillmer S, Robinson DG, Venis MA, Blatt MR, Lazarus CM, Napier RM (2000) Overexpression of auxin-binding protein enhances the sensitivity of guard cells to auxin. Plant Physiol 124: 1229-1238
4. Borel C, Frey A, Marion-Poll A, Simmoneau T, Tardieu F (2001) Does engineering abscisic acid biosynthesis in *Nicotiana plumbaginifolia* modify stomatal response to drought ? Plant, Cell Env 24: 477-489
5. Clark DG, Gubrium EK, Barrett JE, Nell TA, Klee HJ (1999) Root formation in ethylene-insensitive plants. Plant Physiol 121: 53-59
6. Correia MJ, Pereira JS, Chaves MM, Rodrigues ML, Pacheco CA (1995) ABA xylem concentrations determine maximum daily leaf conductance of field-grown *Vitis vinifera* L. plants. Plant, Cell Env 18: 511-521
7. Davies WJ, Wilkinson S, Loveys B (2002) Stomatal control by chemical signalling and the exploitation of this mechanism to increase water use efficiency in agriculture. New Phytol 153: 449-460
8. Davies WJ, Zhang J (1991) Root signals and the regulation of growth and development of plants in drying soil. Ann Rev Plant Physiol Plant Mol Biol 42: 55-76
9. De Silva DLR, Hetherington AM, Mansfield TA (1985) Synergism between calcium ions and abscisic acid in preventing stomatal opening. New Phytol 100: 473-482
10. Dodd IC (2003) Hormonal interactions and stomatal responses. J Plant Growth Reg 22, 32-46
11. Freundl E, Steudle E, Hartung W (1998) Water uptake by roots of maize and sunflower affects the radial transport of abscisic acid and its concentration in the xylem. Planta 207: 8-19
12. Gehring CA, McConchie RM, Venis MA, Parish RW (1998) Auxin-binding-protein antibodies and peptides influence stomatal opening and alter cytoplasmic pH. Planta 205: 581-586
13. Hansen H, Dörffling K (2003) Root-derived trans-zeatin riboside and abscisic acid in drought-stressed and rewatered sunflower plants: interaction in the control of leaf diffusive resistance? Funct Plant Biol 30: 365-375
14. Hartung W, Slovik S (1991) Physicochemical properties of plant growth regulators and plant tissues determine their distribution and redistribution - stomatal regulation by abscisic acid in leaves. New Phytol 119: 361-382
15. Holbrook NM, Shashidhar VR, James RA, Munns R (2002) Stomatal control in tomato with ABA-deficient roots: response of grafted plants to soil drying. J Exp Bot 53: 1503-1514
16. Honour SJ, Webb AAR, Mansfield TA (1995) The responses of stomata to abscisic acid and temperature are interrelated. Proc Royal Soc London B 259: 301-306
17. Hussain A, Black CR, Taylor IB, Roberts JR (1999) Soil compaction. A role for ethylene in regulating leaf expansion and shoot growth in tomato? Plant Physiol 121:1227-1237
18. Incoll LD, Jewer PC (1987) Cytokins and stomata *In* E Zeiger, GD Farquhar , IR Cowan, *eds*, Stomatal Function, Stanford University Press, Stanford, pp 281-292
19. Jackson MB (1993) Are plant hormones involved in root to shoot communication ? Adv Bot Res 19: 103-187.
20. Jia WS, Zhang JH (1997) Comparison of exportation and metabolism of xylem-delivered ABA in maize leaves at different water status and xylem sap pH. Plant Growth Reg 21, 43-49
21. Loveys BR (1984) Diurnal changes in water relations and abscisic acid in field-grown *Vitis vinifera* cultivars. III. The influence of xylem-derived abscisic acid on leaf gas exchange. New Phytol 98: 563-573
22. Mansfield TA, Davies WJ (1981) Stomata and stomatal mechanisms. *In* LG Paleg, D Aspinall, *eds*, The Physiology and Biochemistry of Drought Resistance, Academic Press,

Sydney, pp 315-346
23. McAinsh MR, Brownlee C, Hetherington AM (1990) Abscisic acid-induced elevation of guard cell cytosolic Ca^{2+} precedes stomatal closure. Nature 343: 186-188.
24. Merritt F, Kemper A, Tallman G (2001) Inhibitors of ethylene synthesis inhibit auxin-induced stomatal opening in epidermis detached from leaves of *Vicia faba* L. Plant Cell Physiol 42: 223-230.
25. Morgan PW, He C-J, De Greef JA, De Proft MP (1990) Does water deficit stress promote ethylene synthesis by intact plants ? Plant Physiol 94: 1616-1624.
26. Munns R, Cramer GR (1996) Is coordination of leaf and root growth mediated by abscisic acid? Opinion. Plant Soil 185: 33-49
27. Munns R, King RW (1988) Abscisic acid is not the only stomatal inhibitor in the transpiration stream of wheat plants. Plant Physiol 88: 703-708
28. Netting AG (2000) pH, abscisic acid and the integration of metabolism in plants under stressed and non-stressed conditions: cellular responses to stress and their implication for plant water relations. J Exp Bot 51: 147-158
29. Passioura JB (1977) Grain yield, harvest index and water use of wheat. J Aust Inst Agric Sci 43: 117-121
30. Passioura JB, Condon AG, Richards RA (1993) Water deficits, the development of leaf area and crop productivity *In* JAC Smith & HG Griffiths, *eds*, Water Deficits : Plant responses from cell to community, Bios Scientific Publishers, Oxford, pp 253-264
31. Poorter H, Nagel O (2000) The role of biomass allocation in the growth response to different levels of light, CO_2, nutrients and water : a quantitative review. Aust J Plant Physiol 27: 595-607
32. Pospisilova J, Synkova H, Machackova I, Catsky J (1998) Photosynthesis in different types of transgenic tobacco plants with elevated cytokinin content. Biol Plant 40: 81-89
33. Quarrie SA (1981) Genetic variability and heritability of drought-induced abscisic acid accumulation in spring wheat. Plant, Cell Env 4: 147-151
34. Quarrie SA (1991) Implications of genetic differences in ABA accumulation for crop production. *In* WJ Davies, HG Jones, *eds*, Abscisic Acid : Physiology and Biochemistry, Bios Scientific Publishers, Oxford, pp 137-152
35. Radin JW, Parker LL, Guinn G (1982) Water relations of cotton plants under nitrogen deficiency. V. Environmental control of abscisic acid accumulation and stomatal sensitivity to abscisic acid. Plant Physiol 70: 1066-1070
36. Rodriguez JL, Davies WJ (1982) The effects of temperature and ABA on stomata of *Zea mays* L. J Exp Bot 33: 977-987
37. Saab IN, Sharp RE, Pritchard J, Voetberg GS (1990) Increased endogenous abscisic acid maintains primary root growth and inhibits shoot growth of maize seedlings at low water potentials. Plant Physiol 93: 1329-1336
38. Sanguineti MC, Tuberosa R, Landi P, Salvi DS, Maccaferri M, Casarini E, Conti S (1999) QTL analysis of drought related traits and grain yield in relation to genetic variation for leaf abscisic acid concentration in field-grown maize. J Exp Bot 50: 1289-1297
39. Sauter A, Dietz KJ, Hartung W (2002) A possible stress physiological role of abscisic acid conjugates in root-to-shoot signalling. Plant, Cell Env 25: 223-228
40. Schurr U, Schulze E-D (1995) The concentration of xylem sap constituents in root exudate, and in sap from intact, transpiring castor bean plants (*Ricinus communis* L.). Plant, Cell Env 18, 409-420
41. Sharp RE, Davies WJ (1979) Solute regulation and growth by roots and shoots of water-stressed maize plants. Planta 147: 43-49
42. Sharp RE, Wu Y, Voetberg GS, Saab IN, LeNoble ME (1994) Confirmation that abscisic acid accumulation is required for maize primary root elongation at low water potentials. J Exp Bot 45: 1743-1751
43. Sharp RE, LeNoble ME (2002) ABA, ethylene and the control of shoot and root growth

under water stress. J Exp Bot 53: 33-37
44. Snaith PJ, Mansfield TA (1982a) Control of the CO_2 responses of stomata by indol-3ylacetic acid and abscisic acid. J Exp Bot 33: 360-365
45. Socias X, Correia MJ, Chaves M, Medrano H (1997) The role of abscisic acid and water relations in drought responses of subterranean clover. J Exp Bot 48: 1281-1288.
46. Tardieu F, Davies WJ (1992) Stomatal response to abscisic acid is a function of current plant water status. Plant Physiol 98: 540-545
47. Trejo CL, Davies WJ, Ruiz LMP (1993) Sensitivity of stomata to abscisic acid. An effect of the mesophyll. Plant Physiol 102: 497-502
48. Van Volkenburgh E, Davies WJ (1983) Inhibition of light-stimulated leaf expansion by abscisic acid. J Exp Bot 34: 835-845.
49. Wilkinson S, Clephan AL, Davies WJ (2001) Rapid low temperature-induced stomatal closure occurs in cold-tolerant *Commelina communis* leaves but not in cold-sensitive tobacco leaves, via a mechanism that involves apoplastic calcium but not abscisic acid. Plant Physiol 126: 1566-1578
50. Wilkinson S, Corlett JE, Oger L, Davies WJ (1998) Effects of xylem pH on transpiration from wild-type and *flacca* tomato leaves : a vital role for abscisic acid in preventing excessive water loss even from well-watered plants. Plant Physiol 117: 703-709
51. Wilkinson S, Davies WJ (1997) Xylem sap pH increase: A drought signal received at the apoplastic face of the guard cell which involves the suppression of saturable ABA uptake by the epidermal symplast. Plant Physiol 113: 559-573
52. Wolf O, Jeschke WD, Hartung W (1990) Long distance transport of abscisic acid in NaCl-treated intact plants of *Lupinus albus*. J Exp Bot 41: 593-600
53. Wright STC, Hiron RWP (1969) (+) abscisic acid, the growth inhibitor induced in detached wheat leaves by a period of wilting. Nature 224: 719-720
54. Zhang J, Davies WJ (1989) Sequential response of whole plant water relations to prolonged soil drying and the involvement of xylem sap ABA in the regulation of stomatal behaviour of sunflower plants. New Phytol 113: 167-174
55. Zhang J, Davies WJ (1990) Does ABA in the xylem control the rate of leaf growth in soil-dried maize and sunflower plants ? J Exp Bot 41: 1125-1132
56. Zhang JH, Zhang XP, Liang JS (1995) Exudation rate and hydraulic conductivity of maize roots are enhanced by soil drying and abscisic acid treatment. New Phytol 131: 329-336
57. Zhang SQ, Outlaw WH, Aghoram K (2001) Relationship between changes in the guard cell abscisic acid content and other stress related physiological parameters in intact plants. J Exp Bot 52: 301-308

E4. The Role of Hormones during Seed Development and Germination

Ruth R. Finkelstein

Department of Molecular, Cellular, and Developmental Biology, University of California at Santa Barbara, Santa Barbara, CA 93106 USA.
E-mail: finkelst@lifesci/ucsb.edu

INTRODUCTION

Seed production[1] is an extraordinary adaptation to a terrestrial environment that permits plants to reproduce under dry conditions and broadly disperse their progeny, which can then survive in an arrested state until environmental conditions favor growth of the next generation. Although there are many anecdotal reports of extreme longevity (6), the current record for documented viability is over 1000 years for an Indian Lotus seed, collected from an ancient lake bed in China (51). To accomplish this remarkable feat, seeds contain an embryo and a supply of nutrient reserves, packaged as a dry desiccation-tolerant unit. The features that make seeds an effective means of reproduction (high nutrient content and extended viability during developmental arrest) also make them a convenient food supply, and led to the development of civilizations in cultures that made use of this dependable source of crops to supply themselves with food. Consequently, features such as nutrient content, yield, and germination control are of major agronomic importance and have been the focus of much research.

Seed development contrasts from vegetative growth in that embryogenesis is relatively insulated from environmental effects such that the final product is fairly constant even though the growth rate may vary. This constancy reflects the activity of regulatory mechanisms that lead to an ordered progression through three phases of embryo development: morphogenesis, cell enlargement/reserve accumulation, and desiccation/developmental arrest (Fig. 1). During the morphogenesis phase, the zygote undergoes extensive cell division and differentiation of tissue types to build a miniature plant with a root-shoot axis and the three major tissues of a mature plant: vascular, ground, and dermal. This phase of embryonic development is accompanied by development of the endosperm, which serves as a source of nutrient reserves for the embryo that may continue to grow and persist in the mature seed or be resorbed during the latter phases of seed development,

---

[1]Abbreviations: AtIP5PII, Ins(1,4,5)P35 phosphatase; Em, Early methionine-labeled; LEA, Late embryogenesis abundant; PLC, Phospholipase C.

Hormones in seed development and germination

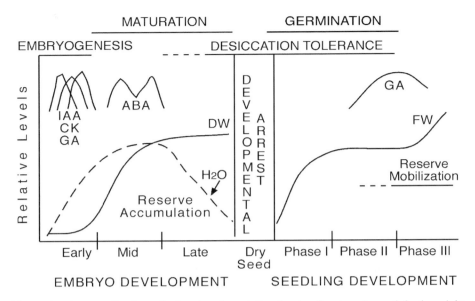

Figure 1. A generalized graph showing the relative levels of water, dry and fresh weights (DW and FW), and hormones during the stages of embryo and seedling development. Time periods associated with each phase of development vary with species and are not included. Developmental arrest may be either quiescence or dormancy, depending on the species. (Adapted from 47)

depending on the species. During the final two phases of seed development, also known as maturation, the embryo and seed prepare for survival when separated from the maternal plant. Toward this end, seeds accumulate reserves to provide nutrients for survival during germination. Depending on the species, storage reserves may accumulate exclusively in the embryo itself, or in both the embryo and endosperm. The developing embryos undergo a transition from growth by cell division to cell enlargement as they begin to accumulate storage reserves. Finally, most seeds prepare for and undergo water loss to achieve a developmentally-arrested, desiccation-tolerant state that permits survival with as little as ~10% free water. The degree of arrest varies among species, ranging from true dormancy (literally, "sleeping") that requires specific environmental cues such as light or chilling to induce germination, to "quiescence" which requires only sufficient water for resumed growth.

Germination involves a reactivation of metabolic activity, a re-differentiation of embryonic tissues to mobilize the reserves they stored, and a shift to growth by meristematic activity. Because completion of germination results in loss of desiccation tolerance, this step constitutes a commitment to growth of the next generation. Consequently, the transition from seed to seedling is highly sensitive to many environmental factors, including light, temperature, and availability of water. Response to many of these environmental signals is mediated by, or interacts with, signaling via one or more hormones.

Regulation of the progression from zygote to seedling has been studied by characterizing the signals that are normally present during these phases, identifying culture conditions that can substitute for the seed environment to maintain "normal" embryonic development, and analyzing mutants with defects in various aspects of seed development or germination (e.g. arrested growth, pattern formation, non-dormant/viviparous, or aberrant reserve accumulation). In addition, early phases of embryogenesis have been studied in somatic embryos, produced by exposure of differentiated tissues to appropriate inducing signals (15, 25). This approach takes advantage of the totipotency of plant cells to produce vast numbers of roughly synchronous genetically identical embryos, free of the many layers of maternal tissue encasing zygotic embryos. Such studies have permitted biochemical characterization of early embryos, and are also relevant to synthetic seed technology (2). Collectively, these studies of both zygotic and somatic embryos have provided evidence for both hormonal and non-hormonal developmental controls, some of which act synergistically. The roles of specific regulators in each of these phases will be addressed in the following sections.

EARLY EMBRYOGENESIS

Two major events of the initial phase of embryogenesis are establishing a body plan for, and ensuring viability of, the developing embryo. Following the double fertilization event that normally initiates seed development in angiosperms, the zygote undergoes an asymmetric division that reflects an already established polarity (25, 53). This polarity is reflected in the asymmetric distribution of a specific cell wall component, an arabinogalactan protein (AGP) whose presence may provide positional information that contributes to cell fate determination. The smaller terminal cell (furthest from the micropyle) will produce the majority of the embryo and the basal cell will give rise to the basal section of the root apical meristem and the suspensor, a small file of cells that attaches the embryo to maternal tissue and serves a nutritive role for the early embryo (Fig. 2). The vastly different fates of embryonic vs. suspensor cells appear to require specific inhibition of embryonic development in the suspensor because either ablation of the embryo proper or mutations in the _SUSPENSOR (SUS)_, _TWIN_, or _RASPBERRY (RSP)_ loci result in a conversion of suspensors to embryonic structures. Over the remaining rounds of cell division in the embryo proper, both radial and apical-basal axes are established to control differentiation of embryonic tissue types and shoot vs. root structures, respectively (25, 53).

Concurrent with the early embryo divisions, the endosperm starts to develop (4). In most species, these divisions go through a syncitial phase to form a liquid endosperm (e.g. coconut milk), then cellularize. The endosperm is often a site of DNA amplification (up to 690C in some maize strains, a 230-fold amplification) and takes over from the suspensor to serve as a source of nutrients for the developing embryo. This tissue may persist and become a site of massive reserve (e.g. starch, lipids, proteins)

Hormones in seed development and germination

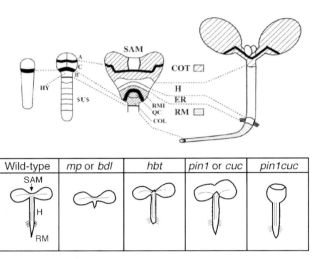

Figure 2. Pattern formation during Arabidopsis embryogenesis. Top, left to right: 2-cell, octant, heart stage embryos and seedling. Thick lines: divisions separating apical (A), central (C) and basal (B) embryo regions. HY, hypophysis. The suspensor is a cell file beneath hypophysis. Cell groups that give rise to seedling structures are indicated in the heart stage embryo. SAM, shoot apical meristem. COT, cotyledons; H, hypocotyl; ER, embryonic root; RM, root meristem; RMI, root meristem initials; QC, quiescent centre; COL, columella root cap. From Willemsen et al. Development 125, 521-531 (1998). Bottom: examples of seedling morphology for selected pattern formation mutants disrupting auxin response (*mp, bdl, hbt*) or transport (*pin1*). Central and basal regions are deleted in the *mp* and *bdl* mutants, root growth is defective in the *hbt* mutants, and cotyledon separation is impaired or blocked in the *pin1* and *cuc* mono and digenic mutants, respectively.

accumulation, as observed in cereals, or be resorbed during the later stages of embryogeny as is common in many dicots.

Measurements of endogenous hormone concentrations have shown that cytokinins, GA and IAA are all high transiently during this period (6, 47) (Fig. 1). Studies manipulating hormone levels or response genetically or in tissue culture suggest that the roles of cytokinins and GAs are primarily nutritive. In contrast, auxin plays a major role in establishing the embryonic body plan via effects on apical-basal polarity/pattern formation and vascular development (Fig. 2)(53, 56).

Growth promotion by GA and cytokinins

Some of the earliest cells to form are the suspensor cells, which remain large and vacuolated (50). In some species, the cell immediately adjacent to the proembryo has highly invaginated walls to increase its surface area for transfer of materials. The transfer-cell-like properties of the suspensor are thought to be important for nutrition of early embryos. Culture studies with *Phaseolus* have shown that exogenous GA can substitute for a detached suspensor in promoting embryonic growth, suggesting that the suspensor may normally provide GAs as well as nutrients to the developing embryo. Cytokinins have also been implicated in promoting suspensor function, but may be even more significant in promoting endosperm growth and grain filling via promotion of cell division (6). Cytokinins may also participate in pattern formation since *WOODEN LEG (WOL)*, a gene required for normal radial patterning of vascular tissue, encodes a cytokinin receptor (57).

Pattern formation

Many lines of evidence support a role for auxin in pattern formation (53, 56). Culture studies with auxin transport inhibitors block the transition from globular to heart stage and prevent cotyledon separation. Characterization of localization of auxin efflux carrier components encoded by the *PIN-FORMED (PIN)* loci shows an evolving pattern for four family members during early embryogenesis (Fig.3A). Immediately following the initial asymmetric division of the zygote, apical localization of the PIN7 protein in the basal cell results in transport of auxin into the apical cell, which gives rise to most of the embryo. In contrast, PIN1 protein and the resulting transport and distribution of auxin, are fairly uniform through the 16-cell stage of embryogenesis. However, beginning around mid-globular (32-cell) stage, the PIN1 protein becomes progressively restricted to the basal ends of procambial cells, resulting in a reversal of auxin flow that correlates with the establishment of distinct apical, central and basal domains of the embryo. This pattern of PIN1 localization would contribute to shoot-to-root polar auxin transport in the developing embryo that could be reinforced by the basal pole-localized expression of other members of the PIN family of auxin

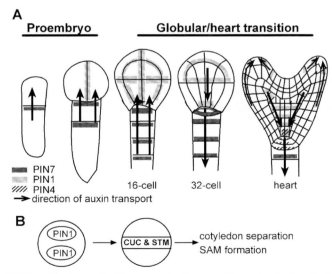

Figure 3. PIN localization and effects on apical pattern formation in developing embryos. (A) Schematic of PIN protein localization and resulting auxin transport in early embryogenesis. (Based on Friml et al. Nature 426, 147-153 (2003) (B) Schematic of cross-section through apical region of heart-stage embryos. PIN1 localization in forming cotyledons, and presumably resulting auxin accumulation, restricts CUC and consequently STM expression to region at boundaries of cotyledons. (Based on 1)

efflux carriers, e.g. PIN4. Localized polar auxin flux is believed to promote vascular development via a positive feedback mechanism (the "canalization" hypothesis) such that the observed procambial localization would promote vascular differentiation in these cells (Chapter. E2). In addition, the polar transport of auxin toward the presumptive root creates a gradient with highest auxin concentration near the tip, as observed in seedling roots. Treatments that disrupt auxin transport and create two auxin concentration maxima produce corresponding duplications of the root apical meristem, suggesting that the polar auxin flux may also contribute to establishment of the root apical meristem in developing embryos (56). Some PIN1 localization is also observed at the apical ends of cotyledon epidermal cells, where it appears to contribute to cotyledon separation by controlling the spatial expression of the *CUP-SHAPED COTYLEDON (CUC)* loci, which are required for cotyledon separation and, via effects on *SHOOT MERISTEMLESS (STM)* expression, meristem formation (1) (Fig. 3B).

Additional evidence for the importance of auxin in establishing polarity comes from the discovery that several apical-basal polarity mutants have defects in auxin signaling factors (56) (Fig. 2). For example, the *MONOPTEROS (MP)* gene encodes an auxin response factor (ARF), a DNA-binding protein that promotes transcription of auxin-inducible genes. Consistent with the *mp* phenotype of reduced vascular, hypocotyl and radicle development, and the frequent failure to develop a second cotyledon (Fig. 2), MP shows a similar pattern of tissue distribution to that of PIN1 in that it becomes progressively restricted to procambial and vascular tissue. The *bodenlos (bdl)* mutant resembles *mp* mutants, apparently because *BDL* encodes an Aux/IAA-class repressor that binds to MP, thereby inhibiting its activation of auxin-inducible genes. The *bdl* mutant produces an abnormally stable repressor, resulting in auxin-resistance. A similar phenotype results from mutations in *AUXIN RESISTANT(AXR)6*, which encodes the CULLIN1 subunit of the SCF (for SKP1, Cullin/CDC53, F-box protein) class of ubiquitin protein ligase complexes (31) that target specific proteins for proteasomal degradation. The net result of all these mutations is a failure to derepress auxin-inducible genes, thereby disrupting pattern formation (Fig. 4). Furthermore, null mutations in the previously-presumed auxin receptor Auxin Binding Protein 1 result in embryo lethality at the globular stage (13). These results indicate that auxin signaling is essential for viability as well as pattern formation in early embryos. However, studies of the *hobbit* mutant illustrate that one must interpret possible pattern formation defects cautiously. Although *HOBBIT (HBT)* was initially described as required for correct orientation of embryonic cell divisions in the embryonic root primordium, recent studies indicate that marker gene expression and patterning occurs normally in *hbt* embryos (7). Instead, cell cycle progression and differentiation of meristematic cells during postembryonic growth is disrupted, possibly due to decreased auxin response. Molecular genetic studies have also provided evidence for regulation of embryonic pattern formation by novel signaling molecules. Two major classes of mutants with conflicting effects on temporal separation of embryonic vs.

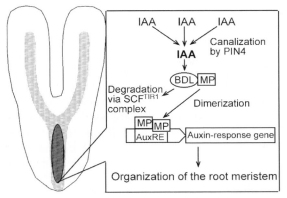

Figure 4. Roles of localized auxin transport, Aux/IAA and ARF proteins, and proteasomal degradation in auxin responses affecting pattern formation in embryonic roots. Auxin is canalized via localized expression of PIN family efflux carriers. When a critical auxin concentration is reached, the repressor BDL is degraded via interaction with the SCF^{TIR1} complex, permitting formation of MP homodimers which can bind to auxin-responsive elements (AuxRE) in the promoters of genes required for root meristem organization. Relative localizations of MP (light gray) and BDL (dark gray), and presumably additional factors, limit this activity to the root tip. (From 57)

vegetative differentiation have been identified: the *leafy cotyledon (lec)* class mutants and those with extra cotyledons (28, 29). Both of these classes of mutants have been described as heterochronic, indicating that they display changes in developmental timing. The *LEC* class loci (*LEC1, LEC2,* and *FUSCA(FUS)3*) are potent promoters of embryonic identity, such that ectopic overexpression of either *LEC* gene results in production of cotyledon-like leaves that produce somatic embryos on their surfaces. All of the *LEC* class loci identified to date encode transcriptional regulators, and at least two of these (*LEC1* and *FUS3*) interact with mediators of ABA signaling to regulate embryo maturation (19). In contrast, the only extra cotyledon class locus molecularly identified so far is *ALTERED MERISTEM PROGRAM (AMP)1*, which encodes a protein with similarity to glutamate carboxypeptidases, leading to the suggestion that it regulates the level of a small as-yet-unidentified signaling peptide (30).

EMBRYO MATURATION

The last two phases of embryo development are sometimes collectively referred to as "maturation" because during this period seeds acquire the ability to survive desiccation and initiate growth of the next generation, independent of the maternal plant. Seed maturation begins when developing embryos cease growth by cell division; this coincides with an increase in seed ABA content, consistent with the fact that ABA induces expression of a cyclin-dependent kinase inhibitor (ICK1) that could lead to cell cycle arrest at the G1/S transition (19). The middle stage of seed development is a period of massive reserve accumulation and cell enlargement as cells fill with protein and lipid bodies (25, 27, 47) (Fig. 1). During this phase, water content (% FW) declines steadily although the total amount of water per embryo is still increasing. The most abundant hormone at this stage is ABA, which reaches peak levels during the period of maximal seed weight gain. In contrast, during the final phase of seed development, embryos become desiccation tolerant, lose water and ABA, and become relatively metabolically inactive.

Although mature seeds are poised for rapid germination, they usually do not germinate until after recognition of signals that favor seedling growth. However, consistent with their progression toward autonomy, embryos removed from the seed from the middle phase onward are capable of "precocious germination" and survival on simple nutrient medium (Fig. 5). The fact that their survival and potential for normal seedling development improves with increasing age at the time of excision from the seed indicates that the reserve accumulation and other events of maturation are important for reproductive success. However, the fact that they can germinate precociously implies that something in the normal seed environment suppresses germination and that the final phase is not absolutely required. In fact, the mangrove (*Rhizopora mangle*) skips this phase as a normal part of its development. As a tropical aquatic plant, temperature and access to water are always favorable to growth, and mangrove seeds progress into seedlings while still attached to the maternal plant, a condition known as vivipary.

Potential regulators of embryo maturation have been identified by determining which conditions or chemicals are present in seeds at these stages and testing their functional significance by physiological and genetic assays. This approach has identified at least two regulatory signals likely to control embryo maturation: ABA and limited water availability (27, 47). In addition to being the most abundant hormone present at this stage, ABA suppresses germination of mature seeds and ABA-deficient mutants are often viviparous. The significance of water availability was suggested by the fact

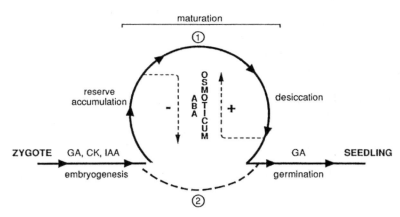

Figure 5. A generalized representation of embryo development in angiosperms, showing three major stages from zygote to seedling: embryogenesis, maturation and germination, and the major hormones regulating growth at these stages. The maturation "loop" (1) can be bypassed in culture by precocious germination of embryos (2), which occurs in the presence of reduced nitrogen (e.g. glutamine) and carbon (e.g. sucrose). In culture, embryos can reversibly enter or leave the maturation loop by application (+) or removal (-), respectively, of high osmoticum or ABA. Maturation is completed by desiccation, and normal germination will occur following imbibition. (From 47)

that water content drops steadily through seed maturation, whereas growth and germination require water uptake. A third "signal" is the developmental context of the embryonic state; this was more difficult to quantify and manipulate before the isolation of mutants defective in this type of regulation.

Physiological studies: Embryos in culture

Physiological studies have tested the significance of potential regulators by determining whether they can mimic the seed environment to maintain embryonic maturation in cultured embryos of different ages. Embryonic development has been defined as the absence of germination (i.e. root growth and production of enzymes required for reserve mobilization or photosynthesis) and the continued or enhanced synthesis of proteins characteristic of maturing embryos. All of these, and subsequent genetic, studies have relied on well-defined sets of molecular markers characteristic of specific stages of embryonic development and germination (25, 27, 47). Initially, these markers were associated with particular processes such as reserve accumulation (e.g. storage proteins, starch, and lipids) or mobilization (e.g. proteases, amylases, or glyoxylate cycle enzymes). Later cataloging efforts identified sets of coordinately expressed genes with expression profiles ranging from constitutive to those with peak expression in early, middle or late embryo development. As anticipated based on the timing of seed filling, the major transcripts in mid-late embryo development encode storage proteins and enzymes associated with other reserve accumulation. In addition, these studies identified a class of genes described as late embryogenesis abundant (LEA). Many of the LEA proteins are unusually hydrophilic, leading to the suggestion that they serve a protective function during desiccation by substituting for water in maintaining hydrophilic interactions needed to prevent denaturation of many cellular components (16). Consistent with this hypothesis, some LEAs or closely related proteins accumulate in vegetative tissues exposed to dehydrating stresses (drought, NaCl or cold) and in maturing pollen, another developmentally regulated desiccation tolerant cell type. Furthermore, expression of at least one LEA confers osmotolerance in yeast (54). Current cataloging efforts are using genomic or proteomic approaches to identify many of the transcripts or proteins accumulating at specific stages in seed development (21, 48).

Some of the major conclusions from the embryo culture studies are that precocious germination is not identical to normal germination, and that the importance of specific regulators changes during development. Although all tested species will germinate precociously, the pattern differs among species (47). Cotton, wheat, and castor bean embryos germinate rapidly and begin expressing germination-specific enzymes within a day after radicle emergence. *Phaseolus* embryos do not germinate immediately upon excision, but undergo a lag inversely proportional to their age at the time of excision, appearing to complete embryonic development before germinating. Soybeans also do not germinate immediately, but require removal of ABA by washing or slow drying to permit germination. Mid-maturation stage rapeseed embryos germinate rapidly, following a brief lag correlated with their ABA content at the time of excision. However, their morphology and physiology is intermediate between embryonic and seedling: new organs are produced from the shoot apical meristem, but these organs resemble cotyledons in both morphology and their continued storage reserve production. Surprisingly these structures can be chimeric, with adjacent sectors of leaf-like and cotyledon-like tissue whose individual cell fates may be determined by short-range signaling (18). In contrast, *lec* class mutants express genes characteristic of both embryonic and vegetative growth within single cells, indicating that these developmental programs are not mutually exclusive, even though they are normally separated temporally (27).

As early as the late 1960s, several researchers found that addition of physiologically relevant levels of ABA (0.1-10 µM) to cultures of immature cotton, corn, or wheat embryos was sufficient to prevent precocious germination (47). Subsequent studies showed that ABA applied in mid-development could induce accumulation of storage reserves characteristic of that stage and even precociously induce expression of some LEAs. Similarly, exposure to high osmoticum (sucrose or a non-metabolized sugar alcohol such as sorbitol) would also prevent germination and maintain embryonic gene expression at this stage. Attempts to determine whether the osmotic effect was mediated by increased endogenous ABA produced mixed results: soybean embryos showed increased ABA levels in response to high sucrose, but rapeseed embryos exposed to high sorbitol had no significant effect on ABA levels. However, whereas osmoticum was still an effective signal in maintaining embryonic growth and promoting maturation of desiccation stage embryos, these older embryos appeared to have lost sensitivity to applied ABA. In combination with the dramatic decrease in endogenous ABA levels at this stage, it seems unlikely that ABA is a major regulator of events in desiccating embryos. Collectively, these studies suggest that ABA is only one of several signals that can maintain embryonic development (27).

Genetic studies of ABA response and seed maturation

The importance of ABA in seed maturation has been demonstrated by analyses of mutants with defects in ABA synthesis or response (19, 27, 40, 47). Many such mutants of maize have been identified on the basis of vivipary, providing strong support for the idea that ABA is required for inhibition of precocious germination. Some of the viviparous mutants (*vp2, vp5, vp7, vp8* and *vp9*) are blocked at an early step in ABA biosynthesis such that they are both ABA and carotenoid deficient, and consequently seedling lethals due to photo-bleaching. Vivipary in these lines requires GA, normally synthesized early in development (19). In contrast, *vp1* mutants have normal ABA levels, but reduced ABA sensitivity in seeds, and display vivipary whether or not they are capable of producing GA. ABA deficient mutants of Arabidopsis were initially identified as non-dormant revertants of GA-deficient mutants that no longer required exogenous GA to promote germination, thereby demonstrating that ABA and GA antagonistically control entry into and escape from dormancy, respectively (19). However, Arabidopsis differs from maize in that ABA deficiency or resistance result only in non-dormancy, not vivipary. Furthermore, null mutants of <u>ABA-insensitive (ABI)3</u>, the Arabidopsis ortholog of the maize *Vp1*, produce green, desiccation-intolerant, non-dormant seeds with reduced levels of many transcripts normally expressed in mid- and late-embryogenesis, but are not normally viviparous. Similarly, phenocopying ABA deficiency by using a seed-specific promoter to express an anti-ABA antibody in transgenic tobacco plants, thereby sequestering any available ABA in the seeds, also results in production of green, desiccation-intolerant seeds with reduced storage reserves.

Several additional ABA response loci of Arabidopsis affect either dormancy (*ABI1* and *ABI2*) or late embryonic gene expression (*ABI4* and *ABI5*) suggesting that these events are controlled by branching pathways (19). Although dozens more ABA response loci have been identified genetically by screens for disruption of ABA response during other phases of the life cycle, relatively few have been specifically tested for effects on seed development (19).

The lack of vivipary in ABA-deficient or -resistant lines of some species might reflect redundancy in ABA biosynthetic genes (Chapter B5) or the presence of additional or redundant control mechanisms. Evidence for the latter comes from the observation that combination of ABA deficiency with even leaky *abi3* mutations leads to a high frequency of vivipary. Growth conditions also appear to play a role: cultivars of some cereals are prone to "preharvest sprouting" under moist conditions and, in some species, this is correlated with altered expression of *Vp1* orthologs (19, 22).

At the opposite end of the germination control spectrum from vivipary, dormancy is an extreme version of the developmental arrest occurring at the end of seed development that delays germination by requiring an external signal or passage of time to indicate that conditions are likely to be favorable for growth. Although all dormancy results in non-germination, the mechanism of dormancy induction and maintenance varies, and is often defined in terms of the treatments required to break dormancy, e.g. disruption of restraint by the seed coat or a change in the physiological state of the embryo.

As discussed above, many ABA-deficient or -resistant mutants are non-dormant and/or viviparous, and increased dormancy is observed in several Arabidopsis ABA-hypersensitive mutants (e.g. *enhanced response to ABA(era)1* and *ABA hypersensitive(abh)1*), suggesting that ABA is an important endogenous inhibitor of germination. During seed maturation in many species, there are two peaks of ABA accumulation (19) (Fig. 1). Elegant genetic studies in Arabidopsis demonstrated that one is of maternal and one of embryonic origin. The first peak is maternally derived, immediately precedes maturation phase, and is important, in conjunction with the *FUS3* and *LEC* genes, for preventing vivipary at the end of the cell division phase of embryogenesis (Fig. 6). Although this ABA peak is

Hormones in seed development and germination

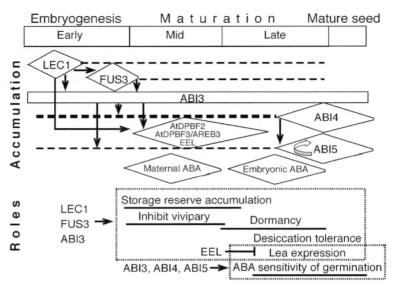

Figure 6. Accumulation and roles of major regulators of embryo maturation. Width of lines/diamonds reflects expression levels, dotted lines indicate weak expression, and position reflects timing. Regulatory roles affecting gene expression and developmental events are based primarily on analyses of single mutants; many additional roles are revealed by analyses of digenic mutants. Arrows and bars indicate documented cross- or auto-regulation, with bars representing inhibitory effects. Dotted Boxes enclose subsets of events regulated by the indicated factors.

reduced three-fold in *fus3* mutants, only the double mutants combining *fus3* with ABA deficiency are highly viviparous. In addition, the *leafy cotyledon* class mutants, which are not themselves defective in ABA response or synthesis, interact with both ABA deficiency and defects in several ABA response loci to produce highly pigmented viviparous seeds that fail to accumulate storage reserves or attain desiccation tolerance. Despite the essentially normal ABA response of the *lec* class mutants, the *abi fus3* and *abi lec1* double mutants are 10- to greater than 30-fold less sensitive to ABA than their monogenic *abi* parents, suggesting that the FUS3 and LEC1 gene products potentiate *ABI*-dependent ABA sensitivity. It is noteworthy that the degree of vivipary and ABA-insensitivity of germination in the *abi fus3* and *abi lec1* mutants are not tightly correlated, supporting the view that suppression of vivipary is not solely dependent on ABA (19).

In wild-type Arabidopsis seeds, embryonic ABA accumulates later and to a lower level than the maternal ABA; however, it is essential for induction of dormancy, which is maintained despite a substantial (~6 fold) decrease in ABA by seed maturity (19). The ABA content of a wild-type mature dry seed is only slightly higher than the peak ABA level in an ABA-deficient mutant, suggesting that dormancy maintenance in mature seeds relies on signals other than residual endogenous ABA. Consistent with this, *de novo* ABA synthesis during imbibition appears required for dormancy maintenance in several species. However, several reduced dormancy mutants (*aberrant testa shape (ats)*, *reduced dormancy(rdo)1*, *rdo2*, *dag(Dof affecting germination)1*, *transparent testa glabrous(ttg)*) have been identified that have wild-type ABA levels and sensitivity to ABA (19, 35). In some of these (*ats* and *ttg*), the decreased dormancy is attributed to defects in the testa, i.e. seed coat. Mutations affecting signaling by other hormones also affect dormancy: constitutive ethylene triple response (*ctr1*) mutants reduce dormancy and enhance ABA resistance, whereas reduced sensitivity to ethylene (*ethylene insensitive(ein)2/era3*), GA (*sleepy, comatose (cts)*) or brassinosteroids (*brassinosteroid insensitive(bri)1, de-etiolated(det)2*) can lead to reduced germination and hyperdormancy. *DAG1* encodes one of a pair of highly homologous Dof (DNA-binding with one finger) transcription factors that have opposite effects on GA sensitivity of germination; in contrast to most of the hormone response mutants, these loci act maternally reflecting their expression in vascular tissue connecting the developing seeds to the maternal plant (26). Additional regulators of dormancy are being identified in Arabidopsis and several cereal species by using quantitative trait loci (QTL) mapping to analyze natural variation among ecotypes and cultivars (22, 35). Collectively, these studies show that although vivipary and dormancy affect the same developmental decision, i.e. whether or not to germinate, these are complex traits and the timing of the decision and the relevant regulatory factors controlling this decision differ both within and among species.

Regulation of seed gene expression by ABA

Molecular studies of seed gene expression began with identification of marker sets of genes expressed at specific stages in seed development, then progressed to identifying specific cis-acting sequences and corresponding trans-acting factors regulating these genes. In recent years, this approach has converged with genetic studies demonstrating that aspects of maturation such as reserve accumulation and *LEA* gene expression are largely controlled by the combinatorial action of transcription factors (Fig. 6). Extensive analyses of promoter sequences for storage protein and *LEA* genes identified elements essential for conferring hormone responsiveness, stage- and tissue-specificity (11, 46, 55). Consistent with this diversity of cis-acting sequences, some of the required trans-acting factors regulate ABA response, while others primarily regulate the transition from embryogenesis to germinative growth, e.g. the *LEAFY COTYLEDON* class genes, or simply confer seed-specific expression.

Studies of cis-acting sequences have made use of fusions combining potential regulatory sequences and a readily assayed reporter gene (e.g. ß-glucuronidase or luciferase) to monitor activity of the promoter in either transiently or stably transformed transgenic tissue. Deletions or specific mutations are used to define the specific sequences that are essential for "normal" regulation. Transient assays usually use either electroporation of protoplasts or biolistic bombardment of intact tissues to introduce the fusion genes into cells that can then be exposed to potential inducing treatments; expression of specific fusions are then compared to co-transformed constitutively expressed control constructs. Stable transformants generally make use of *Agrobacterium* mediated T-DNA transfer to construct transgenic lines that can be analyzed at various developmental stages in intact non-wounded plants, but the dependence on susceptibility to Agrobacterial infection largely limits this approach to dicots. These studies have identified at least 8 classes of regulatory sequences present in the promoters of seed-expressed genes: ABA response elements (ABREs) and the functionally equivalent coupling elements (CEs), RY/Sph elements, MYB-, MYC-, and Dof-class transcription factor recognition sequences, E-boxes (putative recognition sequences for basic helix-loop-helix factors), "A/T" and "WS" motifs, often in multimers or sufficient proximity to permit interactions between trans-acting factors binding discrete sequences in the same promoter (11, 26, 46, 55). The first trans-acting factor isolated on the basis of binding to an ABRE was EmBP1, a basic leucine zipper (bZIP) domain transcription factor that specifically bound ABREs present in the promoter of a *lea* gene, the wheat *Em* (early methionine-labeled) gene. However, *in vitro* binding interactions must be interpreted with caution because they are less specific than *in vivo* interactions, and plant transcription factor families are comprised of dozens to hundreds of related proteins with similar binding specificities (45). Consequently, any potential interactions must be verified by genetic analyses.

Some of the best characterized positive regulators of ABA signaling are the transcription factors encoded by *ABI3*, *ABI4* and *ABI5* (19). These proteins are members of the B3-, APETALA2- (AP2), and bZIP-domain families, respectively, and regulate overlapping subsets of seed-specific and/or ABA-inducible genes. Interestingly, two of the LEC class proteins (LEC2 and FUS3) are also B3-domain family transcription factors. B3-domain proteins bind RY/Sph sequences and bZIP proteins bind ABREs, consistent with direct regulation of many seed-expressed genes by these classes of protein. In fact, the ABREs of the *AtEm* genes are bound and trans-activated *in vivo* by ABI5 and probably other closely related bZIPs, but not by EmBP1 (19, 39). In contrast, the ABI4 binding sequence is found in only a subset of ABI4-regulated genes, suggesting that ABI4 may regulate many genes via effects on other transcription factors.

Studies in which the ABI transcription factors are overexpressed have shown extensive cross-regulation of expression among *ABI3*, *ABI4*, and *ABI5* and that each is sufficient to confer hypersensitivity to ABA for inhibition of root growth (19). Overexpression of *ABI3* or *ABI4* is even sufficient to confer ABA-inducible vegetative expression of several "seed-specific" genes, which is partly dependent on increased *ABI5* expression, but *ABI5* overexpression is less effective, possibly because it is not accompanied by increased expression of the other *ABIs* (9). Taken together, these results suggested that seed-specific or ABA-inducible expression might be at least partially controlled by regulatory complexes containing various combinations of these three transcription factors. Alternatively, it has been suggested that *ABI5* acts downstream of *ABI3* because *ABI5* overexpression can restore ABA sensitivity to a leaky *abi3* mutant (39). Consistent with the combinatorial model, ABI3 (or its monocot ortholog Vp1) and ABI5 (or its rice homolog TRAB1) display direct and synergistic interactions in two-hybrid analyses in yeast and transient reporter activation assays in rice protoplasts (19). However, ABI4 does not appear to interact directly with either ABI3 or ABI5.

In addition to the genetically defined transcriptional regulators discussed above, ABI5 has several close homologs that have been identified on the basis of binding to ABA-responsive (ABRE) promoter fragments or by sequence homology searches. The ABI5-related family in Arabidopsis is comprised of 13 genes variously designated ABFs (ABRE Binding Factors), AREBs (ABA Response Element Binding), AtDPBFs (Arabidopsis thaliana Dc3 Promoter Binding Factor), or bZIP(n), depending on the research groups that identified them; at least 7 of these are expressed in developing seeds. Several of these bZIP factors form heterodimers in some combinations, including pairing with ABI5 (3, 34), suggesting that they may participate in regulation of many of the same target genes. Consistent with this, overexpression of some of these bZIPs in transient or stable transgenic systems is sufficient for increased expression of some ABA-regulated embryonic genes (19, 33). This shared target specificity could provide functional redundancy that would explain the weakly ABA-resistant phenotype of *abi5* null mutants. However, comparison of expression patterns within this subfamily shows differences in developmental timing and regulation by the *ABI* loci that are not consistent with simple functional redundancy (3, 9). Furthermore, at least one locus (*Enhanced Em Level*, [*EEL1/AtDPBF4/bZIP12*]) antagonizes *ABI5* effects on expression of some *LEA* genes, thereby fine-tuning the timing of their expression, but has no effect on ABA sensitivity at germination (3).

Although the *ABI* and *LEC* class loci interact genetically, these transcription factors are unlikely to interact directly because they are expressed at their highest levels at different times in seed development (Fig. 6) and also do not interact in the yeast two-hybrid system (10). However, comparison of *LEC*, *ABI* and related bZIP expression in *abi* and *lec* class mono- and digenic mutants has shown that *LEC1* regulates expression of *FUS3*, *ABI5* and several related bZIPs, possibly indirectly (10). These results suggest that expression of LEC1, which peaks in early embryogenesis, is required to induce expression of FUS3 and many of the bZIPs by mid-development, but it is not clear how many or which signaling intermediates are involved in this regulation. Digenic analyses have also shown that ABI3 protein accumulation is stabilized by *LEC1* and *FUS3*, but these also fail to interact in the yeast two-hybrid system, so the mechanism of this stabilization is still unknown.

Post-transcriptional control has also been demonstrated for accumulation of some ABA-inducible LEA proteins, but the specific regulators involved have not yet been identified. It is possible that some of the recently identified loci encoding putative RNA processing proteins (*abh1*, *hyponastic leaves1*, *supersensitive to ABA and drought1*) might contribute to this level of regulation (17).

Finally, although the ABI ectopic expression studies were initially interpreted to mean that the seed-specificity of embryonic gene expression was due to seed-specific expression of key regulators, most of these regulators are expressed and functional in other processes during vegetative growth (19). An alternate possibility is that seed specificity may be conferred by vegetative repression of embryonic characteristics. Studies of chromatin structure have shown that embryo-specific genes only have an "open" conformation during seed development (38). Possible mediators of such regulation are chromatin remodeling factors, e.g. PICKLE (PKL); *pkl* mutants resemble *LEC* overexpression lines in that they display continued accumulation of embryonic products such as storage proteins and lipids during "vegetative growth". Characterization of *PKL*-regulated gene expression has shown that it is required for mediating GA-repression of embryogenesis-promoting regulators such as the *LEC* class genes (44) as well as genes encoding embryo specific products such as storage proteins (Fig. 7). However, it is not clear which genes are direct targets of PKL repression or how this repressive effect is restricted to vegetative growth since PKL is also expressed throughout embryogenesis.

GERMINATION

Germination occurs when conditions favor growth of the next generation. In quiescent seeds, access to water for hydration is sufficient to permit

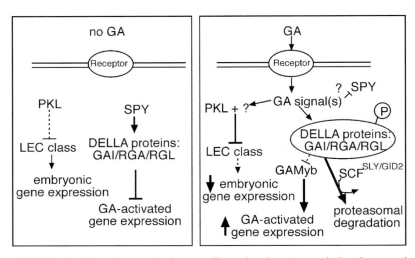

Figure 7. GA signaling in germinating seedlings involves transcriptional repression of embryonic genes and de-repression of GA-activated genes. In the absence of GA, SPY may activate the DELLA proteins by O-linked GlcNAc modification, leading to their repression of GA-activated gene expression. In the presence of GA, at least some of the DELLA proteins are phosphorylated and targeted for proteasomal degradation following ubiquitination mediated by a complex containing the F-box factor SLY/GID2. GA signaling also enhances PKL-dependent repression of embryonic gene expression. Heavy lines represent active regulatory processes; dotted lines represent inactive regulatory processes.

germination. In dormant seeds, the germination inhibition may be due to "coat-imposed dormancy" by physical constraints, such as a hard seed coat that acts as barrier to water uptake, gas exchange, or loss of chemical inhibitors (e.g. ABA or assorted secondary metabolites) that must be leached away to permit germination (5, 6). Alternatively, it may be a physiological state of the embryo requiring a period of dry storage (after-ripening), chilling (stratification), or exposure to light to break dormancy. Stratification is presumed to simulate the passage of the chilling effects of winter, whereas photodormancy is a mechanism for sensing whether a seed is deeply buried, near the surface, or shaded. The latter is particularly important to small seeds that may not have sufficient reserves to support their growth for a long time before gaining access to light. The relative importance of specific requirements for dormancy-breaking vary among species, cultivars, and ecotypes, but these treatments often have additive or synergistic effects. For example, applied GAs can substitute for a light requirement, and light appears to increase both GA synthesis and sensitivity to GA (5, 6, 43).

Germination can be divided into three phases, paralleling the changes in fresh weight of an imbibing seed (5) (Fig. 1). The first phase, imbibition, reflects rapid water uptake and is driven primarily by the matric potential of the dry seed. The second phase shows a lag in weight gain, during which time the seed resumes metabolic activity and prepares for subsequent growth. Seeds remain desiccation tolerant through phase II and will germinate more rapidly and synchronously if re-dried after entering this phase; this is the basis of the agricultural practice of "seed priming", which is just beginning to be extensively characterized molecularly (23). The last phase reflects a resumption of and commitment to growth; the radicle emerges and desiccation tolerance is lost. As one of the most critical developmental decisions in a plant's life, progression to phase III is subject to regulation by diverse signals including light and water availability, salinity, temperature, and nutrient status (e.g. carbon:nitrogen balance and absolute levels of specific metabolites). All of this information must be integrated, and this is mediated by signaling through multiple hormones that either promote (GA, ethylene, brassinosteroids (BR), or cytokinins) or inhibit (ABA and jasmonic acid) germination, and specific regulators that act as "nodes" in a signaling web (19, 24) (Fig. 8).

Genetic studies

Promotion or suppression of germination is a very simple genetic screen and many mutants have been isolated on the basis of intrinsic differences in germinability (e.g. vivipary or altered dormancy) or altered response to specific hormones at this stage (e.g. the *abi* and *era* mutants). Screens for deeply dormant seeds have also led to isolation of Arabidopsis mutants that are severely GA deficient and require GA to germinate. In contrast, semi-dominant *GA insensitive* (*gai*) Arabidopsis mutants were first isolated as dwarfs that resembled GA deficient mutants, but could not be rescued by exogenous GA. Subsequent studies showed that the *GAI* gene belonged to a small family of genetically redundant transcription factors that suppress GA

Hormones in seed development and germination

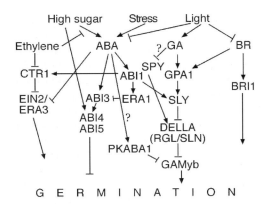

Figure 8. Interactions among some of the hormonal, abiotic and nutrient status signaling pathways affecting germination. Antagonistic effects between GA and ABA signaling are the major regulators of germination, but these pathways interact with those responding to other signals. Genetic studies have implicated numerous loci in mediating response to multiple signals (e.g. *EIN2/ERA3, ABI4, GPA1*), thereby functioning as nodes in a signaling network. Arrows indicate promotion and bars indicate inhibitory interactions.

response in the absence of GA, sometimes referred to as the DELLA family in recognition of the conserved amino acid sequence in its defining domain (43) (Chapter D2). This family of regulators plays an important role in elongation growth and other GA responses, with the <u>*RGA*(*Repressor of ga1-3*)</u>-like genes *RGL1* and *RGL2* acting as repressors of seed germination. Significantly, the *gai*, *Reduced height* (*Rht*) and *dwarf8* (*d8*) mutants of Arabidopsis, wheat, and maize, respectively, contain comparable semi-dominant mutations in their *GAI/RGA* orthologs, resulting in stabilization of these repressors. In contrast to these semi-dominant mutations, recessive mutations in the barley, rice and pea *GAI/RGA* orthologs result in constitutive GA response conferring a "slender" phenotype. Furthermore, the rice and barley *slender* mutants show constitutive GA response in germination, resulting in GA-independent reserve hydrolysis (43) (Chapter. C2). Constitutive GA response is also observed in the <u>*spindly* (*spy*)</u> mutant, which is epistatic to the *gai* mutation; *SPY* encodes a likely O-linked GlcNAc-acyltransferase. Another mutation leading to GA-resistance, <u>*sleepy*</u>, was initially identified as a suppressor of *abi1-1* resulting in extreme dormancy. *SLY*, and its rice ortholog <u>*GA-insensitive dwarf* (*GID*)2</u>, were recently cloned and found to encode F-box proteins that specifically mediate targeting of the *GAI/RGA* class of repressors of GA signaling for proteosomal degradation, thereby permitting activation of GA-inducible genes via transcription factors such as GA-Myb (41, 49) (Fig. 7). Such studies emphasize the difficulty in interpreting genetic interactions mechanistically: in this case, suppression of an ABA signaling mutant is due to defects in de-repression of a GA response pathway. It is possible that the observed interactions with brassinosteroid signaling are similarly indirect.

Evidence for interactions between signaling controlling germination in response to sugar, salt, light, ABA, ethylene, and other hormones has come from physiological and genetic studies (19, 20, 24, 37) (Fig. 8). For example, although low levels of sugars can overcome exogenous ABA inhibition of radicle emergence in a light-dependent manner, greening and subsequent seedling growth is still blocked. In contrast, exposure to high exogenous sugars appears to signal that neither reserve mobilization nor photosynthetic capacity is needed, and leads to developmental arrest even in the absence of ABA. Support for the hypothesis that these signaling interactions involve some common components comes from the observation that the ABA-resistant mutants *abi4* and *abi5* are also resistant to glucose, NaCl, and osmotic inhibition of germination and/or seedling growth (12, 19). Similarly, new alleles of ethylene response genes such as *ctr1* and *ein2* have been identified in screens for suppressors and enhancers of seed sensitivity to ABA, respectively, and it appears that even the single mutants have slightly altered ABA response. Furthermore, ABA-deficient and *ctr1* mutants also display sugar-resistant growth. These results suggest interactions between ABA, ethylene and sugar signaling that are relatively specific because *abi1* and *abi2* mutants, although highly pleiotropic with respect to ABA response, show essentially normal sugar response. The observation that *abi4* or *aba* mutants can ameliorate sugar hypersensitivity conferred by hexokinase overexpression is consistent with models in which ABA and sugar could act in either parallel or intersecting pathways (20). High glucose induces both ABA synthesis and expression of *ABI3*, *ABI4* and *ABI5*, suggesting that altered ABA levels or response may mediate some aspects of sugar signaling (37). However, glucose only induces expression of these *ABIs* when applied relatively early in germination/seedling growth, coinciding with the developmental window of sensitivities to the inhibitory effects of high sugar (19, 20).

Recent studies have emphasized the role of ABA, through ABI5, in mediating "post-germination developmental arrest" (19). When wild-type seeds are incubated on low sucrose and ABA, germination is arrested following radicle emergence. Under these conditions, ABI5 protein accumulates due to protection from proteasomal degradation, and is strongly correlated with maintenance of desiccation tolerance in these arrested seedlings, which appear to be stuck in a state that is a hybrid of phases II and III of germination. Recently, a mutation in a subunit of the regulatory lid for the proteasome was shown to result in ABA hypersensitivity, which is correlated with increased accumulation of ABI5 even in the absence of ABA or stress conditions normally required to stabilize it (52). These studies add ABA to the growing list of hormones subject to regulation by proteosomal degradation of key regulators. It is possible that ABA, the induced ABI5 gene product, and potentially other interacting factors delay or prevent escape from phase II of germination (and the accompanying loss of desiccation tolerance) under conditions of insufficient moisture to support seedling growth. However, the specific targets of ABA/ABI-regulation that are essential to maintaining this state have not yet been identified.

Many additional regulators act at earlier steps than the ABI transcription factors in ABA signaling that control germination (19). *ABI1* and *ABI2*, although initially identified on the basis of dominant ABA-insensitive mutations, encode members of a family of protein phosphatases (PP2Cs), all tested members of which appear to act as negative regulators of ABA signaling. *ERA1* encodes a subunit of farnesyl transferase, and appears to act downstream of *ABI1*, but upstream of the *ABI* transcription factors *ABI3* and *ABI5* (8).

Very early steps in signaling often involve heterotrimeric G-proteins and secondary messengers. Unlike animals, which have multiple variants for each G-protein subunit, plants generally have only one or two genes for each subunit class (32). Arabidopsis has a single Gα subunit gene (*GPA1*), and *gpa1* mutants have pleiotropic but cell/stage-specific defects in hormone response throughout their lives, including reduced sensitivity to GA and BR at germination, normal ABA sensitivity at germination but insensitivity to ABA in stomatal regulation. Consistent with a germination-promoting role for G-proteins in Arabidopsis, seed dormancy is eliminated by overexpression of a G-protein coupled receptor (GCR1) whose ligand has not yet been identified (14), and mutations in a GPA1-interacting protein, AtPirin1, increase dormancy and sensitivity to ABA (36). Although no other interactors have been identified yet for AtPirin1, mammalian pirins interact directly with transcription regulators and comparable interactions could provide a direct link between G-protein activation and transcriptional regulation.

Evidence for secondary messenger involvement in germination regulation has been provided by analyses of mutants or transgenic lines with altered phosphoinositide metabolism ((19). The ABA response at germination is correlated with a rise in inositol trisphosphate (IP$_3$) levels. Transgenic lines that eliminate the ABA-induced rise in IP$_3$ levels, via effects on phospholipase C (AtPLC1) or Ins(1,4,5)P$_3$5 phosphatase (AtIP5PII) activity, display reduced ABA sensitivity in assays of germination, seedling growth and gene expression. Consistent with this, defects in phophoinositide catabolism in *fiery1* mutants increase both IP$_3$ levels and ABA sensitivity. However, ectopic expression of PLC or anti-sense suppression of the Ins(1,4,5)P$_3$5 phosphatase gene expression did not increase IP$_3$ levels or ABA response in the absence of added ABA indicating that PLC1 expression is necessary but not sufficient for ABA response. Secondary messenger participation in reserve mobilization is also well-documented (see below).

Gene expression during germination

Some of the major metabolic activities occurring during germination are the mobilization of stored reserves, production of hydrolases in both the embryo and surrounding tissues that will permit radicle emergence and increases in "growth potential" to promote seedling expansion, and preparation for a transition to photoautotrophic growth (5). Reserve mobilization must be accomplished quickly to fuel the entire process, but it is critical that mobilization not begin prematurely during seed maturation or proceed during germination if conditions are not favorable for continued growth. Consequently, many key enzymes required for mobilization are made *de novo* during germination, and their induction can be blocked by ABA, produced in response to osmotic or other stress. The best-characterized example of reserve mobilization is GA-induced starch mobilization in cereal grains (Chapter. C2).

Starch mobilization is mediated by a GA-induced transcription factor (GAMyb) that specifically binds the GA response elements (GAREs) of responsive α-amylase genes. This induction involves signaling between tissues: GA is produced by the embryo, but much of the response occurs in the aleurone, a cell layer surrounding the endosperm that secretes amylase into the starchy endosperm. Several mechanisms ensure that these genes are not induced during seed maturation. Endogenous GA levels, although transiently high early in seed development, are low in maturing seeds. Mutations in the *GAI/RGA* family of regulators also disrupt GA-induced amylase expression, indicating that this process is repressed in the absence of GA (43). In addition, GA sensitivity is very low in maturing seeds due to a repressor activity of Vp1 (19). Because Vp1 is expressed seed-specifically, this repression is released at germination. However, during germination ABA can inhibit α-amylase expression in the aleurone by inducing expression of a protein kinase, PKABA1, that represses GA-induction of GAMyb (Fig. 8). Early steps in this signaling pathway involve G protein-mediated activation of phospholipase D and signaling by phosphatidic acid and Ca^{2+} as secondary messengers. GA-induced amylase expression in the scutellar epithelium can also be inhibited by the combined effects of ABA and sugars (20).

Recent analyses of hormone levels and transcriptional profiling during Arabidopsis germination have demonstrated some interesting similarities and differences with cereal germination. These studies identified the major active GA correlated with germination (GA_4) and used hierarchical clustering to identify 12 subsets of up- or down-regulated genes with distinct temporal profiles (42). Detailed analyses of some GA regulated genes showed that they were expressed in tissues distinct from the site of GA synthesis, implying signaling between tissues as seen in cereals. However, the GARE motif defined in cereals does not appear to be a major cis-acting component of GA-induction in Arabidopsis. Instead, a surprisingly high proportion of GA-repressed genes had ABREs in their putative promoters, suggesting that down-regulation during germination might actually reflect decreased ABA signaling. Consistent with this, ABI5 transcript levels and endogenous ABA decrease prior to the increased GA levels following imbibition of seeds, and both *ABI3* and *ABI5* are rapidly repressed by GA treatment. Additional hormonal cross-talk suggested by the microarray data includes GA-promotion of ethylene synthesis and response, and auxin synthesis and transport. Finally, as expected based on previous studies in numerous species, GA up-regulates many classes of genes implicated in cell elongation (e.g. wall loosening enzymes, expansins, and aquaporins) and cell division (e.g. cyclins and replication proteins).

SUMMARY

All of the known hormones participate in some aspects of seed development and germination. Early embryos require auxin transport and signaling to mediate pattern formation, whereas GAs and cytokinin are most important for nutritive effects and general growth promotion at this stage. ABA is important for inducing the transition from growth by cell division to growth by cell enlargement during maturation. Seed maturation is subject to overlapping control mechanisms, only some of which are ABA-dependent. Many of the mediators of hormonal signaling during seed development are transcription factors showing complex patterns of cross-regulation, with evidence for both hierarchical and combinatorial interactions. Furthermore, many of these interact genetically with seed-specific regulators that supply a developmental context. Many of these transcription factors are members of families with partially redundant, and sometimes antagonistic, function. In contrast to seed development, germination is acutely sensitive to environmental conditions and there is extensive cross-talk among abiotic, hormonal, and metabolic signals regulating commitment to growth. Signaling intermediates include G-proteins, receptor-kinases, phosphoinositides, phosphatidic acid and Ca^{2+} as secondary messengers, and regulated phosphorylation status of specific proteins, most of which have yet to be identified. In addition, proteasomal degradation and chromatin remodeling appear to play important roles during germination in permitting de-repression of GA-induced responses and decreasing ABA sensitivity in the absence of stress. However, although many of the regulatory components have been identified, the specific mechanisms of most of their interactions are unknown. Finally, many additional gene products of unknown biochemical functions regulate hormonal response and studies of their function are likely to reveal novel signaling mechanisms.

Acknowledgements

I thank Dr. Douglas Bush, Tim Lynch, Emily Garcia, and Dr. Inès Brocard-Gifford for critical review of the manuscript.

References

1. Aida M, Vernoux T, Furutani M, Traas J, Tasaka M (2002) Roles of *PIN-FORMED1* and *MONOPTEROS* in pattern formation of the apical region of the *Arabidopsis* embryo. Development 129: 3965-3974
2. Ara H, Jaiswal U, Jaiswal V (2000) Synthetic seed: Prospects and limitations. Current Sci 78: 1438-1444
3. Bensmihen S, Rippa S, Lambert G, Jublot D, Pautot V, Granier F, Giraudat J, Parcy F (2002) The Homologous ABI5 and EEL Transcription Factors Function Antagonistically to Fine-Tune Gene Expression during Late Embryogenesis. Plant Cell 14: 1391-1403
4. Berger F (2003) Endosperm: the crossroads of seed development. Curr Opin Plant Biol 6: 42-50
5. Bewley JD (1997) Seed germination and dormancy. Plant Cell 9: 1055-1066
6. Bewley JD, Black M (1994) Seeds: Physiology of Development and Germination, 2nd. Plenum Press, New York, 367
7. Blilou I, Frugier F, Folmer S, Serralbo O, Willemsen V, Wolkenfelt H, Eloy N, Ferreira P, Weisbeek P, Scheres B (2002) The *Arabidopsis HOBBIT* gene encodes a CDC27 homolog that links the plant cell cycle to progression of cell differentiation. Genes Dev. 16: 2566-2575
8. Brady S, Sarkar S, Bonetta D, McCourt P (2003) The *ABSCISIC ACID INSENSITIVE 3 (ABI3)* gene is modulated by farnesylation and is involved in auxin signaling and lateral root development in *Arabidopsis*. Plant J. 34: 67-75

9. Brocard I, Lynch T, Finkelstein R (2002) Regulation and role of the Arabidopsis *ABA-insensitive (ABI)5* gene in ABA, sugar and stress response. Plant Physiol. 129: 1533-1543
10. Brocard-Gifford I, Lynch T, Finkelstein R (2003) Regulatory networks in seeds integrating developmental, ABA, sugar and light signaling. Plant Physiol. 131: 78-92
11. Busk PK, Pages M (1998) Regulation of abscisic acid-induced transcription. Plant Molecular Biology 37: 425-435
12. Carles C, Bies-Etheve N, Aspart L, Leon-Kloosterziel KM, Koornneef M, Echeverria M, Delseny M (2002) Regulation of *Arabidopsis thaliana Em* genes: Role of ABI5. Plant J. 30: 373-383
13. Chen J-G, Ullah H, Young J, Sussman M, Jones A (2001) ABP1 is required for organized cell elongation and division in Arabidopsis embryogenesis. Genes Dev. 15: 902-911
14. Colucci G, Apone F, Alyeshmerni N, Chalmers D, Chrispeels M (2002) GCR1, the putative Arabidopsis G protein-coupled receptor gene is cell cycle-regulated, and its overexpression abolishes seed dormancy and shortens time to flowering. Proc. Natl. Acad. Sci. USA 99: 4736-4741
15. Dodeman V, Ducreux G, Kreis M (1997) Zygotic embryogenesis *versus* somatic embryogenesis. J Exp Bot 48: 1493-1509
16. Dure LI (1997) Lea Proteins and the Desiccation Tolerance of Seeds. *In* B. A. Larkins and I. K. Vasil, *eds*, Cellular and Molecular Biology of Plant Seed Development. Kluwer Academic Publishers, Dordrecht, The Netherlands
17. Fedoroff N (2002) RNA-binding proteins in plants: the tip of an iceberg? Curr Opin Plant Biol 5: 452-459
18. Fernandez D (1997) Developmental basis of homeosis in precociously germinating *Brassica napus* embryos: phase change at the shoot apex. Development 124: 1149-1157
19. Finkelstein R, Gampala S, Rock C (2002) Abscisic acid signaling in seeds and seedlings. Plant Cell 14: S15-S45
20. Finkelstein R, Gibson SI (2002) ABA and sugar interactions regulating development: "cross-talk" or "voices in a crowd"? Curr. Opin. Plant Biol. 5
21. Finnie C, Melchior S, Roepstorff P, Svensson B (2002) Proteome analysis of grain filling and seed maturation in barley. Plant Physiol. 129: 1308-1319
22. Foley ME (2001) Seed dormancy: An update on terminology, physiological genetics, and quantitative trait loci regulating germinability. Weed Science 49: 305-317
23. Gallardo K, Job C, Groot SPC, Puype M, Demol H, Vandekerckhove J, Job D (2001) Proteomic analysis of Arabidopsis seed germination and priming. Plant Physiol 126: 835-848
24. Gazzarrini S, McCourt P (2001) Genetic interactions between ABA, ethylene and sugar signaling pathways. Curr. Opin. Plant Biol. 4: 387-391
25. Goldberg R, de Paiva G, Yadegari R (1994) Plant embryogenesis: zygote to seed. Science 266: 605-614
26. Gualberti G, Papi M, Bellucci L, Ricci I, Bouchez D, Camilleri C, Costantino P, Vittorioso P (2002) Mutations in the Dof Zinc Finger Genes DAG2 and DAG1 Influence with Opposite Effects the Germination of Arabidopsis Seeds. Plant Cell 14: 1253-1263
27. Harada J (1997) Seed maturation and control of germination. *In* B. A. Larkins and I. K. Vasil, *eds*, Cellular and Molecular Biology of Plant Seed Development. Kluwer Academic Publishers, Dordrecht, The Netherlands
28. Harada J (1999) Signaling in plant embryogenesis. Curr Opin Plant Biol 2: 23-27
29. Harada J (2001) Role of *Arabidopsis LEAFY COTYLEDON* genes in seed development. J. Plant Physiol. 158: 405-409
30. Helliwell C, Chin-Atkins A, Wilson I, Chapple R, Dennis E, Chaudhury A (2001) The Arabidopsis *AMP1* gene encodes a putative glutamate carboxypeptidase. Plant Cell 13: 2115-2125
31. Hellmann H, Lawrence Hobbie L, Chapman A, Dharmasiri S, Dharmasiri N, del Pozo C, Reinhardt D, Estelle M (2003) Arabidopsis *AXR6* encodes CUL1 implicating SCF E3 ligases in auxin regulation of embryogenesis. EMBO J. 22: 3314-3325

32. Jones A (2002) G-protein-coupled signaling in *Arabidopsis*. Curr Opin Plant Biol 5: 402-407
33. Kang J-y, Choi H-i, Im M-y, Kim SY (2002) Arabidopsis basic leucine zipper proteins that mediate stress-responsive abscisic acid signaling. Plant Cell 14: 343-357
34. Kim S, Ma J, Perret P, Li Z, Thomas T (2002) Arabidopsis ABI5 subfamily members have distinct DNA binding and transcriptional activities. Plant Physiol. 130: 688-697
35. Koornneef M, Bentsink L, Hilhorst H (2002) Seed dormancy and germination. Current Opinion in Plant Biology 5: 33-36
36. Lapik Y, Kaufman L (2003) The Arabidopsis Cupin Domain Protein AtPirin1 Interacts with the G Protein -Subunit GPA1 and Regulates Seed Germination and Early Seedling Development. Plant Cell 15: 1578-1590
37. Leon P, Sheen J (2003) Sugar and hormone connections. Trends Plant Sci. 8: 110-116
38. Li G, Hall T, Holmes-Davis R (2002) Plant chromatin: developmental and gene control. BioEssays 24: 234-243
39. Lopez-Molina L, Mongrand S, McLachlin DT, Chait BT, Chua N-H (2002) ABI5 acts downstream of ABI3 to execute an ABA-dependent growth arrest during germination. Plant J. 32: 317-328
40. McCarty D (1995) Genetic Control and Integration of Maturation and Germination Pathways in Seed Development. Ann. Rev. Plant Phys. Plant Mol. Biol. 46: 71-93
41. McGinnis K, Thomas S, Soule J, Strader L, Zale J, Sun T-p, Steber C (2003) The Arabidopsis *SLEEPY1* gene encodes a putative F-box subunit of an SCF E3 ubiquitin ligase. Plant Cell 15: 1120-1130
42. Ogawa M, Hanada A, Yamauchi Y, Kuwahara A, Kamiya Y, Yamaguchi S (2003) Gibberellin Biosynthesis and Response during Arabidopsis Seed Germination. Plant Cell 15: 1591-1604
43. Peng J, Harberd N (2002) The role of GA-mediated signalling in the control of seed germination. Curr Opin Plant Biol 5: 376-381
44. Rider SJ, Henderson J, Jerome R, Edenberg H, Romero-Severson J, Ogas J (2003) Coordinate repression of regulators of embryonic identity by *PICKLE* during germination in *Arabidopsis*. Plant J. 35: 33-43
45. Riechmann J, Heard J, Martin G, Reuber L, Jiang C-Z, Keddie J, Adam L, Pineda O, Ratcliffe O, Samaha R, Creelman R, Pilgrim M, Broun P, Zhang J, Ghandehari D, Sherman B, Yu G-L (2000) Arabidopsis transcription factors: Genome-wide comparative analysis among eukaryotes. Science 290: 2105-2110
46. Rock C (2000) Pathways to abscisic acid-regulated gene expression. New Phytol. 148: 357-396
47. Rock C, Quatrano R (1995) The role of hormones during seed development. *In* P. J. Davies, eds, Plant hormones: Physiology, biochemistry and molecular biology, 2nd. Kluwer Academic Publishers, Norwell, Massachusetts, USA, 671-697
48. Ruuska S, Girke T, Benning C, Ohlrogge J (2002) Contrapuntal networks of gene expression during Arabidopsis seed filling. Plant Cell 14: 1191-1206
49. Sasaki A, Itoh H, Gomi K, Ueguchi-Tanaka M, Ishiyama K, Kobayashi M, Jeong D-H, An G, Kitano H, Ashikari M, Matsuoka M (2003) Accumulation of phosphorylated repressor for gibberellin signaling in an F-box mutant. Science 299: 1896-1898
50. Schwartz B, Vernon D, Meinke DW (1997) Development of the Suspensor: Differentiation, Communication, and Programmed Cell Death During Plant Embryogenesis. *In* B. A. Larkins and I. K. Vasil, eds, Cellular and Molecular Biology of Plant Seed Development. Kluwer Academic Publishers, Dordrecht, The Netherlands
51. Shen-Miller J, Mudgett M, Schopf J, Clarke S, Berger R (1995) Exceptional seed longevity and robust growth: Ancient Sacred Lotus from China. Am. J. Bot. 82: 1367-1380
52. Smalle J, Kurepa J, Yang P, Emborg TJ, Babiychuk E, Kushnir S, Vierstra RD (2003) The Pleiotropic Role of the 26S Proteasome Subunit RPN10 in Arabidopsis Growth and Development Supports a Substrate-Specific Function in Abscisic Acid Signaling. Plant Cell 15: 965–980
53. Souter M, Lindsey K (2000) Polarity and signalling in plant embryogenesis. J Exp Bot 51: 971-983

54. Swire-Clark GA, Marcotte WR (1999) The wheat LEA protein Em functions as an osmoprotective molecule in *Saccharomyces cerevisiae*. Plant Mol. Biol. 39: 117-128
55. Thomas T (1993) Gene expression during plant embryogenesis and germination: an overview. Plant Cell 5: 1401-1410
56. Vogler H, Kuhlemeier C (2003) Simple hormones but complex signalling. Curr Opin Plant Biol 6: 51-56
57. Yamada H, Suzuki T, Terada K, Takei K, Ishikawa K, Miwa K, Yamashino T, Mizuno T (2001) The Arabidopsis AHK4 histidine kinase is a cytokinin-binding receptor that transduces cytokinin signals across the membrane. Plant Cell Physiol. 42: 1017-1023

E5. Hormonal and Daylength Control of Potato Tuberization.

Salomé Prat
Dpto. de Genética Molecular de Plantas, Centro Nacional de Biotecnología-CSIC, Campus Universidad Autónoma, Cantoblanco, 28049 Madrid, SPAIN.
E-mail: sprat@cnb.uarn.es

INTRODUCTION

Plants utilize light not only as source of energy in photosynthesis but also as a source of information of the environment in which they develop. An incident radiation enriched in FR light, such as that of sunlight filtered through a leaf canopy, is perceived by the plant as the competing presence of other plants for light. This induces an increased elongation response aimed at optimising the capture of incident light. By measuring the relative duration of the day and night plants can also recognise the season of the year in which they are growing. Moving from the equator towards the poles, the days become longer in summer and shorter in winter. The rate at which daylength changes varies during the year, with little change from day to day in mid summer or winter, and more rapid changes as days become longer during spring or shorter during fall. In consequence, plants can detect the season by measuring the relative lengths of day and night, and how they vary in successive days. Characteristic seasonal responses are fall leaf abscission, bud dormancy and cambium activity in trees, flowering, seed germination and potato tuberization. These seasonal responses are synchronized by photoperiod and allow the plant to adapt development to environmental conditions, thus assuring cross-pollination among individuals of the same species or seed survival during adverse winter conditions.

Tuber-bearing potato species have a number of features that make them unique among the *Solanaceae*[1]. They develop specialized underground vegetative propagation organs or tubers which, after a few winter months of rest or dormancy, sprout to form a new plant, thus serving as long lasting perennation organs for these annual tuber-bearing species. Tubers serve a

---

[1]Abbreviations: DFMO, difluoromethylornithine; FR, far red; JA, jasmonic acid; LOX, lipoxygenase; LD, long days; MeJA, methyl jasmonate; NB, night break; ODC, ornithine decarboxylase; PHY, phytochrome; QTLs, quantitative trait loci; R, red light; SAM, shoot apical meristem; SD, short day; SuSy, sucrose synthase; TA, tuberonic acid; TAG, tuberonic acid glucoside.

double function for the plant, as both a storage organ and a vegetative propagation system. Potato tubers accumulate large amounts of starch, and are a rich supply of carotenoids and a balanced source of amino acid and proteins, so they are of great economic importance as a food source (21).

Morphologically, potato tubers are modified stems, with very short swollen internodes and dormant axillary buds referred to as tuber "eyes". Tubers differentiate from the subapical region of specialized underground vegetative shoots, or stolons, that develop at the base of the main stem. Under non-inductive conditions, stolons grow as horizontal stems which, if exposed to sufficient light, grow upward and emerge from the soil to form a new shoot. During this process, the shoot meristem acquires all the characteristics of a shoot apical meristem (SAM), developing new leaves, lateral buds, and eventually flowering. Under favourable conditions however, elongation of the stolon ceases and the tip begins to swell to form a tuber. Cells located in the pith and the cortex of the sub-apical region of the stolon enlarge and divide longitudinally, with subsequent randomly oriented divisions taking place to form the bulk of the tuber (48). Changes in the plane of cell division are accompanied by a switch in the developmental program of the meristem cells of the stolon to form a tuber. As a consequence of this developmental switch meristem growth becomes determinate, and ceases after a few rounds of cell division, the mechanism of sucrose unloading changes from apoplastic to symplastic (46), and the cells begin to accumulate large amounts of starch and to express a new set of proteins (i.e., patatin and proteinase inhibitors) designed to serve as reserve storage for the new developing plant. In its natural environment, the onset of tuberization occurs during autumn or early winter, depending on the potato genotype. Newly formed tubers undergo a period of inactivity or endo-dormancy which is characterized by the absence of bud growth even if tubers are exposed to conditions favorable to sprouting. This period of physiological rest lasts for about three to four months, thereby assuring the survival of the plant during cool winter temperatures. After this period of dormancy bud growth is reactivated and the tuber lateral buds sprout to form a new plant that is genetically identical to the mother plant (6).

THE CONTROL OF TUBER INITIATION

Tuber formation is regulated both by environmental and endogenous factors. Tuberization is favoured by short photoperiods, cool temperatures, and low rates of nitrogen fertilization. Daylength is the environmental factor that has been most investigated as it has the most critical effect on tuberization (24). It is well known that long nights (short days) favour tuberization; potatoes are therefore short day (SD) plants for tuber formation. Although these environmental cues induce tuber formation in all potato species, there is considerable variation in the degree in which different potato varieties respond to a particular environmental stimulus, with tuberization in cultivated potatoes, for example, being relatively independent of daylength.

Genotype

Potato was brought to cultivation in the Andean highlands, where photoperiods are about 12 hours and temperatures are cool. Andean varieties (*Solanum tuberosum* L. ssp. *andigena*) and wild species such as *S. demissum* tuberize very poorly or not at all when grown under the higher temperatures of lowland tropics or the long summer days of temperate zones. Cultivated potatoes (*Solanum tuberosum* L. ssp. *tuberosum*) have been subjected to repeated selection to adapt tuberization to the long summer days of the temperate zones of southern Chile, Europe and North America. *Tuberosum* potatoes give very low yields in the highland tropics because, under the cool temperatures and short day conditions characteristic of these regions, tuber formation is induced very early, when shoot growth is still too reduced to support good tuber yields (9). Such a trend towards prevailing short or longer days for induction is characteristic to all *andigena* or *tuberosum* species, even though there are substantial genetic differences in the daylength response within each group.

Most cultivated potatoes are tetraploid, which adds considerable difficulty to classical genetic experiments. Quantitative trait loci (QTLs) for the ability to tuberize under LD have been identified in reciprocal backcrosses between dihaploid *Solanum tuberosum* and diploid *S. berthaultii*, with eleven distinct loci on seven chromosomes associated with tuberization variation in the offspring (45). Even though most of these loci contribute only with small effects, a major QTL explaining 27% of the variance is found on chromosome 5. Poor tuberization under LD driven by this QTL is also associated with late maturity. The *R1* allele for resistance to *Phytophtora infestans* (potato blight) has been linked to the same RFLP marker on chromosome 5. This may explain the difficulty experienced by breeders who try to obtain early cultivars that are resistant to late-blight. This disease caused in the 19th century starvation of millions of people, and since the 1980's has re-emerged as a major threat to the crop.

Nitrogen supply

High rates of nitrogen fertilization have been reported to reduce the level of tuber induction. Under hydroponic conditions, tuberization can be controlled by altering the nitrogen supply to the plant in a way that a continuous supply of between 1 and 3 mM nitrogen (given either as ammonium or nitrate ions) completely inhibits tuberization, whereas transfer of the plants to a nitrogen-free media produces tuberization in 4 to 6 days (24). Nitrogen addition after the plants have started tuberizing causes reversion of tubers to stolons, and new transfer to nitrogen-free media will cause initiation of a second tuber from each stolon. Repeated cycles of high/low nitrogen then result in the formation of "chain tubers", demonstrating an important role of nitrogen levels in the control of tuber formation. Interestingly, nitrogen supply to the leaves does not inhibit tuberization. Also, a reduction in nitrogen levels in LD or high temperature conditions does not produce tuberization, indicating that nitrogen is not involved in the induction of tuberization but in the repression of tuber formation.

The mechanism by which high levels of nitrogen cause an inhibition of tuberization is not known. There are reports that nitrogen withdrawal causes a reduction in GA and an increase in ABA levels (24) with these hormonal changes then favouring tuber formation. However, changes in nitrogen supply would also affect the ratio of carbohydrate to nitrogen, which also plays a role in the control of tuberization. High levels of carbohydrates in the form of sugars and starch favour photoassimilate export to the storage organs, thus promoting tuberization. High nitrogen levels promote instead shoot and root growth, thereby reducing the amount of available carbohydrate to the tubers. Consistent with this later hypothesis, an inhibitory effect of increased nitrogen levels was observed in *in vitro* tuberization experiments on a low level (2%) sucrose medium (22). Such an inhibitory effect was not observed at higher sucrose concentrations, indicating that at very high carbohydrate levels the effects of other control mechanisms are masked. Interpretation of the results from *in vitro* experiments thus needs to be made with caution.

Temperature

High temperatures are inhibitory for tuberization in both short and long photoperiods, although the inhibitory effect is much greater in long days. Treating different parts of the plant with different temperature regimes established that high temperatures applied to the shoots have the greatest inhibitory effect on the formation of the tuberization stimulus. In addition experimentally-induced high soil temperatures inhibited tuber formation, but did not prevent the production of the tuber inducing signal in the shoots since tuberization occurred after the stolons grew upward and reached cooler air (9). High temperatures cause secondary growth of the tubers, a process known as heat sprouting. Cycles of alternating hot and cool temperatures form "chain tubers" in the same way as observed for successive treatments with high/low nitrogen levels.

There is some evidence that the inhibitory effect of high temperatures is mediated through increased GA levels. Treating plants or cuttings with CCC (chloro-ethyl-trimethyl-ammonium chloride), an inhibitor of GA biosynthesis, was found to overcome the inhibitory effect of high temperatures on tuberization. Such inhibition does not, however, occur if the shoot buds are removed prior to the high temperature treatment, indicating that the synthesis of GAs in response to high temperatures takes place in these organs (28).

Daylength control of tuberization

Short days (long nights) favor tuber formation in all potato species although the critical night length threshold for tuberization and the strength of the photoperiodic response varies with different genotypes (40). Potato species such as *Solanum demissum* and some lines of *Solanum tuberosum* ssp *andigena* are strictly dependent on short days (10h photoperiod) for tuberization, with these species forming tubers in SD conditions, but being unable to tuberize under long days (LD, e.g., 16h photoperiod). Although these plants are referred to as SD-plants it is known that it is the length of the dark period that is critical for the induction of tuberization. When plants grown under short days are given a 30 min light pulse or night break (NB) in the middle of the dark period (SD+NB), tuberization is inhibited, whereas in plants grown under long days conditions, interruption of the light period with 30 min of darkness has no effect on tuber formation (9, 21).

phyB is involved in the photoperiodic response

Red light is the most effective for a night break and the inhibitory effect of a red light pulse (R) can be reverted if a pulse of far red light (FR) is given immediately afterwards (1). These observations implicate the light receptor phytochrome in the perception of the relative lengths of day and night, and thereby in the photoperiodic control of tuber formation. By transformation of potato *andigena* plants with an antisense construct for the *phyB* phytochrome gene, encoding a light stable form of the photoreceptor, we showed that PHYB plays a key role in daylength perception and, subsequently, in activation of the photoperiod-dependent pathway controlling tuber formation (Fig. 1) (20). The *phyB* antisense lines were strongly induced to tuberize, doing so equally well under SD, LD or SD+NB conditions. Thus the perception of non-inductive conditions (LD or SD+NB) by PHYB leads to the activation of a negative regulatory pathway that blocks tuber formation. Such an inhibitory pathway is not activated in the antisense-*phyB* lines, which behave as if they were under SD conditions, even when they are exposed to SD+NB or to LDs (20).

The effect of phytochrome on tuberization appears at least in part to be mediated by GAs. Besides the loss of the photoperiodic control of tuberization, the antisense-*phyB* lines exhibit an elongated phenotype that is reminiscent of plants treated with saturating doses of gibberellins. Similar to *slender* mutants, the antisense-*phyB* lines have paler leaves and very elongated internodes, which suggests that they are altered either in GA biosynthesis, the feed-back regulation of the GA biosynthetic enzymes, or in some step of the GA response pathway. Measurement of endogenous levels of GAs have indeed shown that GA content is elevated in these plants, with 4 to 6-fold higher levels of GA_1 detected in the leaves as compared to non-transformed plants (27).

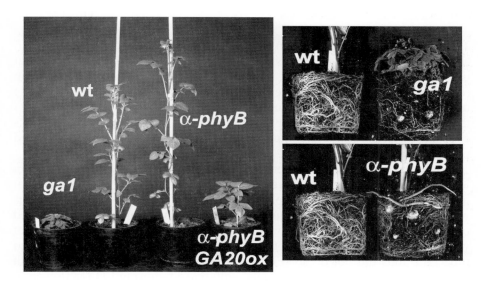

Figure 1 (Color plate page CP14). Phenotype of potato *andigena* plants, the *ga1* mutant, and the *phyB* antisense lines. The *ga1* mutant has a dwarf phenotype, dark green leaves and can tuberize after several month in LD conditions. The *phyB* antisense lines have paler leaves and an elongated phenotype and tuberize equally well in LD, SD or SD + NB conditions. GA 20-oxidase antisense inhibition in the *phyB* genetic background results in shorter plants that tuberize even earlier than the *phyB* mutants.

Another member of the phytochrome family, PHYA, was also shown to be involved in photoperiod sensing via input to the circadian clock. FR or FR+R treatments during the night were found to advance the phase of the circadian movement of the leaves in wild-type plants but not in *phyA* antisense potato plants (49). The circadian movement of leaves was still reset by blue light in these antisense lines, indicating that these plants are not impaired in rhythm entrainment. Extension of the day with FR+R light caused a lesser delay in tuber production in transgenic plants deficient in PHYA. Therefore, it appears that whereas *phyB* mutants of both short-day (potato, sorghum) or long-day plants (Arabidopsis, pea) flower, or in the case of potato tuberize, earlier than wild-type controls, the situation is different for *phyA* where a deficiency in this photoreceptor delays long-day flowering in Arabidopsis and pea, but accelerates short-day tuberization in potato. Hence, these results are indicative of a concerted action of the two phytochromes in photoperiodic control, with PHYB being involved in making the tuberization and flowering responses sensitive to photoperiod, whereas PHYA is involved in sensing the duration of the photoperiod (49).

GRAFTING EXPERIMENTS

The principal site of perception of the photoperiodic signal is in the leaf. Grafting experiments led Gregory and Chapman (16, 6) to hypothesize that the leaves produce some sort of graft transmissible inducing signal in response to the photoperiodic stimulus, which is transported to the underground stolons to induce tuberization. When leaves from potato plants that had been exposed to short photoperiods were grafted onto potato stocks exposed to LDs these were induced to tuberize, even though the grafted plants were kept in non-inducing long day conditions.

Inter-specific grafting experiments showed that the signal produced in leaves of plants induced to flower is similar to or identical to the signal that induces tuberization in potato plants (5). When shoots of tobacco species with various photoperiodic requirements for floral induction were grafted onto *andigena* potato plants it was observed that while tobacco scions that were induced to flower induced tuberization in the potato stock, non-induced tobacco scions did not do so (Fig. 2). Identical results were obtained with LD, SD or day-neutral tobacco plants, with those requiring LD for flowering

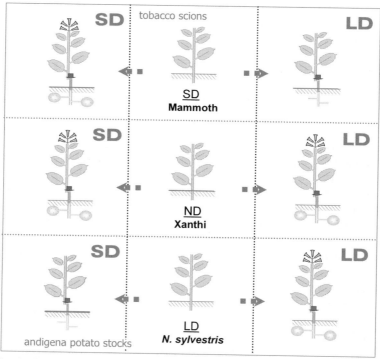

Figure 2: Grafting experiments between tobacco plants requiring SD (*N. tabacum* var. Mammoth), LD (*N. sylvestris*) or being day neutral (*N. tabacum* var. Xanthi) for flowering and potato *andigena* plants showed that the flowering and tuberization stimulus are equivalent. The flowering tobacco scions induced tuberization in the potato stocks independently of being derived from SD, LD, or day neutral varieties for flowering (adapted from. 5).

inducing tuberization on potato only if the tobacco leaves received long days, whereas tobacco that required SD for flowering caused tuberization on potato only when the tobacco leaves were exposed to short days. These observations suggest that the stimulus for flowering in tobacco is the same as the stimulus inducing tuberization in potato, and that the nature of this stimulus is equivalent in day-neutral, SD or LD plants. Similar results were also obtained in Jerusalem artichoke, in which induced sunflower leaves caused tuberization of the Jerusalem artichoke stocks, thus indicating that this phenomenon is not restricted to tobacco and potato plants.

A CONSTANS ortholog plays a role in daylength control of tuberization

Genetic analyses in the LD plant Arabidopsis has led to the identification of four independent pathways affecting flowering time: the autonomous pathway, the vernalization pathway, the GA-dependant pathway, and the LD-dependant pathway (39). Detection and transduction of the light signal by phytochromes and cryptochromes, in combination with the circadian clock oscillator, was shown to mediate activation of the daylength-dependent flowering pathway. The protein CONSTANS (AtCO) has been found to function as a link between the oscillator and the flowering time genes, playing a key regulatory function in daylength flowering control. Loss-of-function *co* mutants flower late in LD conditions but like wild-type plants in SDs. Ectopic over-expression of this gene, on the other hand, promotes flowering both in SD and LD conditions (33).

We have investigated whether a CO-homologue would be involved in photoperiodic control of tuber formation by over-expressing the Arabidopsis *CONSTANS* gene in *andigena* plants. Lines over-expressing the *AtCO* gene (pAtCO lines) were smaller than wild-type plants and in SD conditions tuberized much later than the controls (Fig. 3). While control plants started to form tubers after 2-3 weeks under SD inducing conditions, pAtCO over-expressers required 9 weeks or longer under SDs to tuberize, thereby demonstrating a negative effect of *AtCO* on the photoperiodic control of tuberization.

Analysis of the diurnal rhythm of expression of the gene *StCOL*-1, encoding a zinc finger regulatory protein with homology to CONSTANS, showed that oscillation of this transcript is not affected in the pACO over-expresser lines. Thus the pACO plants are not altered in daylength perception. Instead the negative control of CONSTANS on tuber formation is exerted through a block in the production or transport of the tuberization inducing signal, or an alteration in the stolon response to this inductive signal. When wild-type scions were grafted onto pACO stocks the resulting chimeras tuberized like wild-type plants, whereas when pACO scions were grafted onto wild-type stocks, the resulting chimera tuberized as the pACO plants (33). These experiments demonstrate that function of AtCO is required in the leaves and therefore that AtCO acts upstream of the generation or transport of the tuberization-inducing signal.

A

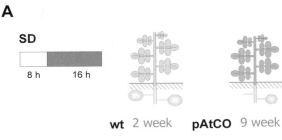

B

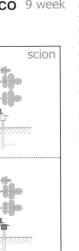

Figure 3: Expression of the Arabidopsis CO protein in potato affects negatively tuberization. While the *andigena* controls have already formed tubers after 2 week entrainment to SD conditions, the potato *CO* over-expressers (pAtCO) require 9 or more weeks under inducing conditions to form tubers (A). Grafting experiments showed that this negative effect is exerted on the leaves since wild-type scions on pAtCO stocks tuberize as wild-type plants, whereas pAtCO scions on wild-type stocks show the same tuberization delay as the pAtCO transformants (B).

Several lines of evidence indicate that *CO* acts as link between the photoperiodic pathway and the developmental pathways that respectively control flowering or tuber formation (Fig. 4). Therefore, it is likely that the genes regulated by *CO* in potato will be different from the *AGL20/SOC1* and *FT* genes, which in Arabidopsis act as floral integrator genes to control floral transition and are directly activated by this regulatory factor. Genes acting downstream of *CO* in potato are currently under study, since they are likely to function as tuberization pathway integrator genes which in response to both endogenous and environmental cues, switch the developmental fate of the cells at the subapical region of the stolon into the tuberization pathway.

HORMONAL CONTROL OF TUBER FORMATION

There are several reasons to believe that flowering and tuberization are under hormonal control. Tuber initiation is accompanied by extensive morphological and biochemical changes above and below ground. Leaves are larger, thinner, and have a flatter angle to the stem. Axillary branching is suppressed, flower buds abort more frequently, and senescence is hastened when tuber formation is induced. The graft transmissible nature of the inducing stimulus and the effect of girdling on movement of the inducing signal are also consistent with a hormonal role in induction. Stem girdling, for example, results in tuber formation on the axillary buds above the site of girdling but not on the stolons, indicating that a disruption of phloem vessels arrests translocation of the mobile signal.

There are reports in the literature describing the importance of gibberellins (GA), cytokinin, jasmonic and tuberonic acid, or abscisic acid (ABA) in tuber induction. Let us consider the evidence for the involvement of each of these particular hormones:

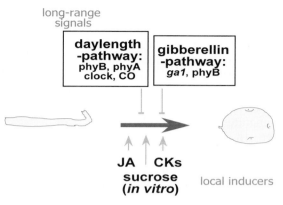

Figure 4. Regulatory pathways controlling tuber transition. Two negative regulatory pathways inhibit tuber formation in non-inducing conditions: a daylength-dependent pathway and a gibberellin-dependent pathway. PHYB, PHYA and CONSTANS act as components of the daylength-dependent pathway, with PHYB modulating the GA-dependent pathway by altering biosynthesis or sensitivity to these hormones. JA, cytokinin and sucrose induce tuberization *in vitro*, thus acting as local inducers.

Gibberellins

The role of gibberellins (GAs) in the induction of tuber formation has been studied mainly by experiments using GA-biosynthesis inhibitors and application of GAs. These experiments substantiated the observation that GAs are inhibitors of tuberization, although how GAs act in relation to tuber formation is still not fully understood (25).

Nitrogen levels, temperature, and light intensity were all shown to have an effect on GA levels. Analysis of GA content in the stolons of plants induced to tuberize showed that levels of GA_1 strongly decrease in the stolons prior to tuber formation (47). Application of the GA biosynthesis inhibitors ancymidol or paclobutrazol also overcomes the strict SD requirement of *andigena* plants for tuberization, suggesting that GAs play a role in the photoperiodic control of tuberization in these species by preventing tuberization in LDs (21). In this respect, the *ga1* dwarf mutant of *S. tuberosum* ssp. *andigena,* which appears to be blocked in the 13-hydroxylation step of the GA biosynthetic pathway, is able to tuberize in LDs although, in contrast to the antisense-*phyB* lines, it requires cultivation for several months under these non-inducing conditions to produce tubers (44). Together, these results indicate that a decrease in GA levels may be involved in the induction of tuberization in potato plants. In line with such a function, GAs are known to promote cell elongation in meristematic tissue, with GA_3

shown to cause microtubules and microfibrils to become transversely oriented to the cell axis, thus resulting in longitudinal cell expansion and stolon elongation. Reducing the levels of GAs in stolons exposed to SD would result in a longitudinal orientation of the microtubules and microfibrils as is observed to occur during treatment with the GA biosynthesis inhibitor uniconazol. This allows lateral cell expansion and division of the meristem cells, with the consequent enlargement of the stolon sub-apical region to generate a tuber (12).

In order to elucidate the role of GAs in the control of tuberization, we have isolated the genes encoding the two last steps for $GA_{1/4}$ biosynthesis: GA 20-oxidase and GA 3-oxidase (3, 2). These enzymes correspond to key regulatory steps in bioactive GA synthesis and are subjected to negative feedback regulation by the end-product GA_1. In addition, they have been shown to be controlled by phytochrome or in response to daylength conditions in several plant species. For example, spinach bolting in response to long days is associated to greater levels of GA 20-oxidase expression, whereas light-dependant germination of Arabidopsis or lettuce seeds is mediated through higher levels of 3β-hydroxylase gene expression (30). We have investigated whether expression of these two GA biosynthetic activities is differentially regulated during transition to tuber formation by analysing gene expression in *andigena* plants induced (SD) and non-induced (SD+NB) to tuberize.

Using degenerate primers we isolated three potato GA 20-oxidase cDNAs, designated as *StGA20ox1*, *StGA20ox2* and *StGA20ox3*, with different patterns of tissue-specific expression (3).Time-course studies in plants entrained to tuber inducing (SD) or non-inducing conditions (LD or SD+NB) did not show significant differences in the levels of accumulation of any of these transcripts. In plants exposed to LD, an extended accumulation of mRNA was detected during the supplementary hours of light, but in contrast to the studies reported in Arabidopsis or spinach, higher levels of transcripts were not detected in LD or SD+NB conditions as compared to SDs. These results would indicate that regulated expression of this biosynthetic activity does not play a principal role in the control of GA synthesis during the transition to tuber formation. Over-expression of the *StGA20ox1* GA 20-oxidase enzyme in *andigena* plants resulted in taller plants that required longer exposures to SD conditions to form tubers. Tuber yields in these plants were smaller than those of controls, thus confirming a negative role of GAs in tuber induction (4).

Two cDNA clones encoding potato 3β-hydroxylase (3-oxidase) (*StGA3ox1* and *StGA3ox2*) were also isolated using degenerated primers, (2). The two cDNAs were expressed in different potato tissues, with the *StGA3ox2* transcript being very abundant in the stolons of non-induced plants and detected in the stem, whereas the *StGA3ox1* message was mainly found in flowers. Expression of gene *StGA3ox2* in the stolons was greatly repressed in induced stolons, suggesting an important role of this gene in regulating stolon GA_1 levels in response to daylength conditions (2).

We have studied the role of potato GA 3-oxidase in the control of tuberization by over-expressing this enzyme in *andigena* potato plants. Transgenic lines that accumulated higher levels of the GA 3-oxidase enzyme had a taller phenotype and increased shoot levels of GAs, which correlated with the levels of expression of the biosynthetic enzyme. However, contrary to what was expected, these plants tuberized earlier in SD conditions and showed higher tuber yields than the untransformed controls. Such a tuberization promoting effect is clearly opposite to the negative effect on tuberization observed in the GA 20-oxidase lines, this result being indicative of different tuberization control activities of the non-3β-hydroxylated GA_{20} precursor and the 3β-hydroxylated GA_1 end-product, with GA_{20} being more

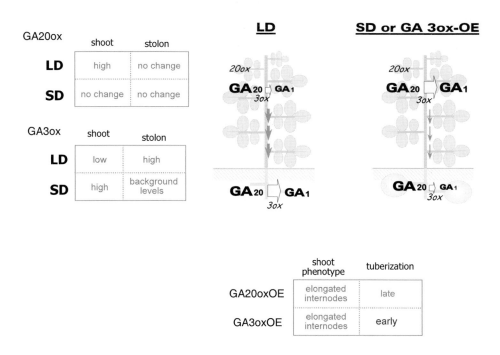

Figure 5: Hypothetical model for GA control of tuberization. GA 20-oxidase is highly expressed in the apex and young leaves both in LD and SD conditions. In LDs 3-oxidase activity is high in the stolons thus catalyzing conversion of GA_{20} to GA_1 and inhibiting tuber formation. Under SD conditions, 3-oxidase activity increases in the nodes and leaves, producing an increase in the rate of GA_{20} to GA_1 conversion in these organs and diverting transport of GA_{20} to the stolons. GA 3-oxidase is, at the same time, reduced to background levels in the stolons, GA_1 production in these organs being strongly reduced under these inducing conditions. Over-expression of 3-oxidase mimics SD conditions by increasing GA_{20} to GA_1 conversion in the leaves, and therefore induces early tuberization in SDs. High rates of GA_{20} to GA_1 conversion in the leaves are also likely to occur in the *phyB* antisense lines, where an elongated shoot phenotype occurs together with a strong tuberization induction.

active in the inhibition of tuberization than GA_1. A reduced flowering activity of 3β-hydroxylated GAs, has also been reported in *Lolium*, where 3β-hydroxylated GA_1 is most active in respect to stem elongation, while non-3β-hydroxylated GAs are more active in promoting flowering (8). Alternatively, a different degree of transport of GA_{20}, the GA_1 precursor, and GA_1 itself, from the leaves to the stolons, would also satisfactorily explain our observations. GA_{20} could be readily transported throughout the plant, whereas GA_1 preferentially remains in the vicinity of the cells where it is produced so that it exerts its action mainly in these and directly adjacent cells.

Several lines of experimental evidence are in support of this second hypothesis (Fig. 5). Independent evidence for a different mobility of these two GA molecules has indeed been also obtained in pea, by using grafting experiments between the *Le* and *Na* mutants and wild-type plants (32). According to this hypothesis, constitutive or leaf-specific over-expression of the GA 3-oxidase activity leads to an increase in the rate of GA_{20} to GA_1 conversion in the aerial tissues of the over-expresser potato plants. This brings about a reduction in the concentration of GA_{20} available for transport to the stolons, thus mimicking the changes in GA content that occur in SD-conditions. Similarly a related diversion of GA_{20} into GA_1 may occur in the *phyB* mutants, in which increased GA_{20} to GA_1 conversion could cause the observed phenotype. In these plants a *slender* phenotype, produced by a 4-6 fold increase in the shoot GA_1 content (27), paradoxically coexists with a strong induction of tuber formation. Therefore, contrary to the general idea that GAs inhibit tuberization, our observations indicate that high levels of GAs in the stolons do indeed inhibit tuberization, whereas a high rate of GA_{20} to GA_1 conversion in the shoot favours tuber formation.

Jasmonates

Ethanolic extracts from induced potato plants led to the identification of the glucoside of 12-OH jasmonic acid (12OHJA), or tuberonic acid (TA), with strong tuber-inducing properties *in vitro*. When added to the agar medium the glucoside of tuberonic acid (TAG) was found to stimulate tuberization at concentrations as low as 3×10^{-8} M. In these bioassays, jasmonic acid (JA), its methyl ester (MeJA), and TA showed all a similar promotive effect on tuberization, although activity of cucurbic acid, which differs from JA by having a hydroxyl group instead of oxygen at C-3, was much lesser.

Nevertheless, subsequent studies in *S. demissum*, which like *andigena* has an absolute SD requirement for tuberization, showed that in leaves of plants grown in SD, 11-OH-jasmonic was more abundant than 12-OH-jasmonic acid and no glycosides of 11-OHJA or 12-OHJA were observed in leaflets of these plants grown in either SD or LDs (19). Furthermore, repeated application of jasmonic acid on non-induced *andigena* leaves had no effect on tuber induction, although jasmonate was shown to be taken up and transported throughout the plant to induce the wound response (21).

Application of salicyl-hydroxamic acid, which inhibits one of the steps in the JA biosynthetic pathway, did not prevent tuberization of *S. demissum* in SD conditions (19). In addition elevated endogenous levels of JA produced by over-expression of the flax allene oxide synthase gene (*AOS*) did not correlate with increased tuber formation (18). This later result needs, however, to be interpreted with caution since constitutive expression of *pin2* or other wound-induced genes was also not observed in the transgenic *AOS* plants, despite the fact that these lines had levels of JA similar or even higher to those found in wounded plants (18).

These observations indicate that differences in the levels of JA themselves do not control tuberization, although it is still possible that tuberonic acid or other JA-related compounds cause tuberization under inductive conditions. In this respect, it has been pointed out that the glucosylated form of tuberonic acid (TAG) would be more easily transported out of the leaf and hence be a better candidate to play a role in tuber initiation. Thus, it was suggested that photoperiod might affect the activity of one or more enzymes controlling the hydroxylation of JA, thus making this compound glucosylatable and therefore susceptible to be transported from the leaf to the stolon to initiate tuber formation (19). In contrast to soil-grown plants, transport of bioactive compounds from the leaves is not required in *in vitro* experiments. In these bioassays, direct application of JA, MeJA, TA and TAG to the target tissues resulted in similar tuber inducing activities. These compounds are likely to promote tuberization by disrupting cortical microtubules of the stolon meristem cells and allowing lateral expansion of these cells, so that they exert their effect by antagonizing the effects of GA on microtubule orientation.

Although the glucoside of 12-OHJA was reported to be present only in leaves of tuberizing species like potato and Jerusalem artichoke, a gene encoding hydroxyjasmonate sulfotransferase has also been recently isolated from Arabidopsis (15). In this plant species 12-OHJA was found to be present at a level about 4-fold lower than JA, with preliminary results indicating that 12-OHJA is also present in barley, tomato, and tobacco leaves. The presence of 12-OHJA in these plant species indicates that its distribution can no longer be considered to be restricted to tuber-producing plants and it raises interesting questions concerning its function in plants that do not produce tubers. Expression of the Arabidopsis sulfotransferase enzyme is strongly induced by the JA precursor, 12-oxophytodienoic acid, JA and 12-OHJA. However, unlike JA or MeJA, 12-OHJA does not induce expression of wound-responsive genes, such as *Thi2.1*, indicating that MeJA and 12-OHJA have different biological activities, with 12-OHJA probably controlling developmental processes other than tuberization in plants (15).

While the debate on whether JA or related compounds are involved in inducing tuberization in soil-grown plants continues, evidence for the involvement of a lipoxygenase (LOX) product in tuberization has also been recently obtained. Transgenic potato plants with reduced levels of expression of a tuber-specific 9-LOX, did exhibit a much lower number of distorted tubers, and leaf bud cuttings of these plants failed to tuberize. Therefore, these results suggest the involvement of a fatty acid hydroperoxide metabolite derived from this LOX enzyme in the induction of tuber formation/tuber enlargement (23).

Cytokinins

Addition of cytokinin (CK) to the high-sucrose *in vitro* tuberization media has also been reported to increase speed of tuber induction. Cell division is one of the early events following tuber initiation, with the cell proliferation promoting effects of CKs being likely to be involved in these early phases of tuber growth. However, attempts to improve whole-plant tuberization by foliar applications of CK have given contradictory or ambiguous results and there are no reports that CK application to non-induced *andigena* plants would induce them to tuberize.

Additional evidence for a role of CKs in tuber formation derives from transgenic potato plants expressing either the *Agrobacterium rhizogenes rolC* gene (11, 36), or the *Agrobacterium* T-DNA-derived *ipt* gene encoding isopentenyltransferase, an enzyme that catalyzes a step of cytokinin biosynthesis that is rate-limiting in plants (14). The rolC protein has been found to hydrolyze *in vitro* conjugates of cytokinins, though its exact mechanism of action in plants is not understood, with some data indicating that *rolC* genes might change sensitivity of cells to CKs. Expression of the *rolC* gene under the control of a leaf-specific promoter in potato produces a slight reduction in plant size and increases number of axillary shoots compared to the controls (11), but does not significantly alter tuber yield. In contrast, plants expressing the *rolC* under the control of the tuber-specific B33 patatin promoter show a significant decrease in tuber yield (to about 50% of controls), these plants exhibiting a 3 to 4-fold increase in the number of tubers, with these tubers being of abnormal shape. Constitutive expression of the *ipt*-gene resulted in highly pleiotropic effects. An advance in the onset of tuberization was, however, observed in the transgenic with lower levels of over-expression of the transgene, application of benzyladenine further promoting tuberization in these plants (14). Formation of tuberized lateral branches, on the other hand, has been observed in transgenic tobacco lines specifically expressing the *ipt* gene in axillary buds. These lines, obtained by transformation with a promoterless *ipt* gene, showed a breakage of axillary bud dormancy during the floral phase, giving rise to very short and swollen lateral branches with short internodes and small narrow scale-leaflets. Starch grains and increased levels of extracellular invertase were observed in the internodes of these swelled lateral branches, thus providing evidence of the tuber-like nature of the morphological alteration (17). These results indicate that localized *ipt* expression can cause tuberization-related processes in a species that normally does not tuberize, by giving sink identity to the new cytokinin accumulating cells. Cytokinins have indeed been reported to have a role in sink creation by regulating the expression of genes implicated in assimilate partitioning and source-sink regulation such as invertase, sucrose synthase and hexose transporter genes (35). Therefore, these observations suggest that cytokinins may play a role in tuber formation by redirecting carbon fluxes towards the stolon cells.

Several recent reports have suggested a link between cytokinins and homeotic genes (38, 13) Increased steady state mRNA levels for the KNOTTED class of homeodomain proteins *KNAT1* and *STM* were observed in the *amp1* mutant of Arabidopsis overproducing cytokinin (38). Overexpression of *KNAT1* in lettuce, on the other hand, produced dramatic alterations in leaf shape which were characterized by a reduction in midvein elongation and altered margins with typical leaf-like structures at the serrations, these alterations being associated with an over-production of cytokinins (13). Likewise, over-expression of *POTH1*, a potato class I *knotted*-like homeobox gene isolated from early-stage tubers, produced dwarf plants with a mouse-ear leaf morphology, less prominent mid-veins and a palmate venation pattern (37). These alterations were associated with decreased levels of GAs and a 2- to 4-fold increase in bioactive cytokinins measured in the SAMs. *In vitro* tuberization was enhanced in the *POTH1* over-expressing lines, with these lines producing more tubers and at a faster rate than controls both under short- and long-day photoperiods (37). These results imply a possible involvement of the *POTH1 knox* gene in triggering tuber formation, by changing stolon cell fate through modification of both GA and cytokinin levels.

Other hormones

Evidence for a role of other classes of growth substances in tuber formation is much less consistent. ABA levels were reported to increase up to 4-fold in SD *andigena* plants compared to plants grown under LD or SD+NB conditions (25). An increase in ABA levels in potato shoots and roots was also observed upon removal of nitrogen supply in hydroponically grown potatoes (24), thus suggesting an association between increased ABA levels and tuberization. However, the potato *droopy* ABA-deficient mutant forms tubers normally, which indicates that ABA is not essential for tuberization, with the promotive effects of ABA being likely to be due to the antagonistic effects to GAs (47). Polyamines were also reported to be needed for tuberization of node explants grown *in vitro*, as treatments with difluoromethylornithine (DFMO), an inhibitor of the polyamine biosynthesis enzyme ornithine decarboxylase (ODC), inhibited tuber formation *in vitro* (31). These growth regulators were postulated to be required in cell division during early events of tuber formation and in agreement with a possible role at early induction; over-expression of the S-adenosyl-Met decarboxylase enzyme in potato has been reported to lead to an increase in tuber number accompanied by a decrease in tuber size (31).

CARBOHYDRATE METABOLISM

Induction of tuberization is difficult to study in plants grown in soil due to the low level of synchrony of tuber development and the need of plant manipulation to observe new forming tubers. Periodic inspection of the underground parts of the plant produces damage to the stolons and roots to a certain extent, therefore having a negative effect on tuber growth. As an alternative to soil-grown plant tuberization studies, *in vitro* methods have been developed, essentially consisting of single stem node explants cultured in high sucrose (8%) medium and incubated in darkness (7). In some instances, high sucrose content is supplemented with the gibberellin biosynthesis inhibitor CCC and a cytokinin, resulting in more rapid induction. Under these conditions axillary buds differentiate into a tuber instead of a leafy shoot, with the ultrastructure and protein content of these tubers being comparable to that of soil-grown tubers. These *in vitro* systems have been very useful for the analysis of the tuber-inducing properties of different substances, but are not suitable to the study of leaf-derived long-range signals, such as the phloem transported florigen/tuberigen. Although daylength is the environmental factor that most influences tuberization, these high-sucrose *in vitro* tuberization systems show little or no response to photoperiod, with tuber formation observed both in explants obtained from plants grown under LD or SD conditions.

Plants grown under conditions that favour tuber induction have a very high percentage of their assimilate directed to the tubers. A high supply of sucrose to developing stolons, or to stem node explants grown *in vitro*, favours tuber induction, so that sugar transport and starch formation have been promoted for some time as part of the inducing stimulus. Results obtained using transgenic approaches have nevertheless indicated that starch formation is not required for tuber induction. The most direct evidence in this direction was obtained from plants with an antisense repression of the ADP-glucose pyrophosphorylase enzyme, responsible for conversion of glucose-1-phosphate into ADP-glucose, which is the first committed precursor for starch synthesis. These plants were shown to exhibit significantly reduced levels of starch, but to display normal tuber formation, with an even larger number of tubers and a decrease in individual tuber weight, indicative of decreased sink competition between the tubers (29). Attempts to increase starch biosynthesis in tubers by increasing the level of ADP-glucose have also been undertaken, either by increasing the level of its immediate precursors in the starch biosynthetic pathway (41), or by increasing the activity of ADP-glucose pyrophosphorylase (42). While the former of these approaches failed, expression of ADP-glucose pyrophosphorylase displaying modified allosteric properties led to substantial increases in starch content. However, this did not correlate with an increased tuber number, which again indicates that the regulation of tuber induction occurs independently of starch accumulation.

Sucrose is the major form of photoassimilate transported to the tuber.

Whether import of sucrose into the tubers occurs via symplastic or apoplastic unloading has been a matter of controversy over many years. Using both fluorescent dies and radiolabelled sugar precursor, clear evidence has been obtained for a change in sucrose unloading from apoplastic to symplastic during tuberization (Fig. 6) (46). Concomitant with the first visible sign of tuber initiation, a change in the unloading mechanism from predominantly apoplastic to basically symplastic could be observed, this latter not requiring an active sucrose transport mechanism but involving transport through the plasmodesmata and subsequent metabolism of sucrose in the cytosol. In parallel with this switch in sucrose unloading, a decline of alkaline and acidic invertases and an increase of sucrose synthase activity could be observed (10), with acid invertase predominating during early stages of tuberization, but sucrose synthase (SuSy) becoming the major activity involved in sucrose breakdown during the storage phase of developing potato tubers.

Such a prevalent role of sucrose synthase with respect to sucrose metabolism in growing potato tubers has also been confirmed using a transgenic antisense approach. Starch accumulation was severely reduced in plants with decreased sucrose synthase activity, indicating that sucrose synthase is the major sucrose cleaving determinant during the storage phase. Tuber number and fresh weight was unaltered in these plants, further indicating that starch accumulation does not play a role in tuber initiation (50).

The impact of increased sucrose breakdown activity on tuber development has also been analyzed by targeting a yeast-derived invertase either to the cytosol or the apoplast of transgenic potato plants under the control of a tuber-specific class I patatin promoter (41). In both cases sucrose content declined whereas hexose content, especially glucose, increased in mature tubers. Simultaneous expression of a heterologous glucokinase led to a reduction of the glucose content, accompanied by decreased starch content (10). Surprisingly, whereas elevated cytosolic invertase was found to cause a severe reduction in tuber yield, overexpression of apoplastic invertase led to an increase in tuber size and a decrease in tuber number per plant. These data demonstrate that, in contrast to the minimal effects of starch synthesis, levels of cytosolic sucrose in the tuber would have a strong impact on tuber development, with a possible preferential use of these hexose substrates depending on the sucrose breakdown system involved at their generation and/or an inefficient uptake of hexose-phosphates into the amyloplast.

In this respect, it is interesting to note that more recent studies have shown that adenylate supply to the plastids is of fundamental importance for starch biosynthesis in storage organs (43). Overexpression or antisense inhibition of the amyloplastidial ATP:ADP translocator resulted, respectively, in increased and reduced starch accumulation. Down-regulation of the plastidial isoform of adenylate kinase had substantial effects on the pool size of the various adenylates leading to a large increase in the starch content of the tuber and increased tuber yield (34).

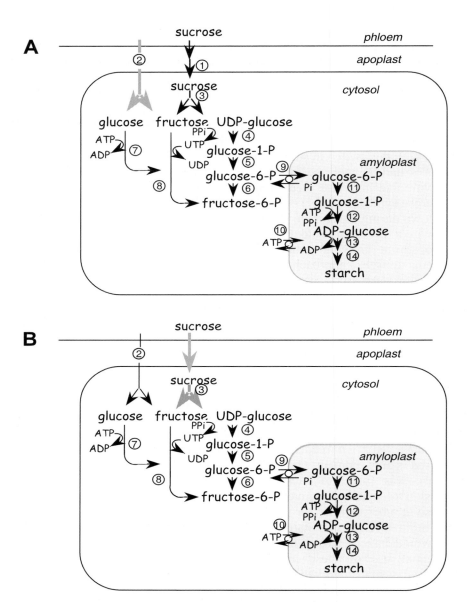

Figure 6: Carbon metabolism and starch synthesis in potato stolons and tubers. (A). In stolons undergoing extension growth sucrose import is predominantly apoplastic. (B). During tuberization there is a switch to symplastic transport. Concomitant with this switch, there is a marked decline in soluble invertase activity and a rise in sucrose synthase activity.
Numerical equivalents: 1, sucrose transporter; 2, invertase; 3, sucrose synthase; 4, UDP-glucose pyrophosphorylase; 5, cytosolic phosphoglucomutase; 6, phosphoglucose isomerase; 7, hexokinase; 8, fructokinase; 9, glucose 6-phosphate transporter; 10, adenylate transporter; 11, plastidial phosphoglucomutase; 12, ADP-glucose pyrophosphorylase; 13, starch synthase; 14, starch branching enzyme.

CONCLUSIONS

Induction of tuberization in response to environmental conditions implies a complex regulatory network involved in signalling stolon cell transition to tuber development. Although many plant growth substances have proved to be active at inducing tuber formation, it is important to make a distinction between the inducing events that take place in the leaves and those that directly affect the response of the stolons. The principal site of perception of the photoperiodic signal is in the leaves, with two independent negative regulatory pathways controlling tuber formation in response to environmental conditions, i.e. a LD-pathway that inhibits tuber formation in non-favorable daylength conditions and is activated by phyB, and a GA-dependent pathway, which is induced by GA_1 and promotes elongation growth of the stolons, thus preventing tuber formation. Cross-regulation between both pathways appears to occur, with, for example, levels of expression of the GA 3-oxidase enzyme regulated by both daylength and PHYB. Evidence has been obtained for a different inhibitory activity on tuberization of GA_{20} as compared to GA_1, likely due to a reduced transport of the GA_1 biosynthetic end-product towards the stolons. Under inductive SD conditions expression of GA 3-oxidase increases in the stem and leaves, but is strongly inhibited in the stolons, changes that result in a reduction in GA transport towards the stolons. A similar increased conversion of GA_{20} to GA_1 in the leaves may also occur in the *phyB* mutants, in which a 4-6 fold increase in the shoot GA_1 content was found to paradoxically coexist with a strong induction of tuber formation.

Reduction in stolon GA levels produce a longitudinal orientation of cell microtubules and microfibrils, thus triggering tuber formation by inducing lateral cell expansion and cell division of the cells at the subapical region of the stolons. Antagonists of GA action like jasmonic acid or ABA were found to promote tuberization *in vitro* by inducing a similar disruption of the cell cortical microtubules. It is possible that in addition to this effect in microtubule orientation, GA levels control carbohydrate metabolism directing sucrose utilization towards starch synthesis. Cytokinins and high sucrose are likely to be required for cell division and starch accumulation once tuber induction has occurred, with CKs reported to have a role in sink creation by regulating the expression of genes implicated in assimilate partitioning and source-sink regulation. In addition, evidence has been obtained for the involvement of a potato *CO* ortholog in tuberization control, which would act as signaling intermediate in the daylength-pathway, thereby regulating tuber formation in response to daylength conditions. Action of this regulatory protein is required in the leaves and appears to control generation of the tuberization inducing signal, thus serving as a connection point between the proper photoperiod-induced pathway and the tuber formation pathway. Defining the primary targets of the CO signaling intermediate will help to elucidate the molecular identity of the tuberization-inducing signal, and also to identify the tuberization pathway integrator genes that in response to both endogenous and environmental cues switch the developmental fate of the meristem cells at the subapical region of the stolon to differentiate into a tuber.

Acknowledgements

I thank Elmer Ewing for his helpful comments and suggestions.

References

1. Batutis EJ, Ewing EE (1982) Far-red reversal of red light effect during long-night induction of potato (*Solanum tuberosum* L.) tuberization. Plant Physiol 69: 672-674.
2. Bou J, Martínez-García J, García-Martínez JL, Prat, S (2003) Role of potato gibberellin 3beta-hydroxylase in the photoperiodic control of tuber induction, submitted for publication.
3. Carrera E, Jackson SD, Prat S (1999) Feedback control and diurnal regulation of gibberellin 20-oxidase transcript levels in potato. Plant Physiol 119: 765-74.
4. Carrera E, Bou J, García-Martínez JL, Prat S (2000) Changes in GA 20-oxidase gene expression strongly affect stem length, tuber induction and tuber yield of potato plants. Plant J 22: 247-256.
5. Chailakhyan MKh, Yanina LI, Devedzhyan AG, Lotova GN (1981) Photoperiodism and tuber formation in grafting of tobacco onto potato. Doklady Akademic Nauk SSSR 257: 1276.
6. Chapman HW (1958) Tuberization in the potato plant. Physiol Plant 11: 215–224.
7. Coleman WK, Donnelly DJ, Coleman SE (2001) Potato microtubers as research tools: a review. Am J Potato Res 78: 47-55.
8. Evans LT, King RW, Mander LN, Pharis RP (1994) The relative significance for stem elongation and flowering in *Lolium temulentum* of 3β-hydroxylation of gibberellins. Planta 192: 130-136.
9. Ewing EE, Struik PC (1992) Tuber formation in potato: induction, initiation and growth. Hortic Rev 14: 89-197.
10. Fernie AR, Willmitzer L (2001) Molecular and biochemical triggers of potato tuber development. Plant Physiol 127: 1459-1465.
11. Fladung M, Ballvora A, Schmülling T (1993) Constitutive or light regulated expression of the *rolC* gene in transgenic potato plants alters yield attributes and tuber carbohydrate composition. Plant Mol Biol 23: 749-757.
12. Fujino K, Koda Y, Kikuta Y (1995) Reorientation of cortical microtubules in the sub-apical region during tuberization in single node stem segments of potato in culture. Plant Cell Physiol 36: 891-895.
13. Frugis G, Giannino D, Mele G, Nicolodi C, Chiappetta A, Bitonti MB, Innocenti AM, Dewitte W, van Onckelen H, Mariotti D (2001) Overexpression of *KNAT1* in lettuce shifts leaf determinate growth to a shoot-like indeterminate growth associated with an accumulation of isopentenyl-type cytokinins. Plant Physiol 126: 1370-1380.
14. Gális I, Macas J, Vlasák J, Ondrej M, van Onckelen H (1995) The effects of an elevated cytokinin level using the *ipt* gene and N6-benzyladenine on single node and intact potato plant tuberization *in vitro*. J Plant Growth Regul 14: 143-150.
15. Gidda SK, Miersch O, Levitin A, Schmidt J, Wasternack C, Varin L (2003) Biochemical and molecular characterization of a hydroxyjasmonate sulfotransferase from *Arabidopsis thaliana*. J Biol Chem 278: 17895-17900.
16. Gregory LE (1956) Some factors for tuberization in the potato. Ann Bot 41: 281-288.
17. Guivarc'h A, Rembur J, Goetz M, Roitsch T, Noin M, Schmülling T, Chriqui D (2002) Local expression of the *ipt* gene in transgenic tobacco (*Nicotiana tabacum* L. cv. SR1) axillary buds establishes a role for cytokinins in tuberization and sink formation. J Exp Bot 53: 621-629.
18. Harms K, Atzorn R, Brash A, Kuhn H, Wasternack C, Willmitzer L, Peña-Cortes H (1995) Expression of a Flax Allene Oxide Synthase cDNA Leads to Increased

Endogenous Jasmonic Acid (JA) Levels in Transgenic Potato Plants but Not to a Corresponding Activation of JA-Responding Genes. Plant Cell 7: 1645-1654.
19. Helder H, Miersch O, Vreugdenhil D, Sambdner G (1993) Occurrence of hydroxylated jasmonic acids in leaflets of *Solanum demissum* plants grown under long- and short-day conditions. Physiol Plant 88: 647-653.
20. Jackson SD, Heyer A, Dietze J, Prat S (1996) Phytochrome B mediates the photoperiodic control of tuber formation in potato. Plant J 9: 159-166.
21. Jackson SD (1999) Multiple signalling pathways control tuber induction in potato. Plant Physiol 119, 1-8.
22. Koda Y, Okazawa Y (1983) Influences of environmental, hormonal and nutritional factors on potato tuberization *in vitro*. Jpn J Crop Sci 52: 582-591.
23. Kolomiets MV, Hannapel DJ, Chen H, Tymeson M, Gladon RJ (2001) Lipoxygenase is involved in the control of potato tuber development. Plant Cell 13: 613-26.
24. Krauss A (1985) Interaction of nitrogen nutrition, phytohormones and tuberization. *In* PH Li, *ed*, Potato Physiology. Academic Press, London, pp 209–231.
25. Machackova I, Konstantinova TN, Seergeva LI, Lozhnikova VN, Golyanovskaya SA, Dudko ND, Eder J, Aksenova NP (1998) Photoperiodic control of growth, development and phytohormone balance in *Solanum tuberosum*. Physiol Plant 102: 272–278.
26. Martínez-García JF, Virgós-Soler A, Prat S (2002) Control of photoperiod-regulated tuberization in potato by the Arabidopsis flowering-time gene *CONSTANS*. Proc Natl Acad Sci USA 99: 15211-15216.
27. Martínez-García J, García-Martínez JL, Bou J, Prat S (2002) The interaction of gibberellins and photoperiod in the control of potato tuberization. J Plant Growth Regul 20: 377-386.
28. Menzel CM (1983) Tuberization in potato at high temperatures: gibberellin content and transport from buds. Ann Bot 52: 697-702.
29. Müller-Röber B, Sonnewald U, Willmitzer L (1992) Inhibition of the ADPglucose pyrophosphorylase in transgenic potatoes leads to sugar storing tubers and influences tuber formation and expression of tuber storage protein genes. EMBO J 11: 1229-1238.
30. Olszewski N, Sun TP, Gubler F (2002) Gibberellin signaling: biosynthesis, catabolism, and response pathways. Plant Cell 14: 61-80.
31. Pedros AR, MacLeod MR, Ross HA, McRae D, Tiburcio AF, Davies HV, Taylor MA (1999) Manipulation of S-adenosylmethionine decarboxylase activity in potato tubers. An increase in activity leads to an increase in tuber number and a change in tuber size distribution. Planta 209: 153-60.
32. Proebsting WM, Hedden P, Lewis ML, Croker SJ, Proebsting LN (1992) Gibberellin concentration and transport in genetic lines of pea. Plant Physiol 100: 1354-1360.
33. Putterill J, Robson F, Lee K, Simon R, Coupland G (1995) The *CONSTANS* gene of Arabidopsis promotes flowering and encodes a protein showing similarities to zinc finger transcription factors. Cell 80: 847-857.
34. Regierer B, Fernie AR, Springer F, Perez-Melis A, Leisse A, Koehl K, Willmitzer L, Geigenberger P, Kossman J (2002) Starch content and yield increase as a result of altering adenylate pools in transgenic plants. Nat Biotech 20: 1256-1260.
35. Roitsch T, Ehneβ R (2000) Regulation of source/sink relations by cytokinins. Plant Growth Regul 32: 359-367.
36. Romanov GA, Aksenova NP, Konstantinova TN, Golyanovskaya SA, Kossman J, Willmitzer L (2000) Effect of indole-3-acetic acid and kinetin on tuberisation parameters of different cultivars and transgenic lines of potato *in vitro*. Plant Growth Regul 32: 245-251.
37. Rosin FM, Hart JK, Horner HT, Davies PJ, Hannapel D (2003) Overexpression of a *Knotted*-like homeobox gene of potato alters vegetative development by decreasing gibberellin accumulation. Plant Physiol 132: 106-117.
38. Rupp HM, Frank M, Werner T, Strnad M, Schmülling T (1999) Increased steady state mRNA levels of the *STM* and *KNAT1* homeobox genes in cytokinin overproducing *Arabidopsis thaliana* indicate a role for cytokinins in the shoot apical meristem. Plant J 18: 557-563.

39. Simpson GG, Dean C (2002) Arabidopsis, the Rosetta Stone of Flowering Time? Science 296: 285-289.
40. Snyder E, Ewing EE (1989) Interactive effects of temperature, photoperiod and cultivar on tuberization of potato cuttings. Hortic Sci 24: 336–338.
41. Sonnewald U, Hajirezaei MR, Kossmann J, Heyer A, Trethewey RN, Willmitzer L (1997) Increased potato tuber size resulting from apoplastic expression of a yeast invertase. Nat Biotechnol 15: 794-797.
42. Stark DM, Timmerman KP, Barry GF, Preis J, Kishore GM (1992) Regulation of the amount of starch in plant tissues by ADP glucose pyrophosphorylase. Science 258: 287-292.
43. Tjaden J, Mohlmann T, Kampfenkel K, Henrichs G, Neuhaus HE (1998) Altered plastidic ATP/ADP-transporter activity influences potato (*Solanum tuberosum* L.) tuber morphology, yield and composition of tuber starch. Plant J 16: 531-540.
44. van den Berg JH, Simko I, Davies PJ, Ewing EE, Halinska A (1995) Morphology and (^{14}C) gibberellin A$_{12}$ aldehyde metabolism in wild-type and dwarf *Solanum tuberosum* ssp. *andigena* grown under long and short photoperiods. J Plant Physiol 146: 467-473.
45. van den Berg JH, Ewing E, Plaisted RL, McMurry S, Bonierbale MW (1996) QTL analysis of potato tuberization. Theor Appl Genet 93: 307-316.
46. Viola R, Roberts AG, Haupt S, Gazzani S, Hancock RD, Marmiroli N, Machray GC, Oparka KJ (2001) Tuberization in potato involves a switch from apoplastic to symplastic phloem unloading. Plant Cell 13: 385-398.
47. Xu X, van Lammeren AAM, Vermeer E, Vreughdenhil D (1998) The role of gibberellin, abscisic acid and sucrose in the regulation of potato tuber formation *in vitro*. Plant Physiol 117: 575-584.
48. Xu X, Vreugdenhil D, van Lammeren AMM (1998) Cell division and cell enlargement during potato tuber formation. J Exp Bot 49: 573-582.
49. Yanovsky MJ, Izaguirre M, Wagmaister JA, Jackson SD, Thomas B, Casal JJ (2000) Phytochrome A resets the circadian clock and delays tuber formation under long days in potato. Plant J 23: 223-232.
50. Zrenner R, Salanoubat M, Willmitzer L, Sonnewald U (1995) Evidence of the crucial role of sucrose synthase for sink strength using transgenic potato plants. Plant J **7**: 97-107.

E6. The Hormonal Regulation of Senescence

Susheng Gan
Department of Horticulture, Cornell University, Ithaca, New York, 14850 USA. E-mail: sg288@cornell.edu

INTRODUCTION

Senescence occurs ubiquitously in living organisms. At the cellular level, a cell's life history consists of two processes: mitotic division and post-mitotic life pattern (6). A mother cell or germ-like cell can undergo a finite number of divisions to produce daughter cells. When the cell ceases dividing, this cell is said to undergo mitotic senescence. In literature concerning yeast and mammalian cells in culture, this type of senescence is often referred to as replicative senescence or replicative aging. Although the mitotically senescent cell can no longer divide, it may live for a certain period before its ultimate attrition/death; the functional degenerative process of the cell is called post-mitotic senescence. The degeneration of a neuron and the dying and peeling of a skin cell represent mitotic senescence in nature. Plants exhibit both mitotic and post-mitotic senescence. A typical example of mitotic senescence in plants is the arrest of shoot apical meristems. Meristem consists of non-differentiated, germline-like cells that can divide for a certain finite number of times to produce cells that will then differentiate to form a primodium of a new organ such a leaf or a flower. The arrest of the apical meristem is also called proliferative senescence. Post-mitotic senescence occurs in plant organs such as a leaf and a floral sepal, petal or carpel. Once formed, cells in these organs rarely divide further but undergo cell expansion and later a massive, genetically-programmed cell degeneration. The process encompassing the period from maturation to ultimate cell death of these organs is called leaf, floral or fruit senescence, respectively (6). This chapter[1] focuses primarily on post-mitotic senescence, leaf senescence in particular.

---

[1] Abbreviations: 2,4-D, 2,4-dichlorophenoxyacetic acid; ACC, 1-aminocyclo propane-1-carboxylic acid; AMP, adenosine 5'-phosphate; AOA, aminooxyacetic acid; ATP, adenosine 5'-triphosphate; BA, benzylaminopurine; F_m, maximal chlorophyll fluorescence; F_v, variable chlorophyll fluorescence; GA, gibberellic acid; HXK, hexokinases; JA, jasmonic acid; MAP kinase, mitogen-activated protein kinase; MeJA, methyl jasmonate; NAA, naphthylacetic acid; NS, nonsenescent; OPR, 12-oxo-phytodienoic acid-10,11-reductase; PA, polyamine; ROS, reactive oxygen species; Rubisco, ribulose bisphosphate carboxylase/oxygenase; SAGs, senescence-associated genes.

Senescence of leaves, flowers and fruits can be regulated by an array of external and internal factors. Many environmental stresses (such as extreme temperatures, drought, nutrient deficiency, insufficient light/shade or total darkness) and biological insults (such as pathogen infection) can induce senescence. Internal factors influencing senescence include age, levels of plant hormones and other growth substances, and developmental processes such as reproductive growth. These factors may act individually or in combination. The control of senescence by some environmental factors may be mediated by plant hormones. For example, ethylene may be responsible for senescence induced by chilling stress, and ABA may serve as a signal for the induction of senescence by drought. Therefore, plant hormones and related growth substances play important roles in regulating senescence. In general, ethylene, ABA, jasmonate and its derivative methyl jasmonate, salicylic acid, brassinosteroids, and sugars promote senescence; sugars have recently been shown to be an important class of plant growth regulators. In contrast, cytokinins, auxins, gibberellins, and polyamines are retardants of senescence.

Three approaches have been frequently employed to investigate the role of a specific plant hormone/growth substance in senescence, including (i) external application of a substance, (ii) correlation studies by measuring hormone levels in organs before and after the onset of senescence, and (iii) molecular genetic approach by which the levels or the perception/signaling of an endogenous hormone are altered in mutant or transgenic plants. This chapter presents evidence, especially that from recent molecular genetic studies, for the regulatory function of some of the plant hormones and growth substances in leaf senescence. The underlying molecular mechanisms will also be discussed. To facilitate the discussions, it is necessary to first review the cellular, biochemical and molecular aspects of leaf senescence.

OVERVIEW OF LEAF SENESCENCE

Leaf Senescence Syndrome

Leaf senescence is the final phase from maturation to attrition in the leaf life history. Leaf senescence also occurs in detached/harvested leaves. The visible sign of leaf senescence is yellowing, although other colors may also develop, as seen in autumn leaves. The yellowing results from preferential degradation of the green pigment chlorophyll but not the yellow carotenoid pigments. The autumn leaf colors also result from the biosynthesis of anthocyanins and other pigments that accompanies leaf senescence in some plant species. Structurally, the chloroplast, the organelle that contains up to 70% of proteins in a leaf mesophyll cell, is broken down first, and the nucleus is the last organelle to be dismantled during senescence. Functionally, because of the loss of chlorophylls, the photosynthetic capacity of a leaf decreases sharply during senescence. The anabolism of macromolecules, such as proteins, lipids, and nucleotides (DNA and RNA), is replaced by catabolism (Fig. 1), and most of the released nutrients are recycled to active growing regions, such as new buds, young leaves, developing fruits and seeds, or stored in the stems of perennial plants for the next growing season (16). The nutrient recycling process is thus considered to be important in evolutionary fitness. The massive operations of catabolism and nutrient mining lead to the ultimate cell death.

Molecular Aspect of Leaf Senescence

The onset of leaf senescence represents a significant developmental transition in a leaf's life history from a functional photosynthetic organ to a degradative phase of senescence. This developmental switch, like many other developmental processes, is genetically controlled. Changes in

Figure 1 (Color plate page CP15). Phenotypic and biochemical changes during leaf senescence in Arabidopsis. A. Stages of leaf senescence. NS, nonsenescent, fully expanded green leaf; S1, first visible signs of senescence (yellowing): chlorophyll loss at leaf tip; S2 – S5 are up to 25%, 25-50%, 50-75%, and greater than 75% loss of chlorophyll, respectively. B. Changes in the relative amount of chlorophyll (Chl), protein and RNA during leaf senescence (5).

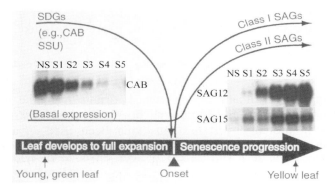

Figure 2. Differential gene expression during leaf senescence. Senescence down-regulated genes (SDGs) include photosynthesis-related genes such as the chlorophyll a/b-binding protein gene (*CAB*) and the Rubisco small subunit gene (*SSU*). Senescence-associated genes (SAGs) are those genes whose expression is up-regulated during leaf senescence. Class I SAGs (e.g., *SAG12*) are senescence specific while Class II SAGs (e.g., *SAG15*) have a basal level of expression during early leaf development (9). Stages (NS, S1–S5) as in Fig. 1.

gene expression associated with leaf senescence have been well document. The vast majority of genes that are expressed in young, photosynthetically active leaves are down regulated at the onset of senescence; these include the chlorophyll a/b binding protein gene (*CAB*) and the Rubisco small subunit gene (*SSU*) (Fig. 2). In contrast, the expression of a subset of genes is up regulated during leaf senescence; these genes have been named senescence-associated genes (SAGs). Some of the SAGs are already expressed at a low level prior to the onset of senescence, but their expression is elevated in senescing cells. Some other SAGs are solely expressed in senescing cells. Many SAGs have been cloned from various plant species, including Arabidopsis, aspen, barley, Brassica, maize, soybean, and tomato. Recent genomic analysis in Arabidopsis (13) has revealed that nearly 2,500 genes (10% of the genes in the Arabidopsis genome) are expressed in senescent leaves. As expected, many of the genes are predicted to encode proteinases, lipases and other enzymes that are involved in the degradation of proteins, lipids, polysaccharides, and other cellular components, while the predicted products of some other genes are related to nutrient recycling. Most strikingly, among the genes are 134 transcription factor genes and 182 genes encoding components of various signal transduction pathways such as MAP kinase cascades. Transcription factors often act as a switch to cause differential gene expression by binding to specific *cis* elements of target gene promoters, resulting in the activation or suppression of the target genes (13).

Although leaf senescence involves both activation and inactivation of different sets of genes, the inactivation *per se* is not sufficient to cause senescence; the activation of SAGs are required in order for a leaf cell to undergo the active degenerative process. This is supported by two major lines of evidence: (i) inhibitors of RNA and protein biosynthesis can block leaf senescence, and (ii) chloroplasts in enucleated protoplasts possess a longer life span than those in nucleated protoplasts (6). In other words, SAG expression is perhaps the driving force of leaf senescence at the molecular level, and as discussed in later sections, gene expression is likely the target for hormonal regulation.

Indices of Leaf Senescence

Leaf yellowing, the hallmark of senescence, results from the preferential degradation of chlorophyll. Chlorophyll degradation (sometimes together with changes in protein level) has thus long been used as an index of senescence in leaves. As shown in Figure 1, the loss of chlorophyll correlates very well with the biochemical degradation of proteins and RNA in Arabidopsis. However, in some mutants, the chlorophyll degradation is slowed down, or chlorophyll biosynthesis is enhanced, rendering leaves green for a much prolonged period, but the leaves have little photosynthetic activity (so called "stay-green" mutants). Therefore, the levels of chlorophyll may not always reflect declines in leaf function. Photosynthetic activity has thus been used as another criterion for leaf senescence. In practice, the ratio of variable to maximal chlorophyll fluorescence (F_v/F_m) is used as an index of senescence. F_v/F_m reflects the photochemical quantum efficiency of photosystem II, as well as the photoreduction efficiency of the primary electron-accepting plastoquinone of photosystem II. Although the F_v/F_m ratio in fully expanded young leaves varies from 0.7 to 0.85, depending on plant species and light conditions, it decreases with the progression of leaf senescence. The third index for senescence is senescence-specific molecular markers. For example, *SAG12*, a gene encoding a cysteine proteinase in Arabidopsis, is highly senescence specific and has been widely used as a molecular marker for leaf senescence. The molecular markers are especially useful to distinguish leaf senescence from other cell death processes such as the hypersensitive response in disease resistant plants (34).

THE PROMOTION OF LEAF SENESCENCE

Ethylene and Leaf Senescence

Ethylene is involved in many aspects of plant growth and development including the promotion of seed germination, formation of adventitious roots, sex determination, abscission and the senescence of fruits, flowers and leaves. Ethylene plays a key role in promoting senescence of climacteric fruits and flowers although it is less effective in stimulating non-climacteric fruits and flowers to senesce. Climacteric refers to the phenomenon of a sudden rise of respiration rate during fruit or floral senescence in certain plant species such as apple, tomato and mango; in contrast, such a respiratory rise is not observed in non-climacteric plants such as grape, lemon and strawberry. An increased level of ethylene production occurs immediately prior to (and overlaps with) the respiratory increase (Fig. 3), and it is this that causes the respiratory surge and the related senescence process. Suppression of ethylene production and/or perception by various means has been an effective approach to the extension of storage and shelf life of many climacteric fruits and flowers in practice. A climacteric-like respiratory surge has also been observed in senescing leaves of several plant species, such as barley, laurel, ivy, oat and tobacco, for which ethylene is

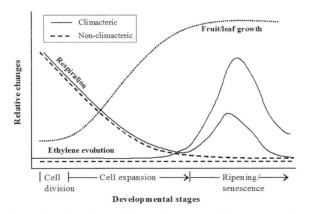

Figure 3. General correlations between ethylene production and respiration rate in climacteric vs. non-climacteric plants during fruit/leaf senescence

believed to be responsible. This correlation between ethylene production and leaf senescence in climacteric species supports a regulatory role of ethylene in senescence.

Another line of evidence comes from experiments involving treatment of leaves with ethylene; most of these experiments show that application of ethylene can promote changes that are characteristic of leaf senescence in many (but not all) plant species, including degradation of chlorophyll, proteins and other macromolecules, increase of proteinase, and other catabolic enzymatic activities (27).

The third line of evidence comes from studies involving the genetic manipulation of ethylene production. The last two steps in the ethylene biosynthesis pathway are catalyzed by ACC synthase and ACC oxidase respectively. Antisense suppression of genes encoding either of the enzymes has been used to block the ethylene biosynthesis pathway in transgenic plants. For example, transgenic tomato plants containing either of the antisense constructs produce much reduced levels of ethylene (11). The onset of leaf senescence in ACC oxidase antisense tomato plants, as assessed by chlorophyll loss, was delayed by 10-14 days compared with that of wild-type plants, but the senescence progressed normally (23). As expected, in addition to leaf senescence, fruit senescence in these transgenic plants was also significantly retarded (11).

Analysis of mutants deficient in ethylene perception and signal transduction further supports a promoting role of ethylene in leaf senescence. The Arabidopsis *ETR* gene encodes an ethylene receptor. A mutation in this gene, *etr1-1*, makes the receptor unable to bind ethylene, thus rendering the mutant plants insensitive to ethylene. Leaves of the *etr1-1* mutant plants live 30% longer than the wild-type leaves, and the increased longevity is primarily due to the delayed onset of senescence; leaf senescence in *etr1-1* progresses at a rate that is essentially identical to that of the wild-type control (Fig. 4) (12). Similarly, leaf senescence of the Arabidopsis mutant *ein2* is also delayed (33). EIN2 is a component in the ethylene signal transduction

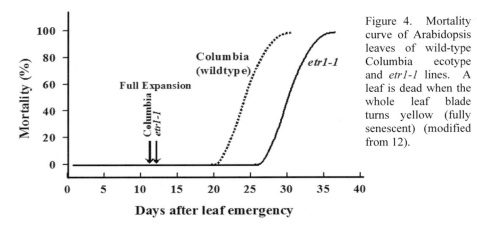

Figure 4. Mortality curve of Arabidopsis leaves of wild-type Columbia ecotype and *etr1-1* lines. A leaf is dead when the whole leaf blade turns yellow (fully senescent) (modified from 12).

pathway, and *ein2* renders the plants insensitive to ethylene.

It should, however, be pointed out that not all experiments support a promoting role for ethylene in leaf senescence. For example, both Arabidopsis plants grown in the continuous presence of ethylene and the Arabidopsis mutant *ctr1*, which displays constitutive ethylene responses, do not show a precocious leaf senescence phenotype. Similarly, no premature leaf senescence is observed in ethylene-overproducing tomato plants that constitutively express the ACC synthase gene (12). Considering the above, and the fact that leaves of plants deficient in either ethylene production or perception undergo senescence (although delayed), it has been concluded that ethylene alone is neither necessary nor sufficient for promoting leaf senescence, at least in Arabidopsis and tomato plants (12). Ethylene may promote senescence only in leaves that are mature or old enough. It also should be noted that ethylene is generally less effective in promoting leaf senescence than in fruit and floral senescence.

Sugars and Leaf Senescence

In plants, sugar status modulates and coordinates internal regulators and environmental cues that govern growth and development (40). It has been generally agreed that sugar levels regulate leaf senescence but the effect of high vs. low sugar levels on leaf senescence has been controversial. Leaves are the primary site where sugars are produced through photosynthesis. It is known that the photosynthetic activity declines sharply during leaf senescence, and that photosynthetic activity is low in leaves in shadow or darkness. Both shadow and darkness induce leaf senescence. This led to the proposal that the low level of sugars induces leaf senescence (18, 35). There are, however, three lines of evidence that a high level of sugars triggers leaf senescence program. First, sugar levels are higher in senescing leaves of Arabidopsis and tobacco plants (26, 42). Second, when yeast invertase is expressed in the extracellular space of leaves of Arabidopsis, tobacco, and tomato plants, sugars are accumulated, and the leaves undergo premature senescence (6). Similarly, a sugar hypersensitive mutant in Arabidopsis causes accumulation of sugars (as well as starch) in leaves (30); the leaves of the mutant plants appear to undergo precocious senescence.

Sugar levels are believed to be sensed by hexokinases (HXKs), the gateway enzyme of the glucose metabolism. When a HXK was overexpressed in tomato plants, the leaves became precociously senescent as if they contained an increased concentration of sugars (although the actual sugar concentration was lower than that of control plants). In contrast, the Arabidopsis *hxk1* knockout/null mutant plants displayed a delayed leaf senescence phenotype in the presence of high levels of sugars that were obtained either by an exogenous sugar application or by increasing light intensity (thus increasing photosynthetic sugar production) (28). The sugar sensing in plants is independent of the catalytic activity of HXKs, because even a mutant HXK1 gene producing a defective ATP binding site or catalytic site restored the *hxk1* null line's ability to sense sugars (28). The decoupling of sugar sensing and metabolism clearly indicates that leaf senescence triggered by accumulating sugars most likely involves mechanisms other than enhanced sugar metabolism and calorie production.

Jasmonic Acid (JA) and Leaf Senescence

JA and its derivative methyl jasmonate (MeJA) are a class of plant growth regulator that plays pervasive roles in several other aspects of plant development including seed germination, wounding response and leaf senescence (19). In fact, the first demonstrated biological function of JA was the promotion of senescence in detached oat leaves (43). Exogenous application of JA at a physiological level of 30 µM causes premature senescence of attached and detached leaves in wild-type Arabidopsis, but fails to induce precocious senescence of JA-insensitive mutant *coi1* plants (Fig. 5), suggesting that JA-signaling pathway is required for JA to promote leaf senescence (15). Genes involved in JA biosynthesis are shown to be

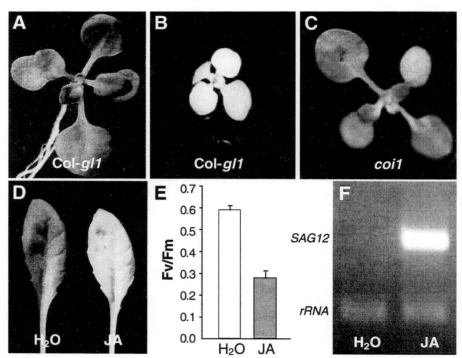

Figure 5 (Color Plate page CP15). The effect of exogenous JA on leaf senescence in Arabidopsis. Wild-type Col-*gl1* (A and B) and JA-insensitive mutant *coi1* (C) were grown on agar containing 0 (A) or 30 (B and C) µM JA for 12 days. D. Detached young wild-type leaves treated with water or 30 µM JA for 4 days under darkness. E. Variable fluorescence (F_v)/maximal fluorescence (F_m) values of leaves in D. F. Expression of the senescence-specific marker gene *SAG12* in leaves shown in D (15).

differentially up-regulated during leaf senescence, and JA levels in senescing leaves are 4-fold higher than in non-senescing leaves (15). The JA-induced leaf yellowing was indeed an active senescence process, because several *SAG*s are expressed in the yellowing leaves (Fig. 5) (16).

However, a promoting role for JA in leaf senescence has not been without controversy. For example, transgenic potato plants constitutively expressing a flax allene oxide synthase gene (*AOS*) overproduce JA (14), but no visible early senescence is observed. Also, the Arabidopsis triple mutant *fad3-2 fad7-2 fad8* and the *OPR3* T-DNA knockout lines produce little JA, but leaf senescence is not significantly delayed (15). Nonetheless, these facts may not necessarily contradict a senescence-promoting role of JA and derivatives. The overproduced JA might be sequestered so that it could not exert its biological action because JA-inducible genes are not expressed in the JA-overproducing potato plants (14). The JA-underproducing Arabidopsis mutants might not display a significantly retarded leaf senescence phenotype because of the plasticity of leaf senescence (9). It is possible that other factors may induce leaf senescence in the absence of JA.

Salicylic Acid (SA) and Leaf Senescence

Although SA has been well studied as a signaling molecule in plant disease defense responses, its role in leaf senescence has not been recognized until recently. There are two lines of evidence for a controlling role of SA in leaf senescence in Arabidopsis. First, levels of endogenous SA in Arabidopsis leaves increase with the progression of senescence: the concentration of SA in fully expanded, mature yet green leaves is 128 ng g^{-1} fresh leaves, whereas the SA level increases to 206 and 494 ng g^{-1} in leaves that are at the onset of senescence or that display visible yellowing, respectively (29). Second, Arabidopsis plants that produce much less SA due to overexpression of a SA-degrading enzyme (salicylate hydroxylase) gene *NahG* (from *Pseudomonas putida*), and Arabidopsis mutants defective in SA-signaling such as *npr1* and *pad4*, display a retarded (by approximately 4 days) leaf senescence phenotype and an altered expression of some *SAG*s (29). More research is needed to confirm the regulatory role of SA in leaf senescence.

Brassinosteroids (BRs) and Leaf Senescence

Plant steroid hormones play an essential role in diverse developmental programs including leaf senescence. Two lines of evidence suggest a leaf senescence-promoting role of BRs. First, external application of 24-*epi*-brassinolide (*e*BR) induces senescence in mung bean leaves and cucumber cotyledons (1). Second, several Arabidopsis mutants that are deficient in either BR biosynthesis (e.g., *det2*) or in the BR signaling pathway (e.g., *bri1*) display a delayed leaf senescence phenotype, in addition to other characteristic changes (1), whereas the *bri1* suppressor mutant exhibits accelerated leaf senescence (47). Reactive oxygen species (ROS) may mediate the BR-induced senescence because after BR treatment the level of malondialdehyde increases markedly while the activities of superoxide dismutase and catalase are inhibited (16). *e*BR also induces a subset of potential *SAG*s in Arabidopsis (17).

Abscisic Acid (ABA) and Leaf Senescence

External application of ABA promotes senescence in detached leaves, but is much less effective *in planta* (3, 20, 31). Levels of ABA increase at the onset of and during senescence in leaves, but decease at the late stage of senescence (31). ABA was believed to be the strongest promoting factor of leaf senescence, second only to ethylene (31). However recent molecular genetic studies of many mutants deficient in either ABA biosynthesis or perception do not reveal specific roles for ABA in promoting leaf senescence (4). The function of ABA in leaf senescence remains to be defined.

THE INHIBITION OF LEAF SENESCENCE

Cytokinins and Leaf Senescence

The cytokinins were first discovered as a factor that promotes cell division in plant cell and tissue culture. Thus cytokinins are inhibitors of mitotic senescence in plants. Intriguingly, kinetin has also been shown to delay mitotic senescence in human cells in culture and prolong the lifespan of fruitflies (41). As discussed elsewhere (Chapters A1, B3, C3 and D3), cytokinins are involved in almost all stages of plant growth and development including the post-mitotic leaf senescence. Among the plant hormones, cytokinins are perhaps the most studied hormone in terms of the regulatory role in leaf senescence. There are three major lines of evidence supporting an inhibitory role of cytokinins in leaf senescence. First, external application of cytokinins to detached and *in planta* leaves can prevent the leaves from senescing. This cytokinin effect was first observed in detached cocklebur leaves by Richmond and Lang (39). External application of cytokinins can even cause a yellowing leaf to rejuvenate or regreen. A large body of literature dealing with cytokinin treatment experiments has developed; in general, these experiments demonstrate that cytokinins can interfere with senescence in detached tissues of both dicotyledons and monocotyledons, and in fact has been developed into a bioassay for measuring cytokinin levels (8). It should be noted that externally applied cytokinins are not always effective in blocking senescence of excised leaves, and that the effect of cytokinins on senescence can also be different under different experimental conditions. For example, treatment with 6-benzyladenine retards senescence in Arabidopsis leaves in the dark but accelerates the loss of chlorophyll in the light (8).

Second, analysis of cytokinin levels in leaves before and after the onset of senescence has revealed an inverse correlation between cytokinin levels and the progression of senescence in a variety of tissues and plant species.

For example, cytokinin levels, as well as the capacity to synthesize cytokinins, drops with the progression of leaf senescence in tobacco plants (8, 44). Although leaves are capable of synthesizing cytokinins, roots are the primary site of cytokinin production, and the cytokinins produced in roots are transported through the xylem to the shoot including leaves. Therefore, there is also an inverse correlation between the cytokinin levels in the xylem sap and the stages of leaf development and senescence (8). For example, cytokinin levels in xylem sap of soybean plants drop sharply with the onset of leaf senescence. Similarly, the amount of zeatin and zeatin riboside in xylem sap in mature plants of the non-senescent 'Tx2817' cultivar of sorghum is higher than that of the 'Tx7000' cultivar, which senesces normally. Another example is rice. Akenohoshi is a late-senescing rice cultivar and Nipponbare is a cultivar that senesces normally. The xylem sap of Akenohoshi contains a higher level of cytokinins than does Nipponbare.

Third, genetic manipulation of cytokinin production in transgenic plants has provided very conclusive evidence for an inhibitory role of cytokinins in leaf senescence. Unlike some other hormones such as ethylene, a hunt for mutants that over- or underproduce cytokinins, or are deficient in cytokinin perception and signaling has not been fruitful. This is perhaps partly due to the pervasive involvement of cytokinins in many aspects of plant growth and development such that a related genetic lesion may cause lethality. To circumvent this a transgenic approach has been employed to manipulate cytokinin biosynthesis in a variety of plant species. The cytokinin biosynthesis pathway begins with the transfer of an isopentenyl group from isopentenyl pyrophosphate to adenosine 5'-phosphate (AMP) to form isopentenyladenine ribotide, which is subsequently converted into various cytokinin forms (8) (Chapter B3). This first step is catalyzed by an isopentenyl transferase (IPT) and appears to be the committed yet rate-limiting step because overexpression of *IPT* almost always leads to the overproduction of cytokinins in transgenic plants. Various gene promoters have been used to direct the IPT expression in different plant species such as Arabidopsis, tobacco, tomato and potato plants. The gene promoters include the CaMV 35S promoter, heatshock-inducible promoters, a light-inducible promoter and a wounding-inducible promoter (for a list of promoters see (8)). The cytokinin levels in these transgenic plants increase up to 500-fold compared with non-transgenic control plants, and the transgenic plants display both a delayed leaf senescence phenotype and other physiological and developmental changes that are characteristic of cytokinin overproduction, including reduced plant and leaf size, less developed vascular and root systems, and weakened apical dominance. The leaky regulation of IPT by the inducible promoters can readily cause the overproduction of cytokinins and the related developmental abnormalities even under non-inducible conditions. The abnormality associated with the transgenic plants complicates the interpretation of the role of cytokinins in the regulation of senescence because leaf senescence is often influenced by developmental processes of other organs/tissues (referred to as the correlative control of leaf senescence) (8).

To avoid the abnormality associated with the transgenic plants, Gan and Amasino (7) developed an autoregulatory senescence-inhibition system in which the highly senescence-specific promoter of *SAG12* is used to direct the

Figure 6. (Color plate page CP15). Inhibition of leaf senescence by autoregulated production of cytokinins. A. The senescence-specific (SS) promoter of *SAG12* directs expression of IPT at the onset of senescence, resulting in the increase of endogenous cytokinin production, which in turn inhibits senescence. The inhibition of senescence results in the attenuation of the SS promoter to prevent from overproduction of cytokinins. B. A significantly delayed leaf senescence phenotype is observed in the 7-week-old transgenic Arabidopsis plants containing the construct in A (left pot) compared with the wild-type *Landsberg erecta* (right pot). C. A 3-week-old transgenic tobacco plant (left) is shown to develop as normally as the wild-type plant (right). D. A significantly delayed leaf senescence in a 20-week-old transgenic tobacco plant (left) is compared with that of an age-matched wild-type plant (right). E. Senescence is also markedly retarded in the detached leaves of the transgenic tobacco plant (left) in comparison with that of wild-type leaves (right) (7).

IPT gene (Fig. 6A). *SAG12* encodes an apparent cysteine proteinase and is the most senescence-specific gene (5). At the onset of senescence, the *SAG12* promoter is activated to direct *IPT* expression, resulting in the biosynthesis of cytokinins. The increase in endogenous cytokinins will in turn inhibit senescence. Because senescence is inhibited, the senescence-specific promoter becomes attenuated or inactivated, which prevents the overproduction of cytokinins. Because the cytokinin production is targeted specifically to the senescing phase, transgenic Arabidopsis (Fig. 6B) and tobacco plants (Fig. 6C) harboring this system develop normally except for a significantly-delayed leaf senescence phenotype (Fig. 6B and 6D) (7). Senescence in detached leaves of the transgenic tobacco plants are also significantly retarded (Fig. 6E). The leaves with regarded senescence are photosynthetically active. The levels of cytokinins, measured as the zeatin riboside (ZR) equivalents, increase with the age of the leaves in transgenics. A similar phenotype has been observed in other transgenic plants such as alfalfa, bok-choy, broccoli, lettuce, petunia, rice, and tomato (6). These transgenic results confirm the antagonistic role of cytokinins in leaf senescence.

In addition, overexpression of components of the cytokinin signal transduction pathway such as CKI1 (a histidine protein kinase that may serve as a cytokinin receptor) and ARR2 (the rate-limiting factor in the response to cytokinins) also delays dark-induced leaf senescence in Arabidopsis (21), further confirming the inhibitory role of cytokinins in leaf senescence.

Polyamines (PAs) and Leaf Senescence

PAs, including putrescine, spermidine and spermine, are ubiquitous cellular components that play an important role in cell proliferation, cell growth and synthesis of protein and nucleic acids in a variety of living organisms, ranging from yeasts, plants and animals (6, 22). In plants, PA levels are high in active dividing cells but low in cells that do not grow and divide (25), suggesting that PAs may have a role in preventing mitotic senescence in plants (45). The role of PAs in preventing post-mitotic senescence in plants are well supported by the fact that exogenous application of PAs can retard leaf senescence by preventing chlorophyll loss, membrane peroxidation and by inhibiting RNase and protease activities in many plant species and experimental systems, although there are a few negative examples (2). PA levels are also higher in green as opposed to senescing leaves (2). The genetic manipulation of PA production in transgenic tobacco plants also supports an inhibitory role of PAs in leaf senescence. *S*-adenosylmethionine is the common substrate for the biosynthesis of PAs and ethylene in plants. When ethylene production is blocked by antisense technology, more *S*-adenosylmethionine is channeled into the PA biosynthesis pathway. PA levels in such transgenic tobacco plants are significantly elevated and stress-induced leaf senescence is retarded (46). However, as discussed above, ethylene promotes leaf senescence and blocking ethylene biosynthesis may at least in part contribute to the delayed leaf senescence phenotype in the transgenic tobacco plants.

Auxin and Leaf Senescence

A retardant role for auxin in leaf senescence has been supported primarily by early studies involving the external application of auxin and an inverse correlation between endogenous auxin and leaf senescence. In general, application of either synthetic or natural auxins delays chlorophyll loss and protein degradation in detached leaves of various plant species, and the levels of endogenous auxins decrease at the onset of and during senescence in leaves (31). However, there are many exceptions to these general observations. More often than not, auxin treatment has no effect on leaf senescence, sometimes even accelerating senescence (31). The use of 2,4-D (a synthetic auxin species) at a high dose level to kill weeds is an example, though this does not mimic the process of natural senescence. It is known that auxin, especially at a high dose, can promote ethylene production, which in turn promotes senescence as discussed above. Some early experiments showed that the endogenous auxin activity increases just prior to or at the onset of leaf senescence (31). A more recent study finds that the levels of free IAA (the biologically active form) increase during Arabidopsis leaf senescence, and reach approximately 2-fold higher in leaves that are 50% yellowing than in fully expanded green leaves (36). Consistent with the findings, two nitrilase genes (*NIT2* and *NIT4*) are expressed during leaf senescence; nitrilase converts IAN to IAA (36). It is thus difficult to conclude that auxin has an antagonistic role in regulating leaf senescence at the present time. Molecular and genetic analysis of auxin mutants should shed some light on this fundamental question.

Gibberellins and Leaf Senescence

Based on external application and correlation studies, GAs appear to be able to inhibit mitotic and post-mitotic senescence in pea apical buds (49), and post-mitotic leaf senescence in many plant species. Most of the early work show application of GA can effectively inhibit degradation of chlorophyll, proteins and/or nucleic acids in leaves (31). Some GA species are more effective than others; for example, GA_4 is 2 orders of magnitude more active in delaying leaf senescence of *Alstroemeria hybrida* than GA_1 (24). GA_{4+7} appears to be very effective in retarding leaf senescence in *Lilium* (38) and many other plant species, and this formula has been commercialized for horticultural applications. An inverse correlation between GA activity and leaf senescence has been observed in several plant species such as lettuce, nasturtium and dandelion (31). Recent literature concerning molecular genetic analysis of GA-related mutants has rarely discussed leaf senescence. Whether these mutants display no leaf senescence-related phenotype or the phenotype is overlooked is unknown.

THE COMBINATORIAL CONTROL OF LEAF SENESCENCE

While each individual plant growth substances may regulate leaf senescence independently as discussed above, these regulators more likely act in various combinations to control the leaf senescence processes and interact with each other via some cross-talking networks.

Such cross-talk has been widely observed among senescence-promoting hormones/growth regulators. One common scenario is that application of one growth substance often alters the endogenous levels of other growth substances. A striking example is the internal interaction between ethylene and ABA. ABA application increases ethylene production and ethylene treatment also elevates the levels of ABA in many experimental systems (31). This may form a stronger force for the promotion of leaf senescence. However, the promotion of senescence by ABA is not mediated through its stimulation of ethylene production, at least in Arabidopsis, because ABA-stimulated ethylene production is inhibited by application of aminooxyacetic acid (AOA) but ABA-promoted leaf senescence is not affected (48). BR treatment enhances ethylene production in mung bean hypocotyls by activating the expression of ACC synthase gene (16). BR also stimulates JA biosynthesis by inducing the expression of OPR3, a key enzyme in the JA biosynthesis pathway (16). JA application has been shown to enhance ethylene production in tomato plants (16). Which of these is the prime inducer of senescence, or whether all three contribute individually or together, is not clear from such application experiments.

Cross-talks have also been observed among senescence retardants. Numerous early studies revealed that exogenous application of kinetin (cytokinin) causes an increase in GA-like activity (32, 44). The beneficial interaction between GA and cytokinin has led to the development of a commercial formula (GA_{4+7} + BA) that has been widely used for delaying leaf and flower senescence in many horticultural crops (37). Both kinetin and IAA inhibit yellowing in detached tobacco leaves, as do BA and NAA (auxin) in intact soybean leaves, but the cytokinins and auxins exert their effect differently. In detached rape and pumpkin leaves, auxin prevents yellowing, and kinetin sustains the levels of endogenous auxin activity. There are also synergistic interactions between auxin and GA in delaying leaf senescence. Little is known about crosstalk between polyamines and other hormones.

The interactions take place not only among the hormones with the same biological effect (either promoting or inhibiting) on leaf senescence, but they can also occur amongst hormones with opposite effects on senescence. Application of cytokinins can lower endogenous ABA levels and may antagonize the senescence-promoting effects of ABA. Conversely, treatments with ABA reduce cytokinin levels (32). Lowering the levels of their antagonists will help inhibit (in the case of cytokinins) or promote (ABA) leaf senescence. Similarly, the levels of endogenous PAs and ethylene change in opposite directions: when ethylene production is reduced, the PA biosynthesis is enhanced (46). This is because, as discussed above, they share the common precursor in their biosynthesis pathway and the precursor is channeled to the PA pathway upon the blockage of ethylene production. In contrast to these inverse relationships between antagonists, there are cases in which one growth substance can increase the levels of its antagonist, which may represent a fine tuning of the senescence processes. For example, exogenous cytokinin, GA, and auxin can all cause ethylene production (32). The biological significance of such interactions is unknown.

Although the above observations are largely from early physiological studies involving hormonal applications, they are most likely to be true in natural situations. Recent molecular genetic analyses of hormonal mutants in various plant species, Arabidopsis in particular, have unraveled the existence of cross-talk networks in plant hormone actions in general (10, 40).

THE MOLECULAR BASIS OF HORMONAL CONTROL OF LEAF SENESCENCE

As discussed above, it is generally accepted that leaf senescence is driven by changes in gene expression. Therefore, the molecular mechanisms by which various plant growth substances regulate leaf senescence are likely involving activation or suppression of the expression of a subset of SAGs. They may also inhibit or enhance activities of some of the senescence-down regulated genes such as the photosynthesis-related genes. Although nearly 2,500 genes are transcribed in senescent Arabidopsis leaves (13), and various DNA microarrays are available, large-scale analysis of the expression of these genes as regulated by different plant growth substances with regard to leaf senescence has not yet been performed. Nonetheless, recent research involving small subsets of SAGs and photosynthesis-related genes support the regulation of these genes by various growth regulators.

Upregulation of SAGs by Senescence-Promoting Growth Substances

A screening of Arabidopsis enhancer trap lines led to the identification of 147 leaf senescence-specific lines in which a reporter gene is expressed in senescing leaves, but not in young, non-senescing leaves (17). Analysis of the regulation of the reporter expression by six senescence-promoting factors, including ethylene, ABA, JA, BR, darkness and dehydration, reveals that each of the growth substances can induce the expression of potential SAGs in a subset of the enhancer trap lines. For example, the reporter gene in 2 lines is up regulated by ABA, 14 lines by JA, 14 lines by ethylene, and 4 lines by BR. The reporter expression in some of the lines can be induced by more than one factor. For example, the reporter expression in 4 lines can be regulated by both ethylene and BR, and one line by JA, ethylene and BR, suggesting cross-talks among the senescence-promoting factors (17).

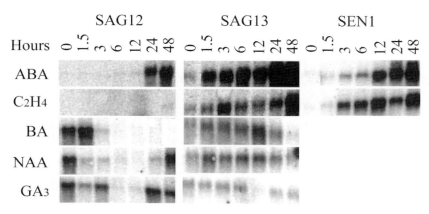

Figure 7. Examples of regulation of SAGs by plant growth substances. For the treatments with ABA and ethylene, fully expanded, nonsenescent leaves of Arabidopsis were incubated for the indicated time periods. For the treatments with BA, NAA or GA, senescing leaves (S2) were incubated for the indicated times. Total RNA was extracted and RNA gel blot analysis was performed.

Ethylene has been widely used to investigate regulation of individual SAGs. It can up-regulate the expression of some, but not all, SAGs. For example, ethylene can readily induce expression of *SAG2*, which encodes a cysteine proteinase, in Arabidopsis leaves after 36 hr of treatment (12). *SAG13* and *SEN1* can also be readily induced by ethylene, whereas *SAG12* is not inducible after the same period of exposure to ethylene, and is barely detectable after 48 hr of treatment (Fig. 7). Like ethylene, ABA can readily induce *SAG13* and *SEN1*, but is less effective for inducing *SAG12* (Fig. 7). JA has been shown to up regulate a number of SAGs in Arabidopsis, including *SEN1*, *SEN4*, *SEN5*, *SAG14*, and *SAG15* (15). SA can also induce a subset of SAGs (29).

Suppression of SAGs by Senescence-inhibiting Growth Substances

It has been repeatedly shown that cytokinins can suppress some SAGs and enhance photosynthesis-related genes (8). For example, both *SAG12* and *SAG13* are expressed in senescing Arabidopsis leaves. When the senescing leaves are detached and placed in BA (cytokinin) solution, *SAG12* transcripts are diminished after 3 hr of incubation, whereas the *SAG13* expression appears to be unaffected (Fig. 7). NAA (auxin) and GA_3 can also down regulate *SAG12*, but *SAG13* expression is not significantly influenced (Fig. 7). Whether PAs have a similar role in suppressing SAG expression needs to be determined.

A Regulatory Model of Leaf Senescence

Leaf senescence is a complex process that is driven by the activation of thousands of SAGs. The down-regulation of the photosynthesis-related

Hormonal regulation of senescence

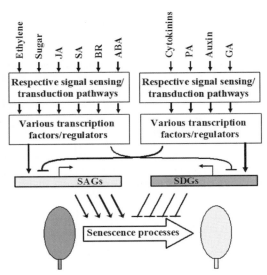

Figure 8. A simple model to illustrate the molecular regulatory mechanisms by which various plant growth substances promote or retard leaf senescence. The senescence-promoting substances are most likely to exert their effect by activating a subset of SAGs and by suppressing the expression of senescence down-regulated genes (SDGs). In contrast, the retardation of leaf senescence by senescence-inhibiting substances is most likely achieved by activating SDGs and possibly by suppressing SAGs.

genes may also contribute to the senescence progression. It has been proposed that there exist multiple regulatory pathways leading to senescence, and that these pathways may interconnect to each other to form a regulatory network. Plant hormones and other growth substances play important regulatory roles in either promoting or inhibiting leaf senescence. These regulations are most likely achieved by controlling the SAG expression (Fig. 8). Specifically, senescence-promoting hormones/substances may activate sets of SAGs and inactivate the expression from photosynthesis-related genes; in contrast, senescence-inhibiting hormones/substances may inhibit expression of some SAGs and enhance photosynthesis-related genes.

Many details in the model need further investigation. Given the availability of various microarrays and the complexity of the senescence process, a systems biology approach should be taken for deciphering the molecular regulatory mechanisms of leaf senescence by plant hormones/substances and other environmental and internal factors. Once senescence is fully understood, it should allow us to devise ways to manipulate leaf senescence for agricultural improvement.

Acknowledgement

I thank Richard Amasino for introducing me to the senescence research field and Peter Davies for offering me this opportunity to write this chapter and for his editing and patience. I also thank members of my laboratory for useful discussions. Our research on leaf senescence has been supported by the US Department of Energy's Energy Biosciences program, by USDA NRI, by the Kentucky Tobacco Research and Development Center at the University of Kentucky, and by a new faculty startup fund from Cornell University.

References

1. Clouse SD (2001) Brassinosteroids. *In* CR Somerville, EM Meyerowitz, eds, The Arabidopsis Book. American Society of Plant Biologists, Rockville, MD,

doi/10.1199/tab.0009, http://www.aspb.org/publications/arabidopsis/
2. Evans PT, Malmberg RL (1989) Do polyamines have roles in plant development? Annu Rev Plant Physiol Plant Mol Biol 40: 235-269
3. Fan L, Zheng S, Wang X (1997) Antisense suppression of phospholipase Da retards abscisic acid- and ethylene-promoted senescence of postharvest Arabidopsis leaves. Plant Cell 9: 2183-2196
4. Finkelstein R, Rock CD (2002) Abscisic acid biosynthesis and response. *In* CR Somerville, EM Meyerowitz, eds, The Arabidopsis Book. American Society of Plant Biologists, Rockville, MD, doi/10.1199/tab.0058, http://www.aspb.org/publications/arabidopsis/
5. Gan S (1995) Molecular characterization and genetic manipulation of plant senescence. Ph.D. Dissertation. University of Wisconsin-Madison, Madison, WI, USA
6. Gan S (2003) Mitotic and post-mitotic senescence in plants. Sci SAGE KE 2003: RE7 http//sageke.sciencemag.org/cgi/content/full/sageke;2003/38/re7
7. Gan S, Amasino RM (1995) Inhibition of leaf senescence by autoregulated production of cytokinin. Science 270: 1986-1988
8. Gan S, Amasino RM (1996) Cytokinins in plant senescence: from spray and pray to clone and play. BioEssays 18: 557-565
9. Gan S, Amasino RM (1997) Making sense of senescence. Molecular genetic regulation and manipulation of leaf senescence. Plant Physiol 113: 313-319
10. Gazzarrini S, McCourt P (2003) Cross-talk in plant hormone signalling: what Arabidopsis mutants are telling us. Ann Bot 91: 605-612
11. Giovannoni J (2001) Molecular biology of fruit maturation and ripening. Annu Rev Plant Physiol Plant Mol Biol 52: 725-749
12. Grbic V, Bleecker AB (1995) Ethylene regulates the timing of leaf senescence in *Arabidopsis*. Plant J. 8: 595-602
13. Guo Y, Cai Z, Gan S (2003) Transcriptome of Arabidopsis leaf senescence. Plant, Cell & Environment 27: 521-549
14. Harms K, Atzorn R, Brash A, Kühn H, Wasternack C, Willmitzer L, Peña-Cortés H (1995) Expression of a flax allene oxide synthase cDNA leads to increased endogenous jasmonic acid (JA) levels in transgenic potato plants but not to a corresponding activation of JA-responding genes. Plant Cell 7: 1645-1654
15. He Y, Fukushige H, Hildebrand DF, Gan S (2002) Evidence supporting a role of jasmonic acid in Arabidopsis leaf senescence. Plant Physiol 128: 876-884
16. He Y, Gan S (2003) Molecular characteristics of leaf senescence. *In* Recent Research Developments in Plant Molecular Biology, Vol 1. Research Signpost, Kerala, India, pp 1-17
17. He Y, Tang W, Swain JD, Green AL, Jack TP, Gan S (2001) Networking senescence-regulating pathways by using Arabidopsis enhancer trap lines. Plant Physiol 126: 707-716
18. Hensel LL, Grbic V, Baumgarten DA, Bleecker AB (1993) Developmental and age-related processes that influence the longevity and senescence of photosynthetic tissues in *Arabidopsis*. Plant Cell 5: 553-564
19. Hildebrand DF, Fukushige H, Afitlhile M, Wang C (1998) Lipoxygenases in plant development and senescence. Chapt. 8. *In* AF Rowley, H Kuhn, T Schewe, eds, Eicosanoids and Related Compounds in Plants and Animals. Portland Press Ltd, London, pp 151-181
20. Hung KT, Kao CH (2003) Nitric oxide counteracts the senescence of rice leaves induced by abscisic acid. J Plant Physiol 160: 871-879
21. Hwang I, Sheen J (2001) Two-component circuitry in Arabidopsis cytokinin signal transduction. Nature 413: 383-389
22. Jeevanandam M, Petersen SR (2001) Clinical role of polyamine analysis: problem and promise. Curr Opin Clin Nutr Metab Care 4: 385-390

23. John I, Drake R, Farrell A, Cooper W, Lee P, Horton P, Grierson D (1995) Delayed leaf senescence in ethylene-deficient ACC-oxidase antisense tomato plants: molecular and physiological analysis. Plant J. 7: 483-490
24. Kappers IF, Jordi W, Tsesmetzis M, Maas FM, Van Der Plas LHW (1998) GA4 does not require conversion into GA1 to delay senescence of Alstroemeria hybrida leaves. *In,*
25. Kaur Sawhney R, Shekhawat NS, Galston AW (1985) Polyamine levels as related to growth, differentiation and senescence in protoplast-derived cultures of Vigna aconitifolia and Avena sativa. Plant Growth Regul 3: 329-337
26. Masclaux C, Valadier M-H, Brugiere N, Morot-Gaudry J-F, Hirel B (2000) Characterization of the sink/source transition in tobacco (*Nicotiana tabacum* L.) shoots in relation to nitrogen management and leaf senescence. Planta 211: 510-518
27. Mattoo AK, Aharoni N (1988) Ethylene and plant senescence. *In* LD Noodén, AC Leopold, eds, Senescence and Aging in Plants. Academic Press, San Diego, pp 242-281
28. Moore B, Zhou L, Rolland F, Hall Q, Cheng WH, Liu YX, Hwang I, Jones T, Sheen J (2003) Role of the Arabidopsis glucose sensor HXK1 in nutrient, light, and hormonal signaling. Science 300: 332-336
29. Morris K, MacKerness SA, Page T, John CF, Murphy AM, Carr JP, Buchanan-Wollaston V (2000) Salicylic acid has a role in regulating gene expression during leaf senescence. Plant J 23: 677-685
30. Nemeth K, Salchert K, Putnoky P, Bhalerao R, Koncz-Kalman Z, Stankovic-Stangeland B, Bako L, Mathur J, Okresz L, Stabel S, Geigenberger P, Stitt M, Redei GP, Schell J, Koncz C (1998) Pleiotropic control of glucose and hormone responses by PRL1, a nuclear WD protein, in Arabidopsis. Genes Dev 12: 3059-3073
31. Noodén LD (1988) Abscisic acid, auxin, and other regulators of senescence. *In* LD Noodén, AC Leopold, eds, Senescence and Aging in Plants. Academic Press, San Diego, pp 329-367
32. Noodén LD (1988) Abscisic acid, auxin, and other regulators of senescence. *In* LD Noodén, AC Leopold, eds, Senescence and aging in plants. Academic Press, San Diego, pp 330-368
33. Oh SA, Park J-H, Lee GI, Paek KH, Park SK, Nam HG (1997) Identification of three genetic loci controlling leaf senescence in *Arabidopsis thaliana*. Plant J 12: 527-535
34. Pontier D, Gan S, Amasino RM, Roby D, Lam E (1999) Markers for hypersensitive response and senescence show distinct patterns of expression. Plant Mol Biol 39: 1243-1255
35. Quirino BF, Noh Y-S, Himelblau E, Amasino RM (2000) Molecular aspects of leaf senescence. Trends in Plant Science 5: 278-282
36. Quirino BF, Normanly J, Amasino RM (1999) Diverse range of gene activity during Arabidopsis thaliana leaf senescence includes pathogen-independent induction of defense-related genes. Plant Mol Biol 40: 267-278
37. Ranwala AP, Miller WB (1999) Timing of gibberellin$_{4+7}$ + benzyladenine sprays influences efficacy against foliar chlorosis and plant height in Easter lilies. HortSci 34: 902-903
38. Ranwala AP, Miller WB (2000) Preventive mechanisms of gibberellin$_{4+7}$ and light on low-temperature-induced leaf senescence in *Lilium* cv. Stargazer. Postharvest Biol Technol 19: 85-92
39. Richmond AE, Lang A (1957) Effect of kinetin on protein content and survival of detached *Xanthium* leaves. Science 125: 650-651
40. Rolland F, Moore B, Sheen J (2002) Sugar sensing and signaling in plants. Plant Cell 14(Suppl): S185-205
41. Sharma SP, Kaur J, Rattan SI (1997) Increased longevity of kinetin-fed Zaprionus fruitflies is accompanied by their reduced fecundity and enhanced catalase activity. Biochem Mol Biol Int 41: 869-875
42. Stessman D, Miller A, Spalding M, Rodermel S (2002) Regulation of photosynthesis

during Arabidopsis leaf development in continuous light. Photosyn Res 72: 27-37
43. Ueda J, Kato J (1980) Isolation and identification of a senescence-promoting substance from wormwood (*Artemisia absinthium* L.). Plant Physiol 66: 246-249
44. Van Staden J, Cook E, Noodén LD (1988) Cytokinins and senescence. *In* LD Noodén, AC Leopold, eds, Senescence and aging in plants. Academic Press, San Diego, pp 281-328
45. Walden R, Cordeiro A, Tiburcio AF (1997) Polyamines: small molecules triggering pathways in plant growth and development. Plant Physiol 113: 1009-1013
46. Wi SJ, Park KY (2002) Antisense expression of carnation cDNA encoding ACC synthase or ACC oxidase enhances polyamine content and abiotic stress tolerance in transgenic tobacco plants. Mol Cells 13: 209-220
47. Yin Y, Wang ZY, Mora-Garcia S, Li J, Yoshida S, Asami T, Chory J (2002) BES1 accumulates in the nucleus in response to brassinosteroids to regulate gene expression and promote stem elongation. Cell 109: 181-191
48. Zacarias L, Reid MS (1990) Role of growth regulators in the senescence of *Arabidopsis thaliana* leaves. Physiol. Plant. 80: 549-554
49. Zhu YX, Davies PJ (1997) The control of apical bud growth and senescence by auxin and gibberellin in genetic lines of peas. Plant Physiol 113: 631-637

E7. Genetic and Transgenic Approaches to Improving Crop Performance

Andy L. Phillips
CPI Division, Rothamsted Research, Harpenden, Hertfordshire, AL5 2JQ, U.K. E-Mail: andy.phillips@bbsrc.ac.uk

INTRODUCTION

As this volume illustrates, plant hormones play important roles in all aspects of the growth and development of plants, and in their interactions with the environment and other organisms. Many of the developmental and physiological processes regulated by plant hormones are of agronomic importance: examples that will be further explored in this chapter include the roles of gibberellins in regulating stem elongation and of ethylene in promoting fruit ripening. As a consequence of the central roles of plant hormones, plant hormone signalling has been a target of genetic modification by various selective processes, from the evolution of land plants 400 million years ago, through the adaptation of species to specific ecological niches, the domestication of crop plants by prehistoric farmers, the directed breeding of improved varieties with the advent of modern genetics 100 years ago and, most recently, the targeted modification of hormone signaling pathways in transgenic plants to improve specific agronomic traits.

Crop performance and quality may also be chemically modified through the application of plant growth regulators (PGRs[1]), produced either by purification from microorganisms or by chemical synthesis. Chemical growth regulators include bioactive plant hormones or synthetic mimics thereof, inhibitors of plant hormone biosynthesis and compounds that interfere with hormone perception or signal transduction. In a previous edition of this book, Thomas Gianfagna pointed out that the use of chemical growth regulators

---

[1] Abbreviations: 1-MCP, 1-methylcyclopropene; 2ODD, 2-oxoglutarate-dependent dioxygenase; ACC, 1-aminocyclopropane-1-carboxylic acid; ACO, ACC oxidase; ACS, ACC synthase; AOS, allene oxide synthase; AVG, aminoethoxyvinylglycine; BR, brassinosteroid; CIMMYT, Centro Internacional de Mejoramiento de Maíz y Trigo; CPS, copalyl diphosphate synthase; DGWG, Dee-geo-woo-gen; EKO, *ent*-kaurene synthase; GGPP, geranygeranyl diphosphate; GM, genetic manipulation; IRRI, International Rice Research Institute; JA, jasmonic acid; KAO, *ent*-kaurenoic acid oxidase; K_m, Michaelis Constant; MeJA, methyl jasmonate; NCED, 9-*cis*-epoxycarotenoid dioxygenase; PCR, polymerase chain reaction; PGR, plant growth regulator; SA, salicylic acid; SAM, S-adenosylmethionine; TILLING, targeting induced local lesions in genomes.

dates back to before the identification of plant hormones themselves (19). He cites the example that the induction of flowering in pineapple by treatment with smoke from slow fires was subsequently found to be due to stimulation by ethylene, produced as a result of incomplete combustion.

Plant growth regulators are now widely used on a variety of agricultural, horticultural and amenity species, and the worldwide value of the PGR market is in the region of a billion US dollars. PGRs are classified as pesticides, and as such are subject to the same rigorous regulatory framework as herbicides and fungicides. The number of PGRs available for use on crops, particularly those for human consumption, decreases yearly. This is partly due to increasing public concern about the use of 'chemicals' in agriculture. For example, a major food scare in 1989 involved the potential carcinogenetic properties of a breakdown product of daminozide ('Alar'), an inhibitor of GA biosynthesis that was used to enhance fruit quality in apples. Despite no conclusive evidence that Alar is a danger to human health, the compound remains barred from food use in many parts of the world. A second restriction on the PGRs currently available results from the high development costs and regulatory charges that, taking into account the relatively small worldwide market compared with other agrochemicals, have reduced the number of new PGRs being developed.

Notable recent exceptions to this include 1-methylcyclopropene (1-MCP), a gaseous antagonist of ethylene, registered in the US for use in prolonging the life of cut flowers and also for food use on fruit crops. Prohexadione-Ca, a relatively new growth retardant, acts as an inhibitor of 2-oxoglutarate-dependent dioxygenases (2ODDs) in the GA biosynthetic pathway and has been commercialized for use on fruit trees, groundnuts and cereals. On apple and pear trees, prohexadione-Ca also affords some protection against fire-blight and other bacterial pathogens. It appears that this is caused by alterations in the flavonoid profile of the treated trees, which accumulate 3-deoxycatechins due to inhibition of key 2ODDs in the flavonoid biosynthetic pathway (49).

Alternative approaches to chemical PGRs include the use of controlled growing conditions, for example the manipulation of light quality through shading in glasshouse crops, the use of mutant alleles in crop species, selected either by agronomic trait or with molecular markers for specific hormone signalling components, and the direct manipulation of genes involved in hormone biosynthesis or transduction in transgenic plants. The genetic approaches also allow more precise targeting of the effect to specific organs and tissues, an effect not generally achievable through chemical treatments. This chapter will discuss genetic

Figure 1: Pre-harvest sprouting in wheat. Courtesy of Dr. J. Lenton.

and transgenic strategies for the modification of crop production and quality, organized in terms of the physiological processes involved.

SEED GERMINATION AND DORMANCY

Seed germination and early seedling establishment are clearly key stages in the growth of crop plants. The interaction of the different plant hormones involved, particularly ABA and GA, is a classic example of the role of hormones in the regulation of plant development and has been studied in depth in both model and crop species. In cereals, ABA promotes maturation and dormancy during late seed development, and there is evidence that GA is antagonistic in this process: for example, seeds of the maize mutant *viviparous-5*, that is deficient in ABA, do not undergo full maturation to dormancy and desiccation-tolerance and germinate precociously while still on the ear, but this can be suppressed by mutations or inhibitors that reduce bioactive GA content (62). GA and ABA are similarly antagonistic in promoting and inhibiting, respectively, the germination of seeds of many species.

In crop plants, the requirements for full seed maturation and rapid germination can be incompatible. For example, a low degree of dormancy of mature wheat seeds is important to ensure rapid and synchronous germination and to allow planting of a new crop soon after harvest. However, under certain environmental conditions such as cool temperature, high humidity or rainfall just prior to harvest, the low embryo dormancy can lead to pre-harvest sprouting (Fig. 1), in which precocious seed germination results in the production of α-amylase by the embryo that has a serious effect on the bread-making quality. Holdsworth and colleagues have studied the molecular processes involved in the regulation of dormancy in wheat by isolating homologues of a gene shown in model species to be important in dormancy. Mutations in the maize *VP1* or Arabidopsis *ABI3* genes confer insensitivity to ABA in the seed and result in precocious seed germination, suggesting that these related genes normally function to promote seed maturation and dormancy. In wild oat (*Avena fatua*), it was shown that the variable level of dormancy of in-bred lines was strongly correlated with the level of expression of *AfVP1*, an oat homologue of maize *VP1* (29). As described above, wheat has a low embryo dormancy; analysis of the expression of three *VP1* homoeologues of wheat (*TaVP1*) showed a high proportion of mis-spliced transcripts, most of which encoded aberrant proteins that would have reduced function, which suggested that the low dormancy of wheat was due to the dysfunction of the *VP1* gene (41). This was supported by overexpression in transgenic wheat of a functional wild oat *AfVP1* cDNA under the maize ubiquitin promoter, which increased the sensitivity of seed germination to ABA and decreased the rate of pre-harvest sprouting under laboratory conditions by up to 50%. Interestingly, the dysfunctional *VP1* genes of modern hexaploid wheat varieties do not appear to been acquired as a result of selection for low dormancy, as the diploid and tetraploid progenitor species of wheat contain similar mis-spliced transcripts (41).

One possible application of increased seed dormancy is in the prevention of volunteers in GM crops, where seed lost before or during harvest germinates in the soil to contaminate the succeeding crop. Seeds modified to remain indefinitely dormant could reduce environmental pollution by GM crops and prevent the accumulation of, for example, weed beet in sugar beet production. Arabidopsis mutants such as *ga1* that are extremely deficient in bioactive GAs fail to germinate unless treated with exogenous GA; however, such mutants invariably have an extreme dwarf phenotype and other GA-related defects. One possible approach would be to reduce endogenous GA levels in the seed through tissue-specific expression of the GA-catabolizing enzyme, GA 2-oxidase (GA2ox): overproduction of GA2ox in vegetative tissues results in a dramatic reduction in bioactive GA content and an extreme dwarf habit (52). Such GA-deficient seeds might, however, require treatment with GA_3 (which lacks a 2β-H and is therefore not inactivated by GA2ox) before sowing to allow germination and to encourage seedling establishment.

PLANT ARCHITECTURE

Excessive stem growth of crops plants directly affects achievable yield by directing photoassimilate away from the harvested storage organ and into unwanted elongation growth. In addition, unchecked growth of stems can make crops vulnerable to damage from wind and rain. As an agrochemical solution to this problem, a wide variety of crops are treated with plant growth retardants, most of which act by inhibiting the biosynthesis of GA, the principal hormone involved in elongation growth. In UK agricultural crops, the largest single use of growth retardants is the application of chlormequat chloride, which inhibits copalyl diphosphate synthase, to cereal crops to minimize losses due to lodging (stem collapse). Growth retardants are similarly used to control elongation growth in cotton, oilseed rape (canola), grapes, apple and many other crops, as well as ornamental and amenity species.

Dwarfing genes in crop plants

There is excellent evidence that a genetic approach to the control of elongation growth can be extremely successful. The 'Green Revolution' of the 1960s and 1970s dramatically increased the global production of wheat and rice, due to a combination of higher-yielding varieties, increased use of fertilizer and the development of effective pesticides. A key trait of the new cereal varieties was a reduction in height, such that the increased growth enabled by the use of quantities of nitrogen fertilizer was directed into grain filling rather than straw production, such that the harvest index – the

Figure 2. Effects of different *Rht* combinations on plant height in wheat cv. Maris Huntsman. From (14).

proportion of total plant weight that is in the grain – rose from 35% in the 1920s to 50-55% in modern varieties (11).

The origin of the new wheat varieties lay in Japan, where in 1917 the semi-dwarf 'Daruma' was successively crossed into American varieties to yield 'Norin 10'. Progeny from a further cross were sent to Norman Borlaug at CIMMYT in 1954, where they were bred with Mexican wheats to yield the new varieties of semi-dwarfs that spread throughout the developing world. Although a number of *Rht* genes yielding a dwarf or semi-dwarf habit have been identified, the *Rht1* and *Rht2* genes from 'Norin 10' proved to be most successful and now are present in 70% of all commercial wheat varieties (11). *Rht1* and *Rht2* are partially-dominant homoeologous alleles located on the B and D genomes of hexaploid wheat and are now named *Rht-B1b* and *Rht-D1b*, respectively, to reflect their chromosomal locations. *Rht-B1b* and *Rht-D1b* confer a relatively mild dwarfing affect of about 14% and 17%, respectively (Fig. 2), and this is additive in lines carrying both alleles, which are reduced in height by about 42% (14). A third homoeologous allele, *Rht-B1c* (*Rht3*) affects height more severely, with a 50% reduction, but is rarely used in commercial varieties because, although the harvest index is increased over *Rh1-B1b/Rht1-D1b*, the overall plant size is diminished and overall yield is thereby reduced.

These naturally-occurring *Rht* dwarfing alleles reduce plant height by affecting response to endogenous GAs and thus have a similar effect to the semi-dominant *gai* mutant of Arabidopsis, which was generated by X-ray

Figure 3. The effect of mutation in *SD1* (*OsGA20ox2*) on height in different rice backgrounds: (1) Taichung 65 (wild-type) and IR8 (*sd1*); (2) Woo-gen (wt) and Dee-geo-woo-gen (*sd1*); (3) Calrose (wt) and Calrose 76 (*sd1*); (4) Fujiminori (wt) and Reimei (*sd1*); (5) Taichung 65 (wt) and Jikkoku (*sd1*). From (1).

irradiation (32). The *GAI* gene was cloned by Harberd and colleagues (43) by insertional mutagenesis; the dwarfing *gai* allele was shown to encode a protein with a 17-amino-acid ('DELLA') deletion near the N-terminus. A search for *GAI*-related sequences in a rice database identified a potential homologue that was used to screen a cDNA library from wheat, which in turn was mapped to genomic fragments from wheat chromosome arms 4A, 4B and 4D, corresponding to the location of the *Rht-1* alleles (44). The DNA sequence of the dwarfing alleles *Rht-B1b* (*Rht1*) and *Rht-D1b* (*Rht2*) were isolated and shown to contain translation stop codons after the 'DELLA' sequence. It is likely that translational re-initiation results in the production of a protein lacking part the N-terminal domain, including the 'DELLA' sequence that is deleted in *gai*. It is suggested that the wild-type GAI/rht protein represses GA responses in the absence of GA and is suppressed by GA, while the truncated or deleted gai/RHT proteins are no longer sensitive to the GA stimulus, and operate constitutively to repress GA downstream functions.

The production of semi-dwarf varieties of rice parallels that of wheat. Rice breeding in Taiwan, under Japanese and later Chinese direction, identified a dwarf 'Dee-geo-woo-gen' (DGWG; Fig. 3), which entered the IRRI breeding programme in 1962 and resulted in the release of the lodging resistant line IR8, known as 'miracle rice' (11). In contrast to the partially-dominant *Rht-1* alleles of wheat, the agriculturally-important dwarfing genes of rice are all recessive alleles of a single gene, *Semidwarf* (*SD1*). Rice lines containing the *sd1* mutation, the main dwarfing allele at this locus, are fully GA-responsive but have a partial block in the biosynthesis of the hormone (1). Analysis of the GA content of *sd1* suggested that the affected step in GA biosynthesis involved the conversion of GA_{53} to GA_{20}, catalysed by GA 20-oxidase. GA 20-oxidase fragments were amplified from rice genomic DNA using degenerate primers and a novel GA 20-oxidase gene fragment isolated that mapped closely to the *SD1* locus (1); the *sd1* allele from DGWG and IR8 was shown to have a deletion in the first exon of (*OsGA20ox2*) that would result in a frame shift and premature translational termination. The *sd1* allele is therefore a knockout of *OsGA20ox1*, and the semi-dwarf stature of *sd1* is likely to be due to the activity of other *GA20ox* genes in rice. Two other groups simultaneously used positional cloning to identify the *SD1* gene (42, 54).

Another dwarfing gene with great historical significance is the *Long internode* (or *Length*, *Le*) gene of pea, an allele of which (*le*) was used by Mendel in his classical experiments on heredity. Dwarf peas are responsive to applied gibberellic acid (GA_3), suggesting that the lesion in *le* is in GA biosynthesis rather than response. Application of radiolabelled GA_{20} to shoots of tall and dwarf peas showed that conversion to GA_1 was much slower in *le*, indicative of a reduction in GA 3β-hydroxylase (GA3ox) activity. Two related approaches were used to clone *Le*: Martin and colleagues (37) used PCR with degenerate primers derived from the Arabidopsis *GA4* (*AtGA3ox1*) gene to isolate *GA3ox* sequences from pea, while Lester used *GA4* as a heterologous probe against a pea cDNA library (35). Mendel's *le* mutation was shown to be due to an alanine-to-threonine substitution near the active site of the enzyme that raised the K_m for the GA substrate by about 100-fold. The *le* allele is still used in the majority of commercial and garden pea varieties as the bushy habit it confers allows cultivation without support.

Transgenic approaches to the manipulation of plant stature

As mentioned above, a wide variety of crop plants are treated with retardants to limit excessive elongation growth, while GA_3, extracted from cultures of the fungus *Gibberella fujikuroi,* is used, for example, to promote germination in malting barley and to stimulate berry expansion in seedless grapes. In the absence of naturally occurring or induced mutations that affect GA biosynthesis or response, alternative approaches are being used to manipulate these pathways in model species and crop plants.

The GA biosynthetic pathway from geranygeranyl diphosphate (GGPP) involves fourteen reactions catalyzed by eight enzymes (Fig. 4). Analysis of the endogenous regulatory mechanisms of the pathway suggests a number of enzymes that might be limiting and therefore suitable targets for manipulation. The first enzyme, copalyl diphosphate synthase (*CPS*) is expressed at very low levels and might act to regulate flux into the pathway. When *CPS* was over-expressed in Arabidopsis, the level of *ent*-kaurene in rosettes increased by up to 1000-fold, and the level of GA_{12} by up to 12-fold, but there was no change in the level of GA_9 or of the major bioactive GA in Arabidopsis, GA_4 (13). One possible explanation for the failure to increase the levels of GA_9 and GA_4 might be the operation of the feedback pathway by which expression of GA 20-oxidase and GA 3β-hydroxylase are down-regulated by GA action; however, no change in *GA20ox1* or *GA3ox1* expression was observed. The authors suggest that in these transgenic plants, other activities, including *ent*-kaurene synthase (*EKO*) and *ent*-kaurenoic acid oxidase (*KAO*) are rate-limiting (13). Significantly, the GA biosynthetic enzymes are localized to specific tissues and cells within the Arabidopsis rosette whereas over-expression of *CPS* and *KS* under the CaMV-35S

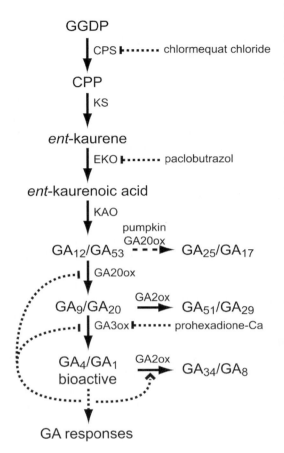

Figure 4. A simplified representation of the GA biosynthetic pathway from GGDP, showing the sites of inhibition by chemical growth retardants and the feedback/feedforward mechanism by which bioactive GAs regulate their biosynthesis and inactivation.

promoter increases *ent*-kaurene biosynthesis in a much wider range of tissues. As *ent*-kaurene is hydrophobic and unlikely to be mobile within plant tissues, most of the *ent*-kaurene produced is likely to accumulate in tissues not competent for further metabolism. Furthermore, it is important to note that the main bioactive GAs, GA_1 and GA_4, are not end-products of the pathway but intermediates (Fig. 4) and their levels may therefore be unaffected by changes in flux.

Studies on the regulation of GA biosynthesis and catabolism in plants suggest that the dioxygenases that catalyze the later steps might be important control points. For example, *GA20ox*, *GA3ox* and *GA2ox* genes have been shown to be regulated by light (39, 65, 67), *GA20ox* and *GA3ox* by negative feedback control (5, 46) and *GA2ox* by positive feed-forward (57). The roles of these genes in regulating endogenous GA levels suggest that they might be suitable targets for genetic manipulation of the pathway to achieve changes in bioactive GA levels. This is supported by the theory of metabolic control analysis which suggests that although enzymes that are controlled by feedback may not be good targets for increasing the overall flux through a pathway, they can effectively and quickly alter the levels of intermediates close to the feedback control point.

Increasing GA levels
Possible strategies for increasing bioactive GA levels in transgenic plants thereby include overexpression of *GA20ox* or *GA3ox* or reduction in catabolism by suppression of *GA2ox* genes. *GA2ox* in many species is

Plant hormones and crop performance

Figure 5. Increased plant height in aspen (*Populus tremuloides*) achieved by overexpression of Arabidopsis *GA20ox1*. From (10).

represented by small multi-gene families with overlapping expression patterns (Woolley, Phillips & Hedden, unpublished) and suppression of a single gene may not be effective; an exception is the *slender* mutant of pea affecting a seed-specific GA 2-oxidase that normally converts GA_{20} to GA_{29}. In the mutant, GA_{20} accumulates in the mature seed and is converted to GA_1 on germination, producing excessive elongation of the first few internodes (38).

In Arabidopsis, overexpression of *GA3ox* did not produce any discernable effect on plant development (Williams, Phillips and Hedden, unpublished). However, over-expression of *GA20ox* genes yielded plants with a GA-overproduction phenotype that had reduced dormancy, flowered early and had increased stem elongation, associated with an increase in bioactive GAs (6, 26). In tobacco, which normally follows the early-13-hydroxlation pathway, overexpression of a citrus *GA20ox* cDNA resulted in shift to the non-13-hydroxylation pathway such that GA_4 was the predominant bioactive GA (60), presumably due to the introduced GA 20-oxidase enzyme effectively competing for substrate with the endogenous GA 13-hydroxylase enzyme.

There is considerable interest in modification of the GA content of trees to modify growth rates and the properties of the wood formed. Eriksson and colleagues (10) over-expressed the Arabidopsis *GA20ox1* cDNA in hybrid aspen (*Pupulus tremula* x *P. tremuloides*) and achieved 20-fold increases in GA_1 and GA_4 levels that led to higher rates of both elongation and diameter growth in the transgenic trees (Fig. 5). Significantly, the trees also showed increases in the number and length of xylem fibres, qualities that are important in paper production.

Decreasing GA levels
Applications for reduction in GA levels to limit elongation growth are more widespread, reflecting the use of growth retardants that inhibit GA biosynthesis in agriculture and horticulture. A number of transgenic strategies for reducing bioactive GA levels in model and crop species have

Figure 6. Control of stem elongation in apple cv. Greensleeves by co-suppression of a GA 20-oxidase gene. Courtesy of Dr. D. James.

been attempted, with varying degrees of success. Suppression of *GA20ox* genes in Arabidopsis by expression of antisense RNA was successful in Arabidopsis, although only in a small proportion of transgenic lines (6). Notably, suppression of different members of the *GA20ox* gene family gave different phenotypes corresponding to their expression pattern: for example, suppression of *AtGA20ox1* resulted in short hypocotyls and a semi-dwarf phenotype in the adult plant (6). This has potential for manipulation of crop plants as tissue-specific traits may be targeted by suppression of the *GA20ox* gene expressed in that tissue without the need for tissue-specific promoters. For example, anti-sense suppression or co-suppression of an apple *GA20ox* gene expressed in shoot tips and leaves yields a reduction in shoot elongation and internode number in the transgenic lines (Bulley, Phillips, Hedden and James, unpublished) (Fig. 6), corresponding to a substantial reduction in GA_1 in these tissues. Promisingly, the semi-dwarf stature of these transgenic apple lines is maintained when grafted onto invigorating rootstocks, which would allow the use of disease- and drought-resistant rootstocks while limiting elongation growth of the scion and reducing the need for chemical retardants.

The level of expression of certain *GA20ox* genes may also be reduced by overexpression of the KNOTTED-1 class of homeobox transcription factors. For example, ectopic expression of *NTH15*, a tobacco *KNOX* gene, in transgenic tobacco decreases expression of an endogenous *GA20ox* gene (*Ntc12*) by binding to a sequence in its first intron and results in reduced GA levels and abnormal leaf and flower morphology (51). However, increased *KNOX* gene expression has many pleiotropic effects including increases in cytokinin production, and thus may not be appropriate for targeted control of elongation growth.

Antisense suppression of *GA3ox* genes is also effective in reducing GA levels in crop plants. The *OsGA3ox2* gene of rice corresponds to the *D18* locus, loss-of-function mutations in which result in severe dwarfism cause by a reduction in GA_1. Itoh and colleagues produced transgenic rice plants expressing an antisense copy of *OsGA3ox2*, that had reduced expression of the target gene and had a semi-dwarf habit that may be useful in rice breeding programmes (27).

The normal activity of GA 20-oxidase is to carry out the sequential oxidation of $GA_{12}/GA_{53} \rightarrow GA_{15}/GA_{44} \rightarrow GA_{24}/GA_{19} \rightarrow GA_9/GA_{20}$; the final step involves the loss of C-20 to form the C_{19} GAs. However, a GA20ox found in developing seeds of pumpkin (CmGA20ox1) is unusual in that it converts most of the GA_{24} to the tricarboxylic acid, GA_{25}, which is not active *per se* and cannot be converted to an active GA (Fig. 4). In theory, over-production of this enzyme in tissues involved in GA biosynthesis should divert the pathway into inactive by products and thereby reduce the levels of bioactive GAs. Several groups have adopted this strategy, with mixed results: over-expression of *CmGA20ox1* in Arabidopsis reduced bioactive GA levels, but stem height was only slightly reduced (66), suggesting that stem growth in Arabidopsis is relatively insensitive to changes in GA levels. In addition, the expression of the endogenous biosynthetic genes *AtGA20ox1* and *AtGA3ox1* were upregulated in the transgenic lines, presumably due to the operation of the feedback pathway, at least partially compensating for the diversion of substrates. In contrast, Curtis and colleagues (7) were successful in producing semi-dwarf lines of the ornamental plant *Solanum dulcamara* by overexpression of *CmGA20ox1*; the stems and leaves of the transgenics were also more heavily pigmented than the control. Although GA_1 levels in leaves of the transgenic lines were reduced by 80%, GA_4 levels were maintained or even increased, possibly indicating that this species is more sensitive to 13-hydroxy-GAs. Over-expression of *CmGA20ox1* in lettuce similarly reduced endogenous GA_1 levels to 16% that of control plants and the transgenic lines had a semi-dwarf habit and flowered was delayed. In conclusion, over-expression of pumpkin GA 20-oxidase may be useful where a relatively mild dwarfing phenotype is desired, but with the disadvantage that the effects on GA levels and plant development are not entirely predictable.

The different strategies above aim to reduce elongation growth by limiting the rate of GA biosynthesis. An alternative approach is to increase the rate of GA turnover through enhanced production of GA catabolizing enzymes. GA 2-oxidase, which inactivates bioactive GAs by hydroxylation at C- 2β, was first cloned from scarlet runner bean (*Phaseolus coccineus*) by functional screening (57). Over-expression of this cDNA (*PcGA2ox1*) under the CaMV-35S promoter in Arabidopsis led to several lines with an extreme dwarf phenotype (Fig. 7), while other transgenic lines were similar to controls (Thomas, Phillips & Hedden, unpublished). Transcript and GA analyses indicated that high level expression of the *PcGA20ox* transgene resulted in low bioactive GA levels and an extreme dwarf phenotype, similar to the GA-deficient *ga1-3* mutant, whereas lower expression of the transgene gave no phenotype due to up-regulation of the biosynthetic genes *GA20ox1* and *GA3ox1* which restored close-to-normal GA levels.

Figure 7. Manipulation of GA levels by increased turnover. Over-expression of the runner bean *GA2ox1* cDNA under the CaMV35S promoter in Arabidopsis generates an extreme dwarf phenotype (front row). Lower levels of expression in some transgenic lines yields no phenotype, partly due to increased expression of endogenous *GA20ox* and *GA3ox* genes through the action of the feedback pathway; wt: wild-type control. From (23).

Over-expression of *GA2ox* genes has been used to demonstrate this strategy for reduced elongation growth in several crop plants. Ectopic expression of a rice *GA2ox* cDNA, *OsGA2ox1*, under the rice actin promoter in transgenic rice plants yielded extreme dwarf lines that were reduced in height by more than 80% (52). The dwarf lines did not flower unless rescued by application of exogenous GA_3. Moreover, the severe reduction in internode elongation prevented emergence of most flower panicles, which senesced in their leaf sheaths; the few flowers that did emerge were completely sterile (53). Over-expression of the runner bean *GA2ox* gene, *PcGA2ox1*, in transgenic wheat (cv. Canon) under the maize ubiquitin promoter generated a range of phenotypes including semi-dwarfs and extreme dwarfs (Evans, Thomas, Wilkinson, Jones, Lenton, Phillips & Hedden, unpublished); the extent of dwarfing correlated with the level of expression of the transgene. At higher levels of transgene expression, there was also a noticeable loss of negative geotropism in the shoots, to the extent that in the extreme dwarfs the tillers grew almost horizontally (Fig. 8). Other characters affected included reduced flower emergence, reduced seed weight and reduced α-amylase expression in germinating seeds. Overexpression of *PcGA2ox1* in sugar beet, with the aim of increasing resistance to vernalisation, resulted in reduced leaf expansion and a dwarf habit; however, the transgenic plants grown under inductive conditions bolted at a similar time to controls, although the flowers were infertile (Mutasa-Göttgens, Thomas, Phillips & Hedden, unpublished).

Figure 8. Production of a range of dwarf phenotypes in wheat by over-production of runner bean GA 2-oxidase. From (22).

Thus, as an approach to reducing overall GA levels, over-production of GA-catabolising enzymes is very effective, but for targeted effects on elongation growth tissue-specific promoters may be necessary. This was elegantly demonstrated by Sakamoto and colleagues (53) who expressed the rice *OsGA2ox1* cDNA at the site of GA biosynthesis by using the promoter from a GA 3-oxidase gene, *OsGA3ox2*, which is mainly expressed in vegetative tissues. The resulting plants were moderately dwarfed but had normal flower and grain development (Fig. 9). This is an interesting approach because the *Os3ox2* promoter is down-regulated by bioactive GA: thus, the *pOsGA3ox2::OsGA2ox1* chimaeric transgene is up-regulated in low GA levels to counteract the endogenous homeostatic mechanisms that promote GA biosynthesis under such conditions.

Manipulation of GA sensitivity
The worldwide success of semi-dwarf varieties of wheat containing *Rht* alleles provides encouragement that modification of GA response in transgenic crop plants would be an effective means for controlling elongation growth. To demonstrate this, Harberd and colleagues expressed *gai*, the Arabidopsis gain-of-function homologue of *Rht*, in transgenic Basmati 370 rice using the maize ubiquitin promoter (17). Even a low level of expression

of the *gai* transgene resulted in clear dwarfing effects and a loss of GA sensitivity. These authors also showed that over-expression of the wild-type *GAI* in rice conferred a GA-insensitive phenotype in an expression-level dependent manner, with high-level expression of *GAI* resulting in extreme dwarfism and a complete loss of GA responses. In contrast to over-expression of *gai*, a low level of expression of wild-type *GAI* yielded plants with relatively mild dwarfism that retained some responsiveness to applied GA. This clearly shows the potential for the introduction of varying levels of dwarfing into elite lines of cereal crops.

A similar strategy has been applied to reduce stem growth in chrysanthemum, an ornamental species that is treated with growth retardants when produced commercially in order to maintain a bushy habit. Transformation of chrysanthemum with the gain-of-function *gai* gene from Arabidopsis generated lines with a range of dwarf phenotypes, correlating with variable levels of expression of the transgene that are presumably related to its location within the genome (45). Such ornamental species are ideal targets for genetic manipulation, as traditional breeding approaches are frequently hampered by polyploidy, incompatibility and lack of genetic variation in breeding material. Furthermore, genetically-manipulated non-food crops may be more acceptable to consumers in parts of the world where suspicion of the GM approach is high.

Figure 9. Production of semi-dwarf rice by over-production of a GA 2-oxidase under the *OsGA3ox2* promoter. From (53).

RESISTANCE TO BIOTIC AND ABIOTIC STRESSES

As sessile life forms, plants are particularly vulnerable to changes in environmental conditions and attack by other organisms and have therefore developed a wide range of biochemical and physical defenses to combat such stresses. However, constitutive expression of these defensive mechanisms has costs in terms of both the resources consumed (for example, the diversion of fixed carbon and nitrogen into secondary metabolism) and possible side-effects on plant growth and development, with consequent effects on fitness as measured by survival and seed production. As a result, plants have evolved signalling systems that allow the defense mechanism to be invoked only when challenged by the particular biotic or abiotic stress. A number of plant hormones, particularly abscisic acid (ABA), ethylene, jasmonic acid (JA) and salicylic acid (SA), but also brassinosteroids (BR) are involved in the signalling pathways between stress perception and response; the specific roles of the hormones in these processes are described in detail elsewhere in this book and will not be discussed further here.

These hormone signalling pathways are attractive targets for the genetic modification of resistance to both biotic and biotic stresses, because they act as an interface between the diverse perception mechanisms and the wide range of response genes. This suggests that manipulation of the biosynthesis or catabolism of these hormones might lead to increased resistance to a range of pathogens or abiotic stresses due to expression of a set, or subset, of resistance genes. However, one serious drawback of this approach is that some components of the resistance are then no longer inducible but constitutive, with associated costs in the absence of the stress. These costs can be determined experimentally by direct application of the hormone to plants: Baldwin (2) manipulated the resistance of leaves of *Nicotiana attenuata* to herbivory by treatment with methyl jasmonate (MeJA), which induces nicotine accumulation. In populations with modest levels of herbivory, MeJA-treated plants produced more seeds than untreated plants, whereas in the absence of herbivores untreated plants were more successful. Thus, plants with constitutive expression of resistance may only have an advantage where the pressure from pathogens or adverse environmental conditions is high. It is likely that crop plants frequently experience such adverse conditions, as the pathogen load is likely to be high in monocultures and crops are often grown outside the ecological niche in which they originally evolved.

Salicylic Acid
Several studies have shown that genetic modification of hormone biosynthetic pathways can lead to increased resistance to pathogens or wounding. Constitutive overproduction of SA in tobacco, achieved by coordinate overexpression of the *E. coli entC* gene encoding isochorismate synthase and the *Pseudomonas fluorescens pmsB* gene encoding isochorismate pyruvate lyase, resulted in increased levels of resistance to both viral and fungal pathogens (59). Although SA levels were high in the transgenic lines, the plants had a normal appearance and development, suggesting that this approach might be successfully applied to crops.

Jasmonic Acid
Attempts to overproduce JA in transgenic plants have generated conflicting results. A key enzyme in the biosynthetic pathway, allene oxide synthase

(AOS) appears to be localized in the plastid. Overproduction of a flax seed AOS, complete with plastid targeting sequence, in transgenic potato plants increased endogenous JA levels by 8-12 fold but did not lead to the induction of JA-responsive genes in the transgenic plants (21), suggesting that JA produced by the action of plastidic AOS may be unavailable for the induction of wound-responsive genes. In contrast, overproduction of the complete Arabidopsis AOS in tobacco did not increase endogenous JA levels in unwounded plants, while wounded plants accumulated 2-3 fold more JA than non-transgenic controls (34). Similar results were obtained when the flax AOS, with plastid transit peptide removed, was expressed in tobacco, increasing the wound-induced JA levels by 2.5-5 fold above wounded plants not expressing the transgene (61). In these latter cases, the wound-induced JA also led to expression of JA-responsive genes. Whether this led to increased resistance to wounding or susceptibility to pathogens was not investigated.

Ethylene
Ethylene similarly plays significant roles in defense against pathogens, wounding and flooding. Increased ethylene levels have been detected in plants challenged with a wide range of biotic and abiotic stresses, and the ethylene produced induces a wide range of responses, which differ between species. In some cases, it has been shown that ethylene responses increase the resistance of plants to pathogen attack: for example, tomato plants pre-treated with ethylene are less susceptible to *Botrytis cinerea* (8). However, the magnitude of the ethylene response is frequently so extreme that processes such as senescence and chlorosis are induced, leading to tissue damage such that the response may be more damaging than the initial stress. Paradoxically, therefore, modifications that limit ethylene biosynthesis or that decrease sensitivity to the hormone (the approaches discussed under Fruit Ripening and Flower Senescence, below) can result in increased tolerance of pathogens. For example, the tomato *never-ripe* mutant, which is impaired in ethylene perception, shows greater tolerance of bacterial infection than wild type controls, with reduced severity of symptoms and increased survival of seedlings to infection (36). Plants with reduced ethylene accumulation due to over-expression of ACC deaminase showed similarly reductions in symptoms of bacterial spot disease (36). This and other evidence suggests that, in many cases, ethylene is essential for the development of symptoms but not for basal resistance. Crops engineered to have reduced ethylene accumulation or response, may, therefore, have usefully enhanced tolerance to pathogens and wounding.

Abscisic Acid
The principal role of ABA in non-seed tissues of plants is in mediating responses to various environmental stresses, particularly drought. The increasing use of marginal lands for crop production and the degradation of arable land by irrigation and climate change result in high levels of drought,

salinity and cold stress that strongly influence plant productivity. Manipulation of ABA biosynthesis to increase ABA levels may, therefore, enhance the endogenous resistance of crops to such stresses. Analysis of intermediates in the ABA biosynthetic pathway suggests that cleavage of 9-*cis*-epoxycarotenoids, catalysed by 9-*cis*-epoxycarotenoid dioxygenase (NCED), is an important regulatory step; this is supported by evidence that drought stress in maize strongly induces the expression of *NCED* (56).

NCED has been over-expressed in a number of species, including tomato, Arabidopsis, tobacco and *Nicotiana plumbaginifolia*. Transgenic tomato plants with constitutive overexpression of tomato *NCED* were wilty, with epinastic leaves and other characteristics of ABA deficiency, suggesting a reduction in endogenous *NCED* expression due to co-suppression (58). Other lines had an over-guttating phenotype, which was interpreted as a symptom of ABA over-production. These plants were shown to have, on average, 2.8-fold higher ABA levels in the leaves and had only half the level of stomatal conductance of control plants. Seeds from these plants also exhibited increased dormancy, reversed by treatment with norflurazon, which inhibits carotenoid biosynthesis and this reduces ABA levels. These authors also showed that transient overexpression of *NCED* in tobacco, using a tetracycline-inducible promoter, was even more successful in transiently increasing ABA levels, by as much as 10-fold over wild-type levels (58). It is likely that ABA levels in plants with sustained over-expression of *NCED* might eventually return closer to normal, as ABA promotes its own inactivation through induction of ABA 8-hydroxylase. Indeed, over-expression of *NCED* in *Nicotiana plumbaginifolia* using the CaMV 35S promoter resulted in 3.5-fold increases in ABA, but 6-fold increases in the ABA catabolite, phaseic acid (PA) (48). Furthermore, inducible expression of *NCED* using a glucocorticoid-regulated promoter showed that ABA levels increased markedly during the first six hours of dexamethasone (DEX) induction but then leveled off, while PA levels continued to increase (48). These plants showed increased seed dormancy, induced by DEX, and enhanced drought tolerance: transgenic plants treated with DEX and transferred to low humidity showed reduced wilting when compared with controls (Fig. 10). This indicates that endogenous ABA production, induced by water loss, is less effective at protecting from drought than increases in ABA induced by DEX treatment before the onset of stress and suggests that plants in the field with enhanced ABA levels might be similarly protected

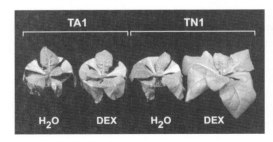

Figure 10. Enhanced drought tolerance in *Nicotiana plumbaginifolia* by expression of bean NCED under a dexamethasone-inducible promoter. TA1: control plant transformed with an empty vector; TN1: line transformed with *PvNCED* under the DEX-inducible promoter; DEX: treated with dexamethasone. From (48).

from early symptoms of drought. However, high ABA levels in the absence of drought may reduce plant productivity, and such transgenic crops may be limited to situations where drought stress is the rule rather than the exception.

An alternative strategy to increasing ABA levels in the protection of plants against stress is the manipulation of candidate ABA signalling components. The Arabidopsis ABI3 gene product is a transcription factor involved in the control of dormancy and cellular maturation; mutations in *ABI3* germinate in the presence of ABA suggesting that ABI3 is a component of, or interacts with, the ABA signalling pathway, although there has been little evidence for a direct role in ABA signalling. Ectopic expression of *ABI3* resulted in increased sensitivity of Arabidopsis seedlings to ABA, as determined by the expression of ABA-induced genes, including those associated with tolerance to low temperatures (55). As a consequence, the transgenic lines showed an increased induction of tolerance to freezing after ABA pre-treatment or after a period of low temperature. This provides further evidence for the involvement of ABI3 in ABA responses outside the seed, and illustrates the advantage in manipulating global signalling regulators, such as hormone signalling components, to induce a large set of target genes, where manipulating individual target genes themselves might have only minimal effect.

FRUIT DEVELOPMENT AND PARTHENOCARPY

The development of fruit is normally dependent on fertilization and in the absence of pollination the ovary will cease growth and senesce or abscise. However, there are several reasons why parthenocarpic growth - the development of fruit in the absence of fertilization – is desirable. In several fruit crops, particularly grapes and citrus, seedless varieties are more popular with consumers, while in aubergine (eggplant) the seeds impart a bitter taste and result in increased browning of the flesh. Difficulty in achieving pollination can also limit productivity: unfavorable environmental conditions can affect pollen production or the abundance of pollinating insects. While this can be circumvented by growing under glass and supplying the insect pollinators, this is an expensive solution.

There is good evidence that in many species continued fruit growth depends upon hormonal signals, principally GA and auxin, originating from the developing seed. Parthenocarpy can thus often be achieved by application of auxin or GA to the unpollinated flower bud. Alternatively, in several crops mutations have been identified that induce parthenocarpy. For example, two non-allelic genetic systems in tomato, *pat-2* and *pat-3/pat-4*, have been identified that allow full development of unfertilized ovaries to form complete fruit. Analysis of *in vitro* translation products from unpollinated ovaries suggests that *pat-2* and *pat3/pat-4* are related in their molecular modes of action, and GA analysis of developing ovaries showed that both mutations result in increased GA_1 and GA_3 content compared with wild-type

Plant hormones and crop performance

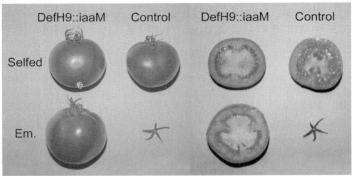

Figure 11. Induction of parthenocarpy in tomato by over-production of auxin in placenta/ovule tissues through transformation with the *DefH9::iaaM* construct. Selfed: self-pollinated flowers; Em.: emasculated flowers. From (12).

tissues (16). In the *pat3/pat-4* system all 13-hydroxy GAs are elevated in the ovary, indicating a stimulation of GA biosynthesis generally, whereas in the *pat-2* mutant only C_{19}-GAs are elevated, suggestive of an increase in GA 20-oxidase activity (16). Parthenocarpy has also been induced in tomato by over-production of auxin: expression of the *iaaM* gene from *Pseudomonas syringae*, under the *DefH9* placenta/ovule-specific promoter from *Antirrhinum*, results in production of indole 3-acetamide which is converted to IAA in plant tissues. The transgenic tomato plants containing *DefH9-iaaM* produced parthenocarpic fruit that were identical in size and weight to fruit from pollinated flowers (Fig. 11) (12). As auxin is known to increase GA biosynthesis in other systems (Chapter E1), it seems likely that at least part of the mechanism of auxin action in both tomato fruit development is to promote GA biosynthesis in the unfertilized ovaries.

The same *DefH9-iaaM* construct has been similarly used to promote parthenocarpy in eggplant: the transgenic plants produced fruit of a similar size as pollinated ovaries (Fig. 12) (50). Furthermore, in winter condition, where control plants failed to set fruit even from hand-pollinated flowers, the *DefH9-iaaM* transgenic lines produced parthenocarpic fruit from emasculated flowers. The transgenic lines were also superior to control lines that were treated with a commercial formulation of β-naphthoxyacetic acid and GA_3, giving a 25% yield increase and an improvement in fruit quality (9)

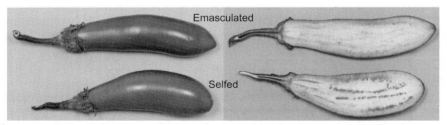

Figure 12 (Colour plate page CP1). Induction of parthenocarpy in eggplant by over-production of auxin in placenta/ovule tissues through transformation with the *defH9::iaaM* construct. Selfed: self-pollinated flowers. From (50).

when grown under glass. With the added advantages of reduced labour and phytohormone costs, it is hopeful that these transgenic eggplants will see commercial production in the near future.

POTATO TUBERIZATION

Tuberisation of potato is variably influenced by photoperiod, with short days (SD) generally promoting tuber formation: the subspecies *Solanum tuberosum andigena* has an obligate requirement for SD in tuberisation, mediated by the phytochrome PHYB. Grafting experiments have shown that a transmissible substance produced in long days (LD) inhibits tuber formation; treatment with bioactive GAs similarly represses tuberisation, suggesting that a gibberellin may be the transmissible substance. The current model for induction of tuber formation (Chapter E5) proposes that LD promotes expression of a GA 20-oxidase gene, *StGA20ox1*, increasing the level of GA_{20} in leaves; the mobile GA_{20} moves to the roots where GA 3β-hydroxylase converts it to GA_1, inhibiting tuberisation. Transgenic *S. tuberosum andigena* plants with reduced leaf *StGA20ox1* mRNA levels, achieved through antisense suppression, have lower concentrations of GA_{20} in the leaves and formed tubers earlier in SD (4). However, the transgenic plants did not form tubers in long days, suggesting either that GA levels were not sufficiently lowered in the non-inductive conditions, or that an additional signal was involved. It also remains to be seen whether a similar approach in commercial varieties that are less sensitive to daylength will be equally successful.

FRUIT RIPENING AND FLOWER SENESCENCE

Ripening in climacteric fruit such as apples, tomatoes and bananas is typically characterised by an increase in respiration, the production of wall-softening enzymes such as polygalacturonase and a change in colour due to the loss of chlorophyll and the accumulation of other pigments (eg. lycopene in tomatoes); thereafter, the fruit continues into senescence and decay. Unfortunately, modern commercial practices involving transport over great distances, with consequent delays between harvest and sale to the consumer, can result in fruit that are past their best on arrival in the supermarkets. Although shelf life can be extended by storage in controlled atmosphere, this method is only effective for a limited time. Many cut flowers similarly deteriorate after harvest, and flower longevity is one of the most important targets of plant breeders and biotechnologists.

Ripening and senescence are promoted by ethylene, which promotes its own biosynthesis through increases in both ACC synthase (ACS) and ACC oxidase (ACO) activities, leading to a rapid rise in levels of the hormone. Chemical approaches to delaying ripening act by blocking ethylene synthesis (eg. aminoethoxyvinylglycine, AVG, which inhibits ACC synthase) or perception (eg. 1-methylcyclopropene, 1-MCP, an ethylene antagonist, and Ag^+ ions, that compete with Cu^+ as cofactor in the ethylene receptor). However, as discussed above, there is growing concern over the use of chemical growth regulators in food production, both in terms of environmental protection and human health. A number of biotechnological approaches have therefore been developed to target this signalling pathway, mainly in tomato, but also other fruit and cut flowers.

Although ACS is considered to be the rate-limiting step in ethylene biosynthesis (Chapter B4), a cDNA encoding ACO was the first to be used in attempts to suppress ethylene biosynthesis in transgenic plants: the pTOM13 cDNA from tomato fruit, whose endogenous expression profile paralleled that of ethylene biosynthesis, was expressed in tomato as antisense under the CaMV-35S promoter (47). The transgenic plants showed reduced ACO activity and reduced ethylene production by wounded leaves and ripening fruit. In progeny homozygous for the transgene, ethylene production in ripening fruit was inhibited by 97%; the effect on the ripening process was less pronounced, although fruit took longer to reach normal levels of lycopene accumulation and were noticeably less prone to shrivelling (Fig. 13) (47). Leaf senescence was also noticeably delayed in these plants (28).

Similar approaches involving suppression of *ACO* genes have been used in other ethylene-sensitive fruit and flower crops. In melon, climacteric ethylene production was completely abolished using an antisense melon *ACO* gene; in this case, ripening of the rind was prevented while pulp ripening was only delayed (15). This suggested that a second melon *ACO* gene not suppressed by the antisense construct might be involved in ethylene production by the pulp, or that the pulp was more sensitive to endogenous

Figure 13 (Colour plate page CP14). Antisense suppression of ACO in tomato retards fruit ripening and senescence. From (47).

ethylene. A similar situation was observed in broccoli, in which an antisense tomato *ACO* cDNA (pTOM13) was successful in reducing ethylene biosynthesis associated with an early phase in senescence, while a later burst of ethylene production was unaffected (24). In carnation, ethylene produced in the gynoecium induces the expression of *ACS* and *ACO* genes in the petals, leading to senescence. Co-suppression of *ACO* in the gynoecium prevented ethylene signalling to the petals and neither *ACS* nor *ACO* were induced in petal tissues, resulting in a two-fold extension in vase life (33).

Ethylene levels have also been reduced by suppression of *ACS* gene expression. In tomato, *ACS* is encoded by a small multi-gene family, at least two members of which are expressed in ripening fruit. Antisense expression of one of these genes, *LeACC2*, succeeded in suppressing transcripts from both fruit *ACS* genes and reduced ethylene levels in fruit by as much as 99.5%. Fruit from these plants failed to ripen, even 90 days post-harvest, unless treated with ethylene. This has been commercially developed by DNA Plant Technology Corporation through the use of partial sense suppression ("Transwitch") of *ACS*; the fruit have been approved for sale in the US under the trade name "Endless Summer", although they are not in commercial production. A similar strategy has been applied to the apple cultivar 'Royal Gala', where plants transformed with an antisense *ACS* construct showed 90-99% reductions in ethylene production; fruit from these plant left at room temperature and humidity for three months showed no sign of senescence (Fig. 14) (25). Transgenic carnations in which *ACS* has been suppressed, doubling the vase life of the cut flowers, are similarly close to commercialisation by Florigene.

Alternative transgenic strategies for reducing ethylene production in plants involve removal of intermediates in the ethylene biosynthetic pathway through the over-expression of metabolising enzymes (Fig. 15). S-adenosylmethionine (SAM) is the immediate precursor of ACC; to reduce the levels of SAM, a gene from bacteriophage T3 encoding SAM hydrolase was expressed in tomato using the fruit-specific *E8* promoter (20). Ethylene production in fruit from the transgenic lines was reduced by about 80%, which was enough to double the time taken to full ripeness and delay senescence by up to three months. The SAM hydrolase technology is being commercialised by Agritope for a variety of cherry tomato. Similarly, a reduction in ACC levels has been achieved by over-expression of an ACC deaminase gene from *Pseudomonas spp.* in tomato under the CaMV-35S

Figure 14 (Colour plate page CP13). Increased shelf life in apple with reduced ethylene biosynthesis, through antisense suppression of an endogenous *ACS* gene. Fruit from control and three transgenic lines were kept for 94 days at room temperature and humidity. From (25).

Plant hormones and crop performance

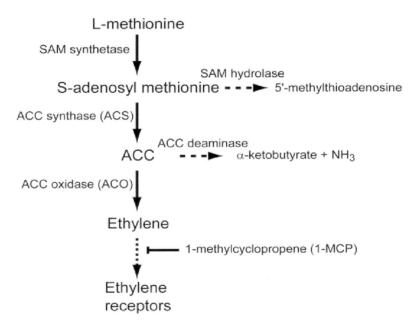

Figure 15. Pathways of ethylene biosynthesis and manipulation.

promoter: leaf ethylene levels were reduced by over 90% and ripening and senescence were retarded in harvested fruit (31). It was noted that transgenic fruit left on the vine ripened much more quickly than detached fruit from the same plant: the detached fruit have lower ethylene levels, possibly due to diffusion of the gas through the stem scar (30).

Mutations in ethylene receptor genes that cause loss of hormone binding confer a dominant ethylene-insensitive phenotype. In tomato, the dominant *Never-ripe* mutant is impaired in ethylene perception due to a lesion in the sensor domain of a protein with homology to ethylene receptors (64). Expression of such dominant mutant ethylene receptors in transgenic plants confers ethylene insensitivity: for example, over-expression of the Arabidopsis *etr1-1* gene under the CaMV-35S promoter in tomato resulted in plants that were highly resistant to ethylene, with a loss of the triple response in seedlings and delays in flower senescence and abscission, while the fruit never fully ripened (63). Similarly, over-expression of *etr1-1* in petunia and carnation bestowed an ethylene-insensitive phenotype that extended the life of flowers by 2-3 fold in each case (3, 63).

Cytokinins also play a role in the control of senescence as revealed by application of cytokinin to detached leaves which delays senescence. Over-production of cytokinins in transgenic plants may be accomplished by expression of the *IPT* gene from *Agrobacterium tumifaciens*, encoding *iso*pentenyl transferase. Excessive accumulation of cytokinins may be deleterious, causing stunted growth, under-developed vascular tissues and root systems and loss of apical dominance. Gan and Amasino (18) avoided these developmental abnormalities by driving expression of *IPT* in tobacco with a leaf-specific, senescence-regulated promoter, thus introducing a feedback control mechanism into expression of the transgene. This autoregulatory system maintained *IPT* expression at a minimum level, such that there were no developmental abnormalities but leaf senescence was prevented and the transgenic plants produced near twice the number of flowers than the controls, due to their extended life-span. Several groups are applying a similar approach to crops species.

CONCLUSIONS

Recent advances in our knowledge of the pathways of hormone biosynthesis and perception in plants has permitted genetic manipulation of these signalling systems. As plant hormones have multiple roles in plant development, approaches that involve global modulation of hormone status or signalling may have limited uses, as beneficial effects in one process may have unintended consequences on other aspects of plant development. This should be avoidable in many cases through the use of tissue-specific and inducible promoters to achieve control of the transgenes.

Antagonism to transgene technology, particularly in Europe, continues to slow the development and commercialization of genetically manipulated crops. It is possible that transgenic non-food crops, such as cut flowers, will be more acceptable and will help to allay fears of the new technology. However, it is likely that, for the foreseeable future, for a transgenic crops to succeed they will have to be clearly superior to those produced by established breeding or agrochemical methods. For example, transgenic tomatoes engineered for increased shelf life by manipulation of ethylene biosynthesis or perception have not yet been commercialized, because this product competes directly with equivalent, non-transgenic varieties containing the *rin* (ripening inhibitor) mutation.

One alternative technology that may bridge the gap between scientific progress and public acceptance is TILLING: targeting induced local lesions in genomes (40). This allows mutations in specific genes to be identified in large natural or mutagenised populations, through gene-specific PCR followed by heteroduplex analysis. For example, tomatoes with reduced ethylene production could be developed by TILLING for mutations in *ACS* or *ACO* genes. However, this approach may not be appropriate for gene families with a high level of redundancy, or in polyploid species such as wheat.

References

1. Ashikari, M., Sasaki, A., Ueguchi-Tanaka, M., Itoh, H., Nishimura, A., Datta, S., Ishiyama, K., Saito, T., Kobayashi, M., Khush, G.S., Kitano, H. and Matsuoka, M. (2002) Loss-of-function of a rice gibberellin biosynthetic gene, GA20 oxidase (*GA20ox-2*), led to the rice 'green revolution'. Breed. Sci., 52, 143-150.
2. Baldwin, I.T. (1998) Jasmonate-induced responses are costly but benefit plants under

attack in native populations. Proc. Natl. Acad. Sci. U.S.A., 95, 8113-8118.
3. Bovy, A.G., Angenent, G.C., Dons, H.J.M. and van Altvorst, A.C. (1999) Heterologous expression of the Arabidopsis *etr1-1* allele inhibits the senescence of carnation flowers. Mol. Breed., 5, 301-308.
4. Carrera, E., Bou, J., Garcia-Martinez, J.L. and Prat, S. (2000) Changes in GA 20-oxidase gene expression strongly affect stem length, tuber induction and tuber yield of potato plants. Plant J., 22, 247-256.
5. Chiang, H.H., Hwang, I. and Goodman, H.M. (1995) Isolation of the Arabidopsis *GA4* locus. Plant Cell, 7, 195-201.
6. Coles, J.P., Phillips, A.L., Croker, S.J., Garcia-Lepe, R., Lewis, M.J. and Hedden, P. (1999) Modification of gibberellin production and plant development in Arabidopsis by sense and antisense expression of gibberellin 20-oxidase genes. Plant J., 17, 547-556.
7. Curtis, I.S., Ward, D.A., Thomas, S.G., Phillips, A.L., Davey, M.R., Power, J.B., Lowe, K.C., Croker, S.J., Lewis, M.J., Magness, S.L. and Hedden, P. (2000) Induction of dwarfism in transgenic *Solanum dulcamara* by over-expression of a gibberellin 20-oxidase cDNA from pumpkin. Plant J., 23, 329-338.
8. Diaz, J., ten Have, A. and van Kan, J.A.L. (2002) The role of ethylene and wound signaling in resistance of tomato to *Botrytis cinerea*. Plant Physiol., 129, 1341-1351.
9. Donzella, G., Spena, A. and Rotino, G.L. (2000) Transgenic parthenocarpic eggplants: superior germplasm for increased winter production. Mol. Breed., 6, 79-86.
10. Eriksson, M.E., Israelsson, M., Olsson, O. and Moritz, T. (2000) Increased gibberellin biosynthesis in transgenic trees promotes growth, biomass production and xylem fiber length. Nature Biotechnol., 18, 784-788.
11. Evans, L.T. (1998) Feeding the Ten Billion: Plant and Population Growth. Cambridge University Press, Cambridge.
12. Ficcadenti, N., Sestili, S., Pandolfini, T., Cirillo, C., Rotino, G.L. and Spena, A. (1999) Genetic engineering of parthenocarpic fruit development in tomato. Mol. Breed., 5, 463-470.
13. Fleet, C.M., Yamaguchi, S., Hanada, A., Kawaide, H., David, C.J., Kamiya, Y. and Sun, T.-p. (2003) Overexpression of *AtCPS* and *AtKS* in *Arabidopsis* confers increased *ent*-kaurene production but no increase in bioactive gibberellins. Plant Physiol., 132, 830-839.
14. Flintham, J.E., Borner, A., Worland, A.J. and Gale, M.D. (1997) Optimizing wheat grain yield: Effects of *Rht* (gibberellin-insensitive) dwarfing genes. J. Agric. Sci., 128, 11-25.
15. Flores, F.B., Martinez-Madrid, M.C., Sanchez-Hidalgo, F.J. and Romojaro, F. (2001) Differential rind and pulp ripening of transgenic antisense ACC oxidase melon. Plant Physiol. Biochem., 39, 37-43.
16. Fos, M., Proano, K., Nuez, F. and Garcia-Martinez, J.L. (2001) Role of gibberellins in parthenocarpic fruit development induced by the genetic system *pat-3/pat-4* in tomato. Physiol. Plant., 111, 545-550.
17. Fu, X.D., Sudhakar, D., Peng, J.R., Richards, D.E., Christou, P. and Harberd, N.P. (2001) Expression of arabidopsis *GAI* in transgenic rice represses multiple gibberellin responses. Plant Cell, 13, 1791-1802.
18. Gan, S.S. and Amasino, R.M. (1995) Inhibition of leaf senescence by autoregulated production of cytokinin. Science, 270, 1986-1988.
19. Gianfagna, T. (1995) Natural and synthetic growth regulators and their use in horticultural and agronomic crops. In: *Plant Hormones*. Davies, P. J. (ed.). Dordrecht: Kluwer, pp. 751-773.
20. Good, X., Kellogg, J.A., Wagoner, W., Langhoff, D., Matsumura, W. and Bestwick, R.K. (1994) Reduced ethylene synthesis by transgenic tomatoes expressing S-adenosylmethionine hydrolase. Plant Mol. Biol., 26, 781-790.
21. Harms, K., Atzorn, R., Brash, A., Kühn, H., Wasternack, C., Willmitzer, L. and Peña-Cortés, H. (1995) Expression of a flax allene oxide synthase cDNA leads to increased

endogenous jasmonic acid (JA) levels in transgenic potato plants but not to a corresponding activation of JA-responding genes. Plant Cell, 7, 1645-1654.
22. Hedden, P. and Phillips, A.L. (2000) Gibberellin metabolism: new insights revealed by the genes. Trends Plant Sci., 5, 523-530.
23. Hedden, P. and Phillips, A.L. (2000) Manipulation of hormone biosynthetic genes in transgenic plants. Curr. Opin. Biotechnol., 11, 130-137.
24. Henzi, M.X., Christey, M.C. and McNeil, D.L. (2000) Morphological characterisation and agronomic evaluation of transgenic broccoli (*Brassica oleracea* L. var. italica) containing an antisense *ACC oxidase* gene. Euphytica, 113, 9-18.
25. Hrazdina, G., Kiss, E., Galli, Z., Rosenfield, C., Norelli, J.L. and Aldwinckle, H.S. (2003) Down Regulation of Ethylene Production in 'Royal Gala' Apples. Acta Horticulturae, 628, in press.
26. Huang, S.S., Raman, A.S., Ream, J.E., Fujiwara, H., Cerny, R.E. and Brown, S.M. (1998) Overexpression of 20-oxidase confers a gibberellin- overproduction phenotype in Arabidopsis. Plant Physiol., 118, 773-781.
27. Itoh, H., Ueguchi-Tanaka, M., Sakamoto, T., Kayano, T., Tanaka, H., Ashikari, M. and Matsuoka, M. (2002) Modification of rice plant height by suppressing the height-controlling gene, *D18*, in rice. Breed. Sci., 52, 215-218.
28. John, I., Drake, R., Farrell, A., Cooper, W., Lee, P., Horton, P. and Grierson, D. (1995) Delayed leaf senescence in ethylene-deficient ACC oxidase antisense tomato plants - molecular and physiological analysis. Plant J., 7, 483-490.
29. Jones, H.D., Peters, N.C.B. and Holdsworth, M.J. (1997) Genotype and environment interact to central dormancy and differential expression of the *VIVIPAROUS-1* homologue in embryos of *Avena fatua*. Plant J., 12, 911-920.
30. Klee, H.J. (1993) Ripening physiology of fruit from transgenic tomato (*Lycopersicon esculentum*) plants with reduced ethylene synthesis. Plant Physiol., 102, 911-916.
31. Klee, H.J., Hayford, M.B., Kretzmer, K.A., Barry, G.F. and Kishore, G.M. (1991) Control of ethylene synthesis by expression of a bacterial enzyme in transgenic tomato plants. Plant Cell, 3, 1187-1193.
32. Koornneef, M., Elgersma, A., Hanhart, C.J., van Loenen-Martinet, E.P., van Rijn, L. and Zeevaart, J.A.D. (1985) A gibberellin insensitive mutant of *Arabidopsis thaliana*. Physiol. Plant., 65, 33-39.
33. Kosugi, Y., Waki, K., Iwazaki, Y., Tsuruno, N., Mochizuki, A., Yoshioka, T., Hashiba, T. and Satoh, S. (2002) Senescence and gene expression of transgenic non-ethylene-producing carnation flowers. J. Jpn. Soc. Hortic. Sci., 71, 638-642.
34. Laudert, D., Schaller, F. and Weiler, E.W. (2000) Transgenic *Nicotiana tabacum* and *Arabidopsis thaliana* plants overexpressing allene oxide synthase. Planta, 211, 163-165.
35. Lester, D.R., Ross, J.J., Davies, P.J. and Reid, J.B. (1997) Mendel's stem length gene (*Le*) encodes a gibberellin 3 beta- hydroxylase. Plant Cell, 9, 1435-1443.
36. Lund, S.T., Stall, R.E. and Klee, H.J. (1998) Ethylene regulates the susceptible response to pathogen infection in tomato. Plant Cell, 10, 371-382.
37. Martin, D.N., Proebsting, W.M. and Hedden, P. (1997) Mendel's dwarfing gene: cDNAs from the *Le* alleles and function of the expressed proteins. Proc. Natl. Acad. Sci. U.S.A., 94, 8907-8911.
38. Martin, D.N., Proebsting, W.M. and Hedden, P. (1999) The SLENDER gene of pea encodes a gibberellin 2-oxidase. Plant Physiol., 121, 775-781.
39. Martinez-Garcia, J.F., Santes, C.M. and Garcia-Martinez, J.L. (2000) The end-of-day far-red irradiation increases gibberellin A_1 content in cowpea (*Vigna sinensis*) epicotyls by reducing its inactivation. Physiol. Plant., 108, 426-434.
40. McCallum, C.M., Comai, L., Greene, E.A. and Henikoff, S. (2000) Targeting induced local lesions in genomes (TILLING) for plant functional genomics. Plant Physiol., 123, 439-442.
41. McKibbin, R.S., Wilkinson, M.D., Bailey, P.C., Flintham, J.E., Andrew, L.M., Lazzeri,

P.A., Gale, M.D., Lenton, J.R. and Holdsworth, M.J. (2002) Transcripts of *Vp-1* homeologues are misspliced in modern wheat and ancestral species. Proc. Natl. Acad. Sci. U.S.A., 99, 10203-10208.
42. Monna, L., Kitazawa, N., Yoshino, R., Suzuki, J., Masuda, H., Maehara, Y., Tanji, M., Sato, M., Nasu, S. and Minobe, Y. (2002) Positional cloning of rice semidwarfing gene, *sd-1*: Rice "Green revolution gene" encodes a mutant enzyme involved in gibberellin synthesis. DNA Res., 9, 11-17.
43. Peng, J.R., Carol, P., Richards, D.E., King, K.E., Cowling, R.J., Murphy, G.P. and Harberd, N.P. (1997) The Arabidopsis *GAI* gene defines a signaling pathway that negatively regulates gibberellin responses. Genes Dev., 11, 3194-3205.
44. Peng, J.R., Richards, D.E., Hartley, N.M., Murphy, G.P., Devos, K.M., Flintham, J.E., Beales, J., Fish, L.J., Worland, A.J., Pelica, F., Sudhakar, D., Christou, P., Snape, J.W., Gale, M.D. and Harberd, N.P. (1999) 'Green revolution' genes encode mutant gibberellin response modulators. Nature, 400, 256-261.
45. Petty, L.M., Harberd, N.P., Carre, I.A., Thomas, B. and Jackson, S.D. (2003) Expression of the Arabidopsis *gai* gene under its own promoter causes a reduction in plant height in chrysanthemum by attenuation of the gibberellin response. Plant Sci., 164, 175-182.
46. Phillips, A.L., Ward, D.A., Uknes, S., Appleford, N.E.J., Lange, T., Huttly, A.K., Gaskin, P., Graebe, J.E. and Hedden, P. (1995) Isolation and expression of three gibberellin 20-oxidase cDNA clones from Arabidopsis. Plant Physiol., 108, 1049-1057.
47. Picton, S., Barton, S.L., Bouzayen, M., Hamilton, A.J. and Grierson, D. (1993) Altered fruit ripening and leaf senescence in tomatoes expressing an antisense ethylene-forming enzyme transgene. Plant J., 3, 469-481.
48. Qin, X.Q. and Zeevaart, J.A.D. (2002) Overexpression of a 9-cis-epoxycarotenoid dioxygenase gene in *Nicotiana plumbaginifolia* increases abscisic acid and phaseic acid levels and enhances drought tolerance. Plant Physiol., 128, 544-551.
49. Rademacher, W. (2000) Growth retardants: Effects on gibberellin biosynthesis and other metabolic pathways. Annu. Rev. Plant Physiol., 51, 501-531.
50. Rotino, G.L., Perri, E., Zottini, M., Sommer, H. and Spena, A. (1997) Genetic engineering of parthenocarpic plants. Nature Biotechnol., 15, 1398-1401.
51. Sakamoto, T., Kamiya, N., Ueguchi-Tanaka, M., Iwahori, S. and Matsuoka, M. (2001) KNOX homeodomain protein directly suppresses the expression of a gibberellin biosynthetic gene in the tobacco shoot apical meristem. Genes Dev., 15, 581-590.
52. Sakamoto, T., Kobayashi, M., Itoh, H., Tagiri, A., Kayano, T., Tanaka, H., Iwahori, S. and Matsuoka, M. (2001) Expression of a gibberellin 2-oxidase gene around the shoot apex is related to phase transition in rice. Plant Physiol., 125, 1508-1516.
53. Sakamoto, T., Morinaka, Y., Ishiyama, K., Kobayashi, M., Itoh, H., Kayano, T., Iwahori, S., Matsuoka, M. and Tanaka, H. (2003) Genetic manipulation of gibberellin metabolism in transgenic rice. Nature Biotechnol., 21, 909-913.
54. Spielmeyer, W., Ellis, M.H. and Chandler, P.M. (2002) Semidwarf (*sd-1*), "green revolution" rice, contains a defective gibberellin 20-oxidase gene. Proc. Natl. Acad. Sci. U.S.A., 99, 9043-9048.
55. Tamminen, I., Makela, P., Heino, P. and Palva, E.T. (2001) Ectopic expression of *ABI3* gene enhances freezing tolerance in response to abscisic acid and low temperature in Arabidopsis thaliana. Plant J., 25, 1-8.
56. Tan, B.C., Schwartz, S.H., Zeevaart, J.A.D. and McCarty, D.R. (1997) Genetic control of abscisic acid biosynthesis in maize. Proc. Natl. Acad. Sci. U.S.A., 94, 12235-12240.
57. Thomas, S.G., Phillips, A.L. and Hedden, P. (1999) Molecular cloning and functional expression of gibberellin 2- oxidases, multifunctional enzymes involved in gibberellin deactivation. Proc. Natl. Acad. Sci. U.S.A., 96, 4698-4703.
58. Thompson, A.J., Jackson, A.C., Symonds, R.C., Mulholland, B.J., Dadswell, A.R., Blake, P.S., Burbidge, A. and Taylor, I.B. (2000) Ectopic expression of a tomato 9-cis-epoxycarotenoid dioxygenase gene causes over-production of abscisic acid. Plant J., 23,

363-374.
59. Verberne, M.C., Verpoorte, R., Bol, J.F., Mercado-Blanco, J. and Linthorst, H.J.M. (2000) Overproduction of salicylic acid in plants by bacterial transgenes enhances pathogen resistance. Nature Biotechnol., 18, 779-783.
60. Vidal, A.M., Gisbert, C., Talon, M., Primo-Millo, E., Lopez-Diaz, I. and Garcia-Martinez, J.L. (2001) The ectopic overexpression of a citrus gibberellin 20-oxidase enhances the non-13-hydroxylation pathway of gibberellin biosynthesis and induces an extremely elongated phenotype in tobacco. Physiol. Plant., 112, 251-260.
61. Wang, C., Avdiushko, S. and Hildebrand, D.F. (1999) Overexpression of a cytoplasm-localized allene oxide synthase promotes the wound-induced accumulation of jasmonic acid in transgenic tobacco. Plant Mol. Biol., 40, 783-793.
62. White, C.N., Proebsting, W.M., Hedden, P. and Rivin, C.J. (2000) Gibberellins and seed development in maize. I. Evidence that gibberellin/abscisic acid balance governs germination versus maturation pathways. Plant Physiol., 122, 1081-1088.
63. Wilkinson, J.Q., Lanahan, M.B., Clark, D.G., Bleecker, A.B., Chang, C., Meyerowitz, E.M. and Klee, H.J. (1997) A dominant mutant receptor from Arabidopsis confers ethylene insensitivity in heterologous plants. Nature Biotechnol., 15, 444-447.
64. Wilkinson, J.Q., Lanahan, M.B., Yen, H.-C., Giovannoni, J.J. and Klee, H.J. (1995) An ethylene-inducible component of signal transduction encoded by Never-ripe. Science, 270, 1807-1809.
65. Xu, Y.L., Gage, D.A. and Zeevaart, J.A.D. (1997) Gibberellins and stem growth in *Arabidopsis thaliana* - effects of photoperiod on expression of the *GA4* and *GA5* loci. Plant Physiol., 114, 1471-1476.
66. Xu, Y.L., Li, L., Gage, D.A. and Zeevaart, J.A.D. (1999) Feedback regulation of *GA5* expression and metabolic engineering of gibberellin levels in Arabidopsis. Plant Cell, 11, 927-935.
67. Yamaguchi, S., Smith, M.W., Brown, R.G.S., Kamiya, Y. and Sun, T.P. (1998) Phytochrome regulation and differential expression of gibberellin 3β-hydroxylase genes in germinating Arabidopsis seeds. Plant Cell, 10, 2115-2126.

F. THE ROLES OF HORMONES IN DEFENSE AGAINST INSECTS AND DISEASE

F1. Jasmonates

Gregg A. Howe
MSU-DOE Plant Research Laboratory, Michigan State University, East Lansing, MI 48824, USA. E-mail: howeg@msu.edu

INTRODUCTION

Jasmonic acid (JA[1]) and its volatile methyl ester, MeJA, are fatty acid-derived cyclopentanones that occur ubiquitously in the plant kingdom. Since the discovery of jasmonates (JAs) in plants over 40 years ago, our understanding of the biosynthesis and physiological function of these compounds has been marked by several major developments. Experiments performed in the 1980s elucidated the JA biosynthetic pathway and demonstrated that exogenous JAs exert effects on a wide range of physiological processes. The discovery in the early 1990s that JAs act as potent signals for the expression of defensive proteinase inhibitors (PIs) aroused intense interest in the function of hormonally active JAs in plant defense. Research in the past decade has led to several key developments, including identification of genes encoding most of the JA biosynthetic enzymes and discovery of novel biologically active JAs. Identification of a large collection of JA biosynthesis and response mutants has provided important tools to assess the role of JAs in plant developmental and defense-related processes. The widespread occurrence of JAs in plants and some lower eukaryotes, together with their capacity to regulate physiological

---

[1] Abbreviations: 13-HPOT, 13-hydroperoxy linolenic acid; EST, Expressed Sequence Tag; PA, Phosphatidic acid; SA, Salicylic acid; ISR, Induced systemic resistance; FAA, Fatty acid amide; JAs, Jasmonates; MeJA, Methyl jasmonate; MAP, Mitogen-activated protein; PPO, Polyphenol oxidase; COX, cyclooxygenase; DOX, dioxygenase; OPDA, 12-oxophytodienoic acid; MFP, Multifunctional protein; COR, Coronatine; PI, Proteinase inhibitor; SCF, Skp1 Cullin F-box.

processes in animals (e.g., insects), reinforces the notion that JAs are of general biological interest.

JASMONATE BIOSYNTHESIS

Oxylipin Metabolism in Plants and Animals

Jasmonates belong to the family of oxygenated fatty acid derivatives, collectively called oxylipins, which are produced via the oxidative metabolism of polyunsaturated fatty acids. In animals, members of the eicosanoid (C_{20}) group of lipid mediators are synthesized from arachidonic acid and function as regulators of cell differentiation, immune responses, and homeostasis. In plants, oxygenated compounds derived mainly from C_{18} α-linolenic acid (18:3) control a similarly broad spectrum of developmental and defense-related processes (4, 13, 19, 30, 47, 48). The biochemical logic underlying the synthesis of oxylipins in plants and animals is remarkably similar (Fig. 1). Species in both kingdoms utilize cytochromes P450, lipoxygenase (LOX), and cyclooxygenase or cyclooxygenase-like (e.g., plant α-dioxygenase) activities to oxidize polyunsaturated fatty acid substrates (3). The resulting oxygenated fatty acids are further metabolized by various enzymatic and non-enzymatic systems to an array of intermediates and end products. These compounds are typically synthesized *de novo* in specific cell types upon activation of lipases that release fatty acids from membrane lipids.

Plant and animal pathways for oxylipin biosynthesis may have evolved from an ancestral lipid-based signaling system (2). Alternatively, oxylipin-based signaling in plants and animals may have evolved independently, with natural selection acting on intrinsic features that allow these compounds to transduce cellular information: (i) large supply of precursors in the form of membrane lipids, (ii) rapid synthesis in response to extracellular cues, (iii) structural diversity that permits functional specificity, (iv) physical properties that facilitate intra- and intercellular transport, and (v) rapid turnover or

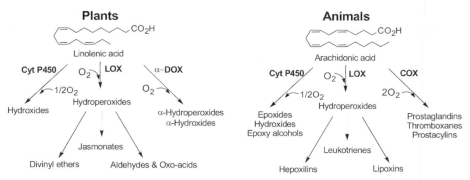

Figure 1. Oxylipin metabolism in plants and animals. LOX; lipoxygenase; COX, cyclo-oxygenase; DOX, dioxygenase.

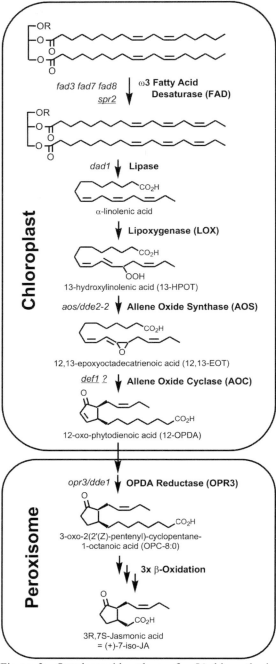

Figure 2. Octadecanoid pathway for JA biosynthesis. Mutations that block JA biosynthesis in Arabidopsis and tomato (underlined) are indicated to the left of the arrows.

inactivation. A greater understanding of JA biosynthesis and action promises to provide insight into the evolutionary origins of fatty acid-based signaling pathways in diverse biological systems.

The Octadecanoid Pathway

Elucidation of the JA biosynthetic pathway was largely the result of research conducted by Vick and Zimmermann in the 1980s (45). As shown in Figure 2, the pathway is initiated in the chloroplast by lipoxygenase (13-LOX)-catalyzed addition of molecular oxygen to the C_{13} position of 18:3. The resulting 13-hydroperoxy derivative (13-HPOT) is converted to an unstable allene oxide by the action of allene oxide synthase (AOS). AOS is a member of the CYP74 family of cytochromes P450 that appears to have evolved specifically for the metabolism of hydroperoxy fatty acids (13, 19). Allene oxide cyclase (AOC) transforms the AOS reaction product to the first cyclic compound in the pathway, namely the cyclopentenone 12-oxo-phytodienoic acid (OPDA). The two chiral centers of OPDA allow for four possible stereoisomers. An important feature of

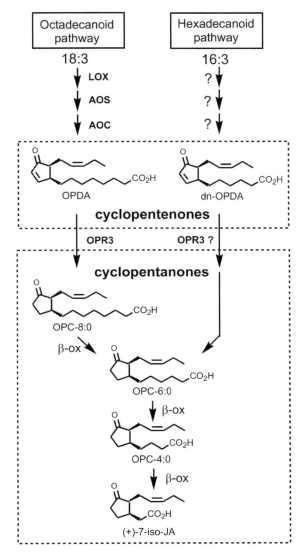

Figure 3. The octadecanoid and hexadecanoid pathways for jasmonate biosynthesis.

AOC is its ability to produce only the 9S,13S isomer of OPDA. This stereochemistry is maintained throughout the remaining biosynthetic steps. The terminal reactions of JA synthesis occur in peroxisomes, which are the site of fatty acid β-oxidation in plants. First, the cyclopentenone ring of OPDA is reduced by OPDA reductase (OPR3) to yield OPC-8:0. Three cycles of β-oxidation remove six carbons from the carboxyl side chain, completing the biosyntheis of JA.

Because JA is derived from 18:3 (octadecatrienoic acid), the series of enzymatic reactions leading to JA biosynthesis is often referred to as the octadecanoid pathway. Recent analysis of the oxylipin content in Arabidopsis leaves revealed the existence of a novel C_{16} cyclopentenone called dn-OPDA (48). The structural similarity of this compound to OPDA led to the proposal that dn-OPDA is produced via a "hexadecanoid pathway" starting from 16:3 (Fig. 3). Assuming that dn-OPDA is a substrate for OPR3, this C_{16} compound could be further metabolized to JA by β-oxidation enzymes in the peroxisome. For the purposes of this discussion, the term "jasmonate" is used to include biologically active cyclopentenones (e.g., OPDA) and cyclopentanones (e.g., JA/MeJA) that are derived from oxygenated polyunsaturated fatty acids. Other authors (47) have used the terms octadecanoids and jasmonates to refer more specifically to C_{18} and C_{12} products, respectively, of the pathway.

Tissue and Cell Type-Specific Location of JA Biosynthesis

Endogenous JA levels can vary over several orders of magnitude depending on the environmental conditions and the specific tissues and cell-types under consideration. In unstressed plants, JA levels are typically higher in young sink tissues than in older vegetative tissues (4, 47). Developing reproductive structures such as fruit and flowers accumulate very high levels of JAs, which is consistent with evidence that JAs play an important role in reproductive development (see below). Additional work is needed to determine how intracellular levels of JAs are regulated in different cell types and tissues.

It is well established that JA levels in healthy leaf tissues increase rapidly and transiently in response to wounding, herbivory, elicitor treatments, and other biotic and abiotic stimuli (4, 47). Genes encoding many JA biosynthetic enzymes are preferentially expressed in vascular bundle cells of leaf tissue. This observation is in agreement with the finding that JA and OPDA accumulation in tomato leaves is enriched in veins compared to the leaf lamina (39). Preferential accumulation of JAs in vascular tissues may account for the cell type-specific expression of JA-regulated vegetative storage proteins in soybean (4). In tomato leaves, JA-regulated PIs are expressed predominantly in leaf mesophyll cells. In this case, restriction of octadecanoid pathway enzymes to vascular bundles suggests that JAs produced in one cell type exert effects on neighboring cells (35, 39).

Identification of Genes Encoding JA Biosynthetic Enzymes

Work performed in many laboratories during the past 10 years has identified genes encoding nearly all enzymes involved in JA biosynthesis. Much of this work has involved the use of traditional biochemical approaches for enzyme purification, followed by use of appropriate molecular techniques to identify the corresponding cDNAs. Heterologous expression systems, usually *E. coli*, have been used to verify the enzymatic activity of cDNA products and to determine the substrate specificity and kinetic parameters of these enzymes. This strategy was successfully employed for the first identification of *AOS*, *AOC*, and *OPR* cDNAs from flax, tomato, and Arabidopsis, respectively (36, 47). The discovery of these "pioneer" genes of the octadecanoid pathway has provided the necessary DNA sequence information to identify homologous genes in a wide variety of plant species.

The recent completion of plant genome sequencing projects has also provided insight into the number of genes that encode various pathway enzymes. For those pathway enzymes encoded by a multigene family, a future challenge will be to determine the physiological function of each isoform in JA homeostasis. As illustrated by an increasing collection of JA biosynthetic mutants (Fig. 2), genetic approaches are expected to play an important role in addressing this question. The following discussion highlights the ways in which combined genetic and biochemical approaches have advanced our understanding of the molecular basis of JA biosynthesis. Additional details about specific enzymes and their role in the regulation of JA biosynthesis can be found in recent reviews (5, 13, 36, 43).

Production of linolenic acid
Important insight into the role of JA in plant growth and development has come from genetic analysis of trienoic fatty acid biosynthesis in Arabidopsis (46). The Arabidopsis genome encodes three ω3 fatty acid desaturases (FAD3, FAD7, and FAD8) that convert dienoic fatty acids (16:2 and 18:2) to trienoic fatty acids (16:3 and 18:3). The *FAD3* gene product is an endoplasmic reticulum-localized enzyme that desaturates 18:2, whereas the *FAD7* and *FAD8* gene products desaturate both 16:2 and 18:2 in the chloroplast. Because of functional redundancy between these enzymes and the trafficking of lipids between the chloroplast and extrachloroplast membranes, mutations in any one of these genes have only modest effects on trienoic fatty acid accumulation and JA levels. However, a *fad3fad7fad8* "triple" mutant produces no trienoic fatty acids and, as a consequence, is completely deficient in JA (46). With the exception of being deficient in JA-mediated defense, this mutant exhibited normal vegetative growth. Interestingly, however, attempts to grow the mutant to maturity revealed that it was male sterile. This observation, together with the ability of exogenous 18:3 and JA to restore fertility to the mutant, provided the first conclusive evidence that JAs play an essential role in male reproductive development. The strict dependence of male fertility on JA has been exploited as a convenient assay to identify additional JA biosynthetic mutants of Arabidopsis (see below).

Additional insight into the role of ω3-FADs in JA biosynthesis has come from the characterization of the *spr2* mutant of tomato (25). This mutant was originally isolated in a genetic screen for plants that are defective in anti-herbivore defenses in response to wounding and the peptide wound signal, systemin. The *Spr2* gene encodes a chloroplast fatty acid desaturase (LeFAD7) that is homologous to the Arabidopsis *FAD7/8* genes. *spr2* mutant plants are severely deficient in 18:3/16:3 and JA accumulation and, as expected, are compromised in defense against insect attack (see below). These results indicate that chloroplast pools of 18:3 are required for wound- and systemin-induced JA synthesis and subsequent activation of defense responses in tomato leaves. The *spr2* mutant exhibits normal growth, development, and reproduction, indicating that LeFAD7 is not required for fertility in tomato. The availability of ω3-FAD mutants in tomato and Arabidopsis should assist future work to determine the function of JAs in diverse plant species.

Release of linolenic acid from membrane lipids
Jasmonate accumulation in plant tissues appears to be regulated at several steps along the octadecanoid pathway (39, 47). Nevertheless, recent evidence indicates that phospholipases (PLs) that release fatty acid precursors from

membrane lipids play a critical role in this regulation. A significant advance in this area came from the characterization of the male sterile *delayed in anther dehiscence1* (*dad1*) mutant of Arabidopsis that is deficient in JA accumulation in floral tissues (20). *DAD1* encodes a PLA_1 that liberates 18:3 from the *sn1* position of membrane lipids (Fig. 2). Localization of DAD1 to the chloroplast is consistent with the notion that it generates fatty acid substrate for the plastid-localized enzymes of the octadecanoid pathway. Analysis of substrate specificity showed that recombinant DAD1 was highly active against phospholipids such as phosphatidylcholine. The enzyme was somewhat less active against galactolipids that are abundant in the chloroplast envelope membrane where the initial steps of JA biosynthesis are thought to occur (47). Nevertheless, the ability of DAD1 to utilize galactolipids is interesting in the context of the recent discovery that a large pool of OPDA is esterified to chloroplast galactolipids (19, 48). This finding raises the possibility that DAD1 may be involved in releasing OPDA from the plastid for subsequent metabolism in the peroxisome. Substantiation of this hypothesis will require experiments aimed at determining the precursor-product relationship between OPDA-containing galactolipids and JA.

The capacity of *dad1* mutant plants to accumulate normal levels of JA in wounded leaves (20) indicates that other lipases are involved in JA biosynthesis in response to insect and pathogen attack. Other members of the *DAD1* gene family, which includes several putative chloroplast isozymes, are candidates for such a lipase. There is also evidence for the involvement of PLA_2 and PLD in wound-induced JA biosynthesis (5, 19, 43). For example, depletion of PLDα1 in Arabidopsis resulted in reduced JA accumulation and reduced expression of JA-regulated genes in response to wounding. Given that PLD is an extrachloroplastic lipase that generates phosphatidic acid (PA) rather than free fatty acids, it appears that PLD plays an indirect role in generating precursor 18:3 for JA synthesis. For example, PA may serve as a substrate for PLA, or as an intracellular regulator of the lipase directly involved in wound-induced JA biosynthesis.

Chloroplast reactions
Genes encoding LOX, AOS, and AOC, which catalyze the three "core" chloroplastic reactions in the octadecanoid pathway, have now been identified in several plant species. Localization studies using biochemical fractionation and immunocytochemical and *in vitro* chloroplast import assays have demonstrated a chloroplast location for these enzymes, and further indicate that LOX and AOS may be associated with the chloroplast envelope (47). A precise understanding of the contribution of the chloroplast reactions to JA homeostasis is complicated by the presence of multiple isozymes for these enzymes. For example, the completed sequence of the Arabidopsis genome indicates the existence of four genes for both 13-LOX and AOC (5, 13, 17, 47). Antisense expression of one 13-LOX isoform (LOX2) resulted in decreased JA accumulation in response to wounding, but did not affect male fertility (4, 13). This phenotype may reflect incomplete loss of function of

LOX2 in the antisense lines, or functional compensation by another 13-LOX. It is possible that individual isoforms of 13-LOX and AOC modulate JA levels in specific tissue or cell types. Further insight into the function of individual isoforms in JA homeostasis will likely come from the characterization of the corresponding T-DNA knock-out lines of Arabidopsis. In contrast to LOX and AOC, Arabidopsis has a single gene encoding AOS. Consistent with this, T-DNA-mediated disruption of *AOS* resulted in decreased JA accumulation and male sterility (33).

Forward genetic analysis has identified three mutants of tomato that are defective in JA biosynthesis. These include the *spr2* ω3-FAD mutant described above and two mutant lines (JL1 and JL5) that were identified in a genetic screen for plants that fail to express PIs in response to wounding (26). Characterization of the latter (renamed *def1*) showed that this mutant is deficient in wound- and systemin-induced JA accumulation and more susceptible to herbivore attack than wild-type plants. As expected for a block in JA synthesis, treatment of *def1* plants with exogenous JAs induced the expression of defense-related genes and restored resistance to herbivory (28). Available data suggest that *Def1* does not correspond to any known octadecanoid pathway gene in tomato. Rather, it appears that this locus is involved in the regulation of JA synthesis, perhaps by affecting the activity of AOC (28, 39). A defect in JA biosynthesis in the JL1 mutant line is indicated by a severe deficiency in wound-induced JA accumulation and the ability of the mutant to respond to exogenous JA (G.I. Lee and G.A. Howe, unpublished). The specific biosynthetic step affected in JL1 remains to be determined.

Peroxisomal reactions
Purification of enzymatically active OPR from cell suspension cultures of *Corydalis sempervirens* led to the isolation of the first OPR-encoding cDNA from Arabidopsis (36). The deduced amino acid sequence of this cDNA showed significant similarity to the Old Yellow Enzyme of yeast. This flavin-containing protein catalyzes the reduction of the olefinic bond of α,β-unsaturated carbonyls, such as the cyclopentenone ring of OPDA. Subsequent characterization of *OPR* genes showed that both Arabidopsis and tomato contain three highly related OPR genes (*OPR1*, *OPR2*, and *OPR3*). Biochemical analysis of recombinant OPR enzymes, however, revealed that only OPR3 is capable of reducing the $9S,13S$ stereoisomer of OPDA, which is the physiologically relevant precursor of JA. This finding is consistent with the demonstration that OPR3 is targeted to peroxisomes, whereas OPR1 and OPR2 are cytosolic isoforms (41). The conclusion that OPR3 is the only OPR isoform capable of initiating the transformation of $9S,13S$-OPDA to JA was confirmed in studies demonstrating that an *opr3* null mutant (also known as *dde1*) lacks JA and is male sterile (46).

An important question raised by the organization of the octadecanoid pathway into two distinct compartments is the mechanism by which OPDA is transported between chloroplasts and peroxisomes. By analogy to what is known about peroxisomal β-oxidation of straight-chain fatty acids, it is conceivable that interorganellar trafficking of OPDA involves formation of an OPDA-CoA intermediate (Fig. 4). Although very little is known about specific transport systems for fatty acyl-CoAs, recent evidence suggests that ABC transporters may play a role in the trafficking of these compounds to the peroxisome (15). Additional work is needed to address this question and to determine whether OPR3 accepts OPDA-CoA as a substrate. Metabolic labeling experiments (45) demonstrated that OPDA is reduced to OPC-8:0 prior to β-oxidation. The three core enzymes of the β-oxidation cycle are acyl-CoA oxidase, the multifunctional protein (MFP, containing 2-*trans*-enoyl-CoA hydratase and L-3-hydroxyacyl-CoA dehydrogenase activities), and 3-ketoacyl-CoA thiolase (Fig. 4). In Arabidopsis, these enzymes are encoded by small gene families, and mutations in some of these genes have been identified (15). In the context of JA biosynthesis, it is noteworthy that isoforms of some β-oxidation enzymes (e.g., acyl-CoA oxidase) exhibit specificity for fatty acids of a particular chain length. Whether or not specific β-oxidation isozymes mediate the conversion of OPC-8:0 to JA remains to be determined. Analysis of JA levels in various β-oxidation mutants will be useful to address this question.

JASMONATE METABOLISM

The balance between the *de novo* formation of JA and its further metabolism is likely to play a critical role in controlling the activity of the hormone. The simplest transformation is epimerization of newly synthesized (+)-7-iso-JA to the more stable (-)-JA isomer in which the side-chains at positions 3 and 7 are in the thermodynamically more stable *trans* configuration (Fig. 5). Commercially available JA typically consists of a mixture (approximately 9:1) of these two isomers. Although the extent to which epimerization occurs *in vivo* is unknown, it is generally assumed that (+)-7-iso-JA is the biologically relevant and more active form of JA. Support for this idea comes from the observation that the phytotoxin coronatine (see below) and synthetic indanoyl conjugates, which exhibit greater activity than JA in some bioassays, possess cyclopentanone ring constituents that are held rigidly within a six-membered ring structure (Fig. 5; 24).

Newly synthesized JA is subject to a variety of enzymatic transformations that profoundly affect the range of signaling activities of the molecule. As illustrated in Figure 6, the major metabolic routes include: (i) methylation of C_1, yielding MeJA; (ii) hydroxylation at C_{11} or C_{12}, yielding tuberonic acid-related derivatives; (iii) conjugation of the carboxy terminus to amino acids or other adducts; (iv) reduction of C_6, yielding cucurbic acid-related derivatives; and (v) degradation of C_1 to form (Z)-jasmone. Products of the reduction and hydroxylation pathways may be glycosylated to give the corresponding O-glucosides. Most of these metabolites have been shown to occur naturally (47).

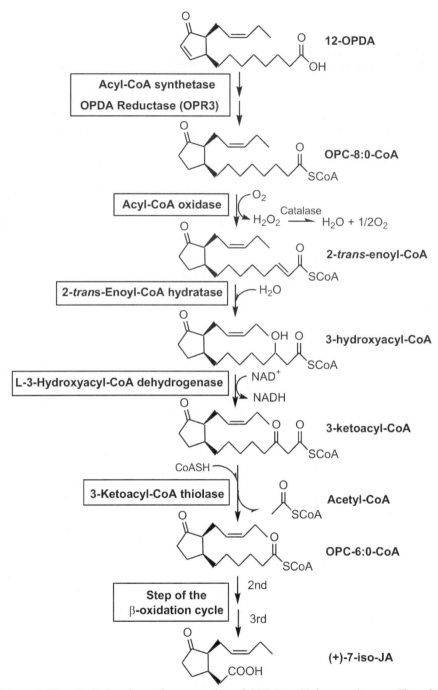

Figure 4. Hypothetical pathway for conversion of OPDA to JA in peroxisomes. The relative order in which Acyl-CoA synthetase and OPR3 act in the pathway is unknown, and may be reversed from that shown. See text for details.

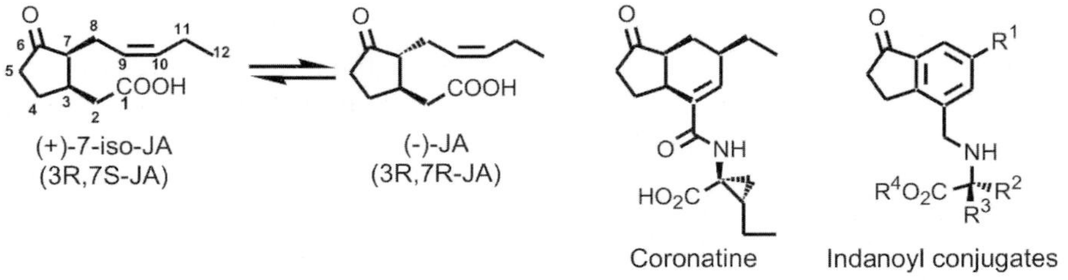

Jasmonates

Figure 5. Chemical structures of JA isomers, coronatine, and indanoyl conjugates.

A significant recent advance in our understanding of JA metabolism was the identification of genes encoding various JA-metabolizing enzymes. Two examples of this are genes encoding a JA-specific methyl transferase (JMT) and a sulfotransferase (ST2a) catalyzing the sulfonation of 12-OH-JA (Fig. 6). Transgenic studies in Arabidopsis have provided insight into the physiological function of these metabolic tranformations. For example, overproduction of JMT resulted in increased levels of MeJA, constitutive expression of JA-responsive genes, and enhanced resistance to a fungal pathogen (37). These results suggest that methylation of JA is an important regulatory step in jasmonate-signaled defenses. Overexpression of *St2a* in Arabidopsis resulted in decreased levels of 12-OH-JA and delayed onset in flowering time (47). This provocative finding suggests that 12-OH-JA may function as a component of the inductive signal for flowering.

Characterization of the Arabidopsis *jar1* mutant that is defective in JA-signaled root growth inhibition led to the discovery that the *JAR1* gene encodes an adenylate-forming enzyme that catalyzes ATP-dependent

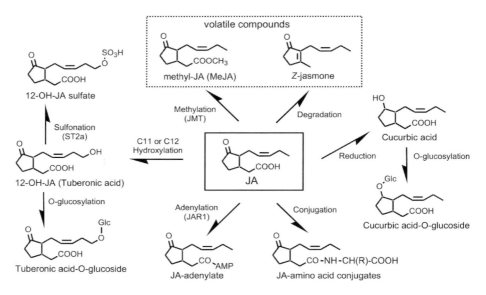

Figure 6. Pathways for JA metabolism.

adenylation of JA (Fig. 6; 38). JAR1 belongs to a superfamily of enzymes that activate a wide array of carboxylic acids. Although the precise biochemical function of JAR1 in JA metabolism remains to be determined, the activity of the enzyme suggests a role in conjugation to amino acids or other adducts, transport of JA to specific cellular locations, or metabolism of JA to an as-yet-unidentified product. Significantly, other members of the JAR1 family of enzymes were shown to catalyze adenylation of salicylic acid (SA) and indole-3-acetic acid (IAA) (38). This observation is important because it suggests that adenylation is a general feature of plant hormone metabolism. In this context, the insensitivity of *jar1* plants to JA indicates that adenylation promotes some JA-signaled responses (e.g., inhibition of root growth). Additional work should clarify the role of JAR1 in the JA signal transduction pathway.

JASMONATE ACTION

Regulation of Gene Expression

The discovery that MeJA and JA activate the expression of wound-inducible PIs in tomato aroused intense interest in the role of these compounds as regulatory signals for plant defense (9, 10). It is now widely recognized that JAs are among the most potent and important signals for the regulation of defense-related genes in species throughout the plant kingdom. Recent developments in technology to assess genome-wide patterns of gene expression have prompted research aimed at defining the size and function of the jasmonate transcriptome (14, 22, 30, 50). These studies have demonstrated that JA production in response to some biotic stress conditions (e.g., wounding) results in large-scale changes in the transcription of genes, many of which function in the reconfiguration of metabolism and the elaboration of diverse defense traits. Interestingly, JAs also regulate gene expression in herbivorous insects (29). Jasmonate perception by insects may allow these herbivores to "eavesdrop" on host plant defensive systems, and to rapidly mount counter-defenses against JA-induced phytotoxins. These findings raise the possibility that components of the JA signaling pathway may be conserved between the plant and animal kingdoms.

Despite extensive knowledge about the effects of JAs on gene expression, our understanding of the signaling events that couple the production of JA and other bioactive oxylipins to the activation of downstream target genes is still in its infancy. Systematic analysis of mutants obtained from forward and reverse genetic approaches is beginning to yield valuable insight into this question. For example, studies of the interaction of the *opr3* mutant of Arabidopsis with insect and fungal pests led to the discovery that OPDA promotes defense responses in the absence of its conversion to JA (40). This finding is consistent with previous studies showing that OPDA, rather than JA, is the physiologically relevant signal for the tendril coiling response of *Bryonia* (47, 48). The general theme that emerges from these studies is that individual cyclopentenone (e.g., OPDA) and cyclopentanone (e.g., JA) signals work together to optimize expression of specific downstream responses.

One of the more remarkable effects of JAs on plant cells is increased production of secondary metabolites (e.g., phytoalexins) that play a known or suspected role in plant defense. Zenk and colleagues demonstrated that JAs are an integral part of the signaling cascade that couples elicitor action to the activation of genes for the biosynthesis of phytoalexins in plant cell cultures (16). This pioneering study of plant metabolism has spurred research aimed at using JAs to enhance the production of various phytochemicals. A survey of the literature shows that the accumulation of compounds belonging to nearly all of the major classes of plant secondary metabolites is enhanced in response to applied JAs (Table 1). The phenomenon of JA-induced secondary metabolism is interesting not only for what it reveals about how metabolism is reconfigured in response to stress signals, but may also have practical importance for the production of economically useful compounds (14). For example, MeJA was shown to significantly enhance production of the anti-cancer drug, taxol, in *Taxus* cell suspension cultures (49).

The ability of JAs to promote the accumulation of diverse defensive phytochemicals implies the existence of JA-responsive regulatory factors that coordinate the expression of numerous biosynthetic genes within a particular pathway. Support for this has come from two lines of investigation. The first is the use of gene expression profiling techniques to simultaneously monitor the expression of hundreds or thousands of genes in response to conditions that induce the JA pathway (14, 22, 50). The second approach has been the investigation of the underlying transcription factors. Analysis of JA-induced secondary metabolism in *Catharanthus roseus* (periwinkle) led to the identification of ORCA3, a jasmonate-responsive member of the AP2/ERF-domain family of plant transcription factors (32). ORCA3 specifically binds to the promoter region of several JA-responsive genes, including those involved in the synthesis of terpenoid indole alkaloids, to regulate their expression. Increasing evidence indicates that regulation of defense-related genes by JAs requires the action of additional signals. For example, several defense responses in Arabidopsis and tomato have been shown to require both JA and ethylene (7, 23). A recent breakthrough in our understanding of the molecular basis of this phenomenon was the identification of ERF1, an AP2-domain transcription factor that regulates ethylene/JA-dependent defense responses in Arabidopsis (31). Transgenic studies support the idea that ERF1 plays a pivotal role in integrating defense responses to both signals. The observation that *ERF1* expression is rapidly up-regulated by JA and ethylene indicates the involvement of transcription factors that act upstream of ERF1.

Table 1. Examples of secondary metabolites induced by JA/MeJA

Class	Metabolite	Plant species
Flavonoids	Anthocyanins	*Glycine max*
	Feruloylsaponarin	*Hordeum vulgare*
	Isobavachalcone	*Crotolaria cobalticola*
Phenylpropanoids	Phlorotannin	*Fucus vesiculosus*
	Coumarins	*Nicotiana tabacum*
	Lignin	*Bryonia dioica*
Alkaloids	Benzophenanthridines	*Eschscholtzia californica*
	Catharanthine	*Catharanthus roseus*
	Nicotine	*Nicotiana* spp.
Monoterpenoids	Camptothecin	*Camptotheca acuminata*
	β-Ocimine	*Arabidopsis thaliana*
	Valepotriates	*Valerianella locusta*
Sesquiterpenoids	Homoterpene I	*Phaseolus lunatus*
	Lettucenin A	*Lactuca sativa*
	Tessaric acid	*Tessaria absinthioides*
Diterpenoids	Levopimaric acid	*Picea abies*
	Momilactone A	Rice
	Taxol	*Taxus* spp.
Triterpenoids	Ginsenosides	*Panax ginseng*
	Oleanolic acid	*Scutellaria baicalensis*
	Soyasaponin	*Glycyrrhiza glabra*
Polyamines	Putrescine	Rice
	Methylputrescine	*Hyoscyamus muticus*
	Caffeoylputrescine	*Nicotiana attenuata*
Acetate-malonate derived compounds	Hypericin	*Hypericum perforatum*
	Rubiadin	*Rubia tinctorum*
	Xanthones	*Centaurium* spp.
Acetylenics	Alkamides	*Echinacea pallida*
	Diacetylenes	*Tanacetum parthenium*
Amino acid derivatives	Indole glucosinolates	Oilseed rape and mustard
	Tocopherol	*Arabidopsis thaliana*
Naphthoquinones	Alkannin	*Alkanna tinctoria*
	Shikonin	*Lithospermum erythrorhizon*

Jasmonates as Regulators of Diverse Defensive Traits

JAs regulate diverse aspects of plant defense (Fig. 7) (7, 22, 43, 46, 47). So-called direct defenses are mediated by JA-regulated phytochemicals that interact directly with the plant invader to negatively

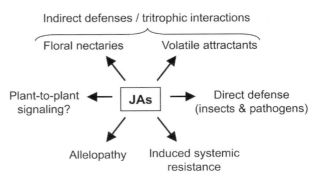

Figure 7. Proposed roles for JA in plant defense.

affect its feeding, growth or reproduction. Examples of JA-regulated anti-herbivore phytochemicals include PIs and polyphenol oxidases (PPOs) that reduce the digestibility of damaged leaf tissue and the toxic alkaloid nicotine produced by tobacco. Several lines of evidence support a central role for JA in the regulation of this type of defensive trait. First, treatment of plants with exogenous JAs results in major re-programming of gene expression, including defense-related genes that are activated by mechanical wounding and herbivore attack. Second, endogenous JA levels increase rapidly in response to wounding and other types of biotic stress. Third, mutants that are defective in either the biosynthesis or perception of JAs are compromised in resistance to herbivores (Fig. 8). Conversely, constitutive activation of JA signaling results in enhanced resistance to herbivores (28). In addition to anti-herbivore defense, genetic studies in Arabidopsis have shown that JA signaling promotes direct defense responses to fungal pathogens such as *Pythium* and *Alternaria* (46). This defensive trait appears to be mediated, at least in part, by low-molecular-weight antimicrobial peptides such as thionins and defensins. In conjunction with ethylene, the JA signaling pathway also participates in rhizobacteria-mediated induced systemic resistance (ISR) against different types of bacterial and fungal pathogens (47). The identity of specific protective compounds involved in ISR has not yet been determined.

JAs also play an important role in orchestrating "indirect" defenses. This fascinating type of self-protection strategy involves the interaction of organisms at three trophic levels: plant host, herbivore, and natural enemy (parasitoid or predator) of the herbivore (7, 22, 47). One of the best examples of this is the production of plant volatiles (e.g., terpenoids) in response to fatty acid amide (FAA) elicitors found in the oral secretions of foraging lepidopteran herbivores. These emitted volatile chemicals enable parasitic wasps to locate host caterpillars. Increasing evidence indicates that production and emission of volatiles in response to FAA elicitors involves the host plant's JA signaling pathway (7, 22, 42). Another remarkable example of JA-mediated indirect defense is the production of plant extrafloral nectar that attracts ants that fend off insect herbivores (18).

Extrafloral nectar production is induced by herbivore attack and can be effectively mimicked by JA application.

Finally, there is some evidence to suggest that JAs mediate plant-to-plant signaling (11). Initial insight into this phenomenon came from experiments showing that MeJA released from sagebrush (*Artemesia tridentate*) induced the expression of PIs in neighboring tomato plants (9). Results of experiments conducted under more natural conditions using sagebrush and wild tobacco (*Nicotiana attenuata*) support the hypothesis that MeJA is a natural wound signal for interplant communication of defense responses (21). A second example of JA-mediated plant-to-plant communication is the allelopathic interaction between *A. tridentate* and *N. attenuata*. Studies performed under both laboratory and field conditions indicate that MeJA released from *A. tridentate* inhibits the germination of nearby *N. attenuata* plants (34). An exciting trend in the study of this and other JA-mediated defense responses is the use of multidisciplinary approaches that exploit modern genetic and biochemical tools to study interactions between organisms in ecologically relevant settings (7, 22).

Antagonism Between Jasmonate and Salicylate Signaling Pathways

Salicylic acid (SA) plays a key role as a signaling molecule for defense responses against pathogens, including the induction of systemic acquired resistance. An important emerging theme in the study of plant-pest interactions is the antagonistic interaction between the JA and SA signaling pathways (7, 12, 23, 43, 50). This form of negative cross-talk appears

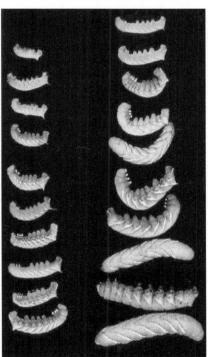

Figure 8. Loss of function of a fatty acid desaturase in the *spr2* JA biosynthetic mutant of tomato results in increased susceptibility to tobacco hornworm. (Top) Representative wild-type and *spr2* plants after the feeding trial. (Bottom) Hornworm larvae recovered at the end of the trial from representative wild-type (left) and *spr2* (right) plants.

Jasmonates

to provide plants with the plasticity to mount a defense response that is appropriate and specific for a particular invader, while minimizing the expression of inappropriate defenses. Intensive research efforts are now underway to understand the physiological and ecological relevance of this phenomenon, as well as the molecular mechanisms involved. Initial clues to the antagonistic effect of SA on JA signaling came from studies of PI expression in tomato plants (23). Numerous studies have now confirmed and extended this observation by showing an antagonistic effect of JA on the SA signaling pathway. At the molecular level, genetic evidence indicates that mitogen-activated protein (MAP) kinases play an important role in integrating the JA and SA signaling pathways. For example, a wound-inducible MAP kinase in tobacco was shown to be involved both in the activation of wound-induced JA biosynthesis and the suppression of SA-dependent signaling. Similarly, the *MPK4* gene product of Arabidopsis simultaneously inhibits SA biosynthesis and promotes JA signaling (23, 43).

Additional insight into the negative interaction between the JA and SA signaling pathways has come from the study of the phytotoxin coronatine (COR) that is secreted by plant pathogenic strains of *Pseudomonas syringae*. COR is a non-host specific, chlorosis-inducing polyketide that enhances bacterial growth and increases the severity of disease symptoms (23). The capacity of COR and JA to effect a variety of the same physiological processes including chlorosis, ethylene emission, inhibition of root elongation, and volatile production suggests that COR and JA may interact with the same or similar macromolecular targets. This hypothesis is supported by the remarkable structural similarity between COR and JA (Figs. 5 and 9). Several lines of evidence indicate that COR promotes disease by simultaneously activating the host's JA signaling pathway and suppressing the SA pathway (23, 50). First, COR produced during *P. syringae* infection of tomato plants caused massive

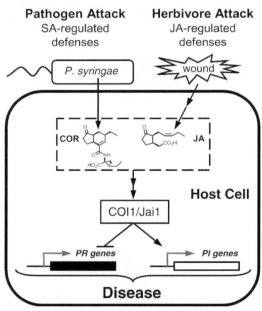

Figure 9. Model depicting the role of the phytotoxin coronatine in the virulence of *Pseudomonas syringae*. Coronatine (COR) produced by *P. syringae* activates the JA signaling cascade leading to expression of anti-herbivore defenses (e.g., PIs) and suppression of SA-regulated anti-pathogen defenses (e.g., PR proteins). The COI1 and JAI1 gene products are required for the action of both COR and JA in Arabidopsis and tomato, respectively.

expression of JA/wound-responsive genes that play a role in anti-herbivore defense. Second, the expression of pathogenesis-related (PR) genes having a putative function in SA-mediated defense was suppressed during this interaction. Third, Arabidopsis *coi1* and tomato *jai1* mutants that are insensitive to both JA and COR exhibit elevated resistance to COR-producing strains of *P. syringae*. Finally, *coi1* Arabidopsis plants inoculated with *P. syringae* exhibited increased levels of SA and hyperexpression of *PR* genes. Taken together, these observations indicate that the virulence effects of COR result from its ability to activate the host JA signaling pathway and simultaneously suppress SA-dependent defenses (Fig. 9). Further study of the molecular basis of COR action promises to provide insight into the mechanisms by which microbial toxins co-opt eukaryotic signaling pathways to promote disease. Discussion of other aspects of the interaction between the JA and SA pathways can be found in recent review articles (7, 12, 22, 23, 43).

Systemic Signaling

Many induced plant defense responses occur both locally at the site of attack and systemically in undamaged tissues. In 1972, Green and Ryan observed that localized wound damage of potato and tomato leaves results in systemic accumulation of defensive PIs (35). This observation implied that signals generated at the wound site travel through the plant and activate anti-herbivore defense responses in undamaged leaves. In the 30 years since this discovery, a wealth of genetic and biochemical evidence has shown that systemin, the first peptide signal discovered in plants, and JA are both involved in this systemic defense response (27, 35). Relatively little is known about how systemin and JA interact with one another to effect cell-to-cell communication over long distances. One attractive hypothesis put forth is that systemin, upon its production at the wound site, is translocated through the phloem to distal undamaged leaves (35). Binding of systemin to target cells in distal leaves would then activate the octadecanoid pathway for JA synthesis and subsequent PI expression.

The availability of tomato mutants that are defective in JA biosynthesis (e.g., *spr2*), JA perception (e.g., *jai1*), and systemin perception (e.g., *spr1*) has provided an opportunity to rigorously test this model (27, 44). Because these mutants are defective in wound-induced systemic PI expression, classical grafting techniques could be employed as a simple means to determine whether a particular mutant is defective in the production of the systemic (i.e., graft-transmissible) wound signal in wounded leaves or the recognition of that signal in undamaged responding leaves. As shown schematically in Figure 10A, analysis of systemic wound signaling in grafts between wild-type and *jai1* plants revealed that JA perception is required for recognition, but not production, of the transmissible wound signal. Conversely, grafts between wild-type and the *spr2* plants indicated that jasmonate synthesis is needed to produce the systemic signal in wounded

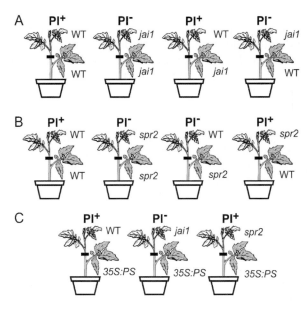

Figure 10. Schematic illustration of grafting experiments used to investigate the role of JA in wound-induced systemic proteinase inhibitor (PI) expression in tomato. Scions and rootstocks (shaded) of the indicated genotype (wild-type, WT; *35S:prosystemin*, 35S:PS) were joined at the graft junction (horizontal bar). For experiments shown in A and B, rootstock leaves were wounded and PI expression in undamaged scions leaves was assessed 8 hours later using RNA blot analysis (PI$^+$, high PI expression; PI$^-$, low PI expression). For the experiments depicted in panel C, no wounds were inflicted because 35S:PS transgenic rootstock constitutively produces a systemic signal.

leaves, but is not required in systemic undamaged leaves (Fig. 10B). The most straightforward interpretation of these results is that a signaling compound derived from the octadecanoid pathway acts as a transmissible wound signal. Such a scenario is analogous to the well-characterized function of prostaglandins and leukotrienes as intercellular signals for defense responses in mammalian cells.

What then is the role of systemin in wound-induced systemic expression of *PI* genes? Because the *spr2* mutant is responsive to JAs but unresponsive to systemin (25), grafting experiments involving this mutant provided important insight into this question. The ability of *spr2* scions to perceive the signal generated in wild-type rootstock leaves (Fig. 10B) strongly suggests that systemin is not the long-distance signal. Rather, systemin appears to function at or near the wound site (i.e., in the rootstock tissues) to amplify jasmonate synthesis to a level that is necessary for the systemic response. This hypothesis is supported by grafting experiments using a transgenic line of tomato that overexpresses prosystemin from the cauliflower mosaic virus 35S promoter and, as a consequence, exhibits constitutive PI expression in the absence of wounding (35). Initial grafting experiments performed with these transgenic plants (called *35S:PS*) showed that they constitutively produce a systemic signal that activates PI expression in wild-type scion leaves (Fig. 10C; 35). Grafts between *35S:PS* rootstock and various mutant scions showed that recognition of the *35S:PS*-derived signal in scion tissue requires JA signaling but not JA biosynthesis. These findings suggest that *35S:PS* constitutively activates the synthesis of JA, which is then mobilized to scion leaves. A proposed role for systemin in the amplification of JA levels in damaged leaves is consistent with the wound response phenotype of tomato mutants that are defective in systemin biosynthesis and systemin perception (35, 44). Although the precise mechanism of systemic wound signaling remains to be determined, one possibility is that jasmonate and systemin both act as mobile signals, and that positive feedback between them serves to propagate the long-distance signal (44).

The Jasmonate Signaling Pathway

Despite extensive knowledge of the effects of JAs on gene expression and defense-related processes, our understanding of the signaling events that couple the production of JAs to the activation of target genes is still in its infancy. Our current understanding of this problem has been significantly enhanced by the identification and characterization of mutants that are insensitive to JAs (1, 5, 43, 47, 48, 50). For example, analysis of the *coi1* mutant of Arabidopsis provided a causal link between JA signaling and selective proteolysis mediated by the ubiquitin-proteasome pathway. The *COI1* gene encodes an F-box protein that is a component of a E3-type ubiquitin ligase complex referred to as SCF^{COI1} (named for the three major proteins of the complex: $\underline{S}$kp1, $\underline{C}$ullin, and $\underline{F}$-box protein). The presumed function of this multiprotein complex is to attach ubiquitin to regulatory proteins that interact with the leucine-rich repeat domain of COI1. By analogy to other ubiquitin-dependent signaling pathways, a model for JA signaling has been proposed (5). This model postulates that, in the absence of JA, JA-responsive genes are repressed by a negative regulator. Increased levels of JA initiate a signaling cascade that results in modification (e.g., phosphorylation) of the hypothetical negative regulator, such that SCF^{COI1} recognizes and targets this protein for degradation. Such a mechanism would theoretically permit rapid de-repression of the JA transcriptome in response to any cue that increases endogenous JA levels.

Identification of proteins that interact with COI1 is beginning to provide insight into the mechanism by which the ubiquitin pathway controls JA responses (6). The existence in Arabidopsis of over 300 F-box proteins, which provide specificity to the SCF complex, suggests that the ubiquitin pathway plays a major role in many aspects of plant growth and development. Indeed, there is evidence to implicate the ubiquitin-proteasome pathway in numerous responses, including those triggered by auxin, ethylene, gibberellic acid, sucrose, light, and pathogen attack. Elucidation of the mechanisms by which regulated proteolysis is controlled by one or more of these signals to effect specific physiological responses will be a major focus of plant hormone research in the years to come.

A significant gap in our understanding of JA signaling is the identification of the primary cellular target (i.e., receptor) of these compounds. Because JA and OPDA regulate distinct (but overlapping) sets of target genes, it is conceivable that different JAs use different receptor systems to "fine-tune" downstream responses (24, 40, 48). The capacity of

COR to mimic many JA- and OPDA-mediated responses allows one to further speculate that the primary target of this phytotoxin is the jasmonate receptor. A potential approach for identifying the macromolecular targets of JA/OPDA/COR involves identification and purification of proteins that bind these molecules. A key requirement for the success of this approach is the availability of chemical probes that bind specifically and tightly to the target molecule. The synthesis of photoaffinity derivatives of indanoyl amino acid conjugates that mimic the activity of endogenous JAs (Fig. 5; 24) may prove to be useful for this purpose.

Analysis of JA-insensitive mutants provides another strategy to identify components of the jasmonate perception apparatus. Hormone-insensitive mutants have been instrumental for the identification of receptors for ethylene, brassinosteroid, and cytokinins. Characterization of existing JA-insensitive mutants, however, has not yet revealed gene products that are obviously involved in JA perception. For example, current models suggest that COI1 functions downstream of the initial JA perception event (5, 43). It is likely that functional redundancy of jasmonate receptors precludes facile identification of mutations that affect this signaling component. The problem of redundancy may be overcome through the use of genetic screens designed to identify weak JA-insensitive alleles. Alternatively, the recent identification of mutations that cause constitutive JA signaling may provide insight into this question (1, 48). A gene defined by the *cev1* mutation was shown to encode a cellulose synthase, indicating a link between cell wall biosynthesis and JA signaling (8). However, because the constitutive JA signaling phenotype of *cev1* plants results from overproduction of JA and ethylene, it appears that CEV1 is not a primary component of the JA signaling pathway. Mutants that exhibit constitutive signaling in the absence of JA may be particularly useful for identifying novel components of the JA response pathway.

The Role of Jasmonates in Plant Development

Exogenous JA/MeJA exerts both inductive and inhibitory effects on a variety of plant developmental processes (4, 47). However, because applied compounds do not target specific cell types and are often used at non-physiological concentrations, caution must be used in extrapolating these effects to the function of endogenous JAs. Mutants that are defective in JA biosynthesis or signaling provide an opportunity to rigorously assess the function of JAs in specific developmental processes, and to discover new roles for these compounds in development. The following section offers a few examples of developmental processes that appear to be regulated in part by endogenous JAs.

One of the most pronounced effects of exogenous JAs is general inhibition of growth. Genetic screens in Arabidopsis have exploited the JA-mediated inhibition of root growth as a phenotype for the isolation of mutants having reduced sensitivity to JAs. Conversely, screens for constitutive JA signaling mutants have led to identification of mutants exhibiting features of JA-treated wild-type plants, including short roots, stunted growth, and anthocyanin accumulation (1). The *cev1* mutant is one that constitutively accumulates JAs (8). Although the short root phenotype of *cev1* plants is consistent with a role for endogenous JAs in growth inhibition, genetic analysis showed that the *coi1* mutation does not suppress this phenotype. Thus, it can be concluded that root growth inhibition does not result simply from increased signaling through COI1.

Another prominent effect of JA/MeJA application in many plants is decreased expression of genes involved in photosynthesis and reduction in chlorophyll content. Jasmonate-treated leaves typically exhibit chlorosis and increased abscission. This phenomenon led to the suggestion that JA may play a role in promoting senescence (4, 47). Recent support for this hypothesis came from the demonstration that senescence of Arabidopsis leaves is correlated with increased expression of JA biosynthetic genes and increased JA levels (17). It was also shown that induction of senescence-like symptoms in JA-treated leaves requires COI1. Similarly, MeJA-induced chlorosis and abscission of tomato leaves is abrogated by the *jai1-1* mutation that disrupts the function of the tomato homolog of COI1 (L. Li and G.A. Howe, unpublished results). Despite these observations, it is noteworthy that JA biosynthesis/signaling mutants of Arabidopsis and tomato do not exhibit obvious delayed-senescence phenotypes; such a phenotype would be predicted if JA functions as a causal signal for senescence. It would thus appear that either JAs are not strictly required for the normal progression of senescence (17), or that senescence-like effects induced by exogenous JA do not accurately reflect the normal senescence program.

As mentioned above, genetic studies have provided convincing evidence that JA is essential for male reproductive development in Arabidopsis (46). Multiple aspects of male fertility appear to be controlled by JA: i) development of viable pollen, ii) timing of anther dehiscence and; iii) elongation of the anther filaments. Jasmonate application experiments further demonstrated that JA is both necessary and sufficient to promote these reproductive processes. Interestingly, OPDA cannot substitute for JA in male gametophyte development, pointing to the existence of cellular mechanisms that can distinguish between these two signals. Based on the temporal and spatial expression pattern of the PLA_1-encoding *DAD1* gene that is required for male fertility, it was proposed that JA-regulated water transport within floral tissues may promote synchronous pollen maturation, anther dehiscence, and flower opening (20). Studies aimed at identifying JA-regulated genes in Arabidopsis reproductive tissues may provide insight into the specific biochemical processes involved in these aspects of reproduction.

In contrast to Arabidopsis, JA biosynthetic mutants of tomato display normal fertility as determined by production of viable seed (25). This phenotype may reflect the fact that wild-type tomato flowers accumulate high levels of various JAs (47), and that tomato JA biosynthetic mutants retain significant levels of JA in floral tissues. It is also possible that male fertility in tomato is not strictly dependent on JA. This idea is consistent with the observation that sterility of the *jai1* mutant of tomato (which is analogous to the *coi1* mutant of Arabidopsis) results mainly from a defect in female, not male, reproductive development (26). Additional studies are needed to determine whether sterility of *jai1* plants is due to a defect in ovule development, embryogenesis, or another maternal process involved in seed maturation. A dysfunction in embryogenesis would be consistent with previous studies implicating JA as an endogenous regulator of embryo development in oilseeds (4). The apparent differences in the roles of JAs in reproductive development in tomato versus Arabidopsis underscore the importance of studying hormone-signaled processes in diverse plant species.

It is becoming increasing clear that the role of JA in plant development is highly dependent on interactions with other growth regulators; a similar theme has emerged for JA-signaled defense processes. The complexity of these interactions notwithstanding, one of the more interesting features of JAs is their capacity to function as dual regulators of development and defense (4, 26). For example, it is often observed that defense genes (e.g., *PI*s in tomato) whose expression is dependent on wound-induced JA synthesis in leaves are constitutively expressed in reproductive tissues that accumulate high JA levels (4, 47). It seems likely that expression of such genes in flowers provides an important preformed deterrent against predators. However, the sterile phenotype of jasmonate mutants indicates that JA-regulated "defense" proteins may also serve a role in plant reproductive development. The coming years are expected to bring exciting advances to our understanding of not only how JA biosynthesis is regulated in different tissues, but also how JAs interact with other signals to modulate essential aspects of the plant life cycle.

Acknowledgments

I would like to thank Dr. Hui Chen for critical reading of the manuscript and for the information provided in Table 1. Jasmonate research in my laboratory is supported by grants from the National Institutes of Health, the U.S. Department of Energy, the Michigan Life Sciences Corridor, and the Michigan Agriculture Experiment Station at MSU.

References

1. Berger S (2002) Jasmonate-related mutants of Arabidopsis as tools for studying stress signaling. Planta 214: 497-504
2. Bergey DR, Howe GA, Ryan CA (1996) Polypeptide signaling for plant defensive genes exhibits analogies to defense signaling in animals. Proc Natl Acad Sci USA 93: 12053-12058
3. Blée E (2002) Impact of phyto-oxylipins in plant defense. Trends Plant Sci 7: 315-322
4. Creelman RA, Mullet JE (1997) Biosynthesis and action of jasmonates in plants. Annu Rev Plant Physiol Plant Mol Biol 48: 355-381
5. Creelman RA, Rao MV (2001) The oxylipin pathway in arabidopsis. In CR Somerville, EM Meyerowitz, eds, The Arabidopsis Book. American Society of Plant Biologists, Rockville, MD. doi/10.1199/tab.0009,
6. Devoto A, Nieto-Rostro M, Xie D, Ellis C, Harmston R, Patrick E, Davis J, Sherratt L, Coleman M, Turner JG (2002) COI1 links jasmonate signalling and fertility to the SCF ubiquitin-ligase complex in *Arabidopsis*. Plant J 32: 457-466

7. Dicke M, van Poecke RMP (2002) Signalling in plant-insect interactions: signal transduction in direct and indirect defence. In D Scheel, C Wasternack, eds, Plant Signal Transduction. Oxford University Press, Oxford, UK
8. Ellis C, Karafyllidis I, Wasternack C, Turner JG (2002) The Arabidopsis mutant *cev1* links cell wall signaling to jasmonate and ethylene responses. Plant Cell 14: 1557-1566
9. Farmer EE, Ryan CA (1990) Interplant communication: Airborne methyl jasmonate induces synthesis of proteinase inhibitors in plant leaves. Proc Natl Acad Sci USA 87: 7713-7718
10. Farmer EE, Ryan CA (1992) Octadecanoid precursors of jasmonic acid activate the synthesis of wound-inducible proteinase inhibitors. Plant Cell 4: 129-134
11. Farmer EE (2001) Surface-to-air signals. Nature 411: 854-856
12. Felton GW, Korth KL (2000) Trade-offs between pathogen and herbivore resistance. Curr Opin Plant Biol 3: 309-314
13. Feussner I, Wasternack C (2002) The lipoxygenase pathway. Annu Rev Plant Biol 53: 275-297
14. Goossens A, Häkkinen ST, Laakso I, Seppänen-Laakso T, Biondi S, De Sutter V, Lammertyn F, Nuutila AM, Söderlund H, Zabeau M, Inzé, Oksman-Caldentey K-M (2003) A functional genomics approach toward the understanding of secondary metabolism in plant cells. Proc Natl Acad Sci USA 100: 8595-8600
15. Graham IA, Eastmond PJ (2002) Pathways of straight and branched chain fatty acid catabolism in higher plants. Prog Lipid Res 41: 156-181
16. Gundlach H, Müller MJ, Kutchan TM, Zenk MH (1992) Jasmonic acid is a signal transducer in elicitor-induced plant cell cultures. Proc Natl Acad Sci USA 89: 2389-2393
17. He Y, Fukushige H, Hildebrand DF, Gan S (2002) Evidence supporting a role of jasmonic acid in Arabidopsis leaf senescence. Plant Physiol 128: 876-884
18. Heil M, Koch T, Hilpert A, Fiala B, Boland W, Linsenmair KE (2001) Extrafloral nectar production of the ant-associated plant, *Macaranga tanarius*, is an induced, indirect, defensive response elicited by jasmonic acid. Proc Natl Acad Sci USA 98: 1083-1088
19. Howe G, Schilmiller AL (2002) Oxylipin metabolism in response to stress. Curr Op Plant Biol. 5: 230-236
20. Ishiguro S, Kawai-Oda A, Ueda K, Nishida I, Okada K (2001) The DEFECTIVE IN ANTHER DEHISCENCE1 gene encodes a novel phospholipase A1 catalyzing the initial step of jasmonic acid biosynthesis, which synchronizes pollen maturation, anther dehiscence, and flower opening in Arabidopsis. Plant Cell 13: 2191-2209
21. Karban R, Baldwin IT, Baxter KJ, Laue G, Felton GW (2000) Communication between plants: induced resistance in wild tobacco plants following clipping of neighboring sagebrush. Oecologia 125: 66-71
22. Kessler A, Baldwin IT (2002) Plant responses to insect herbivory: the emerging molecular analysis. Annu Rev Plant Biol 53: 299-328
23. Kunkel BN, Brooks DM (2002) Cross talk between signaling pathways in pathogen defense. Curr Opin Plant Biol 5: 325-331
24. Lauchli R, Boland W (2003) Indanoyl amino acid conjugates: tunable elicitors of plant secondary metabolism. Chem Record 3: 12-21
25. Li C, Liu G, Xu C, Lee GI, Bauer P, Ling HQ, Ganal MW, Howe GA (2003) The tomato *Superssor of prosystemin-mediated responses2* (*Spr2*) gene encodes a fatty acid desaturase required for the biosynthesis of jasmonic acid and the production of a systemic wound signal. Plant Cell 15: 1646-1661
26. Li L, Li C, Howe GA (2001) Genetic analysis of wound signaling in tomato. Evidence for a dual role of jasmonic acid in defense and female fertility. Plant Physiol 127: 1414-1417
27. Li L, Li C, Lee GI, Howe GA (2002) Distinct roles for jasmonic acid synthesis and action in the systemic wound response of tomato. Proc Natl Acad Sci USA 99: 6416-6421
28. Li C, Williams MM, Loh Y-T, Lee GI, Howe GA (2002) Resistance of cultivated tomato to cell content-feeding herbivores is regulated by the octadecanoid-signaling pathway. Plant Physiol 130: 494-503
29. Li X, Schuler MA, Berenbaum MR (2002) Jasmonate and salicylate induce expression of

herbivore cytochrome P450 genes. Nature 419: 712-715
30. Liechti R, Farmer E E (2002) The jasmonate pathway. Science 296: 1649-1650
31. Lorenzo O, Piqueras R, Sanchez-Serrano JJ, Solano R (2003) ETHYLENE RESPONSE FACTOR1 integrates signals from ethylene and jasmonate pathways in plant defense. Plant Cell 15: 165-78
32. Memelink J, Verpoorte R, Kijne JW (2001) ORCAnization of jasmonate – responsive gene expression in alkaloid metabolism. Trends Plant Sci 6: 212-219
33. Park JH, Halitschke R, Kim HB, Baldwin IT, Feldmann KA, Feyereisen R (2002) A knock-out mutation in allene oxide synthase results in male sterility and defective wound signal transduction in Arabidopsis due to a block in jasmonic acid biosynthesis. Plant J 31: 1-12
34. Preston CA, Betts H, Baldwin IT (2002) Methyl jasmonate as an allelopathic agent: sagebrush inhibits germination of a neighboring tobacco, *Nicotiana attenuata*. J Chem Ecol 28: 2343-2369
35. Ryan CA (2000) The systemin signaling pathway: differential activation of plant defensive genes. Biochim Biophys Acta 1477: 112-121
36. Schaller F (2001) Enzymes of the biosynthesis of octadecanoid-derived signaling molecules. J Exp Bot 52: 11-23
37. Seo HS, Song JT, Cheong JJ, Lee YH, Lee YW, Hwang I, Lee JS, Choi YD (2001) Jasmonic acid carboxyl methyltransferase: a key enzyme for jasmonate-regulated plant responses. Proc Natl Acad Sci USA 98: 4788-4793
38. Staswick PE, Tiryaki I, Rowe ML (2002) Jasmonate response locus JAR1 and several related Arabidopsis genes encode enzymes of the firefly luciferase superfamily that show activity on jasmonic, salicylic, and indole-3-acetic acids in an assay for adenylation. Plant Cell 14: 1405-1415
39. Stenzel I, Hause B, Maucher H, Pitzschke A, Miersch O, Ziegler J, Ryan CA, Wasternack C (2003) Allene oxide cyclase dependence of the wound response and vascular bundle-specific generation of jasmonates in tomato - amplification in wound signalling. Plant J 33: 577-589
40. Stintzi A, Weber H, Reymond P, Browse J, Farmer EE (2001) Plant defense in the absence of jasmonic acid: The role of cyclopentenones. Proc Natl Acad Sci USA 98: 12837-12842
41. Strassner J, Schaller F, Frick U, Howe GA, Weiler EW, Amrhein N, Macheroux P, Schaller A. (2002) Characterization and cDNA-microarray expression analysis of 12-oxophytodienoate reductases reveals differential roles for octadecanoid biosynthesis in the local versus the systemic wound response. Plant J 32: 585-601
42. Thaler JS (1999) Jasmonate-inducible plant defences cause increased parasitism of herbivores. Nature 399: 686-688
43. Turner JG, Ellis C, Devoto A (2002) The jasmonate signal pathway. Plant Cell S153-S164.
44. Stratmann J (2003) Long distance run in the wound response – jasmonic acid pulls ahead. Trends Plant Sci (in press)
45. Vick B, Zimmerman DC (1984) Biosynthesis of jasmonic acid by several plant species. Plant Physiol 75: 458-461
46. Wallis JG, Browse J (2002) Mutants of *Arabidopsis* reveal many roles for membrane lipids. Prog Lipid Res 41: 254-278
47. Wasternack C, Hause B (2002) Jasmonates and octadecanoids: signals in plant stress responses and development. Prog Nucleic Acid Res Mol Biol 72: 165-221
48. Weber H (2002) Fatty acid-derived signals in plants. Trends Plant Sci 7: 217-224
49. Yukimune Y, Tabata H, Higashi Y, Hara Y (1996) Methyl jasmonate-induced overproduction of paclitaxel and baccatin III in *Taxus* cell suspension cultures. Nature Biotechnol 14: 1129-1132
50. Zhao Y, Thilmony R, Bender C, He SY, Howe GA (2003) The Hrp type III secretion system and coronatine of *Pseudomonas syringae* pv. *tomato* coordinately modify host defense by targeting the jasmonate signaling pathway in tomato. Plant J 36: 485-499

F2. Salicylic Acid

Terrence P. Delaney
Department of Botany and Agricultural Biochemistry, University of Vermont, Burlington, VT 05405, USA. E-mail: terrence.delaney@uvm.edu

INTRODUCTION

Plants are the source of many substances used in treating human disease and discomfort. One of the earliest known plant-derived therapeutic compounds originated from the bark of willow trees (*Salix* spp.) (Fig. 1), which in traditional medicine was chewed to provide relief from pain and inflammation, a practice that can be traced to over two thousand years ago. This ancient remedy was described in writings by the Greek physician Hippocrates (Fifth Century B.C.) and the physician and botanist Pedanius Dioscorides (First Century A.D.), who described the ingestion of willow leaves and bark as a means for relieving pain (20). In the 1820s, the

Figure 1. Willow, the source of salicin, the folk remedy against pain and inflammation. Left: *Salix babylonica*. Right: *S. alba* by Prof. Dr. Otto Wilhelm Thomé from: Flora von Deutschland, Österreich und der Schweiz. 1885, Gera, Germany.

Salicylic acid

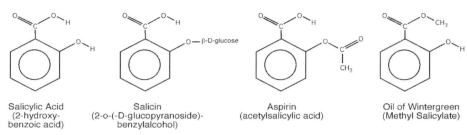

Salicylic Acid
(2-hydroxy-
benzoic acid)

Salicin
(2-o-(-D-glucopyranoside)-
benzylalcohol)

Aspirin
(acetylsalicylic acid)

Oil of Wintergreen
(Methyl Salicylate)

Figure 2. Chemical structure of salicylic acid and related compounds.

predominant active ingredient in this natural product was identified as salicin, and the presence of this compound was discovered in several other plant species including meadowsweet (*Spirea* spp.) and myrtle (*Myrica* spp.). In the following decade, salicin from natural sources was shown to consist of both a sugar, and an aromatic component initially called spirsaure and later salicylic acid (SA[1]), named for the source genera *Spirea* and *Salix*, respectively. In 1852, the first de novo synthesis of SA was described and the chemical structure of SA deduced as 2-hydroxybenzoic acid (Fig. 2). In addition to SA, a number of different SA-related compounds are found in plants, such as salicin, the glycosylated derivative of SA, and the fragrant oil of wintergreen, a methylated form of SA obtained from plants such as teaberry (*Gaultheria* spp.) (Fig. 2). Though salicin from natural sources was used as an analgesic and antipyretic through the 19th century, ingestion of plant materials that contained this compound frequently produced stomach irritation that limited the usefulness of salicin as a pain and fever-reducing drug. The synthesis of a less irritating acetyl-SA derivative (Fig. 2) led to the popularization of this form of SA as a drug and to early growth in the pharmaceutical industry, exemplified by the patenting in 1899 of Aspirin by the Bayer Company in Germany (20).

While humans have exploited the therapeutic powers of willow for centuries, plants have for millions of years employed SA as an endogenous signaling molecule active in defense and other processes. In addition to its role in controlling defense against pathogens, SA signaling in plants can also influence the expression of defenses against insect pests, as well as modifying physiology and reproductive development in some plants (33). This chapter will discuss these roles of SA in plants, the synthesis and storage of SA, and the signal transduction pathways that are responsive to SA signals.

---

[1] Abbreviations: BTH, benzo(1,2,3)thiadiazole-7-carbothioic acid S-methyl ester; ET, ethylene; HR, hypersensitive response; ICS, isochorismate synthase; INA, 2,6-dichloroisonicotinic acid; JA, jasmonic acid; MeSA, methyl salicylate; PR proteins, Pathogenesis-Related proteins; SA , salicylic acid; SABP, SA-binding protein; SAG, salicylic acid β-glucoside; SAR, systemic acquired resistance; SH, salicylate hydroxylase; TMV, Tobacco Mosaic Virus; β-GTase, UDP-glucose:SA glucosyltransferase

SALICYLIC ACID AND ITS ROLE IN THERMOGENESIS

A dramatic response regulated by SA is the production of heat in the spadix of voodoo lilies (*Sauromatum guttatum*), which can elevate in temperature by 14°C during anthesis. Heat production results from a shunt in respiratory electron transport, and volatilizes foul-smelling compounds attractive to some pollinators. The endogenous signal that triggers this intense response was named calorigen by Van Herk in 1937, and in 1987 was identified by Meeuse, Raskin and their collaborators as SA (33). In a survey of a diverse set of plants, high SA levels were observed in several other thermogenic *Arum* lily and *Cycad* species (33). However, heat production and SA accumulation were not correlated in all cases because heat-producing palms and water lilies did not accumulate detectable amounts of SA, while a number of non-thermogenic species did. The role of SA in heat production may be through its effect at inducing expression of an alternative oxidase enzyme; these enzymes have been characterized from *S. guttatum* and several other plant species (47).

A phylogenetically diverse range of plants accumulate SA or related compounds in their tissues in addition to those exploited for their analgesic properties (33). SA levels typically range from 0.05 to 5.0 µg SA per gram of leaf tissue, while a few species tested produce much more: e.g., rice may accumulate over 30 µg SA per gram leaf tissue, whereas inflorescence tissues of the thermogenic cycad species *Dioon edule* have been found to have almost 100 µg SA per gram of male cone tissue. As will be described below, pathogen attack also triggers high levels of SA production in many plants.

SALICYLIC ACID IN PATHOGEN DEFENSE

Even by 1933, dozens of reports had been published suggesting that exposure of plants to pathogens may confer protection from secondary infections, warranting a comprehensive review of the phenomenon by Chester (6). While most of the accounts Chester reviewed were anecdotal, carefully controlled studies performed in 1960 by Ross showed convincingly that infection of a local lesion tobacco host with Tobacco Mosaic Virus (TMV) conferred substantial local and systemic resistance to a secondary TMV inoculation (Fig. 3) (37). Ross called this induced whole-plant resistance response systemic acquired resistance (SAR) because it was evident in leaves distant from the primary inoculation site.

Interestingly, the SAR response could be activated by exposure to many different pathogens that cause tissue necrosis, and the resistance response is effective against a diverse range of pathogens, including bacteria, viruses, fungi and oomycetes (36). Necrosis is not mandatory, however, as pathogens that do not cause tissue damage also can activate SAR. Other studies by several groups showed that induction of SAR is accompanied by production of approximately a dozen families of soluble basic and acidic proteins called Pathogenesis-Related (PR) proteins; the genes encoding these proteins have

Salicylic acid

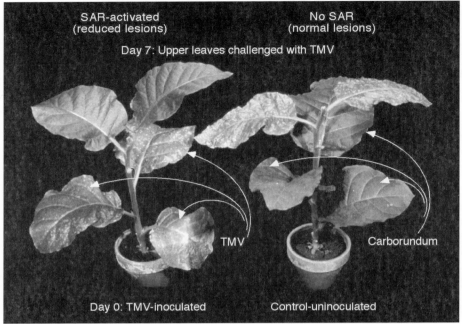

Figure 3 (Color plate page CP16). Induction of systemic acquired resistance in tobacco. Three lower leaves were inoculated with TMV using an abrasive (left plant) or mock-inoculated with just the abrasive (right). Seven days later, two upper leaves were challenge-inoculated with TMV, and symptoms evaluated after a week. TMV lesions on the SAR-activated plant (left) are 20-35% of the size of lesions on the control plant. Adapted from Ross (1961), and the Cornell University archives.

been characterized and most are induced during SAR (26, 36, 46). Though some SAR proteins have modest antimicrobial activity on their own, the key factor or combination of factors responsible for robust SAR-mediated resistance has thus far eluded discovery.

Application of SA to Plants Triggers SAR

The first insight that SA might participate in SAR came from experiments by White, who in 1979 described the induction of TMV resistance by infiltration of tobacco leaves with dilute solutions of acetylsalicylic acid (aspirin) or SA. These treatments also induced production of PR proteins like those that accumulated in tobacco after TMV infection. These observations showed that application of SA could mimic pathogen-induced SAR. Later work showed that SA treatment faithfully mimicked pathogen-activated SAR by inducing accumulation of the same large set of *PR* gene products and a similar broad spectrum resistance profile as that caused by pathogens (22, 36).

Increased SA Levels Accompany the Induction of SAR

In view of prior work showing that phloem tissue may transport the systemic signal in SAR, Métraux and collaborators in 1990 examined phloem exudates from cucumber stems following infection of leaves with the SAR-inducing pathogen *Colletotrichum lagenarium* or Tobacco Necrosis Virus. Following infection, phloem sap was harvested and fractionated by HPLC, which revealed a fluorescent metabolite shown to be SA. The increase in phloem SA accumulation occurred later than leaf symptom formation following *C. lagenarium* or Tobacco Necrosis Virus infections. By contrast, the increase in SA levels preceded expression of SAR, as measured by a reduction in lesion formation following a challenge inoculation with *C. lagenarium*. Also in 1990, in an independent study, Malamy and colleagues examined SA accumulation in two tobacco varieties susceptible to or resistant to TMV. Following TMV inoculation, SA levels increased only in the Xanthi cultivar able to detect and mount a resistance reaction against this virus. This increase in SA preceded induction of the SAR associated *PR-1* gene.

Together, these studies showed that endogenous SA levels increase both in primary infected leaves and systemically in the plant prior to induced *PR* gene expression and expression of SAR, consistent with the hypothesis that SA acts as a pathogen-induced signal that triggers the activation of SAR.

SA-Deficient Plants Fail to Activate SAR, are Hypersusceptible to Pathogens, and Compromised in Expression of Race-Specific Resistance

To test the hypothesis that SA acts as a signal for inducing SAR, Ryals and colleagues created plants that could not accumulate SA, and assessed them for the ability to activate SAR (16). This was achieved by expressing in tobacco plants the bacterial *nahG* gene that encodes salicylate hydroxylase (SH). Salicylate hydroxylase converts SA into catechol (Fig. 4), a compound that does not affect disease susceptibility in tobacco. Transgenic plants expressing SH were tested for SAR by inoculating lower leaves with TMV, and later challenging upper leaves with a secondary inoculation of TMV. NahG tobacco and wild type control plants were examined for SA accumulation seven days after the primary inoculation; wild type plants showed 185-fold greater SA levels than untreated plants, while plants expressing high levels of SH showed just a 2-3-fold increase.

Disease symptom development in secondary inoculated leaves was dramatically different in the two types of plants. Normal tobacco showed a typical SAR response, with lesions on the challenge leaves

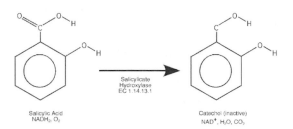

Figure 4. Reaction catalyzed by salicylate hydroxylase from *Pseudomonas putida*.

about 40% the diameter of lesions produced on naive, mock-inoculated plants that did not receive a primary inoculation. However, NahG tobacco plants showed no evidence of SAR, with the upper leaf lesions equal in size on NahG plants that experienced or did not receive a primary inocu-lation. These results, together with the earlier work described above, provide compelling evi-dence that SA accumulation is required for induction of SAR.

In addition to an inability to mount an SAR response, NahG tobacco plants also showed much more severe TMV symptoms than normal tobacco plants, with TMV lesions expanding across the leaf, coalescing, and spreading down the leaf petiole, whereas in wild type plants, infection produced punctate, determinate lesions. The susceptibility phenotype of NahG plants was further examined in tobacco and Arabidopsis after infection with a diverse range of virulent bacterial, fungal and oomycete pathogens, and in each case, NahG plants displayed a general hypersusceptibility to these parasites (9). The hypersusceptibility phenotype of NahG plants to virulent pathogens indicates that SA accumulation is required for expression of basal resistance against disease-causing microbes, and that plants infected with virulent pathogens rely on SA-mediated defenses to keep infection in check. The expression of race-specific or gene-for-gene resistance was also examined in Arabidopsis NahG plants, and, in most cases, SA signaling was required for full expression of resistance (9, 15, 32).

The failure of NahG plants to express SAR, their hypersusceptibility to virulent pathogens, and the breakdown they display in expression of race-specific resistance, indicates that SA plays a central role in expression of plant defense. While these types of defenses share a reliance on SA signaling, genetic studies described below indicate that multiple pathways are responsive to SA signals. Further, while induction of SAR relies upon SA accumulation, expression of race-specific resistance is only partially dependent on SA signaling, as indicated by the relaxed, but not abolished resistance to avirulent pathogens seen in NahG plants (9, 32, 40).

Synthetic Analogues of SA Activate SAR

The ability of SAR to simultaneously produce resistance to multiple pathogens made this system attractive as a means for controlling disease in agriculture. Though SA is effective at inducing SAR, problems with its phytotoxicity and field effectiveness led to efforts to identify more potent SAR-inducing compounds in large-scale screens of chemicals by the then Ciba-Geigy Corporation. These screens identified two types of synthetic compounds that induced resistance in pathogen assays, INA (2,6-dichloroisonicotinic acid), BTH (benzo(1,2,3)thiadiazole-7-carbothioic acid S-methyl ester), and related compounds. Because INA and BTH induce broad spectrum resistance and a profile of *PR* gene expression similar to biological inducers of SAR, and are not themselves toxic to microbes, they are regarded as faithful mimics of SA and biotic inducers of SAR (22). INA and BTH share a similar structure with SA, in that all contain a planar aromatic ring linked to a carboxylic acid group (Fig. 5), suggesting that the

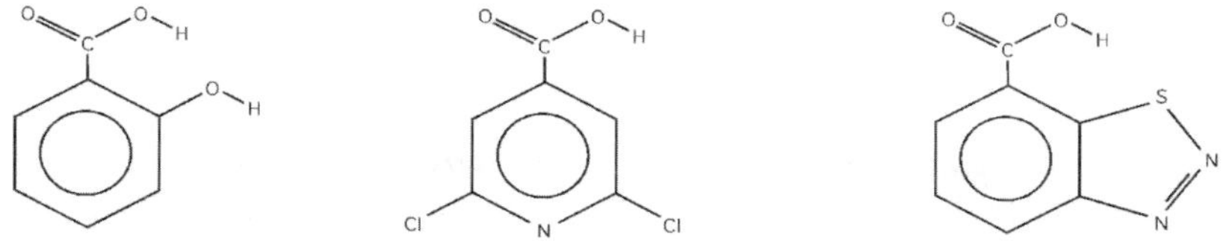

T. P. Delaney

Salicylic Acid	INA	BTH
(2-hydroxy-benzoic acid)	(2,6-dichloroiso-nicotinic acid)	(7-carboxy benzo,1,2,3-thiadiazole)

Figure 5. Structure of salicylic acid and its functional analogues.

three compounds may activate SAR through interaction with a common receptor. The effectiveness of these compounds is specific however, as some very similar compounds fail to activate SAR (22).

INA and BTH appear to interact with the SAR signal transduction pathway either at or downstream of the site of SA action, because both compounds are able to induce *PR* gene expression and SAR in NahG plants. The effectiveness of INA and BTH in NahG plants also indicates that these compounds are not substrates for salicylate hydroxylase activity. An alternative explanation for the effectiveness of these chemicals at inducing resistance in NahG plants is that they function independent of the SAR pathway. However, genetic studies described below show that this is unlikely, as SA, INA and BTH all utilize a common signaling pathway in Arabidopsis defined by the *NIM1/NPR1* gene that is required for expression of SAR.

Is Salicylic Acid the Long Distance Signal in SAR?

The induction of systemic resistance following a localized infection requires some kind of long distance communication within the plant. Following Ross' description of SAR, Kuc and coworkers found that blocking phloem transmission by stem girdling prevented induction of SAR in leaves distal to the block, leading to their conclusion that the long distance signal moved through phloem tissue. Later findings in cucumber and tobacco showed that phloem sap became enriched in SA after infection of leaves that led to SAR (28, 51). Together, these observations suggested that SA may be the long distance signal in SAR. Since then, several types of experiments have been performed to examine this hypothesis.

Kinetic Evidence Indicates that SA is Not the Systemic Signal in SAR
Localized infection of cucumber causes an increase in SA levels within the infected leaf, and later in the whole plant. The infected leaf can be excised from the plant as early as four hours after infection, and distal uninfected leaves still show induction of SA, indicating that the primary leaf had exported the systemic signal within this time period. However, phloem from petioles of the primary infected leaf did not show a detectable increase in SA content until eight hours after infection (34). These results were interpreted to indicate that production of the long distance signal in SAR precedes SA

accumulation, and therefore that SA is not the primary signal of induced resistance in cucumber.

Grafting Studies Indicate that SA is Not the Long Distance Signal in SAR
NahG tobacco plants provided an elegant means for testing whether SA is the systemic signal for SAR in tobacco. Vernooij and coworkers grafted wild type and NahG tobacco plants, and analyzed the chimeric plants in SAR assays by inoculating rootstock leaves with TMV, then later assaying for SAR expression in the scion, using a second TMV inoculation (48). As controls, wild type (Xanthi) scions were grafted onto Xanthi rootstocks, and NahG scions onto NahG stocks. The grafted controls performed like non-grafted plants, with NahG grafted plants showing no SAR, while Xanthi grafted plants showed strong SAR. This confirmed earlier observations by Kuc that the SAR signal was graft transmissible. Comparisons between the NahG-Xanthi and Xanthi-NahG reciprocal grafted plants showed that SAR was induced in the scion leaves only if the scion was from a Xanthi plant, and that SAR could be induced in a wild type scion grafted onto a NahG rootstock. These results showed that SA accumulation was not required in the primary infected (rootstock) leaf for export of an SAR-inducing systemic signal. Further, the response to the systemic signal induced resistance only in (wild type) scion leaves that could accumulate SA, indicating that SA accumulation is required to respond to the long distance signal, but not for its generation. While these experiments are compelling evidence against SA as the systemic signal, this conclusion must be accepted with caution, as a small amount of residual SA is present in NahG plants.

Transport of Labeled SA shows that SA Moves in Phloem
Although the experiments described above are strong evidence that SA is not the systemic signal in SAR, in vivo $^{18}O_2$ labeling experiments show that SA made in an infected leaf can move systemically in tobacco, consistent with SA playing a long distance signaling role. These studies examined SA levels in uninoculated leaves having a strong vascular connection with the infected leaf, and showed that in these leaves, a substantial amount (approx. two-thirds) of the nascent SA originates from infected leaves (42).

Thus, while labeling experiments show that SA movement occurs in plants undergoing pathogen attack, grafting studies indicate that SA accumulation in rootstock leaves is not required for export of the systemic SAR-inducing signal. It is possible that SA may be transported in parallel with another systemic signal, and that SA itself is not sufficient to activate SAR in uninfected leaves. A definitive test of whether SA is the primary long distance inducer of SAR may require grafting experiments with plants unable to manufacture SA, or identification of an authentic long distance signal in SAR through mutant or biochemical analyses. SA-deficient mutants have been found in Arabidopsis, but thus far few workers have tackled grafting experiments in this small rosette species.

BIOSYNTHESIS AND FATE OF SALICYLIC ACID

SA accumulation is a function of its rate of production balanced by the rate of removal through degradation or formation of derivatives of SA. Further, partitioning of SA or its conjugates can effect local concentrations of SA within cells or in plant tissues. Our understanding of the biosynthetic pathways involved in SA production have benefited from both biochemical studies and mutant analysis.

Biosynthetic Pathways for Salicylic Acid Production

Much of the organic carbon on our planet is a product of the shikimic acid pathway, as this is the source of aromatic amino acids that are precursors of lignin, a stable and abundant constituent of wood. Shikimic acid lies at the top of a linear pathway that leads to chorismic acid, which is a common intermediate for several branched pathways that yield tyrosine, tryptophan, phenylalanine, and many other aromatic compounds formed in plants. Of these, two pathways have been implicated to play a role in SA synthesis in plants.

Biochemical labeling studies of TMV-infected tobacco plants showed that SA in these plants is produced from the shikimate pathway's phenylalanine branch through the phenylpropanoid pathway. Phenylalanine is a substrate for phenylalanine ammonia lyase, forming *trans* cinnamic acid, which can be converted to either benzoic acid or *ortho* coumaric acid (Fig. 6). Labeling studies in tobacco leaves undergoing TMV-elicited HR showed that while ^{14}C-phenylalanine, ^{14}C-*trans* cinnamic acid or ^{14}C-benzoic acid are all rapidly converted to labeled SA, ^{14}C-*ortho* coumaric acid is not, indicating that in these plants, SA is derived from benzoic acid rather than *ortho* coumaric acid (Fig. 6) (33). The conversion of benzoic acid to SA is catalyzed by a benzoic acid 2-hydroxylase that is itself induced by virus attack. In other work, evidence was obtained in other plants for SA synthesis through the *ortho* coumarate intermediate (44).

Another pathway from chorismate leads to production of isochorismic acid by action of isochorismate synthase (ICS). In microorganisms, this pathway has been shown to play a role in SA synthesis. Genetic studies in Arabidopsis show that this pathway also functions in plants. In mutant screens to identify plants deficient in pathogen-induced SA accumulation or that displayed enhanced disease susceptibility, the *sid2* (*salicylic acid deficient 2*) and *eds16* (*enhanced disease susceptibility 16*) mutants were isolated, found to be allelic, and to map to the bottom of chromosome 1 (11, 30). Arabidopsis contains two putative *ICS* genes, and one (*ICS1*) maps near the *SID2/EDS16* locus. The equivalence of *ICS1* and *SID2/EDS16* genes was indicated by the presence of mutations in the *ICS1* gene in *sid2-1* and *sid2-2/eds16-1* mutants, pathogen induction of *ICS1*, and the similarity between the Arabidopsis ICS1 and periwinkle ICS that has been shown to have ICS biochemical activity (50). Putative ICS genes have been found in a wide range of plant species, indicating that this pathway is probably a general

Salicylic acid

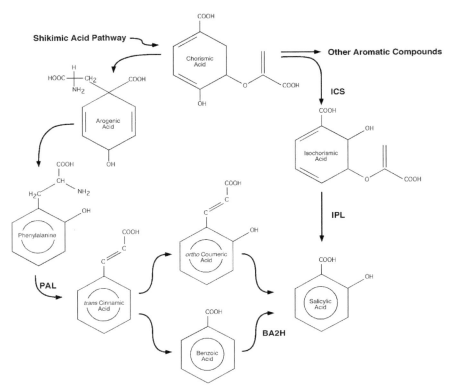

Figure 6. Biosynthetic pathways for salicylic acid production in plants. The shikimic acid pathway produces chorismic acid, which is at the top of several pathways that yield aromatic compounds, including the aromatic essential amino acids. One branch from chorismate produces phenylalanine, which by action of PAL is converted to *trans* cinnamic acid, which can form two compounds implicated in SA synthesis, *ortho* coumeric acid and benzoic acid. The latter compound is hydroxylated by BA2H, to yield SA. Another pathway from chorismate yields isochorismic acid through action of ICS, and the product further modified by IPL to form SA. The ICS gene has been shown in genetic studies to be required for most SA production in parasitized Arabidopsis plants. Abbreviations: BA2H, benzoic acid 2 hydroxylase; ICS, isochorismate synthase; IPL, isochorismate pyruvate lyase; PAL, phenylalanine ammonia lyase; SA, salicylic acid. Most arrows imply action of more than one enzymatic conversion.

mechanism for SA synthesis in most plants (Fig. 6). The plant ICS genes that have been examined have putative chloroplast transit sequences, indicating plastid localization for this branch of the SA biosynthetic pathway.

The reduced SA accumulation found in *sid2/eds16* mutants indicates that an important source of SA in parasitized plants derives from chorismate through action of ICS (Fig. 6). The magnitude of flux through this pathway is suggested by the phenotype of mutants defective in ICS, which have just 5-10% of the SA found in wild type plants after infection, suggesting that the majority of pathogen-elicited SA may be produced through the isochorismate pathway. The reduction of SA in *sid2* plants is significant because the mutants are strongly suppressed for induction of *PR-1* and SAR (11, 30).

Together, these studies show that SA production can occur through multiple pathways. It will be important to discern the relative importance of each pathway for SA synthesis, as well to understand whether these pathways play different roles in plant tissues. Additional studies that combine mutations that disrupt multiple SA biosynthetic pathways will help establish the significance of SA production through each pathway for development of SAR and other SA-dependent processes.

Conjugation and Partitioning of Salicylic Acid

Glucosyl salicylic acid and its localization

The accumulation of SA in leaves following pathogen infection coincides with the appearance of a related compound, salicylic acid β-glucoside (SAG) (18, 33). SAG levels are very low in the absence of infection, but in infected tissues can exceed by several fold the amount of free SA that accumulates. SAG production is catalyzed by a UDP-glucose:SA glucosyltransferase (β-GTase), whose activity is stimulated by elevated SA levels within the infected leaves (13). In naive leaves of tobacco, very little β-GTase activity or SAG is found.

The function of SAG is not well established, but may serve to sequester free SA, thus protecting the cell from potentially toxic effects of high concentrations of SA. Conjugation may also serve as a mechanism for shutting off responses to SA in cells after events have passed that caused the increase in cellular SA. Alternatively, the SAG pool that accumulates after infection may act as a reservoir of SA that can be liberated upon pathogen attack. This possibility was suggested by labeling studies that showed SAG infiltrated into the leaf air space to be quickly hydrolyzed to liberate free SA that is then taken up by leaf cells, where it is converted again into SAG by action of cytoplasmic β-GTase (18). These studies showed SAG hydrolase activity to be located in the apoplast, while SAG accumulation occurs in the cytoplasm or vacuole. The separation of SAG and hydrolase in different cellular compartments may provide a mechanism for the rapid production of SA in tissues undergoing attack, where membrane leakage or cell lysis may allow mingling of substrate and hydrolase. This mechanism may liberate a bolus of SA at the site of attack, providing a means by which the plant can mount a rapid and aggressive response to a secondary attack by pathogen.

Methyl salicylate as a defense-activating signal

The fragrant oil of wintergreen is produced by a number of plant species, including some birches (*Betula* spp.) and the teaberry plant (*Gaultheria* spp.), where it is responsible for the appealing smell of crushed roots and foliage. This volatile oil is the methyl ester of SA (methyl salicylate, MeSA) (Fig. 2), and is widely used as a traditional medicine, especially in traditional Chinese medicine, where it is often the predominant component in many topical formulations. The desirable medicinal effects attributed to oil of wintergreen are likely due to its rapid de-esterification in animal tissues, and to the well-established therapeutic effects of SA liberated by this reaction. However, the pleasant smell and sweet taste of oil of wintergreen, together with the popularity of natural remedies containing MeSA, has led to many cases of salicylate poisoning; one ml of MeSA is the equivalent of over twenty standard (325 mg) aspirin tablets, and ingestion of a teaspoonful can be fatal.

MeSA has also been shown to be produced in some plants in response to attack by phytopathogens, such as in tobacco leaves infected by TMV. MeSA production was most dramatically seen in TMV-infected tobacco plants that express the *N* resistance gene, following a temperature shift that led to systemic necrosis (43). These studies exploited the temperature sensitive nature of the tobacco *N* gene product, which at elevated (32°C) temperatures is nonfunctional, allowing TMV to spread systemically in the plant and form mosaic symptoms that give the virus its name. When systemically infected plants are shifted to a permissive temperature (24°C), the *N* gene is reactivated, TMV is recognized, and a massive whole-plant defense response is triggered culminating in a generalized hypersensitive response (HR); this reaction is aptly called systemic necrosis. Tobacco plants displaying systemic necrosis emit large amounts of MeSA, concurrent with expression of high levels of *PR-1* gene expression. Interestingly, when plants undergoing systemic necrosis are placed in a gas-tight chamber with a healthy tobacco plant, the naive plant shows induction of the SAR associated *PR-1* gene, and displays heightened resistance to TMV infection (43). The authors of this study concluded that infected plants may use gaseous MeSA to communicate with nearby healthy plants, and induce defenses in those plants.

The prospect of plant-to-plant communication through MeSA signaling is an exciting possibility, but one that should be weighed carefully. The amount of MeSA produced in plants that are undergoing a more natural disease process is much less that those undergoing systemic necrosis, and the atmospheric dilution of a volatile signal would be expected to further reduce the amount of gaseous MeSA available for perception by naive plants. These complications are not as severe for another suggested role for MeSA signaling - that the compound is important as a local signal for activating defenses in leaf parts distal to an infection focus or HR lesion. Within a leaf, a necrotic lesion or HR can present a significant insult to the organ, and the concentration of a nascent airborne signal could be preserved within the air space trapped between the leaf lamina. The production of a volatile signal within a leaf may thus provide a means for local defense reactions that are more rapid than responses triggered by symplastically-propagated signals.

SALICYLIC ACID SIGNAL TRANSDUCTION

Genetic Dissection of SA-Mediated Defenses

Because of the important role played by SA in activating defenses against pathogens, the identity of signaling pathway components that respond to SA signals have been sought using genetic and biochemical methods. Several groups looked for Arabidopsis mutants defective in their response to the synthetic SA analog INA, in mutant screens that focused on the failure of INA to induce resistance to *P. parasitica* (8) or activate expression of reporter genes driven by *PR* gene promoters (2, 41). These efforts led to discovery of over a dozen independent mutants called *nim1* (*non-inducible immunity 1*), *npr1* (*non-expresser of PR genes 1*) or *sai1* (*SA insensitive 1*), which fail to respond to INA, SA or BTH in the induction of *PR* gene expression and resistance. Complementation analysis showed all mutants to involve the same gene now called *NIM1* or *NPR1*. In later screens for enhanced disease susceptibility mutants, several other mutant alleles at this locus were isolated (17). The *NIM1/NPR1* gene was positionally cloned and found to encode a protein containing ankyrin repeat and BTB/POZ domains (3, 38), motifs that in many proteins have been shown to mediate interactions with other proteins. The NIM1/NPR1 protein show possible homology to IκB (38), a protein that binds to and inhibits activity of the transcription factor Nf-κB in vertebrates. The Nf-κB pathway is homologous to the Dorsal pathway in Drosophila, pathways that in both animals regulate expression of the innate immune system that is activated by pathogen exposure and controls expression of broad-spectrum disease resistance, analogous to SAR in plants. In plants, a number of genes involved in pathogen defense have been found to be homologous to components of the animal Nf-κB and Dorsal pathways, suggesting that the pathways that regulate innate immunity in animals and plants may be conserved, and thus of ancient origin (1, 7, 38).

Using the NIM1/NPR1 protein as a probe in yeast two-hybrid screens, several TGA proteins, members of the bZIP family of transcription factors, were identified that bind to NIM1/NPR1, implicating these factors in SA-mediated changes in gene expression (10, 23, 52, 54). Elucidating the specific roles played by these factors in defense is complicated by their redundancy (53) and ability to impinge upon defense pathways that function independent of SA or NIM1/NPR1 activity (23). Another two-hybrid screen revealed several related novel proteins that may act to bridge between NIM1/NPR1 and TGA factors, suggesting that a complex of proteins may be involved in mediating SA signal transduction (49).

Additional studies of NIM1/NPR1 action have examined the role of post-translational modifications on the function of this protein in activating defense. Intriguing results from Dong and coworkers have shown that specific cysteine residues in NIM1/NPR1 are important for its function, and can mediate formation of disulfide bonds and multimerization of the protein (29). The authors of this study speculate that inducers of SAR act by reducing the cellular redox state, promoting the liberation of putatively active NIM1/NPR1 monomers that enter the nucleus and modulate SAR gene expression. In other work, a potential role for phosphorylation in the regulation of NIM1/NPR1 activity was implicated by the finding that specific serine and threonine residues are both important for SAR-induction activity in vivo, and also are substrates for phosphorylation of the protein (J-Y. Ko and T. P. Delaney, unpublished).

To fully understand SA responsive signaling pathways, it is also essential to define the molecules that act as receptors for SA, and trigger downstream responses in response to elevated SA levels in plants. The mutant screens described above did not reveal obvious SA receptors, but biochemical approaches to identify SA-binding activities have recently provided clues to the identity of the elusive SA receptor.

Salicylic Acid Binding Proteins

The Search for an SA Receptor
The features expected for an SA receptor are that it binds specifically to SA, that its binding affinity to SA be relevant to concentrations of SA that occur *in planta*, and that its function be required for SA-mediated responses. Also, because biologically active synthetic analogs of SA are known, such as INA and BTH, an SA receptor may also bind to and be activated by these compounds, but not by other chemically similar but inactive compounds.

To identify potential SA receptors, Klessig and colleagues sought proteins that bind in vitro to radiolabeled SA. Several SA-binding proteins (SABPs) that vary in properties and structure were found in tobacco; the two best characterized will be discussed here. SABP1 has a coefficient of SA binding (K_d) of approximately 14 μM (4), and upon cloning was found to be a catalase enzyme whose activity is reduced by SA binding but not by biologically inactive analogs of SA (5). This discovery led to speculation that SA-inactivation of the SABP1 catalase leads to accumulation of H_2O_2, which then acts as a second messenger for SAR activation (5). This hypothesis was examined further and excluded based on the relatively low binding affinity of SABP1 to SA relative to SA amounts found in systemic tissues of infected plants, the lack of a measurable increase in H_2O_2 levels after induction of SAR with SA, and the demonstration that H_2O_2 induction of SAR genes is associated with and requires SA accumulation (19). A role for SABP1 in defense seems more likely to involve local responses, such as hypersensitive cell death, as tissues surrounding an HR can accumulate sufficient SA to inhibit SABP1 activity (19).

The group also characterized another SA-binding protein called SAPB2 (12), which is distinguished from SABP1 by its lower abundance and 150-fold higher affinity for SA (K_d=90 nm). The SABP2 protein was recently cloned and shown to be a lipase with an activity stimulated by SA (24). The importance of SABP2 in defense is indicated by the phenotype of tobacco plants in which the *SABP2* gene was silenced. SABP2 silenced plants displayed greater susceptibility to TMV, and were unable to activate SAR in response to TMV inoculation, similar to NahG plants that are unable to accumulate SA.

The finding that SABP2 is an SA-stimulated lipase suggests that SA accumulation may cause the liberation of lipid-derived compounds that trigger SA-dependent defensive processes (24). The SA-binding affinity of SABP2 is similar to SA levels found in systemic tissues of plants undergoing pathogen attack, suggesting that SABP2 may play a role as an SA receptor in activation of SAR. The authors point out that in Arabidopsis 18 SABP2-like genes are found, and 10 of the encoded products found to bind SA. The efficient functional genomics tools available in this plant will enable these genes to be tested for a role in defense and help address how general is the regulation of defense through SA-stimulated lipase signaling.

The possibility that lipid-derived signals play a general role in regulating defense pathways is also suggested by other defense mutants that have perturbations in genes that appear to be involved in lipid metabolism or transport. These include *EDS1* (*ENHANCED DISEASE SUSCEPTIBILITY 1*) and *PAD4* (*PHYTOALEXIN DEFICIENT 4*), which encode lipase-like proteins and are required for expression of race specific resistance against specific pathogens (14, 21, 31). The *DIR1* (*DEFECTIVE IN INDUCED RESISTANCE 1*) gene encodes a putative lipid transport protein, and *dir1* mutants display an inability to transport the systemic signal in SAR (27).

Influence of Salicylic Acid on Other Signal Transduction Pathways

In response to insect damage, plants may activate other inducible pathways that provide defense against herbivores. For example, insect grazing of tomato leaves leads to systemic production of proteinase inhibitor proteins that reduce plant food quality by limiting the digestibility of plant proteins to the herbivore. Insect-induced defenses have been examined in a diverse range of plant species, and been shown in many cases to be regulated by jasmonic acid (JA) and ethylene (ET) signaling (35). JA and ET activate expression of sets of genes that largely overlap with each other, but are mostly distinct from those induced by SA (39). In general, JA and ET responses promote resistance to pests as well as some necrotrophic plant pathogens, but do not play a large role in defense to biotrophic pathogens (35).

SA pathway signaling has been found in many cases to interfere with JA and ET signaling in defense, indicated by repression of JA or ET-responsive genes. Further, suppression of SA signaling by mutations in *NIM1/NPR1* or expression of salicylate hydroxylase leads to hyperactivity of jasmonate responses (15, 25). Conversely, JA pathway activity can suppress induction of SA responses, including *PR* gene expression and SAR. While many SA and JA responses show this mutual antagonism, some genes are induced by both compounds, revealing complexities in the network of defense pathways. These patterns have been most evident in large-scale gene expression profiling experiments in Arabidopsis, where the transcriptomes of SA and JA treated plants were identified and compared (39).

The evolution of mutually antagonistic SA and JA signaling pathways suggests that this regulation has adaptive value, such as in channeling the investment of resources to the most appropriate mode of defense for a given attack, or through fine tuning of the defense response to be most effective against the particular pathogen elicitor of the response (25, 45).

Summary

The past decade has greatly expanded our understanding of the diverse roles of SA in several of the more experimentally tractable plant species. The current challenge is to better understand the structure, complexity and functions of the signaling pathways that control SA responses, as well as how those pathways interact with other response pathways. An important extension of this work will be to apply this knowledge to other plant species, which will provide new strategies for improving crop traits. Comparative studies across plant species will also facilitate study of the evolution of SA responses and of the pathways that control these responses.

Acknowledgements

I thank Dr. D. F. Klessig for sharing results prior to publication, and Dr. K. Stüber for providing the digital image of *S. alba*. This work was supported by funding from the National Science Foundation (IBN-0241272) and USDA NRICGP (No. 9802134).

References

1. Belvin MP, Anderson KV (1996) A conserved signaling pathway: the Drosophila Toll-Dorsal pathway. Annu Rev Cell Dev Biol 12: 393-416
2. Cao H, Bowling SA, Gordon AS, Dong X (1994) Characterization of an Arabidopsis mutant that is nonresponsive to inducers of systemic acquired resistance. Plant Cell 6: 1583-1592
3. Cao H, Glazebrook J, Clarke JD, Volko S, Dong X (1997) The Arabidopsis *NPR1* gene that controls systemic acquired resistance encodes a novel protein containing ankyrin repeats. Cell 88: 57-63
4. Chen Z, Klessig D (1991) Identification of a soluble salicylic acid-binding protein that may function in signal transduction in the plant disease-resistance response. Proc Natl Acad Sci USA 88: 8179-8183
5. Chen Z, Silva H, Klessig DF (1993) Active oxygen species in the induction of plant systemic acquired resistance by salicylic acid. Science 262: 1883-1886.
6. Chester KS (1933) The problem of acquired physiological immunity in plants. Quart Rev Biol 8: 275-324
7. Dangl JL, Jones JD (2001) Plant pathogens and integrated defence responses to infection. Nature 411: 826-833.
8. Delaney TP, Friedrich L, Ryals JA (1995) Arabidopsis signal transduction mutant defective in chemically and biologically induced disease resistance. Proc Natl Acad Sci USA 92: 6602-6606
9. Delaney TP, Uknes S, Vernooij B, Friedrich L, Weymann K, Negrotto D, Gaffney T, Gut-Rella M, Kessmann H, Ward E, Ryals J (1994) A central role of salicylic acid in plant disease resistance. Science 266: 1247-1250

10. Despres C, DeLong C, Glaze S, Liu E, Fobert PR (2000) The Arabidopsis NPR1/NIM1 protein enhances the DNA binding activity of a subgroup of the TGA family of bZIP transcription factors. Plant Cell 12: 279-290
11. Dewdney J, Reuber TL, Wildermuth MC, Devoto A, Cui J, Stutius LM, Drummond EP, Ausubel FM (2000) Three unique mutants of Arabidopsis identify *eds* loci required for limiting growth of a biotrophic fungal pathogen. Plant J 24: 205-218.
12. Du H, Klessig D (1997) Identification of a soluble, high-affinity salicylic acid-binding protein in tobacco. Plant Physiol 113: 1319-1327
13. Enyedi AJ, Raskin I (1993) Induction of UDP-glucose: salicylic acid glucosyltransferase activity in tobacco mosaic virus-inoculated tobacco (*Nicotiana tabacum*) leaves. Plant Physiol 101: 1375-1380
14. Falk A, Feys BJ, Frost LN, Jones JDG, Daniels MJ, Parker JE (1999) EDS1, an essential component of R gene-mediated disease resistance in Arabidopsis has homology to eukaryotic lipases. Proc Natl Acad Sci USA 96: 3292-3297
15. Feys BJ, Parker JE (2000) Interplay of signaling pathways in plant disease resistance. Trends Genet 16: 449-455.
16. Gaffney T, Friedrich L, Vernooij B, Negrotto D, Nye G, Uknes S, Ward E, Kessmann H, Ryals J (1993) Requirement of salicylic acid for the induction of systemic acquired resistance. Science 261: 754-756
17. Glazebrook J (1999) Genes controlling expression of defense responses in Arabidopsis. Curr Opin Plant Biol 2: 280-286
18. Hennig J, Malamy J, Grynkiewicz G, Indulski J, Klessig DF (1993) Interconversion of the salicylic acid signal and its glucoside in tobacco. Plant J 4: 593-600
19. Hunt MD, Neuenschwander UH, Delaney TP, Weymann KB, Friedrich LB, Lawton KA, Steiner H-Y, Ryals JA (1996) Recent advances in systemic acquired resistance research: a review. Gene 179: 89-95
20. Jack DB (1997) One hundred years of aspirin. Lancet 350: 437-439
21. Jirage D, Tootle TL, Reuber TL, Frost LN, Feys BJ, Parker JE, Ausubel FM, Glazebrook J (1999) *Arabidopsis thaliana PAD4* encodes a lipase-like gene that is important for salicylic acid signaling. Proc Natl Acad Sci USA 96: 13583-13588
22. Kessmann H, Staub T, Hofmann C, Maetzke T, Herzog J, Ward E, Uknes S, Ryals J (1994) Induction of systemic acquired resistance in plants by chemicals. Annu Rev Phytopathol 32: 439-459
23. Kim HS, Delaney TP (2002) Over-expression of *TGA5*, which encodes a bZIP transcription factor that interacts with NIM1/NPR1, confers SAR-independent resistance in *Arabidopsis thaliana* to *Peronospora parasitica*. Plant J 32: 151-163
24. Kumar D, Klessig DF (2003) High-affinity Salicylic Acid-Binding Protein 2 is required for plant innate immunity and has salicylic acid-stimulated lipase activity. Proc Natl Acad Sci USA
25. Kunkel BN, Brooks DM (2002) Cross talk between signaling pathways in pathogen defense. Curr Opin Plant Biol 5: 325-331
26. Linthorst H (1991) Pathogenesis-related proteins of plants. Crit Rev Plant Sci 10: 123-150
27. Maldonado AM, Doerner P, Dixon RA, Lamb CJ, Cameron RK (2002) A putative lipid transfer protein involved in systemic resistance signalling in Arabidopsis. Nature 419: 399-403
28. Métraux J-P, Signer H, Ryals J, Ward E, Wyss-Benz M, Gaudin J, Raschdorf K, Schmid E, Blum W, Inverardi B (1990) Increase in salicylic acid at the onset of systemic acquired resistance in cucumber. Science 250: 1004-1006
29. Mou Z, Fan W, Dong X (2003) Inducers of plant systemic acquired resistance regulate NPR1 function through redox changes. Cell 113: 935-944

30. Nawrath C, Métraux JP (1999) Salicylic acid induction-deficient mutants of Arabidopsis express PR-2 and PR-5 and accumulate high levels of camalexin after pathogen inoculation. Plant Cell 11: 1393-1404.
31. Parker JE, Feys BJ, van der Biezen EA, Noël L, Aarts N, Austin MJ, Botella MA, Frost LN, Daniels MJ, Jones JDG (2000) Unravelling R gene-mediated disease resistance pathways in Arabidopsis. Molec Plant Pathol 1: 17-24
32. Rairdan GJ, Delaney TP (2002) Role of salicylic acid and NIM1/NPR1 in race-specific resistance in Arabidopsis. Genetics 161: 803-811
33. Raskin I (1995) Salicylic Acid. *In* PJ Davies, ed, Plant Hormones, Physiology, Biochemistry and Molecular Biology. Kluwer Academic Publishers, Dordrecht, pp 188-205
34. Rasmussen JB, Hammerschmidt R, Zook MN (1991) Systemic induction of salicylic acid accumulation in cucumber after inoculation with *Pseudomonas syringae* pv *syringae*. Plant Physiol 97: 1342-1347
35. Reymond P, Farmer EE (1998) Jasmonate and salicylate as global signals for defense gene expression. Curr Opin Plant Biol 1: 404-411
36. Ryals J, Neuenschwander UH, Willits MG, Molina A, Steiner H-Y, Hunt MD (1996) Systemic acquired resistance. Plant Cell 8: 1809-1819
37. Ryals J, Uknes S, Ward E (1994) Systemic acquired resistance. Plant Physiol 104: 1109-1112
38. Ryals J, Weymann K, Lawton K, Friedrich L, Ellis D, Steiner HY, Johnson J, Delaney TP, Jesse T, Vos P, Uknes S (1997) The Arabidopsis NIM1 protein shows homology to the mammalian transcription factor inhibitor I kappa B. Plant Cell 9: 425-439
39. Schenk PM, Kazan K, Wilson I, Anderson JP, Richmond T, Somerville SC, Manners JM (2000) Coordinated plant defense responses in Arabidopsis revealed by microarray analysis. Proc Natl Acad Sci USA 97: 11655-11660
40. Shah J (2003) The salicylic acid loop in plant defense. Curr Opin Plant Biol 6: 365-371
41. Shah J, Tsui F, Klessig DF (1997) Characterization of a salicylic acid-insensitive mutant (*sai1*) of *Arabidopsis thaliana*, identified in a selective screen utilizing the SA-inducible expression of the *tms2* gene. Molec Plant-Microbe Interact 10: 69-78
42. Shulaev V, León J, Raskin I (1995) Is salicylic acid a translocated signal of systemic acquired resistance in tobacco? Plant Cell 7: 1691-1701
43. Shulaev V, Silverman P, Raskin I (1997) Airborne signalling by methyl salicylate in plant pathogen resistance. Nature 385: 718-721
44. Sticher L, Mauch-Mani B, Métraux JP (1997) Systemic acquired resistance. *In* RK Webster, ed, Annu Rev Phytopathol, Vol 35. Annual Reviews Inc., Palo Alto, pp 235-270
45. Thomma BP, Penninckx IA, Broekaert WF, Cammue BP (2001) The complexity of disease signaling in Arabidopsis. Curr Opin Immunol 13: 63-68.
46. van Loon LC (1997) Induced resistance in plants and the role of pathogenesis-related proteins. Europ Journal Plant Pathol 103: 753-765
47. Vanlerberghe GC, McIntosh L (1997) Alternative oxidase: From gene to function. Annu Rev Plant Physiol Plant Mol Biol 48: 703-734
48. Vernooij B, Friedrich L, Morse A, Reist R, Kolditz-Jawhar R, Ward E, Uknes S, Kessmann H, Ryals J (1994) Salicylic acid is not the translocated signal responsible for inducing systemic acquired resistance but is required in signal transduction. Plant Cell 6: 959-965.
49. Weigel RR, Bauscher C, Pfitzner AJ, Pfitzner UM (2001) NIMIN-1, NIMIN-2 and NIMIN-3, members of a novel family of proteins from Arabidopsis that interact with NPR1/NIM1, a key regulator of systemic acquired resistance in plants. Plant Mol Biol 46: 143-160.
50. Wildermuth MC, Dewdney J, Wu G, Ausubel FM (2001) Isochorismate synthase is required to synthesize salicylic acid for plant defence. Nature 414: 562-565

51. Yalpani N, Silverman P, Wilson TMA, Kleier DA, Raskin I (1991) Salicylic acid is a systemic signal and an inducer of pathogenesis-related proteins in virus-infected tobacco. Plant Cell 3: 809-818
52. Zhang Y, Fan W, Kinkema M, Li X, Dong X (1999) Interaction of NPR1 with basic leucine zipper protein transcription factors that bind sequences required for salicylic acid induction of the *PR-1* gene. Proc Natl Acad Sci USA 96: 6523-6528
53. Zhang Y, Tessaro MJ, Lassner M, Li X (2003) Knockout analysis of Arabidopsis transcription factors TGA2, TGA5, and TGA6 reveals their redundant and essential roles in systemic acquired resistance. Plant Cell 15: 2647-2653
54. Zhou JM, Trifa Y, Silva H, Pontier D, Lam E, Shah J, Klessig DF (2000) NPR1 differentially interacts with members of the TGA/OBF family of transcription factors that bind an element of the *PR-1* gene required for induction by salicylic acid. Molec Plant-Microbe Interact 13: 191-202

F3. Peptide Hormones for Defense, Growth, Development and Reproduction

Clarence A. Ryan and Gregory Pearce
Institute of Biological Chemistry, Washington State University, Pullman, WA 99164-6340 USA. E-mail: cabudryan@hotmail.com

INTRODUCTION

Peptide hormones are extracellular signaling molecules that are commonly found in animals and plants (38). A broad spectrum of physiological processes are regulated by peptide signals including metabolism, cell division, growth, pain, well being, reproduction and immunity, as examples. The first peptide hormone, insulin, was isolated in 1921 (1), which led to the isolation of hundreds of peptide signals from animals and, more recently, to their isolation from plants (34). The majority of our fundamental knowledge about peptide signals has been from studies in animals, where peptide hormones are synthesized through the secretory pathway and are derived from larger precursor proteins by proteolytic processing enzymes. Two general classes of peptide hormones are found in animals, endocrine hormones (10) and membrane-anchored cytokines and growth factors (21). Endocrine peptide hormone precursors such as pro-insulin are synthesized through the secretory pathway and most are processed while sequestered in vesicles by proteinases of the kexin family (15). The mature peptides are released from the vesicles in response to physiological signals and travel to nearby or distant locations where they interact with membrane receptors to initiate intracellular responses. Precursors of membrane-anchored cytokines and growth factor precursors, such as transforming growth factor (TGF[1]) and tumor necrosis factor alpha (TNFα), are not processed in ER vesicles where they are synthesized, but are anchored in the vesicle membranes (21). The vesicle membranes fuse with cell membranes to present the peptides to extracellular spaces where they can be processed by proteinases activated in response to physiological cues. The proteinases that process these precursors to mobile peptide signals have different specificities than enzymes of the

---

[1] Abbreviations: HypSys, hydroxyproline-rich systemins; MAP(K), mitogen-activated protein (kinase); ORF, open reading frame; PSKα, phytosulfokine-α; SLG, S-Locus Glycoprotein; SCR, S-locus cysteine rich proteins; SRK, S-receptor kinase; SAM, shoot apical meristem.

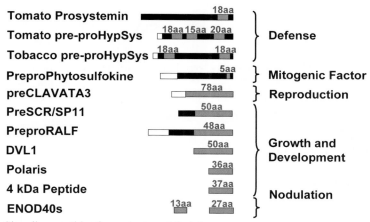

Figure 1. Signaling peptides from plants and their known precursors.
Open boxes, leader peptides; black boxes, pro-sequences; grey boxes, peptide hormones.

kexin family. The processing sites of plant peptide hormone precursors identified so far have not been identified.

Before 1991, plants were thought to utilize small organic molecules, called phytohormones, to regulate all physiological processes. These hormones included auxin, cytokinins, gibberellins, ethylene, abscisic acid, and brassinosteroids (19, 20). In 1991, an 18 amino acid peptide, called systemin (34) was isolated and identified as the systemic wound signal for the regulation of the expression of defensive genes in tomato leaves in response to insect attacks or other severe mechanical wounding. Since then, over a dozen peptide hormones have been isolated from plants or identified by genetic approaches that regulate various processes involved in defense (31, 33), cell division (23), growth and development (4, 32, 46, 47) and reproduction (18). Known peptide signals derived from plants are shown in Fig. 1. Some are synthesized as precursors through the secretory pathway, having pre- pro- or prepro-sequences analogous to those of animal peptide hormones, while others lack these sequences and may be synthesized on free ribosomes in the cytoplasm. Receptors that have been identified for several plant peptide hormones are shown in Fig. 2.

PEPTIDE SIGNALS FOR DEFENSE

Systemins

Tomato Systemin
The tomato systemin peptide $^+$AVQSKPPSKRDPPKMQTD$^-$ (34) is a powerful primary signal that is released into the vascular system of tomato plants at sites of herbivore attacks. Systemin is processed from a 200 amino acid precursor (Fig. 1) called prosystemin (26) in response to wounding and initiates a signaling cascade that activates the synthesis of defense-related

Peptide hormones

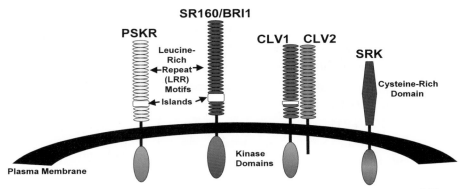

Figure 2. Plant hormone receptors. Systemin receptor, SR160/BRI1; CLV3 receptor; SCR receptor, SRK; Phytosulfokine-α receptor, PSKR.

proteins in leaves throughout the plants (Fig. 3). Early events following the systemin-receptor interaction (Fig. 4) include an increase in cytosolic calcium, the activation of a MAP kinase cascade, and the activation of a phospholipase that cleaves linolenic acid (18:3) from membranes. The linolenic acid (18:3) is converted to the oxylipin jasmonic acid, a powerful inducer of defense genes (37).

The systemin signaling pathway is complex, and in some ways, is analogous to the inflammatory response in animals (2). Jasmonic acid exhibits structural similarities to animal prostaglandins that are derived from membrane-associated arachidonic acid (20:3) in response to injury or pathogen attacks and that mediate inflammation. It is of interest that aspirin inhibits the synthesis of jasmonic acid in plants and of prostaglandins in animals, and that suramin, an inhibitor of receptor function in animals, inhibits systemin binding to its receptor.

Systemin activates the synthesis of over 20 wound-inducible, defense-related genes that encode proteins that have been functionally assigned to three groups: signal pathway components; proteolytic enzymes; and defensive proteins (2, 37). The net results of these genes are to amplify the signaling system to produce large quantities of defensive proteins in leaves that can interfere with the digestive processes of the attacking herbivores or pathogens. Prosystemin is synthesized and sequestered in the cytoplasm of phloem parenchyma cells of

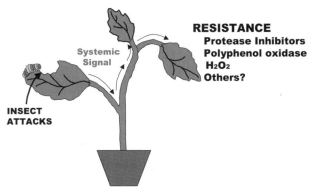

Figure 3. Systemic wound-signaling in tomato plants. Wounding releases systemic wound signals that are amplified throughout the plants to activate defense-related genes.

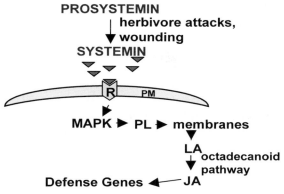

Figure 4. Early signaling events in response to systemin in tomato leaves. Systemin interacts with its receptor to initiate an intracellular cascade in which the early events, including the activation of MAP kinase and phospholipase (PL) activities, lead to the release of linolenic acid (LA) from membranes and its conversion to jasmonic acid.

vascular bundles (29) and is subsequently processed and released in response to wounding. The proteolytic processing of prosystemin does not involve the typical dibasic amino acids that are sites for processing animal secretory prohormones. Processing enzymes that convert prosystemin to systemin have not been identified, nor have the initial cleavage sites.

The potency of systemin is in the nM range, similar to those of animal peptide hormones (34). The function of systemin as a primary signal for the activation of defense genes was confirmed by producing transgenic tomato plants that constitutively express an antisense prosystemin gene (26). The plants do not exhibit a systemic activation of defense genes in response to mechanical damage or to wounding caused by feeding *Manduca sexta* larvae. The larvae rapidly consume the plants and grow very rapidly, in contrast to slowly growing larvae feeding on wild-type plants.

Prosystemin gene orthologs are present in various species of the Solanaceae family tribe Solaneae, but the gene has not been identified in species of any of the other tribes, or in other plant families. Why and how systemin evolved as a systemic wound signal in only one subtribe of the Solanaceae is not known. The mechanism for the release and transport of systemin through the plants is also not understood, but its long range signaling effects in response to wounding by attacking insects is thought to involve a complex system for amplification of the wound signal. The presence of prosystemin in vascular bundle cells suggests that its localization there is important for long distance signaling.

Systemin exhibits only weak secondary and tertiary structural features in aqueous solutions. The structures, determined by NMR spectroscopy, reflect a poly(L-proline) II, 3_1 helix secondary structure (PP II), due to two proline-doublets in the central region of the peptide (44). The PP II structures may be important for recognition by the systemin receptor, which is a 160 kD membrane-bound protein called SR160 (39).

The extracellular domain of the systemin receptor contains 22 leucine-rich repeats (LRRs), a transmembrane domain, and an intracellular receptor kinase domain (40) (Fig. 2). The K_d for the interaction of systemin with the receptor is about 10^{-10} M, which is generally similar to the dissociation

Peptide hormones

constant found with many peptide-LRR receptor kinase interactions found in animals. The receptor is structurally identical to the brassinolide receptor (BRI1) in tomato leaves (27), indicating that the receptor has dual signaling functions, one for defense and one for development. This suggests that receptor sites for systemin and brassinolide occur within the extracellular regions, which result in the activation of different intercellular signaling components, likely associated with the kinase domain of the receptor. This dual function for receptors is unique in the plant kingdom, but receptors that recognize dual ligands have been reported in animals. Research to understand the systemin/brassinolide receptor promises new insights into the understanding of the integration of signaling pathways that regulate both defensive and developmental processes.

Tomato and Tobacco Hydroxyproline-rich Glycosylated Defense Signals.
Peptide signals that are not systemin homologs, but that activate defense genes in response to wounding, have been isolated and characterized from tomato and tobacco leaves. While not homologs, these peptides do have some structural characteristics similar to systemin. Two hydroxyproline-rich glycopeptides were isolated from tobacco leaves (31) and three from tomato leaves (33) (Figs. 1 and 5) that range from 15 to 20 amino acids in length. Their posttranslational modifications infer that their synthesis is through the secretory pathway. However, their sites of synthesis and final localizations have not been established. The peptides, like systemin, are powerful activators of proteinase inhibitor genes when supplied to plants through their cut stems at nM concentrations. Three peptides isolated from tomato leaves activate the same defensive genes as systemin, and the tobacco peptides activate proteinase inhibitors in tobacco plants which are related to the

		Carbohydrate Pentoses/polypeptide
Tom HypSys-I	RTOYKTOOOOTSSSOTHQ	8-17
Tom HypSys-II	GRHDYVASOOOOKPQDEQRQ	12-16
Tom HypSys-III	GRHDSVLPOOSOKTD	10
Tob HypSys-I	RGANLPOOSOASSOOSKE	9
Tob HypSys-II	NRKPLSOOSOKPADGQRP	6
Tom Systemin	AVQSKPPSKRDPPKMQTD	0

Figure 5. A comparison of the HypSys peptide defense hormones from tomato and tobacco leaves. O, hydroxyproline. P, proline, T, threonine, and S, serine are highlighted in grey.

wound-inducible inhibitor II in tomato leaves. The tomato and tobacco peptide precursor genes are both inducible by jasmonic acid. These peptides are not considered to be mobile signals because tomato plants that express an antisense prosystemin gene lack a systemic response, and tobacco plants that lack a prosystemin gene exhibit a very weak systemic response. The hydroxyproline-rich peptides are thought to be involved in wound signaling by being synthesized and released in response to jasmonate to amplify the production of jasmonates, but the peptides are not acting as long range mobile signals. Because of their sizes and their biological activates in signaling the activation of defense genes, they have been assigned to the systemin family, which is defined functionally as a family of proline-rich or hydroxyproline-rich peptides derived from plants that activate defense-related genes (33). The hydroxyproline-rich systemins are called TobHypSys I and II, and TomHypSys I, II, and III.

The two tobacco systemins are derived from a single precursor protein of 165 amino acids that includes a signal sequence, with the TobHypSys I sequence processed from the N-terminus and the TobHypSys II sequence from the C-terminus (Fig. 1). The three tomato peptides, called TomHypSys I, II and III, are also derived from a single precursor of 146 amino acids that exhibits weak amino acid identity to the tobacco precursor (33). In animals, multiple peptide hormones are commonly found to originate from a single precursor, but the tobacco and tomato HypSys precursors are the only examples in plants of precursors harboring multiple peptide signals. No homology is apparent between tomato prosystemin and the two HypSys peptide precursors, but the –PPS- sequence present in systemin is found in the primary translation products of the TomHypSys and TobHypSys mRNAs. Receptors for the tomato and tobacco HypSys peptides have not been identified.

Although wound-inducible defense genes are found in most plant families, systemin family peptides have been found so far only in two subtribes of the Solanaceae family, with systemin found only in the subtribe Solaneae, and the HypSys peptides in both the Solaneae and Nicotianeae subtribes. Considering the complexity of the signaling pathway for tomato systemin, and the similarity of the signaling properties of all systemins, the intracellular signaling pathways may have evolved from a common ancient precursor gene. Because of these similarities, the systemin family peptide defense signals are expected to be in other plant families.

PEPTIDE SIGNALS FOR GROWTH AND DEVELOPMENT.

Phytosulfokine
A sulfated pentapeptide with the structure $Tyr(SO_3H)$-Ile-$Tyr(SO_3H)$-Thr-Gln, called phytosulfokine-α (PSKα) (23) plays a role in cell proliferation of suspension cultured plant cells. When suspension cultured cells are transferred to fresh medium for propagation, they divide slowly for some time and then proliferate rapidly. By adding the culture medium recovered

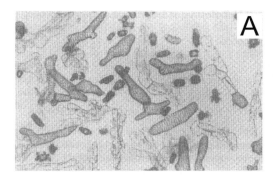

Figure 6. The differentiation of Suspension cultured *Zinnia elegans* mesophyll cells (B) to tracheary elements (A) following the addition of PSKα to the cells (24).

from rapidly growing cells (called conditioned medium) to the newly transferred low-density cells, the cells rapidly divide. The component of the medium causing the response is the PSKα-peptide, which is active at 10^{-9} M (23). In the presence of defined concentrations of auxin and cytokinin, PSKα can stimulate tracheary element differentiation of isolated *Zinnia* mesophyll cells (Fig. 6) and somatic embryogenesis of cultured carrot cells (24). In intact plants, the peptide exhibits hormonal effects, stimulating the growth of adventitious roots of cucumber and bud formation in snapdragon (49). Sulfated tyrosines are often found in animal proteins, but not in peptide hormones, and PSKα is the only example of post-translational sulfated tyrosine residues in plants. The sulfate groups in PSKα are required for activity.

PSKα orthologs are found in many plants, where they are derived from precursor proteins (49). Rice PSKα, for example, is proteolytically cleaved from the C-terminal region of an 89-amino acid precursor that includes a 22 amino acid leader peptide (50) (Fig. 1). Its synthesis, modification, and processing suggests that the precursor is synthesized through the secretory pathway. The rice genome has only a single copy of the *PSKα precursor* gene, while Arabidopsis has four copies (51). Two of the Arabidopsis cDNAs were isolated and characterized, revealing that their encoded PSKα peptides are identical to the rice PSKα. The Arabidopsis PSKα precursors all contain dibasic processing sites near the N-termini of the PSKα sequences, suggesting that kexin-like processing enzymes found in animals and yeast may be involved in processing the precursor. However, in rice, dibasic sequences do not flank the PSKα sequence, so that other processing sites and perhaps other enzymes may be involved in processing events. Overexpression of the two Arabidopsis PSKα precursor genes in transgenic Arabidopsis cells caused calli to grow twice as large as calli from wild type cells, but transforming cells with an antisense PSKα precursor gene had no apparent effect on growth (49).

PSKα-binding proteins of 120 kD and 150 kD were identified in plasma membranes. The larger component is an LRR receptor kinase with a K_d in the low nM range (22). The receptor contains an extracellular LRR domain, a transmembrane domain, and a receptor kinase domain (Fig. 2), similar to the systemin/brassinolide receptor, the CLV3 receptor, and many LRR receptors in animals.

PSKα has been proposed to be secreted from various cells in response to auxin and cytokinin and to act as a growth factor to regulate cellular differentiation and proliferation within various physiological environments. However, downstream signaling pathway components that interact with the PSKα-receptor have not been identified.

CLAVATA3

CLAVATA3 (CLV3) is an extracellular signaling peptide that plays a central role in the shoot apical meristem (SAM) that determines stem cell fate during development (3, 5, 28). The identity and position of all aerial organs of plants is initiated and established in the SAM, where division and differentiation of stem cells is maintained as the plant grows (25). Stem cells are continually dividing, providing cells to the surrounding areas where they multiply and differentiate to form leaf and flower primordia (5, 25). CLV3 is a 79 amino acid peptide in the epidermal layers of the Arabidopsis SAM stem cells that is secreted to the apoplast where it interacts with a receptor complex in underlying cells to coordinate cell growth and expansion (36) (Fig. 7). *CLV3* mRNA is initially translated as a 96 amino acid protein from which a leader peptide of 18 residues is proteolytically removed to produce the active peptide (11). Whether any further processing of CLV3 occurs after removal of the signal peptide is not known.

The Arabidopsis CLV3 receptor appears to exist as a complex composed of CLV1, a 980 amino acid LRR transmembrane receptor kinase, and CLV2, a 720 amino acid LRR receptor-like protein that lacks an intracellular kinase domain (7, 17). Mutations in *CLV1* and *CLV3* lead to identical phenotypes that accumulate undifferentiated stem cells in shoot and flower meristems (8), supporting their associated functions. As CLV3 is secreted from the stem cells, it interacts with the CLV1/CLV2 complex in the underlying cells to activate the signal transduction pathway (36). An intracellular autophosphorylation of CLV1 activates the immediate

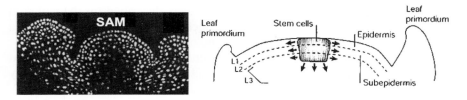

Figure 7. Left panel; a cross-section of an Arabidopsis stem apical meristem (SAM) (25) showing individual layers of cells. Right panel; an illustration showing the localization of interactions among cell layers LI, L2 and L3, to regulate cell fate (6)

Peptide hormones

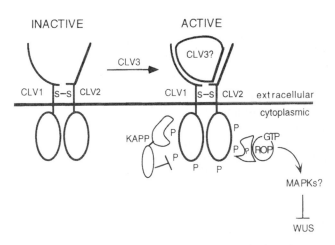

Figure 8. Early events of CLV signaling in Arabidopsis apical meristems. CLV3 interact with the receptor CLV1 and its associated protein CLV2 to initiate a cascade of intracellular events that suppresses WUS gene expression, resulting in cell proliferation and differentiation (45).

downstream signaling components, including a kinase-associated protein phosphatase called KAPP, and a GTPase-related protein called ROP, that appear to initiate signaling through a MAPK cascade (Fig. 8) (45). Individual downstream signaling components have not been identified, but the fate of stem cells as they differentiate is regulated by a transcription factor called WUSCHEL (WUS) (5, 6). While CLV3 and its receptor CLV1 have been studied extensively in *Arabidopsis thaliana*, the *CLV3* gene is a member of a large family of related genes (9) that are found in several plant families in which they may have diverse signaling roles. Further studies of the CLV3/CLV1/CLV2 signaling complex promise new insights into the roles of peptide-receptor interactions leading to signal transduction cascades that regulate early developmental processes of both leaf and flower organ formation.

RALF
RALF is a peptide signal that blocks cell division and elongation of roots when applied to the media of germinating seeds (32). The peptide was discovered and isolated while searching for defense peptide signals in tobacco leaf extracts using the alkalinization assay described for the isolation of HypSys peptides. RALF was named because of its ability to cause a strong, rapid alkalinization of the medium of suspension-cultured cells of tomato, tobacco and alfalfa. RALF is 49 amino acids in length and is derived from the C-terminus of a 115-amino acid precursor that is initially synthesized with a 25 amino acid leader peptide at its N-terminus (Fig. 2). Databases revealed the existence of *RALF* orthologs in more than 15 plant species from nine families, present in a variety of tissues and organs. A family of RALF peptides was isolated from hybrid poplar leaves (16), and

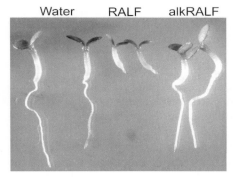

Figure 9 (Color plate page CP16). The inhibition of growth and development of Arabidopsis roots by exposing the germinating seeds to a solution of RALF peptide. Treatment of seedlings with an inactive form of RALF having modified cysteine residues (alkRALF) had no effect on root growth (32).

two unique *RALF* cDNAs were isolated from a poplar cDNA library. While both genes were constitutively expressed in poplar saplings, the expression of one paralog in suspension cultures was transiently suppressed by methyl jasmonate. The roles(s) of the RALF peptides in poplar is not known.

Two binding proteins, 120 kD and 25 kD, were identified by photoaffinity labeling (JM Scheer and CA Ryan, submitted). The proteins are thought to be part of a receptor complex that recognizes RALF to initiate a signaling pathway. However, neither binding proteins have been isolated, and little else is known of their possible relationships to signaling pathways.

The addition of µM concentrations of tomato RALF to germinating tomato and Arabidopsis seedlings caused root cell division and elongation to be arrested, and root growth to cease (DS Moura, G Pearce and CA Ryan submitted) (Fig. 9). Nine *RALF* genes are present in the Arabidopsis genome that code for highly conserved RALF sequences. The N-terminal amino acid sequences of each RALF precursor are much less conserved than the C-terminal sequences encoding the RALF peptide. The presence of tandem arginine residues just upstream from the putative N-termini may be a recognition site for processing enzymes, since mutation within this dibasic pair blocks processing (DS Moura and CA Ryan, unpublished). The region just upstream of the dibasic site is rich in polar residues that are highly conserved among RALF precursors from different species (32) and, as found in animals, may be part of a recognition site for a processing proteinase.

The functional role of RALF in growth and development of roots and other organs is not yet known. The presence of several paralogs in Arabidopsis, and the identity of *RALF* orthologs throughout the plant kingdom in various tissues and organs, suggests that the peptide may have an important fundamental physiological role in plants in regulating growth of various organs and tissues. The powerful effects of RALF on root growth reinforces the anticipation that additional peptide signals will be found that play central roles in plant growth and development.

POLARIS

Polaris is a peptide of 36 amino acids in length that was identified in Arabidopsis seedlings and root tips where it appears to be required for the modulation of root growth and leaf patterning in the presence of auxin and cytokinin (4). The peptide is synthesized without a leader peptide and its initiating methionine codon is immediately preceded by a stop codon for an 8 amino acid open reading frame that has no known function, but that also may be a peptide signal.

The C-terminus of POLARIS has a predicted α-helix structure containing a repeat motif of –KLFKLFK-. The helix is leucine rich, suggesting that it may have a role in protein-protein interactions. Whether POLARIS is processed to smaller, active peptides after synthesis is not known. An -RRR- motif near the center of the peptide suggests that it may be a likely site for processing to produce either smaller, active peptides, or to degrade the peptide, as found with the internal dibasic pair found at the center of systemin.

Arabidopsis plants having a mutated *pls* gene exhibit shortened roots and defective vasculature. The *POLARIS* gene is up-regulated by auxin and down-regulated by cytokinin. The interplay between *POLARIS* and cytokinins and auxins may be similar to the interactions of *PSKα precursor* gene with these phytohormones, and suggests that peptide cross-interactions with other hormones may be more widespread than previously thought. The further understanding of the specific cellular localization of synthesis and mode of signaling of POLARIS remain to be determined. The full understanding of both POLARIS and PSK should provide novel insights into the interplay between peptide signals and phytohormones in regulating growth and development.

DVL1

DVL1 is the most recent member of the growing list of small peptides that appear to regulate developmental processes. The peptide was identified by activation tagging in Arabidopsis (47). The name DVL1 was derived from the abnormal devil's fork-like structure of the siliques of flowers of plants that overexpress the gene (Fig. 10) (47). The peptide is 51 amino acids in length and is synthesized without a leader peptide. Other characteristics of plants that overexpress the gene are shortened stems, rounder leaf rosettes than wild-type, and various modifications of flowers and fruit. Like systemin and POLARIS, DVL1 lacks a leader peptide. DVL1 shows no sequence identity with any known protein, but twenty-two members of the *DVL* gene family were found in Arabidopsis, and 24 members in rice. Orthologs were found only in seed bearing plant species. Among all of the genes, similarities were found within the C-terminal regions. Five Arabidopsis genes with

Figure 10. A phenotype of Arabidopsis overexpressing the DVL1 peptide gene. The plants exhibit an abnormal development of siliques resembling a Devil's fork (right) as compared to wild type siliques (left) (47)

the closest sequence identities were individually overexpressed and found to produce similar phenotypes, but with identifiable variations, especially in siliques.

The peptide is associated with plasma membranes, but it is not known to be secreted or if it interacts with a receptor to cause the abnormal structures in the phenotypes. The N-terminal region of the peptide is highly charged, similar to prosystemin, and all paralogs contain dibasic groups, with 100% conservation of an RR pair adjacent to one of two cysteine groups found within the C-terminal region that are also totally conserved. Whether the conserved dibasic pair is a processing site or a site for degradation is not known. It is not known if the synthesis of the peptide is associated with cytokinin or auxin functions, as found with POLARIS and PSK, or if the peptide has a unique signaling role that produces the phenotypes resulting from its overexpression.

PEPTIDE SIGNALS FOR SELF-INCOMPATABILITY

SCR/SP11

Self-incompatibility (SI) is commonly found in plants and has been studied extensively in the Brassicaceae family, where one type is the result of a communication system that occurs between the pollen and the stigma. The stigma membrane receptor recognizes small peptides secreted from pollen, triggering a signal transduction system that results in the hydration of the pollen grain and the interruption of pollen tube development (18).

The secreted pollen peptides consist of a family of at least 28 members that are 47-60 amino acids in length, cumulatively called S-locus Cysteine Rich proteins (SCR) (30) or S-locus Protein 11 (SP11) (12, 41). SCR/SP11 peptides are derived from cysteine-rich preproteins of 74 to 83 amino acids in length that include leader peptides. The processed SCR/SP11 peptides are about 50 amino acids in length and contain four disulfide bridges that are also conserved among family members. In contrast to most animal and plant leader peptides, the amino acid sequences of the SCR/SP11 leader peptides are highly conserved with respect to the processed peptide regions, indicating that the peptides are mutating much faster than the leader sequence. On the other hand, the four disulfide bridges are conserved (41, 43), indicating that the secondary structure of the peptides is important for receptor recognition, but the individual amino acids other than the cysteine can vary considerably without affecting function.

The S-receptor kinase for SCR/SP11, called SRK, is composed of 858 amino acids, including a leader peptide, a cysteine-rich extracellular domain, a transmembrane domain and an intracellular protein kinase domain (Fig. 2). The receptor is synthesized specifically in the stigma epidermis and is localized in the external membranes associated with the cell walls of stigma papillae (18). SRK has a functional serine/threonine kinase at its C-terminus with a transmembrane region in between it and the extracellular N-terminal

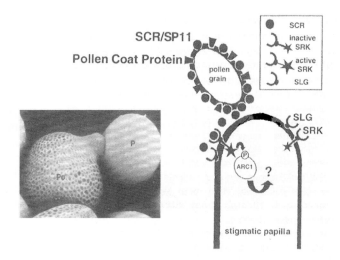

Figure 11. Early events in self-incompatibility signaling in Brassica species. The electron microscope photo on the left shows the interaction of a pollen grain (Po) with a stigma papilla (P) (24). At right is a model showing the signal transduction cascade initiated by the interaction of the SCR/SP11 peptide secreted by pollen with its receptor SRK on the stigma papilla. This interaction leads to unknown biochemical events that result in dehydration of the pollen and a cessation of the fertilization process (12).

cysteine-rich domain. An abundant soluble 60 kD S-Locus Glycoprotein (SLG) appears to be associated with SRK in some Brassica species but not in others, and is often present at a 100-fold excess where it appears to facilitate a functional SCR/SRK/SLG complex (30). A downstream signaling "arm repeat"-containing protein called ARC1, is a stigma-specific protein that interacts with the kinase domain of SRK, and is phosphorylated (14) (Fig. 11). Other downstream signaling components for events leading to dehydration and incompatibility have not been identified.

Legume Peptides

ENOD40
Nodulation in legume roots results in the establishment of symbiotic nitrogen-fixing bacteria to supply atmospheric nitrogen to plants for growth and development. Nodules develop on roots in response to signals from the bacteria that alter the cell division and development of plant cells that are destined to become nodules to harbor the nitrogen-fixing bacteria (13). ENOD40 is a gene found in legume species whose expression has been associated with nodule formation in response to signals (Nod factors) and cytokinin. Plants constitutively expressing the *ENOD40* gene show accelerated nodule formation, while plants with suppressed expression abort nodule development (42). The *ENOD40* gene does not contain any long open reading frames (ORFs), but does harbor several short ORFs. Two ORFs of 13 and 27 amino acids have been deduced from the *ENOD40* gene from alfalfa (46), while 12 and 24 amino acid peptides (Fig. 2) were deduced from

the soybean *ENOD40* gene (35). The peptides lack leader sequences and are likely synthesized in the cytosol. The expression of each alfalfa peptide separately mimics the effects of the complete genes. Evidence that the peptides are expressed has been difficult to establish. A role for the ENOD40 peptides was proposed based on the ability of the soybean peptides to bind strongly to sucrose synthase to alter its activity. The peptides have been proposed to act not as hormones in the sense that they are secreted, but to act as modulators of metabolic activity (35), which would be a novel role for peptides in plants.

4 kD Legume Peptide
A family of highly conserved peptides of 38 amino acids containing six half cysteines is present in several legume species and its presence in callus cells enhances growth and cell proliferation (48). The peptides were isolated by affinity chromatography using an immobilized 43 kD protein found in soybean seeds that has structural characteristics of a transmembrane receptor similar to the insulin receptor of animals. Transfection of carrot and bird's foot trefoil cells with a plasmid containing the peptide gene caused the cells to grow rapidly, compared to untransformed cells. However, transgenic plants grown from the transformed cells did not show any effects of the expression of the *4 kD* gene. The high conservation of amino acid sequences among different legumes species suggests that the peptide may have an important physiological role, but its identity as a signaling peptide remains to be elucidated.

PERSPECTIVES FOR THE FUTURE

Over a dozen peptide hormones have now been identified in plants, involved in processes of growth, development, reproduction, and defense. The similarities in the synthesis, processing and receptor-mediation of some plant peptide hormones suggest that they may have evolved from common ancestral progenitors. On the other hand, because of the diversity of plant peptide hormone synthesis and processing, it is not clear that they all trace back to a common ancestral precursor, or if peptide signals in plants are the result of convergent evolution. It is clear that peptide signaling confers distinct advantages to plants, and has been a mode of signaling that evolved by natural selection from ancestral systems. Peptide signals provide plants with infinite possibilities for evolving structures for specific interactions with membrane receptors. They also provide structures that can be processed by proteases to regulate activity in response to physiological cues and to control their temporal existence in tissues and organs. The recent discovery of several peptide signals in plants has placed to final rest the long-held hypothesis that plant processes are regulated only by small organic non-peptide phytohormones.

To date, the known receptors for peptide hormones are limited to LRR receptor kinases and cysteine-rich receptor kinases. However, other types of receptors may be found, since only a small number of plant peptide hormones and their receptors have been identified to date. Over 100 putative receptor kinases are found in the Arabidopsis genome, and it is likely that many of these receptors interact with peptide ligands that are yet to be discovered. Peptide signaling is a new frontier for research on plant signaling systems. Such knowledge holds promise to reveal fundamental information about the biology and evolution of plants and for the application of this knowledge to agriculture and medicine.

References

1. Banting FG, Best BA (1922) The internal secretion of the pancreas. J. Lab. Clin. Med. 7: 251-266
2. Bergey D, Howe G, Ryan CA (1996) Polypeptide signaling for plant defensive genes exhibits analogies to defense signaling in animals. Proc. Natl. Acad. Sci. (USA) 93: 12053-12058
3. Brand U, Hobe M, Simon R (2001) Functional domains in plant shoot meristems. Bioessays 23: 131-141
4. Casson SA, Chilley PM, Topping JF, Evans IM, Souter MA, Lindsey K (2002) The POLARIS gene of Arabidopsis encodes a predicted peptide required for correct root growth and leaf vascular patterning. Plant Cell 14: 1705-1721
5. Clark SE (1997) Organ formation at the vegetative shoot meristem. Plant Cell 9: 1067-1076
6. Clark SE (2001) Cell signaling at the shoot merirstem. Nature Rev. Mol. Cell Biol. 2: 276-284
7. Clark SE, Williams RW, Meyerowitz EM (1997) The CLAVATA1 gene encodes a putative receptor kinase that controls shoot and floral meristem size in Arabidopsis. Cell 89: 575-585
8. Clark SE, Running MP, Meyerowitz EM (1995) CLAVATA3 is a specific regulator of shoot and floral meristem development affecting the same processes as CLAVATA1. Development 122: 1567-1575
9. Cock JM, McCormick S (2001). A large family of genes that share homology with CLAVATA3. Plant Physiol. 126: 939-942
10. Douglass J, Civelli O, Herbert E (1984) Polyprotein gene expression: Generation of diversity of neuroendocrine peptides. Annu. Rev. Biochem. 53: 665-715
11. Fletcher JC, Brand U, Running MP, Simon R, Meyerowitz EM (1999) Signaling of cell fate decisions by CLAVATA3 in Arabidopsis shoot meristems. Science 283: 1911-1914
12. Franklin-Tong VE, Franklin, FCH (2000) Self-incompatibility in Brassica: The elusive pollen S gene is identified. Plant Cell 12: 305-308
13. Geurts R, Bisseling T (2002) *Rhizobium* nod factor perception and signaling. Plant Cell 14: S239-S249
14. Gu T, Mazzurco M, Sulaman W, Matias DD, Goring DR (1998) Binding of an arm repeat protein to the kinase domain of the S-locus receptor kinase. Proc. Natl. Acad. Sci. USA 95: 382-387
15. Harris RB (1989) Processing of pro-hormone precursor proteins. Arch. Biochem. Biophys. 275: 315-333
16. Haruta M and Constabel CP (2003) Rapid alkalinization factors in poplar cell cultures. Peptide isolation, cDNA cloning, and differential expression in leaves and methyl jasmonate-treated cells. Plant Physiol. 131: 814-823

17. Jeong S, Trotochaud AE, Clark SE (1999) The Arabidopsis CLAVATA2 gene encodes a receptor-like protein required for the stability of the CLAVATA1 receptor-like kinase. Plant Cell 11: 1925-1935
18. Kachroo A, Nasrallah ME, Nasrallah JB (2002) Self-incompatibility in the *Brassicaceae*: Receptor-ligand signaling and cell-to-cell communication. Plant Cell 14: S227-S238
19. Kende H, Zeevaart JAD (1997) The five "classical" hormones. Plant Cell 9:1197-1210
20. Mandava NB (1988) Plant growth-promoting brassinosteroids. Annu. Rev. Plant Physiol. Mol. Biol. 39, 23-52
21. Massague J, Pandiella A (1993) Membrane-anchored growth factors. Annu. Rev. Biochem. 62: 515-541
22. Matsubayashi Y, Ohawa M, Morita A, Sakagami Y (2002) An LRR receptor kinase involved in perception of a peptide plant hormone, phytosulfokine. Science 296: 1470-1472
23. Matsubayashi Y, Sakagami Y (1996) Phytosulfokine, sulfated peptides that induce the proliferation of single mesophyll cells of *Asparagus officinalis* L. Proc. Natl. Acad. Sci. USA 93: 7623-7627
24. Matsubayashi Y, Takagi L, Omura N, Morita A, Sakagami Y (1999) The endogenous mitogenic peptide, phytosulfokine-α stimulates the tracheary element differentiation of isolated mesophyll cells of *Zinnia*. Plant Physiol 120: 1043-1048
25. Meyerowitz EM (1997) Genetic control of cell division patterns in developing plants. Cell 88: 299-308
26. McGurl B, Pearce G, Orozco-Cardenas M, Ryan CA (1992) Structure, expression and antisense inhibition of the systemin precursor gene. Science 255: 1570-1573
27. Montoya T, Nomura T, Farrar K, Kaneta T, Yokota T, Bishop GJ (2002) Cloning the tomato *curl3* gene highlights the putative dual role of the leucine-rich receptor kinase tBRI1/SR160 in plant steroid hormone and peptide hormone signaling. Plant Cell 14: 3163-3176
28. Nakajima K, Benfey PN (2002) Signaling in and out: Control of cell division and differentiation in the shoot and root. Plant Cell 14: S265-S276
29. Narvaez-Vasquez J, Ryan CA The cellular localization of prosystemin: A functional role for phloem parenchyma in systemic wound signaling. In Review
30. Nasrallah JB, Kao TH, Chen CH, Goldberg ML, Nasrallah ME. (1987) Amino acid sequence of glycoproteins encoded by three alleles of the S locus of *Brassica oleracea*. Nature 326: 617-619
31. Pearce G, Moura DS, Stratmann J, and Ryan CA (2001) Production of multiple plant hormones from a single polyprotein precursor. Nature 411: 817-820
32. Pearce G, Moura DS, Stratmann J, Ryan CA (2001) RALF, a 49 amino acid polypeptide signal arrests root growth and development. Proc. Natl. Acad. Sci. USA 98: 12843-12847
33. Pearce G, Ryan CA (2003) Systemic signaling in tomato plants for defense against herbivores: A sinle precursor contains thee defense-signaling pepides J. Biol. Chem. In press
34. Pearce G, Strydom D, Johnson S, Ryan CA (1991) A polypeptide from tomato leaves activates the expression of proteinase inhibitor genes. Science 253: 895-897
35. Röhrig H, Schmidt J, Miklashevichs E, Schell J, Hohn M (2002) Soybean ENOD40 encodes two peptides that bind to sucrose synthase. Proc. Natl. Acad. Sci. USA 99: 1915-1920
36. Rojo E, Sharma VK, Kovleva V, Raikel NV, Fletcher JC (2002) CLV3 is localized to the extracellular space, where it activates the Arabidopsis CLAVATA stem cell signaling pathway. Plant Cell, 14: 969-977
37. Ryan CA (2000) The systemin signaling pathway: differential activation of plant defensive genes. Biochem. Biophys. Acta. 1477: 112-121
38. Ryan CA, Pearce G, Moura DS, Scheer JM (2002) Polypeptide hormones. Plant Cell 14: S251-S264

39. Scheer JM, Ryan CA (1999) A 160 kDa systemin cell surface receptor on *Lycopersicon peruvanium* cultured cells. The Plant Cell 11: 1525-1535
40. Scheer JM, Ryan CA (2002) The systemin receptor SR160 from *Lycopersicon esculentum* is a member of the LRR receptor kinase family. Proc. Natl. Acad. Sci. in press.
41. Schopfer CR, Nasrallah ME, Nasrallah JB (1999) The male determinant of self-incompatabilidty in *Brassica*. Science 286: 1697-1700
42. Staehelin C, Charon, C, Boller T, Crespi M, Konderosi A (2001) *Medicago truncatula* plants overexpressing the early nodulin gene ENOD40 exhibit accelerated mycorrhizal colonization and enhanced formation of arbuscles. Proc. Natl. Acad. Sci. USA 98: 15366-15371
43. Takayama S, Shiba H, Iwano M, Shimasato H, Che, F-S, Kai N, Watanabe M, Suzuki KG, Hinata K, Isogai A (2000) The pollen determinant of self-incompatability in *Brassica campestris*. Proc. Natl. Acad. Sci. USA 97: 1920-1925
44. Toumadje A, Johnson WC, Jr. (1995) Systemin has the characteristics of a poly(L-proline) II type helix. J. Am. Chem. Soc. 117: 7023-7024
45. Trotochaud AE, Hao T, Wu G, Yang S, Clark SE (1999). The CLAVATA1 receptor-like kinase requires CLAVATA3 for its assembly into a signaling complex that includes KAPP and a Rho-related protein. Plant Cell 11, 393-405
46. van de Sande K, Pawlowski K, Czaja I, Wieneke U, Schell J, Schmidt J, Walden R, Matvienko M, Wellink J, van Kammen A, Frannsen H, Bisseling T (1996) Modification of phytohormone response by a peptide encoded by ENOD40 of legumes and a nonlegume. Science 273: 370-373
47. Wen J. Lease KA, Walker JC (2003) The DVL family of small polypeptides regulates Arabidopsis development. Proc. Natl. Acad. Sci. USA in review
48. Yamazaki T, Takaoka M, Katoh E, Hanada K, Sakita M, Sakata K, Nishiuchi Y, Hirano H. (2003) A possible physiological function and the tertiary structure of a 4-kDa peptide in legumes. Eur. J. Biochem. 270: 1269-1276
49. Yang H, Matsubayashi Y, Hanai H, Sakagami Y (2000) Phytosulfokine-□, a peptide growth factor found in higher plants: Its structure, functions, precursor and receptors. Plant Cell Physiol. 41: 825-830
50. Yang H, Matsubayashi Y, Nakamura K, Sakagami Y (1999) *Oryza sativa* PSK gene encodes a precursor of phytosulfokine-□, a sulfated peptide growth factor found in plants. Proc. Natl. Acad. Sci. USA 96: 13560-13565
51. Yang H, Matsubayashi Y, Nakamura K, Sakagami Y (2001) Diversity of Arabidopsis genes encoding precursors for phytosulfokines, a peptide growth factor. Plant Physiol. 127: 842-851

G. HORMONE ANALYSIS

G1. Methods of Plant Hormone Analysis

Karin Ljung, Göran Sandberg and Thomas Moritz
Umeå Plant Science Centre, Department of Forest Genetics and Plant Physiology, Swedish University of Agricultural Sciences, SE-901 83 Umeå, Sweden. E-mail: Karin.Ljung@genfys.slu.se

INTRODUCTION

The development of sensitive analytical methods[1] for determining hormone levels in plant tissues is essential for elucidating the role and function of plant hormones in growth and development. During the last decade the trend has been to use mass spectrometry as one of the principal tools in plant hormone analysis, thereby increasing the quality of the analyses dramatically. The development of user-friendly bench-top mass spectrometers has revolutionized analytical chemistry, enabling many laboratories to switch from fairly unspecific bioassays and immunoassays to a methodology that can be both sensitive and accurate. Analysis of plant hormones is an essential component of studies of plant development, so the demands made on the methods used will often be linked to the questions asked by developmental biologists. Clearly, many of the mechanisms controlling plant development operate in specific tissues, in many cases at the cellular level. It is within this context that the analytical technology of the future has to be developed. Cellular markers for hormone concentrations have aroused a

---

[1] Abbreviations: ACC, 1-aminocyclopropane-1-carboxylic acid; AdoMet, S-adenosylmethionine; APCI, atmospheric pressure chemical ionization; ARFs, auxin-response transcription factors; C_{18}, 18-carbon straight-chain hydrocarbon sorbent; CI, chemical ionization; D_2O, deuterated water; DEA, diethylaminopropyl sorbent; DEDTCA, diethyldithiocarbamic acid; DPM, disintegrations per minute; EI, electron ionization; ESI, electrospray ionization; FID, flame ionization detection; frit-FAB, frit fast atom bombardment; GC, gas chromatography; GFP, green fluorescent protein; GUS, β-glucuronidase; HPLC, high performance liquid chromatography; IS, internal standard; LC, liquid chromatography; MCX, mixed-mode polymeric C_{18}/SCX sorbent; Me, methylation; MRM, multiple reaction monitoring; MS, mass spectrometry; MS-MS, tandem mass spectrometry; m/z, mass to charge; NCI, negative chemical ionization, NMR, nuclear magnetic resonance spectroscopy; PFB, pentafluorobenzylation; QTOF, quadrupole time-of-flight; RC, radioactive counting; SAX, trimethylaminopropyl sorbent; SCX, propylbenzenesulfonyl sorbent; Si , unbonded, activated silica sorbent; SPE, solid phase extraction; SRM, selected reaction monitoring; TMS, trimethylsilylation.

great deal of interest lately. A combination of this technique and mass spectrometry for verification and turnover studies is likely to be the most potent analytical approach in the future. However, cellular marker systems have not yet developed to a level where they can be used as general tools to monitor hormone concentrations. The attraction of this technology is obvious, but in its development we should critically evaluate the emerging methodologies as rigorously as we have scrutinised current techniques!

Plant extracts are complex, multi-component mixtures. The major problem associated with plant hormone analysis is that the amount of hormones is very low, usually in the range of 0.1 to 50 ng g^{-1} fresh weight. As very large numbers of components are found in plant extracts at low concentrations, many compounds can interfere with hormone analyses: many more than when compounds present at high concentrations are being analysed (13). Therefore, accurate and precise analyses can only be ensured by thorough knowledge of the analytical principles involved. Even with the use of modern mass spectrometers a number of errors can be made, leading to inaccurate and imprecise quantifications of plant hormones. The aim of this chapter is to explain the principles of hormone analysis, with the focus on extraction, purification and final analysis by mass spectrometry.

EXTRACTION AND PURIFICATION

In order to identify and quantify plant hormones the plant tissue must first be homogenized and extracted with a suitable solvent. Interfering substances (e.g. proteins, carbohydrates, pigments and lipids) then have to be removed from the crude extract to obtain a sufficiently pure sample for the final analysis. The choice of extraction and purification method depends not only on the analyte, but also on the type of analysis to be performed and the analytical equipment available.

Homogenisation of large amounts of plant material can be performed either by grinding fresh or frozen plant material in cold extraction medium using knife homogenisers, or by grinding "dry" frozen plant material with a mortar and pestle, then adding an appropriate solvent to the ground material. Small amounts of plant material can also be homogenised directly in plastic vials using a pestle mounted on an electric drill, or with ball grinders. An important consideration, whichever method is used, is that the plant material must be kept cold during homogenisation to avoid enzymatic induction of metabolic changes or chemical degradation of the compounds of interest. The extraction procedure is an important, but sometimes insufficiently carefully considered step in the analytical method. Addition of internal standards (compounds labelled with appropriate radioactive or stable isotopes) to the extract makes it possible to compensate for analyte losses during the sample purification procedure, but they cannot be used to monitor extraction efficiency. The solvent used has to extract the analyte efficiently without removing too much interfering substances from the tissue. Solvents such as methanol, methanol/water mixtures or neutral pH buffers are often used for extraction of plant hormones. The extraction period must be long enough to extract most of the analyte into the medium and to allow for isotope equilibrium between the endogenous compound and the added internal standard. However, if the extraction periods used are too long, the risks of the analyte(s) breaking down, or of hormone conjugates hydrolysing increase, thereby liberating free hormone into the medium. The risk of enzymatic or chemical breakdown of the analyte can be minimised by performing the extraction at low temperatures, e.g. +4°C (or even lower for cytokinins) with suitable solvents. Antioxidants such as DEDTCA (diethyldithiocarbamic acid) can be added to the medium to prevent non-enzymatic breakdown of certain other compounds, e.g. indole metabolites to IAA (46).

Specific purification methods must usually be developed, depending on the type of final analysis to be performed (quantitative or qualitative). The sensitivity of modern plant hormone analysis allows the quantification of many plant hormones and their metabolites in very small amounts of tissue (35). For definitive qualitative analysis (the identification of putative plant hormone metabolites), the amount of plant material extracted has to be much larger: from several grams to kilograms of tissue. Purification of plant extracts for identification of unknown metabolites is usually done by a combination of solvent partitioning and purification on silica or polymer-based extraction columns (SPE-columns). This removes the bulk of interfering substances from the extract. The extract usually needs to be purified further by immunoaffinity chromatography or preparative HPLC (high performance liquid chromatography) before final analysis by mass spectrometry. An extensive overview of extraction and purification methods developed for the major classes of plant hormones can be found in different textbooks (30, 44). The degree of purification needed is dependent, to a large degree, on the physico-chemical detection method to be employed, i.e. the analytical equipment available (see the identification of plant hormone metabolites section, below).

Known plant hormones and their metabolites can be quantified in much lower amounts of plant material than needed for qualitative analysis. The development of highly selective and very sensitive methods of analysis using combined gas chromatography or liquid chromatography tandem mass spectrometry (GC/MS-MS or LC/MS-MS) has made it possible for some time now to quantify these substances in mg amounts of plant tissue, or even less (18, 34). This means that homogenisation, extraction and purification methods can be miniaturised to a much higher degree than before. Homogenisation and extraction can be performed directly in small, disposable plastic vials, and purification methods such as liquid/liquid extraction and HPLC can often be replaced by solid phase extraction (SPE) methods. SPE-columns are manufactured by many companies, and a variety of columns are available with silica-, carbon- or polymer- based packing material. Examples of silica-based products are C_{18}, SAX, SCX and Si columns. The analyte is bound to the column by chemical interaction with the solid phase, and interfering substances are removed by washing the column with a suitable solvent. The analyte is then eluted from the column using a stronger solvent. The most common retention mechanisms in SPE are based on Van der Waals forces (non-polar interactions), hydrogen bonding, dipole-dipole forces (polar interactions) and cation-anion interactions (ionic interactions). An efficient clean-up method can only be developed if different separating mechanisms are used in each of the purification steps. This will give the highest recovery of the analyte, combined with the most efficient reduction in interfering compounds. A combination of SPE columns of differing selectivity is often used to enhance the purification. Mixed-mode SPE columns have also been developed for specific groups of compounds. Sample purification methods based on SPE are very well suited to the use of purification robots to speed up sample preparation, leading to higher throughput in the analyses.

To monitor analyte recovery during purification, radioactively labelled standards can be added to the extract after homogenisation of the tissue. The recovery of the analyte can then be calculated by measuring the ratio of DPM (disintegrations per minute) added to the extract to the DPM recovered after purification, using liquid scintillation counting. In trace analysis of plant hormones, the amount of "cold" hormone in the added radioactive tracer can be so great that it interferes with the final analysis, making it impossible to obtain an accurate quantification. This can usually be avoided by the use of radioactive standards with very high specific activity (e.g. ^{3}H-labelled compounds), but it is very important to check the purity of all standards, labelled with either radioactive or stable isotopes. The amount of native hormone originating from the added standard can then be measured, and the derived value corrected accordingly.

IDENTIFICATION OF PLANT HORMONES

Identification of plant hormones has two major purposes. One is to identify new hormones and their metabolites. The other is to confirm the occurrence of a specific substance in the tissue of interest. Identification of organic compounds requires a physico-chemical detector that can conclusively distinguish structurally similar compounds from each other. The two most common techniques for identification are nuclear magnetic resonance spectroscopy (NMR) and mass spectrometry (MS). Although NMR is extremely useful for identification it has relatively poor sensitivity and it cannot be readily used to analyse complex mixtures of samples. These problems are especially pronounced when analysing plant hormones, which generally occur in very low concentrations. The revolutionizing development of mass spectrometry in the last 10-15 years has greatly simplified the identification of organic compounds, and therefore mass spectrometry has become the analytical method of choice for the identification and quantification of plant hormones. A mass spectrometer is an instrument in which molecules are introduced into an ion source and

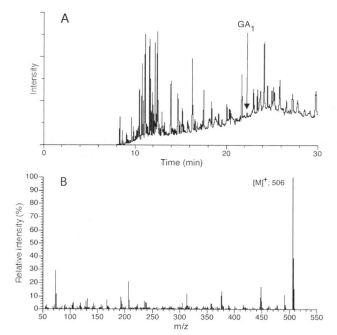

Figure 1. Identification of GA_1 from an extract of hybrid aspen by GC/MS. The purified extract was analysed following derivatisation with diazomethane and TMS (trimethylsilylation). (A) total ion chromatogram (TIC) with peak corresponding to GA_1-Me-TMS indicated by an arrow. (B) the mass spectrum and retention index obtained for the putative GA_1 were compared with corresponding data for a GA_1-Me-TMS standard.

ionized by different kinds of energies, depending on the type of instrument. The masses of the ions formed are then detected and the resulting mass spectrum provides a fingerprint of the compound. Samples can be introduced in either liquid phase, as in LC/MS; ESI (electrospray ionization), APCI (atmospheric pressure chemical ionization) or frit-FAB (frit fast atom bombardment), or alternatively in gas phase, as in GC/MS; EI (electron ionization), or CI (chemical ionization). The mass spectrum obtained (Fig. 1) is compared with a spectrum from a standard. Alternatively, if no standard is available the identity of the compound can be obtained by interpretation of the mass spectra. The latter approach demands an experienced mass spectroscopist and it is still usually necessary to synthesize the compound chemically for conclusive identification. Tandem mass spectrometry (MS-MS) improves the selectivity of the analysis, since the first MS-analyser removes most of the potentially interfering compounds by selecting a parent ion derived from the compound of interest. The second mass analyser then detects ions formed by the fragmentation of the parent ion. Examples of tandem mass spectrometers are triple quadrupole, ion trap, quadrupole time-of-flight (QTOF) and other hybrid instruments. The principles and applications of mass spectrometry are described in various textbooks (9, 27).

Although mass spectrometry is the method of choice for identification of plant hormones it must be emphasized that there are occasions when NMR must be used to obtain complete structural information, e.g. for determining sugar positions in hormone conjugate identification (38).

METABOLIC STUDIES

The pool size of a specific plant hormone is influenced by diverse metabolic processes, such as *de novo* synthesis, transport, conjugation, catabolism and compartmentation. All these processes help maintain the hormone content in plant cells at optimal levels for growth and development. It is not possible to estimate the relative contribution to the total pool of these processes merely by measuring hormone concentrations. Methods involving isotope labelling in combination with mass spectrometry makes it possible to investigate the processes involved, and these methods have greatly increased our knowledge of both hormone synthesis pathways and the rates of hormone synthesis and turnover. The synthesis of catabolic end products, reversible and irreversible conjugation and transport can also be investigated using these methods.

With the use of stable isotope feedings it is possible to measure synthesis and turnover rates for different compounds, giving an estimate of metabolic fluxes (11, 45). The analytical methods used for these types of measurements are quite similar to ordinary quantification procedures. If the synthesis pathway(s) for a specific substance is known, feedings can be performed with identified precursors. The precursors are labelled with "heavy" isotopes such as ^2H, ^{13}C, ^{15}N or ^{18}O. Incorporation of label from the precursor into the hormone can be monitored by mass spectrometry, and relative synthesis rates calculated (37, 41). This method has also been used to establish precursor-product relationships in specific synthesis pathways (2). Feedings are usually performed as "pulse-labelling" experiments, and the method of feeding has to be adapted to the chemical properties of the precursor and the plant species under study. The feeding can be performed with plants growing in liquid culture, and in this case the label will enter all tissues of the plant simultaneously. Substances of interest can also be fed to plants through the root system, injected into specific tissues of the plant, or administered through the surface of appropriate organs, e.g. leaves. The concentration of the precursor in the tissue should be in the same range as the endogenous concentration, otherwise the normal synthesis pathway(s) may be perturbed. Potential sources of error in theses studies can also arise from differences in uptake of the labelled compound between tissues or from differences in compartmentation of the labelled and endogenous compound.

In cases where the synthesis pathways are not known, or where there is redundancy in the synthesis pathways, feedings can be conducted with general precursors such as deuterium oxide (deuterated water) or $^{13}CO_2$. The degree of incorporation into different putative precursors and the hormone itself can then be measured and synthesis rates calculated (Fig. 2) (3, 32).

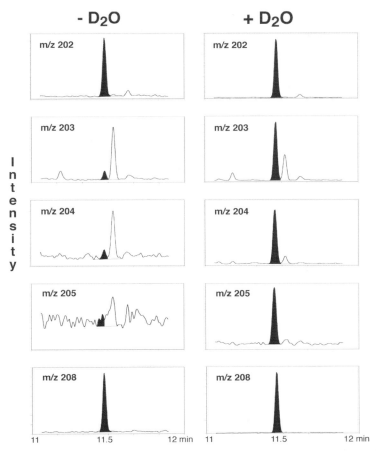

Figure 2. Determination of IAA biosynthesis rates by GC/MS. Ion chromatograms from Arabidopsis control samples (-D$_2$O) and samples from seedlings fed with deuterated water (+D$_2$O), showing an increase in the abundance of the isotopomers m/z 203-205 after *de novo* IAA synthesis. The base peak of methylated and trimethylsilylated IAA (m/z 202-205) and ^{13}C$_6$-IAA (m/z 208) were used as diagnostic ions.

Feeding with labelled isotopes of the plant hormone itself can be conducted in order to measure "turnover rates" and metabolic products of the hormone, giving estimates of the rates of conjugation and catabolic events in the plant (19, 31). Feeding experiments with radioactive (usually ^{14}C or ^{3}H labelled) precursors is a powerful technique for detecting unknown hormone metabolites. With the help of HPLC-RC (radioactive counting), putative metabolites that are formed from the fed precursor can be discovered, and later identified by mass spectrometry (31, 39). Using this technique metabolic profiles from interesting materials, such as different mutants, can be compared, giving valuable information on differences in hormone synthesis, catabolism and conjugation (5). Another approach for identifying unknown metabolites of specific plant hormones, which does not

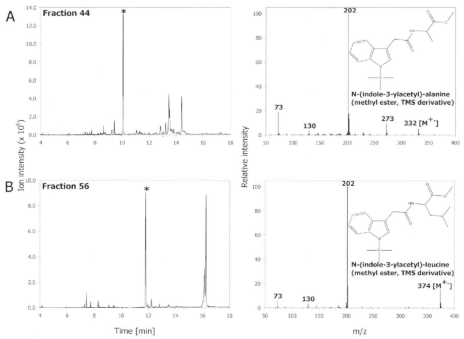

Figure 3. Ion chromatograms for m/z 202 (left) and mass spectra (right) of endogenous amide conjugates from Arabidopsis identified during a screen for indolic compounds in vegetative tissues. (A) N-(indole-3-acetyl)-α-Ala from HPLC fraction 44. (B) N-(indole-3-acetyl)-α-Leu from HPLC fraction 56: the asterisk on the ion chromatogram indicates the peak from which the mass spectrum was taken. (From 29)

involve feeding labelled compounds, is to use full scan or precursor ion scan GC/MS or LC/MS analysis, looking for characteristic fragment ions. This approach has been used to screen for IAA metabolites via preparative HPLC fractionation and GC/MS analysis (Fig. 3) (29).

QUANTIFICATION OF PLANT HORMONES

The most widely used method for quantification of plant hormones today is isotope dilution analysis by mass spectrometry. Correctly used, this method gives quantitative results of high accuracy and precision. A known amount of an analogue of the analyte of interest labelled with a stable isotope (an "internal standard", or IS) is added to the sample as early in the purification process as possible. During the purification procedure, the ratio of unlabelled (endogenous) to labelled analyte should remain unchanged, irrespective of sample losses. The intensity of specific ions formed from the unlabelled and labelled compound can then be measured by mass spectrometry, and the ratio between the ion intensities can be used to estimate the concentration of the endogenous compound in the sample (Fig. 4).

Amount of endogenous compound in sample =

$$\frac{\text{Peak area of endogenous ion}}{\text{Peak area of internal standard ion}} \times \text{Amount of internal standard added}$$

Internal standards are often labelled with deuterium (^{2}H) or ^{13}C atoms, but sometimes ^{15}N or ^{18}O labelled compounds are available. These heavy-labelled compounds can be easily distinguished from the endogenous compound in the mass spectrometer due to the difference in masses. The accepted analytical procedure is to produce a standard curve by analysing standards containing known concentrations of unlabelled and labelled analytes. The standards should cover the normal concentration range of the analyte in the samples. A linear (or near-linear) standard curve can be produced if the IS is labelled with three or more deuterium or ^{13}C atoms per molecule (Fig. 4b). If there are fewer labelled atoms in the IS, corrections have to be made for the abundance of natural isotopes in the spectrum of the endogenous compound, since these ions will contribute to the measured area of the IS. Alternatively, a non-linear curve can be fitted and used for quantification. Examples of internal standards used in quantitative analysis of plant hormones are listed in (26).

However, a number of specific problems are associated with trace analyses of plant hormones in very small amounts of plant tissue. Clearly, it is important to avoid losses of the analyte during sample purification. Such losses

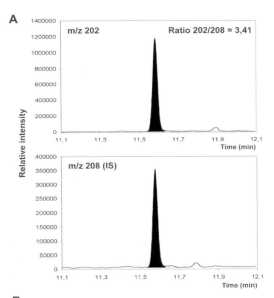

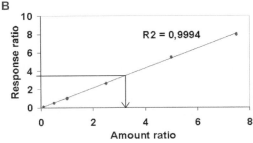

Figure 4. Isotope dilution mass spectrometry quantification of IAA. (A) ion chromatograms from the base peak of methylated and trimethylsilylated IAA (m/z 202) and ^{13}C$_6$-IAA (m/z 208). (B) the ratio between the areas for m/z 202 and m/z 208 (response ratio) is inserted into the calibration curve to calculate the content of IAA in the sample.

could arise either from low recovery of the analyte in the different purification steps, or from adsorption of the analyte to vials, pipettes or other equipment used in the cleaning up procedure. It is also important to analyse blank samples (samples without plant material, purified in exactly the same way as ordinary samples) to monitor possible contamination of solvents, glassware or other laboratory equipment. In addition, drying of the sample for prolonged times and at high temperatures can cause large losses of the analyte and should be avoided (43). The amount of purification required depends on the type of mass spectrometric detector to be used. If the final analysis is to be done by tandem mass spectrometry (MS-MS), extensive purification of the sample is usually not necessary, due to the high selectivity of the MS-MS analysis. The number of purification steps can thus be minimised, improving the recovery of the analyte. However, it must be emphasized that analysis of non-purified extracts can result in suppression of the ionization in the ion source of the mass spectrometer, thereby reducing sensitivity or reproducibility. With the method called "successive approximation" (42) it is possible to determine if a purification procedure is sufficient to give an accurate estimation of the analyte concentration in the sample. If the quantification is accurate, further purification of the sample should not alter the estimated value.

The final separation of the sample before it is introduced into the mass spectrometer is usually performed either by capillary gas chromatography (GC) or by high performance liquid chromatography (HPLC). IAA, ABA and gibberellins are usually analysed by GC/MS, but LC/MS methods are preferred for analysis of cytokinins and other substances that are very polar or thermo-labile (35, 40).

For a long time GC/MS has been the most widely used hyphenated technique in mass spectrometry. It is robust, sensitive, reproducible and can be easily automated. Derivatisation of plant hormones prior to GC/MS analysis injection is usually needed to increase the volatility of the analyte (Table 1). However, derivatisation prior to LC/MS analysis can also be advantageous since, for instance, it can increase the selectivity of the analysis by giving a useful fragmentation pattern with strong diagnostic ions (1). The

Table 1. Methods of derivatisation and MS-analysis for the major plant hormones
The methods shown are either the most commonly used or the most sensitive reported for the different classes of hormone. (More information can be found in 8, 30, 44).

Hormone class	Derivatisation	MS-analysis	References
IAA	Me + TMS; Me	GC/MS	18; 43
Cytokinins	Propionylation	LC/MS	1
GAs	Me + TMS	GC/MS	34
ACC	Propylation + benzylation; PFB	GC/MS; GC/NCI-MS	25; 49
ABA	Me; PFB	GC/MS; GC/NCI-MS	44; 17
Brassinosteroids	Methaneboronation + TMS	GC/MS	48

increase in analyte mass after derivatisation usually improves the signal to noise ratio in the analysis of low molecular weight compounds such as 1-aminocyclopropane-1-carboxylic acid (ACC). Derivatisation can also improve peak shape in LC-MS analysis, as it can reduce the polarity of the compound.

Whatever analytical method is used, it is very important to validate it for the species, as well as for both the amount and type of tissue on which the analysis is to be performed, in order to establish the precision and accuracy of the analysis. Systematic errors affect the accuracy of the analysis, i.e. the closeness of the estimated value to the true value. The precision of the analysis, on the other hand, is affected by random errors due to biological variation in the plant material, sampling and weighing, the purification processes and the instrumental analysis. The magnitude and significance of random errors can be estimated by making repeated measurements and by applying appropriate statistical tests. Readers are referred to suitable textbooks (33) for more information.

Auxins

GC/MS is the most commonly used method for IAA analysis. This hormone is usually methylated and often also trimethylsilylated in order to increase its volatility. Such derivatisation also greatly improves its chromatographic behaviour and the sensitivity of the analysis. The sensitivity of ordinary bench-top single mass spectrometers is not usually high enough to allow analysis of IAA in low milligram amounts of plant tissues. Extensive purification of the extract, optimisation of analyte recovery and modification of the mass spectrometer are needed to ensure acceptable sensitivity (43). However, the need for extensive purification can be avoided by the use of tandem mass spectrometry, or high-resolution mass spectrometry: both of these techniques offer much higher selectivity for the analyte(s) of interest.

We have developed a method (18) that allows IAA to be quantified in milligram amounts of plant tissue with relatively little purification using a double focusing magnetic sector instrument with reversed geometry allowing selected reaction monitoring (SRM) GC/MS. This scanning technique involves the detection of daughter ions originating from specific metastable parent ions, giving extremely high sensitivity and selectivity. The method has been further improved since the cited study was published, and currently allows quantitative analysis in samples down to 0.05-0.1 mg. The usefulness of the method has been proven for analysis of tissue-specific IAA distribution in Arabidopsis seedlings (32) and for measurements of IAA content and IAA turnover in one-millimetre sections of Arabidopsis seedling roots (21, 50). In fact, the technology has now reached a stage at which contamination and sample preparation are equal or sometimes even more limiting than the sensitivity of the mass spectrometer!

Tandem mass spectrometry can also be performed with other types of mass spectrometers, such as ion trap MS-MS instruments or triple

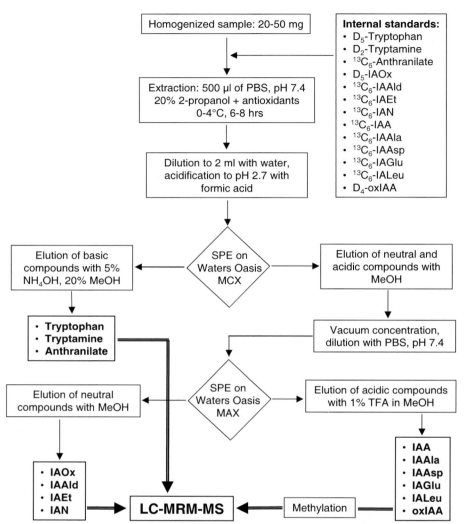

Figure 5. Purification protocol for known precursors, conjugates and catabolites of IAA (Kowalczyk, unpublished). The purification is done by solid phase extraction, separating the metabolites into basic, neutral and acidic compounds. Final analysis is by tandem mass spectrometry.

quadrupoles. For instance, a method for analysis of IAA and other hormones has been described using an ion trap MS-MS instrument (36), but it requires 20-200 mg of plant material for quantification. Methods using LC-electrospray tandem mass spectrometry for analysis of IAA and IAA metabolites have also been developed (Figs. 5, 6) (29, 40). These methods show both very high sensitivity and selectivity and are therefore very useful for quantitative work.

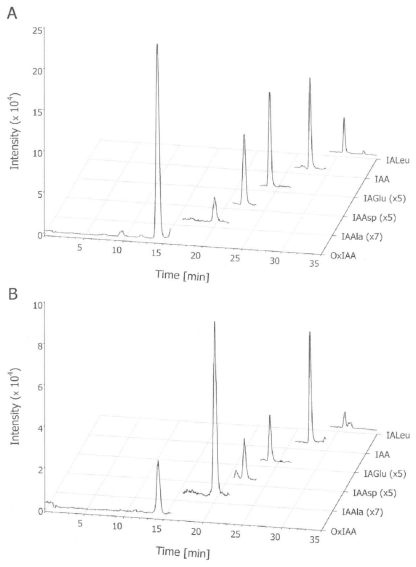

Figure 6. LC/MRM-MS measurements of IAA, OxIAA and IAA conjugates in 100 mg Arabidopsis tissue. Endogenous metabolites of IAA in roots (A) and in the 3rd and 4th leaves (B) 10 d after germination (DAG). Traces for IAAla, IAAsp, and IAGlu are magnified to allow the same scaling for all chromatograms. (From 29)

Cytokinins

Analysis of cytokinins is associated with a number of specific problems related to the chemical nature of the compounds. For instance, in order to determine the amount of cytokinin nucleotides present in a sample

Methods of plant hormone analysis

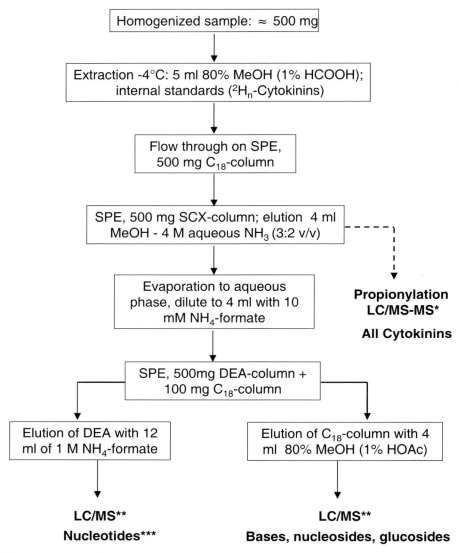

Figure 7. Examples of purification protocols for cytokinins. *With a LC/MS-MS instrument, cytokinins can be analysed without extensive purification after derivatisation. **Single LC/MS can be used in SIM mode, but tandem mass spectrometry is preferred. For LC/MS, extracts should be purified by immunoaffinity chromatography. ***The usual strategy is to convert nucleotides to corresponding nucleosides by alkaline phosphatase treatment prior to LC/MS analysis.

accurately, care must be taken during extraction, as phosphatases may be activated during the extraction procedure. A common method is to extract the sample in Bielesky buffer (60% MeOH, 25% $CHCl_3$, 10% HCOOH and 5% H_2O) at $-20°C$ (7). Although other buffers, such as 80% MeOH can be used, it is important to add a suitable internal standard that can be used to

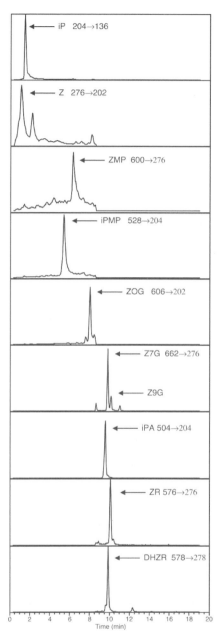

Figure 8. Quantification of endogenous cytokinins in 100 mg Arabidopsis tissue (36a). The extract was derivatised by propionylation prior to LC/MS-MS analysis. MRM parent/daughter ion transitions are indicated on the chromatograms.

monitor the possible degradation of cytokinin nucleotides, or convert the nucleotides to corresponding cytokinin ribosides, before measuring the combined nucleotide/riboside content. Extracts can be further purified using strong cation ion exchange SPE (solid phase extraction) columns, e.g. SPE-SCX, and anion chromatography (SPE-DEA; SPE-SAX). It is then possible to separate nucleotides from bases and ribosides (Fig. 7). A protocol involving an SPE column with both C_{18} and SCX material (MCX) has been used for separating the different cytokinins (16). Immunoaffinity column purification has also been used extensively in cytokinin analysis (20). Although immunoaffinity chromatography is slow, the advantage is that the extracts become very clean, thus simplifying the final detection by mass spectrometry.

There are several methods for derivatising cytokinins for GC/MS analysis (35), but recently LC/MS has been the preferred approach for cytokinin analysis. A number of publications have described cytokinin analyses by frit-FAB LC/MS: a technique that can give high sensitivity (1). However, it is also complex and today electrospray ionisation (ESI) LC/MS is the analytical tool of choice for cytokinin analysis (40). Although the structure of cytokinins is suitable for analysis by ESI, further improvement can be achieved by derivatisation, which not only increases the absolute ESI ionisation, but also improves the chromatographic properties of the cytokinins. With derivatisation it is possible to separate all of the major cytokinins, including nucleotides, and achieve low femtomol detection limits (Fig. 8).

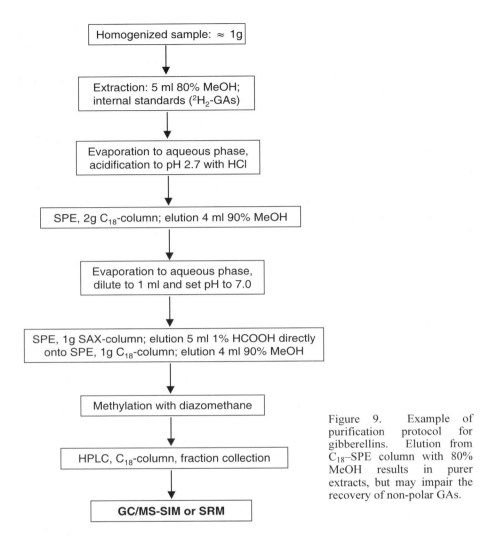

Figure 9. Example of purification protocol for gibberellins. Elution from C_{18}–SPE column with 80% MeOH results in purer extracts, but may impair the recovery of non-polar GAs.

Gibberellins

Gibberellins are primarily analysed as methyl ester trimethylsilyl-ether derivatives by GC/MS[2]. As the number of GAs in each tissue can be very high, it is very important to optimise the GC-system for separating both isomeric GAs, e.g. GA_1 and GA_{29}, and GAs with common mass spectra fragmentation patterns. This is especially important when different GAs are not separated prior to GC/MS analysis, as co-eluting GAs can cause inaccuracy. Although there are a few examples of GAs being quantified in small amounts of tissue without extensive purification using highly selective SRM detection (28, 34) it is important to note that in many plant materials,

---

[2]Mass spectra for all identified GAs can be found at www.plant-hormones.info/ga1info.htm

e.g. Arabidopsis leaves, there are a number of structurally related compounds that can interfere with the GC/MS analysis of GAs. Therefore, more extensive purification by HPLC is usually necessary prior to GC/MS analysis than in, for instance, IAA analysis. With HPLC it is possible to separate a large number of GAs, but by combining the fractions it is possible to reduce the number of GA fractions that need to be analysed. SPE and HPLC are the methods of choice for purification, and GC/MS for final analysis (Fig. 9).

LC/MS has mainly been used for analysis of GA-conjugates (47). Although there are cases where GAs have been analysed by LC/ESI-MS-MS on a triple quadrupole instrument (10), trace analysis of GAs in specific tissues is not (as yet) possible by LC/MS. Instead, GC/SRM-MS was used in the only published study in which small amounts of plant tissue were analysed for GA content. In this case, levels of GA_1, GA_4 and GA_5 were followed in apices of *Lolium* during the long-day induction of flowering (28). This analysis demanded detection limits lower than 0.5 pg, and high recovery during extraction. The analysis was therefore performed without any HPLC purification of the extract, and to confirm the identity of the analysed GAs some extracts were spiked with non-labelled GAs. This methodology is very useful for validating methods and can be very helpful for distinguishing compounds of interest from interfering substances.

Ethylene, Abscisic Acid and Brassinosteroids

The gaseous nature of the small ethylene molecule makes it easy to measure the low amounts that evolve from whole plants or plant organs using gas chromatographs coupled to appropriate detection systems such as flame ionisation, photoionisation or photoacoustic laser detectors. Very sensitive measurements can be obtained by photoacoustic spectroscopy, but the greatest problem is to analyse ethylene in specific tissues, due to its gaseous nature. However, levels of its precursor, 1-aminocyclopropane-1-carboxylic acid (ACC) may reflect ethylene levels since it has been suggested that ACC biosynthesis from S-adenosylmethionine (AdoMet) is a rate-limiting step in ethylene biosynthesis, and ACC levels in unstressed vegetative plant tissues are low. Therefore, instead of analysing the active hormone, ACC is sometimes quantified instead. ACC has not often been analysed in a routine fashion, presumably due to the relatively complex derivatisation protocols required for its analysis by GC/MS and the rather low sensitivity of the published LC/MS methods. ACC can be determined by GC/MS using the *N*-benzoyl *n*-propyl derivative (25). However, the chemical nature of ACC can cause chromatographic problems, which can sometimes complicate GC/MS analysis. An alternative option is LC/MS. Derivatisation is needed for this, since ACC is such a small compound that it is difficult to distinguish it from background noise, resulting in poor sensitivity. An example of ACC analysis by LC/APCI-MS-MS from Arabidopsis is shown in Figure 10. Derivatisation increases the molecular weight sufficiently to ensure both chromatographic integrity and reduction of background noise in the MS-analysis.

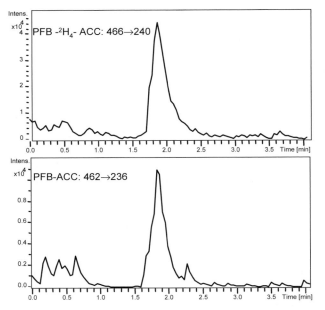

Fig. 10 Negative-APCI LC/MS-MS analysis of ACC (below), including the internal standard $^2H_4$-ACC (above), from an extract of 200 mg Arabidopsis tissue. ACC was derivatised with pentafluorobenzylbromide.

ABA is a weak acid and can therefore often be analysed together with other analytes, such as GAs or IAA, using similar or even identical purification protocols. ABA is usually analysed by GC/MS as its methyl ester derivative. Using high resolution selected ion monitoring (SIM), detection limits can be reduced to less than 10 pg, but with low resolution SIM the detection limits are higher. A sensitive method for quantifying ABA extracted from different plant species has been described, in which ABA is analysed as its pentafluorobenzyl derivative using GC/MS-SIM in negative chemical ionisation mode with methane as reagent gas (17). A method using LC/ESI-MS-MS has also recently been described (23), in which samples obtained after extraction with methanol:water (80:20) or water were injected without any further purification or derivatisation.

Brassinosteroids are a group of more than 40 different plant sterols, some of which are considered to be important growth regulating factors. The endogenous concentrations of brassinosteroids (BRs) range from 0.01-0.1 ng/g fresh weight in shoots and leaves to 1-100 ng/g fresh weight in pollen and immature seeds (12). Due to the very low amounts of the active brassinolide very few methods have been described for the analysis of BRs. Methods for purification of BRs usually involve solvent partitioning, column chromatography and preparative HPLC (15). Analysis of endogenous concentrations of BRs has usually been performed by GC/MS, in either full-scan or SIM mode, using gram amounts of plant material (22, 48). The BRs are often converted to methaneboronates or methaneboronate-trimethylsilyl ethers before GC/MS analysis. A method for analysing brassinosteroids by LC/APCI-MS has been presented, but not used in routine analysis. For tissue-specific analysis of BRs in milligram amounts of plant material, much more sensitive and selective methods of analysis have to be developed, probably using LC or GC tandem mass spectrometry.

HORMONE PROFILING

Most studies of plant hormones have dealt with only one group of hormones at a time. However, it has been known for a long time that mutual interactions amongst the hormones profoundly affect their biosynthesis, metabolism and catabolism, as well as the sensitivity of specific tissues to them. Therefore, increasing interest has arisen in analysing several hormones in the same sample. There are two different strategies for this. The first is to divide a sample into different parts, and then analyse each part for different hormones. However, this is not a good choice when the amount of plant material is limited and specific tissues are the targets for analysis. In such cases a method must be developed that enables extraction, purification and detection of all the major hormones in the same extract. A GC/MS-ion trap approach has been used for the combined analysis of IAA, jasmonate, salicylic acid and ABA in Arabidopsis (36). The GC/MS was operated in chemical ionisation multiple reaction monitoring (MRM) mode using methane as reagent gas. With 20-200 mg plant tissue, microextraction, and solid-phase extraction it was possible to analyse routinely up to 60 samples per day. Similarly, an LC/MS-MS approach for profiling cytokinins, IAA, ABA, and GAs in lettuce seeds was used in a study of thermo-dormancy (10). In the cited study, extracts were only purified with a C_{18}-SPE column, and all hormones were analysed in 40 min. Recently, a new technique for the combined analysis of several carboxylated plant hormones (IAA, ABA, JA and SA) has been described (46a). With the use of vapor phase extraction (VPE) coupled to GC-MS, these plant hormones could be analysed in a single run. The plant extract was first methylated to make the analytes more volatile. The sample was then heated, and volatile compounds were collected on a polymeric adsorbent. After elution with methylene chloride, the sample was analysed using chemical-ionization-GC-MS. Although this made it possible to analyse several hormones in single samples, there are still problems associated with these methods. The major drawback is that the sensitivity of the analysis is not yet high enough to determine hormone levels in small amounts of plant tissue, weighing perhaps 1-10 mg fresh weight.

As some hormones are preferentially analysed by GC/MS and others by LC/MS, an alternative to the LC/MS approach is to separate different classes of hormones by SPE-chromatography, and then apply the most sensitive MS-method for each, specific compound. It was showed (16) that by using an MCX-column (C_{18} and SCX) it is possible to separate acidic hormones like IAA and ABA from cytokinins. Although the cited authors did not analyse IAA and ABA, but only cytokinins, from real plant extracts, this method could also be used to analyse GAs, salicylic acid, jasmonate, brassinolide and ACC (Nordström and Moritz unpublished). One fraction can then be analysed by GC-MS and the other(s) by LC/MS. However, it is too early to claim that this method can be used for hormone profiling, since it has not been validated against other methods. Therefore, it must be emphasized that hormone profiling is still not suitable for routine analyses. The development of highly sensitive and user-friendly mass spectrometers will simplify the development of profiling methods. However, it should be recognised that hormone profiling will never be straightforward, as plant extracts are complex, and the various hormones occur in very low amounts, and have widely differing chemical properties.

REPORTER GENES AND IMMUNOLOCALISATION AS TOOLS FOR DETECTING HORMONE LEVELS

The high cost of analytical instruments and the lack of methods sensitive enough to perform tissue- and cell- specific analysis of plant hormone concentrations have raised obstacles for many researchers. Alternatives to analytical methods, based on reporter genes and immunolocalisation, have therefore been developed. The use of reporter gene constructs to visualise promoter activity in plants is a well-known technique that has mainly been used for IAA so far, and to a lesser extent for the other hormone classes such as cytokinins. A number of IAA-inducible genes have been investigated (24) and constructs with either GUS (β-glucuronidase) or GFP (green fluorescent protein) fused to the promoters of these genes are now available in several plant species. The DR5::GUS construct in Arabidopsis, where synthetic auxin response elements are fused to GUS, has been widely used as a marker for high IAA concentrations and for changes in IAA concentration following different treatments. Although this approach seems to be valid for the root tip, where DR5::GUS activity and IAA concentration appear to be correlated (50), reliable relationships have not been established for other parts of the plant. This is well illustrated by the fact that the large difference in IAA concentration found between young expanding and fully developed leaves (32, 36) appears to be entirely uncorrelated to differences in DR5::GUS staining. It has also been observed that seedlings of the auxin-overproducing mutant *sur2* have increased endogenous IAA levels in all parts of the plant (5), although they do not show any marked differences in DR5::GUS staining compared to wild-type plants (Fig. 11). These and other examples show that it is important to use reporter genes with caution, and to verify detected differences in reporter gene signals with chemical analysis. The promoter activity and the GUS-signal can be

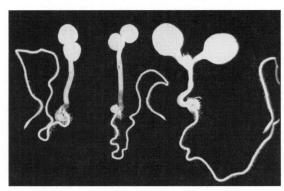

Figure 11 (Colour plate page CP16). Photograph of one-week-old Arabidopsis wild type (right) and mutant *sur2* (left) seedlings containing the construct DR5::GUS (Catherine Bellini and Celine Sorin, unpublished).

attenuated or amplified by numerous factors such as substrate availability, transcription factors and protein stability. The developmental stage of the plant and the type of tissue under study also affect the signal.

Immunolocalisation has also been used to monitor the distribution of plant hormones such as IAA and cytokinins. Since plant hormones are small, easily-diffusible molecules they have to be chemically cross-linked to proteins within the cell before this technique can be used. After cross-linking, the tissue is sectioned and incubated first with monoclonal antibodies raised against the hormone, and then with anti-antibodies, usually coupled to an enzyme that generates a signal to visualise the distribution of the hormone. This method has been used, together with DR5::GUS staining, to monitor IAA distribution in developing organs of Arabidopsis seedlings (4, 6). It has also been used in cytokinin research (14). The efficiency of the cross-linking and the specificity of the antibodies used are very important parameters in these types of studies. Furthermore, there is always a risk that cross-reactions may occur between the antibody and compounds that are chemically similar to the hormone under study, such as conjugates and different metabolites. Reporter genes and immunolocalisation are both qualitative techniques, and cannot provide accurate measurements of hormone concentrations. The results are also highly dependent on the protocol used, making it difficult to compare results from different laboratories. It is thus important to verify the results obtained using these techniques with quantifications of endogenous levels of plant hormones.

FUTURE PROSPECTS

The rapid technical development of mass spectrometers has provided us with tools enabling plant hormones to be detected in mg amounts of plant tissues. This is especially true for IAA, the level of which can now be determined in samples as small as 0.05-0.1 mg of tissue. Thus, for IAA we have reached a technical level where losses of reproducibility due to factors such as contamination and precision in tissue sampling can pose equal or even larger problems than the analytical sensitivity *per se*. This is not the case for several of the other hormones, for which improvements in instrumentation and sample purification would greatly facilitate attempts to address questions being asked in developmental biology. However, in the future we will not focus solely on the concentration of the active hormone. The importance of the regulation of the metabolic machinery, (i.e. synthesis, catabolism and conjugation) on hormone action is becoming increasingly apparent. It will therefore be essential to develop techniques allowing measurements of synthesis and turnover rates, as well as analysis of metabolic markers in very low amounts of tissue, in order to perform tissue-specific analysis of these processes. It is clear that future progress in studies of plant development will require both sensitive mass spectrometry-based techniques and reliable cellular marker systems. This is one of the most challenging tasks for the future!

Acknowledgements

We would like to thank our colleagues Mariusz Kowalczyk, Anders Nordström, Catherine Bellini and Celine Sorin for sharing unpublished results (Figures 5, 8 and 11).

References

1. Åstot C, Dolezal K, Moritz T, Sandberg G (1998) Precolumn derivatisation and capillary liquid chromatographic/frit-fast atom bombardment mass spectrometry analysis of cytokinins in *Arabidopsis thaliana*. J Mass Spectrom 33: 892-902
2. Åstot C, Dolezal K, Nordström A, Wang Q, Kunkel T, Moritz T, Chua NH, Sandberg G (2000) An alternative cytokinin biosynthesis pathway. Proc Natl Acad Sci USA 97: 14778-14783
3. Åstot C, Dolezal K, Moritz T, Sandberg G (2000) Deuterium *in vivo* labelling of cytokinins in *Arabidopsis thaliana* analysed by capillary liquid chromatography/frit-fast atom bombardment mass spectrometry. J Mass Spectrom 35: 13-22
4. Avsian-Kretchmer O, Cheng J-C, Chen L, Moctezuma E, Sung, ZR (2002). Indole acetic acid distribution coincides with vascular differentiation pattern during *Arabidopsis* leaf ontogeny. Plant Physiol 139: 199-209
5. Barlier I, Kowalczyk M, Marchant A, Ljung K, Bhalerao RP, Bennett M, Sandberg G, Bellini C (2000) The SUR2 gene of A*rabidopsis thaliana* encodes the cytochrome P450 CYP83B1, a modulator of auxin homeostasis. Proc Natl Sci USA 97: 14819-14824
6. Benkova E, Michniewicz M, Sauer M, Teichmann T, Seifertova D, Jurgens G. Friml J (2003) Local, efflux-dependent auxin gradients as a common module for plant organ formation. Cell 115: 591-602
7. Bielesky RL (1964) The problem of halting enzyme action when extracting plant tissues. Anal Biochem 9: 431-442
8. Birkemeyer C, Kolasa A, Kopka J (2003) Comprehensive chemical derivatisation for gas chromatography-mass spectrometry-based multi-targeted profiling of the major phytohormones. J Chromatogr A 993: 89-102
9. Chapman JR (1993) Practical Organic Mass Spectrometry – A Guide for Chemical and Biochemical Analysis, Ed 2. John Wiley & Sons, Chichester, 330 pp
10. Chiwocha SD, Abrams SR, Ambrose SJ, Cutler AJ, Loewen M, Ross AR, Kermode AR (2003) A method for profiling classes of plant hormones and their metabolites using liquid chromatography-electrospray ionization tandem mass spectrometry: an analysis of hormone regulation of thermodormancy of lettuce (*Lactuca sativa* L.) seeds. Plant J 35: 405-417
11. Christensen B, Nielsen J (1999) Isotopomer analysis using GC/MS. Metabolic Engineering 1: 282-290
12. Clouse SD, Sasse JM (1998) Brassinosteroids: essential regulators of plant growth and development. Annu Rev Plant Physiol Plant Mol Biol 49: 427-451
13. Crozier A, Moritz T (1999) Physico-chemical methods of plant hormone analysis. *In* PJJ Hooykaas, MA Hall, KR Libbenga, *eds*, Biochemistry and Molecular Biology of Plant Hormones. Elsevier, Amsterdam, pp 23-60
14. Dewitte W, Chiappetta A, Azmi A, Witters E, Strnad M, Rembur J, Noin M, Chriqui D, Van Onckelen H (1999) Dynamics of cytokinins in apical shoot meristems of a day-neutral tobacco during floral transition and flower formation. Plant Physiol 119:111-122
15. Dinan L, Harmatha J, Lafont R (2001) Chromatographic procedures for the isolation of plant steroids. J Chromatogr A 935: 105-123
16. Dobrev P, Kaminek M (2002) Fast and efficient separation of cytokinins from auxin and abscisic acid and their purification using mixed-mode solid-phase extraction. J Chromatogr A 950: 21-29
17. Duffield PH, Netting AG (2001) Methods for the quantitation of abscisic acid and its p recursors from plant tissues. Anal Biochem 289: 251-259

18. Edlund A, Eklöf S, Sundberg B, Moritz T, Sandberg G (1995) A microscale technique for gas chromatography-mass spectrometry measurements of picogram amounts of indole-3-acetic acid in plant tissues. Plant Physiol 108: 1043-1047
19. Epstein E, Cohen JD, Bandurski RS (1980) Concentration and metabolic turnover of indoles in germinating kernels of *Zea mays* L. Plant Physiol 65: 415-421
20. Faiss M, Zalibulova J, Strnad M, Schmuelling, T (1997) Conditional transgenic expression of the ipt gene indicates a function for cytokinins in paracrine signalling in whole tobacco plants. Plant J 12: 401-415
21. Friml J, Benková E, Blilou I, Wisniewska J, Hamann T, Ljung K, Woody S, Sandberg G, Scheres B, Jürgens G, Palme K (2002) AtPIN4 mediated sink-driven auxin gradients and root patterning in *Arabidopsis*. Cell 108: 661-673
22. Fujioka S, Noguchi T, Yokota T, Takatsuto S, Yoshida S (1998). Brassinosteroids in *Arabidopsis thaliana*. Phytochemistry 48: 595-599
23. Gomez-Cadenas A, Pozo OJ, Garcia-Augustin P, Sancho JV (2002) Direct analysis of abscisic acid in crude plant extracts by liquid chromatography-electrospray/tandem mass spectrometry. Phytochem Anal 13: 228-34.
24. Hagen G, Guilfoyle T (2002) Auxin-responsive gene expression: genes, promoters and regulatory factors. Plant Mol Biol 49: 373-385
25. Hall KC, Else MA, Jackson MB (1993) Determination of 1-aminocyclopropane-1-carboxylic acid (ACC) in leaf tissue and xylem sap using capillary column gas chromatography and a nitrogen/phosphorus detector. Plant Growth Regul 13: 225-230
26. Hedden P (1993) Modern methods for the quantitative analysis of plant hormones. Annu Rev Plant Physiol Plant Mol Biol 44: 107-129
27. Hoffman ED, Stroobant V (2002) Mass Spectrometry: Principles and Applications, Ed 2. John Wiley & Sons, Chichester, 420 pp
28. King RW, Moritz T, Evans LT, Junttila O, Herlt AJ (2001) Long-day induction of flowering in *Lolium temulentum* involves sequential increases in specific gibberellins at the shoot apex. Plant Physiol 127: 624-632
29. Kowalczyk M, Sandberg G (2001) Quantitative analysis of indole-3-acetic acid metabolites in *Arabidopsis*. Plant Physiol 127: 1845-1853
30. Linskens HF, Jackson JF (*eds*) (1986) Modern Methods of Plant Analysis Vol 3 – Gas Chromatography/Mass Spectrometry. Springer-Verlag, Berlin, 304 pp
31. Ljung K, Östin A, Lioussanne L, Sandberg G (2001) Developmental regulation of indole-3-acetic acid turnover in Scots pine seedlings. Plant Physiol 125: 464-475
32. Ljung K, Bhalerao RP, Sandberg G (2001) Sites and homeostatic control of auxin biosynthesis in *Arabidopsis* during vegetative growth. Plant Journal 28: 465-474
33. Miller J, Miller JN (2000) Statistics and Chemometrics for Analytical Chemistry, Ed 4. Prentice Hall, New York, 288 pp
34. Moritz T, Olsen JE (1995) Comparison between high-resolution selected ion monitoring, selected reaction monitoring, and four-sector tandem mass spectrometry in quantitative analysis of gibberellins in milligram amounts of plant tissue. Anal Chem 67: 1711-1716
35. Moritz T (1996) Mass spectrometry of plant hormones. *In* RP Newton, TJ Walton, *eds*, Applications of Modern Mass Spectrometry in Plant Science Research. Clarendon Press, Oxford, pp 139-158
36. Müller A, Düchting P, Weiler EW (2002) A multiplex GC/MS/MS technique for the sensitive and quantitative single-run analysis of acidic phytohormones and related compounds, and its applications to *Arabidopsis thaliana*. Planta 216: 44-56
36a. Nordstrom A, Tarkowski P, Tarkowska D, Dolezal K, Astot C, Sandberg G, Moritz T (2004) Derivatization for LC-electrospray ionization-MS: A tool for improving reversed-phase separation and ESI responses of bases, ribosides, and intact nucleotides. Anal Chem 76: 2869-2877
37. Normanly J, Cohen JD, Fink GF (1993) *Arabidopsis thaliana* auxotrophs reveal a tryptophan-independent biosynthetic pathway for indole-3-acetic acid. Proc Natl Acad Sci USA 90: 10355-10359
38. Östin A, Catala C, Chamarro J, Sandberg G (1995) Identification of glucopyranosyl-beta-1, 4-glucopyranosyl-beta-1-N-oxindole-3-acetyl-N-aspartic acid, a new IAA

catabolite, by liquid chromatography tandem mass spectrometry. J Mass Spectrom 30: 1007-1017
39 Östin A, Kowalczyk M, Bhalerao RP, Sandberg G (1998) Metabolism of indole-3-acetic acid in *Arabidopsis*. Plant Physiol 118: 285-296
40 Prinsen E, Van Dongen W, Esmans EL, Van Onckelen HA (1998) Micro and capillary liquid chromatography-tandem mass spectrometry: a new dimension in phytohormone research. J Chromatogr A 826: 25-37
41 Rapparini F, Cohen JD, Slovin JP (1999) Indole-3-acetic acid biosynthesis in *Lemna gibba* studied using stable isotope labeled anthranilate and tryptophan. Plant Growth Regulation 27: 139-144
42 Reeve DR, Crozier A (1980) *In* J MacMillan, *ed*, Encyclopedia of Plant Physiology, Vol 9. Springer, Heidelberg, p.203
43 Ribnicky DM, Cooke TJ, Cohen JD (1998) A microtechnique for the analysis of free and conjugated indole-3-acetic acid in milligram amounts of plant tissue using a benchtop gas chromatograph-mass spectrometer. Planta 204: 1-7
44 Rivier L, Crozier A. (*eds*) (1987) Principles and Practice of Plant Hormone Analysis, Vol 1 and 2. Academic Press, London, 401 pp
45 Rosenblatt J, Chinkes D, Wolfe M, Wolfe RR (1992) Stable isotope tracer analysis by GC/MS, including quantification of isotopomer effects. Am J Physiol 263: E584-E596
46 Sandberg G, Crozier A, Ernstsen A (1987) Indole-3-acetic acid and related compounds. *In* L Rivier, A Crozier, *eds*, Principles and Practice of Plant Hormone Analysis, Vol 2. Academic Press, London, pp 169-302
46a. Schmelz EA, Engelberth J, Alborn HT, O'Donnell P, Sammons M, Toshima H, Tumlinson JH 3rd (2003) Simultaneous analysis of phytohormones, phytotoxins, and volatile organic compounds in plants. Proc Natl Acad Sci USA 100: 10552-10557
47 Schneider G, Schmidt J (1996) Liquid chromatography-electrospray ionization mass spectrometry for analysing plant hormone conjugates. J Chromatogr A 728: 371-375
48 Shimada Y, Goad H, Nakamura A, Takatsuto S, Fujioka S, Yoshida S (2003) Organ-specific expression of brassinosteroid-biosynthetic genes and distribution of endogenous brassinosteroids in *Arabidopsis*. Plant Physiol 131: 287-297
49 Smets R, Claes V, Van Onckelen HA, Prinsen E (2003) Extraction and quantitative analysis of 1-aminocyclopropane-1-carboxylic acid in plant tissue by gas chromatography coupled to mass spectrometry. J Chromatogr A 993: 79-87
50 Swarup R, Friml J, Marchant A, Ljung K, Sandberg G, Palme K, Bennett M (2001). Localization of the auxin permease AUX1 suggests two functionally distinct hormone transport pathways operate in the *Arabidopsis* root apex. Genes Dev 15: 2648-2653

Table of Genes

Where duplicate names or alleles exist they are listed in the gene name column. Note that abbreviations in different species are sometimes totally different genes. For compactness, numbers at the end of a gene name are sometimes omitted unless the gene is known to encode a protein of different function from the other numbers. In general the genes as listed as the wild-types in capital letters; mutants would be in lower case letters. In the species column will be found the main species of reference; if several the one in which it was found first is usually also listed, though some derive from e.g., yeast. For the sake of avoiding duplication the initials of the species name are omitted before the gene name in the table below, though sometimes included they are within the gene name The species are listed in the species column:

A, animals; Ac, Actinidia chinensis (Kiwifruit); Agt, Agrobacterium tumefaciens; Ar, A. rhizogenes; At, Arabidopsis thaliana; Bsp, Brassica species; Bo, B. oleracea (Broccoli); Cr, Catharanthus rosea (Madagascar periwinkle); Cm, Cucurbita maxima (pumpkin); Cme, Cucumis melo (melon); Cs, Cucumis sativus (cucumber); Cp, Cucurbita pepo (zucchini); Dc, Dianthus caryophyllus (carnation); Dm, Drosophila melanogaster; Ec, E. coli; Fa, Fragaria ananassa (strawberry); Gm, Glycine max (soybean); H, human; Hv, Hordeum vulgare (barley); Le, Lycopersicon esculentum (tomato); M, Mammals; Md, Malus domestica (apple); mi, microbial; Mt, Medicago truncatula (alfalfa); Np, Nicotiana plumbaginifolia; Nt, N. tabacum (tobacco); Os, Oryza sativa (rice); Pd, Prunus domestica (plum); Ph, Petunia hybrida; Pi, P. inflata; Ph, Phalenopsis sp.; Pa, Phaseolus aureus (mung bean); Pc, P. coccineus (runner bean); Pv, P. vulgaris (bush bean); Pl, P. lunatus (lima bean); Ps, Pisum sativum (pea); Pp, Pseudomonas putida; Psa, P. savastanoi; Psp, Pseudomonas sp.; Sc, Saccharomyces cerevisiae (yeast); Sp, Schizosaccharomyces pombe; Ssp, Solanum sp.; St, Solanum tuberosum (potato); So, Spinacea oleracea (spinach); Ta, Triticum aestivum (wheat); Vf, Vicia faba (faba bean); Vv, Vitis vinifera (grape); Zm, Zea mays (maize); Ze, Zinnia elegans.

Other abbreviations: eh, eukaryotic homologues; plh, plant homologues S, several.

Gene symbol	Gene name meaning	Species	Function	Chap.
AAO3	Abscisic Aldehyde Oxidase	At	Oxidation of abscisic aldehyde to abscisic acid	B5
AAPK	ABA-Activated Protein Kinase	Vf	Serine threonine protein kinase activated by ABA	D6
ABA1	Abscisic Acid Deficient 1	At; Np	Zeaxanthin epoxidase; epoxidation of zeaxanthin to violaxanthin; the first step of ABA synthesis	B5,D6 E4
ABA2	Abscisic Acid-Deficient	At; Np	Oxidation of xanthoxin to abscisic aldehyde	B5
ABA3	Abscisic Acid-Deficient	At	Sulfurylation of molybdenum cofactor	B5
ABC	ATP-Binding Cassette	A plants yeasts	Large family of membrane transporters possessing wide substrate specificity	E1
ABH1	ABA Hypersensitive 1	At	RNA cap-binding protein; mutation of which confers an ABA hypersensitive phenotype	D6,E4
ABI1,2	ABA-insensitive	At	Protein phosphatases (PP2Cs); dominant mutations confer ABA-insensitivity	D6,E4 E7

695

Table of Genes

ABI3	ABA-insensitive	At	Promotes embryonic development; B3 domain transcription factor; orthologous to maize Vp1	D6,E4 E7
ABI4	ABA-insensitive	At	Promotes embryo maturation and seedling stress response; AP2 domain transcription factor	D6,E4 E7
ABI5	ABA-insensitive	At	Promotes embryo maturation and seedling stress response; bZIP domain transcription	D6,E4 E7
ABP1	Auxin Binding Protein 1	many	Binds auxins; putative auxin receptor	D1
ABRK	ABA-Related Kinase	Vf	Serine threonine protein kinase activated by ABA, probably identical to AAPK	D6
ACC deaminase	ACC deaminase	Psp	Conversion of ACC into α-ketoglutaric acid	B4
ACO	ACC Oxidase	Le; Cme; Pi; M; Dc;Bo; Ph; many	Oxidation of ACC to ethylene	B4,E7 D4,D5
ACS	ACC Synthase	Le;Cp; Ph;Md Pd;At; many	Conversion of S-adenosyl methionine to ACC in ethylene biosynthesis	D4,B4 D5,E7 D3
AGL20 SOC1	Agamous-Like20/ Suppressor of Overexpression of Constans1	At	Flowering time genes	E5
AGPase	ADP-Glucose Pyrophosphorylase	many	Starch synthesis	E5
AGR1	Agravitropic1	At	Identical to EIR1 and PIN2 auxin efflux regulators	E1
AHK1	At Histidine Kinase	At	Putative osmosensing histidine kinase	D3
AHK2,3,4	At Histidine Kinase 2/3/4	At	Membrane-bound histidine kinases; putative cytokinin receptors, ahk4 is allelic to cre1-1 and wol	C3,D3
AHK5	At Histidine Kinase 5	At	Unknown	D3
AHP	At Histidine Phosphotransfer Protein	At	Phosphorelay from AHKS to ARRS	D3
AIR3	Auxin Induced Root3	At	Putative subtilisin-like protease	D1
AKIP1	AAPK-Interacting Protein 1	Vf	Single-stranded RNA binding protein that is a substrate for AAPK	D6
AKT1	At K$^+$ Transporter 1	At	Shaker-like inwardly rectifying K$^+$ channel	D6
AKT2/3	At K$^+$ Transporter 2	At	K$^+$ channel	D6

Table of Genes

ALH1	*ACC-Related Long Hypocotyl 1*	*At*	Ethylene and auxin crosstalk	D4
AMP	*Altered Meristem Program*	*At*	Regulates number of cotyledons formed; similar to glutamate carboxypeptidases	E4
ANT	*Aintegumenta*	*At*	Transcription factor modulating ovule and lateral organ development and embryogenesis	C4
AOC	*Allene Oxide Cyclase*	many	Conversion of 12,13-EOT to 12-OPDA in JA biosynthesis	F1
AOS	*Allene Oxide Synthase*	*At*; flax; many	Conversion of 13-hydroperoxylinolenic acid to 12,13-epoxy-octadecatrienoic acid, the first step specific to the octadecanoid pathway; JA biosynthesis	E5,E6 G1
AOX	*Amine Oxidase*	*mi*	Conversion of tryptamine to indole-3-acetaldehyde	B1
AP1,2	*Apetala 1,2*	*At*	Floral homeotic gene, specifies floral meristem and sepal identity; transcription factor	B2,D3
APRR 3,4,5	*At Pseudo-Response Regulator 3,4,5*	*At*	Circadianly regulated putative transcription factors	D3
APRR1/ TOC1	*At Pseudo-Response Regulator/Timing of CAB 1*	*At*	Regulator of circadian rhythm timing	D3
APRR2	*At Pseudo-Response Regulator 2*	*At*	Putative MYB transcription factor	D3
APT	*Adenine Phospho-ribosyltransferase*	*At*	Cytokinin nucleotide formation via salvage pathway	B3
ARATH CDKA;1	*At Cyclin-Dependent Protein Kinase A;1*	*At*, plh	Cell cycle regulator, interacts with cyclin in cell cycle dependent manner to form serine/threonine specific protein kinase complex	C3
ARATH CYCD3;1	*At Cyclin D3;1*	*At*, plh	Cell cycle regulator, interacts with cyclin-dependent protein kinase in a cell cycle-dependent manner to form a serine/threonine specific protein kinase complex	C3
ArcB	*Aerobic Respiration Control B*	*Ec*	Response regulator; negative transcriptional regulator of genes in aerobic pathways	D3
ARF	*Auxin Response Factor*	*At*; many	Transcription factor for auxin-dependent gene expression that binds to Auxin Response Elements	D1,E1 E4
ARF-GEF	*Guanine Nucleotide Exchange Factor for GTPases of the ARF Type*	many	Guanine nucleotide exchange factor for small ADP-ribosylation factor-type (ARF-type); GTPases involved in the regulation of intracellular vesicle trafficking	E1

Table of Genes

ARR (other)	At Response Regulator	At	Type-As are negative regulators of cytokinin signaling; type-Bs are putative transcriptional activators	D3
ARR 1,2,11,10	At Response Regulator 1, 2, 11, 10	At	Transcription factor type response regulators	D3
ARR2	At Response Regulator 2	At	A type-B response regulator mediating cytokinin-induced expression of type-A ARRs	E6
ARR4	At Response Regulator 4	At	Response regulator; stabilizes PHYB-fr and negatively regulates cytokinin signaling	C3,D3
ARR7	At Response Regulator 7	At	Response regulator; negatively regulates cytokinin signaling	C3
ASK1	At SKP1-Like 1	Sc	Subunit of SCF (E3 ligase)	D1
ATR1	Altered Tryptophan Regulation	At	Myb transcription factor	B1
ATS	Aberrant Testa Shape	At	Maternally inherited effects on seed shape	E4
AUX/IAA	Auxin/Indole Acetic Acid	many	Transcription factor involved in auxin signaling	D1,E4
AUX1	Auxin-Resistant 1	At	Similar to amino acid permeases; putative auxin influx carrier	D4,E1 E2
AXR1	Auxin Resistant 1	At	Subunit of the RUB activating enzyme Similar to the ubiquitin-activating enzyme E1 Involved in auxin action	D1,D4
AXR 2 IAA7	Auxin Resistant 2,	At	See AUX/IAA	D1
AXR 4	Auxin Resistant 4	At	Unknown	D1
AXR 6 CUL1	Auxin Resistant 6	At	Cullin subunit of SCF ubiquitin ligase complex	D1,E1 E4
AZ34 NAR2A		Hv	Conversion of ABA aldehyde to ABA	E3
BAK1	Bri1-Associated Receptor Kinase	At	Brassinosteroid signal transduction	D7
BAS1-D	PhyB Activation-Tagged Suppressor 1 – Dominant	At	Brassinosteroid-26-hydroxylase	B6
b-CHI, ATHCHIB	At Basic Chitinase	At	Class1 chitinase involved in ethylene and jasmonic acid signaling during systemic acquired resistance	D4
BDL BODENLOS IAA12	Bodenlos (Bottomless)	At	AUX/IAA-class transcriptional repressor (IAA12) for ARF5; involved in regulation of embryo patterning	D1,E4 E1
BES1	bri1-EMS-Suppressor 1	At	Brassinosteroid signal transduction; can be nuclear localized	D7

Table of Genes

BIG	*Big*	plants *A*	Identical to *DOC1* and *TIR3*, homologous to the *Drosophila* Calossin (calO)/Pushover protein. Involved in vesicle trafficking	E1
BIN2	*Brassinosteroid-Insensitive 2*	*At*	Shaggy-like kinase A negative regulator of brassinosteroid signaling	D7
BR22ox CYP90B1 DWF4	*Brassinosteroid 22-Oxidase*	*At*	C22 hydroxylation of campestanol	B7
BR23ox CYP90A1 CPD	*Brassinosteroid 23-Oxidase*	*At*	C23α-hydroxylation of cathasterone and 6 deoxocathasterone	B7
BR24red DIM/DWF1 LKB	*Brassinosteroid 24-reductase*	*At; Ps*	Isomerization and reduction of the $\Delta^{24(28)}$ bond of 24-methylenecholesterol during BR biosynthesis	B7
BR5red DET2; LK	*Brassinosteroid 5-Reductase*	*At Ps*	C5α-reduction of (24R)-ergost-4-en-3-one during BR biosynthesis	B7
BR6ox CYP85A	*Brassinosteroid-6-Oxidase*	*Le; At*	C6 oxidation of 6-deoxo intermediates in brassinosteroid biosynthesis	B6,B7
BRI1 LKA(in Ps)	*Brassinosteroid Insensitive*	*At; Ps;* many	Brassinosteroid receptor kinase involved in the perception of brassinosteroids; the *bri1* null mutants are extreme dwarfs with multiple developmental defects	B7,D7 E6,E4
BRS1	*Bri1-5 Suppressor 1*	*At*	Carboxypeptidase putatively involved in brassinosteroid signaling	D7
BRZ1	*Brassinazole Resistant 1*	*At*	Brassinosteroid signal transduction; can be nuclear localized	D7
BSAS	*ß-Substituted Alanine Synthase*	*At; So*	A family of genes with some of them capable of encoding ß-cyanoalanine synthase	B4
BTB	*Broad-Complex, Trimtrack And Bric-A-Brac*	*A*	Transcription factor	C2
BX1	*Benzoxazineless*	*Zm*	Indole synthase, tryptophan synthase alpha paralog	B1
CAB	*Chlorophyll a/b Binding Protein*	many	Platform of light-harvesting chlorophyll a/b	E6
CAS	*ß-Cyanoalanine Synthase*	*At; So*	See *BSAS*	B4
CBP	*Cytokinin Binding Protein*	*At*	Unknown	D3
CCA1	*Circadian Clock Associated1*	*At*	MYB transcription factor; negative regulator of TOC1	D3
CCD	*Carotenoid Cleavage Dioxygenase*	*At*	Oxidative cleavage of carotenoids	B5
CDC2	*Cell Division Cycle 2*	*Sp,* eh	Mitotic cyclin dependent protein kinase	C3

Table of Genes

CDC25	Cell Division Cycle 25	Sp, eh	Protein phosphatase	C3
CEV1	Constitutive Expression of VSP1	At	Cellulose synthase A3 (CESA3)	F1
CheY	Chemotaxis Y	Ec	Response regulator; regulates direction of flagellar rotation	D3
CHK	Cytokinin Hypersensitive	At	Unknown	D3
CHL1	Chlorate Resistant 1	At	Mutant of *atnrt1*; confers a chlorate-resistant phenotype	D6
CIM1	Cytokinin-Induced Message1, B-Expansin	Gm	Cell-wall loosening	C4
CIN	Cytokinin Insensitive	At	Unknown	D3
cisZOG1,2	Cis-Zeatin O-Glucosyltransferase	Zm	O-glucosylation of *cis*-zeatin	B3
CKI	Cytokinin independent 1-2 (gain of function mutants)	At	Histidine protein kinase required for female gametophyte development	D3,E6
CKX	Cytokinin Oxidase/ Dehydrogenase	many	Cytokinin degradation	B3
CLA1	Cloroplastos Alterados (Altered Chloroplasts)	At	1-deoxy-D-xylulose 5-phosphate synthase (DXS)	B2
CLV1	Clavata Receptor Kinase	At	CLV3 receptor; helps determine apical meristem cell fate	F3
CLV2	Clavata Receptor-Like Protein	At	Associates with CLV1 to produce an active receptor	F3
CLV3	Clavata (club-like)	At	Signaling peptide; 96AA ligand for CLV1/CLV2 receptor kinase; restricts SAM size	D3,F3
CO	Constans	At	Transcription factor; B-box type zinc finger protein; serves as link between the clock oscillator and flowering time genes; required for flowering in response to long days	B2,E5
COI1	Coronatine Insensitive 1 (Phytotoxin)	At	Required for response to jasmonates; protein contains 16 leucine-rich repeats and an F-box motif; component of E3 ubiquitin ligase; involved in wounding and parthenogenesis	D4,E6 F1
COL	Constans-Like	St	Unknown	E5
COP1	Constitutive Photo-morphogenesis 1	At	Light-dependent regulator of HY5 protein stability; repressor of photomorphogenic development	D1,D3
COP9	Constitutive Photo-morphogenesis 9	At	Subunit of the signalosome; regulates the 26S proteasome	D3

Table of Genes

CP1	Cysteine Proteinase1	At	Protein turnover	B2
CPD BR23ox CYP90A1	Constitutive Photomorphogenesis and Dwarfism	At	C-23α-steroid hydroxylase of cathasterone and 6 deoxocathasterone involved in brassinosteroid biosynthesis	B7,D7 B6
CPH, ORC; SMT	Cephalopod	At	Sterol methyltransferase; plant sterol biosynthesis	E1
CPR5,6	Constitutive Expression of Pr Genes5,6	At	Regulator of expression of pathogenesis-related (PR) genes	D4
CPS GA1(in At) LS (in Ps)	ent-Copalyl Diphosphate Synthase	At; Ps S	Converts geranylgeranyl diphosphate to ent-copalyl diphosphate	B2,B7
CRE1	Cytokinin Response (resistant) 1	At	Membrane-bound histidine kinase cytokinin receptor; cre1-1 is allelic to ahk4 and wol	C3,D3
CSBP	Cytokinin Specific Binding Protein	Pa	Unknown	D3
CTR1	Constitutive Triple Response 1	At; Le	Mitogen-activated protein kinase	D4,D5 E6,E4
CTS	Comatose	At	ATP binding cassette (ABC) transporter regulating transport of acyl-coAs into the peroxisome; promotes germination and represses embryo dormancy	E4
CU3	Curl 3	Le sp	Encodes tomato BRI1	D7
CUC1	Cup-Shaped Cotyledon 1	At	No apical meristem (NAM) domain protein; functions redundantly with CUC2 to promote embryonic apical meristem formation, cotyledon separation and expression of STM	E4
CUC2	Cup-Shaped Cotyledon 2	At	Transcriptional activator of the NAC gene family modulating shoot apical meristem and cotyledon; see CUC1	C4,E4
CVP1	Cotyledon Vascular Pattern 1	many	Transferring a methyl group to C-24[1] position to form C29 sterols	B6
CYCD3	Cyclin D3	gene family many	Control of the cell cycle at the G1 to S transition; however tobacco CycD3;1 (Nicta CycD3;1) may have a role at mitosis; interacts with cyclin-dependent protein kinase in cell cycle dependent manner to form serine/ threonine specific protein kinase complex	C3,D3
CYP79B2,3	Cytochrome P450	At	Conversion of tryptophan to indole-3-acetaldoxime	B1
CYP79F1	Cytochrome P450	At	Synthesis of short-chain methionine-derived aliphatic glucosinolates abolished in mutant allele	B1

Table of Genes

CYP83B1	Cytochrome P450	At	Synthesis of indole glucosinolates; N-hydroxylation of indole-3-acetaldoxime in vitro	B1
CYR1	Cytokinin Resistant1	At	Unknown	D3
D1	Dwarf1	Os	Putative α subunit of heterotrimeric G protein	D2
D8	Dwarf8	Zm	DELLA-class repressor of GA-inducible gene expression	D2,E4
DAD1	Delayed Anther Dehiscence1	At	Phospholipase A1 involved in JA biosynthesis in Arabidopsis anthers	F1
DAG1,2	DOF Affecting Germination (see DOF)	At	Highly homologous zinc finger transcription factors with opposing effects on germination	E4
DBP	At DNA Binding Protein	At	Auxin-inducible DNA binding protein	D3
DCT1	Divalent Cation Transporter 1	M	Metal ion transporter, NRAMP2, similar to $Smf1$ and MVL	D4
DDE1	Delayed Dehiscence1	At	12-OPDA reductase in JA biosynthesis; see OPR	F1
DDE2	Delayed Dehiscence2	At	Allene oxide synthase See AOS	F1
DEF1	Defenseless 1	Le	Unidentified gene involved in wound-inducible JA synthesis	F1
DET2 BR5red	De-Etiolated	At, Ps	A steroid 5α-reductase involved in the formation of campestanol from campesterol in BR biosynthesis	B7,D7 E6,E4 B6
DFL1 GH3-6	Dwarf in Light	At	Adenylate-forming enzyme	D1
DIM/DWF1 BR24red	Diminutive/ Dwarf	At	Isomerization and reduction of the $\Delta^{24(28)}$ bond of 24-methylenecholesterol	B7
DIR1	Defective In Induced Resistance	At	Putative lipid transport protein involved in SAR signaling	F2
DOC1	Dark Overexpression of CAB	plants A	Identical to TIR3 (BIG)	E1
DOF	DNA-Binding with One Finger	many	Transcription factor	C2
DVL1	Devil 1	At	Unknown	F3
DWF1 BR24red LKB	Dwarf 1	At, Ps	Δ^5-sterol-Δ^{24}-oxidoreductase involved in sterol biosynthesis	B6,D7
DWF4 BR22ox CYP90B1	Dwarf 4	At	C-22α-steroid hydroxylation of campestanol in brassinosteroid biosynthesis	B6,B7 D7
DWF5	Dwarf 5	Sc, At, H	$\Delta^{5,7}$-sterol-Δ^7-reductase involved in sterol biosynthesis	B6,D7
DWF7	Dwarf 7	At	Desaturase involved in sterol biosynthesis	B6,D7

Table of Genes

E2F	E2 Promoter Binding Factor	HeLa cells-H; eh	Transcription factor; originally isolated in human HeLa cells as binding promoter of adenovirus E2 protein	C3
E8	Ethylene induced 8	Le	Unknown; Fe(II) dioxygenase family Negative feedback regulation of ethylene biosynthesis	B4
EBF1,2	Ein3-Binding F Box Protein 1,2	At	F box proteins that interact with EIN3	D4
ECR1	E1 C-Terminal Related 1	At	Subunit of the RUB activating enzyme	D1
EDS1	Enhanced Disease Susceptibility 1	At	Lipase-like protein involved in race-specific resistance	F2
EDS16	Enhanced Disease Susceptibility 16	At	See SID2 Also known as SID2	F2
EEL	Elevated Em Levels	At	bZIP class transcription factor; same clade as ABI5	E4
EER1	Enhanced Ethylene Response 1	At	See RCN1	D4
EFR1	Ethylene Response Factor 1	At	AP2-domain transcription factor	F1
EIL1,2	Ethylene-Insensitive 3-Like 1,2	At	Ethylene signal transduction	D4
EIN2	Ethylene Insensitive 2	At; Le	Ethylene signal transduction; allelic to ERA3	E4,D4 D5,E6
EIN3	Ethylene Insensitive 3	At	Transcription factor involved in ethylene signal transduction	D4,D5
EIN4	Ethylene Insensitive4	At	Ethylene receptor	D4,D5
EIN5,7	Ethylene Insensitive 5,7	At	Ethylene signal transduction	D4
EIN6/EEN	Ethylene Insensitive 6/Enhancer of Ethylene Insensitivity	At	Ethylene insensitive double mutant associated with mechanical stimuli pathway and ethylene signal transduction	D4
EIR1 PIN2	Ethylene Insensitive Root 1	At	Identical to AGR1 and PIN2 auxin efflux regulators	E1
EKO	See KO			
ENOD40	ENOD40 Nod Factor Precursor	Legumes	Role in establishing symbiotic N-fixation	F3
ERA1	Enhanced Response to ABA 1	At	β subunit of farnesyl transferase; ABA-hypersensitive phenotype	D6,E4
ERA3	Enhanced Response To ABA	At	Signal transduction for multiple hormones; allelic to EIN2	E4
ERF1	Ethylene Response Factor 1	At	Transcription factor mediating expression of ethylene-inducible genes; EREBP like protein that binds CGG box of ethylene regulated promoters	D4,D5 F1

Table of Genes

ERS1, 2	Ethylene Response Sensor	At	Ethylene receptors	D4,D5
ETO1	Ethylene Overproducing	At	A protein that interacts with the C-terminal end of AtACS5 and increases its stability	B4,D4
ETO2/3	Ethylene Overproducing 2/3	At	Forms of AtACS5/9 (respectively) mutated within the C-terminal domain	B4,D4
ETR1,2	Ethylene Response 1,2	At; Cme; Le	Ethylene receptor histidine kinase; mutant form confers dominant ethylene insensitivity	D3,D4 D5,E6 E7
EXP	α-Expansin	Cs; At, Le,Os; many	Cell-wall loosening	C4
EXPB	β-Expansin	many	Cell-wall loosening	C4
EXPL	Expansin-Like	many	Unknown	C4
EXPR	Expansin-Related	many	Unknown	C4
FAD	Fatty Acid Desaturase	At; many	Introduction of double bonds into fatty acyl chain; the triple mutant (genes 3, 7 & 8 produces little linolenic acid (thus is unable to accumulate jasmonates)	F1,E6
FK	Fackel (torch, flare)	many	$\Delta^{8,14}$-sterol-Δ^{14}-reductase	B6
FLC	Flacca	Le	Sulfurylation of molybdenum cofactor; conversion of ABA aldehyde to ABA	B5,E3
FliM	Flagella M	Ec	Subunit of the flagellar motor complex	D3
FRY1	Fiery	At	Phosphoinositide catabolism	E4
FT	Flowering Locus T	At	Control of floral transition	E5
FUS3	Fusca (brown, dusky)	At	Promotes embryonic development; B3 domain transcription factor	E4
FUS9 COP10	Fusca9/Constitutive Photomorphogenic10	At	Similar to E2 ubiquitin-conjugating enzyme; interacts with COP1 and COP9	D3
FZY	Floozy	Ph	Flavin monooxygenase, overexpression results in IAA accumulation	B1
GA1; CPS	GA-Deficient-1	At	GA biosynthesis (ent-CPP synthase)	E7
GA20ox GA5 (in At)	Gibberellin 20-Oxidase	Cm;At St; Ps; many	Converts GA_{12} to GA_9 and GA_{53} to GA_{20} Converts GA_{12} to GA_{25} (in Cm)	B2,B7 E5,E7
GA2ox GA 2β-Hydroxylase SLN (in Ps)	Gibberellin 2-Oxidase	Pc; Ps; At; many	GA catabolism; converts C19-GAs to biologically inactive 2β-hydroxy analogs and to 2-oxo analogs (GA-catabolites); converts C20-GAs to 2β-hydroxy analogs	B2,B7 E7
GA3ox GA 3β-hydroxylase GA4; GA4H (in At); Le (in Ps)	Gibberellin 3-Oxidase GA 3-hydroxylase	many At	Converts GA_9 to GA_4 and GA_{20} to GA_1	A2,B1 B2,B7 E7

Table of Genes

GAI	GA-Insensitive	At; Vv	DELLA protein, a negative regulator of GA signaling	C2,D2 E4,E7
GAMYB	GA Regulated MYB	Hv	MYB transcription factor	C2,E4
GCA2	Growth Controlled By ABA 2	At	Unknown, but gca2 mutant has an ABA-insensitive phenotype	D6
GCR1	G-Protein Coupled Receptor	At	Promotes germination	E4
GH3	(Isolated by) Gretchen Hagen 3	Gm many	Auxin-responsive gene, JAR1-like; adenylate-forming enzyme	B1,D1 E2
GH45	Glycosyl Hydrolase Family-45	many fungi	Hydrolysis of glycosidic bond	C4
GID2	GA-Insensitive Dwarf 2	Os	F-box factor that targets DELLA proteins for proteasomal degradation; a positive regulator of GA signaling; orthologous to SLY	C2,D2 E4
GIN1,5	Glucose Insensitive	At	See ABA2/3 respectively	A2,B5
GL2	Glabra2	At	Homeodomain transcription factor modulating hair formation	C4
GLUT4	Glucose Transporter4	M	Insulin-regulated glucose transporter	E1
GMPOZ	GAMYB Associated POZ	Hv	Transcription factor	C2
GN	Gnom	many	ARF-GEF	E1
GORK1	Guard cell Outwardly Rectifying K^+ Channel 1	At	Outwardly rectifying K^+ channel	D6
GPA1	G Protein α Subunit 1	At; many	α subunit of heterotrimeric G protein; gpa1 mutants show ABA-insensitivity in certain guard cell responses; promotes germination	D6,E4
GRD2	GA-Responsive Dwarf 2	Hv	Putative GA3ox, a GA biosynthetic enzyme	D2
GSE	GA-Sensitivity	Hv	A positive regulator of GA signaling	D2
Gα	Gα - A Subunit of Heterotrimeric G Proteins	many	Signal transduction	B2
HB	Homeobox HD-ZIP gene	At	Encoding transcriptional regulator expressed early in procambial (or provascular) cells	E2
HBT	Hobbit	At	Homolog of the CDC27 subunit of the anaphase-promoting complex (APC); required for cell division and cell differentiation in meristems	E4
HK1	Histidine Kinase 1	Zm	Cytokinin receptor	B3

Table of Genes

HLS1	Hookless 1	At	Ethylene regulated apical hook development, putative N-acetyltransferase	D4
HOG1	High Osmolarity Glycerol Response 1	Sc	MAP kinase central to the high-osmolarity signaling pathway	D3
HRT	Hordeum Repressor of Transcription	Hv	Transcription factor	C2
HXK1	Hexokinase 1	At	Conversion of glucose to glucose-6-phosphate and sense of sugar level	E6
HYD1	Hydra 1	Sc, At	Sterol Δ^8-Δ^7 isomerase	B6
HYL1	HYponastic Leaves 1	At	Double-stranded RNA binding protein; *hyl1* mutant shows reduced sensitivity to auxin and cytokinin and ABA hypersensitivity (although guard cell ABA responses are wild-type)	D6,E4
IAA1-3	Indole Acetic Acid 1-3	At	Auxin-inducible nuclear-localized proteins	D3
IAA17 AXR3	Indole Acetic Acid 17/Auxin Resistant 3	At	Auxin-inducible nuclear-localized protein	D3
IAA3 SHY2	Indole Acetic Acid 3/ Short Hypocotyl	At	Auxin-inducible nuclear-localized protein (suppressor of *HY2*)	D3
iaaL	IAA-lysine synthetase	Psa	over-expression results in extremely low IAA concentrations	E2
iaaM	tryptophan mono-oxygenase	Agt	Produces indoleacetamide from tryptophan; over-expression results in high IAA concentrations	E2
IaaspH	IAA-Aspartic Acid Hydrolase	mi	Hydrolysis of IAA-Asp	B1
IAGLU	IAA-Glucose Synthase	Zm	UDP-glucosyl transferase specific to IAA-Glucose formation	B1
IAH	IndoleAcetamide Hydrolase	mi	Conversion of indole-3-acetamide to IAA	B1
IAO	Indole-3-Acetaldehyde Oxidase	mi	Conversion of indole-3-acetaldehyde to IAA	B1
IAP1	IAA-Modified Protein	Pv	IAA-modified protein	B1
IBC6	Induced By Cytokinin 6 (Same as ARR5)	At	Negative regulator of cytokinin signaling	D3
IBC7	Induced By Cytokinin 7 (Same as ARR4)	At	See *ARR4* above	D3
ICK1	Cyclin-Dependent Kinase Inhibitor	At	Suppress cell division	E4
IGL	Indole-3-Glycerol Phosphate Lyase	Zm	Indole synthase, tryptophan synthase alpha paralog	B1

Table of Genes

IPDC	Indole Pyruvate Decarboxylase	mi	Conversion of indole-3-pyruvic acid to indole-3-acetaldehyde	B1
IPT1,3-8	Isopentenyl Transferase 1,3-8	Agt; At; many	Cytokinin biosynthesis; catalyses the rate-limiting step of cytokinin biosynthesis: the condensation of 2-Δ-isopentenyl PPi with AMP to form isopentenylAMP	B3,C3 E3,E5, E6
IPT2,9	Isopentenyl-transferase 2,9	At	Isopentenylation of tRNA	B3
ISI4	Impaired Sucrose Induction	At	See ABA2	B5
JAI1 COI1	Jasmonate-Insensitive 1	Le	Regulator of JA signaling; see COI1	F1
JAR1 FIN219 GH3-11	Jasmonic Acid Resistant 1; Far Red Insensitive 219	At	Adenylate-forming enzyme; acyl adenylate-forming firefly luciferase superfamily; adenylation of JA	B1,D1 F1
JMT	Jasmonate Methyl Transferase	At	Conversion of JA to methyl-JA	F1
KAO1,2 NA (in Ps) CYP88A6,7	Ent-Kaurenoic Acid Oxidase	Hv; Cm; Ps; At; many	GA biosynthesis; oxidation of *ent*-kaurenoic acid to GA_{12}	B2,E7 B7
KAT1,2	Voltage-Gated K^+ Channel of At1/2	At	Inwardly rectifying K^+ channel	D6
KGM	Kinase Associated With GAMYB	Hv	Protein kinase	C2
KIP	Kinase Inhibitory Protein	H; eh	Inhibition of cyclin/CDK complexes, homologs in plants known as Kip-related proteins (KRP)	C3
kn1	Knotted 1	Zm	Homeobox transcription factor	D3
KNAT1-2	Knotted-Like From At 1-2	At	Homeobox transcription factor; class I KNOX gene	D3,E5
KNOX	Knotted1-Like Homeobox	many	Transcription factors involved in establishing organ identity	B2
KO GA3 (inAt) LH (in Ps) CYP701A10	ent-Kaurene Oxidase	S At; Ps	Oxidizes *ent*-kaurene to *ent*-kaurenoic acid	B2,B7 E7
KS GA2(inAt)	ent-Kaurene Synthase	Cm;At; S	Converts *ent*-copalyl diphosphate to *ent*-kaurene	B2,E7
LAX	Like AUX1	At	Homologue of *AUX1*; putative auxin uptake carrier	E1
LE GA3ox1	Length. Pea length genes (L-) are named in alphabetical order	Ps	GA_{20} 3-oxidation	B7,E5 E7
LEC1	Leafy Cotyledon	At	Promotes embryonic development; CCAAT-box binding factor	E4

Table of Genes

LEC2	Leafy Cotyledon	At	Promotes embryonic development; B3 domain transcription factor	E4
LFY	Leafy	At	Transcription factor modulating floral organs; promotes transition from inflorescence to floral meristem	B2,C4
LH; KO1 CYP701A10	See LE	Ps	Ent-kaurene oxidation	B7
LHY	Late Elongated Hypocotyl	At	MYB transcription factor; negative regulator of TOC1	D3
LK BR5red	See LE	Ps	C5α-duction during BR biosynthesis (see DET2 below)	B7
LKA; BRI1	See LE	Ps	BR receptor	B7
LKB BR24red	See LE	Ps	C24 reduction during BR biosynthesis (see DIM/DWF1)	B7
LKC	See LE	Ps	Unknown BR mutant	B7
LKD	See LE	Ps	Unknown BR mutant	B7
LOS5,6	Low Expression of Osmotic Stress-Responsive Genes	At	See ABA3,1 respectively	B5
LOX	Lipoxygenase	St; many	Oxygenation of polyunsaturated fatty acids	E5,F1
LS; CPS	See LE	Ps	Synthesis of copalyl diphosphate	B7
LUC	Luciferase	Firefly	Catalyzes the oxidation of luciferin producing light	D3
MAN2	Endo-B-Mannanase	Le	Hydrolysis of mannan	C3
MDR1	Multidrug Resistance 1	many	Multi-drug resistance sub-family of ABC transporters. Some members involved in auxin transport	E1
MNK	Menkes Copper-Transporting ATPase	M	Substrate-regulated efflux transporter for Cu ions	E1
MP ARF5	Monopteros	At	Auxin response factor (ARF) transcription factor; regulator of embryo patterning; putative interactor with BDL	D1,E1 E2,E4
MPK4,6,13	Mitogen Activated Protein Kinase 4,6,13	At	Protein kinase (note: kinases of kinases repeat the K in the abbreviation)	D4,F1
MRP5	Multidrug Resistance-Related Protein 5	At	Closely related to MDR5	E1
MSG2 IAA19	Massugu	At	See AUX/IAA	D1
MSR1	Mitochondrial-Specific Arginyl-tRNA Synthetase 1	Sc	Arginyl-tRNA synthetase	D3
MVL MALVOLIO	Malvolio	Dm	NRAMP Metal-ion transporter similar to Smf1 and DCT1	D4

Table of Genes

NA; KAO1 CYP88A6	Nana	Ps	Ent-kaurenoic acid oxidase	A2,B7 E5
NAC1	NAM,ATAF1,CUC2	Ph	Transcription factor	D1
NAHG	Salicylate Hydroxylase	Pp	Converts salicylic acid to catechol	E6,F2
NAR2A	Molybdenum Cofactor	Hv	Molybdenum cofactor synthesis	B5
NCED	Nine-Cis-Epoxy-Carotenoid Dioxygenase	Zm; Le; many	Cleavage of 9-cis-epoxy-carotenoids to xanthoxin in ABA biosynthesis	B5, E7
NIA1,2	Nitrate Assimilation 1,2	At	Cytokinin-inducible nitrate reductase	C3,D6
NIM1	Non-Inducible Immunity 1	At	Ankyrin repeat protein that transduces the SA signal that activates SAR Also known as NPR1, SAI1	F2
NIT1-,4	Nitrilase 1-4	At	Conversion of IAN (indole-3-acetonitrile) to IAA in vitro, null allele is resistant to inhibitory effects of IAN	B1,E6
NOT NCED	Notabilis	Le	See NCED	B5
NPH4 MSG1 ARF7	Non-Phototrophic Hypocotyl 4 Massugu 1	At	Affects blue light and gravitropic and auxin mediated growth responses; see ARF	D1,D4
NPQ2	Non-Photochemical Quenching	At	See ABA1	B5
NPR1 NIM1, SAI1	Nonexpresser of Pr Genes 1 SA Insensitive	At	Controls systemic acquired resistance (SAR) Confers resistance to pathogens; see NIM1	D4,F2
NR	Never-Ripe	Le	Ethylene receptor	D4,D5 E7
NRT1	Nitrate/Chlorate Transporter 1	At	Dual affinity nitrate transporter	D6
ODC	Ornithine Decarboxylase	many	Polyamine biosynthesis	E5
OPR	12-Oxo-Phytodienoic Acid Reductase	At; many	Conversion of 12-oxo-PDA to 3-oxo-2-(2'(Z)-pentenyl)-cyclopentane-1-octanoic acid (OPC-8:0) in JA biosynthesis	E6;F1
ORCA3	Octadecanoid-Responsive AP2-Domain Protein	Cr	ERF/AP2-domain transcription factor	F1
ORE12	Oresara 12 (delayed senescence)	At	Gain of function mutation in AHK3	D3
ORP	Orange Pericarp	Zm	Tryptophan synthase beta	B1
OST1	Open Stomata 1	At	ABA-activated serine-threonine protein kinase; probably ortholog of AAPK ost1 mutants show guard cell insensitivity to ABA	D6

Table of Genes

PAD4	Phytoalexin Deficient 4	At	Lipase-like protein involved in race-specific resistance	F2
PAS	Pasticcino (tartlet)	At	Mutants show uncontrolled cell division	D3
PAT	Parthenocarpic	Le	Unknown; mutation promotes parthenocarpy	E7
PBF	Prolamin Box Binding Factor	Cereals	Transcription factor	C2
PDF1	Protodermal Factor 1	At	Encodes a putative extracellular proline-rich protein	B2
PDF1.2	Plant Defensin 1.2	At	Encodes an ethylene- and jasmonate-responsive plant defensin	D4
PEP	Pepino (Same as Pasticcino 2)	At	Unknown	D3
PGP1	P-Glycoprotein 1	At	Member of a sub-group of MDR proteins	E1
PHOR1	Photoperiod Responsive 1	St	U-box arm-repeat protein, a positive regulator of GA signaling	D2
PHYA	Phytochrome A	At; many	Light labile red/far-red absorbing photoreceptor; serine-threonine kinase	D3,E5
PHYB	Phytochrome B	At; many	Light stable red/far-red absorbing photoreceptor; putative histidine kinase	D3,E5
PID	Pinoid	At	Serine/threonine protein kinase involved in auxin transport and/or signaling	D1,E1
PIF3	Phytochrome Interacting Factor 3	At	Putative helix-loop-helix transcription factor	D3
PIN	Pin-Formed	At	Family of auxin efflux regulators	A2,E1 E2,E4
PIN2	Proteinase Inhibitor II	Le	Defense-related	E5,G1
PIRIN1	Pirin1	At	Interacts with α subunit of G-protein; promotes germination	E4
PIS1	Polar Auxin Transport Inhibitor Sensitive 1	At	Putative negative regulator of polar auxin transport	E1
PKABA1	Protein Kinase Responsive To ABA 1	Hv	Protein kinase; suppresses GA-inducible gene expression in aleurone	C2,E4
PKL	Pickle	At	Chromatin remodeling factor; suppresses embryonic development; a positive regulator of GA signaling?	D2,E4
PLD	Phospholipase D	At	Catalyzes hydrolysis of phosphatidyl-choline to phosphatidic acid and choline	D3
PLS	Polaris	At	36AA peptide of unknown function	D3
POTH1	Potato Homeodomain 1	St	KNOX gene	E5
POZ	Poxvirus Zinc Finger	Virus; A	Transcription factor	C2

Table of Genes

PRO-SYSTEMIN	Systemin Precursor	Ssp	Systemic wound signaling	F1,F3
ProTomHysSys	Hydroxyproline-Rich Glycopeptide Precursor	Nt; Le	Precursor of two hydroxyproline-rich peptide defense signals	F3
Ps-IAA4/5	Ps Indole Acetic Acid 4/5	Ps	See AUX/IAA	D1,C4
PSKα precursor	Phytosulfokine-α Precursor	many	Regulates cellular de-differentiation and proliferation	F3
PSY	Phytoene Synthase	many	Converts geranylgeranyl diphosphate to phytoene	B2
RAC1	ras-related C3 botulinum toxin substrate	At	RHO-like small GTPase; synonymous with ROP6; negative regulator of guard cell ABA response	D6
RALF precursor	Rapid Alkalinization Peptide Precursor	many	Unknown	F3
RAN1	Responsive-To-Antagonist 1	At	ATP dependent copper transporter vital for ethylene response pathway	D4
RB1	Retinoblastoma-Like Protein 1	Nt, eh	Cell cycle regulator, hyperphosphorylated by cyclin-dependent protein kinase complex	C3
RBX1	Ring Box 1	Sc	SCF subunit	D1
RCE1	Rub Conjugating Enzyme 1	At	Conjugation of RUB to substrates	D1
RCN1	Roots Curl In NPA 1	At	Serine/Threonine protein phosphatase type 2A regulatory subunit; rcn1 mutant shows impaired guard cell response to ABA	D4,D6 E1
RcsC	Regulator of capsule synthesis C	Ec	Sensor histidine kinase; regulates genes encoding envelope proteins	D3
RDO	Reduced Dormancy	At	Four loci of unknown cellular/molecular function	E4
RGA	Repressor of ga1-3	At	DELLA protein, a negative regulator of GA signaling	C2,D2 E4
RGL1,2,3	RGA-LIKE 1-3	At	DELLA proteins, negative regulators of GA signaling	D2,E4
RHD6	Root Hair Defective6	At	Unknown	C4
RHT1	Reduced Height 1	Ta	DELLA-class repressor of GA-inducible gene expression	D2,E4 E7
RIN	Ripening Inhibited	Le	MADS box transcription factor Loss-of-function mutant fruits fail to ripen	E7,D4 D5
ROLC		Ar	Hydrolysis of CK conjugates	E5
ROP6,10 RAC1	RHO of Plants 6,10	At	RHO-like small GTPase; rop10 null mutants are ABA hypersensitive	D6
RPN12	Regulatory Particle Non-ATPase 12	Sc; At; eh	Regulatory component of the 26S proteasome complex	C3,D3

Table of Genes

RSP1,2 RAS	Raspberry	At	Promote embryonic morphogenesis; suppress embryonic development of suspensor	E4
RUB1 NEDD8	Related to Ubiquitin	Sc	Modifier of CUL1; regulates SCF activity	D1
SABP1	SA Binding Protein 1	Nt	SA-sensitive catalase	F2
SABP2	SA Binding Protein 2	Nt	SA-stimulated lipase and putative SA receptor	F2
SAD	Scutellum and Aleurone Expressed DOF	Hv	Transcription factor	C2
SAD1	Supersensitive To ABA And Drought 1	At	Sm-like small nuclear ribonucleoprotein; sad1 mutants are ABA and drought hypersensitive	D6,E4
SAG12	Leaf Senescence-Specific Gene 12	At	Vacuole-targeted cysteine proteinase	E6
SAG13	Leaf Senescence-Specific Gene 13	At	Short-chain alcohol dehydrogenase	E6
SAI1;NIM1, NPR1	SA Insensitive	At	See NIM	F2
SAMase	S-Adenosyl Methionine Hydrolase	T3 bacteriophage	Conversion of S-adenosyl methionine into methylthioadenosine	B4
SAMS	SAM Synthase	At; Le; Ac	ATP:L-methionine S-adenosyltransferase involved in the transfer of the adenosyl moiety from ATP to methionine	B4
SAUR	Small Auxin-Up RNA	Gm;At	Unknown	D1
SAUR-AC1	Small Auxin-Up RNA-Arabidopsis Columbia 1	At	Unknown	D3
SCR precursor	Brassica S-Locus Cysteine- Rich Peptide Precursor	Bsp	Produces S-locus peptide signal for self incompatibility	F3
SD1	Semidwarf-1	Os	GA biosynthesis (GA 20-oxidase)	E7
SDGs	Senescence-Down-Regulated Genes	At	Expression is down-regulated during leaf senescence	E6
SENs	Senescence-Associated Genes	At	Expressed during leaf senescence in At	E6
SHI	Short Internodes	At	Zinc finger protein, a negative regulator of GA signaling?	D2
SHO	Shooting	Ph	See IPT1	B3
SHY1,6 IAA6,3	Short Hypocotyl	At	See AUX/IAA	D1
SID2	SA-Deficient 2	At	Encodes isochorismate synthase involved in SA synthesis Also known as EDS16	F2

Table of Genes

SIMKK	Salt Stress-Induced MAPKK	Mt	Salt stress- and pathogen-induced *Medicago* MAPKK	D4
SIS4	Sugar Insensitive 4	At	See *ABA1*	B5
SIT	Sitiens	Le	Conversion of ABA-aldehyde to ABA	B5,E3
SLG	Brassica S-Locus Glycoprotein	Bsp	Activates SRK to trigger an incompatibility signaling cascade	F3
SLN GA2ox1	Slender	Ps	GA_{20} 2-oxidation, GA_1 2-oxidation	B7
SLN1	Synthetic Lethal of N-End Rule 1	Sc	Two-component histidine kinase involved in osmosensing	D3,D4 D5
SLN	Slender	Hv	DELLA protein, a negative regulator of GA signaling	C2,D2 E4
SLR1	Slender Rice 1	Os	DELLA protein, a negative regulator of GA signaling	B2,C2 D2
SLR1;IAA14	Solitary Root 1	At	See *AUX/IAA*	D1
SLY1	Sleepy 1	At	F-box factor that targets DELLA proteins for proteasomal degradation; a positive regulator of GA signaling orthologous to *GID2*	D2,E4
SMF1	Suppressor of MIF 1	Sc	NRAMP Metal-ion transporter similar to *MVL* and *DCT1*	D4
SMT1	Sterol C-24 Methyl Transferase 1	At; many	Synthesis of membrane sterols; transferring a methyl group to C-24 position of sterols, allelic to *ORC*, *CPH*	B6,E1
SPR2 FAD7	Suppressor of Prosystemin-Mediated Responses2	many	Omega-3 Fatty acid desaturase involved in the production of linolenic acid for JA biosynthesis	F1
SPY	Spindly	At	O-linked GlcNAc transferase; a negative regulator of GA signaling	C2,D2 E4
SR 160	Tomato Systemin Receptor	Le	Interacts with systemin to initiate defense signaling	F3
SRK	Brassica S-Locus Receptor Kinase	Bsp	Interacts with *SLG* and *SCR* to initiate self incompatibility	F3
SRK2E	SNF1-Related Protein Kinase 2E	At	Snrk2-type protein kinase; synonymous with *OST1*	D6
SSI1	Suppressor of Salicylic Acid Insensitive 1	At	Activator of defense response gene expression and lesion formation	D4
SSK1	Suppressor of Sensor Kinase 1	Sc	Response regulator; negatively regulates the HOG1 pathway	D3
SSU	Small Subunit Gene	many	Component of ribulose-1,6-bisphosphate carboxylase/oxygenase (rubisco)	E6
ST2A	Sulfotransferase	At	Hydroxyjasmonate sulfotransferase	F1
STE,DWF7	Sterol/Dwarf 7	At	Steroid C-5 desaturase	B6

Table of Genes

STM	Shoot Meristemless	At	KNOX homeobox transcription factor; regulator of shoot meristem formation and maintenance	B2,D3 E4,E5
STP1	Stunted Plant 1	At	Monosaccharide/H^+ symporter	D3
SUS1	Abnormal Suspensor 1	At	MicroRNA processing; allelic to Short integuments1 and Carpel factory	E4
SUS2	Abnormal Suspensor	At	Putative pre-mRNA splicing factor	E4
SUSY	Sucrose Synthase	many	Sucrose breakdown	E5
SYR1	Syntaxin-Related 1	Nt	Syntaxin	D6
TAT	Tryptophan Aminotransferase	mi	Conversion of tryptophan to indole-3-pyruvic acid	B1
TCH3	Touch 3	At	Calmodulin-like protein, expression induced by touch and darkness	D4
TCH4	Touch 4	At	Xyloglucan endotransglucosylase/hydrolase	D7
TCP10	TB1/CYC/PCF10	At	Putative transcription factor	D3
TDC	Tryptophan Decarboxylase	mi, Cr	Conversion of tryptophan to tryptamine	B1
TED	Tracheary Element Differentiation	Ze	Expressed during early stages of vessel element differentiation	E2
THI2.1	Thionin	At	Defense-related	E5
TIR1	Transport Inhibitor Response1	At	F-box protein; related to yeast GRR1P and human SKP2 proteins, involved in ubiquitin-mediated processes	D1,D4
TIR3	Transport Inhibitor Response3	At	Identical to *BIG* and *DOC1*. Putative NPA-binding protein	E1
TMO	Tryptophan Monooxygenase	mi	Conversion of tryptophan to indole-3-acetamide	B1
TMR	Tumor Morphology of Roots	Agt	See *IPT*	B3
TRAB1	Transcription Factor Responsible for ABA Regulation 1	Os	bZIP domain transcription factor; likely *ABI5* ortholog	E4
TRIP1	Transforming Growth Factor-Beta Receptor Interacting Protein 1	At, Pv	WD-domain protein, subunit of eif3 translation initiation factor Homolog of mammalian signaling protein	D7
TT4	Transparent Testa 4	At	Chalcone synthase, a key enzyme in flavonoid biosynthesis	E1
TTG	Transparent Testa Glabrous	At	WD40 repeat protein, binds transcription factors modulating the fate of root epidermal cells and testa structure	C4,E4
TWN1	Twin 1	At	Required for suppressing embryogenic development in suspensor cells	E4
TWN2	Twin 2	At	Valyl-tRNA synthetase; required for proper proliferation of basal cells	E4

Table of Genes

TZS	Trans-Zeatin Secretion	Agt	See *IPT*	B3
UBA2a	UBP1-Associated Protein 2a	At	Single-stranded RNA binding protein	D6
UCU1	Ultracurvata 1	At	Allelic to *BIN2*	D7
UGT84B1	UDP-Glucosyl-ltransferase	At	UDP-glucosyl transferase specific to IAA-Glucose formation	B1
VH	Vascular highway	At	A leucine-rich receptor kinase, expressed in provascular/procambium cells	E2
VP1	Viviparous	Zm	B3 domain transcription factor; *ABI3* ortholog, Embryo maturation	E4,E7
VP14 NCED1	Viviparous	Zm	See *NCED*	B5
Vp2,7,8 and 9	Viviparous	Zm	ABA biosynthesis	E4
VP5	Viviparous	Zm	Defect in carotenoid biosynthesis; causing phytoene accumulation and ABA deficiency	E3,E4 E7
WEE1	Wee 1 (i.e., small)	Sp, eh	Protein kinase, cell cycle regulator active at G_2/M transition	C3
WEI2,3,4	Weak Ethylene Insensitive	At	Ethylene signal transduction	D4
WOL	Wooden Leg	At	Membrane-bound histidine kinase, cytokinin receptor, wol mutant allele is impaired in cytokinin binding, see also *AHK4* and *CRE1*	C3,D3 E4
WUS	Wuschel (ruffled, disheveled)	At	Homeobox transcription factor for stem cell identity; shoot and floral meristem organization	D3,F3
XET4	Xyloglucan Endo-transglycosylase	Le	Endotransglycosylation of xyloglucan	C4
XTH5	Xyloglucan Endo-transglycosylase/ Hydrolase	At	Cell wall modification	B2
YojN	Regulator of Capsule Synthesis (Same as RCSD)	Ec	Sensor histidine kinase; regulates colanic capsule synthesis	D3
YPD1	Tyrosine Phos-phatase Dependent 1	Sc	His-phosphotransfer protein in the HOG1 pathway	D3
YUCCA	Yucca	At	Flavin monooxygenase, conversion of tryptamine to N-hydroxyl tryptamine in vitro, overexpression results in IAA accumulation	B1
ZEA3	Zeatin Resistant 3	At	Unknown	D3
ZmHP2	Zm Histidine Phosphotransfer Protein 2	Zm	His-phosphotransfer protein	D3

Table of Genes

Zmp Zm-p60.1		Zm	Cytokinin-glucoside specific β-glucosidase; releasing free cytokinins from cytokinin-O-glucosides	C3,E3
ZmRR1,2	Zm Response Regulator 1,2	Zm	Response regulator involved in nitrogen signaling	D3
ZOG1	Trans-Zeatin O-Glucosyltransferase1	Pl	O-glucosylation of *trans*-zeatin	B3
ZOX1	Trans-Zeatin O-Xylosyltransferase 1	Pv	O-xylosylation of *trans*-zeatin	B3

INDEX

AAO3 143, 147, 149
AAPK 405
ABA - see Abscisic acid
ABA genes 30, 31, 143, 147, 149, 530, 533
ABC transporters 452, 618
abh1 527
ABI 14-15, 398, 406, 524, 526-528, 530-531
Abiotic stress 116,137, 360
ABP1 283-284
Abscisic acid **137-155, 391-412, 493-537**
9-*cis*-epoxycarotenoid dioxygenase see NCED
ABI3 584
aba1 143
ABI5 531, 533
ABRE 526-527
abscission 9, 137
AC1 397
acclimation 391
accumulation 505, 523
actin 397, 399
aldehyde 141-2, 149
aldehyde oxidase 143
aleurone 236
analysis 680, 687-688
antagonism 584
antibody 523
Arabidopsis 397, 399, 523, 524, 689
biosynthesis **137-155**, 496, 570
  AAO3 147
  aba3 146, 147
  abscisic aldehyde 140, 146
  aldehyde oxidase 146
  Arabidopsis 145, 147, 148
  capsanthin-capsorubin synthase 148
  carotenoid 138, 141, 144-145 148
  cellular dehydration 394
  cowpea 145, 148
  dehydration 394, 496
  dimethylallyl diphosphate 138
  direct pathway 139
  dormancy 524
  early steps 142

enzymes 145
epoxy-carotenoid 148
farnesyl diphosphate 138
ferredoxin 144
flacca 147
flavoprotein monooxygenases 143
gene expression 148
genes 142
geranyl diphosphate 138
geranylgeranyl diphosphate 138
gibberellin 146
green fluorescent protein 145
HPLC 144
indirect pathway 140
inhibition 524
isopentenyl diphosphate 138, 140
isoprenoids 138
labeling pattern 141-142
later steps 146
lignostilbene dioxygenases 144
lycopene 147
lycopene cyclase 148
mass spectrometry 141
methylerythritol phosphate 138, 140
MoCo synthesis 146
mutants 142-143, 146, 570
NADPH 144
nar2a 146
NCED 144-145, 147, 598
neoxanthin 147, 148
neoxanthin synthase 148
nitrate reductase 146
notabilis 145
nutrient deficiency 496
osmotic stress 148
pea 145
phenotypes 142
photo-isomerization 147
phytoene 138
plastids 145
potato 148

precocious germination 143
prolycopene 147
Pseudomonas 144
regulation 148
ring modifications 146
root 394
seed germination 140
sites 9
sitiens 147
stomatal conductance 143
stroma 145
sucrose 143
temperature 496
tomato 145, 147, 148
transcription 148
translation 148
violaxanthin 140, 143, 147, 148
viviparous 144
wilting 143
xanthine dehydrogenase 146
xanthoxin 140, 146
zeaxanthin 143
zeaxanthin epoxidase 148
ZEP 143, 148
binding protein 394
bud dormancy 9
calcium 394-395, 398-399, 495
calmodulin 230
carbon dioxide 391
carotenoid 145
catabolism 151, 498
catabolites 142, 151, 152
cell wall extensibility 506
cereal 10
channels 399
chloroplast 146, 496
cold 391, 407, 598
Commelina. communis 397
conjugates 496
ctr1 531
cytokinin 575
cytoskeleton 397
Dc3 promoter binding factor 527
defense 11
deficiency 523
deficient mutants 30, 145,

717

497, 499, 520, 522, 553
  embryos 145
  phenotypes 145
  viviparous 520
degradation 137
dehydration 393
dehydrin 407
depolymerization 397
derivatisation 680, 688
desiccation 137
development 391
developmental arrest 531
discovery 3, 137
diphosphate 9
dormancy 9-10, 137, 236, 524, 584, 598
drought 391, 407, 598
effects 10
ein2 531
elongation 391
embryo 522, 524
embryo maturation 236
embryogenesis 137
endosperm hydrolases 227
enhanced response 523
ethylene 531, 575
ester 152
expansin 274
farnesylation 397
flux 498
G protein 397
ga1 143
GC/MS 688
GCA2 407
gene expression 137, 405, 406
gene transcription 11
germination 10, 522-523, 529, 530-532, 533
gibberellin 10, 143, 221-222, 230, 236, 238, 553, 584
  α-amylase 10, 236
  antagonism 236
glucose 30-31
glucose ester 152, 496
glyceraldehyde-3-phosphate 10
GPA1 532
growth 30, 391, 505
guard cell 391, 394, 407
H^+ ATPases 395
H_2O_2 400
hypersensitive 407, 523
immunoaffinity column 498
immunoassay 497

immunolocalization 397
inactivation 151
inhibitor 9
inositol phosphates 399
insect 11
insensitive 30, 398, 523
isoprenoids 139
kinase 232, 236
LEA 522
leaf 10, 497
lettuce 689
maize 151
manipulation 598
maternal 524
maturation 584
membrane potential 395
metabolism 137
methyl ester 688
microtubule 397
MS-analysis 680
mutants
  aba1 393
  abi1-1 393
  abi2-1 393
  assimilate partitioning 505
  biosynthesis 142-143, 146, 570
  ethylene 507
  growth analysis 505
  hypersensitive 407, 523
  insensitive 30, 398, 523
  leaf growth 506
  shoot growth 506
NAD 146
nature 9
NCED 144-145, 147, 150, 598
negative regulator 397
nitrogen 541, 553
null mutants 497
oscillations 398, 399
OST1 405
oxygen 400
pathogens 236
perception 394
pH 399, 394, 495, 496, 401
phaseic acid 151
phloem 499
plasma membrane 394 397
potassium 394, 399
potato 553
PP2A 407
precocious germination 522
precursor 141, 148, 496

pre-harvest sprouting 523, 584
proteasome 531
protein 10
protein kinase 404, 407
receptor 495
redistribution 496, 498
regulated genes 527
regulation 401
response 397, 407, 525
response elements 526-527
rest 11
rhizosphere 499
RNA 407
root growth 10, 397, 505
ROP 397
SAG 577
salinity 391, 407, 598
secondary messengers 400, 406, 409
seed 10
  development 137
  dormancy 391
  germination 391, 397
  maturation 523
sensitivity 34, 532
separation 689
shoot growth 10
signal transduction 393-395, 398, 402-404, 408-409, 495, 599
  ABI3 599
  blue light 394
  CO_2 394
  cold tolerance 599
  ectopic expression 599
  fusicoccin 394, 395
  G protein 402, 409
  humidity 395
  inositol 1,4,5 triphosphate 403
  kinases 398
  light 395
  lipids 403
  myo-inositol-hexakisphosphate 403
  phosphatidylinositol-phosphate 403
  phospholipase 403-404, 409
  phospatidylinisitol-3-phosphate 404
  PLC 404
  protein kinases 404
  receptors 408
  red light 394
  sphingosine-1-

Index

phosphate 402
sphingosine kinase 402
stomata 404
soybean 521
sphingosine phosphate 399
stomata 33, 392, 497-499, 502, 504
  closure 9, 34, 137, 392, 394, 397
  calcium 394
  cytokinin antagonism 504
  potassium 394
  oxygen 394
  opening 392, 393
storage protein 9
storage reserves 522
stress 598, 599
  tolerance 137
structure 10, 137, 139
sugar 531
tobacco 598
tomato seed 274
transcription 391, 526
transpiration 392-393
transport 10, 33
tuberization 553
tyrosine phosphatases 406
vesicle trafficking 397
Vicia. faba 394, 395, 397, 399
viviparous 520, 524, 584
Vp 143-145, 523, 584
water 10, 137, 393, 401
xylem 10, 33, 394, 497, 509
  photosynthesis 509
  sap 498
  sap pH 394
  stomatal conductance 509
Abscisin II 10
Abscission 137, 377, 538
  zone 384
  abscisic acid 137
  leaf 538
ACC-oxidase **120-122, 129-130**
accumulation 131
Agrobacterium rhizogenes 132
anther 130
antisense 131
apoplast 122
apple 122
ascorbate 121

biotechnology 131
cantaloupe 131
carnation 387
catalytic cycle 121
cell fractionation studies 122
climacteric 129
CO_2 121
corolla 130
cucurbits 130
cytosol 122
ethylene-forming enzyme 120
expression 129, 386
flavanone-3-hydroxylase gene 120
flower 130, 386
fruit ripening 120, 131
gene 126
gene family 129
GUS 130
hypersensitive response 130
immuno-cytolocalization studies 122
inactivity 121
leaf senescence 129, 132
localization 122
melon 120
morphological changes 132
negative feedback 133
orchid 384
ovule development 132
peduncle 131
petals 130
petunia 130
Phalenopsis 384
pigment 131
pollination 130, 384
polygalacturonase 131
pTOM13 120
radiochemical studies 122
Ralstonia 130
regulation 126, 127
resistance 132
rind yellowing 131
ripening 129
Saccharomyces 120
senescence 130, 386
sense 131
stigma 130
style 130
tobacco 130, 132
tomato 120-122, 126, 130-131
Xenopus oocytes 120
ACC-synthase **119-120,**

123-129
abscission 383
ACC 119
ACS gene 124, 335, 345
activity 127, 384
aminoethoxyvinylglycine 357
analog 119
antisense 131, 383
Arabidopsis 122-125, 127
auxin 124
AVG 357
biotechnology 131
calyculin 128
cycloheximide 124
cytokinin 128
dimerization 119
divergence 123
eto 128, 129
expression 124, 125, 385-386
flowers 385-386
fruit ripening 124
gene 123, 126, 127
inhibition 119, 124
inhibitor 357
LE-ACS, 124-125
methylthioadenosine 119
orchid 384
Phalenopsis 384
phosphorylation 123, 128-129
phylogenetic tree 124
promoter 124
protein stability 127
regulation 123
regulation 125-128, 129
ripening 383
S-adenosyl methionine 119
senescence 383, 386
sense 131
stability 129
stimulation 124
tomato 119, 123, 124, 126, 128
transcription 123
turnover 129
zucchini 119
β-γ elimination 128
Acclimation 391
ACD9 127
Acetate-malonate derived compounds 623
Acetic acid 198
Acetylenics 623
Acid
  growth hypothesis 265

719

Index

invertase 555
phosphatase 228
AcSAM3 123
Actin 397, 399, 444, 446
Active oxygen species 131
Acyl-CoA oxidase 618
Acylcyclohexanediones 89
Adenylate 555, 556
ADP-glucose 554, 556
ADP-ribosyl cyclase 399
AGL20/SOC1 546
Agrobacterium 21, 552
Agrobacterium rhizogenes 132, 552
Agrobacterium tumefaciens 100
AHKs 254, 517, 324-330, 338, 340, 342
AHPs 26, 330, 332, 336-341, 346
AKIP1 407
Al21 protease 234
Alcohol dehydrogenase 143
Aldehyde oxidase 143
Aleurone
  ABA 236
  amylase 81
  barley 227
  calcium 230
  cell walls 225
  cells 225
  enzymes 228
  GAMYB 234
  gene expression 227
  germination 227
  gibberellin 221-224, 228, 235-236, 533
  hormonal control 228
  hydrolases 227-228
  layer 81, 222
  plasma membrane, gibberellin 229
  protein synthesis 228
  regulation 228
  reserve hydrolysis 227
  starch 225
  vacuolation 225
Algae 115
ALH1 364-365
Alkaloids 623
Alkylation 164
Allene oxide cyclase 614, 616
Allene oxide synthase 551, 569, 614, 616-617
  jasmonate accumulation 617
  tuber formation 551

Allergens 264
Alstroemeria hybrida 574
Altered meristem program 519
Alternaria 624
Alternative oxidase 637
Amines 11
Amino acid 539, 623, 644
1-Aminocyclopropane-1-carbocylic acid (ACC) **117-133**
  β-cyanoalanine 118
  β-cyanoalanine synthase 122
  β-ketobutyric acid 133
  γ-glutamyltranspeptidase 122
  γ-glutamyl-β-cyanoalanine 118
  ammonia 133
  analog 118
  analysis 681, 687
  application 361-362, 364
  Arabidopsis 687, 688
  arsenic inhibition 133
  ATP 118
  biosynthesis 116-7, 119, 384, 687
  biotechnology 133
  catalytic conversion 120
  conjugate 118
  cytosol 118
  deaminase 133
  derivatisation 680-681, 687
  distribution 118
  enzymes 122
  ethylene 9, 116-118, 132-133
  flooding 133
  flower 386
  GC/MS 687
  glucose 31
  heavy metals 133
  LC/MS 687
  long hypocotyl 364-365
  MACC-hydrolase 118
  malonyl ACC 117-118
  malonyltransferase 122
  metabolism 118, 122
  MS-analysis 680
  orchid 384
  oxidase: see ACC oxidase
  oxidation 384
  pathogen 133
  petunia 383
  Phalenopsis 384
  phloem 118

  senescence 386
  SPE 689
  synthase: see ACC synthase
  tobacco 383
  translocation 384
  transport 118
  vacuole 118
  xylem 118
Aminoethoxyacetic acid 386
Aminoethoxyvinylglycine (AVG) 357, 386, 502
Aminopeptidases 226
Ammonium 540
amp1 553
α-Amylase 7, **221-237**
  ABA 533
  activator 233
  aleurone 222-223, 225
  arabinogalactan proteins 230
  cell wall 225
  endoplasmic reticulum 225
  expression 223, 229, 232, 233, 235
  GAI/RGA 533
  gene expression 222, 224, 232
  germination 223, 227, 228
  gibberellin 7, 221, 224, 228-229, 236
  Golgi 225
  green fluorescent protein 223
  high-pI 226
  HvGAMYB 237
  localization 225
  promoter 230, 233-235, 237
  repressor 233
  scutellum 223
  sugar 533
  vacuolation 232
Amylopectin 226
Amyloplastidial ATP:ADP translocator 555
Amylose 226
Analysis **671-694**
Antagonism 649
Andigena 540, 546-547, 550
Angiosperms 157
Anion
  channel 452
  efflux carriers 442-443

Anther dehiscence1 (*DAD1*)
  mutant 616
Anthers 130, 487
Anthesis 637
Anthocyanins 562
Anthranilate 682
Antioxidants 673
APCI LC/MS-MS 688
Apical 438
  bud 33
    IAA 29
  dominance 297, 321,474
    cytokinin 321
    leaf 474
  hook 163, 453
  meristem 351, 518
Apoplast 122, 230, 393, 495-497
  calcium 230
Apoplastic 408, 498, 555-556
Apple 33, 116, 118, 122, 142, 371
  ABA 33, 142
  ethylene 118
  fruits 371
Aquaporin 504, 533
Arabidopsis
  12-oxophytodienoic acid 551
  ABA 523, 527
    activated protein kinase 407
    biosynthesis 143, 145-147, 149
    dormancy 523
    mutants 523
    stomata 392, 397, 399-400, 404-407
    abscisic aldehyde 147
  ACC 687, 688
    synthase 123
  ACS 127
  AKIP1 407
  aldehyde oxidase gene 147
  ARF genes 291
  Aux/IAA 286
  auxin
    ABP 283-284
    distribution 464
    DR5 478
    ethylene 458
    influx 447
    PIN 450-451, 453, 457, 461, 463
    proteasome 296-297, 300

  response 290-291
  signal transduction 285-288, 290-292
  transport 292, 446, 452-454, 457, 459, 464, 481
  BAK1 423, 426
  big/tir3/doc1 453
  brassinosteroid mutants 414, 418
  BRI1 418, 421, 423, 426
  cell polarity 461
  coi1 mutant 629
  CPS 82
  CTR1 380
  cytokinin 26, 100-102, 685
  Dc3 Promoter Binding Factor 527
  DNA insertional null mutants 401
  DR5 480, 690
  dwarf mutants 65
  embryogenesis 466
  ent-kaurene 73
  ethylene 371, 372, 374, 622
    biosynthesis 122, 125
    receptor 374, 376-377
    signal transduction 387
    expression 480
  flower abscission 382-383
  flowering 545
  GA2ox 76, 83
  GA3ox 74, 76, 83
  GA20ox 17, 76, 83-84
  ga1 mutant 65, 71, 79
  ga2 mutant 65, 71
  ga3 mutant 65, 71
  ga4 mutant 65, 84
  ga5 mutant 65, 84
  GAI/RGA 235
  GC/MS 687
  genetic screens 416
  genome 90
  GGPP synthase 70
  gibberellin
    action 79
    mutants 66, 83
    response 79
    signalling 25, 87, 236
  glucose 30, 31
  GPA1 401, 402
  IAA 691
  ICS genes 643
  jasmonate 13, 622, 624
  kinases 405
  LD plant 545

leaf 473, 486
LRR-receptor-like kinases 423
MEP pathway 69-70
mutant screens 5, 647
MYB 87
NAA 481
NahG plants 640
NR 378
photoperiod 86
PIN mutants 22
potassium channels 395
proteasome 292-293, 296-297, 300
RGA 26
SABP2-like genes 649
SAGs 564
Salicyclic acid-deficient mutants 642
senescence 416, 563-569, 571-573
shoot apex 82
SPINDLY 236
stomata 397, 399-400, 404-407
sulfotransferase 551
sur2 690
TIBA 481
TRIP1 426
UBA2a 407
ubiquitin 292-293, 296-297
vascular differentiation 471
vein pattern formation 471
β-cyanoalanine synthase gene 130
Arabinofuranosidase 228
Arabinogalactan 229-230, 237, 515
Arabinoxylans 226
Arachidonic acid 611
ARF 285, 290, 297-300
  GEF 454-455
  genes 289-291
  binding protein 290
Arginine 11
ARRs 253, 328-336, 338-343, 345
Artemesia tridentate 625
Arum 13, 637
ASK gene family 293
Aspen 564
Aspirin 346, 636
Assimilate partitioning 552-553, 557
ATP 116, 118

721

Index

Aux/IAA
  Arabidopsis 286
  auxin 289, 299-300
    binding 284
    signal transduction
      285-287, 300, 461
    degradation 293-295,
      298-299
    dimerization 290, 298
    DNA binding 287
    hormonal regulation 294
    instability 294
    pea 286
    phosphorylation 296
    post-translational
      modification 296
    proteins 287, 290, 294-
      296, 299, 300
    proteolysis 294, 296,
      298-299
    repressors 298, 299, 518
    SCF complex 294-295
    soybean 286
    stability 295, 297
    TIR1 298
    turn over 299
    ubiquitin 294, 299
AUX 364, 447-448, 455,
  462-463
Auxin 36-62, 204-220, 282-
    303, 437-492
  2,3,5-triiodobenzoic acid
    216
  2,4-D 445
  4-Cl-IAA 37
  ABP1 283, 284
  abscission 383, 384
  accumulation 450, 456,
    476
  acid growth theory 210,
    211
  action 4
  activated gene 24
  adventitious root 479
  affinity chromatography
    283
  amino acids 39
  analysis 681
  anion efflux carriers 442
  anti-IAA antibodies 456
  apical hook formation
    453
  apical-basal polarity 516
  apoplastic pH, wall
    loosening 209
  Arabidopsis see
    Arabidopsis: auxin
  ARF 286-287, 291, 300,
    455, 519
  assimilate partitioning 6
  Aux/IAA 286, 289, 300,
    519
    proteins 293, 299, 461
  AUX1 447
  Avena see maize
  AXR 295, 518
  basipetal stream 462
  bdl 465
  bean 458
  binding activity 283
  binding proteins 282-283,
    518
  biosynthesis 5, 40-47,
    384
  Brassica 465
  brefeldin 445
  bundle sheaths 479
  calcium 285
  calmodulin 285, 288
  cambium 478, 483
  canalization 480, 481,
    518-519
  carrier 23, 444-446, 460,
    517
  cell division 6, 7, 241,
    248, 282, 284, 459, 460
  cell elongation 4, 6, 282,
    459
    2,4-dinitrophenol 207
    acid-growth theory 209
    apoplastic pH 211
    ATP 207
    ATPase 207
    azide 207
    cellulose 215
    extension 215
    growth lag 205
    growth parameters 208
    inhibitors 207
    intact plants 215
    KCl 215
    KCN 207
    long-term 211
    maintenance 214
    osmoregulation 214
    pea stems 216
    polarity 460
    protein synthesis 207
    roots 217
    stem 216
    sucrose 215
    time course 205
    turgor 207
    wall extensibility 208,
      265
    wall loosening 209
    wall yield threshold
      208
  cell expansion 284
  chemiosmotic polar
    diffusion 442
  chloroplast 18, 473
  coleoptile 206, 441
  compartmentation 17
  concentration 17, 462,
    482
  conjugates 38, 289
  constitutive over-
    expression 284
  cop 288, 297
  cotton 458
  cycloheximide 443
  cytokinin 241, 334, 341,
    343, 344
  cytoskeleton 455
  DefH9::iaaM 600
  density 482
  differentiation 284, 464
  diffusible 17
  diffusion 442
  discovery 2-3
  distribution 466
  DR5 290, 450, 476
  early 286
  effect 6, 284
  efflux 17, 444, 446, 449,
    452, 455, 457
    PIN 286, 449, 452, 517
    GN 455
    inhibitor 446, 449, 450,
      452
    machinery 449
    Menkes copper-
      transporting ATPase
      446
    protein synthesis 446
    secretory system 446
    transparent testa 444
  elongation 200
  embryogenesis 284, 466,
    516-519
  epidermis 476, 480
  ethylene 6, 384, 385, 458,
    574
  exocytosis 456
  expansin 265-271
  expression 297-298, 299
  extractable 17
  fin219 288
  floral abscission 384
  flowering 6
  flowers 384
  flows 476
  flux 450

Index

free 473, 474, 489
fruit 6
function 437
G protein 285
GC/MS 681
gene expression 284
gene family 447 Kinetics 286
GH3 285-286, 288
gibberellin 20, 28, 30, 38, 84, 197-200, 217, 300, 533, 575, 600
GN 465
Golgi- 461
gradient 474, 482, 489
gravitropism 23, 216
growth 20, 33
  active site 17, 18
  kinetics 20
  rate 32
  responses 284, 289
GUS 288, 290, 293, 450, 464, 476
homeostasis 36-37, 437, 465
hydathodes 473
hydrolysis 39
hypocotyl 206, 453
iaaM 600
IBA 38, 51
inducible 288
inducible transcription 285
inflorescence 479
influx 447
  carriers 284
  inhibitors 447, 458
inhibitory effects 6
initial growth response 20
insensitive mutants 285
interaction 334
jar1 288
jasmonic acid 288-289
kinetics 286
lateral distribution 24
lateral organ formation 268
lateral root 479
LE gene 200
leaf 472
leaves 6, 473
level 36-37
lipid 461
location 21
luciferase 293
maize 206, 213, 215, 283, 288, 441-458
MAPK cascade 285

mass spectrometry 681
mechanisms 439
metabolism 284
metaxylem 479
methylated 681
microsomal 283
mitogen activated protein kinase 285
monopteros (mp) 454, 465
morphogen 464, 465
morphogenetic signal 482
movement 439, 476, 478-479, 481
mutants 289
mutations 291
NAA 445
NAC1 gene 300
natural 37
nature 5
Nicotiana tabacum 445
non-polar movement 439, 476
NPA 406, 457
onion 288
organ identity genes 269
orientation 460
overexpression 288
overproducing mutants 690
parenchyma 478
parthenocarpy 600
pattern formation 53, 517
patterns of cell division 464
pea 32, 83, 206, 445
pericycle 479
pH gradient 442
phellogen 478, 480
phenotypes 288, 447
phenylacetic acid 37, 39
phloem-translocated 440
phospholipases 285
phosphorylation 296
phototropism 2, 33, 464
phytotropins 457
PIN 447, 456, 461, 465
pine 464
pinoid mutants 453, 457
plasma membrane 284, 441, 459
pls mutants 343-344
polar flux 518
  secretion model 442
  transport 440, 458, 488
polarity 450, 459, 518
pollen 383
pollination 383

precursors 5
procambium 478
production sites 472-473
  hydathodes 473
  primary sites 473
prolonged growth response 20
promoters 289
proteasome 291, 294
protein 39
  degradation 291
  phosphatase 456-457
  phosphorylation 285
proton excretion, ATP 212-213
protoxylem vessels 479
RCN1 453, 457
reactivation 483
receptor 17, 282-284, 295, 518
redistribution 23
regulation 37, 266, 299
regulator 293
reporter gene 285, 290
resistant 6, 447, 465, 518
response 284, 285, 291, 293, 516
  elements 124, 289 298-299
  factor 287, 289, 454, 516, 518
  genes 23, 285-286, 289, 297-298, 300
  promoters 289
response genes 286
retention 456
root 6, 217, 387
root agravitropic 447
SAURs 286
SCF ubiquitin-ligase 297
secondary messenger 285
seedling morphology 516
senescence 574
sensitivity 384
sieve tubes 471
signal transduction 4, 282, 288, 292
signaling 297, 299-301, 384, 518
sink 476
sites 476
soybean 287
stem 217
stipules 473
stomata 502
storage 18
sugar 39
superoxide radicals 271

723

Index

synthetic 5
TIBA 441
tobacco 284, 445, 457
tomato 600
transcription factors 285, 454
transcriptional regulation 293
transcriptionally induced 288
translocated 439
transport **437-492**
  ABC transporters 452
  abscission 384
  apical 439
  Arabidopsis 446
  aux1 458
  axiality 438
  basal 439
  basipetal 23, 364, 445
  brefeldin 455
  BIG 456
  calossin 456
  cell differentiation 459
  cell-to-cell polar transport 439
  cycling 455
  cytokinin 458
  development 459
  efflux inhibitors 455
  embryo patterning 517, 519
  ethylene 458
  flavonoids 444
  fluxes 455
  genetics 446
  GN 454
  gravity 46
  growth 33
  guanine nucleotide 454
  histogenesis 459
  hypocotyl 365
  inhibitors 364, 441, 443, 444, 461, 517
  inverted umbrella 23
  lateral organs 269
  lateral root initiation 32
  localization 444, 450-451
  MDR1 452
  morphogen 464
  multidrug resistance 452
  mutants 461
  naphthylphthalamic acid 441, 443-444, 453
  nomenclature 439
  NPA-binding protein 456
  organogenesis 459
  pathway 440, 472, 475-478, 489
  pattern formation 438, 459
  PGP1 452
  phosphorylation 453, 456, 457
  phytotropin 443-444
  PIN 450
  PINOID 456
  polar 438, 446, 459
  polarity 438, 473
  protein phosphatase 2A 456
  pyrenoylbenzoic acid 443
  quercetin 443
  RCN1 gene 456
  receptor 455
  regulation 456
  root elongation 452-453
  seedling morphology 516
  sieve tube 489
  signal transduction 455
  stem elongation 32
  stimulus 463
  subcellular dynamics 454
  tir 292, 452-453
  triodobenzoic acid 443
  vascular development 32, 461
  vascular tissue 489
  vesicle trafficking 454-455, 458
  trees 479, 483
  trimethylsilylated 681
  Tropaeolum 39
  tropistic responses 6, 21
  tryptophan 439
  ubiquitin 287, 291-292, 297, 518
  unloading 447
  uptake 442
  vacuole 18
  vascular bundle 480
  vascular cambium 478
  vascular development 516, 518
  vascular tissues 6, 473, 478
  vessel 471, 474 478-479, 482-483
  wall loosening 213, 214
(1-3,1-4)-β-glucan 214
  apoplastic pH 210
  cellulose 214
  enzymes 209, 213
  hydroxyl radicles 211
  in vitro wall extension assay 214
  mechanism 213
  pectic chains 214
  pH optimum 214
  proteins 215
  protons 209
  stored growth 207
  target 207
  wall calcium 214
  wall loosening factors (WLF) 209
  wall loosening proteins 214
  xyloglucan 214
xylem 478-479
Zea mays 441
zucchini 442
Avena 205-206, 210, 215
Avena fatua 584
AVG 360
Avocado 149, 371
AXR 285, 293, 295-296, 364
Az34 506
β-alanine 116
β-amylase 226, 227
β-catenin 308
β-cyanoalanine synthase 122, 130
β-glucan 223
β-glucanase 228
β-glucuronidase (GUS) 23, 79, 293
β-oxidation 613
B3-domain proteins 526
Baeyer-Villiger 160
Banana 380
Barley (*Hordeum vulgare*) 143, 234, 237, 499, 539
  aleurone 226
  β-glucan 226
  calcium 230
  GA_1 222
  GAI 530
  GAMYB 234
  gibberellin 221, 224
    deficient dwarfs 72
    response complexes 232
    signalling 235
  kao mutant 72
  SAGs 564

Index

Basal 438
BASIC CHITINASE (b-CHI) 361, 365
Botrytis 365
BDL 465, 516, 518
Bean 5, 50, 458
Begonias 387
Benzo(1,2,3)thiadiazole-- 636, 640-641
Benzoic acid 2-hydrolase 643
Benzyladenine 503, 504, 552, 570, 577
Bicollateral bundle 479
Bielesky buffer 684
BIG 453, 456
Bioassays 19, 671
Biomass allocation 509
Biotechnology 131, 407
Biotic stresses 116
Blue light 364
BODENLOS 461
Bolting 163
BPBF transcription factor 234
Branching enzyme, starch 556
Branching 21
Brassica 465, 564
Brassica chinensis 414
Brassica napus 156
Brassica pollen 4, 11, 156
Brassin 156
Brassinolide 11, 156, 159, 160, 266, 689
  growth effect 266
  LKB 30
  structure 12
Brassinosteroid **156-178, 186-197, 413-436**
  ABA 142
  action 413
  active compound 192
  analysis 687
  application 569
  Arabidopsis 196, 433
  biosynthesis **158-159**, 569, 160-161, 190-191, 197
    C22 oxidation 159
    C6 oxidation 159-160, 187, 197
    campestanol 160
    campesterol 160
    CPD 159
    DDWF1 159
    DET2 159
    dwarf mutants 162

DWF4 159
  feedback regulation 191
  LK 190
  pathway 5, 156, 159-160, 162
  phenotype 162
  rate-limiting step 191
  regulation 179
  differentiation 415
  division 12, 414
  elongation 12, 216, 414
  length 190
  wall loosening 12, 414
cholesterol 156
concentrations 11, 688
crosstalk 5
cyclins 430, 432-433
dark 197
de-etiolated phenotype 196
derivatisation 680, 688
discovery 3, 4
dwarf 162-163
effects 12
epinasty 12
ethylene 12, 195, 575
fertility 12
GC/MS 688
gene expression 430, 432-433
germination 529-530
GPA1 532
hormones 3, 14, 459
jasmonate 575
leaf development 413
levels 197
light 197, 413
male fertility 413
metabolism 158
microarray analysis 432
MS-analysis 680
mode of action 413
mutants 5, 189-190, 196
nature 11
negative regulators 196
organ elongation 413
pea 29, 179, 186-193, 195-197
purification 688
receptor 5, 192
reproductive biology 415
root 12
seed germination 415
senescence 12, 413, 415
shoot elongation 190
signal transduction 413,
  Arabidopsis 417, 419

autophosphorylation 422, 424, 426
BAK 419, 420, 424-426, 428, 430
BES 430
BIN 429-431
receptor 421
binding activity 421
BRI 416-418, 420, 422, 425-426, 428, 430
BRS1-D 422
BZR 430, 434
CRR-receptor kinase 417
cu3 417, 418
d61 417, 418
downstream components 428
GSK3/shaggy kinases 429, 431
heterodimerization 420, 424-425
mass spectrometry 422
mechanistic parallels 431
mutant screens 417
negative regulators 429
nuclear steroid receptor 416
phosphorylation 420, 424, 425, 428-430
proteasome 429-431
radiolabelled 421
receptor-like kinases 420
rice 417, 419
SR160 422
steroid-binding protein 421
tBRI1 422
TGF-β receptor kinbase 424-425, 526
tomato 417, 419, 422
TRIP-1 424-425, 428
wingless/wnt 429-431
signaling 161
steroid hormones 156
structures 157
tomato 196
vascular differentiation 12, 413
xyloglucanases 12
Brasssinazole 196
Brefeldin 23, 445, 451
BRI1 530, 569
Broccoli 132
Brown algae 483
Bryonia 621

Index

Bud growth 539
1-butanol 404
Bundle 484, 486-487
BY-2 cell culture 247
bZIP proteins 526
C_{18} HPLC 673
Calcium
  ABA 394-396, 398-406, 495
  aleurone 230
  apoplast 230
  barley 230
  cytokinin 343
  cytosol 495
  channel 400
  GA 238
  intracellular 495
  oscillations 399, 403
  protein kinase 343 404
  wheat 230
Calmodulin 230, 232
Calorigen 13, 637
Calossin 456
Calyculin 128
Cambium 538, 489
cAMP 230
Campestanol 159-160
Campesterol 160
Canalization 461
Cantaloupe 131
Capsanthin 148
α-Capthesin-like protease promoter 234
Carbohydrate 541, 554
  nitrogen ratio 541
Carbon dioxide 391
Carbon metabolism 556
Carnation 132
  ethylene 381-383-387
  pollination 385
  SAMS 123
Carotenoid 139, 141, 148, 539
  abscisic acid 10, 145, 142, 144
  apocarotenoid 145
  biosynthesis 70
  cleavage reaction 144
  deficient 522
  indirect pathway 139
  metabolism 141
  photo-bleaching 522
Carrot 50
Casasterone 197
Castor bean 521
Catalase 648
Catharanthus roseus 160
Caulonemata 268

CBPs 345
CCA1 335
CCC 541, 554
cdc2 kinases 237
Cell
  cycle 242-243, 247-248, 340
  division 284, 328, 341, 414, 460
    naphthylphthalamic acid 460
    auxin 7
    brassinosteroids 414
    cytokinin 328, 341
    SAM 344
  elongation/enlargement 163, 204-218, 209, 460, 539
    acid-induced growth 210
    apoplastic pH, neutral buffers 211
    auxin 460
    Avena coleoptile 207, 211
    brassinosteroids 163, 204, 217, 413, 414
    cell wall 460
    cytokinin 204
    dwarf mutants 163
    ethylene 204
    growth parameters 205
    microtubules 460
    plasma membrane 209
    receptor 209
    yieldin 208
Cercospora rosicola 139
Cereal
  aleurone 237
  grains 221
  gibberellin 81, 221
  half-seeds 81
  starch mobilization 532
CEV1 630
cGMP 231, 238
Chain tubers 540
CHD3 308
Chemical ionization 675, 689
Chemiosmotic hypothesis 442, 450
CheY 322, 329, 332
chk 344
Chloro-ethyl-trimethyl-ammonium chloride (CCC) 541, 554
3-Chloro-4-hydroxyphenylacetic

acid 458
4-Chloroindole-3-acetic acid 37-38, 83-84, 197
4-Chloro-IAAsp 37
Chloronemata 268
Chloroplast 145, 146, 156, 321, 473
5α Cholestane steroids 156
Cholesterol 156
Cholodny-Went hypothesis 462
Chorismic acid 643
cim1 277
CIMMYT 586
Circadian rhythms, *ARR* 331, 543, 545
CKI1 26, 324, 326, 328, 340
  ectopically expresses 326
  function loss 326
  histidine kinase 326
CKI2 330
  over-expression 330
CLA1 gene 69, 70
CLAVATA3 (CLV3) 661, 662
Climacteric 129, 379, 565
CO_2 116
COI1 358, 365, 626-627, 631
Cold 391, 407
Coleoptile, *Avena* 211
Coleus blumei 483
Collateral bundle 479
Collettotrichum lagenarium 639
Commelina beghalensis 503
Commelina communis 393, 397, 404, 495, 499, 502, 503
Communication 641
CONSTANS (CO) 86, 545-546, 557-558
Constitutive photomorphogenesis (cop)/de-etiolated (det/fusca (fus) 163
CONSTITUTIVE TRIPLE RESPONSE1 355, 357
COP1 288, 345
COP9 signalosome complex 297
  auxin-related phenotypes 297
  ubiquitin mediated proteolysis 297
COP9/signalosome 345

Index

cytokinin 345
Copalyl diphosphate synthase 71, 79, 88
Copalyl pyrophosphate 89
Copper transporter 355
Coronatine 618-619, 626-627, 630
Corolla 130
Cotton 89, 392, 458, 521
Cotyledon 161, 517, 522
  Cup-shaped (CUC) 517, 518
Cowpea 145, 148
cpd/dwf3 161
CP1 80
CPR 366
cre1, see AHK4
CRE1 324
Crops **582-609**
Cross-talk 576
Cryptochromes 545
CSBP 345
CTR1 31, 354-356, 358, 360, 361, 371, 380, 530-531
Cucumber 66, 569, 639
Cucumis sativus 473
Cucurbic acid 550, 618
Cucurbita maxima 74, 414, 483
Cucurbita pepo 445
Cucurbitaceae 65
Cucurbits 130
CUL1 293, 296, 297, 299
Cyanide 122
Cycad 637
CycD3 250, 252-253, 341, 342
Cyclic ADP ribose 399, 404
Cyclins 24, 242, 252-253, 533
Cyclin/CDK-complexes 242-244, 252-253
Cycloartanol 164
Cycloheximide 124, 287, 443, 445, 455
Cytoskeleton 446
Cyanide, respiration 122
CYPS1 158
CYR1 345
Cysteine endopeptidases 227
Cysteine protease 234
Cysteine synthase 130
Cytochrome oxidase 122
Cytochrome P450-dependent mono-

oxygenase 72
Cytokinin **95-114, 241-261, 321-349**
  ABA 142, 503, 575
  absence 248
  accumulation 552
  action 4
  active forms 96, 97
  activity 103
  adenine derivatives 7
  Agrobacterium 21
  Agrobacterium rhizogenes rolC 552
  AHPs 336
  alfalfa 573
  alternative pathway 103, 104
  analysis 683, 685
  anthocyanin production 345
  apical dominance 321
  Arabidopsis 26, 103, 573, 685
  aromatic 95
  ARRs, 330, 573
  autoregulation 572
  auxin 8, 241, 334, 341, 343, 344
  benzyladenine (BA) 97
  binding 327
    proteins 345
  bioassay 96, 98, 103, 570
  biosynthesis 8, **95-114**, 340, 342, 552, 571
    gene 21
    senescence 571
    sites of 8
  bok-choy 573
  branching 21
  broccoli 573
  BY-2 cell cycle 247
  calcium 343, 346
  carbon 553
    flux 553
  CDPK 343
  cell 8
    cycle 240, 242-243, 247-248, 340-342
    CycD3 250, 252-253, 341
    division 8, 241, 244, 328, 341, 517
    enlargement 8
    expansion 345
    proliferation 244, 342, 552
  cells 552
  chickpea 103

chloroplast 8, 321
cis-type 103
cis-zeatin 97, 103
cis-zeatin riboside 5'-monophosphate 97
cis-zeatin riboside 97
CKI1 573
conjugates 552
COP9/signalosome 345
cre1-1 327
crown gall tumors 8
CycD3 342
decapitation 102
defined 96
degradation 98, 685
derivatisation 680, 685
detection limits 685
developmental fate 241
differentiation processes 244
dihydrozeatin 97
dihydrozeatin riboside 97
discovery 3, 95
distribution 244
embryogenesis 514
effects 8
endogenous 244
ethylene 127-128, 335, 345, 357
exogenous 244
expansin 277
extraction 684
flower 342
foliar applications 552
free-base 96, 97
fruits 8
GC/MS 685
gene induction 331
gene responsive 26
genechip 334
germination 529
glucose 32
glucosides 684
histidine kinase 324, 325, 327
homeotic 340, 342, 553
identification 244
immunolocalisation 244, 691
in vitro binding assay 96
inactive forms 98
induced genes 330
interaction 334
ipt gene 552
isopentenyl transferase 604
isopentenyladenine 32, 97, 100

727

isopentenyladenine riboside 97
isopentenyladenine riboside 5' phosphate 97, 103
isopentenyltransferase (*IPT*) 98-104, 503, 552, 571-572, 604
isoprenoid 95, 99
KNOTTED 553
lateral root 102
LC/MS 685
leaf expansion 8
senescence 8
lettuce 573, 689
levels 247
light 334, 341
long-day 102
maize 103
manipulation 604
mass spectrometry 244
measurement 244
meristem 321, 341, 342
methoxytopolin 97
methoxytopolin riboside 97
mitosis 242
mode of action 242
morphogenesis 8
moss 8
MS-analysis 680
mutant 344, 571
N-alanyl 98
natural 95-97
nature 7
negative regulators 343
N-glucosides 98
nitrate 340, 343, 344
nitrogen 102, 334
nucleoside/tide 97, 683-685
nucleus 26
O-glucosides 97, 98
overexpression 98
overproduction 553, 571
oxidase 97, 342
pattern formation 517, 533
perception 4
petunia 573
phenylurea 95, 96
phloem 97, 102
phospholipase D 343
phosphorelay 321, 323, 346
phosphorylation 339
phytochrome 346
potato 103

primary response 326, 328, 330, 332, 334, 340
promoters 334
proteasome 335
protein degradation 334
protein kinase 26
protein stability 346
protein turnover 345
purification 684
quantification 685
receptor 96, 103, 321, 324, 326-328, 517
 CHASE domain 324
 gene 690
response mutants 343
responses 344
riboside 8, 685
ribotides 8
rice 571, 573
root 8, 21, 327, 336, 344, 345
seeds 8, 514
senescence 321, 328, 336, 342, 570, 572, 604
sensitivity 552
separation 689
shoot apical meristems 342, 344
shoot development 342
shoot initiation 8, 333, 345
side chain 98
signal transduction 4, 26
signaling 321, 333, 339, 340, 343, 346
 calcium 343
 model 339, 340
 negative feedback 340
 phosphorylation 339
 transcriptional activation 339
sink 552, 557
sorghum 571
source-sink 553
soybean 571
stomata 8, 503-504
storage forms 96, 97, 103
stress 341
structural diversity 96
structure 97
suspensor 516, 517
synthetic 95
tissue culture 7, 26
tobacco 21, 571, 573
tomato 573
topolin 97
transgenic tissues 26
translocation forms 96, 97

transport 8
trans-zeatin 97
trans-zeatin riboside 5'-phosphate 97
trans-zeatin riboside 97
tRNA 103
tuber 547, 552
tuberization 552-553
vascular morphogenesis 330
vascular stele 102
xylem 97, 503, 571
Cytoskeleton 397, 455
Cytosol 118, 122, 363, 555-556
D18 gene 81, 309
DAD1 616
DAG1 525
Daminozide 89
Darwin, 2
Daylength 538-540, 545
DCT1 357
Decapitation 199, 474
Decreased sucrose synthase activity 555
Deepwater rice 272
De-etiolated 163, 196
Defensins 624, 656
Dehydration 393
Dehydrin 407
DELLA 309-314, 528, 530
6-Deoxocasasterone 197
Deoxyxylulose phosphate 67-69
Desiccation 137
 tolerance 521, 524, 531
det 161, 163
Deuterium 676-677, 679
DFL1 288, 289
Diacylglycerol 403
Diastase 226
Diazomethane 675
Dicots 123
Dichloroisonicotinic acid 636, 640, 641, 647-648
2,4-Dichlorophenoxyacetic acid (2,4-D) 440, 441, 445, 447, 574
Dictyostelium discoideum 99-100
Diethyldithiocarbamic acid 673
Difluoromethylornithine 553
Digitalis 381
Dihydrophaseic acid 151,152

Dimethylallyldiphosphate 70, 99
Dioxindole acetic acid 57-58
Dioxygenase 73, 89
Diphosphocytidel- 68-69
DIR1 649
Disease resistance 4, 116, 365
Diterpenoids 623
Divinyl ethers 611
Dof transcription factor 526
Dormancy 514, 532, 538, 552, 585
 abscisic acid 11, 137, 524
 after-ripening 529
 Avena fatua 584
 axillary buds 539, 552
 brassinosteroids 525
 bud 538, 539
 coat-imposed 529
 de-etiolated (det) 525
 desiccation tolerance 524
 dormancy-breaking 529
 endo-dormancy 539
 ethylene 525
 GA 523, 525, 585
 maintenance 524
 mutants 524-525
 natural variation 525
 photodormancy 529
 pre-harvest sprouting 584
 quantitative trait loci 525
 seed germination 584
 stratification 529
 testa 525
 wheat 584
Dormin 10
DR5 280, 473-474, 451, 478, 690, 691
DR5::GUS 280, 475, 478
Droopy 553
Drought 391, 407, 494, 499, 504, 507, 598
DVL1 664-665
Dwarf 156, 161, 163, 180-187, 230, 272, 305-309, 569,
DWF 158, 161, 163
Deoxyxylulose 69-70
E. coli 76, 119
E3 ubiquitin ligase 307
E8 133, 370
EBF1 358
Ecdysone 157
EDS1 643, 649
EEL 524, 527
EER1 361, 356, 364

Efflux carrier proteins 454
Efflux inhibitors 453
EIL1 358, 362
eil1/wei5 362
EIN2 31, 354, 356-358, 360, 365, 371, 381, 383, 530-531
EIN3 356, 358, 366, 372, 380, 381, 384
EIN4/5/6/7 357, 360, 372
EIR1 364
Electron ionization 675
Electron transport 637
Electrospray ionization 675
Elongation 162, 195, 199, 391
Embryo 221, 222
 ABA 522
 apical-basal axis 515
 basal cell 515
 castor bean 521
 cell division 514
 cell enlargement 514
 cotton 521
 cultured 521
 development 513, 632
 embryogenesis 520
 germination 520, 522
 gibberellin 221, 222, 533
 maturation 519-520, 522, 524
 patterning 466, 517
 Phaseolus 521
 radial axis 515
 rapeseed 522
 somatic 515
 wheat 521
 zygotic 515
Embryogenesis 513, 515-517, 525, 632
Embryogenic 622
Embryonic 284, 519, 521, 528
Endo-(1-3,1-4)-β-glucanase 226
Endo-(1-4)-β-xylanase 226
Endoplasmic reticulum 225, 283, 355
Endosperm 226, 273, 513-515
 amylase 533
 breakdown 223
 development 513
 endo-β-mannanase 274
 germination 223, 227
 gibberellin 222-223
 hydrolases 227
 hydrolysis, starch 225

LeEXP4 expression 274
liquid 515
nutrients 515
reserve 515
source 515
starch 225, 226
wheat 226
xyloglucan endotransglycosylase 274
ent 63
ent-6α, 7α-dihydroxykaurenoic acid 72, 73
ent-7α-hydroxy-kaurenoic acid 71, 72
ent-copalyl diphosphate 70
ent-gibberellane 6, 65, 71
ent-kaura-6, 16-dienoic acid 72-73
ent-kaurenal 72
ent-kaurene 66-67, 70-73, 79, 86, 89, 589
ent-kaurene oxidase 71-72
ent-kaurene synthase 71, 588
ent-kaurenoic acid 71-72
ent-kaurenoic acid oxidase 71-72, 588
ent-kaurenoids 70
ent-kaurenol 72
Environment 193, 333, 538, 540, 557
Epidermis, IAA 480
Epinasty 378
ERA1 397-398, 530, 531
Erectoides mutants 187
ERF domain 360
ERF1 356, 358, 365, 372, 622
ERS 352-355, 372-374
ESTs 283
24α-ethyl (homobrassinolide) 157
Ethylene **115-134, 350-388**
 1-aminocyclopropane-1-carbocylic acid see that entry
 1-methylcyclopropene (1-MCP) 583, 604
 2-oxo-4-methylthiobutyric acid 115
 5'methylthioadenosine (MTA) 116
 5-methylthioribose (MTR) 116
 ABA 531, 575
 abiotic stresses 116

Index

abscission 9, 369, 387
ACC See ACC, ACC synthase, ACC oxidase
algae 115
aminoethoxyvinylglycine (AVG) 601
analysis 687
antagonists 360
anther 130
antisense 132
apical hook 9
apical mutants 364
apple 116, 118, 604
applications 134
Arabidopsis 122, 371, 372, 374
ATP 116
auxin 124, 385, 458, 574, 575
  activity 384
  cross talk 365
  sensitivity 384
binding 372, 374, 377, 387
biosynthesis 9, **115-134**
  accumulation 505
  autocatalytic synthesis 370, 382, 386
  AVG 357, 360,
  control 386-388
  cytokinin 345
  discovery 116
  enzymes 119, 382; see ACC oxidase, ACC synthase
  fruit 380
  history 116
  1-MCP 387
  pathogenesis 378
  pathway 116, 119, 133
  pollination 384
  receptor synthesis 378-379
  sites 9
  Yang's cycle 117
biotechnology 131, 603
biotic stresses 116
brassinosteroid 575
broccoli 132
carnation 132, 382-384, 386, 604
catalytic conversion 120
climacteric 124, 129, 131, 371, 379, 381, 565
CO_2 116
commercial applications 132
copper 373

corolla 130
CTR1 31, 351, 355, 371, 380, 531, 567
cucurbits 130
cytochrome oxidase 122
cytokinin 335, 345, 575
deepwater rice 274
defense 9, 375, 597
deficient 9
development 383
discovery 3
disease resistance 9, 116
dissociation 377
domain 373
dormancy 9
drought 507
E8 370
effects 9
EIN2 31, 351, 354, 356-358, 360, 365, 371-372, 381, 383, 530-531
EIN3 356, 358, 366, 372, 380, 381, 384
EIN4/5/6/7 357, 360, 372
elongation 508
enhanced responsiveness 360, 364
environmental factors 123
ERF1 356, 358, 372, 622
ERS 352-355, 372-374
Eto 351
ETR1 324, 338, 352-355, 357, 360, 363, 372-374, 376-377, 383, 387, 604
expression 385
femaleness 9
ferns 115
flacca 143, 147
flower 9, 130, 369, 374, 382-386
flowering 583
fruit 9, 123, 369-371, 375, 378, 379, 380
  immature 375
  ripening 9, 116, 123, 370-371,374-375, 378, 380, 565, 602
GA 557, 575
GAF domain 373
gene expression 375
gene regulation 123
genetic engineering 380
germination 529, 530
glucose 31
GUS 130
histidine kinase 324, 325, 376

hormone 14
hypersensitivity 130, 358, 374, 378
IAA 27
inactivation 379
inducible genes 370, 374
injury 9
insensitivity 354, 363-364, 372-373, 376, 508
jasmonate 370, 575
jasmonic acid 575
kinase 378
lateral growth 9
leaf 9, 129
levels 27
linoleic acid 116
lipid peroxidation 115-116
liverworts 115
malonyl ACC 133
mango 379
MAP kinase 371
1-MCP 387
melon 132
metabolism 604
methionine 9, 115-116
molecular clock 379
mosses 115
muskmelon 379
mutants 31, 351, 567, 597, 604-605
negative-feedback 358
never-ripe 125, 127, 375-379, 597 604
olefin 115, 350
orchid 128, 383-385
overproducing 567
oxygen 131
passion fruit 379
pea 116
peach 379
pear 379
perception 370, 372, 379, 380, 386, 388
petal 130, 382-383, 386
petunia 130, 376, 383
petunia 382
Phalaenopsis 128
phosphorylation 133
plums 132
pollen 383
pollination 116, 130, 382, 383, 386
polyamines 11
powder mildew 131
precursors 133
primary response element 358

730

Index

product quality 386
production 34, 131, 363, 382-383
propanal 116
protein phosphatase 128
Ralstonia solanacearum 130
receiver 373
receptor 27, 324, 353, 371-378, 381, 384, 386-388, 604
  abscission 604
  action 377
  activity 324
  bound ethylene 376
  dissociation 376
  domain 373
  EIN2 31, 351, 354, 356-358, 360, 365, 371-372, 381, 383, 530-531
  EIN3 356, 358, 366, 372, 380, 381, 384
  EIN4/5/6/7 357, 360, 372
  ERS 352-355, 372-374
  ETR1 324, 338, 352-355, 357, 360, 363, 372-374, 376-377, 379, 383, 387, 604
  expression 374, 375, 378, 379
  family 352
  fruit 375
  gene 386
  histidine 378
  inactivation 379
  kinase 373-375, 377, 378
  level 378
  mutants 374, 381
  negative regulators 374
  over-expression 604
  pathogen 375
  pollination 384
  receiver 373
  replenishment 379
  senescence 604
  suppression 378
  synthesis 379, 388
  tomato 604
  type 353, 373
regulation 127, 133, 134
respiration 122, 131
response 376-378, 380, 381, 383
  element binding proteins 358, 360
  ERF1 356, 358, 372,

622
  transcription 360
  genes 531
  mutants 383-384
response element binding protein 356, 359-360, 364,366
responsiveness 376
Rin 127
ripening 9, 129, 370, 379-381, 388, 601
root 9, 508
Rumex 274
S-adenosylmethionine 116, 118, 133
SAG 576-577
salicylate 370
SAM hydrolase 604
seed germination 371
seedlings 116
SEN1 577
senescence 9, 128, 130, 132, 369, 382-383, 386-388, 565-567, 577, 601, 602-604
sensitivity 27, 116, 374-375, 378, 370, 381, 384, 386, 388
serine/threonine kinase 373
shoot 9
signal transduction 357-358, 369, 371, 384
  flowers 369
  fruits 369
  pathway 357
  positive regulator 357-358
  EIN2 357-358
  pollination 384
signaling 355, 358, 370-372, 375, 380, 385-386, 649
silver 6
stem 9
stigma 130, 384
storage containers 386
stress 9, 351, 369, 596
style 130
sycamore figs 116
system 1/2 370, 379, 382
tobacco 130
tomato 116, 118-124, 126-134, 374, 376, 378, 602, 604, 622
transcription factor 370, 372
transport 9

triple response 9, 127, 351, 360, 371, 376
two component regulators 323, 372
wilting 132
wounding 129, 597
Yang cycle 118
α-ketoglutaric acid 115
β-alanine 116
24-ethylidene (homodolichnosterone) 157
Etiolated phenotype 195
etn2 365
ETO 127, 129
ETR1 324, 338, 352-355, 357, 360, 363, 372-374, 376-377, 379, 383, 387, 604
Evaporative demand 493
Evapotranspiration 393
Evolution 483
Exocytosis 456
Expansin **262-279**
  α-expansins 262
  β-expansins 263, 264
  abscisic acid 278
  acid-induced growth 262-265
  adhesion 265
  Arabidopsis 276
  auxin 565-571
  cell growth 271
  cell proliferation 277
  cell walls 262
  cellulose 265
  creep tests 264
  cucumber 265
  cytokinin 277
  dicot 264
  sensitivities 264
  domains 263
  drought 278
  seed germination 273
  elasticity 264
  ethylene 274
  extensometer 262
  family-45 263
  ferns 274
  forces 270
  fruit ripening 276
  gene 266, 276
  gibberellin 271, 533
  grass 264
  hormone action 262
  hypocotyl 268
  hypocotyl 265
  leaf morphogenesis 270

Index

leaf phyllotaxis 269
leaf primordium 268, 270
localized induction 270
maize 268
moss 268
pH 262-265
pine 268
plasticity 264
polysaccharides 265
proteins 268
root hair 276
rooting 268
seed germination 271
strawberries 276
stress relaxation 263
tobacco 270
tomato 266-270, 274-275, 278, 414
wall extension 263
wall loosening 264
Expansion 284
Extensometer 263
Extra cotyledon 519
Extrafloral nectar 624
FACKEL 158, 163
fad 615
Far red light 538
Fatty acid 612, 624
F-box 314
protein 292, 296, 307, 358, 629
Feedback: GA pathway 186 189, 192
Feed forwards GA pathway 186
Ferns 115
Ferredoxin 144
Fertility 161
Fery1 532
FIN219 285, 288
Firefly luciferase 22, 26
Flacca 143, 147, 499
Flavanone-3-hydroxylase 120
Flavonoid biosynthetic pathway 445
Flavonoids 457, 623
flavoprotein monooxygenases 143
Flax 551, 569
FliM 322
Floriculture crops 387
Florigen 87
Flowering
ethylene 583
Lolium 687
long-day 86
pathway 545

photoperiod 86, 687
gibberellin 86-87, 687
pineapple 583
time 545
genes 545
tobacco 544
Flowers **381-388**, 546
abscission 381-384, 386
carnation 381
CONSTANS 546
development 130
Digitalis 381
ethylene 27, 374, 376-377, 381-388
GA_5 87
IAA 474
induction 87
initiation 381
loss 378
orchids 381
organ 162, 474, 487
Pelargonium 381
petunia 381
pollination 381, 385
senescence 38, 377, 381, 385-386, 388, 601
vascular differentiation 474
Fluoridone 278
Fluoroindole 449
Fosmidomycin 70, 87
Frit fast atom bombardment 675
Fructokinase 556
Fruit 370, 374, 377, 379, 388
aubergine 599
banana 380
citrus 599
climacteric 379-381
development 599
GA 599
grapes 565, 599
ethylene 27, 374, 379, 380
production 380
softening 276
expansins 276
lemon 565
parthenocarpy 599
pectinases 276
ripening 27
cell-wall hydrolases 276
climacteric 276
ethylene 116, **120-134**, **370-371, 374-381, 387-388**, 602

manipulation 601, 602
1-methylcyclopropene 602
non-climacteric 276, 565
senescence 602
tomato 601, 602
shelf life 381
softening 276
strawberry 565
synthesis 379
tomato 380
FT 546
Fujenal 73
Funaria hygrometrica 96
FUS 519, 523-524, 526
Fusicoccin 215, 217
acid growth 265
expansins 265
H-excretion 211
gibberellin 204
hydraulic conductivity 205
maximum growth rate 205
oat, coleoptile 208
polarity 461
roots 217
sensitivity to auxin 205
turgor pressure 205
Vigna 208
wall 191, 460
wall extensibility 205, 208-209
hydrolysis 226
loosening, model 209
relaxation 191
yield threshold 205
water potential gradient 205
G protein 229, 237, 397, 401, 402, 409, 532
GA see also Gibberellin
GA1 gene 71, 79
ga1 mutant 25, 65, 71, 83, 86-87, 305, 309-312, 316
GA_1
analysis 675, 686-687
auxin-induced 197-200
barley 222
bioassay 19
biological activity 64
biosynthesis 20, 74-75
bolting 18
brassinosteroid 180
content 19, 27, 28
dark 27

Index

de-etiolation 27, 85
embryos 222
endosperm 222
flowering 66
GA_{20} 19, 74, 180-181, 550
graft transmission 22
heights 19
2β-hydroxylation 66
IAA 197-200
identification 675
in plants 6
internode length 19
level 6, 19, 547
light 27, 85
location 29
MeTMS 675
mutant 71, 543, 547
pea 27, 200
precursor 7, 550
response 25
senescence 574
shoot 550
speed of action 29
spinach 18
stem elongation 7, 200
tallness 18, 22, 181
transport 7
tuber formation 547
tuberization 547-550, 557
GA_{103} 63
GA_{12} 63, 65-66, 70-75
aldehyde 71-72
GA_{14} 73
GA_{15} 75
GA_{17} 75
GA_{19} 75, 181
GA2 gene 71 / *ga2* mutant 65, 71
GA_{20} 17, 21, 74-75, 180-181, 549-550
bioassay 19
graft transmission 22
stolons 549
GA_1 19, 74, 180-181, 550
pea 180-181
transport 20, 550
tuberization 550, 557
GA 20-oxidase see under gibberellin
GA_{24} 75
GA_{25} 75
GA_{29} 17, 71, 75, 222, 549-550, 557, 686
GA_3: see gibberellic acid
GA 3-oxidase/hydroxylase see under gibberellin
GA3 gene 79; *ga3* mutant 65, 71
GA_4 64-66, 74, 79-81, 87, 548, 574, 687
GA4 gene 79, 84; *ga4* mutant 65, 76
GA_{44} 75, 181
GA5 gene 76, 84; *ga5* mutant 65, 76
GA_5 64, 66, 74-75, 86, 687
GA_{51} 75
GA_{53} 6, 12, 21, 66, 71-72, 74-75, 181
GA_6 64, 66, 86
GA_7 64, 66, 75
GA_8 75, 180-181
GA_9 63, 65-66, 74-75
GA_{4+7} 574
GAI 25-26, 161, **309-313, 315-316**, 530, 533, 587, 594-595
Galactolipids 616
Gas chromatography 673, 680
Gaultheria 636
GC/MS 673, 675, 678, 686-689
GC/SRM-MS 687
gca2 398
GCC box 356, 358, 360, 364
GC-MS, plant hormones 3
Gene **695-716** and see individual genes and topics
Gene-for-gene resistance 640
engineering **582-609**
redundancy 185
screens 285
Geraniums 387
Geranyl diphosphate 70
Geranylgeranyl diphosphate 66, 68-70
Germination **513-537**
α-amylase 227, 228
α-glucosidase 227
ABA 149, 523, 529, 530, 532, 533
aleurone 227
aminopeptidases 226
amylases 521
β-amylase 227
β-glucan 223, 226
β-glucanase 226, 228
brassinosteroids 529, 530
carboxypeptidases 226
cereal grains 222
cysteine endopeptidases 227
cytokinin 529
desiccation tolerance 514
developmental arrest 514
endosperm 223, 227
ethylene 529, 530
ethylene 530
gibberellin 221-224, 227-228, 238, 523, 530, 533
glutamine 520
glyoxylate cycle 521
hydrolases 532
inhibition 529
inositol trisphosphate 532
jasmonic acid 529
light 514, 530
mangrove 520
meristematic activity 514
metabolic activity 514
nuclease 226
nucleic acid 223
nucleosidases 226
phases 529
precocious 520, 521
production 532
proteases 521
protein 223
proteolysis 227
radicle emergence 532
reserve mobilization 222, 227, 514, 521, 532
ribonuclease 226
root growth 521
salt 530
scutellum 223
starch 223
storage protein 226
sugar 149, 530
temperature 514
transcriptional profiling 533
water 222, 514, 521
GH3 24, 288-289, 480
Gibberella fujikuroi 3, 64-65, 71, 73, 304, 588
Gibberellane 63
Gibberellic acid (GA_3) 6
α-amylase 228
acid phosphatase 228
aleurone hydrolases 228
arabinofuranosidase 228
β-glucanase 228
Bakanae (foolish seedling) 304
biological activity 64
biosynthesis 66, 74-75
expansin 271
extensibility 272

Index

formation 74
fungal 6, 304
Gibberella fujikuroi 304
2β-hydroxylase 66
occurrence 65
phospholipid 228
protease 228
ribonuclease 228
Rumex palustris 272
senescence 577
xylopyranosidase 228
Gibberellin **63-94, 179-186, 193-195, 197-200, 221-240, 304-320**; see also individual GAs
13-hydroxylases 73
13-hydroxylation pathway 72
16, 17-dihydro-GAs 89
20-oxidase 21-22, 38, 74-75, 81-82, 86, 543, 548-549, 587-589, 591-592
 location 22
 tuberization 549
20-oxidation 183
2-oxidase 74-76, 85, 195, 316, 589, 592-594
2-oxidation 65, 74, 77, 183
2β-hydroxylation see 2-oxidation
3-oxidase 19, 22, 75, 85, 223, 548-550, 557, 588-589, 591, 594
3β-hydroxylase see 3-oxidase
3β-hydroxylation 65-66, 74
4-Cl-IAA 37-38
α-amylase 7, 221, 225-229 231-232, 234, 236, 309
A numbers 63
ABI 533
abscisic acid 221, 142, 146, 230, 236, 238, 547, 584
acetylglucosamine 309
action 79, 557
activity 19, 574
akadaic acid 231
aleurone 21, 223, 229, 235-237, 223, 316, 533
analysis 680, 686, 689
antagonism 584
antagonists 557
antisense 543
apical meristem 553

Arabidopsis 86, 235, 236, 304, 308
arabinogalactan 229, 237
auxin 28-30, 197-200, 300, 533, 600
bacteria 64
barley 21, 221, 232, 235, 316
bioactivity 18, 65-66
 2β-hydroxylation 65
 3β-hydroxylation 65
bioassay 19, 66
biosynthesis **63-94**
 20-oxidase 38, 74-75, 81-82, 587-589, 591-592
 2-oxidase 74-76, 85, 589, 592-594
 2β-hydroxylation 74
 3-oxidase 22, 75, 85, 223, 588-589, 591, 594
 3β-hydroxylase see 3-oxidase
 3β-hydroxylase inhibitors 88
 3β-hydroxylation 74
 acylcyclohexanedione 88-89
 AMO-1618 88
 ancymidol 88, 547
 andigena plants 547
 antisense 590
 apple 583, 591
 Arabidopsis 66, 591-592
 auxin 83, 197-200
 C_{19}-GAs 74-75
 C_{20}-GAs 74
 catabolism 589
 cell-free systems 78
 cellular localization of 80
 chemical control 87
 chlormequat chloride 88
 copalyl diphosphate synthase 82, 585, 588
 cucurbits 78
 daminozide 88
 decreasing 587
 de-etiolation 85
 deoxyxylulose 69-70
 developmental regulation 78
 dihydro-GA$_5$-13-acetate 88
 dioxygenases 77
 dwarf 592, 593

ent-kaurene 66-67, 70-73, 79, 86, 89, 589
ent-kaurene synthase 70, 588
ent-kaurene oxidase 71-72
ent-kaurenoic acid oxidase 71-73, 588
enzyme 549
feedback 588, 589, 592-594
flavonoid 583
flux 589
Gibberella fujikuroi 77
growth retardants 87
IAA 197-200
inhibitors 26, 78, 79, 87, 89, 224, 541, 547-548, 554
KNOTTED 82, 591
lettuce 592
light 529
location 21
manipulation 588-590
membrane 7
mepiquat chloride 88
mutants 66, 587
negative regulation 78
NTH15 591
onium inhibitors 88, 589
P450 monooxygenase 77
paclobutrazol 88-89, 308, 547
pathway 547
pathway 75, 181
pea 28, 74, 180-184, 588
Phaseolus coccineus 592
PHYB 542
positive regulation 78
prohexadione calcium 88-89, 583
pumpkin 74, 592
regulation 78, 589
response 78
rice 80, 590, 591, 593
seed germination 84
seeds 78
semi-dwarf 587, 590-592
sites 7, 78, 79
Solanum dulcamara 592
stem growth 592
sterol biosynthesis 89

Index

stimuli 78
sugar beet 593
suppression 589, 591
tetcyclacis 88
tobacco 591
triazoles 89
tricarboxylic acid 75
trinexapac 88-89
tuberization 547-550, 601
uniconazol 88-89, 548
wheat 593
bolting 7
brassinosteroids 162
C_{19}-GAs 63, 65-66, 74
C_{20}-GAs 63, 65-66, 76
Ca^{2+} 230-232, 238
calmodulin 230
cAMP 230
carbohydrate metabolism 557
catabolism 77, 78
catabolites 17, 76
cell 547
  division 533
  elongation 216, 533, 547
  length 29
  number 29
cereal grains 221, 237
cGMP 231, 238
chlorophyll 574
chloroplast 7, 67-70
chromatin 308
chrysanthemum 595
concentration 64
conjugates 77, 687
CONSTANS 547
cross-talk 90
cysteine endopeptidases 227, 234
dandelion 574
deactivation 66
deficient mutants 18-20, 25, 28, 65, 71, 83, 86-87, 180-186, 305, 309-312, 316, 522
degradation 21
DELLA 308-314, 587
dependent pathway 547, 557
derivatisation 680
dioxygenases 76
discovery 3
dormancy 523, 585
dwarf 65, 180-187, 230, 305-309; see also deficient mutants

effects 7
elongation growth 530
embryo 533
endogenous levels 542
endosperm 221, 223, 227
ethylene 533
expansin 272, 304
extensibility 272
flower 86, 316
flowering 64, 66
fruit 7, 64
fungi 21, 64, 77, 304
G-protein 229, 237, 306
GA1(ga1) gene 25, 65, 71, 79, 83, 86-87, 305, 309-312, 316
GAI 25-26, 161, 309-313, 315-316, 530, 533, 587, 594-595
GAMYB 233, 234, 236, 237, 316, 530, 532, 533
  (1-3, 1-4)-β-glucanase 234
  ABA 236
  aleurone 234, 316
  barley 234, 316
  binding 234, 237
  expression 233, 236
  factors 87, 530
  GA 236, 237, 530
  inhibitor 236-237
  KGM 237
  mutation 233
  SLN1 236
  transcription 236
  ubiquitin-proteasome pathway 236
  α-amylase 234, 237
  α-capthesin 234
GC/MS 686-687
gene expression 528
genes 81
genetic studies 229
geotropism 593
germination 7, 64, 221, 223, 227-228, 238, 308, 523, 530, 533
Gibberella fujikuroi 3, 63-65, 71, 73, 304, 588
glucosamine 235
glucosyl conjugates 77
GPA protein 532
growth 26
  promotion 533
  inhibition 26
  -active 18
homeostasis 76, 83, 90, 317, 459

HPLC 687
HvSPY 238
hydrolase 224, 229, 232, 237-238
IAA 28-30, 197-200
inactivation 63, 77
inducible gene 234
insensitive 83, 529
ion channels 231
KGM kinase 237-238
kinases 231, 237-238
kinetin 575
KNOX gene 82
leaves 542
lettuce 574, 689
level 27, 533, 547, 557
light 27, 194
LKB 29
LFY 87
Lolium 66
long day 7, 86, 547
long distance transport 79
maize 18, 232-233, 235
maleness 7
mass spectra 680, 686
meristematic tissue 547
metabolism 17, 79, 90
methyl ester
trimethylsilyl-ether 686
microtubules 557
mutants **179-186, 304-317**, 574; see also deficient mutants
nasturtium 574
nature 6
negative feedback 548
negative regulator 229, 235
nitrogen 541
non-13-hydroxylation pathway 72
occurrence 64
pat 600
pathway 186, 199
pea 18, 20, 25, 27-28, 72-73, 76, 179-200, 550, 588
perception 229
pH 231
phosphatases 231
phosphorylation 231, 235-236
photoperiod 86, 547, 60
PHYA 542, 547
PHYB 542
plant growth retardants 588
plasma membrane 237
plastid 67-70

Index

pollen development 64
potato 547-550, 601
proteasome 236, 313, 315
protein degradation 235-236
protein kinase 304
purification 686
pumpkin 73
pyrimidine box 234
quantification 686
receptor 237
regulated genes 80
reporter genes 79
repressors 235
reproductive development 86
response 229, 305, 530, 542
  complex 232, 233-235, 238
  elements 532
  mutants 229
  pathways 237
  constitutively active 305
responsiveness 27, 200
RGA 26, 309-316
RHT 587
rice 55, 64, 66, 81-82, 87, 89, 224, 230, 235, 530, 594, 595
SCF complex 314
second messengers 230
seeds 7, 17, 183, 316, 514, 533
senescence 574
sensitivity 28, 529, 533
shoot 7, 549
short day 220, 557
signaling 25, 81, 90, 229-232, 235-236, 304-318, 528, 587, 594
  degradation 312-315
sites of action 81
SLENDER 183, 235, 237, 309
sorghum 18
SPE 687
Spinacia oleracea 86
SPINDLY 25-26, 236-237, 308-309, 315-316, 528, 530
starch 532
stem elongation 7, 25, 28, 64, 216-217, 304, 310, 316
stolon 57, 547, 458
structure 6, 7, 63, 65

sucrose 557
suspensor 516
tomato 600
transport 7, 550
treatment 86
tuberization 547-550, 601
ubiquitin 236, 314
vascular plants 64
vivipary 522
VP1 533
wheat 232
xyloglucan 304
GID 306, 315
gin 30-32
Girdling, 547
Glucan 226
Glucanase 226, 228, 234
Glucokinase 555
Glucose-1-phosphate 554
Glucose-6-phosphate 30, 556
Glucose 30-32, 555
Glucosidases 226-227, 496
Glyceraldehyde-3-phosphate 7, 10, 69
Glycosyl hydrolase 263
GMPOZ 234, 235
GNOM 451, 454-455, 465
Golgi 446
GPA 402, 403, 532
Grafting 186, 499, 505, 544, 627
Grape 497
GRAS proteins 309-310, 312
Gravitropism 266-267, 291, 462-463
grd2 305
Green fluorescent protein 73, 79, 145, 397, 690
Green Revolution 26, 235, 305, 312, 585
Growth : see individual listings
GTPases 397, 454
GTP-hydrolysis 229
Guard cell
  ABA 391, 394, 403, 405-407, 495-497, 501
  antiporter 395
  apoplast 395
  Arabidopsis 392
  Auxin 502
  cADPR 399
  calcium 394, 398, 401
  cytosol 394
  epidermal peels 392
  G protein 401

  isolation 392
  lipids 403
  nitrate 395
  NO_3^- 395
  osmotiderm 394
  potassium 395
  protein kinase 405
  protoplasts 393
  RNA 406-407
  signaling 401
  sugar 395
  symplast 394
  symporter 395
  tonoplast 399
  turgor 395
  Vicia 403, 405
  volume 394
GUS 24, 130, 288, 290, 464, 690
Gymnosperms 157
Gynoecium 162, 163, 487
Harvest index 507, 509, 585, 586
Haustoria 277
Hbt 516
Heat 541
Herbivores 656
Heterodimeric transcription factor 244
Heterotrimeric G protein 401
Heterotrimeric GTP-binding proteins 306
Hexadecanoid pathway 613
Hexokinase 30, 32, 556, 567
Hexose 555
Histidine
  kinase 323-327, 330, 374
  protein kinase 322
  phosphotransfer protein (see also *AHPs*) 322-323, 336
  sensor kinase 346
HLS 364
HOBBIT 518
HOG 330
Homeobox genes 340
Homeodomain 342, 553
HOOKLESS 364
Hormone (see also plant hormone, and individual hormones)
  analysis 671, 689
  application 461
  biosynthesis 21, 197, 200
    antisense 20
    gene 20
    intracellular site 17

Index

light regime 200
marker genes 4
regulation 197
transgenic plants 4
discovery 2
changes 541
flowering 546
role induction 547
sensitivity 689
tuberization 546
brassinosteroids 459
catabolism 691
concentration 16, 18, 20, 691
conjugation 691
contents 17
conversion 21
cross-talk 575
definition 1, 34
degradation 21
extractable 16
function 32
gibberellins 459
Greek meaning 2
medicine 2
inactivation 379
integrated effects 27
interactions 27, 200
level 27, 690
location 21, 23
multiple 27
origin 2
overexpression 21
profiling 689, 690
quantitation 16
receptor 17
redistribution 32
regulation 16
response 22
responsiveness 24, 25, 27, 369
sensitivity 24, 369
signaling systems 200
sugar 30
synthesis 691
transgenic plants 20
transport 23, 32, 34
HPLC 144, 673, 677-678, 680, 687
HPLC-MS, plant hormones 3
HRT 233
Hydathode 485
HYDRA 158, 163
Hydrodictyon reticulatum 157
Hydrogen peroxide 400, 405

2-hydroxybenzoic acid 636
Hydroxyjasmonate sulfotransferase 551
7β-hydroxykaurenolide 72
Hydroxyl radicals 271
Hydroxymethyl-butenyl 4-diphosphate 68-69
Hydroxyproline-rich glycopeptides 658-659
Jyl 407
Hypersensitive response 122, 130, 378, 548, 636, 646
Hypocotyl 161, 266, 442, 453
Hyponastic leaves 1 527
Hypophysis 516
IAA see Indole 3 acetic acid
IAA genes 285, 334, 343
Illuminating gas 3
Immunoaffinity 283, 685, 673, 685
Immunoassay 4, 671
Immuno-cytolocalization studies 122
Immunolocalisation 122, 397, 690, 691
In vitro 541, 554, 557
Incipient leaf initiation 269
Indanoyl conjugates 618, 619
Indole 46
Indoleacetaldehyde 5, 682
oxidase 53
Indole-3-ethanol 38
Indole-3-lactic acid 38
Indole-3-pyruvic decarboxylase 53
Indoleacetamide hydrolase 53
Indoleacetonitrile (IAN) 44, 682
Indoleacetyl aspartate 48-49, 51, 53-54, 682, 683
Indole-3-acetaldoxime 682
oxidation 58
Indole-3-acetic acid (IAA) (see also auxin)
adenylation 49, 52, 288, 621
Agrobacterium 21
amide conjugates 51
amine oxides 53
amino acids 49
analysis 680, 682, 689, 691
apical bud 29

Arabidopsis 683, 691
AUX1 480
auxin 5, 288, 437, 445
basipetal polar movement 440, 473
biosynthesis 21, 22, **36-55**, 677
location 5, 22
rates 677
microbial 54-55
tryptophan 41-45
bound 473
bundle sheath 477
cambium 22, 477
carriers 22
catabolism 49, 55-58
catabolites 682
cell length 29
concentration 439, 483, 690
conjugate 39, 48-50, 53, 54, 288, 682, 683
amide 51-52
amidohydrolases 53
amino acid 39
Arabidopsis 53
biosynthesis 48-52
degradation
intermediates 39
developmental controls 48-50
dicots 39
ester conjugates 50-51
evolution 48
glycosylated 50
growth-inducing activity 54
high molecular weight conjugates 48
hydrolysis 39, 48-50, 52-54
IAA-aspartate 48, 49, 51, 53-54, 682, 683
IAA-glucose 48-53
IAA glucan 48
IAA-inositol 49-50
IAA-inositol-arabinoside 50
IAA-inositol-galactoside 50
IAA-*myo*-inositol 50
IAA-protein 48
maize 48-49
mobile forms 39
monocots 39
peptide conjugates 39, 48
rooting of cuttings 39

Index

seed 48
soybean 53
storage 39
synthesis 48-50
tissue culture 39
transacylation 50
transport 49
^{13}C 677, 679
content 21
cork 480
de novo biosynthesis 48
discovery 2-3
distribution 691
efflux carriers 22, 441
embryo development 514
epidermis 477
ethylene 27
extracellular wall 443
flow 479, 489
formation 52
free 473, 475, 477-478, 489
function 40
GA biosynthesis 20, 28, 30, 38, 84, 197-200
GA 2-oxidase expression 84
GFP 690
gibberellin 198-200
glucan 39
glucose 39, 49, 51, 682, 683
  synthetase 52
  hydrolase 52
glucosyl transferase gene 51
glycoprotein 39
growth 33
GUS 23, 690
height 20
homeostasis 40-41
hydathodes 474
iaaL gene 483
iaaM gene 483
immunolocalisation 691
influx carriers 442-443, 447
inositol 39, 49-50
intact plants 20
internal route 476
inverted umbrella 22
ion chromatography 677, 679
JAR1 family 52
kernel 50
kinetin 575
lettuce 689
LKB 30

location effect 29
long distance transport 443
lower plants 48
main routes 476
maize shoot 50
metabolic conversion 38
metabolites 678
metabolism 37
methylated 677, 679
microbial biosynthesis 53-54
microsomal 283
morphogen 459
morphological distortions 21
MS-analysis 680
nature 5
nitrilase genes 574
pea 18, 28, 440
peptide 49
pericycle 477
phellogen 477
phloem 479
phototropism 32
polar 479, 480
precursors 682
procambium 22, 477
production 474
purification 682
quantification 682
reporter gene 690
root 22
routes 476
salicylic acid 689
senescence 574
separation 689
sieve tubes 477
slender 20
speed of action 29
stem elongation 20, 28
stem height 21
stem segments 20
stomatal 502
storage 17
structure 5, 37-38, 40
suberin 48
tallness 20
transacylation 49
transgenic plants 21
transport 6, 22, 33, **437-490**
trimethylsilylated 677, 679
tropistic responses 23, 480
tryptophan 5, 40-48, 53
vascular bundles 440, 477, 479
veins 474
vessel size 483
wounding 480
xylem 477, 479
Indole-3-acetyl Ala 678, 682-683
Indole-3-acetyl Aspartate 48, 49, 51, 53-54, 682, 683
Indole-3-acetyl Leu 678, 682-683
Indolebutyric acid 38-39, 51
Indolic compounds 47, 50
Induced systemic resistance (ISR) 624
Induction 540, 549, 554
Inhibitors of GA biosynthesis 78, 79, 87
Inositol phosphates 399, 403
Insect resistance 4, 649
Internal standard 678, 679
Internodal growth 272
Invertase 555, 556
Ion channels 229, 231
Ion chromatograms 678
Ion trap MS-MS 681
Irrigation 509
Isochorismate synthase 636
Isopentenyl adenine 32, 97, 100
Isopentenyladenine riboside 97
Isopentenyladenine riboside 5' phosphate 97, 99, 103
Isopentenyl diphosphate 7, 10, 66, 68-69, 70, 141
Isopentenyltransferase (IPT) 98-102, 503, 604
4-hydroxy-3-methyl-2-(*E*)-butenyl diphosphate 104
  activity 100
  Agrobacterium tumefaciens 100, 101, 104
  Arabidopsis 100, 101, 104, 571
    cytokinin (CK) 571
  Dictyostelium discoideum, 101
  dimethylallopyrophosphate 104
  expression 102
  gene 552

IAA 503
Km 101
localized 552
nitrate 102
overexpression 100
Petunia hybrida 100
phosphate 102
potato 571
SAG12 572
senescence 572
spatial expression 101, 102
substrate preference 101
sulfate 102
tmr 100
tobacco 100, 571
tomato 571
transgenic plants 104
tRNA-IPT 100
tzs 100
Isoprenoids 67, 142
Isotope dilution 678, 679
JAI 626, 627
JAR 288, 621
Jasmine 12
Jasmonates **610-634**
  abscission 13, 631
  action 621-622
  allelopathy 624
  allene oxidase synthase 597
  analysis 689
  anther dehiscence 631
  anthocyanin accumulation 631
  application 551
  Arabidopsis 615, 631, 689
  auxin 288
  biosynthesis 551, 568, 611-612, 614-619
    allene oxide 612
    chloroplast 612, 616
    enzymes 614
    fatty acid desaturases 615
    hexadecanoid pathway 613
    linolenic acid 515
    lipids 615
    lipoxygenase 612
    location 614
    mutants 625, 631
    octadecanoid pathway 612
    oxo-phytodienoic acid 612
    peroxisomal reactions

617, 619
    phospholipases 615
    systemin 615, 617
    tomato 615, 617, 627
    vascular tissues 614
    wound-induced 616, 617
  brassinosteroid 575
  catharanthus roseus 622
  cev1 630
  chemical structures 620
  chlorophyll 631
  COI1 365, 568, 627
  constitutive 630
  coronatine 626
  defense 12, 596, 624
  defensins 624
  degradation 618
  development 630
  discovery 4, 610
  effects 12
  embryo development 632
  endogenous levels 551
  epimerization 618
  ERF1 365, 622
  ethylene 366, 575
  fad 569
  fatty acid amide 624
  fruit ripening 13
  GA 551
  gene expression 621-622
  germination 529
  glycosides 550-551
  grafting 627
  growth 13, 630
  herbivore 625
  hormones 3
  hydroxylation 550-551, 618
  11-OH-jasmonic acid 550
  12-OH-jasmonic acid 550-551
  indirect defenses 624
  induced plant defense response 627
  induced systemic resistance 624
  insect 12, 621
  iso-jasmonate 618
  isomers 620
  jai1 mutants 627
  leaf senescence 568
  levels 614
  linolenic acid 611
  local 627
  long-distance signal 628
  male reproductive development 631

manipulation 597
MAP kinase 626
metabolism 618, 620, 621
methyl ester 12, 551, 610, 613, 618, 620-623, 625, 630-631, 663
microtubule orientation 551
mutants 568, 569, 630
nature 12
nectaries 624
negative regulator 629
occurrence 610
O-glucosides 618
OPR3 569
ORCA3 622
overproduction 569
pathogen 365
perception 627, 630
photoperiod 551
photosynthesis 631
PI expression 626
pigment 13
plant-to-plant signaling 624, 625
pollen 631
potato 597
precursor 551
promotive effect on tuberization 550
prosystemin 628
proteinase inhibitor 12, 628
receptor 630
reconfiguration of metabolism 621
related compounds 551
reproductive development 13, 631
root growth 630
SAG 577
salicylate 625
SCF 629
seed 13, 568, 631
SEN 577
senescence 13, 568, 631
signal peptides 13
signaling 289, 618, 628-630, 649-650
structure 12
synthesized 12
systemic 627
systemin 627, 628
tendril 13
thionins 624
tobacco 597
tomato 626, 631
transcription factors 622

739

Index

tritropic interactions 624
tuberization 13, 547, 550-551, 557
tuberonic acid 12, 551
ubiquitin-proteasome pathway 629
volatile attractants 624
wounding 551, 568, 627
(Z)-jasmone 618
Jerusalem artichoke 545
JL mutant, 617
K^+ channel 394, 399-401, 403-404
Kaurene oxidase 72-73, 79, 89
Kaurene synthase 88
Kaurene oxidation 79
Kaurenolides 72, 73
4 kD legume peptide 667
Ketoacyl-CoA thiolase 618
α-Ketoglutaric acid 115
KGM 236-238
Kinase-inactivated domain 354
Kinases 4, 5, 229, 231
Kinetin 95, 241, 503-504, 575
Kiwi 123
KNAT 342, 553
KNOLLE 451
Knotted-like homeobox 553
KNOX 82, 342
La cry 20
Landsberg erecta 65, 572
Lanosterol 89
Late blight 540
Late embryogenesis abundant 521-5225-526
Lateral organs 268
Lateral root 102, 297
LC/MS 675, 680, 682, 684-685, 687- 689
LE 19, 181, 198, 301, 550
Leaf
  ABA 496, 506-508
  apoplastic pH 506
  Arabidopsis 473
  barley 565
  blades 162
  brassinosteroid 162
  dehydration 496, 507
  drought 506
  development 473
  elongation zone 506
  expansion 493
  growth 505, 506, 508

IAA 472
ivy 565
jasmonic acid 568
laurel 565
morphogenesis 485
oat 565
ontogeny 486
photosynthesis 507
procambium 485
primordium 473, 486
promoter 552
senescence 129, 358, 568
size 191
tobacco 565
vein pattern 472
water potential 496
xylem 506
Leaflet size 200
Leafy cotyledon 519, 522-528
Legumes 14
LEI proteins 527
Lettuce 66, 84, 689
Leucine heptad repeats 310, 312
Leukotrienes 611, 628
LFY 87
LH 182
LHY 335
Life span 161
Light 161, 529,
  *ARR*4 334, 339
  auxin signaling 285
  cytokinin (CK) 334, 341
  de-etiolation 85
  elongation 538
  environmental signals 333
  FR 538
  GA_1 194
  germination 530
  histidine kinase 325
  PhyB 339
  phytochrome 542
  receptor 542
  signal transduction 163
  source 538
  stem elongation 194
  two-component 323
Lignin 123
Lignostilbene dioxygenases 144
Lilium 574
Limit dextrinase 226
Linoleic acid 12, 116, 615
Lipase-like proteins 649
Lipid 528
  metabolites 403

peroxidation 116
pathway 115
rafts 461
signaling 401, 611
evolution 611
Lipoxygenase 551
Liquid chromatography 673
Liverworts 115
LKA 189
LKB 29-30, 161, 188
Lolium 86-87, 89, 687
Long days 557
Long-distance signaling 102, 641
Low-pI α-amylase 226
LOX 552, 616-617
LS 29, 182
LS/MS 673
Luciferase 21, 26
Luffa cylindrica 484
Lupinis cosentinii 504
Lycopene 147-148
MADS box 127
Maize (*Zea mays*)
  ABA 142-144, 151,392, 497-498, 500-501, 505-506
  ABP1 284
  auxin 206, 213, 215, 283-284, 287-288, 458
  biosynthesis 522
  coleoptile 206, 212-213, 284
  cytokinin 95, 504
  GA_1 64
  GA bioassay 66
  GA-response complexes 232
  germination 142-144
  gibberellin 224
  IPT 100
  PIN1 449
  SAGs 564
  stem elongation 64
  viviparous (*vp*) 142-144, 522-523
Male reproductive development 615, 631
Male sterile 616, 617
Malonyl ACC 118, 133
Malvolio 357
Manduca sexta larvae 657
MAP kinase 237, 355-356, 371,404, 626
MAPK cascade 285, 354 356-357, 361
Marsilea quadrifolia 274
Mass spectrometry 671-676,

Index

678-679, 681-682
Maturation 514
Mechanical stimuli 363
Medicago 357
Melon 120, 133
Mendel 19, 180
Meristem 321, 341-342, 516, 539
Metconazole 89
Methaneboronates 688
Methionine 115-116
24α-Methyl (brassinolide) 157
24β-Methyl (24-epi-brassinolide) 157
Methyl jasmonate 12, 551, 610, 613, 618, 620-623, 625, 630-631, 663
Methyl salicylate 645
Methylation 164
Methylcyclopropene (1-MCP) 276, 387
24-Methylene (docholide) 157
24-Methylene-25-methyl (25 methyldolocholide) 157
24-Methylenecycloartenol 164
Methylerythritol phosphate pathway 67-70, 104
5'Methylthioadenosine 116
5-Methylthioribose 116
Mevalonate pathway 90, 138, 160
　24-methylene 158
　carotenoid 68
　cycloartanol 158
　diterpenes 67
　ergosterol 155
　fungi 67
　GAs 67, 68
　isopentenyl pyrophosphata 67
　lanocherol 158
　occurrence 67
　squalene 158
　squitonine 158
　sterols 67
　terpenoids 67-68
MDR1 452
Microarray 196
Microfibrils 548
Microtubule 397, 460, 548, 557
Midvein 473
Millet 224
Mitochondria 130

Mitogen-activated protein kinase (MAPK) 237, 355-356, 371,404, 626
Mitosis 241-242
MoCo sulfurase 143, 149
Molybdenum 143
Monocots 123
Monooxygenases 89
monopteros 454, 461, 465, 488, 516, 518
Monoterpenoids 70, 623
Mosses 115
MPK4 626
MRP5 452
MS-MS 675, 680, 682
Mung bean 569
Muskmelon 379
Mutant: see respective genes
　DNA insertional null mutants 401
MYB transcription factor 526, 532, 533
　(1-3, 1-4)-β-glucanase 234
　ABA 236
　aleurone 234, 316
　barley 234, 316
　binding 234-235, 237
　expression 233, 236
　factors 87, 530
　GA 236, 237, 530
　inhibitor 236-237
　KGM kinase 237
　mutation 233
　SLN1 236
　transcription 236
　ubiquitin-proteasome pathway 236
　α-amylase 234, 237
　α-capthesin 234
MYC transcription factor 526
Myo-inositol-hexakisphosphate 403
N 646
NA 183, 185, 550
NAD 146, 399
NADPH 144, 400
NahG 569, 639-642
Naphthoquinones 623
Naphthalene acetic acid 284, 440, 445, 447, 481, 577
Naphthoxyacetic acid 458
Naphthylphthalamic acid (NPA) 267-268, 406, 440-441, 444, 453, 457, 460
cell division 460
Nar 143
Naringenin 445
N-benzoyl *n*-propyl 687
NPA-binding protein 444, 446, 453, 456
NCED 143-145, 147, 149
Necrosis 637
Necrotrophic fungi 365
NEDD8 protein 296
Negative regulators 24-25
Neopa 152
Neoxanthin 147-148
Never-ripe 125, 127, 374-379, 597 604
nia1 400
Nicotiana attenuata 625
Nicotiana plumbaginifolia 143
Nicotiana. sylvestris 544
Nicotiana. tabacum see tobacco
Nicotine 624
Night break 542
NIM1 (NPR1, SAI1) 365, 647-649
Nine *cis*-epoxy-carotenoid dioxygenase 143-145, 147, 149
Nitrate 333, 340, 343-344, 540
Nitrate reductase 400
Nitric oxide 400-401
Nitrogen 540, 541
N-malonyl ACC 118
NMR 675
Nodulation 14, 666
Non-hair cells 276
Non-heme oxygenases 121
Notabilis 143, 145
nph4 364, 384
N-phenyl-N'-(1,2,3,-thidiazol-4-yl)urea (thidiazuron, TDZ) 96
N-phenyl-N'-(2-chloro-4-pyridyl)urea (CPPU) 96
NPR1 (NIM1, SAI1) 365-366, 647-649
NR 125, 127, 374-379, 597 604
Nramp 356
Nuclear localization signal 311, 312
Nuclear magnetic resonance spectroscopy (NMR) 674

Index

Nuclease 226
Nucleic acid 223
Nucleosidases 226
Nucleus 26, 235, 286
Octadecanoid pathway 612-613, 616
O-linked N-acetylglucosamine transferases 309, 315
Oil of wintergreen 636, 645
Okadaic acid 128
Olefine 115
Onion epidermal cells 288
OPR 614, 617, 621
ORCA 622
Orchid 128, 381-383-385
Ornithine 11
Ornithine decarboxylase 552-553
Oryza sativa see rice
Os*aba1* 143
Osmosensing 328, 330
Osmotic stress 148, 322
OST 405
Ovary 385
Ovule 481, 487, 632
Oxalis stricta 484
Oxidase 385
Oxidation 161
Oxindole acetic acid 57-58, 682, 683
2-oxo-4-methylthiobutyric acid pathway 115
2-oxoglutarate 89
2-oxoglutarate-dependent dioxygenases 19, 75
Oxylipin 611
P450 monooxygenase 71
Paclobutrazol 26, 89, 182
PAD 649
Papillae 163
Parthenocarpy 599, 600
****
PAS1 344
Patatin and proteinase inhibitors 539
Pathogen 378
Pathogenesis-related 626, 637, 638
Pattern formation 519, 533
PCR 4
PDF1 80
PDF1.2 365, 366
Pea **179-200**
  ABA 145
  apical bud 33
  Aux/IAA 286, 289
  auxin 32, 197-200, 440, 445
  brassinosteroid 29, 179, 186-193, 195-197
  darkness 25, 27
  decapitation 28
  de-etiolation 27, 85-86
  dwarf 18-20, 28, 161
  elongation 18, 20, 29-30, 32-33, 64, 206, 208, 216-217, 362
  embryo 17
  ent-6α, 7α-dihydroxykaurenoic acid 73
  epicotyl 206, 362
  ethylene 116, 362
  GA 3β-hydroxylase 588
  GA_{20}, 22
  GAI/RGA orthologs 530
  growth see elongation
  height 18-20, 22
  IAA
    location 18
    GA biosynthesis 83-84, 197-200
    growth 18, 20, 29-30, 32-33, 206, 208, 216-217
    height 28-29,
    transport 32-33, 440, 444-445, 480
  gibberellin
    deficient dwarfs 72
    biosynthesis 18-20, 28, 72-73, 83-86, 179-187, 193-195, 197-200, 588
    catabolites 76
    content 20, 27
    height 18-20, 22, 28, growth 29
    mobility 550
    sensitivity 25, 27-28, 85
    signal transduction 28
  growth 33
  height 20, 28
  kao mutant 72
  kaurenolides 73
  LE mutants 550
  light 25, 27
  LKA 161
  LKB 29
  ls 29
  Mendel 19, 588
  metabolism 17
  Na 22, 25, 28, 550
  nitric oxide 400
  phloem transport 440
  seed coat 17
  seedlings 362
  slender 20, 28
  stem elongation 25, 64
  stem segments 18
  tall 20
  tallness 22
  wild type 190, 194, 546
Pedicels 162
Pelargonium 381, 382
Pentafluorobenzylbromide 688
Peppermint 69
PEPINO/PAS2 344
Pepper 69, 148
Peppermint 69
Peptide hormones 4, **654-670**
Peptide signal 13, 627, 665
Perennation 538
Perfume 12
Perianth 385
Peroxidase 57, 271
Peroximal β-oxidation 618
Peroxisomes 613
Petal 130, 381, 385-386, 481
Petunia 130, 376, 381-383, 388
PGP1 452
pH 231, 398, 399, 495-496, 499
Phaeosphaeria 71
Phalaenopsis 128
Phaseic acid 151-152, 575, 576
Phaseolus 521, 592
Phellogen 480
Phenylacetic acid (PAA) 5, 36, 38
Phenylalanine ammonia lyase 643
Phenylpropanoids 623
Phenylurea 96
Phloem 118, 439
  anastomoses 484, 490
  axial organs 483
  companion cells 102
  differentiation 483, 489
  discontinuities 484, 487
  florigen 554
  foliar organs 483
  girdling 641
  pattern 484, 487
  phloem-only bundles 484
  salicylic acid 639
  sieve tubes 471
  transport 554
  tuberigen 554

Index

PHOR 308
Phosphatases 398, 404-406, 684
(4,5)P35 phosphatase 532
Phosphatidic acid 404, 616
Phosphatidylinositol-phosphate 403
Phosphoglucomutase 556
Phosphoglucose isomerase 556
Phosphoinositol 404
Phospholipase 343, 403-404, 409, 532, 616
Phospholipid 228
Phosphorylation 123, 133, 242-244, 231, 322, 407
  cytokinin (CK) 339
  gibberellin 236
  KGM 237
  polar auxin transport 453, 456-457
  two-component phosphorelay 322
  ubiquitin 244, 296
Photoaffinity labelling 283
Photoassimilate 554
Photo-isomerization 147
Photoperiod 538, 542, 544, 551, 601
Photoperiodic
  adapt 538
  control 543
  environmental conditions 538
  far red light 542
  floral induction 544
  GA 601
  jasmonate hydroxylation 551
  leaves 544, 557
  long days 542
  perception 557
  phyB 542
  phytochrome 543
  potato 544, 601
  red light 542
  requirements 544
  response 542
  short days 542
  signal 544, 557
  stimulus 544
  tobacco 544
Photoreceptor 542
Photosynthesis 507, 538
Phototropism 2, 32, 291, 464,
PhyA 195, 296, 324, 543
PhyB

photoperiod 542
  antisense 542-543, 547, 549
  ARR4 interaction 334
  daylength 542
  flowering 543
  light 339
  mutants 543, 550, 557
  tuberization 542-543, 550, 557
Phyllotaxy 268
Phylogenetic tree 124
Phytium 624
Phytoalexins 622
Phytochrome 84, 545
  circadian clock 543
  cytokinin 346
  ETR1 373
  GA 542
  gene 542
  histidine kinase 324, 325
  photoperiod sensing 542-543
  phyA see PhyA
  phyB see PhyB
  tuberization 542
Phytoene 70
Phytol 70
Phytophtora infestans (potato blight) 540
Phytosulfokine-α 659-661
Phytotropins 444, 457, 458
PI kinase 404
PICKLE 528
PIN **448-457, 461-466**
  actin cytoskeleton 463
  auxin 448-457, 461-466
    efflux carriers 23, 454
  cycling 463
  gene product 23
  localization 456
  mutants 23
  proteins 23, 461, 463
  transport activity 449
PIN1 447-449, 451, 457, 465, 516-518
PIN2 448, 451, 462
PIN3 455, 462-464
PIN4 448, 464-465, 518
PIN7 517
Pine seed 50
PINOID (PID) 285, 453, 456-457
Pisum sativum L. see pea
PKL 307-308, 316, 528
PLA 616
Plane of cell division 539
Plant defense 596, 597

Plant development 179, 194, 630
Plant evolution 479
Plant growth 2, 192, 582-583, 585, 588
Plant hormone 156, 241, 300, 655, 689; see also individual hormones
  abbreviations 1
  amount 16
  analysis **671-694**
    interfering substances 672
    losses 680
  biosynthesis 16, 677
  books on 14-15
  catabolism 676, 677
  cellular markers 671
  characterization 3
  compartmentation 676
  concentration 672, 678-679
  concept 2
  conjugation 16, 676, 677
  criteria 14
  definition 1
  degradation 16
  derivatisation 680
  discovery 2
  early work 2
  effects 4
  extraction 672, 673
  functions 1
  GC-MS 3
  heavy-isotope-labelled 679
  HPLC-MS 3
  identification 2, 674
  interactions 689
  internal standards 672, 679
  isotope labelling 676
  labelled 676-677, 679
  levels 671
  location 16
  mass spectrometry 580, 677
  metabolic products 677
  metabolic studies 676
  metabolites 673, 677
  mixtures 672
  movement 16
  nature 1, 4
  occurrence 1, 4
  purification 3, 672-673, 680
  qualitative analysis 673
  quantification 678

743

Index

quantitation 2
radioactively labelled
  standards 674
receptors 16
responsiveness 16
sensitivity 16
signal-transduction chain 16
synthesis 676
transport 1, 16, 676
transported 14
turnover rates 677
Plant-to-plant signaling 625, 646
Plasma membrane 237, 398, 446, 461
  ABA 394, 397
  ABP1 283
  actin cytoskeleton 444
  arabinogalactan proteins 230
  auxin 441, 459
    receptor 284
  cytoplasm 441
  guard cell 398
  lipophilic 441
  MDR1 452
  PGP1 452
  pH 441
  surface 284
Plasmodesmata 225
Plastics 555
Plastids 70, 145, 556
Plastoquinone 70
Plectosphaerella
  cucumerina 365
PLS 343, 344
POLARIS 663, 664
Pollen 11, 156, 383-384
Pollination 116, 130, 381-382, 637
Polyamines 11, 14, 119, 123, 553, 573, 623
Polygalacturonase 131
Polygonum convolvulus 484
Polyphenol oxidase 624
Post-mitotic first gap (G1 phase) 242
Post-synthetic second gap (G2 phase) 242
Potassium channel 394, 399-401, 403-404
Potato **538-552**
  3β-hydroxylase 548
  Andean varieties 540
  andigena 543, 544
  CONSTANS 546, 557

cultivated 540
GA 20-oxidase 548
GA 3-oxidase 550
genotype 539
gibberellin 20, 547-550, 601
leaves 544
neoxanthin 148
rolC gene 552
short day 539, 542
species 542
tetraploid 540
tissue-specific expression 548
tuber-bearing species 538
tuberization 20, 540-545
tuberosum 540
tuber 539
varieties 540
POTH1 553
Powder mildew 131
PP2A 457
PR 366, 639, 641
Precocious germination 143, 522
Pre-harvest sprouting 584
Prenyl transferases 70
Procambium 485-486, 488
Progesterone 157
Prohexadione-calcium 89
Prokaryotes 140
Proline 296, 358
Prolonged growth response 20
Prolycopene 147
Promoter 124, 269
  analyses 276
  auxin-responsive 289
  CaMV 35S 571
  heatshock 571
  leaf-specific 552
  light 571
  tetracycline-inducible 269
  tuber-specific 552, 555
  wounding 571
Propanal 116
Prostacytes 611
Prostaglandins 611, 628
Prosystemin 628
Protease, GA3 228
Proteasome 307, 313
  auxin 291-294, 296-298, 300
  cytokinin 334, 335
  function 24, 291-292, 358
  EIN3 358

jasmonate 629
protein degradation 334
Protein
  breakdown 24
  degradation 25, 291-292, 236, 334
  germination 233
  kinases 237, 404
  localization 454
  patatin 539
  phosphatase 128. 456
  rolC 552
  traffic 457
Proteinase inhibitors 539, 610, 649
Proteolysis 227
Proteolytic enzymes 656
Proton excretion 207
Proton ATPase 461
Provascular 461, 485, 489
Pseudomonas paucimobilis 144
Pseudomonas putida 569
Pseudomonas syringae 365, 626
Pseudo-response regulator 329
PSKα precursor 664
Pteridophyte *Equisetum arvense* 157
pTOM13 120
Pumpkin 73, 74
Putrescine 11
Pyrimidine box 233-234
Pyruvate 69
Quantitative trait loci 540
Quiescence 514
RAC1 397
Race-specific resistance 640
Rachis 276
Radial polar transport 462
Radiochemical studies 122
Raf serine/threonine protein kinase family 355
RALF 662, 663
Ralstonia solanacearum 130
RAN1 353, 355
Rapeseed 522
Raspberry 515
Ratio of carbohydrate to nitrogen 541
RBX1 297
RCN1 364, 398, 406, 453, 457
Reactive oxygen species 271, 400, 569

Index

Receptors 352, 354-355, 408
Reciprocal backcrosses 540
Red light 542
Regnellidium diphyllum 274
Related to Ubiquitin (RUB1) 296
Replication proteins 533
Reporter gene 289, 690-691
Resistance 540, 636, 640, 649
Respiration 122
Respiratory climacteric 131
Response regulator (see also *ARR*) 322, 324, 329-331, 335, 346
Rest 11
Retinoblastoma protein 244
RFLP 540
RGA 25, 26, **309-316**
rhd6 276
Rhizopus 3
Rhodamine-phalloidin 397
Rht 26, 309, 586
RI allele 540
Ribonuclease 226, 228
Rice (*Oryza sativa*) 82, 89, 143, 224, 291
  PIN1 449
  brassinosteroid 417-418, 421
  BRI1 418, 421
  cytokinin 571
  embryos 223
  F-box 236
  GAI/RGA orthologs 530
  Gibberella 55, 64
  gibberellin 224, 230, 236
    assay 66
    GA 2-oxidase 81-82, 87
    GA 20-oxidase 80-81
    GA 3β-hydroxylase 81
    signalling 81, 235
  growth retardants 89
  salicylate levels 637
  semi-dwarf 587
  ubiquitin-proteasome pathway 236
Rin 125, 127, 381
Ripening see fruit ripening
RNA 182, 406-407
RNAi 20
rolC gene 552
Root
  ABA-deficient 505
  adventitious 387

allocation 504
auxin 277, 463
  transport 22-23, 452
brassinosteroids 162
curl in NPA (*rcn1*) 453
cytokinin 21
development 200
drying 509
elongation 183, 397, 452, 504
ethylene 276-277, 505
expansins 276-278
GA 183, 200
gravitropism 463
growth 278, 494, 504
hair 276, 277
hydraulic conductivity 504
initiation 276
pattern formation 519
water potential 278
uptake 504
ROP10 397
RPN12 334, 335
RUB 296, 299
Rumex acetosa 272
Rumex palustris 272
RY/Sph 526
Ryanodine receptor 399
SABP 648
Saccharomyces cerevisiae 120
Sachs, . 2
SAD1 407
S-adenosylmethionine 11, 116, 118-119, 133, 174, 687
  decarboxylase 11, 553
  hydrolase 133
  sterol-C-methyltransferases1 164
  synthetase 119, 123
SAGs 564, 572-573, 576-578
SAI1 (NIM1, NPR1) 647
Salicin 635, 636
Salicylate hydrolase 569, 636
Salicyl-hydroxamic acid 551
Salicylate hydroxylase 639, 641, 649
Salicylic acid **635-653**
  accumulation 637, 639, 640, 643
  adenylation 621
  analysis 689
  binding protein 636, 648

biosynthesis 13, 636, 643-644
  isochorismate synthase 643
  pathway 643, 649
  phenylpropanoid pathway 643
calorigenic substance 13
conjugation 645
coronatine 626
defense 596, 637, 640, 647, 649
deficient mutants 642
dependant 365
discovery 4
effects 13
ein2 366
glucoside 645
herbivores 649
hormones 3
IAA 689
jasmonate 625
level 569
lipase 649
movement 642
mutants 569
NIM1 (NPR1, SAI1) 365, 647
pathogen-related proteins 13, 638
pathogens 13, 637, 640, 644
phenylalanine 13
phloem 641
PI expression 626
poisoning 645
production 637, 645
Pseudomonas syringae 626
receptors 648, 649
related compounds 636
resistance 640
SAR 638
senescence 569
sequestration 645
signal transduction 647
signaling pathway 650
structure 636, 640, 641
synthetic analogues 640
systemic acquired resistance 13, 638
systemic signal 641
thermogenesis 637
tomato 626
β-glucoside 636, 645
Salinity 391
Salix spp. 635-636
Salt 407, 530

745

Index

SAM (shoot apical meristem) 133, 268, 270, 342, 344, 539
SAUR 287, 334
Sauromatum guttatum 637
SAX 673
SCF 292-296, 307, 313-315
Scions 544-546
Scots pine stem 464
SCR/SP11 665
SCX 673
SD1 587
SDGs 578
Season 538
Secondary messengers 406
Secondary metabolites 622, 623
Secretory pathway 654
Seed **513-537**
  ABA 391, 397, 519
  abortion 182, 186
  abscisic acid 137
  cell cycle arrest 519
  cell enlargement 519
  cell walls 274
  coat 17, 525
  desiccation tolerant 513, 514, 520, 529,
  development 186, 514, 521
    arrest 513, 514
  dormancy 391, 525, 529, 532, 584, 623
  GA 186
  gene expression 525
  germination 140, 274, 391, 397, 530, 584
  jasmonate (JA) 632
  LEA gene expression 525
  LH-2 182, 186
  maturation 519, 521, 534, 632
  nutrient reserve 513
  priming 529
  production 513
  quiescent 528
  reserve 514, 519, 525
  storage tolerance 524
  viability 513
  water content 519, 521
Seedling 116, 149, 350, 454, 515
Self-incompatibility 14, 665
SEN1 577
Senescence **561-581**
  ACC oxidase 129-130 566, 603

ACC synthase 566, 602
AHK3 328
anthocyanins 562
Arabidopsis 563-569, 571-573
auxin 574
benzyladenine 570
brassinosteroids 163, 569
BRI1 569
broccoli 603
chlorophyll 563-565 570
colors 562
cytokinin 321, 328, 336, 342, 570-573, 604
darkness 562
delayed 357
developmental processes 562
drought 562
ein2 566
enhancer trap 576
environmental stress 562
ethylene 128, 132, 565-567, 597, 601-604
etr1 566
evolutionary 563
floral 561
flowers 562, 565
fluorescence 565
fruit 561, 562, 565, 566, 602
gene 21, 563
hormones 562, 578
indices 564-565
invertase 567
isopentenyl transferase 571-573, 604
jasmonate 631
leaf 561, 562, 566, 570
life span 163
light 562
lipids 562
manipulation 578, 604
measurement 564-565
microarrays 576
mitotic 561
molecular markers 565, 568
molecular mechanisms 576
mutants 566
nucleotides 562
nutrient 562-563
onset 563
oxygen 569
photoreduction 565
post-mitotic 561, 570
potato 569

proteins 562
pTOM13 603
rice 571
SAG12 565, 568, 572
SAGs 563, 564
salicylic acid 569
stay-green 564
sugar 567
syndrome 562
temperature 562
tobacco 567, 572-573
tomato 566, 602
transition 563
yellowing 562
zeatin riboside 21
Ser/Thr *O*-linked N-acetylglucosamine transferases 309, 315
Serine threonine protein kinases 404, 405
Sesquiterpenoids 623
SHI 312, 316
Shikimic acid pathway 643, 644
Sho 100
Shoot
  GA_1 550
  decapitation 28
  apical meristem (SAM) 133, 268, 270, 342, 344, 539
    cytokinin 342
    height 200
    brassinosteroid mutants 200
    GA mutants 200
    meristemless 517, 518
Short days 20, 539
Shy 285
Silica 673
Sid2 643, 645
Signal peptides 13, 283, 627, 665
Signal transduction see individual hormones
  ABA 391-412
  auxin 282-303
  brassinosteroid 413-436
  cytokinin 321-349
  ethylene 350-390
  GA_3-action 222
  gibberellin 229, 304-320
  mutants 20
  stomata 391
Siliques 162
Silver 353, 386
SIM (selected ion monitoring) 688

SIMKK kinase 356, 357
SINAT5 300
Sink 554
Sitiens 143, 147, 505
Sitosterol 160, 193
Skotomorphogenesis 161
Sleepy 530
Slender 183, 550
SLN 184, 198, 235-237, 309-310, 313-314, 322-324, 326, 328, 330, 373
SLR1 81, 309, 310, 313, 315
SLY 306, 316, 528, 530
Small Auxin Up RNAs ; see *SAUR*
Small nuclear ribonucleoprotein 407
SMf1p 357
smt1 164
Soil 494, 499, 504
Solanaceae 538
Solanum demissum 540, 550, 551
Solanum tuberosum L 540
Solid phase extraction see SPE
Sorghum 224
Source-sink 553, 557
Soybean 206, 286-289, 449, 521
SPE 673-674, 682, 684-686, 689
  C18 686, 689
  columns 673, 674
  DEA 684, 685
  MAX 682
  MCX 682, 685, 689
  SAX 685, 686
  SCX 684, 685, 689
Spermidine 11
Spermine 11
Sphingolipid 401
Sphingosine 399, 402
Sphingosine-1-phosphate 402
Spinacea oleracea see spinach
Spinach 18, 73, 130
Spirea 636
Spirsaure 636
SPINDLY 25-26, 236-237, 308-309, 315-316, 528, 530
SRM (selected reaction monitoring) 681
SSK 322, 323
SSI1 366

Stamen 487
Starch
  accumulation 539
  aleurone 225
  amylopectin 226
  biosynthesis 554-557
  branching enzyme 556
  degradation 223, 225
  diastase 226
  endosperm 225
  formation 554
  germination 223
  hydrolysis 223, 225-226
  limit dextrinase 226
  mobilization 532
  plasmodesmata 225
  potato 554-556
  precursor 554
  scutellum 225
  sheath 463
  starch 225
  synthase 556
  tuber 554-556
  α-amylase 226
  α-glucosidase 226
  β-amylase 226
Staurosporine 128
COL 545
Stem 547
  elongation 28, 64, 180, 182
  growth 272
  girdling 657
Sterility 163
Steroid hormone 157
Sterol 156, **158**, 160, 163, 164, 189, 192, 488
Stevia rebaudiana 71
Stevioside 71
Stigma 130, 385, 474, 481
Stigmasterol 193
STM 342
Stolon 20, 539, 548, 557
Stomata **391-409**, **493-504**
  ABA 393, 397, 404, 494-504, 509
  aperture 494
  auxin 502
  behaviour 495
  calcium , 395, 399, 400, 403-405, 407
  closure 394, 397, 401-402, 404, 406-407
  CO_2 502
  conductance 143, 494, 497
  cytokinins 503-504
  differentiation 494

ethylene 502
frequency 494
GPA1 401
IAA 502
opening 393, 401, 404
nitric oxide 401-402
pH 499
potassium 394, 399-401, 403-404
responses 391
signal transduction 391, 404
water deficits 504
water supply 493
Stop codon 184
Storage
  organs 539, 541, 555
  proteins 521, 525, 528
  reserves 524
STP1 345
Streptomyces lavedulae 70
Stress
  ABA 598, 599
  cold 598
  cytokinin 341
  plant defense 596
  salinity 598
  signaling 599
  tolerance 137
Striga asiatica 277
Style 130
Submergence 272
Sucrose 143, 539, 541, 554-555, 557
  tuber 555, 557
  unloading 539, 555
  density centrifugation 355
  synthase 555-556
  transporter 556
Sugar 357
  ABA 531
  germination 530
  hexokinase 531
  hormone interactions 30-32
  level 567
  mutant 567
  sensing 137, 567-568
  signaling 531
Sunflower 392, 545
Superoxide radicals 271
sur2 690
Suspensor 515, 516
Sycamore figs 116
Symplastic 408, 495, 555, 556
Synthesis phase (S phase)

Index

242
Systemic acquired resistance (SAR) 636-639, 640-642
  activation 637, 639
  salicylic acid 13, 638-639, 641
  induction 640-641
  signaling pathway 641-642
  systemic signal 639, 642, 649
Systemic necrosis 646
Systemic signaling 627
Systemin 627-629
Talin 397
Tallness 20, 22
Tanginbozu 272
Tapetum 81
Taxol 360, 622
Temperature 495, 541
Tendril coiling response 621
Termination 248
Terpenoid indole alkaloids 622
Terpenoid pathway 160
Testa 525
Tetcyclasis 498
Tetraploid 540
Thermogenesis 13, 637
Thi2.1 551
Thiamin diphosphate 69
Thigmomorphogenesis 363
Thimann, K.V. 1
Thionins 624
Thromboxanes 611
TIBA 444, 481
TILLING 605
TIR1 285, 292-293, 295, 298
Tmr 100, 101
TMV 640, 642
Tobacco
  ACO 130
  auxin 23, 84, 284, 445, 457, 460
  cytokinin 21, 100, 571-573
  day-neutral 544
  ethylene 373, 383
  flowering 544, 545
  gibberellin 84
  guard cells 392
  jasmonate 624
  kinase inhibitors 457
  KNOX genes 82
  kaurene oxidase 73

leaves 545
long days 544-545
MAP kinase 626
mosaic virus 636-637, 643
necrosis virus 639
nicotine 624
polyamines 11
salicylate 648
scions 544
senescence 567, 571-573
short days 544-545
suspension-cultured cells 445, 460
transgenic 132
Tomato (*Lycopersicon esculentum*)
  ABA 143, 145, 147-148, 278
  ACC oxidase 131
  ACS 124
  brassinosteroids 414, 417-419, 42-422
  BRI1 418-419, 421-422
  carotenoid biosynthesis 69
  CTR1 380
  cu3 418
  defense 622, 627
  EIN2 380
  ethylene 371,622
    antisense 602-603
    biosynthesis 116, 118-124, 126-134, 602-603
    insensitive 376, 604
    receptor 374-376, 378-379
    response 378
    signal transduction 380-381
  expansins 266-270, 274-275, 278, 414
  fertility 632
  fruit ripening , 370-371
    ethylene 116, 118-124, 126-134, 275, 374-376, 378-381, 387, 602, 604
    expansins 275
    gene 126
  germination 274, 278
  homolog of *COI1* 631
  IAA 269
  jai1 mutant 632
  jasmonate 612, 614-615, 617, 622, 625-627, 632
  long-distance wound signal 629
  1-MCP 387

NPA 268-269
never ripe 374
reproductive development 632
rin 381
ripening 381
SAGs 564
seeds 274
stomata 499
systemin 422, 655-658
wound signal 621, 625-626, 628
Tonoplast 399
Torenia 387
Touch sensitive genes 360
Tracheary element 488
Transcription
  factors 222, 285, 526, 529-530, 534
  ABA 391
Transgenic 525-526, 551
  Agrobacterium 526
  cis-acting sequences 525
  crops **582-609**
  plants 639
  potato 551
  salicylate hydroxylase (SH) 639
Transition metal 352
Translocation pathway 440
Transparent testa 444
Transpiration 392, 493, 498
Transporter 553, 556
Triadimenol 89
Trichomes 475
Trifolium 480
Triiodo benzoic acid 444, 481
Trimethylsilylation 675
Triphysaria versicolor 277
Triple response 9, 127
Triterpenoids 160, 623
tRNA 99, 100, 103
Tropisms 23, 462
Tryptamine 682
Tryptophan 5, **40-48**, 439, 682
Tuber / Tuberization **538-560**
  bearing species 538
  carbohydrates 541, 554-556
  control 548
  critical night length 542
  cytokinin 547, 552-553
  daylength 539, 547
  development 557
  dormancy 538-539

Index

environmental conditions 557
eyes 539
formation 539- 542, 546, 547, 553, 557
gibberellin 547-550
growth 554
hormonal control 546
in vitro 554
induction 541, 544, 547, 549, 554, 557
inhibition 541-542, 547
initiation 539, 555
jasmonate 12, 547, 550-552
long days 542
LOX 551-552
night length 539, 542
nitrogen 539-541
patatin 552, 555
pathway 546, 557
  integrator gene 546, 557
photoperiods 539
phyB 542
potato 12, 20, 538-560
production 543
promotion 541, 549
rest 539
senescence 546
short days 539-540, 542
signal 557
starch 539
storage organ 539
temperature 539-541
unloading 555
vegetative propagation 538
Tuberonic acid 3, 12, 550-551, 618
Turgor pressure 191
TWIN 515
Two-component relay 322-324, 340, 346
Tyrosine phosphatases 406
UBA2a 407
Ubiquitin
  conjugation 291-292
  function 24-25
  proteasome pathway 24-25, 285, 296, **307**, 358, auxin 287, **292-294**, 518, 629
  ethylene 629
  gibberellin 236, 314-315, 629
  jasmonate 629
  light 629

pathogen attack 629
rice F-box 236
SCF complex 629
protein ligase (E3) 291, 518
UDP-glucose 556, 645
UMFT1 158
Uniconazol 548
Vacuole 118
Vapor phase extraction 689
Vascular 485, 487- 489
  auxin 471-472, 481, 488
  development 485
  bundle 473, 480-481
  cambium 440
  cells 488
  differentiation 161, 461, 485, 487
    Arabidopsis 481
    auxin 481, 488
    cytokinin 471
    ethylene 471
    gibberellin 471
    NAA 481
    organ development 461
    provascular strands 461
    sterol 488
    cell polarity 488
    TIBA 481
  discontinuities 485, 487, elements 440
  maturation 485
  meristem 485
  networks 488
  regeneration 489
  stele 102
  tissue 481-482
    auxin 471-472
    formation 481
    induction 472
    regeneration 478, 479
    wounding 461
Vegetative 522
  propagation 538, 539
  storage proteins 614
Vein 472, 486-487, 490
Vesicle trafficking 397, 454-455, 458
Vessel 481-482, 484-487
Vicia faba 392-395, 397, 399-400, 403-404, 502-503
Violaxanthin 140, 143, 147-148
Vitamin A 146
Viviparous mutants 143-145, 505-506, 522-523
Vivipary 520, 522-525

Vp 143-145, 505-506, 522-523
Voodoo lily 637
Waito-C rice 19
Wall loosening 214, 533
Wall-loosening factor 265
Wall-yield threshold 190
Water **493-512**
  ABA 34, 495-502
  loss 393
  potential 494
  status 493, 508
    leaf 508
    plant 493
    soil 493
  stress 137, 377
  transpiration 507
  use efficiency 507-508
  uptake 493, 504
Wei 356, 361, 362
Went, F.W. 1, 2, 3
Wheat
  amylase 226, 232
  Ca^{2+} 230
  dormancy 584
  dwarfing gene 26
  endosperm 227
  GA-response complexes 232
  GAMYB 234
  germination 521
  gibberellin 222-224, 231-232
  lodging prevention 89
  Norin 10 586
  reserve mobilization 222
  Rht 586
  scutellum 223
  starch 226
  semi-dwarfs 586
  transpiration 499
  VP1 584
Wild oat 229, 517
Willow 635, 636
Wilting 143
WOODEN LEG (wol, AHK4) 254, 324, 326-330, 340, 342, 517
Wound 478, 551
  signal 627
  induced genes 551
  inducible PIs 621
Wounding 129, 377, 461, 478
  abscisic acid 11
  ethylene 129
  vascular development 461

Index

water stress 377
Xanthomonas campestris 378
Xanthium strumarium 137, 141
X-ray diffraction 156
Xanthoxin 140, 149
Xenopus oocytes 120
XTH5 80
Xyloglucan endotransglycosylase 304
Xylem 33, 118, 439, 471, 472, 475, 483- 488, 503
  ABA 33
  axial organs 483
  bundles 475
  cytokinin 503
  development pattern 485, 486
  differentiation 483
  discontinuities 484-486, 488
  fan formation 475
  foliar organs 483
  IAA 475
  initiation 487
  maturation 472, 484, 485, 487
  monopteros 488
  ovules 475
  parenchyma 440
  phloem anastomoses 483
  regeneration 478, 485
  sap 496, 498, 499
    ABA 498, 499
    alkalisation 499
  stigma 475
  vascular bundles 483
  veinlets 475
  vessels 440, 471
Xylopyranosidase 228
Yang cycle 117-118
Yeast 183
  invertase 555
  patatin promoter 555
  two-hybrid screens 647
Yield 493, 507-509, 585
Yieldins 208
YPD1 322-323, 326-337
(Z)-jasmone 618
Zea mays see maize
ZEA3 344
Zeatin see also cytokinin
  BY-2 248
  biosynthesis **95-101**
  cell cycle 242, 247-248
  cis-zeatin 103, 111
  degration 110-111
  glucose 107-109
  hydroxylase 108-109
  in plants 8
  isomerization, 103, 108
  mitosis 247
  O-glucoside 98, 247
  oxidase 110-111
  reductase 108
  riboside 21, 96, 103, 504
  stomatal response 503-504
  structure 8, 95
  trans-zeatin 95-100, 103-105, 107-111, 242
  xylose 109-110
Zeaxanthin 143
Zinnia elegans 483
HK1 103
HP2 333
RR 333
Zucchini 119, 442
Zygote 515, 517

DATE DUE
